Herpetological Osteopathology

Bruce M. Rothschild
Hans-Peter Schultze
Rodrigo Pellegrini

Herpetological Osteopathology

Annotated Bibliography of Amphibians and Reptiles

Springer

Bruce M. Rothschild
University of Kansas
Lawrence, Kansas, USA
bmr@ku.edu

Rodrigo Pellegrini
New Jersey State Museum
Trenton, New Jersey, USA

Hans-Peter Schultze
University of Kansas
Lawrence, Kansas, USA

ISBN 978-1-4614-0823-9 e-ISBN 978-1-4614-0824-6
DOI 10.1007/978-1-4614-0824-6
Springer New York Dordrecht Heidelberg London

Library of Congress Control Number: 2011941386

Printed on acid-free paper

Springer is part of Springer Science+Business Media (www.springer.com)

Preface

Sir William Osler, a famous nineteenth-century physician observed that diagnosis without knowing the literature is like going to sea without charts. The purpose of the current text is to provide the charts: The delineation of available literature on the subject. This approach reconciles a very disparate literature and facilitates access by herpetologists and other professionals interested in amphibian and reptile disease. It should be of interest to anyone working on animal models of disease and their behavioral implications. The book is designed to also be accessible to non-herpetologists with specific information available by common and scientific name and by disease or disease category, with a full glossary of terms provided. The book is annotated to identify mistaken/superseded diagnoses and changes in taxonomic classification. The book provides a "library" for macroscopic osseous manifestations of a variety of diseases in higher vertebrates facilitating understanding of literature-based diagnoses.

One of the challenges of research is that even the existence of articles related to a particular subject may be hidden by the rarity and obscurity of the publication. The literature of osseous pathology is rife with disparate citation, such that a whole set of topical publications has not previously been available in a single source. For example, shell disease in contemporary turtles has been a source of confusion, partially fueled by a disparate literature. The critical studies on turtle shell pathogenesis were published in literature sources not cited in the standard studies of structural pathology. This is an example of one of the important contributions of this annotated bibliography. It makes the whole span of the literature available to all, such that cross-communication and interdisciplinary efforts can be initiated. The current effort delineates the information by publication, characterizes the contents of the publication, establishes the validity of the conclusions and provides a unique indexing system for identifying the pertinent literature by common and scientific names. It represents an exhaustive literature compilation organized searching by subject, examination of articles for additional, apparently, pertinent citations, and review of journal tables of contents. We would be honored to consider this volume as a companion text to Elliot Jacobson's 2007 tome on reptile biology, anatomy, histology, and pathology.

Acknowlegdments

We wish to express appreciation to the many individuals who have assisted us with this study. Assistance with literature access was provided by Shiela Orth, Tina Spray, Lars Leon of the KU Library System and Spencer Research library, Donna Remely and Shaaron Warne of the New Jersey State Library, Anne Kemp from Brisbane, Queensland, Australia, Eduardo Torres of Santiago, Chile, Frank Rüli of the University of Zürich, Switzerland, Ignacio Cerda from Buenos Aires, Argentina, Macarena Pellegrini of Copiapó, Chile and Van Wallach of Cambridge, Massachussetts. Assistance with translation was provided for Hebrew from Israel Herskovitz of Sackler Medical School, Tel Aviv University, Tel Aviv, Israel, for Czech by Jiri Zidek of Praha, Czech Republic, of Russian by Oleg Pisman, Olga Pavlova and Irina Haller, Japanese by Kenshu Shimada from Chicago, Illinois and "Casey" Seto from Ottawa, Kansas, and Japanese by Desui Miao from the University of Kansas, Lawrence, Kansas. Luke Welton assisted with clarification of phylogeny. Specimen access was provided by Laura Abraczinskas (MSU), Alvin Braswell (NCMNH), Judith M. Chupasko (MCZ), Carla Cicero (MVZ), Ted Daeschler (ANSP), Robert Ferenac (NYSM), Darrel R. Frost (AMNH), Michel Gosselin (CMN), James Hanken (NMNH), Johann Hegelbach (UZNH), W. Ronald Heyer (NMNH), Janet Hinshaw (UMMZ), James L. Knight (SCMNH), Kenneth Krysko (FLMNH), F Wayne King (FLMNH), Ross D. MacCulloch (ROM), Larry Martin (KU), Max Nickerson (FLMNH), Ronald A. Nussbaum (UMMZ), WR Pauwels (HU), Alan Resetar (FMNH), Nate H. Rice (ANSP), Stephen P. Rogers (CM), José Rosado (MCZ), Greg Schneider (UMMZ), John E. Simmons (formerly KU), Linda Trueb (KU), Judit Varos (HNHM), Greg Watkins-Colwell (YPM), Yehudah L. Werner (HU) and Kristof Zyskowski (YPM).

About the Authors

Macroscopic Osteopathology: Bibliography of Amphibian and Reptile Pathology

Bruce M. Rothschild, M.D., graduated from New Jersey College of Medicine in 1973. He is a fellow of the American College of Physicians, American College of Rheumatology, and Society of Skeletal Radiology and was elected to the International Skeletal Society. He has been recognized for his work in rheumatology and skeletal pathology where his special interests focus on clinical-anatomic-radiologic correlation, data-based paleopathology, evolution of inflammatory arthritis and tuberculosis, and management of inflammatory arthritis. He is widely recognized for his contributions to understanding radiologic manifestations of rheumatologic disease.

Dr. Rothschild has been a visiting professor at universities in the USA, Canada, the Caribbean, South America, Europe, the Middle East, South Africa, Asia, and Australia and has been an invited lecturer at universities, hospitals, and museums throughout the world. He has published over 600 papers and abstracts, including authoritative papers on the origins of rheumatoid arthritis, spondyloarthropathy, syphilis, and tuberculosis, and character of bone changes in metastatic cancer, myeloma, leukemia, tuberculosis, fungal disease, renal disease, treponemal disease, rheumatoid arthritis, spondyloarthropathy, gout, calcium pyrophosphate deposition disease, and hypertrophic osteoarthropathy. He is the author of four books and has participated in eight Discovery Channel/BBC documentaries on origins of diseases and ancient reptiles.

Since 1986, Dr. Rothschild has been professor of medicine at Northeast Ohio Medical University, Rootstown, OH. He is also adjuvant professor of anthropology at The University of Kansas and of biomedical engineering at The University of Akron, Ohio, and holds research associateships at the Carnegie Museum and Biodiversity Institute of the University of Kansas. He was the first director of the Rheumatology Division at The Chicago Medical School and a prime force behind the resurgence of data-based paleorheumatology and comparative osseous pathology.

Hans-Peter Schultze, Ph.D., graduated from the Universität Tübingen, Germany, in 1965. After 2 years as postdoc at the Naturhistoriska Riksmuseet Stockholm, Sweden, and 11 years at the Universität Göttingen as assistant and associate professor, he moved to the University of Kansas in 1978. He then worked as a curator at the Museum of Natural History (lower vertebrates) and as a professor in the Department of Systematics and Ecology. He served as chairman of the department between 1988 and 1990. After opting for early retirement in 1994, he moved to the Museum für Naturkunde, Humboldt-Universität, Berlin, Germany, as head of the Department of Paleontology to rebuild the department after the reunification of Germany. He also worked as a professor in the Department of Biology and as a director of the museum from 2000 to 2004. He returned to Lawrence, Kansas, in 2004, to continue research at the university.

Rodrigo Pellegrini, M.S., graduated from the University of Kansas in 1998, obtained an M.S. in vertebrate paleontology in 2003 and an M.A. in museum studies in 2006. After serving as curatorial assistant in the Division of Vertebrate Paleontology at the University of Kansas and Core Facility Curator at the Kansas Geological Survey, he became Natural History Bureau Registrar at the New Jersey State Museum. His duties include primarily collections management and scientific research, with occasional fieldwork and exhibit design.

Contents

Introduction

As scientific analysis of testable hypotheses has replaced the speculative approach to the study of bone disease in recent and fossil amphibians and reptiles, the field has advanced from simply reporting observations to analyzing their implications. This process is predicated upon a reproducible database, which explains/diagnoses the nature of bony alterations and a secure review of the literature. Thereby hangs the rub. The herpetological literature extends far back to Aristotle and other authors, whose works are difficult to access (let alone read) and are scattered through many prominent and eclectic journals and in the lay literature. While older diagnoses often have not stood the test of time, the clarity of report descriptions usually allows confident identification of the underlying pathology.

Sir William Osler, a famous nineteenth-century physician, observed that diagnosis without knowing the literature is like going to sea without charts. The purpose of the current text is to provide the charts. It reconciles this information into one resource of validated annotative citations, facilitating its accessibility to students of amphibian and reptile disease. By accessibility is meant that specific information can be located by common and scientific name and by disease or disease category. By validation is meant that the previous mistaken or superceded diagnoses are identified. Given the development of a "library" of macroscopic osseous manifestations of a variety of diseases in higher vertebrates, literature-based diagnoses are now possible.

Much of past confusion has been related to shared terminologies in zoology, paleontology, and medicine (both veterinary and human), totally disparate in meaning. Therefore a full glossary of terms is provided. Additionally, such semantic sources of confusion are identified in specific annotations.

The publications are listed by last name of author, using what appears to be the most common spelling of their names. As some articles were published in languages with unique letter forms/usage (e.g., Cyrillic, Latin, Greek, Hebrew) or even pronunciations (e.g., German and Russian "W" pronounced as "V"), transliteration in and of the literature is sometimes challenging. Thus, "V" and "W" have been used interchangeably, as have "J" and "I." Searches for articles by specific authors must consider the possibility that the source of the citation may suffer this "affliction" and search for possible alternative spellings. Second authors are listed in alphabetical order. Three or more authors are listed by year of publication.

Reports of pathology in ancient amphibians have been limited to congenital disease and isolated fractures, also found in contemporary amphibians. They often provide a window to evolution of environmental conditions, especially toxic contamination. The amphibian is to the external environment as canaries were at one time to mines. They provide an early warning system that we ignore at our peril. Additionally, infectious disease, gout, neoplasia, and even a

B.M. Rothschild et al., *Herpetological Osteopathology: Annotated Bibliography of Amphibians and Reptiles*, DOI 10.1007/978-1-4614-0824-6_1, © Springer Science+Business Media, LLC 2012

naturally occurring model of neuropathic bone disease have been identified in contemporary amphibians.

Reports of pathology in ancient reptiles reveal congenital disease, trauma, and infections (also found in contemporary reptiles), permitting insights to environmental interactions. They also revealed the presence of decompression syndrome (bends/diving disease) in ancient marine reptiles (revealing habitat and interspecies interactions) and the evolutionary development of mechanisms for protection. Osteoarthritis and neoplasia were rare, suggesting that natural environments are not conducive to development of mechanical forms of arthritis and that neoplasia affecting bone (e.g., cancer) is a more modern disease, at least from the point of view of population penetrance. Also documented is the phylogenetic antiquity of a specific form of back arthritis called spondyloarthropathy. The latter is a newly recognized (but actually ancient) phenomenon in reptiles.

Shell disease in contemporary turtles has been a source of confusion, partially fueled by discrepant literature. The critical studies on pathogenesis were published in literature sources not cited in studies of structural pathology. This is an example of one of the important contributions of this annotated bibliography. It makes that literature available to all, such that cross-communication and interdisciplinary efforts can be initiated.

Osteoarthritis and metabolic disease were equally rare in noncaptive reptiles, as in amphibians. Osteoarthritis has often been mistakenly diagnosed in the presence of the specific form of back arthritis mentioned above. Crystalline joint diseases, in the form of gout and pseudogout, and neoplasia have been recognized as isolated phenomenon in contemporary reptiles.

This is the spectrum of disease categories represented in the literature. The current effort delineates the information by publication, characterizes the contents of the publication, establishes the validity of the conclusions, and provides a unique indexing system for identifying the pertinent literature by common and scientific names. While it categorizes the components of each report, there is overlap. Congenital phenomenon may be directly transmitted or may be the result of environmental exposures or lack thereof, or a mixture of both. Authors' perspectives are used as the guideline, updated according to most recent perspectives.

Glossary

Aberrant feature/condition different from usual or normal.

Abscess cavity containing localized pus collection.

Acheiria = acheiropodia = apodia absent paw.

Acheiropodia = apodia = acheiria.

Achondroplasia abnormal cartilage conversion to bone, producing short, abnormally limbed dwarfism.

Acromegaly disproportionate increase in bone size related to overproduction of growth hormone.

ACTH adrenocorticotrophic hormone.

Adenoma benign neoplasm of glandular cells.

Adrenocorticotrophic hormone Anterior pituitary hormone that stimulates the adrenal cortex.

Agenesis failure of formation of all or part of structure.

Adactyly absence of digits.

Amely congenital absence of limb(s).

Amyloidosis disease associated with tissue deposition of abnormal protein or abnormal immunoglobulin.

Anacatadidymus two heads and two tails (Fischer 1868).

Anadidymus a two-tailed (Fischer 1868).

Anakatamesodidymus separated at the anterior and posterior ends and also in the middle of the trunk (joined only along body).

Ankylodactyly fused digit.

Ankylose growing together.

Ankylosis Rigid union.

Anulus fibrosus often misspelled term for the outer layers of intervertebral disks.

Anurans frogs and toads.

Aphalangia absence of some phalanges or finger bones.

Aplasia congenital absence, total failure of development of specific element.

Apodia = acheiropodia = acheiria.

Arthrogryphosis flexed positioning due to muscle fibrosis.

Atlodymus complete cranial duplication, with a single enlarged or duplicated atlas vertebra.

Batrachian amphibian.

Bends nitrogen bubble-induced complication of decompression syndrome (caisson disease).

Bicephalic = dicephalic having two heads.

Bipartite ossification unfused ossification centers as opposed to bipartite bones.

Bone bridge spans space between two margins of bent bone. Linear rays of bone extending filling margins between bone margins.

Brachydactyly abnormally short fingers or toes.

Brachymely = micromelia = acromelia = nanomely shorter limb.

Brachygnathia mandibular micrognathia; abnormal shortness of mandible.

Brachyuria short tail.

Caecilians third order of Lissamphibia.

Calcium pyrophosphate deposition disease crystals of calcium pyrophosphate dihydrate accumulate in the hyaline articular cartilage or fibrocartilage (pseudogout).

B.M. Rothschild et al., *Herpetological Osteopathology: Annotated Bibliography of Amphibians and Reptiles*, DOI 10.1007/978-1-4614-0824-6_2, © Springer Science+Business Media, LLC 2012

Callus reactive bone which forms a splint during fracture healing.

Carcinoma malignant neoplasm derived from epithelial cells. Tissue of origin can be skin, lungs, stomach, breast, cervix, and prostate.

Catadidymus two-headed (Fischer 1868).

Caudata salamanders and newts (= urodeles).

Cebocephaly reduced distance between orbital cavities.

Cephaloderopagus fusion by cranium and cervical vertebrae.

Cephalomegaly additional extremity(ies) in the head.

Cheigagra gout affecting the hand.

Cheilognathopalatoschisis = cheilognathouranoschisis cleft lip, jaw, and palate.

Cheilognathoschisis split (cleft) anterior jaw.

Cheilognathouranoschisis = Cheilognathopalatoschisis.

Chiloschisis hairlip.

Chirodactyly finger curvature.

Chondrodysplasia = chondrodystrophy abnormal cartilage development in long bones, especially at epiphyseal plates, resulting in short-limbed individuals with normal axial skeleton.

Chondrodystrophy = Chondrodysplasia.

Chondrogenesis cartilage formation.

Chondroid cartilage.

Chondromyxoma benign mesenchymal cell tumor additionally containing cartilage differentiated/recognizable cells.

Chondro-osteofibroma benign fibrous neoplasm containing chondroid and osseous elements.

Chondrosarcoma malignant cartilage neoplasm.

Chordoma notochord-derived neoplasm.

Clinodactyly curvature of one or more digits.

Cortical bone component of bone between periosteum and medullary region.

Corticosteroid adrenal hormones of variable forms, with corticosterone dominant in amphibians and reptiles, contrasted with cortisol in metamorphic ranid tadpoles, *Xenopus laevis*, and some urodeles, which are the major endocrine portion of the stress response.

Craniophagus fusion by cranium.

Craniosynostosis premature cranial suture closure.

Dactylomegaly long digit (= macrodactyly).

Derodidymus vertebral column bifurcated in the cervical region, double-headed (two complete heads and necks).

Deropodymus incomplete cranial duplication and complete vertebro-cervical duplication.

Diagnosis formal character state description distinguishing one taxon from another (herpetology definition), contrasted with use in this annotated bibliography as identified disease state.

Diarthrodial synovial-lined joint, at which motion occurs.

Dicephalic = bicephalic having two heads.

Dichocephalic two-headed ribs.

Dipygus caudal duplication below leg.

Dolichocephaly = macrocephaly large or long head.

Dorsoventral vertebral column curvatures commonly attributed to metabolic bone disease; usually refers to thoracic spine (hump-like arching of tortoise carapace).

Dysplasia any abnormality of tissue development. Conventionally used for nonmalignant pathology.

Ectodactyly = ectrodactyly in Rostand absence of one or more digits.

Ectomely absence of one or more limbs or incomplete limb with missing lower portion; the term encompasses amely, hemimely, and meromely.

Ectromelia = phocomelia.

Enchondroma benign cartilage neoplasia, typically within bone. Perhaps representing cartilage remnant.

Endosteal medullary surface of cortical bone.

Enthesis area of muscle, tendon, or joint capsule attachment.

Erosive biologic process-derived bone disruption in living tissue.

Ethmocephaly cyclopi with small eyes and snout.

Etiology cause of the phenomenon/disease.

Exostosis surface bone growth (at muscle attachments, referred to as entheses).

Fibrodysplasia replacement of bone tissue by fibrous tissue (= fibrous dysplasia).

Fibroma (ossifying) mass composed of fibrous tissue or connective tissue.

Fibroma (nonossifying) benign fibroblastic mass, also called fibroxanthoma, nonneoplastic lesions from faulty ossification at the growth plate. Allegedly present at some time in a third of all children. May have a bubbly appearance and sclerotic margin and may expand cortex but has no matrix calcification.

Fibrous dysplasia replacement of bone by fibrous tissue.

Fracture bone broken into one or more pieces.

Gangrene death of tissue.

Gastromely additional extremity(ies) between thorax and pelvic.

Giantism-gigantism overgrowth of body in whole or part.

Gonagra-gonatagra gout affecting the knee/stifle.

Gout a metabolic disorder in which sodium urate crystals deposit in joints (referred to as articular gout) or internal organs (referred to as visceral gout).

Greenstick fracture herpetologic use (e.g., Lane et al. 1984) was to describe the "folding fracture" of metabolic bone disease. This contrasts with medical use to describe a form of incomplete fracture in immature bone.

Hamartoma overgrowth of tissue normally located in the area.

Hemimely defective limbs, especially distal components.

Heterotopic transplantation to abnormal location or occurring in many habitats (herpetology definition), contrasted with use in this annotated bibliography to indicate spontaneous occurrence in locations usually lacking the anlage.

Hump deformed kyphotic spine.

Hypermely additional extremities (= melomely).

Hyperparathyroidism disorder caused by overactivity of the named glands, producing osteitis fibrosa cystica and other bone changes.

Hyperphalangy fingers or toes with supernumerary phalanges.

Hyperplasia nonneoplastic increase in cells of any body tissue.

Hypoplasia underdevelopment incomplete development of a tissue or organ.

Hypertrophy nonneoplastic increase in tissue bulk but not number of component cells. Term is often misused since hyperplasia is the more accurate term.

Ileopolymely multiple limbs coming off ilium.

Interstitial extracellular, extravascular tissue space.

Ischiomely extra limb coming off ischium.

Keratoconjunctivitis conjunctival inflammation, which may be a component of the immunologic disorder Sjögren's syndrome or associated with eye inflammation in reactive arthritis.

Kypholordosis combination of kyphosis and lordosis.

Kyphoscoliosis combination of kyphosis and scoliosis.

Kyphosis/kyphotic curvature of axial skeleton, with accentuated posterior apical apex.

Lesion pathologic area.

Lordosis curvature of axial skeleton, with accentuated anterior apical apex. Usually refers to lumbar spine.

Lymphoid anatomic system of vessels and glands which drain and filter, respectively, interstitial fluids.

Lymphoma malignant neoplasia of lymphoid tissues.

Lymphosarcoma variety of malignant neoplastic tumor of lymphoid origin.

Lytic destruction of structures (herpetologic definition), contrasted with use in this annotated bibliography to holes in bone.

Macrocephaly large or long head (= dolichocephaly).

Macrodactyly = dactylomegaly.

Macromelia long limb.

Medullary marrow space.

Melanoma malignant neoplasm of the variety of skin cells that are capable of producing the pigment melanin.

Melomely form of hypermely with additional extremity(ies) at the base of the normal extremity.

Meromely absence of digits (= adactyly).

Mesenchymal those mesodermally derived cells which form the musculoskeletal, vascular, lymphatic, and urogenital systems.

Mesoderm embryonic germ layer between ecto- and endoderm.

Metabolic bone disease nonspecific term which includes many diseases (e.g., osteoporosis, fibrous osteodystrophy, osteomalacy, rickets). According to Lillewhite, disease related to inadequate dietary calcium or UV light exposure. Actually, probably a renal osteodystrophy.

Microcephaly abnormally small head, snout blunted.

Microcheiria small paw.

Microdactyly = brachydactyly.

Micromely small or short limb (= nanomely).

Museum/Collection Abbreviations:

AMNH American Museum of Natural History, New York City.

AUMP Auburn University Museum of Paleontology, Auburn, Alabama.

BMS Buffalo Museum of Science, Buffalo, New York.

CM Carnegie Museum, Pittsburgh, Pennsylvania.

IVPP Institute of Vertebrate Paleontology and Paleoanthropology, Beijing, China.

KU University of Kansas, Lawrence, Kansas.

MNCN Natural History Museum of Madrid, Spain.

MSU Michigan State University, East Lansing, Michigan.

NMNH National Museum of Natural History, Smithsonian, Washington, D.C.

ROM Royal Ontario Museum, Toronto, Canada

UCMP University of California Berkeley Museum of Paleontology.

UFMNH University of Florida Museum of Natural History, Gainesville, Florida.

UMMZ University of Michigan Museum of Zoology, Ann Arbor, Michigan.

USNM National Museum of Natural History, Smithsonian, Washington, D.C.

YPM Yale Peabody Museum, New Haven, Connecticut.

Mutation heritable genetic alteration.

Mycobacteria rod-shaped, acid-fast gram-positive bacteria. One variety causes tuberculosis.

Mycoplasma A species-dependent pathologic microorganism lacking a cell wall.

Nanomely small or short limb (= micromely).

Necrosis tissue death.

Neoplasm new, abnormal tissue growth.

Notomely additional extremity(ies) on the back (dorsum).

Oligodactyly less than normal number of digits.

Opisthodichotomy axial duplication with two complete bodies and single cranium.

Opodidymus cranium bifurcated two broadly joined heads (mostly three-eyed).

Osteitis inflammation of bone (herpetologic definition), contrasted with use in this annotated bibliography to identify bone infection, not discriminating use from osteomyelitis.

Osteitis fibrosa cystica fibrous tissue replacement of bone secondary to exaggerated osteoclastic resorption in hyperparathyroidism.

Osteoarthritis = osteoarthropathy overgrowth of diarthrodial joint margins, producing osteophyte.

Osteoarthropathy = osteoarthritis.

Osteochondritis dessicans detachment of an articular bone fragment. May complicate intra-articular fracture or avascular necrosis. Often mistakenly called osteochondrosis.

Osteochondrodystrophy variation on chondrodystrophy, wherein the axial skeleton is also affected, often producing flattened or wedged vertebrae.

Osteochondroma benign neoplasm producing a cartilage cap on an exostosis.

Osteochondrosis failure of cartilage region to ossify (transform into bone). Often mistakenly referred to as degenerative cartilage.

Osteochondrosarcoma malignant neoplasm derived from both bone and cartilage.

Osteoclastic cells responsible for resorption component of bone remodeling.

Osteodystrophy defective bone formation.

Osteoid the framework for calcification secreted by bone cells.

Osteolysis resorption or destruction of bone.

Osteoma benign tumor of bone cells.

Osteomalacia vitamin D deficiency-related softening of bone with failure of mineralization, usually refers to the disease in adults (in children, referred to as rickets).

Osteomyelitis infection of bone. Inflammation of marrow cavity (herpetologic definition), contrasted with use in this annotated bibliography to identify bone infection not limited to the external surface.

Osteopathy combination of osteomalacia and rickets.

Osteopenia reduced bone ossification/density.

Osteoperiostitis inflammation/reaction of/at both the outer layer (periosteum) and underlying component (cortex) of bone.

Osteophyte abnormal bony overgrowth extending from margins of articular surfaces of diarthrodial joints. They are the identifier for the condition called osteoarthritis. When affecting vertebral bodies, they are indicative of spondylosis deformans, not osteoarthritis.

Osteoporosis reduction in quantity and quality (e.g., thickness) of trabecular components of bone. Deficient bone mineral content, associated with loss of structural integrity.

Osteosarcoma malignant bone neoplasm.

Osteosclerosis increased bone density.

Osteopetrosis failure of endosteal bone resorption during growth, resulting in much diminished or absent medullary cavity.

Otocephaly absent or underdeveloped lower jaw.

Pachyostosis benign bone thickening.

Paedomorphosis adult retention of juvenile characteristics.

Panostitis reaction of all bone layers (typically sclerosis).

Pathogenesis development of a disease.

Pathognomonic definitive for a specific disease diagnosis.

Pedomorphosis adult retention of juvenile characteristics.

Periosteopathy any disorder of the outer layer (periosteum) of bone.

Periostitis inflammation/reaction of/at outer layer of bone.

Phocomely incomplete limb with missing proximal portion.

Polyarthritis multiple joint involvement by arthritis. Medical convention limits use of the term to individuals with affliction of 5 or more joints. This convention is not necessarily followed.

Polydactyly increased number of metatarsals (= hyperdactyly).

Polymely additional limbs.

Polypody a limb with two or more hands or feet.

Polyphalangy duplicate phalangeal sets.

Podagra pedal gout.

Porosity volume of openings in rock/soil (herpetologic definition), contrasted with use in this annotated bibliography to identify presence of minute surface holes in bone.

Prodichotomous duplication of head and neck.

Pseudoarthrosis false joint related to failure of fracture components to unite.

Pseudogout acute attacks of calcium pyrophosphate deposition disease-crystals of calcium pyrophosphate dihydrate accumulate in the hyaline articular cartilage or fibrocartilage.

Psodidymus parasacral bifurcation.

Psodymus vertebral column bifurcated near sacrum.

Pygomely additional extremity(ies) behind or within the pelvic region.

Pygopagus fused at pelvis.

Rachitis = rickets.

Reactive arthritis form of inflammatory arthritis characterized by erosions and new bone formation, affecting peripheral and/or axial skeleton. It is a form of spondyloarthropathy.

Renal disease disease of the kidney.

Renal osteodystrophy combination of vitamin D deficiency (osteomalacia) and hyperparathyroidism.

Rhinocephaly proboscis-like nose overlying partial or complete eye fusion.

Rhinodymus minimum degree of duplication in mouth and nose area, double-nosed (two snouts).

Rhoecosis vertebral displacement.

Rickets failure of bone osteoid to calcify. Related to deficiency of active form of vitamin D. Usually refers to individuals in whom/which epiphyses have not fused (in adults, called osteomalacia).

Sarcoma malignant tumor of mesenchymal cell origin.

Scoliosis lateral curvature of the vertebral column.

Scurvy vitamin C deficiency.

Septic joint infected joint.

Sesamoid normal intratendinous bone. It provides mechanical advantage to the muscles whose tendons transgress the area.

Shunting redirecting.

Sirenomelia side to side fusion of lower extremities, often associated with pelvic reduction.

Sjögren's syndrome disorder characterized by a complex of symptoms including dry eyes (keratoconjunctivitis), dry mouth, and arthritis.

Spondyloarthropathy inflammatory arthritis characterized by erosions and new bone formation in animals, affecting peripheral and/or axial skeleton.

Spondylosis (actually spondylosis deformans) term accurately utilized to describe vertebral body osteophytes. This is not a sign of osteoarthritis. It is sometimes inaccurately utilized to describe vertebral ankylosis. The latter actually is properly termed a syndesmophyte and is a sign of spondyloarthropathy.

Spongiform bone expansion of cancellous bone at distal tip of ectromeliac limbs.

Symmelia fused limb.

Symodia fused paw.

Syndactyly fusion or failure of separation of fingers or toes. Fusion of two or more pedal elements.

Syndesmophyte calcification/ossification of outer layer of annulus fibrosus. A sign of spondyloarthopathy.

Synovial referring to diarthrodial articulation.

Taumely long bone bent back on itself, forming >90° angle.

Teratogeny concept (precept, knowledge) of causes of formation of anomalies.

Teratology concept (precept, knowledge) of formation of anomalies. Study of embryologic malformations.

Teratodymus an individual with part of body doubled.

Teratopagus independent axial skeletons (e.g., Siamese twins).

Thoracodymus vertebral column bifurcated in the thoracic region.

Ulcerative disease lesion resulting from disruption of surface.

Urodele salamanders.

Uranoschisis cleft palate.

Uveitis inflammation of the middle coat of the eye, which may be a component of reactive arthritis.

Vasculitis inflammation of blood vessels.

Zygodactylus fusion of digits in bundles of two or three.

Classification

Congenital:

Congenital – Malformations/homeobox anomalies/growth:

Amely – congenital absence of limb(s).

Brachydactyly – short digits.

Brachygnathy – mandibular micrognathia – abnormal shortness of mandible.

Bone bridge – spans space between two margins of bent bone. Linear rays of bone extending filling margins between bone margins.

Ectodactyly – absence of one or more digits (from aphalangia, in which some of the phalanges or finger bones are missing), to adactyly (the absence of a digit).

Ectomely – absence of one or more limbs or incomplete limb with missing lower portion; the term includes amely, hemimely, and meromely.

Hemimely – short bone (absence or short fibula or tibia) with normal distal elements.

Kyphosis – convex thoracic spine (spinal deformity).

Meromely – absence of digits (= adactyly).

Microcephaly – abnormally small head, snout blunted.

Micromely – small or short limb (= nanomely).

Phocomely – incomplete limb with missing proximal portion.

Polydactyly – increased number of metatarsals (= hyperdactyly).

Polymely – additional limbs.

Polypody – a limb with two or more hands or feet.

Polyphalangy – duplicate phalangeal sets.

Scoliosis – lateral deviation (curvature) of vertebrae.

Taumely – a limb element oriented at 90° to the long axis of the limb.

All from Taruffi (1884/5)

Gastromely – additional extremity(ies) between thorax and pelvic.

Cephalomely – additional extremity(ies) an the head.

Ischiomely –

Ileopolymely –

Notomely – additional extremity(ies) on the back (dorsum).

Melomely – additional extremity(ies) at the base of the normal extremity.

Pygomely – additional extremity(ies) behind or between the pelvic.

Trauma:

Fracture – bone broken into one or more pieces.

Infection:

Osteomyelitis – acute or chronic bone infection (caused by bacteria).

Septic joint – infected joint.

Metabolic:

Vitamin

D deficiency – osteomalacia/rickets.

C deficiency – scurvy.

A excess – retinoids.

Hyperparathyroidism – disorder of the parathyroid glands.

Amyloidosis – insoluble protein fibers are deposited in tissues and organs, impairing their function.

Renal disease – disease of the kidney.

B.M. Rothschild et al., *Herpetological Osteopathology: Annotated Bibliography of Amphibians and Reptiles*,
DOI 10.1007/978-1-4614-0824-6_3, © Springer Science+Business Media, LLC 2012

Crystal:

Gout – metabolic disease marked by uric acid deposits in the joints.

Calcium pyrophosphate deposition disease – crystals of calcium pyrophosphate dihydrate accumulate in the hyaline articular cartilage or fibrocartilage (pseudogout).

Arthritis – Osteoarthritis – joint disease that mostly affects the cartilage, but which is diagnosed on the basis of diarthrodial joint spurs (osteophytes).

Spondyloarthropathy – a form of inflammatory, erosive arthritis which predominantly produces diarthrodial joint surface erosions and bridging and which causes osseous bridging of vertebral centra.

Neoplasm:

Osteochondroma – tumor that consists of bone and cartilage (neoplasm of the skeleton).

Osteoma – bony outgrowth of membranous bones.

Vertebral – spondyloarthropathy.

Shell/scute disease

Summary of Osseous Pathology in Amphibians and Reptiles

The summary review of turtles is much more extensive than that currently possible for other reptiles and amphibians. The literature on the latter does not lend itself to such organization at this time. It consists mostly of isolated reports, with few epidemiological studies, predominantly representing a delineation of what has been noticed. Much of the discussed reptile analysis is derived from ongoing personal research.

A vocabulary exists for some of the congenital lesions. Except for describing duplication anomalies, it is currently insufficient to allow comparative frequency analyses of osteopathology in amphibians beyond the individual articles. The published literature on non-chelonian reptilian osteopathology often does not contain the information required for application of current criteria for disease recognition. Thus, it is premature to attempt the extensive analysis which turtle studies have allowed.

The sections on amphibians and non-chelonian reptiles are presented to basically identify what information is missing, the information that would be valuable to obtain prospective. Guided by the philosophical advice of Santayana, the purpose of its presentation is to expose the history of recognized macroscopic osteopathology in these groups. So, the sections on amphibians and non-chelonian reptiles, short as they are, appear to represent the state of the art!

Pathology in Contemporary Amphibians

Reported congenital anomalies/pathologies include supernumerary limbs, polydactyly, amelia, brachydactyly, clinodactyly, and syndactyly (Blaustein and Johnson 2003; Bohl 1997; Cavanna 1877; Dubois, 974; Green and Harshbarger 2001). They segregate into isolated cases and environmentally induced outbreaks (e.g., agricultural products) (Alvarez et al. 1995; Amaro and Sena 1968; Burkhart 1998; Guderyan 2006; Kovalenko 2000; Kusrini et al. 2004). Pugener and Maglia (2009) review variation in sacro-urostylic anatomy and its implications (Fig. 6).

There are only isolated reports of injuries, such as fractures (Fig. 1) or traumatic amputations (e.g., by *Erpobdella*, a leech) (Dubois 1977; Korschelt and Stock 1928). Homeobox limb deformities have been attributed to parasitic infection by *Ribeiroia ondatrae* (Dubois 1977). *Ochrobactrum anthropi* was isolated from damaged *Bufo marinus* vertebral endplates by Shilton et al. (2008). Metabolic disease is equally rare,

B.M. Rothschild et al., *Herpetological Osteopathology: Annotated Bibliography of Amphibians and Reptiles*,
DOI 10.1007/978-1-4614-0824-6_4, © Springer Science+Business Media, LLC 2012

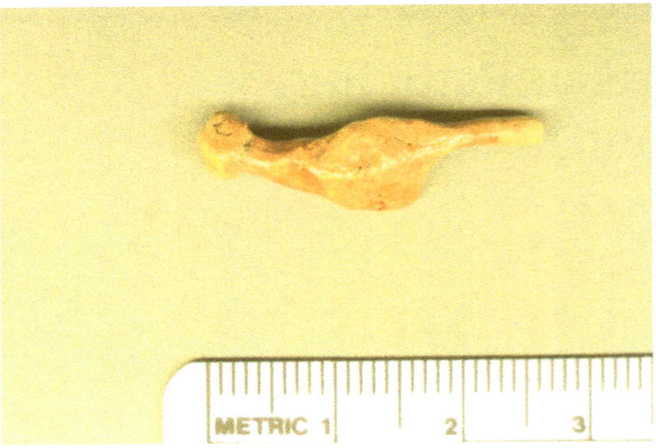

Fig. 1 Displaced fracture in fossil frog BMS F9W-133; middle Holocene, Hiscock Site, Byron, western New York, USA (courtesy of Michael Grenier and Richard Laub)

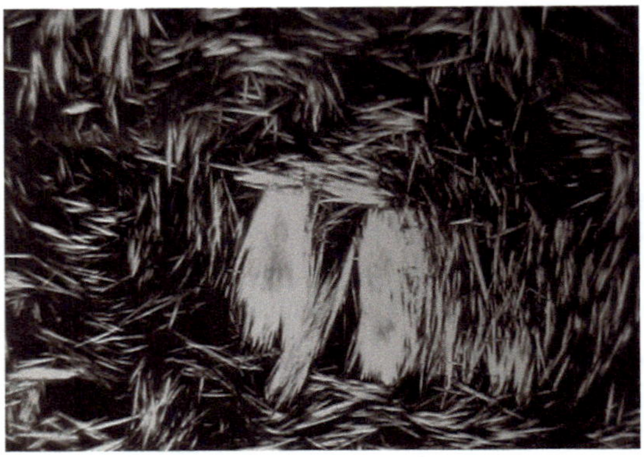

Fig. 2 Polarizing microscopic view of uric acid crystals responsible for gout

with reports limited to from uric acid crystal deposition (Fig. 2) in an African bullfrog *Pyxicephalus* (Crawshaw 1993).

Fusion (of unknown significance) between vertebrae (Fig. 2) has been reported in *Pipa*, *Dactyletra*, *Systoma*, *Ceratophrys*, *Brachycephalus*, *Palaeobatrachus*, *Brachycephalus*, *Rana esculenta*, *Bufo cinereus*, and with the urostyle, in *Bufo variabilis*, *Pipa*, *Dactyletra*, *Systoma*, *Phyllomedusa*, and *Pelobates* (Adolphi 1893, 1895, 1898). Reports of spinal pathology have since been limited to scoliosis (Bacon et al. 2006; Ouellet 2000), with the exception of the observations of Shilton et al. (2008). Their notation of intervertebral space loss in cane toad *Bufo marinus* invading Australia is associated with irregular hyaline cartilage islands and new bone formation, characteristic of neuropathic arthritis (Rothschild and Martin 2006).

Neoplasia is equally rare. Frequent (11.5%) Inkiapo frog *Rana chensinensis* enchondromas (and a single chondrosarcoma) were attributed by Green and Harshbarger (2001) to paper factory waste. Lathyrism-induced chondymyxomas in Pacific chorus frogs *Pseudacris regilla* and *Rana fusca* and chordoma in hybrid toads were attributed to ingestion of sweet pea (*Lathyrus odorata*) by Green and Harshbarger (2001).

Pathology in Extinct Amphibians

Reports in ancient amphibians are limited to congenital (homeobox) phenomenon and isolated fractures. Other pathologies are rarely observed but have not been sufficiently described and analyzed for etiology to be confidently recognized.

Midline skull ossifications or atypical sutures (Table 1) were reported among nasals, frontals, parietals, and postparietals in 4–12% (Gubin and Novikov, 1999). The pineal foramen was absent in *Thoosuchus yakovlevi.* Frontals were laterally expanded in *Thoosuchus yakovlevi* and included in orbit borders in *Platyoposaurus watsoni*, *Konzhukovia vetusta*, *Benthosuchus korobkovi*, and *Melosaurus platyrhinus.*

Reported post-cranial anomalies among extinct amphibians have been limited to the axial skeleton. Sacral anomaly and fusion with adjacent vertebrae or urostyle (Figs. 3 and 4) were noted in Holocene frogs (Böhme 1982) and unilateral sacral wing enlargement, in *Palaeobatrachus*

Table 1 Skull ossification and suture anomalies in fossil amphibians

Apateon	*Eryops megacephalus*	*Osteophorus roemeri*
Aphaneramma	*Gonioglyptus*	*Platyoposaurus*
Batrachosuchoides	*Kestrosaurus*	*Schoenfelderpeton*
Benthosuchus	*Konzhukovia*	*Sclerocephalus*
	Lydekkerina	*Thoosuchus*
Branchierpeton	*Melanerpeton*	*Trematosuchus*
Chelyderpeton	*Micropholis*	*Volgasaurus*
Cochleosaurus	*Mordex calliprepes*	*Wetlugasaurus*
Edops	*Onchiodon*	

[a]Broili (1917); Gubin and Novikov (1999); Gubin et al. (2000)

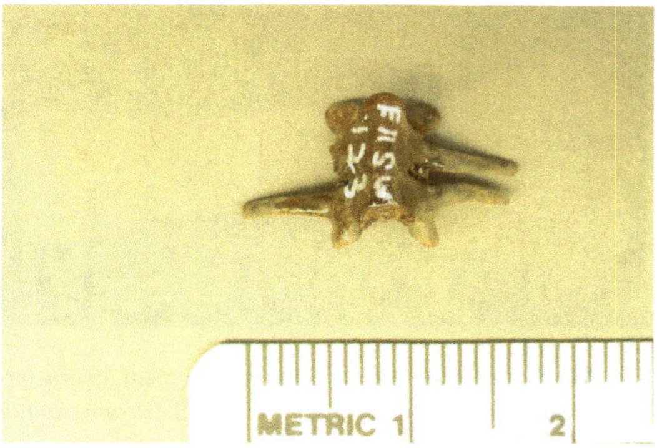

Fig. 3 Congenital fusion of vertebrae in fossil frog BMS E11SW-123; middle Holocene, Hiscock Site, Byron, western New York, USA (courtesy of Michael Grenier and Richard Laub)

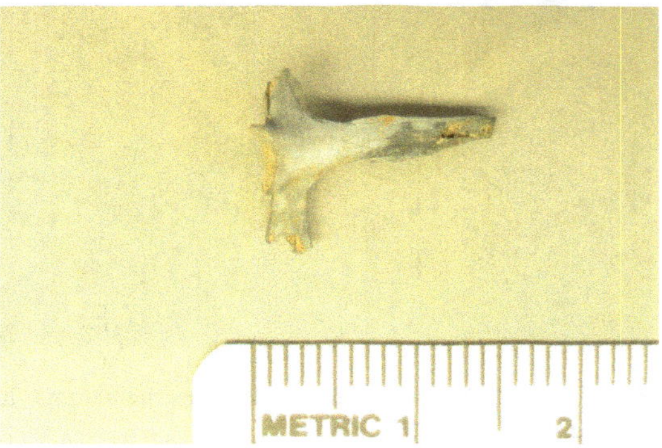

Fig. 4 Pygostyle-vertebral fusion with periosteal reaction along pygostyle in fossil frog BMS F6SW-115; middle Holocene, Hiscock Site, Byron, western New York, USA (courtesy of Michael Grenier and Richard Laub)

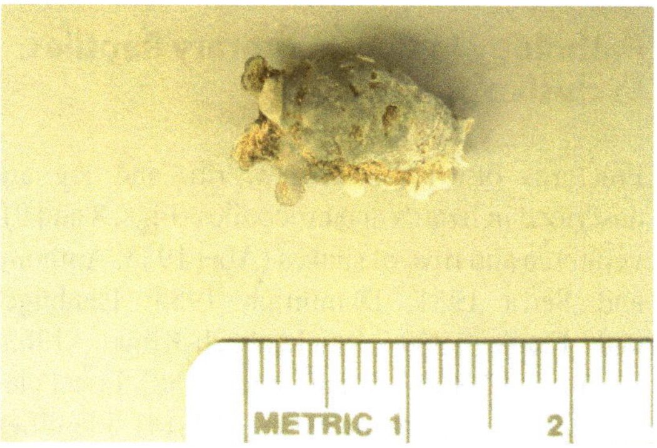

Fig. 5 Infectious vertebral fusion in fossil frog BMS G8SW-182; middle Holocene, Hiscock Site, Byron, western New York, USA (courtesy of Michael Grenier and Richard Laub)

Fig. 6 Congenital anomaly in sacrum of fossil frog BMS F8NE-241; middle Holocene, Hiscock Site, Byron, western New York, USA (courtesy of Michael Grenier and Richard Laub)

luedeckei and *diluvianus* (Špinar 1972). Witzmann (2007) reported a Middle Triassic temnospondyl with a hemivertebrae, restricted dorsal aspect of second pleurocentrum and larger alternate side intercentrum.

Reported femoral and tibial fractures affected Holocene *Rana* sp. and *Bufo* cf. *bufo* (Ippen and Heinrich 1977). The latter were infected (Fig. 5). Fractured, bowed fibulare was fused to tibiale in *Palaeobatrachus* (Špinar 1972). An exostosis which would have hosted a cartilaginous cap (osteochondroma) in a fossil frog is illustrated in Fig. 7.

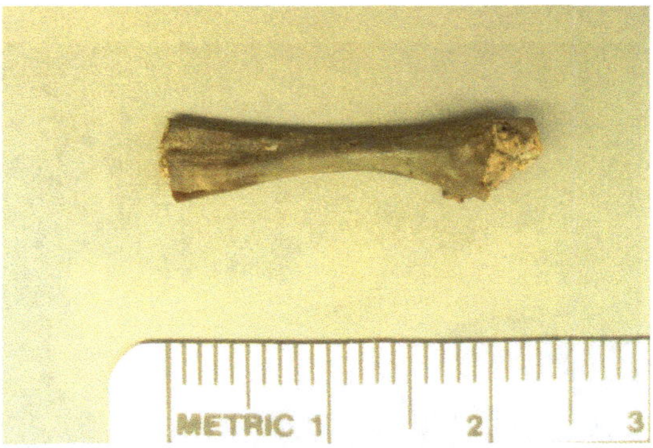

Fig. 7 Osteochondroma in fossil frog BMS F5SE-162; middle Holocene, Hiscock Site, Byron, western New York, USA (courtesy of Michael Grenier and Richard Laub)

Pathology in Contemporary Reptiles, Exclusive of Turtles

Fractures of the lower jaw, rib, and leg are described in lizards and crocodiles (Figs. 8 and 9); vertebrae and ribs, of snakes (Abel 1935; Anthony and Serra 1951; Dämmrich 1985; Isenbügel and Frank 1985; Reichenbach-Klinke 1963; Rothschild 2008; Schlumberger 1958). Focal elevation of periosteum (Figs. 10–14) identifies hematoma secondary to trauma.

Bicephalism is the most commonly reported congenital pathology in snakes and has occasionally been noted in lizards (Acharji 1945; Amaral 1927; Cunningham 1937; Matz 1989; Payen 1991; Wallach 2004). Their congenital origin contrasts with tail duplications, which appear to be post-traumatic events (Blair 1960). Isolated reports of conjoined twins include the slow worm (*Anguis fragilis*), sand lizard (*Lacerta agilis*), and blue-tongued skink (*Tiliqua scincoides*) (Barten 1996; Bellairs 1965; Frye 1991; Hildebrand 1938).

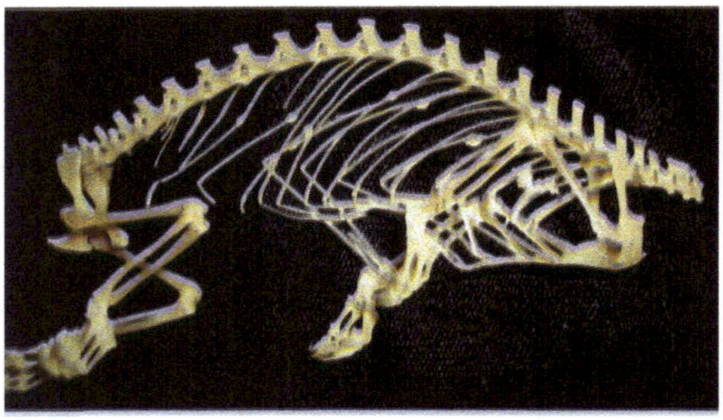

Fig. 8 Lateral view of panther chameleon *Furcifer pardalis* (YPM 13662) skeleton with multiple rib fractures

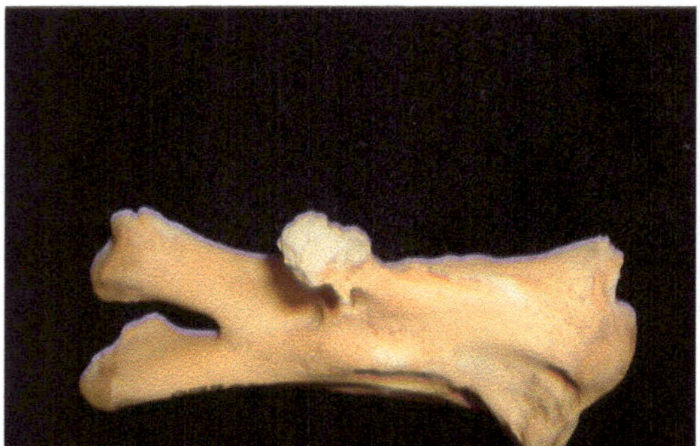

Fig. 9 Healed fracture with exostoses, with apparent cartilage cap, characteristic of osteochondroma; ventral view of tibia and fibula of *Crocodylus niloticus* (NMNH 42150)

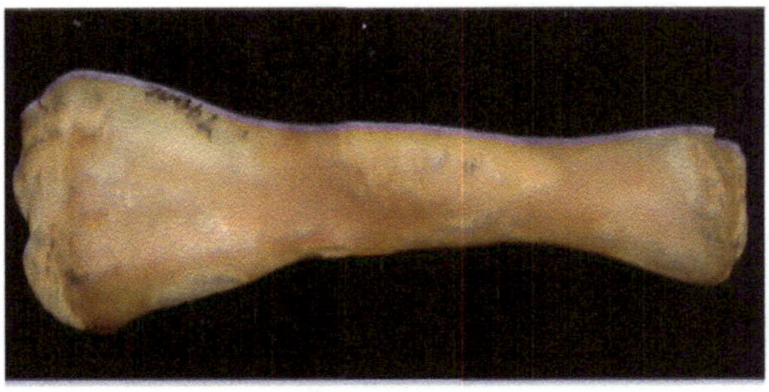

Fig. 10 Large smooth focal bone enlargement, characteristic of hematomas; lateral view of tibia of *Varanus komodoensis* (AMNH 74606)

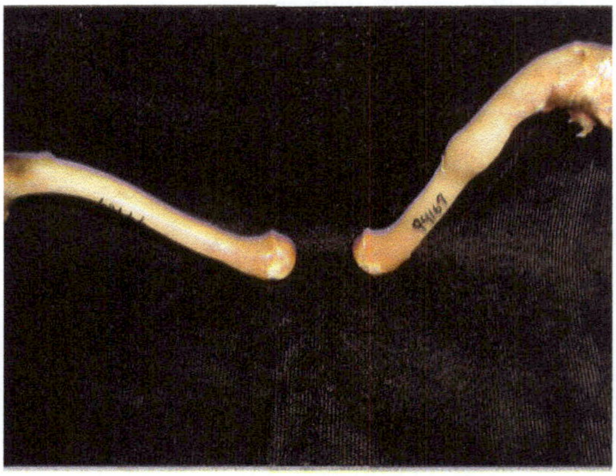

Fig. 11 Isolated bone bump identifies hematoma in *Iguana iguana* (AMNH 94167)

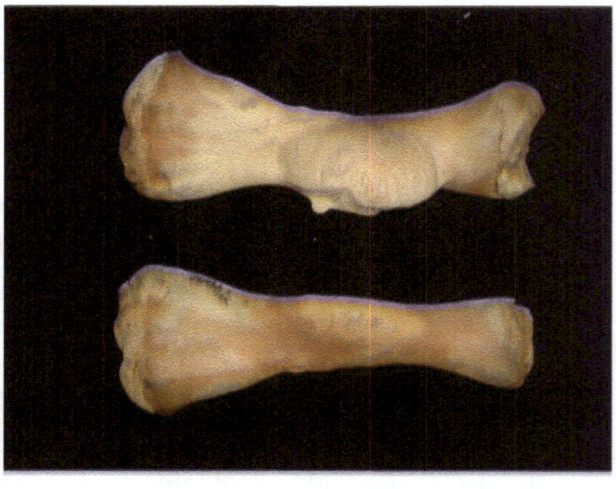

Fig. 12 Hematoma; face view of *Varanus komodiensis* (AMNH 79606)

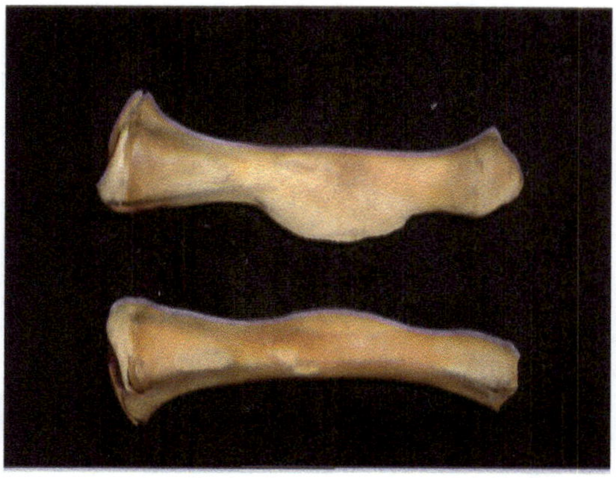

Fig. 13 Hematoma; lateral view of *Varanus komodiensis* (AMNH 79606)

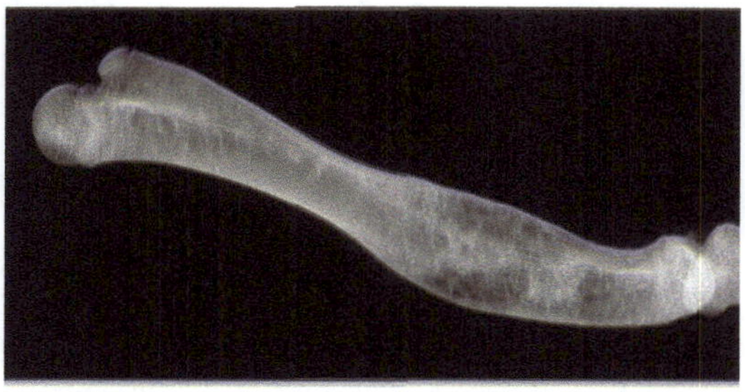

Fig. 14 Lateral x-ray view of *Iguana iguana* ROM 42979 humerus. Periosteal expansion around hematoma

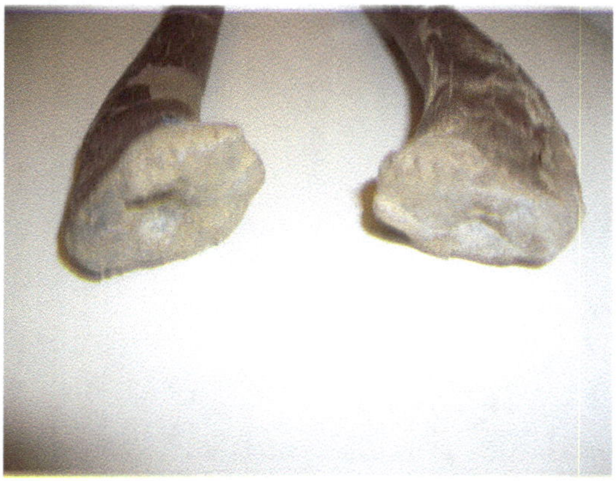

Fig. 15 Divot in articular surfaces, characteristic of osteochondrosis; face view of articular surfaces of *Crocodylus* sp. (NMNH 15r5?82)

Other congenital anomalies (Figs. 11, 15–17) (Table 2) are usually reported as isolated phenomenon (Kälin 1937), but some (e.g., polymely,

Fig. 16 Congenital vertebral fusion producing a spiral pattern; dorsal view of vertebrae of *Iguana iguana* (ROM 42980)

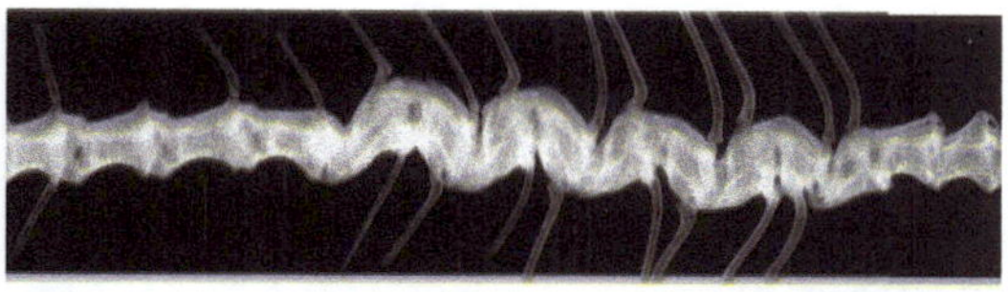

Fig. 17 Dorsal-ventral x-ray view of *Iguana iguana* ROM 42980 vertebral column in Fig. 16. Congenital deformity

Table 2 Anomalies in reptiles[a]

Brachycephaly – New Guinea crocodiles, *Alligator*, *Crocodylus niloticus*, *Crocodylus porosus*
Cleft palate – *C. porosus* *Lacerta lepida*, *Lacerta vivipara*, *Varanus exanthematicus* *Vipera berus*
Protruding mandible – *Crocodylus porosus*, *C. johnsonii*
Snout curvature – gharial
False nostrils – *C. johnsonii*
Tail shortening/absence – *Crocodylus niloticus*, *Osteolaemus tetraspis*
Supernumerary limb – crocodile
Rib/vertebral duplication – *Aelodon priscus*, *Rhacheosaurus gracilis*. *Alligator*, *Crocodylus acutus*, *Heloderma horridum* and *suspectum*
Hemivertebrae – African rock python *Python sebae*
Intercalated half-vertebra[b] – African rock python *Python sebae*, Burmese, Indian or Ceylon *Python molurus*, *Tropidonotus*, *Pelamis platura*, *Gavialis*, *Coluber*, *Diadophis punctatus*, *Natrix natrix*, *Elaphe*, *Lampropeltis*, *Rhadinacea*, and *Thamnophis*
Ectromelia – *Crocodylus moreletii* and caiman

[a] Albrecht (1883); Bateson (1894); Baur (1886, 1891); Bellairs (1965); Huchzermeyer (2003); Kälin (1937); King (1959)
[b] With ankylosis and variable rib duplication

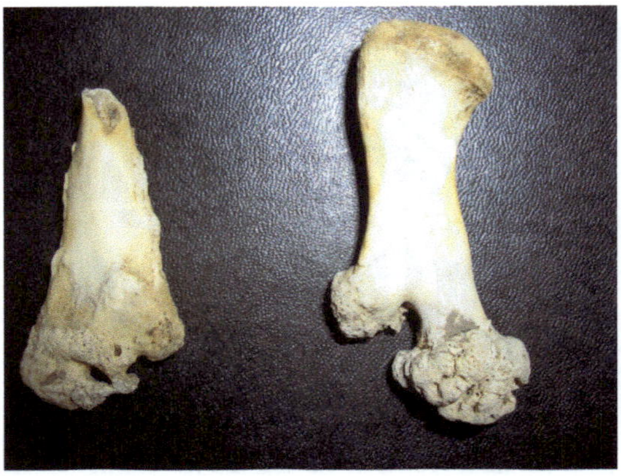

Fig. 18 Destructive changes with new bone formation, characteristic of infection; ventral view of femur and tibia of *Alligator mississippiensis* (UFMNH 34788)

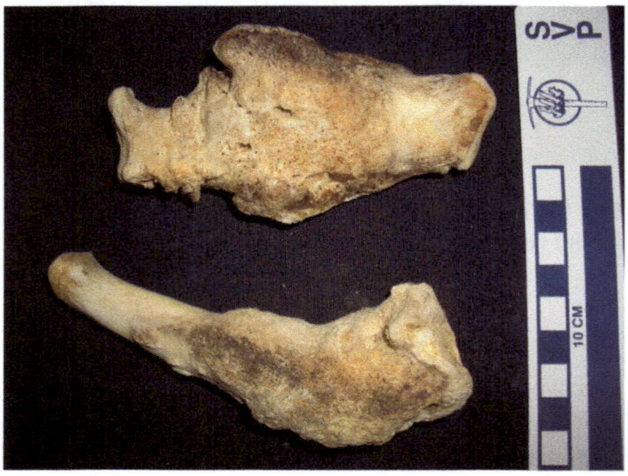

Fig. 19 Gross bone enlargement and distortion with filigree reaction by infectious process; lateral view of limb bones of *Alligator mississippiensis* (KU 19538)

ectomely, and ectodactyly) have been attributed to environmental factors (Johnson et al. 2003; Ouellet et al. 1997; Rostand 1951, 1960).

Vertebral osteomyelitis is the most common infection in snakes (Ippen 1965; Isaza et al. 2000). Multifocal granulomatous spondylitis was noted in the hybrid *Crotalus mitchellii/willardi* (Frye and Kiel 1985). Infection, although less common, is often multifocal (Figs. 18–28) in

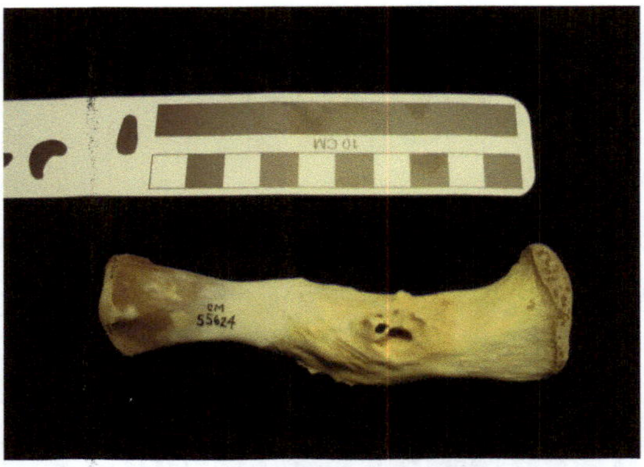

Fig. 20 Draining sinus of osteomyelitis; lateral view of tibia of black caiman *Melanosuchus niger* (CM 55624)

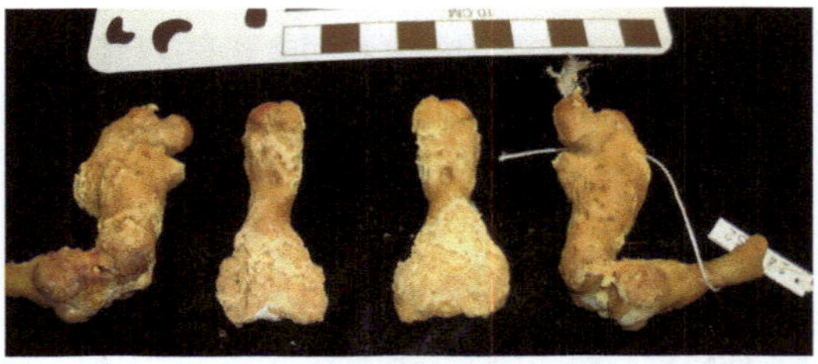

Fig. 21 Surface disruption characteristic of infection; anterior view of femora and humeri of *Varanus exanthematicus* (CM 14052)

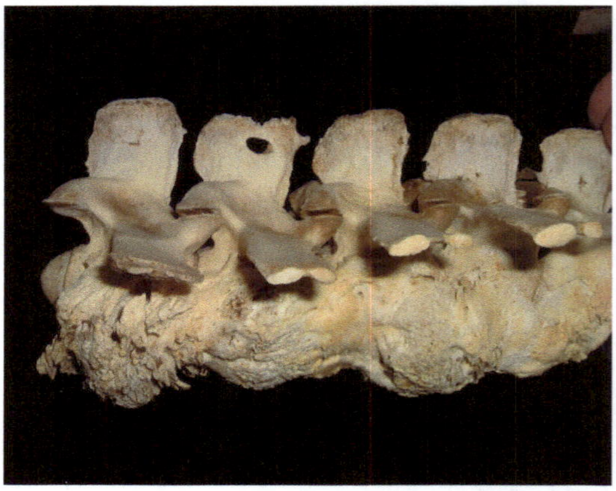

Fig. 22 Reactive new bone with draining sinuses, characteristic of infectious spondylitis; lateral view of vertebrae of *Alligator mississippiensis* (KU 19538)

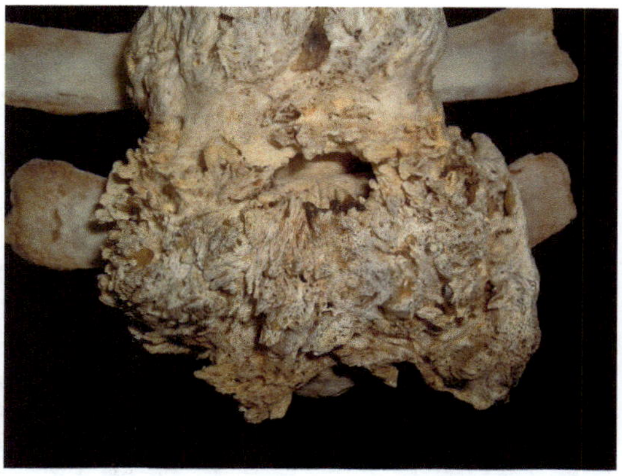

Fig. 23 Reactive new bone with draining sinuses, characteristic of infectious spondylitis; ventral view of vertebrae of *Alligator mississippiensis* (KU 19538)

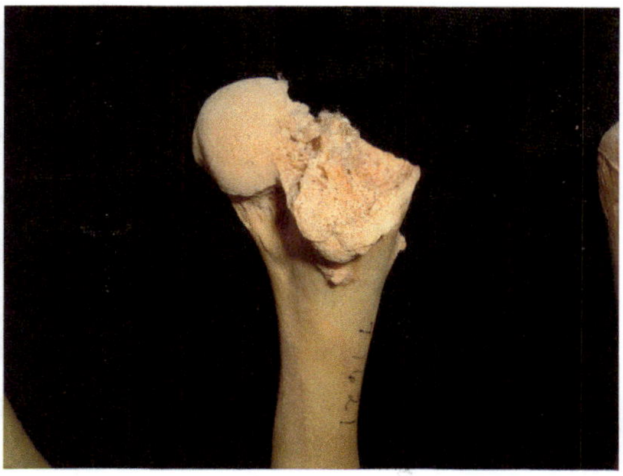

Fig. 24 Severe destructive changes characteristic of infection; lateral view of proximal femur of *Alligator mississipiensis* (AMNH 71621)

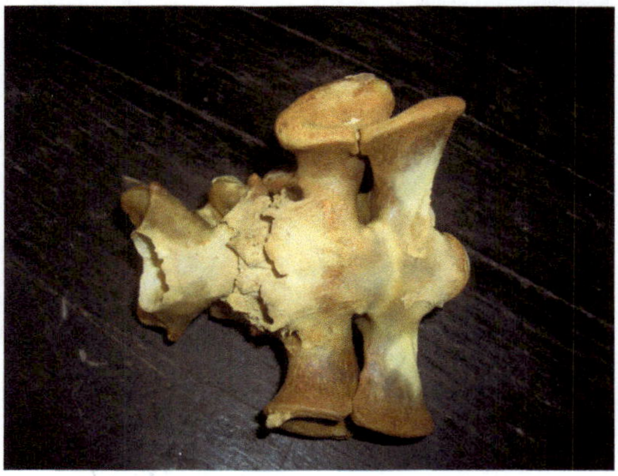

Fig. 25 Bony overgrowth around infectious process; ventral view of sacrum and lumbar vertebrae of *Varanus komodoensis* (UMMZ 236581, with 11,876 = Field number)

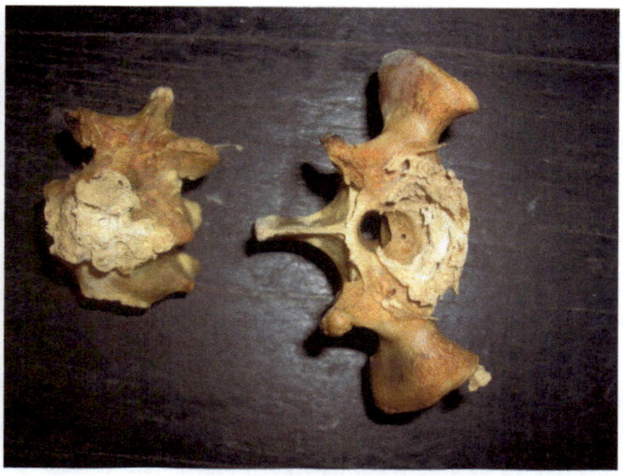

Fig. 26 Destruction by infectious process; en face view of sacrum and lumbar vertebra of *Varanus komodoensis* (UMMZ 236581)

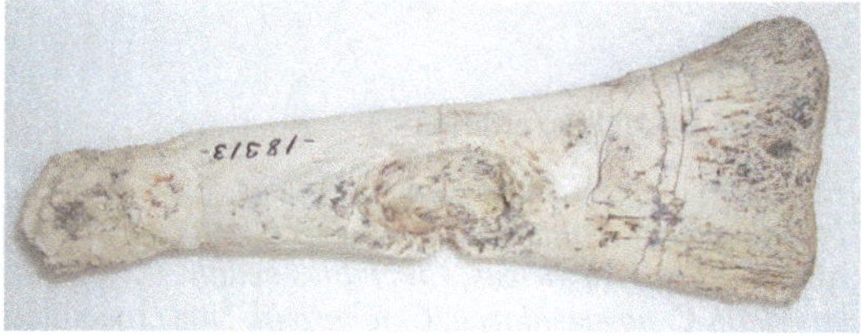

Fig. 27 Surface defect probable from draining sinus; ventral view of phytosaurid (NMNH 18313)

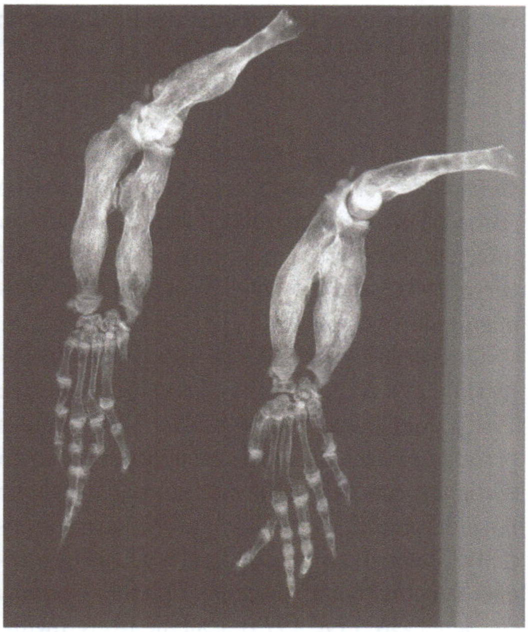

Fig. 28 Oblique x-ray view of Gray's water skink *Tropidophorus grayi* ROM 42982 upper extremity. Disorganized architecture reveals osteomyelitis (Figs. 18–27)

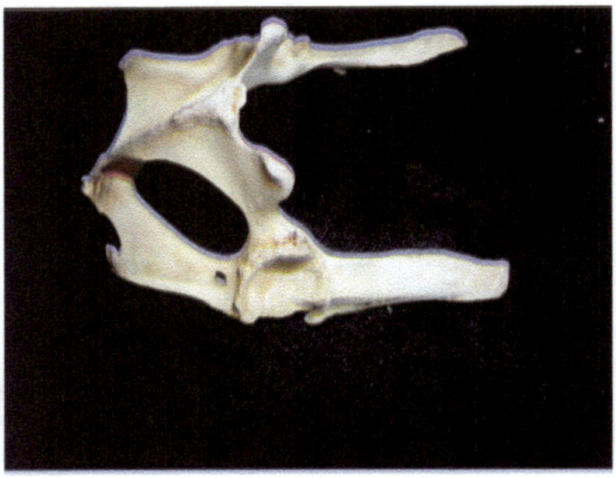

Fig. 29 Overgrowth of bone (osteophytes) around left acetabular joint, characteristic of osteoarthritis; ventral-oblique view of pelvis of *Varanus salvator* (UFMNH 33136)

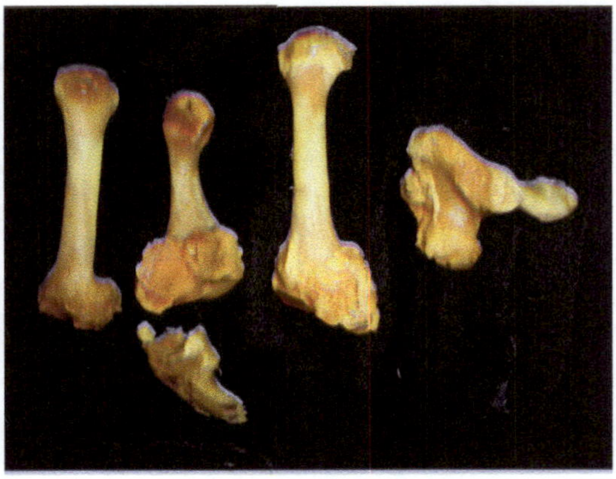

Fig. 30 Calcium deposits on and about joint surfaces; lateral view of metacarpals and carpals of *Varanus komodoensis* (YPM 11626)

Table 3 Crystalline arthritis

Articular Gout – *Varanus exanthematicus, Tupinambis, Iguana iguana, Caiman sclerops, spectacled caiman, Alligator mississippiensis, Crocodilus acutus, C. moreletii, C. novaeguineae, C. johnsonii,* Nile crocodiles, Indo-Pacific crocodile, false gharial and gharial, *Tomistoma schlegelii, Gavialis gangeticus*
Pseudogout – Egyptian spiny-tailed lizard *Uromastyx* – *Varanus komodiensis*

[a] Appleby and Siller (1960); Barten (1996); Frank (1965); Frye (1976, 1984); Huchzermeyer (2003); López del Castillo (1998); Rothschild (2009)

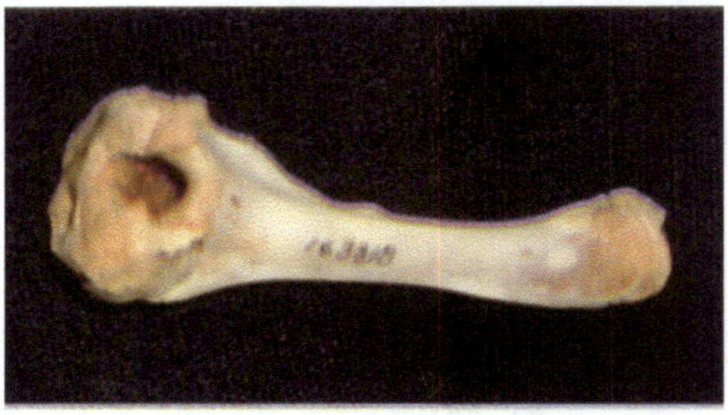

Fig. 31 Ventral view of *Varanus salvator* (NMNH 163810). Hole in bone cause by gout

other reptiles (e.g., *Alligator mississippiensis, Caiman latirostris,* and *Crocodylus siamensis*) (Brown 1971). Other isolated reports include an infected shoulder in *Alligator mississippiensis* (Frye 1991) and *Erysipelothrix* in *Crocodylus acutus* (Jacobson 2007).

Osteoarthritis is essentially, a disease of captivity (Fig. 29), and extremely rare in the wild (Frye 1991). Metabolic disease (Table 3) in reptiles presents as articular gout, pseudogout (calcium pyrophosphate deposition disease), and metabolic bone disease (Figs. 30 and 31). The latter, essentially limited to captive reptiles (Frye 1991), presents as pathologic fractures and folding fractures (Barten 1996; Frye 1973). Calcific overgrowths, noted in boa constrictors, rattlesnakes, a Burmese python (*Python molurus bivittatus*) (Frye 1991), and in a varanid (Figs. 32–39), are of unclear diagnosis.

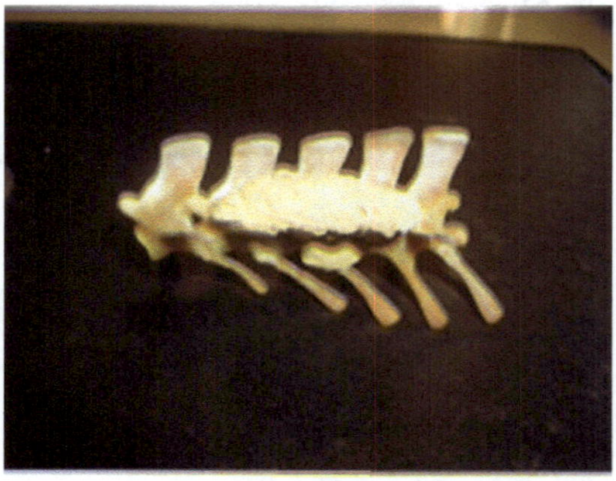

Fig. 32 Mass of bony tissue adjacent to ribs; lateral view of string of vertebrae of *Varanus dorianus* (UMMZ 238879)

While they consist of wormian bone, the diagnosis of osteitis deformans or Paget's disease has been discarded (Jacobson 2007).

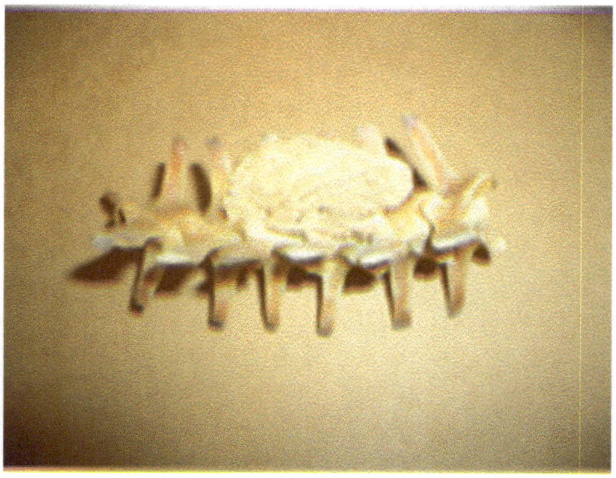

Fig. 33 Mass of bony tissue adjacent to ribs; dorsal view of vertebrae of *Varanus dorianus* (UMMZ 238879)

Fig. 34 Mass of bony tissue adjacent to ribs, with varying degrees of separation; computerized tomographic view of vertebrae of *Varanus dorianus* (UMMZ 238879)

Fig. 35 Mass of bony tissue adjacent to ribs, with varying degrees of separation; computerized tomographic view of vertebrae of *Varanus dorianus* (UMMZ 238879)

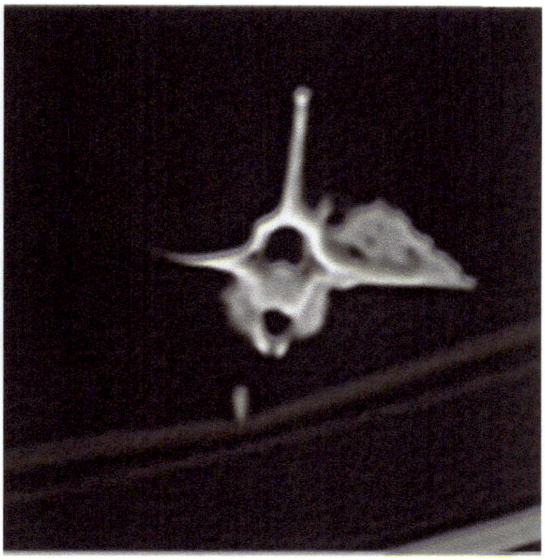

Fig. 36 Mass of bony tissue adjacent to ribs, with varying degrees of separation; computerized tomographic view of vertebrae of *Varanus dorianus* (UMMZ 238879)

Examination of a number of reports of exostoses in *Iguana iguana* (Chiodini and Nielsen 1983) or Paget's disease in a boa constrictor, *Bitis nasicornis*, *Crotalus*, and southern copperheads (Isenbügel and Frank 1985) has revealed a new diagnosis of the inflammatory arthritis of the spondyloarthropathy variety (Rothschild 2009) and infection. Diffuse periosteal reaction may represent infection or hypertrophic osteoarthropathy (Figs. 40 and 41). A form of inflammatory arthritis associated with vertebral fusion and peripheral and axial joint erosions with reactive new bone and joint fusion (Figs. 42–53) has been

Fig. 37 Mass of bony tissue adjacent to ribs, with varying degrees of separation; computerized tomographic view of vertebrae of *Varanus dorianus* (UMMZ 238879)

Fig. 38 Mass of bony tissue adjacent to ribs with varying degrees of separation; computerized tomographic vertebrae view of *Varanus dorianus* (UMMZ 238879)

documented in 2–20% of *Varanus*, dependent upon species (Rothschild 2009), representing the disease category termed spondyloarthropathy.

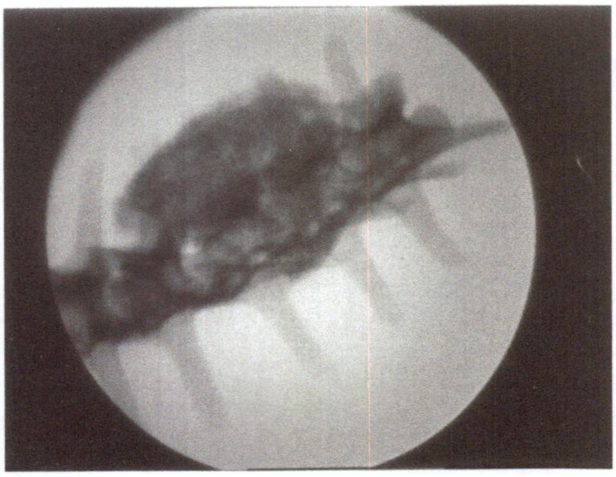

Fig. 39 Dorsal-ventral x-ray view of *Varanus dorianus* (UMMZ 238879) vertebrae. Mass of bony tissue adjacent to ribs. Disruption of adjacent internal vertebral centra architecture, characteristic of osteomyelitis

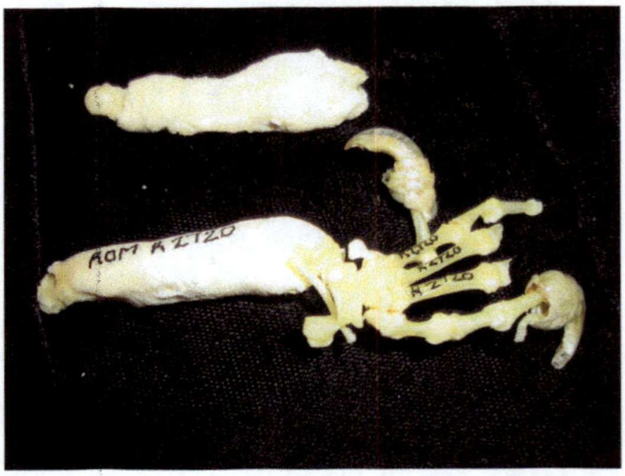

Fig. 40 Diffuse periosteal reaction; ventral (*upper*) and dorsal (*lower*) view of distal limbs of *Iguana iguana* (ROM R2720)

A crystal arthritis, related to deposition of calcium, has been noted in varanids (Figs. 54–61) (Rothschild 2010).

Neoplasia has been directly observed in the form of benign growths such as enchondroma, osteoma (Fig. 63), osteochondroma (Figs. 9, 63–65), and chondro-osteofibromas and malignant growths (Fig. 66), as well as indirectly identified because of their induction of hypertrophic osteoarthropathy (Figs. 41 and 67) (Table 4).

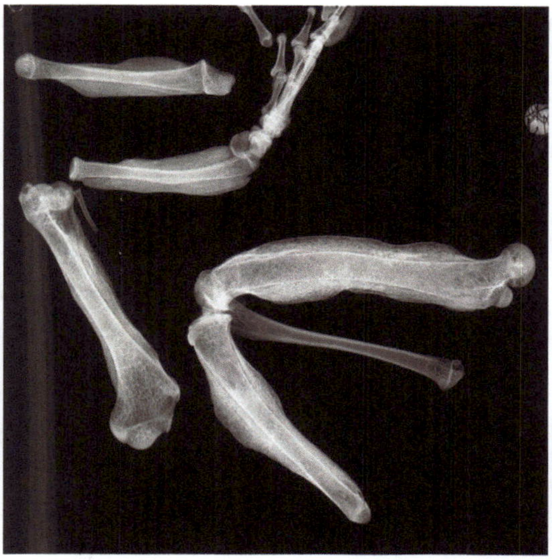

Fig. 41 Lateral x-ray view of *Iguana iguana* ROM 42981 extremities. Generalized periosteal reaction from hypertrophic osteoarthropathy

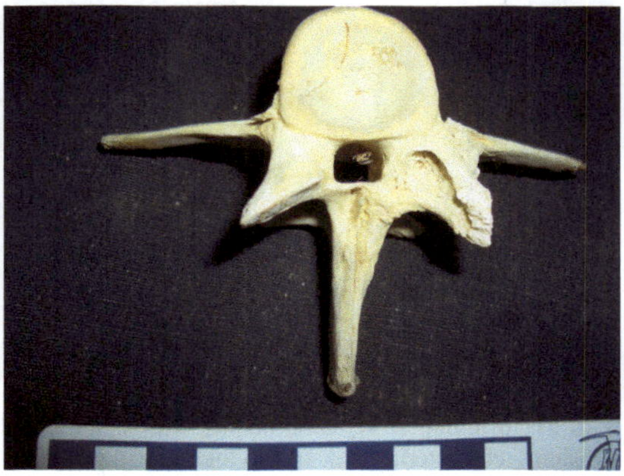

Fig. 42 Destruction of zygapophyseal (facet) joint by erosive process; anterior view of vertebra of *Alligator mississippiensis* (KU 19538)

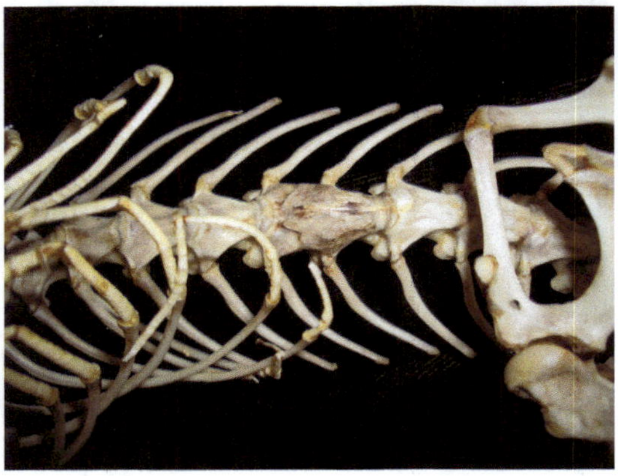

Fig. 43 Smooth vertebral bridging by syndesmophyte, characteristic of spondyloarthropathy; ventral view of caiman lizard *Dracaena guianensis* (YPM 10224) skeleton

Fig. 44 Smooth vertebral bridging by syndesmophyte, characteristic of spondyloarthropathy; lateral view of caiman lizard *Dracaera guianensis* (YPM 10224) skeleton

Fig. 45 Smooth bridging of vertebral centra, characteristic of spondyloarthropathy; ventral view of thoracic vertebrae of *Varanus albigularis* (YPM R 13717)

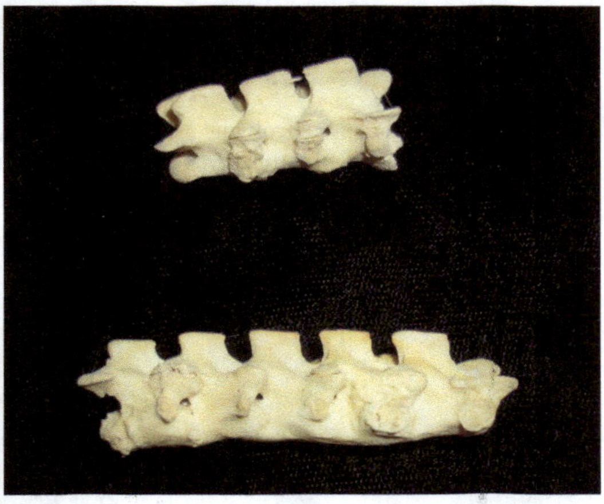

Fig. 46 Smooth bridging of vertebral centra, characteristic of spondyloarthropathy; lateral view of thoracic vertebrae of *Varanus albigularis* (YPM R 13717)

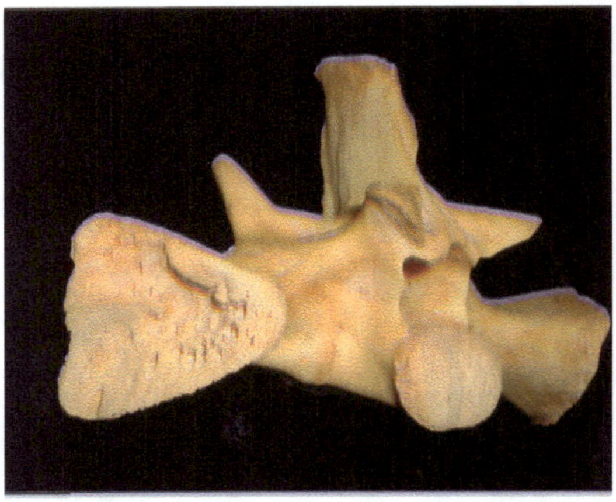

Fig. 47 Erosive process characteristic of spondyloarthropathy; face view of facet joint of *Varanus olivaceus* (UMMZ 218977)

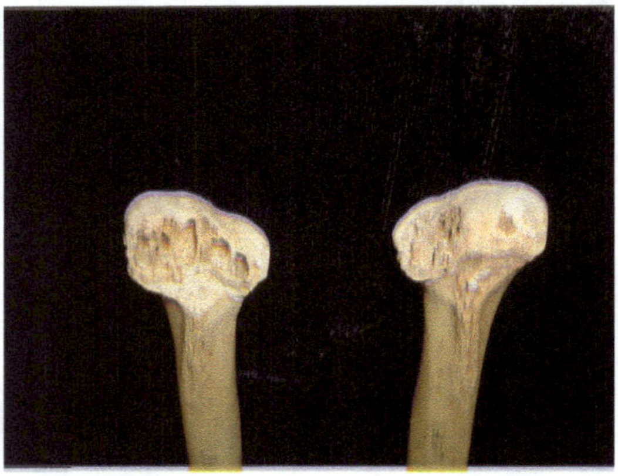

Fig. 48 Erosive process involving subchondral bone, characteristic of spondyloarthropathy; ventral view of two femora of *Varanus olivaceus* (UMMZ 218977)

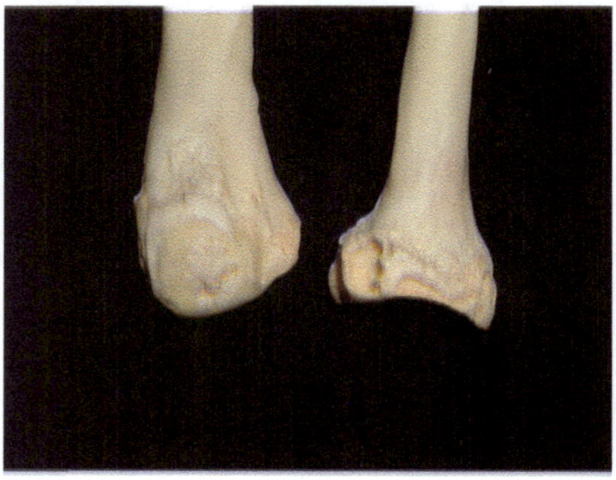

Fig. 49 Subchondral and large marginal erosions with reactive new bone formation, characteristic of spondyloarthropathy; lateral view of distal femur and distal tibia of *Crocodilus acutus* (AMNH 7121)

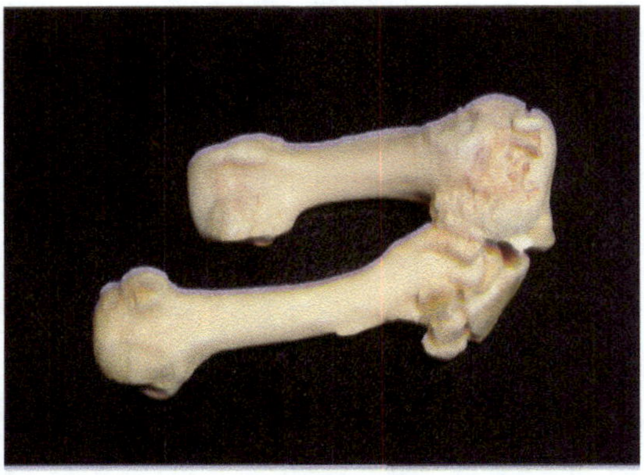

Fig. 50 Subchondral erosions characteristic of spondyloarthropathy; superior-ventral view of metatarsals of *Varanus komodoensis* (ROM 7565)

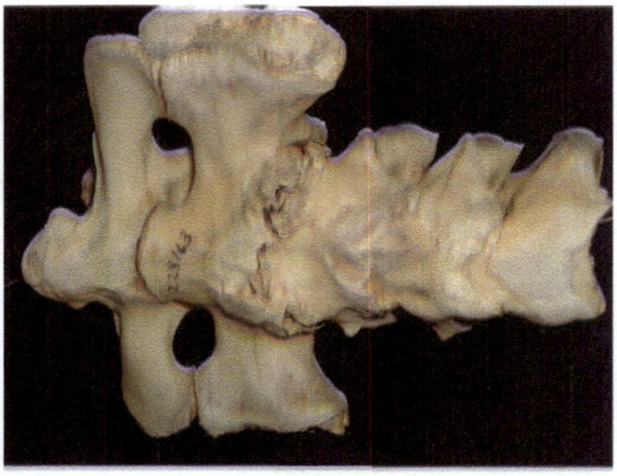

Fig. 51 Bridging of vertebral bodies, with subtle sacral erosions, characteristic of spondyloarthropathy; ventral view of vertebrae and sacrum of *Varanus komodoensis* (NMNH 228163)

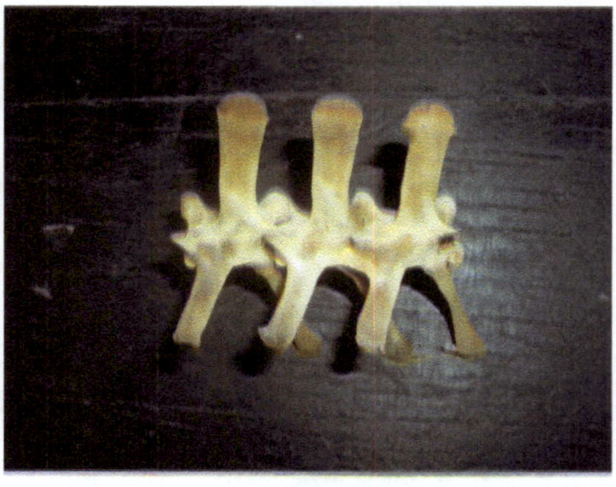

Fig. 52 Dorsal view of *Varanus komodoensis* (UMMZ 236581) vertebrae. Zygapophyseal joint fusion characteristic of spondyloarthropathy

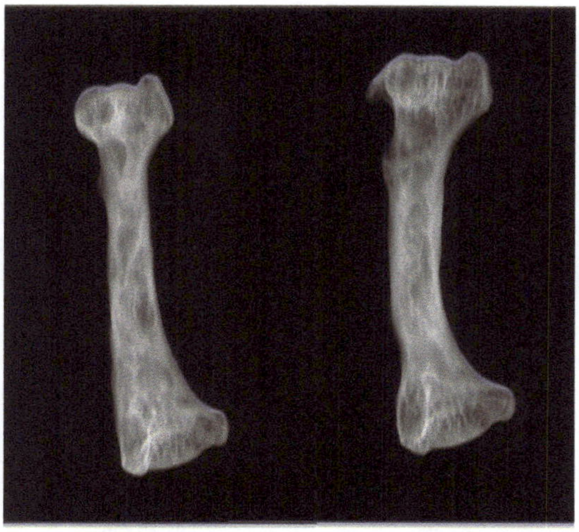

Fig. 53 Dorsal-ventral x-ray view of stump-tailed skink *Trachydosaurus rugosus* ROM 23461 humerus. Erosion with sclerotic margin

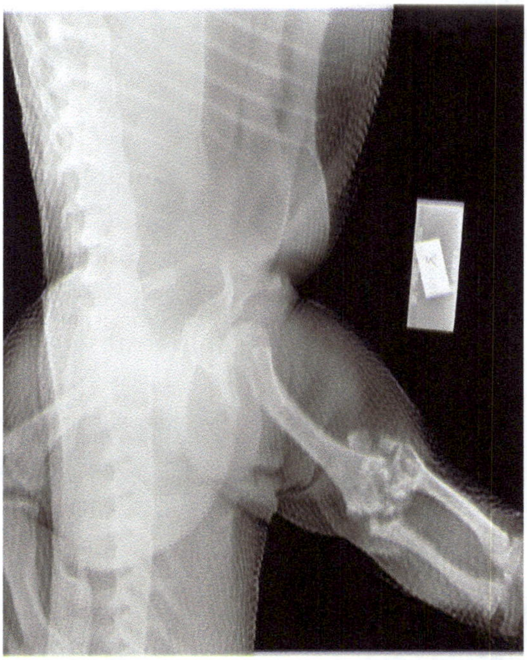

Fig. 54 Articular crumbling with calcium deposits, identifying calcium pyrophosphate deposition disease; dorsoventral x-ray view of right knee of *Varanus komodoensis* (Sedgwick County Zoo 6668)

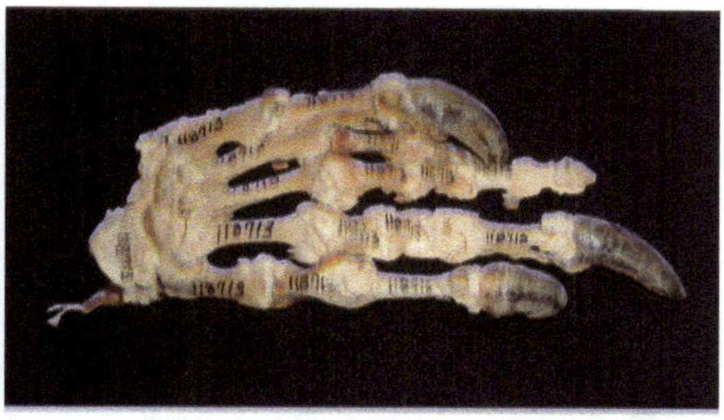

Fig. 55 Calcium deposition around joints, characteristic of calcium pyrophosphate deposition disease; dorsal view of manus of *Varanus bengalensis* (AMNH 118713)

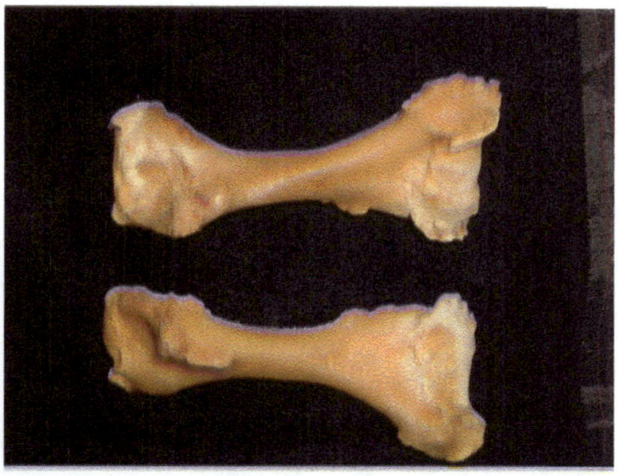

Fig. 56 Calcium deposition around joints, characteristic of calcium pyrophosphate deposition disease; posterior view of humeri of *Varanus olivaceus* (UMMZ 218977)

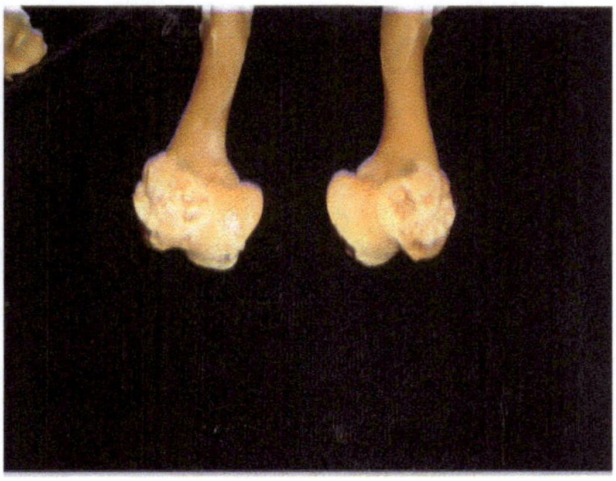

Fig. 57 Ill-defined, crumbling articular surface erosions with calcium deposition, characteristic of calcium pyrophosphate deposition disease; posterior view of distal ends of humeri of *Varanus olivaceus* (UMMZ 218977)

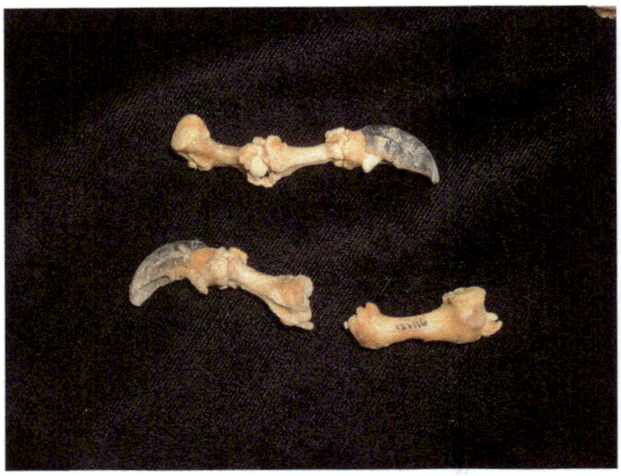

Fig. 58 Calcium deposition around joints, characteristic of calcium pyrophosphate deposition disease; lateral view of digits of *Varanus niloticus* (AMNH R 137116)

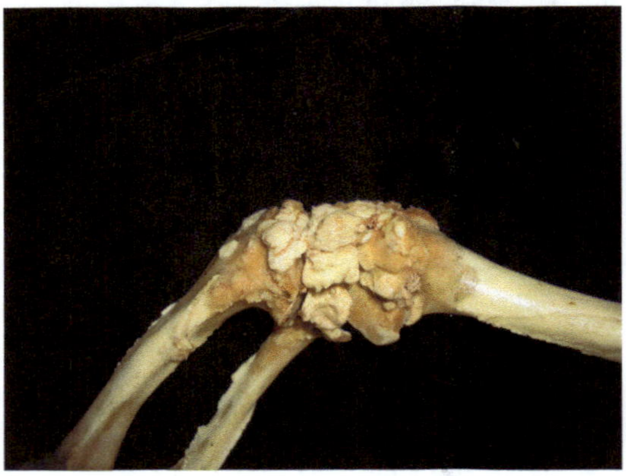

Fig. 59 Calcium deposition around joints, characteristic of calcium pyrophosphate deposition disease; lateral view of the knee of *Varanus salvator* (ROM 23468)

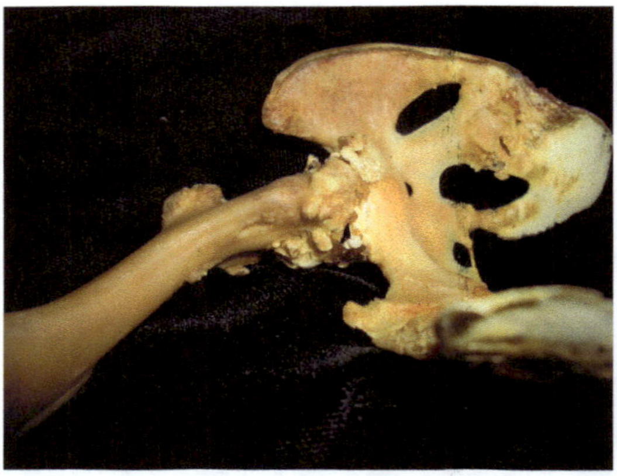

Fig. 60 Calcium deposition around joints, characteristic of calcium pyrophosphate deposition disease; inferior-oblique view of pelvis and femur of *Varanus salvator* (ROM 23468)

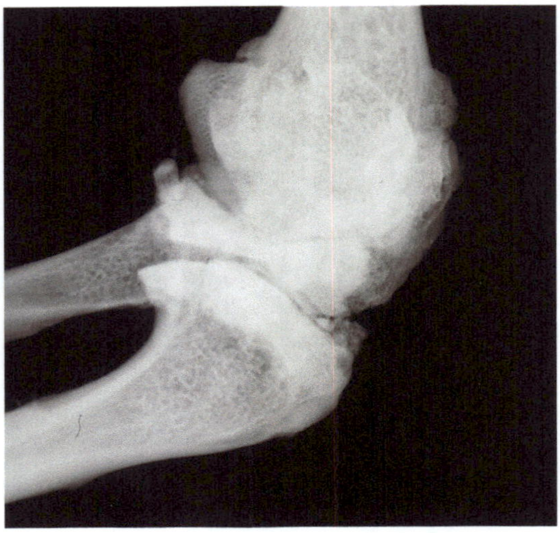

Fig. 61 Lateral x-ray view of *Varanus salvator* ROM 23468 knee. Cartilage calcification from calcium pyrophosphate deposition disease

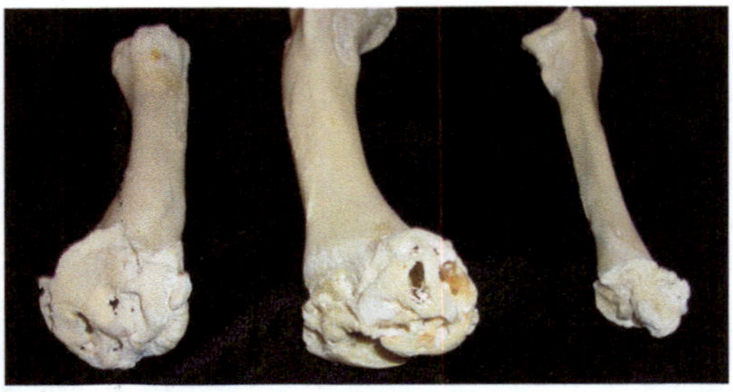

Fig. 62 Ill-defined erosions on subchondral surface with subchondral geodes; oblique view of proximal portion of metacarpals of *Varanus komodoensis* (ROM 7565) (Figs. 30 and 31, and to 54–62)

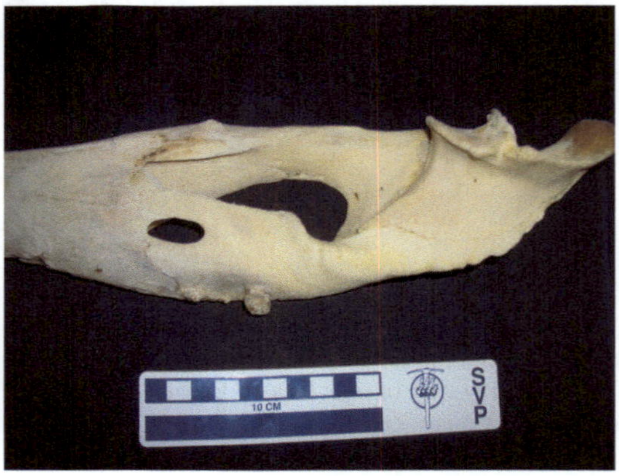

Fig. 63 Small bone overgrowth identifies osteoma; medial view of right mandible of *Alligator mississippiensis* (KU 19538)

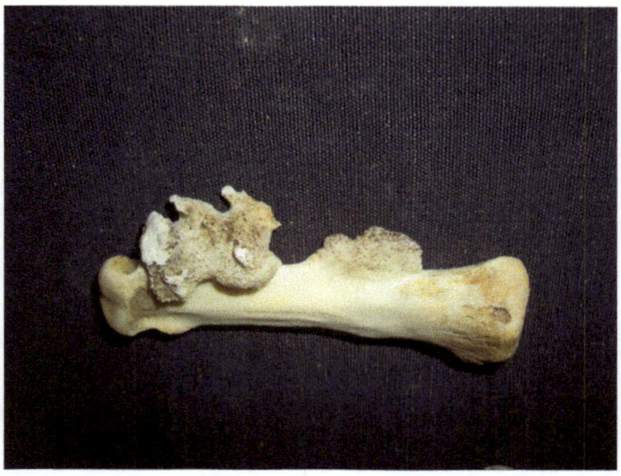

Fig. 64 Exostotic bone overgrowth with cartilage cap (osteochondroma); oblique view of metatarsal of *Alligator mississippiensis* (KU 19538) (Figs. 9, 63–65)

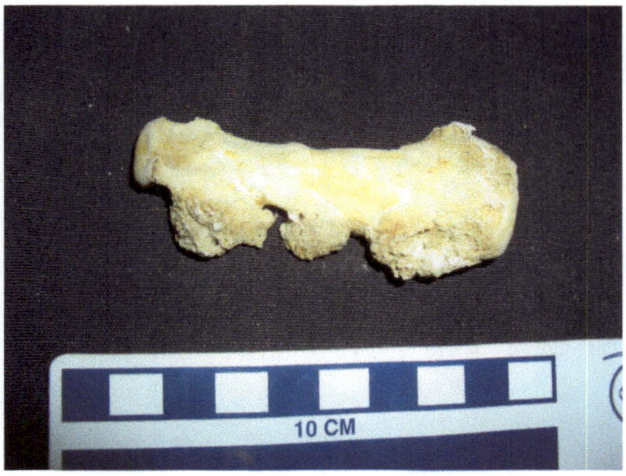

Fig. 65 Exostotic bone overgrowth with cartilage cap (osteochondroma); medial view of metatarsal of *Alligator mississippiensis* (KU 19538)

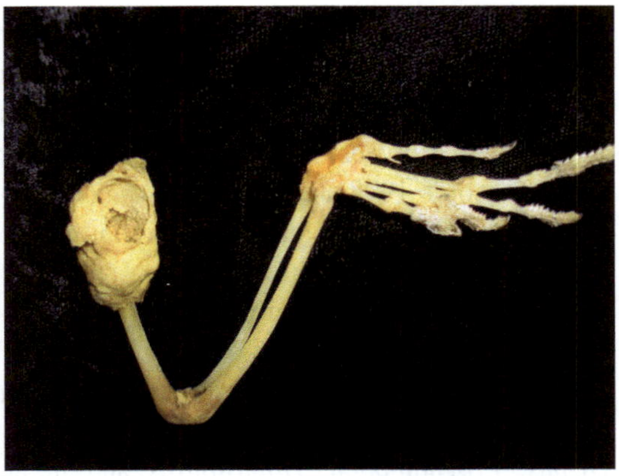

Fig. 66 Large proximal humeral mass with large spherical proximal defect; lateral view of lower extremity of *Hypsilurus dilophus* (YPM R 11404)

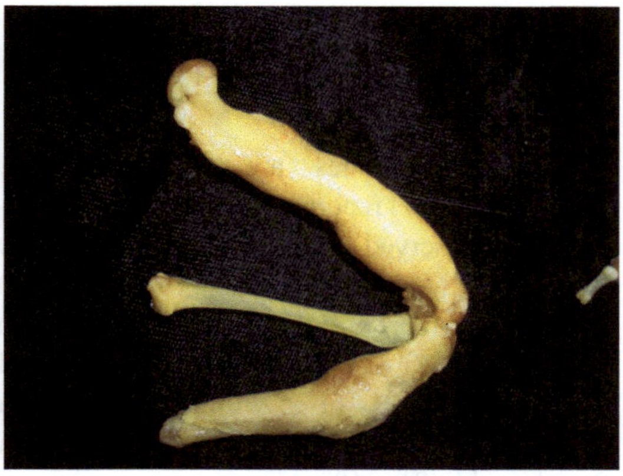

Fig. 67 Diffuse bone enlargement by hypertrophic osteoarthropathy; lateral view of *Iguana iguana* (ROM 42981) lower extremity

Table 4 Neoplasia in reptiles[a]

Benign	
Enchondroma	*Varanus dracaena*, *Varanus indicus*
Osteoma	*Varanus salvator*, *Lacerta viridis*, *Uromastyx hardwickii*
Osteochondroma	*Varanus bengalensis*, *V. niloticus*, *Iguana iguana*, *Cyclura nubila*, *Uromastyx hardwicki*, *Naja nigricollis*
Chondro-osteofibroma	*Cyclura cornuta*, *Lacerta viridis*, African rock python *Python sebae*
Malignant	
Chondrosarcoma	*Elaphe guttata*
Osteosarcoma	*Lampropeltis zonata*, *Molurus bivittatus*, *Rhamphiophis rostratus*, *Lacerta viridis*, *Iguana iguana*
Osteochondrosarcoma	*Naja nigricollis*
Metastatic	
Melanoma	*Pituophis melanoleucus*
Squamous cell	*Tupinambis teguixin*
Hypertrophic	
osteoarthropathy	*Iguana iguana*

[a] Ball et al. (1999); Dawe et al. (1980); Done (1996); Frye and Dutra (1973); Ippen (165,1985); Jacobson (1980); Latimer et al. (2000)

Extinct Reptile Pathology, Exclusive of Turtles and Dinosaurs

Congenital anomalies are rarely reported. Witzmann (2007) reported a congenital hemivertebra was observed in a Jurassic *Cimoliasaurus plicatus* (Witzmann 2007), and Hopley (2001) reported Schmorl's nodes (vertebral endplate holes) in a plesiosaur. MacGregor (1906) noted that the odontoid is fused with axis centrum in Phytosauria. Tooth malposition is present in Pseudosuchia (Fig. 68).

Injuries have been reported predominately as isolated observations. These include fractures and exostoses (Figs. 69–71). The latter are found especially on lower jaws, vertebrae (especially caudal), and paddles (especially digits). Williston (1898) reported distal caudal vertebral fractures with pseudoarthroses. The exception is Sawyer and Erickson's 1998 study of *Leidyosuchus formidabilis*, which documented gastralia as the most common fracture site.

Decompression syndrome (bends) is universally recognized on the basis of linear patterns of bone loss or articular surface subsidence (Fig. 72) in *Platecarpus*, *Tylosaurus*, *Mosasaurus*, *Plioplatecarpus*, *Prognathodon*, *Hainosaurus* and in an Antarctic mosasaur, and invariably absent from *Clidastes*, *Ectenosaurus*, *Globidens*, *Halisaurus*, and *Kolposaurus* (Rothschild and Martin 1987,

Fig. 68 Malignment of teeth; dorsal view of mandible of Pseudosuchia

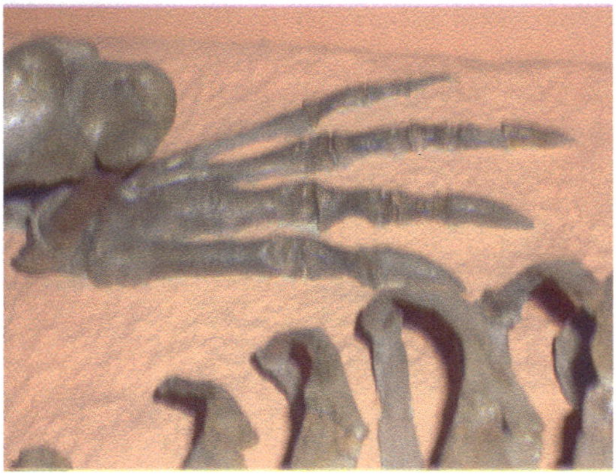

Fig. 69 Healed second metatarsal; dorsal view of hindfoot of *Crocodylus clarus* (NMNH 12719)

Fig. 70 Fractured neural spines; lateral view of Permian *Dimetrodon grandis* (NMNH 8635)

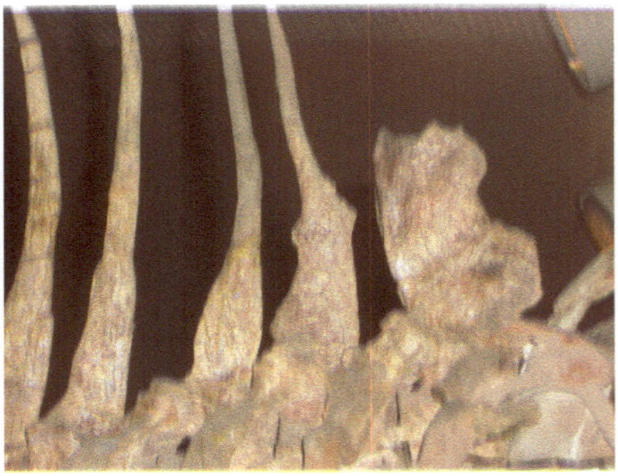

Fig. 71 Lateral view of Permian *Dimetrodon grandis* (NMNH 8635) spinous processes. Healed fracture with callus formation

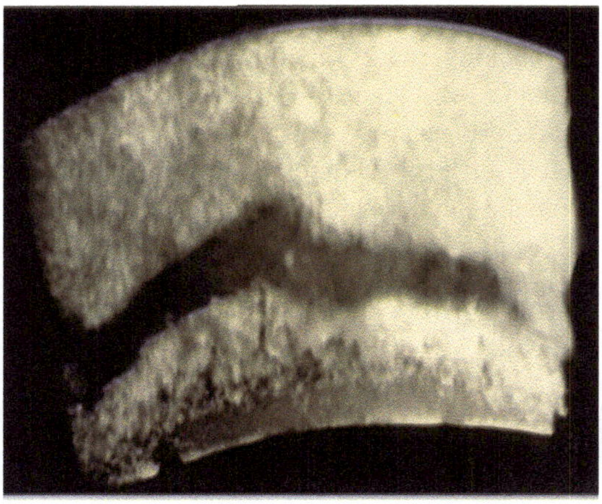

Fig. 72 Radiolucent zone transects vertebra, identifying avascular necrosis; dorsoventral x-ray view of Late Cretaceous *Platecarpus* (KU 152143) vertebra

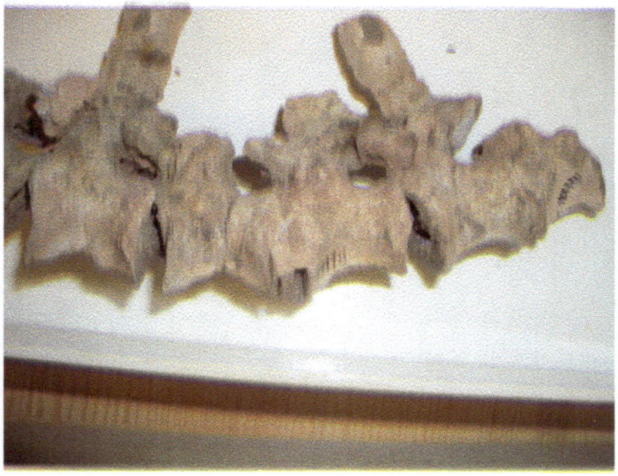

Fig. 73 Syndesmophyte bridging the centermost vertebrae, characteristic of spondyloarthropathy; lateral view of vertebrae of Early Permian *Ctenorhachis jacksoni* (NMNH 437711)

2006). With the exception of the large percentage of the skeleton afflicted in *Platecarpus*, the frequencies of vertebral involvement in the other affected genera were indistinguishable ($p < 0.02$). Among the basal portion of the family Sauropterygia, avascular necrosis was found in only a single genus, *Neusticosaurus* (Rothschild and Storrs 2003). It is seen in both early and late Plesiosauridae, Elasmosauridae. Large intergeneric variation in frequency was noted among ichthyosaurs (Montani and Rothschild, 1999). *Ophthalmosaurus*, *Ichthyosaurus*, *Temnodontosaurus*, and *Platypterygius* equally affected; *Leptonectes*, less so (chi-square = 5.36, $p < 0.025$) (Montani and Rothschild 1999), while *Stenopterygius* was spared.

Infections are noted, predominantly as isolated phenomenon (Fig. 27). Auer (1909) suggested that changes in a *Metriorhynchus* cf. *moreli* femur represented fungal infection.

The only report of osteoarthritis in a fossil reptile was in the wrist and metacarpal phalangeal joints of pterodactyloid pterosaurs (Bennett 2003). Spondyloarthropathy has been observed in *Dimetrodon*, *Diadectes*, and *Ctenorhachis* (Figs. 73–75).

Moodie's 1918 diagnosis of an osteoma in a mosasaur has been reassessed as another form of benign tumor, a hamartoma. The latter is actually overgrowth of normal tissue and not truly a neoplasia. Sawey and Erickson (1998) are credited with the first report of an actual osteoma and also

Fig. 74 Bridging of vertebral centra and facet joints, characteristic of spondyloarthropathy; lateral view of vertebral column of Early Permian *Dimetrodon incisivus* (NMNH 6723)

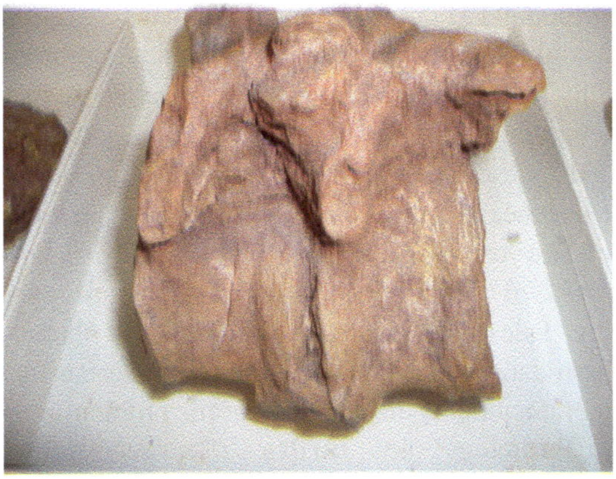

Fig. 75 Note anterior bony bridging indicative of spondyloarthropathy; lateral view of vertebrae of Early Permian *Diadectes* (NMNH)

Fig. 76 Lateral view of fused Late Cretaceous mosasaur (AUMP PR 164) vertebrae

Fig. 77 Cross section of fused Late Cretaceous mosasaur (AUMP PR 164) vertebrae illustrated in Fig. 76

described a probable chondroma in Eocene *Crocodylus* sp.

Vertebral fusion is well recognized in the fossil record. While the pathology in some mosasaurs is clearly related to the trauma of shark bites (Moodie 1917), other mosasaurs have vertebral bridging more characteristic of spondyloarthropathy (Figs. 76 and 77), as has been noted in contemporary varanids (Rothschild 2009). Ruffer's 1917 description and illustrations of a Lower Miocene crocodile *Tomistoma dowsoni* and those of Sawyer and Erickson (1998) of *Leidyosuchus formidabilis* clearly document spondyloarthropathy. Pales' 1930 description of anulus fibrosus ossification in a Pleistocene Cuban saurian further documents the antiquity of the problem.

Pathology in turtles (Reprinted from Rothschild BM, Schultze H-P, Pellegrini R. in press. Osseous and articular pathologies in turtles and abnormalities of mineral deposition. In: Morphology and Evolution of Turtles, D Brinkman, J Gardner, P Holroyd (eds.). New York: Springer):

It is difficult at times to determine whether anomalies observed in the embryo or neonate are the result of genetic alterations or of exposure to adverse environmental conditions, or to distinguish anomalies that have been inherited versus those occurring secondary to a new mutation. One does not necessarily need to invoke an environmental toxin. Environmental conditions may be just as important, either as a direct or indirect effect. Lynn and Ullrich (1950) examined turtle embryos from eggs exposed to partial drying during gestation interval days 35–50. Embryo pathologies ranged from "almost unrecognizable carapaces and plastrons, distorted limbs, and eyeless or jawless heads." Newman (1923) suggested temperature effect, whereas Coker (1910) considered this as an effect of egg distension. Packard et al. (1977) noted that prolonged exposure to temperatures several degrees below the optimum produced developmental abnormalities. Incubation of the California desert tortoise, *Gopherus agassizii*, above the optimum temperature of 32 °C was associated with upper jaw foreshortening and lack of forelimbs (Frye 1991c), whereas incubation at 33.5°C was associated with anophthalmy, maxillofacial clefts, harelips, forelimb and partial hind limb agenesis, coccygeal hypoplasia and scute and shell plate duplication (Frye 1989, 1991c). However, organochlorine compounds (e.g., PCBs) induce missing claws and eyes, deformed carapaces, tails, limbs and crania, deformed upper and lower jaws in snapping turtles (Bishop et al. 1989). Greatest incidence of deformities was noted in Cranberry Marsh. Frequency in Lake Ontario was greater than in Lake Erie or Algonquin Park, correlated with organochlorine levels.

Whereas most studies represent isolated reports of deformities (e.g., Tables 5–9), several report statistical results. Bellairs (1981) and Yntema (1960) reported that 15% of the common snapping turtle, *Chelydra serpentina*, had abnormal tail, hind limb, or carapace. Carswell and Lewis (2003) reported 0.32% abnormalities in

Table 5 Dicephalic turtles

Species	Common name	Kind of duplication	References
Cryptodira Cheloniidae			
Caretta caretta	Loggerhead	Dicephalic	Anon. (2002)
Chelonia mydas	Green turtle	Duplication of head and forelimbs Second, atrophic head	Haft (1994) Coquelet (1983)
Chelydridae			
Chelydra serpentina	Common snapping turtle	Dicephalic twins	Cederstrom (1931); Canella (1932); Campbell (1967); Feldman (1983); Sims (1989); Anon. (1995, 2001); Yntema (1970, 1971)
Geoemydidae			
Chinemys reevesii	Reeve's turtle	Dicephalic derodymus	Khosatzky (1991) Nakamura (1938)
Emydidae			
Chrysemys picta	Northern painted turtle	Dicephalic derodymus duplication of head and forelimbs Siamese twins attached at the posterior ends	Barbour (1888, 1896a, b); Anon. (1889); Girard (1891–1892); Hildebrand (1938) Barbour (1896c); Derickson (1927); Townsend (1928) Bishop (1908) Hildebrand (1938)
Chrysemys sp.	Painted trutle	Derodymus	Schmidt and Inger (1957); Bellairs (1981)
Emys orbicularis	European pond turtle	Duplication anterior head	Epure and Pogorevici (1940)
Emys sp.	European pond turtle	Dicephalic	Bateson (1894); Przibram (1909); Newman (1923)
Graptemys pseudogeographica	False map terrapin	Dicephalic	Clement (1967); Sims (1989)
Malaclemys sp.	Diamondback terrapin	Dicephalic, derodymus	Brogard (1987)
Malaclemys terrapin	Diamondback terrapin	Dicephalic Derodymus	Anon. (1975); Wright (2005) Schmidt and Inger (1957); Frye (1991a)
Malaclemys terrapin centrata	Carolina diamond-back terrapin	Duplication of head and forelimbs, anakatamesodidymus	Hildebrand (1938)
Mauremys leprosa	Mediterranean pond turtle	Dicephalic	Martins d'Alte (1937)
Pseudemys sp.	River cooter turtle	Dicephalic	Anon. (2000)
Pseudemys concinna floridana	Florida cooter	Dicephalic Siamese twins Plastron to plastron	Hildebrand (1938) Reichenbach-Klinke and Elkan (1965); Reichenbach-Klinke (1977) Hildebrand (1938)
Pseudemys nelsoni	Florida redbelly turtle	Dicephalic Derodymus	Ippen (1985) Bellairs (1981); Ippen (1982)
Terrapene carolina	Eastern box turtle	Twins Partial skull duplication	Cohen (1986) Edwards (1751)
Terrapene carolina triunguis	Eastern box turtle	Twins	Crooks and Smith (1958); Cohen (1986)
Terrapene sp.	Box turtle	Dicephalic	Mitchell (1994); Anon. (2006a)
Trachemys scripta elegans	Red-eared slider	Dicephalic Derodymus Duplication of head and forelimbs Siamese twins Siamese twins fused at pelvis Thoraco-omphalopagus	Clement (1967); Sims (1989); Frye (1991c); Chapple (1999) Frye (1991a); Hasel (1992a, b) Chapple (1999); Anon. (2007a, b) Anon. (1967, 2007a); Chapple (1999) Mader (2006a) Anon. (1967)

(continued)

Table 5 (continued)

Species	Common name	Kind of duplication	References
Trachemys scripta scripta	Yellowbelly slider	Dicephalic Siamese twins	Kritzer (2002); Perkins (2002); Reaves (2004); Anon. (2007c) Reichenbach-Klinke and Elkan (1965); Reichenbach-Klinke (1977)
Trachemys scripta troostii	Cumberland slider	Dicephalic	Byrd (1939)
Trachemys sp.	Slider turtles	Conjoined	Frye (1991a)
Kinosternidae			
Kinosternon flavescens	Yellow mud turtle	Dicephalic	Porras and Beraducci (1980); Frye (1991c)
Testudinidae			
Chersina angulata	Angulate tortoise	Dicephalic	Swarts (2003)
Gopherus agassizii	California desert tortoise	Dicephalic Siames twins attached to caudal shell	Frye (1991c) Frye (1991a, c); Rothschild (2009a)
Gopherus polyphemus	Gopher tortoise	Twins	Hundsacker (1968)
Testudo graeca	Spur-thighed or Greek tortoise	Dicephalic	Stojanov (2000)
Testudo graeca ibera	Turkish tortoise	Dicephalic Duplication of head and forelimbs	Vellard and Penteado (1931) Caullery (1931)
Testudo hermanni	Hermann's tortoise	Dicephalic	Reichenbach-Klinke and Elkan (1965)
Testudo hermanni boettgeri	Eastern Hermann's tortoise	Siamese twins	Sailer et al. (1997)
Trionychidae			
Apalone ferox	Florida softshell	Siamese twins Siamese twins fused at plastron	Bellairs (1981) Hildebrandt (1938); Ippen (1982)
Unspecified	Turtle/tortoise	Duplication of head and forelimbs Dicephalic Derodymus	Anon. (1931) Anon. (1888a, b, 1895); Kuvano (1902); (Hildebrand (1930); Brogard (1987); Brady (1991); Benest (1994); Eigner (1994); Kleinberg (1994); Meisel (1994); Hamilton (1995); Walker (1995); Maddux (1996); Martin (1996); Wilcox and Perrin (1996); Cisneros (1997); Thomason (1997); Wilkins (1998); Gumbel (1999); Nicholson (1999); Anon. (1999a, b, c); Mason (2000); Miller (2002); Reeder (2002); Brown (2003); Hall (2003); Holahan (2003); Montgomery (2003); Smith (2003); Wechsler (2003); Anon. (2003a, b, c, 2004, 2005, 2006b); Feuer (2005); Lee (2005); Rothschild (2009a) Anon. (1897)
Pleurodira Podocnemididae			
Podocnemis unifilis	Yellow-spotted Amazon river turtle	Derodymus	Ferreira (1923)

Anakatamesodidymus – two heads, two tails, and rudimentary fifth leg.
Derodymus – vertebral column bifurcated in the cervical region, double-headed; two complete heads and necks.
Dicephalic – bicephalic, having two heads.
Siamese twins – conjoined twins.
Thoraco-omphalopagus type – twins fused from the upper chest to the lower chest.

Table 6 Limb abnormalities in turtles

Species	Common name	Abnormality	Reference
Cryptodira			
Protostegidae			
Archelon ischyros		Shortened fibula and tibia	Tasnádi-Kubacska (1962)
Cheloniidae			
Caretta caretta	Loggerhead	Complete absence of limb Supernumerary limb Partial absence of limb	Ewert (1979); Ippen (1982); Brogard (1987) Mader (2006a) Ewert (1979); Bellairs (1981)
Syllomus crispatus aegyptiacus		Bony ridge located on the femur between the torchanters and the caput	Weems (1974)
Chelydridae			
Chelydra serpentina	Common snapping turtle	Hindlimb	Yntema (1960); Bellairs (1981)
		Dwarfism	Bishop et al. (1989)
Emydidae			
Emydidae	Pond turtles	Partial absence of limb	Dutta (1931); Bellairs (1981); Ippen (1982)
Emys orbicularis	European pond turtle	Complete absence of limb	Dürigen (1897)
Malaclemys sp.	Diamondback terrapin	Supernumerary limb	Brogard (1987)
Malaclemys terrapin centrata	Carolina diamondback terrapin	Partial absence of limb	Hay (1904)
Terrapene carolina	Eastern box turtle	Partial absence of limb, congenital tarsal absence	Wilkinson et al. (2004)
Trachemys scripta	Slider	Achondroplasy	Frye and Carney (1974); Brogard (1987(; Arvy and Fertard (2002)
Trachemys scripta elegans	Red-eared slider	Achondroplasy	Frye and Carney (1974); Marcus (1981); Frye (1981, 1991a, c); Rothschild (2009a)
Testudinidae			
Gopherus agassizii	Desert tortoise	Syndactyly Achondroplasy	Good (1987) Frye and Carney (1974); Marcus (1981); Frye (1981, 1991a, c); Brogard (1987); Arvy and Fertard (2002); Rothschild (2009a)
Testudo hermanni	Hermann's tortoise	Partial absence of limb (= oligodactyly) Polydactyly	Wermuth (1961); Fojtl (1989) Martínez-Silvestre et al. (1998)
Trionychidae			
Trionyx sp.	Softshell turtle	Complete absence of limb	Dutta (1931); Bellairs (1981); Ippen (1982)
Unspecified	Turtle	Ectromely Achondroplastic dwarfism	Hildebrand (1930); Brogard (1987) Gillespie (1994)

Achondroplasy – abnormal cartilage conversion to bone, producing short, abnormally limbed dwarfism.
Ectromely – absence of one or more limbs or incomplete limb with missing lower portion.
Polydactyly – increased number of metatarsals (= hyperdactyly), supernumerary digit(s).
Syndadctyly – fusion or failure of separation of fingers or toes.

Table 7 Skull defects in turtles

Species	Common name	Defects	Reference
Cryptodira			
Cheloniidae			
Caretta caretta caretta	Loggerhead	Cyclopia	Adelmann (1936); Ewert (1979); Bellairs (1983)
Chelonia mydas	Green turtle	Cleft palate + shortened upper jaw	Ippen (1982)
Chelydridae			
Chelydra serpentina serpentina	Common snapping turtle	Eyeless	Lynn and Ullrich (1950)
Emydidae			
Chrysemys picta	Northern painted turtle	Cyclopia	Adelmann (1936)
Pseudemys concinna floridana	Florida cooter	Cyclopia	Hildebrand (1938)
Malaclemys terrapin	Diamondback terrapin	Variation in size and shape of skull	Bangs (1896)
Terrapene sp.	Box turtle	Absent mouth and jaw	Ewert (1979)
Trachemys scripta	Pond slider	Cleft palate	Ippen (1982)
Trachemys scripta elegans	Red-eared slider	Cyclopia Shortened upper jaw	Adelmann (1936); Ewert (1979) Frye and Carney (1974)
Testudinidae			
Gopherus agassizii	Desert tortoise	Cleft palate Shortened upper jaw	Bellairs (1983); Frye (1989) Frye (1991c)
Pleurodira			
Bothremydidae			
Ummulisani rutgersensis		Loss of frontals	Gaffney and Tong (2008)

14,361 loggerhead *Caretta caretta* turtle eggs in 1992 and 2001, manifest as small size, carapace deformities, folded or forshortened flippers, misshapened mouthparts and twinning. The most complex of the congenital anomalies in turtles is perhaps the Siamese, conjoined or parasitic twin (Anonymous 2007a; Kuvano 1902; Maddux 1996; Nakamura 1938). Most have been reported as isolated phenomena (Table 5). A variety of duplication patterns have been observed: plastron to plastron, side to side, or posterior to posterior. Noting Hildebrand's (1938) tally of 27 cases, Table 5 suggests the rarity of this phenomenon. Only 45 cases appear to have been reported to date, and no epidemiologic studies are available. Complete or partial absence of a limb has been reported in isolated occurrences (Table 6). Not all are congenital. Jackson (1980) noted one instance in which a foreleg was gnawed off during hibernation by a rat, and Frye (1994) reported digit necrosis in red-eared slider turtle, *Trachemys scripta elegans*.

Anomalies of the head (Table 7) may be as "simple" as variation in size and shape of skull, as noted in the diamondback terrapin, *Malaclemys terrapin* (Bangs 1896), and the Florida scooter, *Pseudemys floridana* (Hildebrand 1938). Lack of frontal bones, as noted in a bothremydid side-necked turtle from the Eocene of Morocco *Ummulisani rutgersensis* (Gaffney and Tong 2008) and absent mouth and jaw in a box turtle (Ewert 1979), are examples of limited defects. Anomalies may present with protruding lower jaw and eyeless upper jaw (e.g., common snapping turtle, *Chelydra serpentina*), cyclopia (red-eared slider, *Trachemys scripta elegans*, the loggerhead, *Caretta caretta caretta*, Florida cooter, *Pseudemys concinna floridana* [Table 7]), or as an eyeless stump in common snapping turtle *Chelydra serpentina* (Lynn and Ullrich 1950).

Table 8 Deformities of vertebral column in turtles

Species	Common name	Deformity	Reference
Cryptodira			
Cheloniidae			
Caretta caretta	Loggerhead	Kyphosis in 0.2–0.5%	Coker (1910); Plymale et al. (1978)
Chelonia mydas	Green turtle	Kyphosis in 0.08% Lordosis/scoliosis in 0.03%	Rhodin et al. (1984)
Lepidochelys olivacea	Olive ridley	Kyphosis in 0.1–0.4% Lordosis/scoliosis in 0.3%	Rhodin et al. (1984)
Chelydridae			
Chelydra serpentina	Common snapping turtle	Kyphosis Caudal agenesis Kinked tail	Chan (1937); Plymale et al. (1978); Wilhoft (1980); Willft (1980) Finkler and Claussen (1997) Feldmann (1983)
Macroclemys temminckii	Alligator snapping turtle	Curly tail deformity	McCallum and Trauth (2000)
Dermatochelyidae			
Dermatochelys coracia	Leatherback	Kyphosis in 0.07% Lordosis/scoliosis in 0.07–0.11%	Rhodin et al. (1984)
Geoemydidae			
Melanochelys trijuga	Indian pond terrapin	Kyphosis	Plymale et al. (1978)
Melanochelys trijuga thermalis	Sri Lanka pond terrapin	Kyphosis	Deraniyagala (1939)
Emydidae			
Chrysemys sp.	Painted turtle	Kyphosis	Wilhoft (1980)
Chrysemys picta	Northern painted turtle	Kyphosis	Ernst (1971); Plymale et al. (1978); Pavalko (1986)
Chrysemys picta marginata	Midland painted turtle	Kyphosis	Werner (1959); Ippen (1982)
Chrysemys picta belli	Western painted turtle	Kyphosis in 1/429	Stuart (1996)
Chrysemys picta bellii x marginata	Painted turtle cross	Kyphosis	Necker (1940); MacCulloch (1981)
Clemmys guttata	Spotted turtle	Kyphosis	Ernst (1971); Plymale et al. (1978)
Deirochelys reticularia	Chicken turtle	Kyphosis	Mertens (1940); Plymale et al. (1978)
Graptemys pseudogeographica	False map turtle	Kyphosis	Plymale et al. (1978)
Graptemys ouachitensis	Ouachita map turtle	Kyphosis	Carpenter (1958)
Malaclemys terrapin centrata	Carolina diamondback terrapin	Scoliosis kyphosis	Hildebrand (1930) Plymale et al. (1978)
Pseudemys concinna floridana	Florida cooter	Kyphosis	Plymale et al. (1978)
Terrapene carolina	Eastern box turtle	Kyphosis	Plymale et al. (1978)
Terrapene carolina triunguis	Three-toed box turtle	Kyphosis	Black (1976)
Terrapene ornata	Ornate box turtle	Kyphosis kyphoscoliosis	Plymale et al. (1978); Fox (1941)
Trachemys scripta	Slider	Kyphosis	Plymale et al. (1978)

(continued)

Table 8 (continued)

Species	Common name	Deformity	Reference
Trachemys scripta scripta	Yellowbelly slider	Kyphosis	Carr (1952)
Trachemys scripta yaquia	Yaqui slider	Kyphosis	Plymale et al. (1978)
Kinosternidae			
Sternotherus odoratus	Common musk turtle	Kyphosis	Nixon and Smith (1949); Plymale et al. (1978); Saumure (2001)
Testudinidae			
Gopherus agassizii	Desert tortoise	Kyphosis in <0.04%	Rhodin et al. (1984)
		caudal agenesis	Frye (1989, 1991a)
Testudo hermanni	Hermann's tortoise	Kyphosis	Wandolleck (1904); Plymale et al. (1978)
Unindentified	Tortoise	Kyphosis	Dämmrich (1967); Jarofke and Lange (1993)
Trionychidae			
Apalone ferox	Florida softshell	Kyphosis	Nixon and Smith (1949(; Plymale et al. (1978(; Barber (1991)
Apalone mutica	Smooth softshell	Kyphosis	Smith (1947); Plymale et al. (1978)
Apalone spinifera	Spiny softshell	Kyphosis	Chan (1937); Smith (1947); White and Murphy (1972); Burke (1994)
Apalone spinifera emoryi	Texas spiny softshell	Kyphosis	Smith (1947)
		Kyphosis in 1/20	Stuart (1996)
Lissemys sp.	Flap-shelled turtle	Kyphosis	Wilhoft (1980)
Lissemys punctata punctata	Indian flap-shelled turtle	Kyphosis	Duda and Gupta (1977)
Palea steindachneri	Wattle-necked softshell	Kyphosis	Gressitt (1936, 1937); Smith (1947); Plymale et al. (1978)
Pelodiscus sinensis	Chinese softshell	Kyphosis	Vogt (1922); Gressitt (1936); Plymale et al. (1978)
Trionychidae	Softshell turtles	Kyphosis	Smith (1947); Bellairs (1981)
Trionyx triunguis	African softshell	Kyphosis	Mertens (1940); Plymale et al. (1978); Ippen (1985)
		Scoliosis	Reichenbach-Klinke (1977)
Unidentified		Kyphosis	Williams (1957); Schlumberger (1958)

Kyphosis – curvature of axial skeleton, with accentuated posterior apical apex.
Dorsoventral vertebral column curvatures – commonly attributed to metabolic bone disease; usually refers to thoracic spine (hump-like arching of tortoise carapace).
Kyphoscoliosis – combination of kyphosis and scoliosis.
Lordosis – curvature of axial skeleton, with accentuated anterior apical apex. Usual refers to lumbar spine.
Scoliosis lateral curvature of the vertebral column.

Shortening of the upper jaw, as Frye (1989) noted in the desert tortoise, *Gopherus agassizii*, and cleft palate (Bellairs 1981) has been reported. Cleft palate also has been noted in the pond slider, *Trachemys scripta* (Ippen 1982), green turtle, *Chelonia mydas* (Ippen 1982), and desert tortoise, *Gopherus agassizii* (Frye 1989).

Duplications are the most commonly reported anomalies. Edwards (1751) reported partial skull duplication in a Carolina tortoise. Isolated, sometimes redundant reports of dicephalism (Figs. 78 and 79) of unnamed and named species abound (Table 5). It is difficult to assess the frequency of dicephalism, even with named species, given the same isolated reports and tendency of redundancy in those reports. Extrapolating from Hildebrand's (1938) findings of only two dicephalic individuals among 100,000 Carolina diamondback terrapins, *Malaclemys terrapin centrata*, hatched over 25 years, this would appear to be an extremely

Table 9 Carapace and plastron anomalies in turtles

Species	Common name	Anomaly	Reference
Cryptodira			
Cheloniidae			
Caretta caretta	Loggerhead	Lengthwise division of neural plates Carapace deformities	Zangerl and Turnbull (1955) Carswell and Lewis (2003)
Lepidochelys olivacea	Ridley	Division of neural plate into three parts	Zangerl and Turnbull (1955)
Procolpochelys grandaeva	Miocene carettine sea turtle	Supernumerary peripheral carapace element, division of neural plate into three parts	Zangerl and Turnbull (1955); Weems (1974)
Syllomus aegyptiacus		Splitting of first neural and extra postneurals and costals	Weems (1974)
Chelydridae			
Chelydra serpentina serpentina	Common snapping turtle	Shell anomalies, deeply indented carapace Abnormal carapace Deformed carapace	Lynn and Ullrich (1950); Frye (1991a, c) Yntema (1960); Bellairs (1981); Brooks et al. (1989) Hutchinson and Simmonds (1991)
Emydidae			
Chrysemys picta	Northern painted turtle	One or more carapace or plastron anomalies Fifth neural plate reduced and triangular Abnormally broad carapace with extra plates Fusion of plastron pectorals and abdominals, plastron bulges Shell abnormalities Extra right costal plate, division of first vertebral plate by a suture and composition of fifth by four irregular plates	MacCulloch (1981) Derickson (1927) Hildebrand (1938); Lynch and Ullrich (1950); Lynch (1965) Lynch and Ullrich (1950) Zangerl and Johnson (1957); MacCulloch (1981) Bateson (1894)
Chrysemys picta marginata	Midland painted turtle	Shell abnormalities Supernumerary or deficient bony plates	Whillans and Crossman (1977); MacCulloch (1981) Newman (1906)
Chrysemys picta picta	Eastern painted turtle	Shell anomalies Seven right and three left marginals and unusually broad neurals	Frye (1991a, c) Lynn and Ullrich (1950)
Emys orbicularis	European pond turtle	Anomalies related to number of neurals and costals, peripherals Double nuchal plate	Cherepanov (1994) Epure and Pogorevici (1940)
Glyptemys insculpta	Wood turtle	Irregularities and reduction in number of bony plates of the carapace	Parker (1901)
Graptemys geographica	Common map turtle	Supernumerary or deficient bony plates	Newman (1906)
Malaclemys terrapin	Diamondback terrapin	Supernumerary plates, dwarfed specimen with broad short shell, twisted to one side	Hay (1904)
Pseudemys concinna	Eastern river cooter	Disfigured carapace and plastron	Jacobson (2007a)
Terrapene carolina carolina	Eastern box turtle	Rigid plastron without hinge	Holman (1984)
Trachemys scripta elegans	Red-eared slider	Carapace anomaly Asymmetrical shells, concave plastron Anomalous vertebral plates, plastral axial deviation	Bouwhuis (1972) Frye (1973) Frye (1991a)
Trachemys scripta troostii	Cumberland slider	Plastron and carapace anomalies	Cagle (1950)
Geoemydidae			
Cuora amboinensis	Southeast Asian box turtle	Shell deformity	Chou (1979)

(continued)

Table 9 (continued)

Species	Common name	Anomaly	Reference
Kinosternidae			
Sternotherus odoratus	Common musk turtle	180° torsion	Frye (1991a)
Testudinidae			
Chelonoidis chilensis	Chaco tortoise	Abnormal carapace	Matz (1977)
Geochelone carbonaria	South American red-footed tortoise	Asymmetrical shells Wedge-shaped vertebral plates Soft, deformed carapace and plastron, caused by hyperparathyroidism Carapace or plastron anomalies	Frye (1973) Frye (1991a) Frye and Carney (1975); Rivera and Lock (2008) Vieira de Dandrade and Shinya (1993)
Geochelone nigra	Galapagos giant tortoise	Shell anomaly Congenital carapace anomaly	Hayes and Beaman (1985) Dollinger et al. (1997)
Geochelone sulcata	African spurred tortoise	Deformed carapace	Donoghue (2006)
		Carapace asymmetries	Cloudsley-Thompson (1970)
Gopherus agassizii	Desert tortoise	Carapacial and plastral plate abnormalities, supernumerary vertebral and marginal or reduced carapacial plates Plastral axial deviation Wavy carapace Shell hump	Frye (1989, 1991a) Frye (1991a, c) Rosskopf et al. (1982) Mader (1990a)
Gopherus berlandieri	Texas tortoise	Supernumerary carapace bones	McEwan (1982)
Gopherus polyphemus	Gopher tortoise	Asymmetrical shells Supernumerary carapace bones	Frye (1973) McEwan (1982)
Gopherus sp.		10% abnormal (hard, brittle shells)	Grant (1937)
Mauremys caspica	Caspian terrapin	Anomalies related to number of neurals and costals, peripherals	Cherepanov (1994)
Testudo graeca ibera	Turkish tortoise	Carapace anomalies	Kabisch (1989)
Testudo graeca	Spur-tighed tortoise	Anomalies related to number of neurals and costals, peripherals Distorted carapace 1 cm high humps on vertebralia and costalia	Cherepanov (1994) Lynch and Ullrich (1965) Weichmann (1989)
Testudo hermanni	Hermann's tortoise	Humped carapace	Jackson and Cooper (1981)
Testudo horsfieldi	Horsfield's tortoise	Anomalies related to number of neurals and costals, peripherals	Cherepanov (1994)
Testudo sp.	Tortoise	Carapace hump	Highfield (1990)
Trionychidae			
Apalone mutica	Florida softshell	Prominent curved hump (5 times normal shell height) change in numbers of neurals and costals	Smith (1947)
Apalone spinifera	Spiny softshell	Sharp hump, change in numbers of neurals and costals	Smith (1947)
Apalone spinifera emoryi	Texas spiny softshell	Sharp hump, change in numbers of neurals and costals	Smith (1947)
Palea steindachneri	Wattle-necked softshell turtle	Carapace hump	Gressitt (1937)
Unidentified			
	Turtles	Carapace deformities	Brogard (1987)
	Tortoise	Carapace defects	Frank (1976)

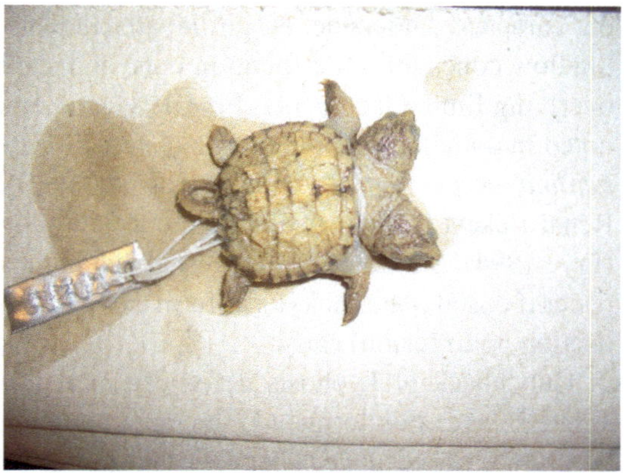

Fig. 78 Dicephalic turtle, *Chelydra serpentina* (UMMZ 119985)

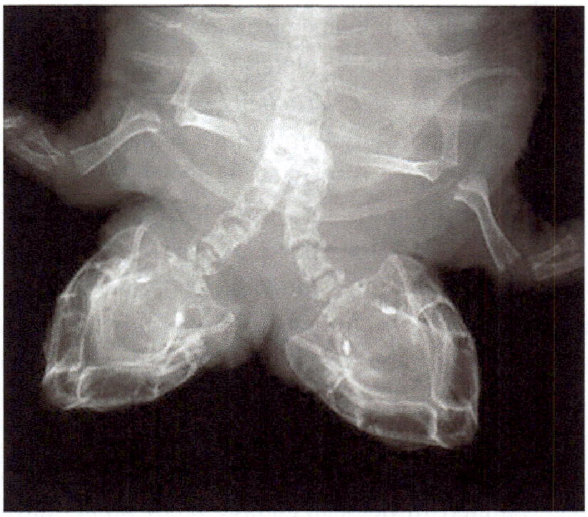

Fig. 79 Dorsal-ventral x-ray view of *Chelydra serpentina* (UMMZ 119985). Dicephalic turtle

uncommon phenomenon. The same appears true for the more complicated duplication, referred to as derodymus (Anonymous 1888a, b, 1897, 2007c; Epure and Pogorevici 1940), which describes duplication of the neck as well as the head (Table 5). Newman (1923) reported dichotomous fission of a rib in association with such doubling. Duplication of the head and forelimbs (Anonymous 1931) has been reported in red-eared slider, *Trachemys scripta elegans* (Anonymous 2007a, b; Chapple 1999), northern painted turtle, *Chrysemys picta* (Bishop 1908), Carolina diamondback terrapin, *Malaclemys terrapin centrata* (Hildebrand 1938), and in the Greek tortoise, *Testudo graeca ibera* (Caullery 1931). Hildebrand (1938) reported a diamondback terrapin, *Malaclemys terrapin centrata*, with two heads, two tails, and rudimentary fifth leg, and Brogard (1987) a bifid tail in *Malaclemys*. A duplication of the vertebral column has been mentioned (Anonymous 1967).

Supernumerary limbs (Table 6) have been reported in diamondback terrapin, *Malaclemys* (Brogard 1987), and *Caretta caretta* (Mader 2006a), polydactyly in the Hermann's tortoise, *Testudo hermanni* (Martínez-Silvestre et al. 1998), and syndactyly (fused toes), in the California desert tortoise, *Gopherus agassizii* (Good 1987). Abnormal shortened limbs have been reported in turtles, in the form of achondroplasy (Gillespie 1994). Such long-bone shortening, cartilage hypertrophy with ossification spicules has been noted in dwarf red-eared slider, *Trachemys scripta*, and in an inbred California desert tortoise, *Gopherus agassizii*, with long-bone shortening, cartilage hypertrophy with ossification spicules. Other reported anomalies include a Cretaceous *Archelon ischyros* with shortened fibula and tibia (Tasnádi-Kubacska 1962) and a *Syllomus aegyptiacus* USNM 24872 with a bony ridge located on the femur between the trochanters and the caput (Weems 1974). Dwarfism, while apparently maintaining normal body proportions, was reported in snapping turtles by Bishop et al. (1989). Absence of limbs was reported by Dutta (1931), oligodactyly by Fojtl (1989), reduction of toes by Wermuth (1961), and leg anomalies in the European pond turtle *Emys* by Dürigen (1897).

Most reports on kyphosis (Table 8) simply relate isolated occurrences. Frequencies in epidemiologic studies (Rhodin et al. 1984; Stuart 1996) have been significantly under 1%, with the exception of Stuart's (1996) report of one instance among 20 Texas spiny softshell turtles, *Apalone spiniferus emoryi*. This may just represent an artifact of small numbers and would not be statistically significant, as the frequency in marine turtles (loggerhead, *Caretta caretta*, was 0.2–0.5%; green turtle, *Chelonia mydas*, 0.08%; leatherback turtle, *Dermochelys coracia*, 0.07% and in the Pacific ridley turtle, *Lepidochelys olivacea*, 0.1–0.4%), the western painted turtle *Chrysemys picta belli*, 0.03%; and the California desert tortoise, *Gopherus agassizii*, 0.04%. Scoliosis may occur in isolation (Table 4; Hildebrand 1930) or be associated with kyphosis, as Fox (1941) observed in *Terrapene ornata*.

Caudal agenesis (= failure of formation of all or part of structure; Table 8) has been noted in California desert tortoise, *Gopherus agassizii* (Frye 1989, 1991a), and common snapping turtle, *Chelydra serpentina* (Finkler and Claussen 1997). McCallum and Trauth (2000) reported a curly tail deformity in alligator snapping turtle, *Macroclemys temminckii*, hatchlings, presenting as tight spiral or coil, and Feldman (1983) reported kinked tail in common snapping turtle, *Chelydra serpentina*.

Distorted turtle shells are often reported as isolated observations (Table 9). Broad (e.g., eastern painted turtle, *Chrysemys picta*) and wedge-shaped plates (e.g., red-footed tortoise, *Geochelone carbonaria*) have been reported, as well as 180° torsion in a common musk turtle, *Sternotherus odoratus* (Frye 1991a, c). There are few statistical statements. Grant (1937) reported that 10% of gopher tortoise specimens (*Gopherus* specimens with hard, brittle shells) are abnormal. McEwan (1982) noted that 50% of gopher tortoise specimens, *Gopherus polyphemus*, have supernumerary carapace bones compared with 68% of the Texas tortoise, *Gopherus berlandieri*. Cherepanov (1994) reported anomalies of the number of neurals, costals, and peripherals in 121 of 510 Caspian pond turtles (*Mauremys caspica*), European pond turtles (*Emys orbicularis*), Mediterranean spur-thighed tortoise (*Testudo graeca*), and Russian tortoise (*Agrionemys horsfieldi*). Congenital carapace anomalies were reported by Dollinger (1997).

Carapace and plastron anomalies are known in fossil sea turtles (Table 9), the Miocene *Caretta* (Zangerl and Turnbull 1955) and *Procolpochelys grandaeva* (Weems 1974; Zangerl and Turnbull 1955), and *Syllomus* (Fig. 3; Weems 1974). They occur only rarely in the extant Ridley, *Lepidochelys* (Zangerl and Turnbull 1955). Frye (1981) suggested that deformation was caused from downward tension on carapace underside at girdle attachments. Shallow concavities are found in bone and shell overlying limb attachments (Frye 1981), as was noted in common snapping turtle, *Chelydra serpentina serpentina* (Lynn and Ullrich 1950). Renal rickets is another etiologic consideration (Frye 1994). Smith (1947) suggested that humps occur if costal plates ankylose (form a close connection up to fusion) early.

Hunchback or kyphosis deformity (see also Table 8) takes two forms: (1) the more common deformed, raised plate, allegedly due to excess dietary protein (McArthur 1996; Rosskopf 1986), or (2) thick, porous, raised plates from renal osteodystrophy/nutritional osteodystrophy (Highfield 1990a, b; Jackson 1987; Scott 1992) or calcium deficiency (Rosskopf 1986; White 1989). Cagle (1950) related the deformity in the Cumberland slider, *Trachemys scripta troostii*, to an abnormal large yolk mass causing distension of carapace. Zwart et al. (1997) reported a thickened (up to 2 cm), loose-textured carapace with large fat-filled spaces in a Hermann's tortoise, *Testudo hermanni*, and reported slightly elevated rims in another specimen (Zwart et al. 1994). Amorphous carapace lumps were reported on carapace of *Syllomus aegyptiacus* USNM 24876 (Weems 1974).

Plastron bulges with wavy border (e.g., eastern painted turtle, *Chrysemys picta picta*) have also been noted (Lynn and Ullrich 1950), as well as a wavy carapace in California desert tortoise, *Gopherus agassizii* (Rosskopf et al. 1982). Similar changes in Galapagos turtle, *Geochelone nigra*, have been attributed to calcium deficiency. A concave plastron was reported in a red-eared slider, *Trachemys scripta elegans* (Frye 1973). Holman (1984) reported an eastern box turtle, *Terrapene carolina carolina*, with a rigid plastron, lacking a hinge.

The record on trauma in turtles is essentially idiosyncratic (Abou-Madi et al. 2004; Arvy and Fertard 2002; Bourdeau 1988b; Devaux 1992; Jacobson 1994; Mautino and Page 1993), although pathologic fractures from nutritional secondary hyperparathyroidism (Avery 1990) and osteomalacia (e.g., Spanish terrapin, *Mauremys leprosa*; Wallis 1927) must also be

Fig. 80 Traumatic scute disruption; dorsal view of *Chelydra serpentina* (Kansas City Zoo)

considered. Injuries (Fig. 80) are attributed to falling, being dropped or stepped on; Vella (2007) stated that turtles are avid climbers. Galois and Ouellet (2007) reported traumatic injury in spiny softshell turtle *Apalone spinifera*, without specifying its nature. A "shark's tooth pertaining to a scavenger species related to *Lamna* was found with the type" of the leatherback turtle, *Dermochelys* (Wieland 1909, p. 120). Wieland (1909) reported an *Archelon ischyros* with obliquely bitten, healed tibia and fibula. Parker (1901) reported sculptured tortoise marginal carapace fractures with irregularity of scutes or reduction in number. Korschelt (1927) reported a 2.5 cm long opening with bony callus on the inner side of left costal plates of the carapace of a European pond turtle, *Emys orbicularis*, and Frye (1991d) a limb fracture in soft shell turtle, apparently caused by a falling rock.

Healed breaks of the carapace were reported in the green turtle, *Chelonia mydas*; the Mediterranean pond turtle, *Mauremys leprosa*; the European pond turtle, *Emys orbicularis*; the Mediterranean spur-thighed tortoise, *Testudo graeca*; Galapagos giant tortoise *Geochelone nigra* (Korschelt 1932); the three-toed box turtle, *Terrapene carolina triunguis* (Barten 1996); the ornate box turtle, *Terrapene ornata* (Bennett 1989); the Gopher tortoise, *Gopherus* sp. (Frye 1973); and unspecified turtles (Dämmrich 1985). Fractures that cross the dorsal midline of the carapace may be associated with spinal fractures (Isaza and Jacobson 1995). Carrick and Reddacliffe (1980) reported automobile-related traumatic shell fractures in Australian side-necked turtles, the common snake-neck turtle, *Chelodina longicollis*, and the Australian short-neck turtle, *Emydura macquarii krefftii*. Epidemiologic information is generally lacking, with exception of Meek and Inskeep's (1981) report of 4% frequency of carapace injuries in Hermann's tortoise, *Testudo hermanni*. Cracks in the rear of the carapace of female Galapagos turtle *Geochelone nigra* occur during mating with the much more massive males.

The literature record also includes a self-amputated hind limb and fractured (shattered) tibia and fibula in an eastern box turtle, *Terrapene carolina carolina* (Boylan 2003), and a fractured humerus in an Aldabra giant tortoise, *Aldabrachelys gigantea* (Crane et al. 1980). Raidal et al. (2006) reported a healed humeral fracture (encompassing a wire) in a loggerhead turtle, *Caretta caretta*. McArthur (2004) noted rat bite trauma to limbs in post-hybernation animals, which may explain that the Carolina diamondback terrapin, *Malaclemys terrapin centrata*, occasionally misses feet and shows irregular stumps or irregular growth (Hay 1904). One individual had lost the posterior half of its body, including one leg.

Arthritis is not common in turtles and tortoises. Shoulder remodeling and bony proliferation in a loggerhead turtle, *Caretta caretta*, was called osteoarthritis (Raidal et al. 2006), but there seems to be subchondral damage. Given negative cultures, the possibility of a spondyloarthropathy must be entertained (Rothschild 2009a). Dämmrich (1985) has used the terms spondylopathia deformans and spondyloarthropathia deformans, but the meaning is unclear. He did describe ankylosis, lending credence to the possibility of spondyloarthropathy. Fitzgerald and Vera (2006) reported proliferative spinal osteopathy, which could represent an infectious process or spondyloarthropathy. Septic arthritis was reported in a leatherback turtle, *Dermochelys coriacea*, by Ogden et al. (1981).

Metabolic abnormalities in turtles predominantly relate to urate-related disease, calcium crystal-related disease, and an entity variously referred to as metabolic bone disease, hypoplastic osteoporosis (reduction in quantity and quality of trabecular components of bone), osteomalacia, rickets osteodystrophy, nutritional osteodystrophy, osteodystrophy fibrosa, osteogenesis imperfecta (an inborn error of metabolism), cage paralysis (complete loss of muscle function), and renal osteodystrophy (defective bone formation from compromised metabolism of vitamin D to the active form producing calcium deficiency and secondary hyperparathyroidism) (Jarofke and Lange 1993; Table 10). The latter may be present as accentuated "interplates" and growth lines (Frye 1994).

Nutritional osteodystrophy is most common in captive animals. Keymer (1978a, b) noted osteodystrophy in 12 of 144 chelonians reported by the Zoological Society of London. Barten (1982, 1983) also reported nutritional shell defects. Coquelet (1983) cites Keymer's (1978a) reports of osteodystrophy in 8.3% of terrestial and 9.8% of freshwater aquatic turtles, whereas Glazebrook (1980) noted its extreme rarity in marine turtles. Literature descriptions of frequencies among captive animals are difficult to assess. Most reports represent isolated occurrences (Table 10). Epidemiologic or facility reports are sometimes also difficult to interpret. Twenty-three (19%) of 122 terrapins autopsied at Zoologic Society of London between 1965 and 1975 had "lesions of the skeletal system" (Keymer 1978a). This included the category of miscellaneous osteopathies, which included nutritional and infectious origins of disease (Table 6). While nutritional and metabolic disorders were reported in 19.7% of studied animals, actual nutritional osteodystrophies were listed in 9.8%. This is difficult to reconcile with Keymer's (1978b) report of "miscellaneous osteopathies" in 4.2% and nutritional osteodystrophies in 0.3% of turtles autopsied at Zoologic Society of London in the same period. Reported frequency differences are even more pronounced in freshwater turtles that were not raised in the London Zoo. Barrows (2004) reported that one-third of freshwater turtles seen in a Leeds/London clinic had nutritional osteodystrophy.

Nutritional osteodystrophy produces shell (e.g., soft, lumps, humps, abnormally large plastron fontanels), maxillary (e.g., overgrown beak), and mandibular (e.g., shortened) deformities (Table 10), as well as deformed, inadequately calcified, misshapened or swollen extremities, kyphoscoliosis and pathologic (e.g., greenstick) fractures (Frye 1973; Jackson and Cooper 1981). Pliable mandible or maxilla, fractures of long bones, fibrodysplasia (= replacement of bone tissue by fibrous tissue) of long bones or jaw, kyphosis (= dorsoventral vertebral column curvatures), lordosis (= curvature of axial skeleton, with accentuated anterior apical apex), scoliosis (= lateral curvature of the vertebral column), widened radiolucent ends of long bones are described in box turtles and desert tortoises, *Gopherus agassizii* (Boyer 1996b; Mader 1990b, 2006d). Widened calcific rings occur near epiphyseal junctions (Glazebrook 1980). Curling of carapace edges (parrot beak), overgrowth of beak, net-like porosity of and overgrowth of the bridge between carapace and plastron have been noted (Boyer 1996b), thick, spongy bone may also occur (Gerlach 2004). Highly porous bone was especially noted at rib ends and peripheral plates (Stojanov 2005). Paradoxically, increased mineralization from secondary renal hyperparathyroidism (= disorder caused by overactivity of the named glands, producing osteitis fibrosa cystica and other bone changes) can also occur (Wilkinson et al. 2004). Cystic calculi (renal stones) may result (Gillespie 1994; Mader et al. 1999). Osteopenia (reduced bone ossification/density) and metastatic calcification (including joints), however, can also occur from hypervitaminosis (= excess vitamin) D (Gillespie 1994). The term osteodystrophy includes secondary hyperparathyroidism (Avery 1990), which can be caused by dietary calcium or vitamin D deficiency, negative $Ca:PO_4$ ratio, or by lack of UV-light exposure (Boyer 1996b). It can occur if vitamin D metabolism is altered secondary to kidney, liver, intestinal, thyroid or parathyroid disease, and hypocalcemia (= presence of low serum calcium levels) (Boyer 1996b).

One component of metabolic bone disease, hyperparathyroidism, can occur as a primary

Table 10 Metabolic bone diseases in turtles

Species	Common name	Disease	Reference
Cryptodira			
Cheloniidae			
Chelonia mydas	Green turtle	Nutritional osteodystrophy	Glazebrook (1980); Coquelet (1983)
Chelydridae			
Chelydra serpentina	Common snapping turtle	Demineralization due to secondary nutritional hyperparathyroidism Osteopathy	Wilkinson et al. (2004) Keymer (1978a)
Geoemydidae			
Heosemys annandalii	Yellow-headed temple turtle	Shell rot	Guirley (2003)
Mauremys caspica rivulata	Western Caspian turtle	Osteopathy and nutritional osteodystrophies	Keymer (1978a)
Mauremys leprosa	Mediterrean pond turtle	Osteopathy and nutritional osteodystrophy Calcioprive osteopathy = osteomalacia + rachitis Gout	Keymer (1978a) Wallis (1927); Dämmrich (1967); Wallis (1927) Figureres (1997)
Emydidae			
Chrysemys ornata calirostris	Ornate slider	Nutritional osteodystrophy	Keymer (1978a)
Chrysemys picta picta	Eastern painted terrapin	Osteopathy	Keymer (1978a)
Graptemys pseudogeographica ouachitensis	Ouachita map turtle	Osteopathy	Keymer (1978a)
Malaclemys sp.	Terrapin	Nutritional osteodystrophy in 31 of 100	Jackson (1980)
Pseudemys sp.	Scooter	Osteoporotic fracture and misshapened carapace (osteomalacy) Calcium pyrophosphate deposition	Frye (1973) Brogard (1987)
Terrapene sp.	Box turtle	Ulcerative shell disease, osteoporosis	Bourdeau (1988a)
Terrapene carolina	Eastern box turtle	Congenital tarsal absence	Wilkinson et al. (2004)
Terrapene carolina triunguis	Three-toed box turtle	Bridge between carapace and plastron	Boyer (1996b)
Terrapene ornata	Ornate box turtle	Curling of carapace edges (parrot beak), overgrowth of beak, and broken legs	Boyer (1996b)
Trachemys calirostris	Ornate terrapin	Metabolic bone disease	Keymer (1978a)
Trachemys scripta	Slider	Articular gout Osteodystrophy with deformed, non-calcified extremities Rachitis	López del Castillo (1998) Bourdeau (1988a) Reichenbach-Klinke (1963, 1977)
Trachemys scripta elegans	Red-eared slider	Osteopathy, nutritional osteodystrophy Articular gout Infected shoulder Secondary hyperparathyroidism producing osteodystrophy fibrosa with residual lacy calcification Osteodystrophy fibrosa with residual lacy calcification Pseudogout = calcium pyrophosphate deposition	Keymer (1978a); Jackson and Sainsbury (1992) Marcus (1981); McArthur (1996) McArthur (2004) Gabrisch and Zwart (1992) Bourdeau (1988a) Frye and Dutra (1976); Frye (1991, 1994); Rothschild and Bruno (2009)
Trachemys scripta scripta	Yellow-bellied slider	Osteopathy Gout	Keymer (1978a) López del Castillo (1998)

(continued)

Table 10 (continued)

Species	Common name	Disease	Reference
Kinosternidae			
Staurotypus triporcatus	Three-keeled terrapin	Osteopathy	Keymer (1978a)
Testudinidae			
Chelonoides carbonaria	South American red-legged tortoise	Soft, deformed carapace and plastron, caused by hyperparathyroidism	Frye and Carney (1975); Rivera and Lock (2008)
Dipsochelys (Aldabrachelys) gigantea	Aldabra tortoise	Osteodystrophia fibrosa and rickets Hyperparathyroidism	Hauser et al. (1977) Gerlach (2004)
Geochelone nigra	Galapagos tortoises	Metabolic bone disease (Zürich zoo)	Dollinger et al. (1997)
Geochelone pardalis pardalis	Leopard tortoise	Secondary nutritional metabolic bone disease Hyperparathyroidism	Raiti and Haramati (1997) Zwart et al. (1994)
Geochelone pardalis babcocki	African leopard tortoise	Ossification of liver and other internal organs	Frye (1991b)
Geochelone radiata	Radiated tortoise	Osteoporosis Osteopathy Articular gout	Isenbügel and Frank (1985) Keymer (1978b) Appleby and Siller (1960)
Geochelone sulcata	African spurred tortoise	Increased mineralization from secondary renal hyperparathyroidism Articular gout	Wilkinson et al., 2004 Appleby and Siller (1960)
Gopherus sp.		Nutritional osteodystrophy	Bourdeau (1988b)
Gopherus agassizii	California desert tortoise	Uncalcified Gout, osteomalacy Osteoclastic resorption, osteopenia, articular gout Nutritional bone disease Calcium pyrophosphate deposition	Boyer (1996b); Mader (2006d); Mader (1990a) Frye (1984); Homer et al. (1998) Homer et al. (1998) Mader (2006c) Rothschild and Bruno (2009)
Kinixys belliana	Bell's hinged tortoise	Osteopathy	Keymer (1978b)
Testudo denticulata	Jaboty or rain forest tortoises	Nutritional osteodystrophy articular gout	Keymer (1978b) Appleby and Siller (1960)
Testudo elegans	Starred tortoise	Osteopathy and nutritional osteodystrophy	Keymer (1978b); López del Castillo (1998)
Testudo graeca	Mediterrenaean spur-thighed tortoise	Nutritional osteodystrophy Demineralization due to secondary nutritional hyperpareathyroidism Osteopathy Articular gout Articular pseudogout = calcium pyrophosphate deposition	Jackson and Duff Fasal (1981) Wilkinson et al. (2004) Keymer (1978b) Appleby and Siller (1960) López del Castillo (1998); Rothschild and Bruno (2009)
Testudo graeca ibera	Turkish tortoise	Metabolic bone disease Shell "infractions" and osteolysis from suppurated infections Hyperparathyroidism	McArthur (2004) Stojanov (2005) Gerlach (2004)

Testudo hermanni	Hermann's tortoise	Osteodystrophia fibrosa and rickets	Dämmrich (1967, 1979)
		Short mandible, misshapened limb bones, and humped carapace shields of nutritional osteodystrophy	Jackson and Cooper (1981)
			Keymer (1978b)
		Osteopathy	Wilkinson et al. (2004)
		Carapace bone loss from secondary hyperparathyroidism	Frye (1981)
		Elbow osteolysis with osteochondromatosis	Appleby and Siller (1960); López del Castillo (1998)
		Articular gout	
Testudo horsfieldi	Horsfield's tortoise	Osteopathy	Keymer (1978b)
Testudo marginatta	Marginated tortoise	Hypoplastic osteoporosis	Häfeli and Zwart (2000)
		Osteopathy and osteodystrophy	Keymer (1978b)
		Metabolic bone disease	McArthur (2004)
Trionychidae			
Apalone mutica	Smooth softshell	Carapace curling related to new born mouse diet	Barten (1982, 1983)
Pleurodira			
Chelidae			
Emydura subglobosa	Red-bellied short-neck turtle	Periarticular hydroxyapatite deposition	Wenker et al. (1999)
Pelomedusidae			
Pelusios subniger	Blackish terrapin	Osteopathy	Keymer (1978a)
Unidentified	Turtles	Nutritional osteodystrophy	Cowie (1976); Jackson (1987)
	tortoises	Osteomalacy	Bourdeau (1988a)
		Gout	Marcus (1968)
		Gout	Frye (1981); Messonnier (1996); McArthur (1997); Arvy and Fertard (2002); Mader (1996)
		Metabolic bone disease	Mader (1990b)
		Hyperparathyoridism	Frye (1981); O'Malley (2008)
		Osteoporosis, rachitis, osteomalacy, osteodystrophia fibrosa	Jarofke and Lange (1993)

Articular pseudogout – calcium pyrophosphate deposition disease.
Gout – a metabolic disorder in which sodium uric acid crystals deposit in joints (referred to as articular gout).
Hyperparathyroidism – disorder caused by overactivity of the named glands, producing osteitis fibrosa cystica (= fibrous tissue replacement of bone secondary) and other bone changes.
Osteodystrophy – defective bone formation.
Osteomalacia – softening of bone due to a lack of vitamin D.
Osteopathy – combination of osteomalacia (= vitamin D deficiency-related softening of bone) and rickets.
Osteopenia – bone mineral density lower than normal.
Osteoporosis – reduction in quantity and quality (e.g., thickness) of trabecular components of bone.
Rickets – failure of bone osteoid to calcify (= rachitis).

event (Frye 1981), causing a soft, deformed carapace and plastron (Frye and Carney 1975). Parathyroid adenomas (a benign neoplasm of glandular cells) have been reported in the South American red-footed tortoise, *Geochelone carbonaria*; the Mediterranean spur-thighed tortoise, *Testudo graeca*; and the California desert tortoise, *Gopherus agassizii* (Frye 1994; Frye and Carney 1975). The other component, osteomalacia (also called rickets prior to epiphyseal closure), is caused by vitamin D deficiency. In addition to softened, malformed bones, enlarged joints and soft shells, widened calcitic rings occur near epiphyseal junctions (Glazebrook 1980; Wallach 1971). Osteomalacia has been noted in the painted turtle, *Chrysemys picta* (Reichenbach-Klinke 1977), ornamental turtles, *Terrapene ornata* (Reichenbach-Klinke and Elkan 1965), and unspecified captive turtles (Hunt 1957).

Most reptiles excrete nitrogen predominantly as uric acid; aquatic turtles, as urea; and sea turtles, as ammonia (Allen and Oftedal 1994). Urate disease occurs from buildup of uric acid in blood or focal locations. Hyperuricemia (elevated uric acid level) results from water deprivation (e.g., post-hibernation dehydration), excess protein or shrimp (high in uric acid) intake, or renal disease (Frye 1981; Marcus 1981; McArthur 1996). Vitamin A deficiency produces renal dysfunction, which produces secondary gout (a metabolic disorder in which sodium urate crystals deposit in joints or internal organs). The latter condition was reported in a red-eared terrapene, *Trachemys scripta elegans* (McArthur 1996). Uric acid may be deposited in the kidneys (in the form of stones, discussed in section on stones below), in joints (referred to as articular gout), and around internal organs (referred to as visceral gout) (Rothschild 2009b). The latter is beyond the scope of this review and will not be discussed further. Articular gout is well recognized in turtles (Table 10), either by presence of classic negatively birefringent (retard light transmission, as visualized by polarizing microscopy) crystals in synovial fluid or by the presence of characteristic bone erosions with sclerotic margins and overhanging edges (Casimire-Etzioni et al. 2004; Rothschild 2009b; Rothschild and Heathcote 1995), typically affecting joints of the forearm, shoulder, hip, and stifle (Homer et al. 1998; López del Castillo 1998; McArthur 2004). Mader (2006c) reported destruction of cervical vertebrae.

Calcium pyrophosphate deposition disease (= articular pseudogout) has been reported in turtles, specifically the red-eared slider, *Trachemys scripta elegans* (Frye and Dutra 1976; Frye 1994; Rothschild and Bruno 2009), *Pseudemys* (Brogard 1987), California desert tortoise, *Gopherus agassizii* (Rothschild and Bruno 2009), and Mediterranean spur-thighed tortoise, *Testudo graeca* (López del Castillo 1998; Rothschild and Bruno 2009). Hydroxyapatite deposition disease in red-belly short-necked turtle, *Emydura albertisii*, is recorded by Wenker et al. (1999). Frye (1991b) reported ossification of liver and other internal organs of African leopard tortoise *Geochelone pardalis babcocki*.

Bladder stones are well recognized in Chelonia (Jackson and Cooper 1981). Kölle et al. (2001) reported calculi in 4% of necropsied tortoises. Keymer (1978b) reported 4.2% in tortoises. Osborne et al. (2008) suggested that 94% of stones were urate, 3% calcium carbonate, and 3% mixed in tortoises. This contrasts with 58% urate, 17% calcium phosphate, 17% calcium carbonate, and 8% mixed in turtles.

Salts (ammonium, sodium, calcium, and potassium) of uric acid also deposit in renal tubules or as bladder stones or calculi (Kölle et al. 2001; Mebs 1965). Whereas isolated reports are delineated in Table 11, frequencies are more difficult to assess. Homer et al. (1998) found 3 California desert tortoises, *Gopherus agassizii*, with urate stones among 24 ill or dead. Kölle et al. (2001) reported presence of calculi in 4% of necropsied tortoises and lizards, noting that the tortoises, *Testudo* and *Geochelone*, were most often affected among chelonians, whereas Grünberg et al. (1977) claimed that uric acid/urate calculi only occur in carnivorous species. Mangone and Johnson (1998) reported that calculi are frequently encountered in many species of captive tortoise. Kölle and Hoffmann (2002) reported that 64% of European tortoises (especially spur-thighed tortoises – as more sensitive to high-protein diet than Afghan tortoises) had "renal alterations" on necropsy. Sixteen percent had gouty tophi in the kidney (Kölle and Hoffmann, 2002). They usually occur as single stones. Mader et al. (1999) reported single calculi in 73; two, in 2; three, in 5; and one Californian desert tortoise, *Gopherus agassizii*, with 5 stones, ranging in size from 0.4 to 14 cm (average = 5.4 cm). Calcium carbonate, calcium phosphate, calcium sulfate, and magnesium phosphate stones have been found in marine turtles (Grünberg 1963/64; Virchow 1878). Basic calcium phosphate, calcium apatite, and struvite (magnesium ammonium phosphate hexahydrare) crystals (Table 11) have been noted. A pseudo-calculus (actually egg) was repored by Mader (2006b) in a California desert tortoise, *Gopherus agassizii.* Low-calcium diet produces caliculi (Wallach 1969).

Osteomyelitis in turtles produces carapace and plastron necrosis. It can be caused by "ordinary" bacteria (e.g., *Pseudomonas*, *Citrobacter*, *Klebsiella*), mycobacteria, fungi (especially Mucor), and algae (Arvy and Fertard 2002; Barnett 2003; Table 12). Shell rot has been attributed to *Mucorales*, *Rusarium*, *Geotrichum*, *Trichosporon*, *Caniothyrium*, algae, and *Beneckea chitinovora* (Garner et al. 1997). The forearm abscess in a Burmese mountain tortoise, *Manouria emys*, reported by Jacobson (1994) was associated with a focal ulcerative lesion of the plastron and linked to a coelomic cavity of a bacterial abscess. Septic arthritis describes a joint infection (Fig. 81). Chelonian infections have been attributed to "ordinary" bacteria, mycobacteria, and fungi (Table 12; Rhodin et al. 1990). Erosion of the distal humerus and replacement of elbow joint with fibrous ankylosis with reactive overgrowth and sclerosis of proximal radius and ulna reported in leatherback turtle, *Dermochelys coriacea* (Ogden et al. 1981), suggests infection. Plastron lesions and swollen left elbow with osteolytic radial and ulnar lesions with sclerotic margins in an Atlantic ridley sea turtle, *Lepidochelys kempii*, were attributed to *Mycobacterium chelonae* (Greer et al. 2003). *Nocardia* and unidentified fungus were isolated from carpal swelling with proximal first metacarpal radiolucency, obliteration of distal carpal row, ulnare, pisiform, and metacarpals II and III in an Atlantic Ridley sea turtle, *Lepidochelys kempii*, by Harms et al. (2002).

Neoplasms (literally, new abnormal tissue) are rare in chelonians (Arvy and Fertard 2002). They are caused by abnormal growth of cells and may be benign or malignant. Garner et al. (2004) reported prevalence referred to his clinic of 2.7–3.2% in turtles and 1.4% in tortoises, without comment on bony involvement. He did note that metastasis was uncommon. Sykes and Trupkiewicz (2006) reported that neoplasia was found only in turtles (not tortoises) in the Philadelphia zoo from 1901 to 2002. The frequency ranged from 0.3% from 1901 to 1967, 0 from 1968 to 1979, 2.4% from 1980 to 1991, and 0 from 1992 to 2002, for a total of 6 afflicted among 511 turtles. Zwart and Harshbarger (1991) reported a Hermann's tortoise, *Testudo hermanni*, with squamous cell carcinoma that caused pathologic fracture of femur and necrosis of the adjoining carapace. A neurilemmal sarcoma was found in *Testudo hermanni* (Cooper et al. 1983). Shell neoplasia is extremely rare. One case in a California desert tortoise, *Gopherus agassizii*, was classified as a chondroma (Rosskopf 1986). Plastron nodules in the loggerhead, *Caretta caretta*, were caused by a lymphoblastic lymphoma (Mauldin and Done 2006; Orós et al. 2001). Weems (1974) described a "tumorous bony growth" in *Syllomus aegyptiacus* USNM 24872, which the senior author recognized as an osteoma or osteoblastoma Rothschild et al, 2011. The specimen had been divided, revealing the characteristic fine trabecular pattern, characteristic of that benign tumor (Resnick 2002).

Frye (1981, 1994) reported osteochondromatosis (= benign neoplasm producing a cartilage cap on an exostosis) and osteolysis (= resorption or destruction of bone) in the elbow of a Hermann's tortoise, *Testudo hermanni.* While this could be the primary disease of numerous osteochondromatous nodule production, a more likely explanation is that it occurred secondary to the osteolytic process and actually represents a neuropathic process. Frye's (1981, 1994) reported osteochondromatosis in Hermann's tortoise, *Testudo hermanni*, that appears to be the result of denervation. Such dramatic osteolysis and new bone formation, with

Table 11 Stone disease in turtles

Species	Common name	Kind of stone	Reference
Cryptodira			
Cheloniidae			
Chelonia mydas	Green turtle	Calcium carbonate, calcium phosphate, calcium sulfate, magnesium phosphate	Virchow (1878)
Emydidae			
Trachemys scripta elegans	Red-eared slider	Uric acid Calcium phosphate	Mebs (1965) Mader (2006b)
Testudinidae			
Chersina angulata	South African bowsprit tortoise	Uric acid	Mebs (1965)
Dipsochelys (Aldabrachelys) gigantea	Aldabaran giant tortoise	Not specified Uric acid	Keymer (1978b) Hamerton (1934, 1938); Grünberg (1963/64)
Geochelone elegans	Indian starred tortoise	Not specified	Keymer (1978b)
Geochelone nigra	Galapagos tortoise	Uric acid	Grünberg (1963/64); Wallach (1971)
Geochelone pardalis	Leopard tortoise	Calcium phosphate	Mader (2006b)
Geochelone ?platynota	Burmese star tortoise	Uric acid	Fife (2007)
Geochelone sp.		Uric acid + other substances	Kölle et al. (2001)
Geochelona sulcata	African spurred tortoise	Uric acid Calcium phosphate (urolith)	Raiti (2004) Mader (2006b)
Gopherus agassizii	Desert tortoise	Bladder stone Uric acid	Long Beach Animal Hospital (2007) Frye (1972); Mader et al. (1999); Mangone and Johnson (1998); Homer et al. (1998)
Kinixys belliana	Bell's hingeback tortoise	Uric acid	Appleby and Siller (1960)
Kinixys homeana	Kuhl's tortoise	Not specified	Keymer (1978b)
	Desert tortoise	Bladder stone	Long Beach Animal Hospital (2007)
Testudo graeca	Spur-tighed tortoise	Not specified Uric acid/urat Tophi gout	Keymer (1978b) Grünberg (1963/64); Ebstein (1899) Kölle and Hoffmann (2002)
Testudo hermanni	Hermann's tortoise	Not specified	Keymer (1978b)
Testudo sp.		Uric acid+ other substances	Kölle et al. (2001)
Trionychidae			
Apalone spinifera hartwegi	Western spiny softshell	Calcium phosphate apatite and struvite	Mader (2006b)

Table 12 Infectious diseases (osteomyelitis) in turtles

Species	Common name	Site affected	Organism	Reference
Cryptodira				
Carettochelyidae				
Carettochelys insculpta	Fly river turtle		*Paecilomyces lilacinus*	Lafortune et al. (2005); Paré and Jacobson (2007)
Cheloniidae				
Caretta caretta	Loggerhead	Shell shell	*Fusarium solani* Barnacles	Rebell et al. (1971); Austwick and Keymer (1981); Rothschild (2009a) Jacobson (2007b)
Lepidochelys kempii	Atlantic ridley sea turtle	Plastron, elbow	*Mycobacterium chelonei*	Greer et al. (2003); Jacobson (2007a)
		Carpals, metacarpals	*Nocardia* and unidentified fungus	Harmes et al. (2002); Jacobson (2007a); Rothschild (2009a)
Chelydridae				
Chelydra serpentina	Common snapping turtle	Not listed	*Erysipothrix*	Jacobson (2007a)
		Plastron	Fungi	Rothschild (2009a)
Dermochelyidae				
Dermochelys coriacea	Leatherback	Erosion of distal humerus and replacement of elbow with fibrous ankylosis with reactive overgrowth of proximal radius and ulna (arthritis)	Not listed	Ogden et al. (1981) Coquelet (1983); Brogard (1987)
Emydidae				
Chrysemys sp.	Painted turtle	Shell	*Beneckea chitinovora*	Brogard (1980); Wallach (1975); Jacobson (2007a); Rothschild (2009a)
Pseudomys concinna	Eastern river cooter	Shell	*Mucorales, Rusarium, Geotrichum, Trichosporon, Coniothyrium*, algae, and *Beneckea chitinovora*	Garner et al. (1997)
Trachemys scripta scripta	Yellow-bellied slider	Shell	*Mucorales, Rusarium, Geotrichum, Trichosporon, Coniothyrium*, algae, and *Beneckea chitinovora*	Garner et al. (1997)
Trachemys scripta elegans	Red-eared slider	Loosened plates around sutures	*Beneckea chitinovora*	Wallach (1975); Jacobson (2007a)
		carapace	*Trichosporon*	Schildger et al. (1991)
Trachemys sp.	Slider	Loosened plates around sutures	*Beneckea chitinovora*	Wallach (1975); Jacobson (2007a); Rothschild (2009a)
Emydide indet		Shell	*Spirorchis*	Jacobson (2007b)
Geoemydidae				
Heosemys annandalii	Yellow-headed temple turtle	Shell rot	*Beneckea chitinovora*	Guirley (2003)

(continued)

Table 12 (continued)

Species	Common name	Site affected	Organism	Reference
Kinosternidae				
Sternotherus sp.	Musk turtle	Carapace	*Beneckea chitinovora*	Wallach (1975); Brogard (1980); Coquelet (1983); Jacobson (2007a); Rothschild (2009a)
Testudinidae				
Astrochelys radiata	Radiated tortoise	Mandible	Chromomycosis	Keymer (1978b); Jacobson (1994); Rothschild (2009a)
Dipsochelys (Aldabrachelys) gigantea	Aldabra tortoise	Plastron	Fungi	Hammerton (1935); Austwick and Keymer (1981)
		Carapace	*Exophiala oligosperma*	Stringer et al. (2009)
Geochelone elegans	Indian star tortoise	Temporomandibular joints and knees	Mycotic infection	McArthur (2004)
Gopherus agassizii	Desert tortoise	Leg	Mycobacteria	Frye (1994d)
		Shell (plastron)	food: *Astragalus*, *Stanleya*, and *Xylorrhiza*	Jacobson et al. (1994)
Manouria emys	Asian brown tortoise	Forelimb	Bacterial abscess	Jacobson (1994)
Testudo sp.		Distal tibia, fibula, and tarsals	Not listed	Jackson and Sainsbury (1992)
Testudo graeca	Spur-thighed tortoise	Shell	*Coniothyrium fuckelianum*	Goodwin (1976); Austwick and Keymer (1981); Rothschild (2009a)
Testudo hermanni	Hermann's tortoise	Carapace	Bacterial infection	Schildger et al. (1991)
Trionychidae				
Trionyx sp.	Softshell turtle	Shell	*Beneckea chitinovora*	Wallach (1975); Brogard (1980); Jacobson (2007a)
Trionychidae indet.		Caseated crateriform ulcers plastron	*Citrobacter freundii*	Boyer (1996a)
		Shell (necrosis)	*Mucorales*	Jacobson (1980)
Cryptodira indet.		Shell	*Beneckea chitinovora*	Marcus (1980)
Pleurodira Chelidae				
Acanthochelys macrocephala	Pantanal swamp turtle	Septic arthritis	Not listed	Rhodin et al. (1990)
Chelodina longicollis	Long-necked terrapin	Plastron	Fungi	Hammerton (1939); Austwick and Keymer (1981); Rothschild (2009a)
Chelus fimbriatus	Matamata terrapin	Plastron	Fungi	Hammerton (1934); Austwick and Keymer (1981); Rothschild (2009a)
Emydura subglobosa	Red-bellied short-necked turtle	Hydroxyapatite deposition disease	Not listed	Wenker et al. (1999)
Podocnemidae				
Podocnemis sp.	South American river turtle	Shell	*Beneckea chitinovora*	Brogard (1980); Arvy and Fertard (2002)
Pleurodira indet.		Loosened plates around sutures	*Beneckea chitinovora*	Wallach (1975); Rothschild (2009a)

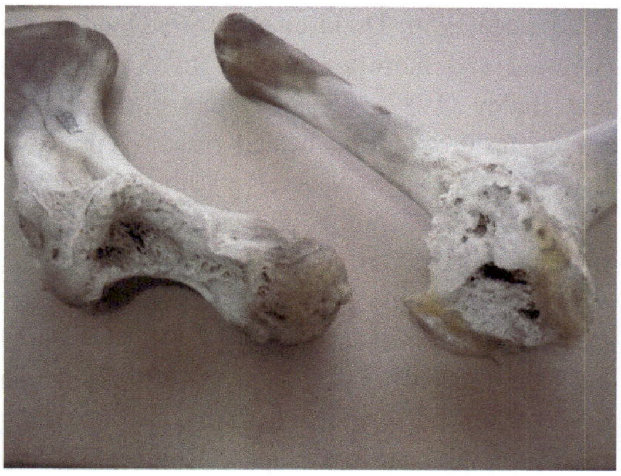

Fig. 81 Disruption from infectious process; face view of pelvic girdle of *Caretta caretta* (IVPP 1206)

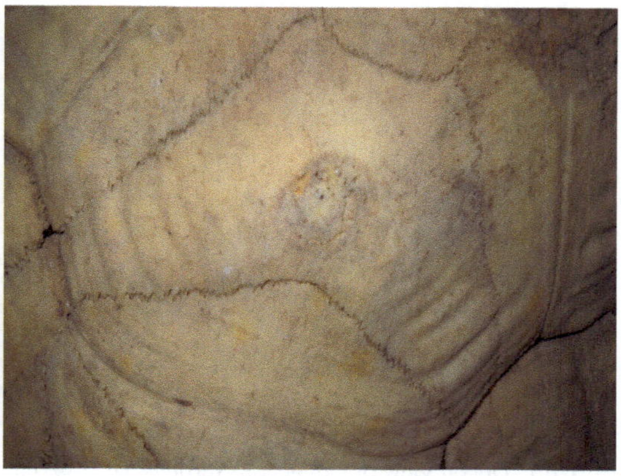

Fig. 83 Superficial deposit on carapace surface; dorsal view of carapace of *Geochelone nigra* (IVPP 1445)

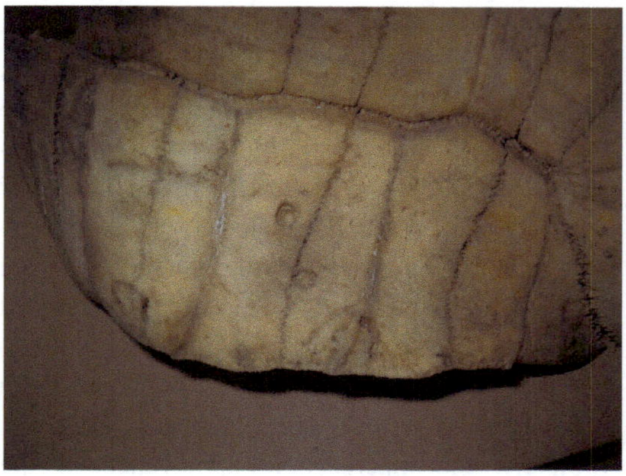

Fig. 82 Multiple defects appear to contain multiple concentric pits; dorsal view of carapace of *Geochelone nigra* (IVPP 1445)

Fig. 84 Dorsal view of Late Jurassic *Plesiochelys* carapace. Multiple surface defects

formation of multiple osteochondral nodules (correctly referred to as osteochondromatosis), is characteristic of "neuropathic disease" (Resnick 2002; Rothschild 2011, p. 57: "partial or complete loss of intervertebral space associated with new bone and irregular hyaline cartilage island formation"), not previously recognized in turtles.

Carapace and plastron damage, also known as shell disease (Figs. 82–85), has been a controversial subject, with pathology variously attributed to trauma, infection, and metabolic disease. Trauma may be direct or contributory. Subsequent infected carapace and plastron injuries may occur from surf injuries from contact with rocks (Balazs 1980; Wright et al. 1977). Cooper and Jackson (1981) reported rostral abrasions, bites and miscellaneous lacerations, bites by cage mates, and crushing. Puncture wounds on posterior and posterodorsal carapace surfaces with thickened bone suggest predator interactions in *Protochelydra* and a fossil trionychid (Erickson 1984). While some damage has been attributed to specific infectious agents (Jacobson 1994; Rosskopf 1986), simple descriptions of the

Fig. 85 Dorsal view of Cretaceous turtle (KU 2249) carapace fragment. Holes reveal bite marks by mosasaur

pathology seem inadequate to allow à priori etiologic assignment. It is unclear if better descriptions of pathology will identify which agent is responsible or if the limited manner in which bone can respond to insults precludes identification of the specific infection from gross examination, unless the organism is seen, components isolated or actual organism recovered (cultured). Perhaps attention to such details as elevated margins and their sharpness or curved nature (as noted by Sowerby and Lear 1872) and the oval shell lesions in Chinese softshell turtle, *Trionyx sinensis*, reported by Wu (1994) might provide clues to specific agents. Most of the discussion of shell disease revolves around pitting, ulcerative shell disease (Rebell et al. 1971), and necrosis. However, Hutchison and Frye (1989) described bone sclerosis (associated with osteolysis) in a fossil Emydidae, which they interpreted as osteomyelitis. There is inadequate description or illustration to further pursue perspectives of shell sclerosis.

Disfiguring shell disease with dermal bone remodeling was reported in eastern river cooter, *Pseudemys concinna*, and yellow-belly turtles, *Trachemys scripta scripta* (Garner et al. 1997; Jacobson 2007a; Lovich et al. 1996). Jacobson (2007a) referred to the damage as segmental necrosis and remodeling. Much of the literature simply reports isolated observations. Epidemiologic studies have revealed 6.5% of three-toed box turtle, *Terrapene carolina triunguis*, and Florida box turtle, *Terrapene carolina bauri*, had carapace pits (Carpenter 1956; Dodd et al. 1997; Dodd 2001; Schwarz and Schwartz 1974). Rosskopf (1986) and Bailey (1987) suggested that pitting represented residua of old, healed osteomyelitis, but that is only one of several possible diagnoses. Rosskopf (1986) suggested that shell ulcerations in diamondback terrapin, *Malaclemys terrapin*, resulted from too much time in freshwater (Rosskopf 1986). Jacobson et al. (1994) suggested possible relationship of shell necrosis in California desert tortoise *Gopherus agassizii* to the spring food, locoweed *Astragalus*, princesplume *Stanleya* and woody aster *Xylorrhiza*, which produces aliphatic nitro compounds. The situation may be even more complex. Garner et al. (1997) reported shell necrosis among turtles with pancreatitis and iron deposition (possible hemochromatosis); inflammation of the pancreas can produce bone necrosis in humans (Resnick 2002). Hutchison and Frye (1989) raised the possibility that the high frequency of pitting in painted turtle, *Chrysemys picta*, females might be related to calcium/phosphate withdrawal for egg laying. Shell disease is a common problem, as Lovich et al. (1996) reported 74% of eastern river cooter, *Pseudemys concinna*, and 35% of slider, *Trachemys scripta*, from Lake Blackshear, Georgia had carapace necrosis. Typically, the macroscopic appearance of the pathology is not fully described; only the putative responsible infectious agent.

Predation injury has also been considered for the origins of pits. Holes, matching predator (mosasaur teeth size, shape, and distribution), have been observed in *Protostega* sp. *gigas* (Rothschild and Martin 2006; Rothschild 2009a). Those in *Echmatemys* UCMP 128283 and *Hadrianus corsoni* UCMP 128418 were attributed to crocodilian bites (Hutchison and Frye 1989; Rothschild 2009a). The latter appeared as conical perforations from predation. They were divided into circular to ovoid with flat bottoms, versus rounded bottoms, irregular pits with discrete margins, track (linear) and rot, the latter represented by large areas of irregular depression or dead lamellar bone. Scratches may imply failed predation.

Carapace necrosis and osteomyelitis have been attributed to microorganisms (Barnett 2003), specifically *Beneckea chitinovora* (Table 12; Marcus 1980), ordinary bacteria (e.g., *Aeromonas, Arizona, Bacteroides, Citrobacter, Enterobacter, Escherichia coli* and *freundii, Klebsiella, Pasteurella, Proteus, Providencia, Pseudomonas, Salmonella typhimurium, regent* and *amrina, Serratia, Staphylococcus aureus*, alpha-hemolytic *Streptococcus*; Arvy and Fertard 2002; Highfield 1990b; López del Castillo 1996; Rosskopf 1986), acid fast organisms (e.g., *Mycobacterium* and *Nocardia*; López del Castillo 1996), spirochetes (Jacobson 2007b), fungi (e.g., *Aspergillus, Basidolobus, Candida albicans, Coniothyrium, Dermatophyton, Fusarium solanae, Geotrichum, Mucor, Mucorales, Paecilomyces, Penicillium, Rusarium, Trichosporon, Trichoderma, Trichophyton*; Garner et al. 1997; López del Castillo 1996; Rosskopf 1986), algae (Garner et al. 1997; Marcus 1971), or even barnacles and parasites (Jacobson 2007b; Weems 1974 on fossil turtle). Barnett (2003) attributed dry forms to fungus and wet forms to *Pseudomonas* or coliform bacilli, *Citrobacter*, and *Klebsiella*. Shell necrosis in fungal infections especially affects the plastron. Shell infections are often of mixed bacterial and fungal origin (Lafortune et al. 2005; Schumacher 2003). Alleged fungal agents reportedly involved in shell damage include *Aspergillus, Basidobolus ranarum, Beauvaria bassiana, Candida albicans, Cladosporium, Fusarium, Geotrichum, Microsporum, Mucor, Paecilomyces, Penicillium*, and *Rhodotolura* (Lafortune et al. 2005; Schumacher 2003). Turtles are more susceptible to these agents when subjected to suboptimal environmental conditions of temperature or humidity, compromise of hygiene or chronic stressors such as overcrowding. Fifty percent of loggerhead and 25% of Kemp's ridley turtles had dermatomycotic scutes, sometimes ulcerating underlying bone (Duguy et al. 1998; Paré and Jacobson 2007). Irregular lytic carapace lesions were attributed to granulomatous disease (usually used to describe the effects of acid fast or fungal infection) in eastern long-necked turtle, *Chelodina longicollis* (Gabrisch and Zwart 1992). While Rosskopf (1986) attributed a hole in the carapace of a Galapagos tortoise, *Geochelone nigra galapagoensis*, as caused by maggots, it was probably a secondary phenomenon. The situation is more complex, and mixed infections may also be responsible. Rosskopf (1986) suggested that *Beneckea chitinovora*–induced shell disease in freshwater turtles requires intermediary crustacean hosts.

Interestingly, pathology and taxonomy have crossed paths in the form of the species name applied to a pond turtle from Europe, North Africa, and Western Asia. Elkan (1983) identifies Pritchard (1967) as attributing the name *Mauremys leprosa* to shell flaking related to a fungal disease. They thought it mimicked leprosy.

Evidence of genetic malformations/anomalies is rare in the fossil record. Gaffney and Tong (2008) noted lack of frontal bones (and septum orbitotemporale) in *Ummulisani rutgersensis* AMNH 30569, a bothremydid side-necked turtle from the Eocene of Morocco.

Tasnádi-Kubacska (1962) reported a shortened fibula and tibia in a Cretaceous *Archelon ischyros*, and Weems (1974) described *Syllomus crispatus aegyptiacus* (USNM 24872) with a bony ridge located between the trochanter and the head of the femur. Zangerl and Turnbull (1955) reported a supernumerary peripheral carapace element in Miocene *Procolpochelys grandaeva*.

Trauma has also rarely been reported. Wieland (1909) reported an *Archelon ischyros* with obliquely bitten, healed tibia and fibula. Hutchison and Frye (1989) reported linear excavations (probable scratch marks) in *Baptemys garmanii* and *Chisternon*. A "shark tooth (from a species related to *Lamna*) was found with the type" of *Dermochelys* (Wieland 1909, p. 120). Erickson (1984, p. 4) reported puncture wounds on posterior and posterodorsal carapace surfaces with thickened bone, affecting *Protochelydra*, a trionychid and "at least two other forms." Shallow, well-rounded pits were noted on outer shell surface of *Syllomus aegyptiacus* (Weems 1974). Hutchison and Frye (1989) noted circular to irregular shaped pits with well-defined margins in Wyoming Eocene *Echmatemys*,

Fig. 86 Multilobular tumor; dorsal view of carapace fragment of *Syllomus aegyptiacus* (USNM 24876)

Fig. 87 Multilobular tumor; close-up dorsal view of carapace fragment of Miocene *Syllomus aegyptiacus* (USNM 24876) from Egypt

Baptemys, and *Hadrianus corsoni* with conical perforations from predation. They divided pits into circular to ovoid with flat bottoms, versus rounded bottoms, irregular pits with discrete margins, track (linear) and rot, represented by large areas of irregular depression or dead lamellar bone. *Echmatemys euthenta* had more than 40 1–3 mm punctate pitting lesions. 9 mm × 9 mm shallow pits and full-thickness partially remodeled pits in a specimen of *Echmatemys* (UCMP 128283) are attributed to crocodilian bite. Weems (1974, p. 299) noted that circular depressions on the right third and fourth costals of *Procolpochelys grandaeva* USNM 24889 were similar to "the basal configuration of small specimens of *Balanus concavus*." As already mentioned, Hutchison and Frye (1989) also suggested the possibility that high frequency of pitting in the northern painted turtle, *Chrysemys picta*, females might be related to calcium/phosphate withdrawal for egg laying.

Shell rot has been reported in Eocene (Bridgerian) trionychids, 10/17 emydids, and 1/7 *Dermatemys* (Hutchison and Frye 1989; Williams and Bartels 1994). All five *Echmatemys septaria* had extensive rot, compared with three of seven *E. wyomingensis*. Emydide specimen UCMP 128418 had near-half-thickness osteolysis crossing sutures and associated with sclerotic bone. Hutchison and Frye (1989, 2001) interpreted the changes as osteomyelitis. Amorphous carapace lumps were reported on the carapace (Figs. 86–88) of USNM 24876 (Weems 1974). Weems (1974,

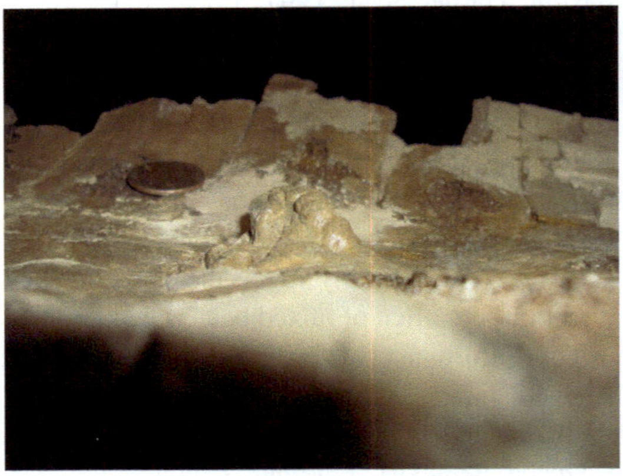

Fig. 88 Multilobular tumor; oblique dorsal view of carapace fragment of Miocene *Syllomus aegyptiacus* (USNM 24876) from Egypt

Fig. 89 Fine trabecular pattern identifies an osteoblastoma; cross-section of mass in carapace of Miocene *Syllomus aegyptiacus* (USNM 24874) from Egypt

Fig. 90 Tumor identified as osteoblastoma; dorsal view of carapace fragment of Miocene *Syllomus aegyptiacus* (USNM 24874) from Egypt

p. 292) reported *Syllomus aegyptiacus* USNM 24874 with a "tumorous bony growth." The senior author recognized it (Figs. 89 and 90) as an osteoblastoma Rothschild et al. 2011.

Presence of avascular necrosis (= blocked blood vessel-related connected tissue death from bends/decompression syndrome) allowed recognition of diving behavior in Cretaceous mosasaurs and their prey, turtles (Martin and Rothschild 1989). Avascular necrosis was limited to marine turtles, with no cases in terrestrial and one instance in a fossil Mediterranean *Trionyx*, which lived in marine habitat at that time (Rothschild 1988). Avascular necrosis was and is present in 8 families of marine/aquatic turtles from the Cretaceous to Recent (Table 13; Rothschild 1987, 1988, 1991). Desmatochelyidae, Toxochelyidae, Protostegidae, and Pleurosternidae were especially afflicted in the Cretaceous. Frequency of avascular necrosis diminished from 41% in the Cretaceous to 9% in the Eocene, 5% in the Oligocene, and 0.3% in the Holocene (Rothschild 1991). Reduction in frequency of avascular necrosis in the early Eocene was followed by near disappearance, subsequent to the Oligocene. Contemporary representation is limited to its extremely infrequent occurrence in Cheloniidae (Ridley's sea turtle, *Lepidochelys olivacea*; Eckert et al. 1986, 1989), Chelydridae (snapping turtles, *Chelydra* and *Macroclemys*), and Kinosternidae (mud turtle, *Kinosternon*) (Rothschild 1987).

Table 13 Avascular necrosis of affected turtle families (Rothschild 1991)

	Number	Avascular necrosis	
Family	evaluated	Number	Proportion
Baenidae	14	1	.07
Protostegidae	6	4	.67
Toxochelyidae	17	5	.29
Cheloniidae	170	12	.07
Desmatochelyidae	6	2	.33
Dermochelyidae	74	4	.05
Pleurosternidae	4	1	.25
Trionychidae	204	2	.001

Annotated Bibliography A-C

Abate AL, Coke R, Ferguson GW, Reavill D. 2003. Chameleons and vitamin A. Journal of Herpetological Medicine and Surgery 13:23–31.

Metabolic – Hypervitaminosis A, more common in Jackson's chameleon *Chamaeleo jacksonii*, mountain chameleon *Chamaeleo montium*, and Johnston's chameleons *Chamaeleo johnstoni*, as manifested by osteomalacia, soft tissue calcification, tail tip necrosis and vertebral kinking, and poorly formed, irregular cement lines.

"Nutritionally related lesions" were reported in Jackson's, panther, and veiled chameleons *Chamaeleo calyptratus*.

Abe Y. 1952. Duplicitas anterior in *Elaphe climacophora*. Annotationes Zoologicae Japonenses 25(1–2):69–70.

Congenital – Derodymous *Elaphe climacophora*, noting previous reports of dicephalic *Elaphe dione* (Nishimura 1938), red-backed ratsnake *Elaphe rufodorsata* (Kuroda 1919), and dicephalic and derodymous Japanese pit viper *Agkistrodon blomhoffii*, respectively (Minobe 1980; Yoshinaga 1901).

Abel O. 1912. Grundzüge der Paläobiologie der Wirbeltiere. [Essentials of paleobiology of vertebrates. Stuttgart: E. Schweizerbart'sche Verlagsbuchhandlung,] 708 pp. [German].

Trauma – Healed fractures in Liassic ichthyosaurs (ribs), mosasaurs *Plioplatecarpus marshi* (ribs), and *Mosasaurus giganteus* (right lower jaw).

Infection – Fungal lesion (or possible actinomycosis) in left femur.

Vertebral – Pachyostosis of second sacral vertebra in the sauropterygian *Proneusticosaurus silesiacus* and overgrowth on sacral vertebra in parasuchian *Metriorhynchus* cf. *moreli* (Auer 1909).

Dental – Caries in *Mosasaurus giganteus*.

Healed fractures in Liassic ichthyosaurs (ribs), mosasaurs *Plioplatecarpus marshi* (ribs), and *Mosasaurus giganteus* (right lower jaw).

Infection – Fungal lesion (or possible actinomycosis) in left femur.

Pachyostosis of second sacral vertebra in the sauropterygian *Proneusticosaurus silesiacus* and overgrowth on sacral vertebra in parasuchian *Metriorhynchus* cf. *moreli* (Auer 1909).

Caries in *Mosasaurus giganteus*.

Abel O. 1922. Schnauzenverletzungen der Parasuchier und ihre biologische Bedeutung [Mouth injuries of parasuchians and their biological importance]. Paläontologische Zeitschrift 5:26–57 [German].

Trauma – Luxurious growth on the snout of parasuchians *Mystriosuchus plieningeri* and *M. planirostris, Phytosaurs kapffi*, and *P. buceros* explained as callus formation after injury during sexual competition fights. Describes injuries and callus formation on jaws and toes in *Alligator mississippiensis* and in whales (*Choneziphius planirotris* from Miocene).

Fossil – Luxurious growth on the snout of parasuchians *Mystriosuchus plieningeri* and *M. planirostris, Phytosaurs kapffi*, and *P. buceros* explained as callus formation after injury during sexual competition fights. Describes injuries and callus formation on jaws in whales (*Choneziphius planirotris* from Miocene).

B.M. Rothschild et al., *Herpetological Osteopathology: Annotated Bibliography of Amphibians and Reptiles*, DOI 10.1007/978-1-4614-0824-6_5, © Springer Science+Business Media, LLC 2012

Abel O. 1924. Neuere Studien über Krankheiten fossiler Wirbeltiere; and Diskussion von H. Joseph, Nopcsa, Wirth. [New studies of sicknesses of fossil vertebrates; and discussion by H. Joseph, Nopsca, D. Wirth] Verhandlungen der zoologisch-botanischen Gesellschaft Wien 73:98–107 [German].

Cites Moodie's (1923) claims:

Trauma – Exostosis, callus formations around parasuchian jaw fracture.

Vertebral – Spondylitis deformans in a Permian *Dimetrodon* vertebra.

Infection – Osteomyelitis in a Permian *Dimetrodon* vertebra, osteoperiostitis in humerus of mosasaur.

Arthritis – "Multiple arthritis" in Cretaceous mosasaur "toe."

Tumor (benign) – Osteoma in dorsal vertebra of mosasaur, hemangioma (actually diffuse idiopathic skeletal hyperostosis) in tail vertebrae of dinosaur.

Other – Necrosis with hyperplasia of radius of mosasaur.

Fossil – Exostosis, callus formations around parasuchian jaw fracture.

Spondylitis deformans in a Permian *Dimetrodon* vertebra, osteomyelitis in a Permian *Dimetrodon* vertebra, and osteoperiostitis in humerus of mosasaur.

"Multiple arthritis" in Cretaceous mosasaur "toe."

Osteoma in dorsal vertebra of mosasaur, haemangioma (actually diffuse idiopathic skeletal hyperostosis) in tail vertebrae of dinosaur.

Necrosis with hyperplasia of radius of mosasaur.

Abel O. 1929. Paläobiologie und Stammesgeschichte. [Paleobiology and phylogeny] X + 423 pp.; Jena: Gustav Fischer. [German]

Trauma – Broken and healed rib fractures in ichthyosaurs and mosasaurs (e.g., *Plioplatecarpus*), broken and healed lower jaw fracture in *Mosasaurus giganteus.*

Infection – Deformation of femoral and sacral vertebra in *Metriorhynchus* cf. *moreli* (reported by Auer and Jaffer, (1909)) considered festering process or necrosis.

Vertebral – Pachyostosis in sauropterygian *Proneusticosaurus.*

Dental – Caries in *Mosasaurus.*

Fossil – Broken and healed rib fractures in ichthyosaurs and mosasaurs (e.g., *Plioplatecarpus*), broken and healed lower jaw fracture in *Mosasaurus giganteus.*

Deformation of femoral and sacral vertebra in *Metriorhynchus* cf. *moreli* (reported by Auer (1909) considered festering process or necrosis.

Pachyostosis in sauropterygian *Proneusticosaurus.*

Caries in *Mosasaurus.*

Abel O. 1935. Vorzeitliche Lebensspuren. [Prehistoric life] XV + 644 pp.; Jena: Gustav Fischer. [German].

Trauma – Broken and healed rib fractures in ichthyosaurs and mosasaurs (e.g., *Plioplatecarpus*), broken and healed lower jaw fracture in *Mosasaurus giganteus.*

Infection – Deformation of femoral and sacral vertebra in *Metriorhynchus* cf. *moreli* (reported by Auer (1909) considered festering process or necrosis.

Vertebral – Pachyostosis in sauropterygian *Proneusticosaurus.*

Dental – Caries in *Mosasaurus.*

Fossil – Broken and healed rib fractures in ichthyosaurs and mosasaurs (e.g., *Plioplatecarpus*), broken and healed lower jaw fracture in *Mosasaurus giganteus.*

Deformation of femoral and sacral vertebra in *Metriorhynchus* cf. *moreli* (reported by Auer (1909) considered festering process or necrosis.

Pachyostosis in sauropterygian *Proneusticosaurus.*

Caries in *Mosasaurus.*

Abou-Madi N, Serivani PV, Kollias GV, Hernandez-Divers SM. 2004. Diagnosis of skeletal injuries in chelonians using computed tomography. Journal of Zoo and Wildlife Medicine 35:226–231.

Trauma – Computed tomographic recognition of fractures (not seen on routine X-ray) in snapping turtles.

Acharji MN. 1945. Axial bifurcation in *Vipera russellii* (Shaw) from Bengal. Journal of the Bombay Natural History Society 45(5):611–613.

Congenital – Anterior axial bifurcation in *Vipera russellii.* Also reports note in Madras Times in 1897 of bicephalic American hog-nosed snake *Heterodon simus.*

Ackerman LJ, Kishimoto RA, Emerson JS. 1971. Nonpigmented *Serratia marcescens* arthritis in a Teju (*Tupinambis teguixin*). American Journal of Veterinary Research 32:823–826.

Infection – Teju *Tupinambis teguixin* swollen left stifle with diffuse lytic lesions of knee with sclerotic reaction caused by *Serratia marcescens* infection.

Adelmann HB. 1936. The problem of cyclopia. Quarterly Review of Biology 11:161–221; 284–304.

Congenital – Cyclopia in *C. picta.*? and red-eared slider *Trachemys scripta elegans.*

Adkesson MHJ, Travis EK, Weber MA, Kirby JP, Junge RE. 2007. Vacuum-assisted closure for treatment of a deep shell abscess and osteomyelitis in a tortoise. Journal of the American Veterinary Medical Association 231:1249–1254.

Shell disease – Focal carapace necrosis in *Geochelone gigantea* with osteomyelitis.

Adkins E, Driggers T, Ferguson G, Gehrmann W, Gyimesi Z, May E, Ogle M, Owens T. 2003. Ultraviolet light and reptiles, amphibians. Journal of Herpetological Medicine and Surgery 13(4):27–37.

Metabolic – Variable UV light – requirements with respect to Vitamin D3 UV lightexposure not required for snakes or nocturnal lizards, but is by most diurnal species; minimally required by crocodilians. Documented UV light deficiency-related osteomalacia reported in bearded dragon *Pogona vitticeps*, brown basilisk *Basiliscus basiliscus*, green-crested basilisk *Basiliscus plumifrons*, Komodo dragons *Varanus komodoensis*, crocodile monitor *Varanus salvadorii*, panther chameleon *Furcifer pardalis*, Western chuckwalla *Sauromalus obesus*, sail-finned lizard *Hydrosaurus* sp., prehensile-tailed skink *Corucia zebrata*, and great-plated lizard *Gerrhosaurus major*.

Adler KK. 1958. A five-legged *Rana* from Ohio. Ohio Herpetological Society. Trimonthly Report 1958(1):21.

Congenital – Supernumerary hind limb in *Rana*.

Adolphi H. 1893. Über Variationen der Spinalnerven und der Wirbelsäule anurer Amphibien. I. (*Bufo variabilis* Pall.). [On the variations of spinal nerves and of the vertebral column in anuran amphibians. I.] Gegenbaurs Morphologisches Jahrbuch 19:313–375 [German].

Congenital – Fusion of vertebrae 1 and 2 in *Pippa, Dactylatra, Systema, Ceratophrys, Brachycephalus, and Palaeobatrachus*; fusion in the middle part of the vertebral column in *Brachycephalus*; fusion of vertebrae 7 to 9 in *Palaeobatrachus*; and fusion of vertebra 9 with urostyle of *Bufo variabilis*, *Pippa*, *Dactylatra*, *Systema*, *Phyllomedusa*, and *Pelobates*.

Adolphi H. 1895. Über Variationen der Spinalnerven und der Wirbelsäule anurer Amphibien. II. (*Pelobates fuscus* Wagl. und *Rana esculenta* L.). [On the variation of spinal nerves and of the vertebral column in anuran amphibians. II.] Gegenbaurs Morphologisches Jahrbuch 22:449–490 [German].

Congenital – Fusion and asymmetry of *Pelobates fuscus* vertebrae; fusion of *Rana esculenta* vertebrae.

Adolphi H. 1898a. Über Variationen der Spinalnerven und der Wirbelsäule anurer Amphibien. III. (*Bufo cinereus* Schneid). [On the variation of spinal nerves and of the vertebral column in anuran amphibians. III. (*Bufo cinereus* Schneid)]. Gegenbaurs Morphologisches Jahrbuch 25:115–142 [German].

Congenital – Fusion of *Bufo cinereus* vertebrae.

Adolphi H. 1898b. Über das Wandern des Extremitätenplexus und des Sacrum bei *Triton taeniatus*. [On the drift of extremities and sacrum of *Triton taeniatus*]. Morphologisches Jahrbuch 25:115–142 [German].

Congenital – Unilateral or bilateral enlargement or loss of *Triton taeniatus* sacral ribs on vertebrae 14 and 15.

Aelianus C. (aka Aelion C.) 3rd Century. De natura animalium. [Natural Animals]. Londini: excudit Gulielmus Bowyer, sumptibus Societatis ad Literas Promovendas institutae: prostant venales apud I. Whiston, I. Osborn, et Fran. Changvion: anno, 1744. [On the Characteristics of Animals – English translation side by side with the original Latin and Greek by AF Scholfield Cambridge: Harvard University Press. Book 16, Chapter 62] [Latin].

Congenital – Cites Pammenes in book 16, chapter 42 (entitled "De serpentibus bicipitus"), in section entitled "concerning wild animals" reporting "two-headed" snakes in the fifth century BC.

Ahl E. 1927. Doppelschwänzige Zauneidechse. [Double-tailed fence lizard]. Aquarium Berlin 1927:70–71 [German].

Trauma – Fence lizard with bifid tail.

Alberch P. 1980. Ontogenesis and morphological diversification. American Zoologist 20:653–666.

Congenital – Six percent of *Bolitoglossa rostrata* and *subpalmata* have segmentation of tarsals 4 and 5, while 8% of *B. mexicana* has conjoined 3 and 5, citing Wake (1966). Also cites Van Valen (1974) on supernumerary legs and digits in *Rana*.

Albrecht P. 1883. Note sur une hémivertèbre gauche surnuméraire de *Python sebae* Duméril. [Note on a left supernumerary hemivertebra in *Python sebae* Duméril]. Bulletin Muséum d'Histoire Naturelle Belgique 2:21–34 [French].

Congenital – Hemivertebrae in African rock python *Python sebae*.

Alderton D. 1988. Turtles & Tortoises of the world. Blandford Press. London, 191 pp.

Congenital – Malformed eggs in leatherback turtle *Dermochelys coriacea*.

Metabolic – Chilling produces abnormal egg thickness.

Temperature effects on gender: Painted turtle *Chrysemys picta* produces predominantly males at 28°C, predominantly females at 30; common snapping turtle *Chelydra serpentina*, males at 22–28, predominantly females at other temperature; Mediterranean spur-thighed turtle *Testudo graeca*, predominantly female at 31–32 and predominantly male at temperatures below 30.

Shell disease – Focal shell indentation in *Gopherus*.

Aldrovandi U. 1637. De quadrupedibus oviparis [Of egg-laying quadripedal animals]. Bononiae Typ. Yo Bapt. Ferronij. [Latin]

Trauma – Bifid tails in *Lacerta viridis* on p. 635 and *Lacerta viridis exsicata* on p. 636.

Aldrovandi U. 1640. Historia Serpentum et Draconum. [History of serpents and dragons]. Liber Primus. VII + 427 pp.; Bononiae: ?Ferroni?, M. Antonij Berni Bibliopole. [Latin]

Congenital - Dicephalic snake on p. 60.

Aldrovandi U. (aka Ambrosini) 1642. Historia Monstrorum. [According to Ercolani 1882 and Cantoni 1921, the title is Monstrorum Historia] [History of Monsters]. Bonniae: Bartolomäus Ambrosinus [Latin].

Congenital – Stated that contrary to Aristotle's suggestion, dicephalism is not rare among serpents, citing cases reported by Aristotle, Aelianus, and Porta, and that a dicephalic snake is preserved dry in the Museum of Bologna [Chap. XI, page 427]. P. 527 (after Borgert 1896) also cites Arneldo's claim that a snake with three heads was killed in the Pyrenees.

Trauma – Lizards with double, triple, and quadruple tail [Chap. VIII pp. 593–596]. Illustrated double-tailed lizards in pages 594–596 and reported encountering a lizard with a very long tail, from "which grew three small rami."

Aldrovandi U. 1645. De qvadrvpedib[us] digitatis viviparis libri tres et de qvadrvpedib[us] digitatis oviparis libri duo. [Of 3 books on live-bearing quadrupedal and 2 books on egg-laying quadrupedal animals].VIII + 718 + 16 pp.; Bononiae: Tebaldinus. [Latin]

Trauma – Lizards with double, triple, and quadruple tail [pp. 630–631, 635–636].

Aldrovandi U. 1657. Paralipomena historiae omnium animalium [Supplement to History of All Animals]. Bononiae Typ. Yo Bapt. Ferronij [Latin].

Congenital – Cites and provides a drawing of a marine amphisbaena that has two heads.

Trauma – Cites four-tailed lizard.

Alekperov A.M. 1954. [The occurrence of two-headed serpents] Zoologicheskii Zhurnal 33(3):716–717 [Russian].

Congenital – Dicephalic *Elaphe quatuorlineata sauromates.*

Alessandrini A. 1854. Catalogo degli oggetti e preparati più interessanti del gabinetto d' anatomia comparata della Università di Bologna [Catalogue of the most interesting objects and prepared specimens of the cabinet of comparative anatomy of the University of Bologna]. Bologna. [Italian] (quoted from Parona 1883) [Italian].

Congenital – Three cases of polymely in *Rana*: One had at its abdominal center a bilobate "tumor" with a rudimentary limb (p. 552, No. 1085); another had two left limbs near the head (No. 3248); and the last had two limbs to the left of the pubis (No. *4/116*). Additionally reported were anomalous *Rana* in page 552 of the catalog (No. 3248), a *Rana esculenta* with an additional scapula, and another specimen that had in the center of the abdomen, over the bilateral symmetry line, a bilobate tumor with an in complete limb (missing the foot).

Alexander AA, Gans C. 1966. The pattern of dermal-vertebral correlation in snakes and amphisbaenians. Zoologische Mededelingen 41:171–190.

Congenital – Variation in body vertebral count.

Species	Count
Amphisbaenidae	
Amphisbaena alba	102–107
carvalhoi	119–122
darwini heterozonata	97–98
dubia	107–112
fuliginosa	104–108
leeseri	111–113
manni	112–117
munoai	104–106
Amps kingi	113–114
Bipes biporus	129–130
Blanus cinereus	115–127
strauchi	92–122
Cadea blanoides	105–107
pallrostrata	136–145
Chirindia e. ewerbecki	131–135
Leposternon affine	100–106
boulengeri	99–102
crassum	98–99
phocaena	84–100
polystegum	125–134
scutigerum	120–121
wuchereri	119–122

(continued)

Species	Count
Loveridgea ionidesii	123–126
Monopeltis c. capensis	87–90
guentheri	123–127
Trogonophidae	
Agamodon a. anguliceps	64–66
Pachycalamus brevus	86–88
Trogonophis wiegmanni elegans	74–77
Typhlopidae	
Albirostris	205–233
Liotyphlops albirostris	234–252
	207–208
Aluensis	175–181
taeniatus	297–307[a]
bituberculatus	219–300
braminus	174–182
congestus	180–182
c. cuneirostris	115–126[a]
cuneirostris calabresii	130–143
polygrammicus	193–206
proximus	160–173
reticulates	137–155
schlegeli	189–202
scortecci	226–255[a]

(continued)

Species	Count
salomonis	219–222
unitaeniatus	315–334[a]
Leptotyphlopidae	
Leptotyphlops conjuncta	191–223
longicauda	208–215
phillipsi	335–343
Uropeltidae	
Brachyophidium rhodogaster	138–143
Platyplecturus madurensis	166–167
Rhinophis blythi	152–157
philippinus	150–166
Uropeltis ocellatus	191–198
pulneyensis	167–173
rubrolineatus	139–169
woodmasoni	166–179
Aniliidae	
Cylindrophis maculatus	195–209

(continued)

Species	Count
Xenopeltidae	
Xenopeltis unicolor	187–190
Boidae	
Acanthophis madagascariensis	236–256
Chondropython viridis	238–242
Epicrates striatus	291–296
colubrinus	187–190
jaculus	174–185
Liasis amethystinus	313–327
Trachyboa boulengeri	131–134
Tropidophis melanurus	195–199
Acrochordidae	
Acrochordus granulatus	201–205
Elachistodon westermanni	210–215
Thrasops flavigularis	205–209
j. jacksoni	195–205
occidentalis	182–189
Hydrophiidae	
Lapemis hardwickii	127–143

[a] body and tail vertebral count

Dermal scale – Body vertebral ratio
1:1 both ventral and dorsal:
Leptotyphlopidae *Leptotyphlops*
Uropeltidae *Brachyophidium, Melanophidium, Platyplecturus, Plectrurus, Rhinophis, Uropeltis*
Aniliidae *Anilius, Cylindrophis*
Xenopeltidae *Xenopeltis*
Boidae *Clavaria, Enygrus, Eryx, Eunectes, Lichanura, Sanzinia, Trachyboa, Tropidophis*
Colubridae *Ahaetulla, Dasypeltis, Duberria, Elachistodon, Malpolon, Psammophylax, Thrasops*
Hydrophiidae *Lapemis?, Laticauda*
Elaphidae *Micrurus*
Viperidae *Crotalus*
Amphisbaenidae *Blanus*
2:1 both ventral and dorsal:
Typhlopidae *Helminthopsis, Typhlops*
Amphisbaenidae *Amphisbaena, Ancyclocranium, Anops, Aulura, Baikia, Bipes canaliculatus, Bronia, Cadea blanoides, Cadea palirostrata, Chirindia, Cynisca, Geocalamus, Leposternon, Loveridgea, Mesobaena, Monopeltis, Rhineura, Tomuropeltis, Zygaspis*
Trophonophidae *Agamodon, Diplometopon, Pachycolamus, Trogonophis*
2:1 dorsal, unknown re ventral:
Typhlopidae *Anomalepis, Liotyphlops*
1:1 ventral, 2:1 dorsal:
Boidae *Acrantophis, Aspidiotes, Boa, Bothrochilus, Charina, Corallus, Epicrates, Liasis, Python*
Bolyeridae *Casarea*
Amphisbaenidae *Bipes biporus*
5:1 ventral, 3.5:1 dorsal to vertebral
Acrochordidae *Acrochordus*

Al-Hussaini AH. 1953. A case of polymely in the Egyptian toad, *Bufo regularis* Reuss. Bulletin of the Zoological Society of Egypt (11):48–51.

Congenital – Polymely in Egyptian toad *Bufo regularis.*

Ali SM. 1941. Studies on the comparative anatomy of the tail in Sauria and Rhynchocephalia. I. *Sphenodon punctatus* Gray. Proceedings of the Indian Academy of Sciences. Section B 13:171–192.

Trauma – Regeneration of difficult to break tail in *Sphenodon punctatus* produces cartilaginous tube, calcified on outer and inner surfaces and divided by union of those surfaces in one or two places.

Ali SM. 1950. Studies on the anatomy of the tail in Sauria and Rhynchocephalia. IV. *Anguis fragilis*. Proceedings of the Indian Academy of Sciences 32(2):87–95.

Trauma – Regenerated tails in *Anguis fragilis.* Perforated cartilaginous regenerated tails in *Pygopus* (Woodland 1920).

Alibardi L. 1995. Muscle differentiation and morphogenesis in the regenerating tail of lizards. Journal of Anatomy 186:143–151.
Trauma – Autotomy in *Anolis* and *Lampropholis.*
Alibardi L. 2010. Ultrastructural features of the process of wound healing after tail and limb amputation in lizard. Acta Zoologica (Stockholm) 91:306–318.
Trauma – Tail regeneration in *Podarcis sicula.*
Alibardi L, Thompson MB. 1999. Morphogenesis of shell and scutes in the turtle *Emydura macquarii.* Australian Journal of Zoology 47:245–260.
Congenital – Discussion of fundamentals of *Emydura macquarii* shell and scute embryology.
Shell disease – Discussion of fundamentals of *Emydura macquarii* shell and scute embryology.
Allen ME. 1989. Dietary induction and prevention of osteodystrophy in an insectivorous reptile *Eublepharis macularius*: Characterization by radiography and histopathology. 3rd International Colloquium on Pathology of Amphibians and Reptiles Orlando, Abstracts:83.
Metabolic – Leopard geckos *Eublepharis macularius* fed crickets developed fractures related to osteomalacia.
Allen A. 1990. Two-headed snakes. Aquarist & Pondkeeper 1990 (August): 20.
Congenital – Derdomous adder.
Environmental – Grass snakes *Natrix natrix* incubate their eggs in manure heaps, where temperature varies from 30° to 33°, which fluctuates up to 40–45°C. One percent of new born snakes in that environment are dicephalic.
Allen ME. 1997. From blackbirds and thrushes … to the gut-loaded cricket: A new approach to zoo animal nutrition. British Journal of Nutrition 78 Suppl 2:S135–S143.
Metabolic – Feeding Leopard gecko *Eublepharis macularius* non-supplemented crickets and Madagascar giant day gecko *Phelsuma madagascariensis* and even calcium-supplemented crickets resulted in bone demineralization and secondary hyperparathyroidism.
Allen ME, Oftedal OT. 1994. The nutrition of carnivorous reptiles. In Murphy JB, Adler K, Collins JT. eds. Captive Management and Conservation of Amphibians and Reptiles. St. Louis, MO: Society for the Study of Amphibians and Reptiles. pp. 71–82.
Metabolic – Most reptiles excrete nitrogen predominantly as uric acid; aqueous reptiles, urea (aquatic turtles) or ammonia (sea turtles and crocodilians).
Allen ME, Bush M, Oftedal OT, Rosscoe R, Walsh T, Holick, M. 1994. Update on vitamin-D and ultraviolet-light in basking lizards. American Association of Zoo Veterinarians and Association of Reptilian and Amphibian Veterinarians Annual Conference Proceedings 1994:314–316.
Trauma – Fractures in Komodo dragons *Varanus komodoensis* may be related to vitamin D deficiency.
Metabolic – Fractures in Komodo dragons *Varanus komodoensis* may be related to vitamin D deficiency.
Alonso L, Alonso L, Jiménez E. 2006. Análisis de varios casos de zoopaleopatología del Eoceno medio de Zamora (España). [Analysis of several cases of zoopalaeopathology from the Middle Eocene in Zamora (Spain)]. Studia Geologica Salmanticensia 42:97–112. [Spanish]
Congenital - *Allaeochelys jimenezi* (Chelonia: Carettochelyidae) from the Middle Eocene of Zamora, Spain with exaggerated pygal plate suture lines, reducing free edge, producing a square, rather than trapezoidal shape.
Trauma –Focal femoral thickening in *Iberosuchus macrodon* (Crocodylia: Sebecidae) femur (STUS 14.108) from the Middle Eocene of Zamora, Spain, probably post-traumatic.
?Iberosuchus clavicle (ITUS 14.115) with gouges and pits.
Shell - *Neochelys* sp. (Chelonia: Pelomedusidae) from the Middle Eocene of Zamora, Spain with gouges in suprapygal (STUS 14.112) and xiphiplastron (STUS 14.109), puncture in mesoplastron (STUS14.114) and healed puncture in pleural (STUS 14.113) attributed to crocodile bites, and multiple irregular defects with irregular margins (STUS 14.114) in first pleural.
Allaeochelys jimenezi (Chelonia: Carettochelyidae) from the Middle Eocene of Zamora, Spain with exaggerated pygal plate suture lines, reducing free edge, producing a square, rather than trapezoidal shape.
Fossil - *Allaeochelys jimenezi* (Chelonia: Carettochelyidae) from the Middle Eocene of Zamora, Spain with exaggerated pygal plate suture lines, reducing free edge, producing a square, rather than trapezoidal shape.
Focal femoral thickening in Iberosuchus macrodon (Crocodylia: Sebecidae) femur (STUS 14.108) from the Middle Eocene of Zamora, Spain, probably post-traumatic.
?Iberosuchus clavicle (ITUS 14.115) with gouges and pits.
Neochelys sp. (Chelonia: Pelomedusidae) from the Middle Eocene of Zamora, Spain with gouges in suprapygal (STUS 14.112) and xiphiplastron (STUS 14.109), puncture in mesoplastron (STUS14.114) and healed puncture in pleural (STUS 14.113) attributed to crocodile bites, and multiple irregular defects with irregular margins (STUS 14.114) in first pleural.
Allaeochelys jimenezi (Chelonia: Carettochelyidae) from the Middle Eocene of Zamora, Spain with exaggerated pygal plate suture lines, reducing free edge, producing a square, rather than trapezoidal shape.
Althoff DM, Thompson JN. 1994. The effects of tail autotomy on survivorship and body growth of *Uta stansburiana* under conditions of high mortality. Oecologia 100:250–255.

Trauma – No effect of tail autotomy on survivorship and body growth of *Uta stansburiana* under conditions of high mortality.

Altmann H. 1980. Erfolgreiche Behandlung der Knochenerweichung bei einer Segelechse. [Successful treatment of osteomalacia of a lizard]. Deutsche Aquarien- und Terrarien-Zeitschrift 33(2):67–70 [German].

Metabolic – Osteomalacia in Philippine lizard, *Hydrosaurus pustulatus* responding to vitamin B and ultraviolet radiation.

Alvarez MI, Herraez I, Herraez P. 1992. Skeletal malformations in hatchery reared *Rana perezi* tadpoles. Anatomical Record 233:314–320.

Congenital – Scoliosis, kyphosis in *Rana perezi.*

Metabolic – Aplasia related to captive *Rana perezi* dietary deficiencies.

Alvarez R, Honrubia MP, Herráez MP. 1995. Skeletal malformations induced by the insecticides ZZ-Aphox and Folidol during larval development of *Rana perezi*. Archive of Environmental contamination and toxicology 28:349–356.

Toxicology – Carbamate insecticides produce scoliosis, short and thick long bones and twisted epiphyses in *Rana perezi*. Organophosphate Folidol and carbamate insecticide ZZ-Aphox produce scoliosis, joint rigidity, and shortening of rear limbs. Cartilage staining with alcian blue was slight or absent with irregular small stained spots, contrasted with normal intensive blue staining. Long bones were shortened, thickened, with twisted epiphyses (mimics rickets); ilium and coccyx, twisted. Vertebral pathology included variable orientation, length and thickness of lateral apophyses and variable dorsal arch width, nonossified vertebral closure. External bone surface was porous, with striations and small cavities. He states that spinal lesions were similar to those of lathyrism.

Amaro A, Sena S. 1968. Ocorrências de anomalias em anfíbios. [Occurrences of anomalies in amphibians]. Atas da Sociedade de Biologia do Rio de Janeiro 12(2):95–96 [Portuguese].

Congenital – Two of 12 specimens of *Leptodactylus ocellatus* (L.) collected in Manguinhos, Rio de Janeiro, Brazil, in 1966, had anomalous limbs. One possessed an extra right forelimb, and the other lacked its right forelimb. Both specimens were still young, and neither appeared impaired due to its anomalies. Amaro and Sena remarked that they decided to report this because of the apparent rarity of the occurrence of anomalous limbs in representatives of the Leptodactylae, Hylidae, Bufonidae, and Brachycephalidae from several regions of Brazil. In a previous study, they examined more than a million individuals of these families in Brazil and found no other anomalous specimens. They also searched the literature and found only one reference (Miller 1968). The specimens described were stored in alcohol in Amaro and Sena's personal collection. Photographs of the two specimens were provided in page 95.

Amrakh B. 1944. [A two headed snake.] Gaz. Azerbaijan Pioneri Nov. 19 [Russian].

Congenital – Dicephalic *Elaphe quatuorlineata sauromates.*

Anan'eva NB, Orlov NL. 1994. Caudal autotomy in colubrid snake *Xenochrophis piscator* from Vietnam. Russian Journal of Herpetology 1:169–171.

Trauma – Intravertebral autotomy is present in Gekkonidae, Pygopodidae, Scincidae, Lacertidae, Teiidae, Anguidae, Cordylidae, Xantusiidae, Dibamidae, and many Iguanidae. Cartilaginous septum is lacking in Chamaeleonidae, Agamidae, and Platynota (Varanidae, Lanthanotidae, and Helodermatidae). Snakes have intervertebral, not intravertebral fracturing, with exception of *Pliocercus* with a caudal intravertebral weakness plane. Caudal autotomy in snakes has been reported in *Scaphiodontophis, Sibynophis, Pliocercus, Natriciteres, Rhadinaea decorata, Xenochrophis piscator*, and *Amphiesma stolatum* (Arnold 1984; Bellairs and Bryant 1985; Mendelson 1993; Sharma 1980), and especially colubrid snakes in museum collections (Mendelson 1993).

Ancel P. 1950. La Chimiotératogenèse. Réalisation des Monstruosités par des Substances Chimiques chez les Vertébrés. [Chemical teratogenicity. Production of monstrosities in vertebrates by chemical substances]. 397 pp.; Paris: Doin & Cie [French].

Toxicology – Sodium chloride-induced pseudo-spina bifida in *Rana fusca.*

Congenital – Cyclops in amphibian – without details as to how produced.

Andersen K Th. 1930. Doppelbildung und Hemmungsmißbildung bei *Lacerta agilis* Embryonen. [Ventro-lateral thoracopagus and eventration in embryonic *Lacerta agilis*]. Zeitschrift für Anatomie und Entwicklungsgeschichte 92:239–257 [German].

Congenital – Description of ventrolateral thoracopagus of *Lacerta agilis* and a tailless embryo.

Anderson MP, Capen CC. 1976a. Ultrastructural evaluation of parathyroid and ultimobranchial glands in iguanas with experimental nutritional osteodystrophy. General and comparative Endocrinology 30:209–222.

Metabolic – Osteopenia with fibrous osteodystrophy, osteoporosis, osteomalacia, and pathologic fractures in iguanas.

Anderson MP, Capen CC. 1976b. Fine structural changes of bone cells in experimental nutritional osteodystrophy of green iguanas. Virchows Archive – B: Cell Pathology 20:169–184.

Metabolic – Low calcium diet produces secondary hyperparathyroidism with osteoclastic and osteocytic osteolysis in *Iguana iguana.*

Anderson MP, Capen CC. 1976c. Nutritional osteodystrophy in captive green iguanas (*Iguana iguana*). Virchows Archive – B: Cell Pathology 21:229–247.

Metabolic – Low calcium diet producing nutritional osteodystrophy with osteopenia and diaphyseal enlargement of long bones and fractures in *Iguana iguana*.

Andrade DV. de, Abe AB. 1993. Malformações em ninhadas de caiçaca, *Bothrops moojeni* (Serpentes: Viviperidae) [Malformations in hatchlings of caiçaca, *Bothrops moojeni* (Serpentes: Viviperidae)]. Memórias do Instituto Butantan 54(1992):61–67 [Portuguese].

Congenital – Dicephalism and kinks in 7.7% of *Bothrops moojeni*.

Andreacchio A, Miller F. 2000. *Salmonella* osteomyelitis transmitted from an iguana. Orthopedics 23:1201–1202.

Infection – *Salmonella* transmitted to human from an iguana.

Andreadis PT, Burghardt GM. 1993. Feeding behavior and an oropharyngeal component of satiety in a two-headed snake. Physiology and Behavior 54(4):649–658.

Congenital – Dicephalic black rat snake *Elaphe o. obsoleta*.

Andrei M. 1985. Un cas de polymélie dans le complex *Rana "esculenta"* Linné (Anura). A case of polymelia in « *Rana "esculenta"* complex Linné (Anura). Travaux du Museum d'Histoire Naturelle "Grigore Antipa" (27):267–268 [French].

Congenital – Supernumerary limb in *Rana esculenta*.

Andrews CW. 1906. A Descriptive Catalogue of the Tertiary Vertebrata of the Fayum, Egypt. Based on the Collection of the Egyptian Government in the Geological Museum, Cairo, and on the Collection in the British Museum (Natural History), London. London: Longmans & Co., 319 pp.

Trauma – Mid-maxilla of Upper Eocene crocodilian *Crocodilus anticeps* BMNH C10036 with expanded smooth excavated area in tooth row and bite marks on superior surface. Unilateral mid-mandibular reactive bone in Middle Eocene *Tomistoma africanum*.

Fossil – Mid-maxilla of Upper Eocene crocodilian *Crocodilus anticeps* BMNH C10036 with expanded smooth excavated area in tooth row and bite marks on superior surface. Unilateral mid-mandibular reactive bone in Middle Eocene *Tomistoma africanum*.

Angel F. 1950. Vie et moeurs des Serpents. [Life and Death of Serpents]. 319 pp.; Paris: Payot [French].

Congenital – Derodymus *Lampropeltis triangulumites*. Cites (though does not reference) dicephalic *Coronella, Coluber, Heterodon, Lycodon*, boa *Epicrates, Natrix, Vipera, Zamenis, Homalopsis, Lachesis, Naja, Distira, Hydrus, Hydrophis, Elaphe, Bothrops, Crotalus*, and *Lampropeltis*.

Trauma – Regeneration of triton, urodeles, and axolotl. Cartilaginous replacement of bone in gecko regeneration.

Ankley GT, Tietge JE, DeFoe DL, Jensen KM, Holcombe GD, Durhan, EJ, Diamond SA. 1998. Effects of ultraviolet light and methoprene on survival and development of *Rana pipiens*. Environmental Toxicology and Chemistry 17:2530–2542.

Environmental – Supernumerary limbs and missing segments or digits of leopard frog *Rana pipiens* induced by UV light exposure, but not by pesticide methoprene.

Ankley GT, Tietge JE, Holcombe GW, DeFoe DL, Diamond SA, Jensen KM, Degitz SJ. 2000. Effects of laboratory ultraviolet radiation and sunlight on survival and development of *Rana pipiens*. Canadian Journal of Zoology 78:1092–1100.

Environmental – Bilateral, symmetrical missing/reduced digits (ectomely and ectodactyly) produced by ultraviolet radiation at stage 25–26, just prior to hind limb development in *Rana pipiens*.

Ankley GT, Degitz SJ, Diamond SA, Tietge JE. 2004. Assessment of environmental stressors potentially responsible for malformations in North American anuran amphibians. Ectotoxicology and Environmental Safety 58:7–16.

Congenital – Suggested digenetic trematodes as responsible for malformations.

Infection – Suggested digenetic trematodes as responsible for malformations.

Environmental – Suggested digenetic trematodes as responsible for malformations.

Annandale N. 1904. Contributions to oriental herpetology. I. The lizards of the Andamans, with the description of a new gecko and a note on the reproduced tail in *Ptychozoon homalocephalum*. Journal of the Asiatic Society of Bengal 73:12–22.

Trauma – Distal tail regeneration in *Ptychozoon homalocephalum*.

Annandale N. 1905. On abnormal ranid larvae from North Eastern India. Proceedings of the Zoological Society of London 1:58–61.

Congenital – *Rana alticola* lacking hind limbs, with only trace of right pelvic arch with sickle-shaped left ilium and asymmetrical forelimbs.

Anon. 1631. A two-headed snake. Stow's Annales 1631:247.

Congenital – Dicephalic snake.

Anon. 1788. Norwich, September 25. Herald of Freedom and Federal Advertiser 6 October 1788;1(7):26.

Congenital – Anacatadidymus rattlesnake.

Anon. 1793. Domestic occurrences. Double headed snake. Massachusetts 5(9):574.

Congenital – Dicephalic rattlesnake.

Anon. 1819. Remarkable. Adams Sentinel, Gettysburg, PA (13 October 1819):1.
Congenital – Dicephalic black snake.
Anon. 1839. A two-headed snake. Indiana Journal 9 November 1839.
Congenital – Dicephalic snake.
Anon. 1858. 2-headed snake. The Cleveland Herald 26 January 1858:1.
Congenital – Dicephalic milk snake.
Anon. 1865. Zweiköpfige Schlangen. [Two-headed snake]. Das Ausland. Eine Wochenschrift für Kunde des geistigen und sittlichen Lebens 38:744.
Congenital – Dicephalic *Tropidonotus sipedon.*
Anon. 1867. Two headed snake. Daily News and Herald (Atlanta, GA) 2 October 1867:1.
Congenital – Dicephalic snake.
Anon. 1872. The Nevada, Ga. Herald. Evening Bulletin (SanFrancisco) 21 January 1872:1.
Congenital – Cites dicephalic snake by Newman in the Georgia Herald.
Anon. 1873a. A two-headed snake. Cleveland Daily Record 27 October 1873:2.
Congenital – Water snake.
Anon. 1873b. A two-headed snake. Cleveland Daily Record 28 October 1873:1.
Congenital – Derodymus water snake.
Anon. 1873c. A two-headed snake. Lowell Daily Citizen & News (Massachusetts) 28 October 1873:1.
Congenital – Derodymus water snake (but rattlesnake more likely?).
Anon. 1877. Two-headed rattlesnake. Field and Stream 9(6):105.
Congenital – Dicephalic rattlesnake.
Anon. 1878a. (Scientific News). American Naturalist 12:264.
Congenital – Reference to publications of Wright (1878: two-headed gopher snake) and Wyman (1863: two-headed common striped snake).
Anon. 1878b. An ounce of mirth. The Butte Miner (Butte, MT) 29 January 1878:1.
Congenital – Two-headed snake.
Anon. 1879a. A snake with two heads. The Daily Inter-Ocean 10 October 1879:10.
Congenital – Dicephalic king snake.
Anon. 1879b. A snake with two heads. The Weekly Inter-Ocean 16 October 1879:6.
Congenital – Dicephalic king snake.
Anon. 1879c. A snake with two heads. The Daily Inter-Ocean 18 October 1879:10.
Congenital – Dicephalic king snake.
Anon. 1879d. A snake with two heads. The Daily Inter-Ocean 19 October 1879:7.
Congenital – Dicephalic king snake.
Anon. 1883. A two-headed snake. Little Rock Daily Republican 27 October 1883:1.
Congenital – Dicephalic water snake.
Anon. 1888a. Sir Rat and a two-headed snake. The New York Times, New York, N.Y. 2 October 1888:7.
Congenital – Derodymus copperhead.
Anon. 1888b. A sacred white crocodile. The animal has two tails – A good time for the city to get a museum. Oregonian 10 August 1888:8.
Trauma – Crocodile with two tails.
Anon. 1888c. A turtle with two heads. Milwaukee Daily Journal 20 October 1888:1.
Congenital – Derodymus turtle.
Anon. 1888d. Two heads that do not agree. The North American (Philadelphia) 6 June 1888:1.
Congenital – Derodymus turtle.
Anon. 1889a. A two-headed snake. The Daily News (Denver) 1 October 1889:10.
Congenital – Derodymus garter snake.
Anon. 1889b. A queer pet. Milwaukee Daily Journal 10 June 1889:1.
Congenital – Dicephalic *Chrysemys picta.*
Anon. 1889c. The snake catcher. Old Zachary Archer of the Storm King Mountain. The Bucks County Gazette (Bristol, PA) 14 March 1889:1.
Congenital – Snakes with two heads found three times per year.
Trauma – Snakes with two tails found three times per year.
Anon. 1889d. The local department: dashes here and there. Denton Journal (Denton, MD) 7 September 1889:1.
Congenital – Dicephalic snake.
Anon. 1890. A two-headed moccasin; an encounter with a venomous reptile double armed. Chicago Daily Tribune, Chicago, IL (19 July 1890):12.
Congenital – Derodymus water moccasin.
Anon. 1891. A two-headed snake. The Atchison Daily Champion 21 November 1891:3.
Congenital – Dicephalic *Pituophis catenifer* (variously referred to as gopher, blow, pine, and bull snake).

Anon. 1893. A two-headed snake. Oregonian (Portland) 10 September 1893:11.
Congenital – Dicephalic snake.
Anon. 1895. A turtle with two heads. The Daily Picayune-New Orleans 21 May1895:8.
Congenital – Dicephalic turtle.
Anon. 1896a. Snakes with two heads. The Atchison Daily Globe 20 April 1896:3.
Congenital – Pseudo-duplication – snake tail mistaken for second head.
Mythology – Snake tail mistaken for second head.
Anon. 1896b. An Oregon snake den. Various kinds there, including the two headed double-ender. The Daily News and Herald (Macon Telegraph) 1 August 1896:2.
Congenital – Alleged head on either end of snake.
Mythology – Alleged head on either end of snake.
Anon. 1897. Two-headed tortoise. The Daily News (Denver) 7 February 1897:21.
Congenital – Derodymus tortoise.
Anon. 1899a. A cotton mouth moccasin. A specimen of our most poisonous serpent – A two-headed snake. The News and Observer 30 September 1899:6.
Congenital – Dicephalic cottonmouth.
Anon. 1899b. Two-headed snake: each of its heads and tails seem to belong to a different variety of reptile. The Lima News (Lima, OH) 21 October 1899:1.
Congenital – Dicephalic snake with two tails. "One head and one tail were from 'an ordinary black snake.' The other was from a variety known as cow snake."
Mythology – Dicephalic snake with two tails. "One head and one tail were from 'an ordinary black snake.' The other was from a variety known as cow snake."
Anon. 1901. Twenty-four pages. The Galveston Daily News (Galveston, TX) 27 October 1901:16.
Congenital – Derodymus rattlesnake.
Anon. 1908. Two-headed snake killed. Daily Gazette and Bulletin (Williamsport, PN) 14 May 1908:4.
Mythology – Black snake with "head on each end."
Anon. 1915. Two headed snake overate. The Washington Post (Washington, DC) 9 March 1915:6.
Congenital – Dicephalic rattlesnake.
Anon. 1919. Two-headed snake can crawl either way. The Syracuse Herald (Syracuse, NY):7 July 1919:2.
Congenital – "Two-headed snake…able to crawl either way." Unclear from this description what kind of anomaly is present.
Mythology – "Two-headed snake…able to crawl either way." Unclear from this description what kind of anomaly is present.
Anon. 1921. Adventurers see two-headed lizard. Reptile with feet that function in opposite directions is shown at club dinner. New York Times 15 May 1921:19.
Mythology – Australian lizard whose tail mimics head.
Anon. 1924. Miss de La Motte inaugurates new rattlesnake fad. The Oakland Tribune (Oakland, CA) 19 October 1924: W-1.
Congenital – Dicephalic rattlesnake.
Anon. 1926. Snake with two heads on view at U.C.'s zoo. Reptile queer brought down here from sunny glen to frighten men. The Oakland Tribune (Oakland, CA) 14 November 1926:A16.
Congenital – Derodymus garter snake.
Anon. 1931. Two headed turtle. Time Magazine 28 September 1931:2
Congenital – Duplication of head and forelimbs in turtle.
Anon. 1933. Snake's two heads lead tail merry chase. The Olean Times Evening Herald, Olean, NY (5 August 1933):7.
Congenital – Dicephalic cornsnake.
Anon. 1933a. Two-headed snake. Columbia Telescope 16 July 1833:B29.
Mythology – Pseudo-dicephalism in *Amphisbaena.*
Anon. 1933b. Freak-snake found by Broken Bow man. McCurtain Gazette (Idabel, McCurtain County Oklahoma) 2 December 1933, 27(70):1.
Congenital – Dicephalic rattlesnake.
Anon. 1935. Two-headed snake captured by youth. The Sheboygan Press (Sheboygan, Wisconsin) 23 October 1935: Parade of Youth p1.
Congenital – Dicephalic water moccasin.
Anon. 1945. More multi-legged frogs. Turtox News 23:86–87.
Congenital – Nine legged *Rana clamitans* tadpole reported as part of contest, with Normal Stewart reporting 7-legged (? mink) frog and Beatrice Mintz, 9-legged.
Anon. 1946. Snake with 2 heads found; eats beefsteak. Lincoln Sunday Star (Lincoln, Nebraska) 15 September 1946:A4.
Congenital – Derodymous water moccasin.
Anon. 1948. Serpentine split personality. The Daily Times-News. Burlington, NC (2 December 1948):1.
Congenital – Derodymus "added" snake.

Anon. 1954. Many-legged frogs. Science News Letter 66:327.
Congenital – *Rana catesbeiana* with 3 to 8 supernumerary legs.
Anon. 1955. [no title]. The Charleroi Mail and Mirror (Charleroi, PN) 21 March 1955:4.
Congenital – Dicephalic snake.
Anon. 1961a. Birmingham News (April 30, 1961).
Congenital – Twelve-legged *Rana catesbeiana* in Alabama.
Anon. 1961b. Two-headed skink of Australia. Great Bend Daily Tribune 23 July 1961.
Congenital – Dicephalic Australian skink.
Anon. 1962a. Alabama biologist has 12-legged bullfrog. Bulletin of the Philadelphia. Herpetological Society 10:24.
Congenital – Bullfrog *Rana catesbeiana* with 12 legs, three of supernumerary with normal 4 digits, one with 1 digit, 2 with 3 digits and 2 with 7 digits, citing Birmingham News April 30, 1961.
Anon. 1962b. Reptiles of Saskatchewan's ancient seas. Saskatchewan Museum of Natural History, popular series 1:1–12.
Trauma – Reported mixture of bones of a large and a small mosasaur, suggesting ingestion of the one by the other.
Fossil – Reported mixture of bones of a large and a small mosasaur, suggesting ingestion of the one by the other.
Anon. 1964. Frogs with 5 legs and more found in pond in Jersey. New York Times 113 (September 5):21.
Congenital – Five-, six- and eight-legged frogs (unidentified) found in pond in Nutley, New Jersey; cause: probably pollution.
Toxicology – Five-, six- and eight-legged frogs (unidentified) found in pond in Nutley, New Jersey; cause: probably pollution.
Anon. 1967a. A case of 'Siamese' twins in the turtle (*Pseudemys scripta elegans*) Texas Journal of Science 19:232.
Congenital – Thoraco-omphalopagus type of conjoined (*Pseudemys scripta elegans*).
Anon. 1967b. Two-headed rattlesnake. The Baltimore Sun (Baltimore, Maryland) 21 September 1967.
Anon. 1975. Turtle with a twist. The Daily Review (Hayward, CA) 2 November:9.
Congenital – Dicephalic diamondback terrapin. Dicephalic snakes.
Anon. 1978. Untitled. The Hitavada, India (25 July 1978) 62:1.
Congenital – Derodymous python.
Anon. 1984. Two-headed *Elaphe g. guttata*. Chicago Herpetological Society Newsletter Sept. (reprinted from Ft. Myers News-Press, August 4, 1984).
Congenital – Dicephalic *Elaphe g. guttata*.
Anon. 1991a. Amphibien am Scheideweg. [Amphibians from Scheideweg]. Natur und Umwelt, Bonn 71:B4–B5 [German].
Congenital – Toads with shortened legs or without leg are figured.
Anon. 1991b. A couple of double-headers! Herpetile 16:175.
Congenital – Dicephalic African house snake and cites J Cole (Snake Breeders magazine, August 1991) report of a dicephalic California kingsnake.
Anon. 1995a. Two-headed rattler found in Lagrange. Birmingham News 9 September 1995:A5.
Congenital – Dicephalic rattlesnake.
Anon. 1995b. Snapping turtles thrive in Chesapeake's salt water. The Washington Times 4 December 1995:C2.
Congenital – Dicephalic snapping turtle.
Anon. 1996. Two-headed snake sees dentist. Durban Daily News (Durban, South Africa) 7 February 1996):3.
Congenital – Derodymus Herald snake.
Anon. 1997. Roving Brief. The week in mice and men. 23 November 1997:1.
Congenital – Dicephalic python.
Anon. 1999a. Stuff – A piece of the action. Providence Journal-Bulletin (Rhode Island) 28 July 1999:1.
Congenital – Dicephalic rattlesnake and turtles.
Anon. 1999b. Asides. Pittsburgh Post-Gazette (Pennsylvania) 25 July 1999:Editorial 1.
Congenital – Dicephalic rattlesnake and turtles.
Anon. 1999c. Two heads are better than one. Telegraph Herald (Dubuque, IA) 24 July 1999.
Congenital – Dicephalic rattlesnake and turtles.
Anon. 2000. Drug king snitches; out 88 years early – A U.S. attorney cites extraordinary help. Florida Times-Union 28 August 2000:B2.
Congenital – Dicephalic snake.
Anon. 2000. A rare find. Richmond Times Dispatch (Virginia) 14 September 2000:1.
Congenital – Dicephalic River Cooter turtle.
Anon. 2001. Strange snapper; Two-headed hatchling gets chance at long life with Mr. Turtle. Grand Rapid Press (Michigan) 20 October 2001:A3.
Congenital – Dicephalic snapping turtle.
Anon. 2002a. Science notebook. Washington Post 18 March 2002:A07.
Congenital – Dicephalic ladder snake.
Anon. 2002b. Double trouble Hobart Mercury 5 April 2002:12.
Congenital – Dicephalic ladder snake.

Anon. 2002c. So scary. Mirror 19 October 2002:35.
Congenital – Dicephalic false smooth snake.
Anon. 2002d. Kicker: Seeing twice as well. Greenwire, September 13, 2002 Friday, KICKER;10(9):1.
Congenital – Dicephalic loggerhead.
Anon. 2003a. Rare two-headed snake attracts attention of buyers. Daily News 19 September 2003:Domestic 1.
Congenital – Dicephalic boa constrictor.
Anon. 2003b. Reptile fan has double slithering surprise. Seattle Times 19 September Rap Zone 1.
Congenital – Dicephalic boa constrictor.
Anon. 2003c. Life! What's On. The Straits Times (Singapore) 6 February 2003:1.
Congenital – Dicephalic turtle.
Anon. 2003d. Senseless crime in Wisconsin. Saint Paul Pioneer Press (Minnesota) 3 June 3, 2003:E2.
Congenital – Dicephalic turtle.
Anon. 2003e. Don't lump prairie dogs in with pets or pet peeves. Pittsburgh Post-Gazette (Pennsylvania) 12 June 2003: Lifestyle.
Congenital – Dicephalic turtle.
Anon. 2003f. Two-headed snake. Kankakee Daily Journal, Kankakee, IL (23 October 2003).
Congenital – Dicephalic snake.
Anon. 2004a. Carlsbad Caverns National Park: Awed, one and all. Los Angeles Times (16 May 2004): Travel 2.
Congenital – Two-headed turtle.
Anon. 2004b. Boy's body found in well. The Hindu 22 April 2004: 1.
Congenital – Attempt to find a two-headed snake.
Anon. 2005a. Two-headed python shocks snakecatcher. South-East Advertiser, Australia 2 February 2005: 1.
Congenital – Dicephalic python.
Anon. 2005b. SOUTH. South China Daily 27 January 2005: 1.
Congenital – Dicephalic turtle.
Anon. 2005c. Two-headed python shocks snakecatcher. South-East Advertiser, Australia 2 February 2005:1.
Congenital – Dicephalic python.
Anon. 2006a. Amazing animals. National Geographic 364:16–17.
Congenital – Dicephalic black snake.
Anon. 2006b. Animal news of the weird. Grand Rapid Press 11 January 2006:A3.
Congenital – Dicephalic rat snake.
Anon. 2006c. Briefly: Two-headed snake. Glasgow Herald 18 January 2006:15.
Congenital – Dicephalic *Elaphe obsoleta.*
Anon. 2006d. Tiny two-headed terrapin faces a fight for survival. Post and Courier 27 October 2006:B8.
Congenital – Dicephalic box turtle.
Anon. 2006e. 2 headed snake picks up corporate sponsorship. St. Louis Post Dispatch 9 November 2006:D6.
Congenital – Dicephalic black rat snake.
Anon. 2006f. Two-headed beast from 100 m BC. London Times 20 December 2006:31.
Congenital – Dicephalic turtle, black rat snake, and derodymus fossil reptile.
Fossil – Derodymus fossil reptile *Sinohydrosaurus lingyuanensis* reported by Eric Buffetaut.
Anon. 2006g. Palaeontolgists discover two-headed lizard fossil. Hindustan Times 20 December 2006:1.
Congenital – Axial bifurcation in choristodere reptile from the Early Cretaceous Yixian rock formation in northeast China.
Fossil – Axial bifurcation in choristodere reptile from the Early Cretaceous Yixian rock formation in northeast China.
Anon. 2006h. Paleontology (A Two-headed fossil). Economist 23 December 2006:1.
Congenital – Dicephalism in snakes and turtles and axial duplication in fossil reptile.
Fossil – Axial duplication in fossil reptile.
Anon. 2006i. Two-headed lizard spied in a fossil. New Scientist 20 December 2006:1.
Congenital – Derodymus choristodera family, Early Cretaceous reptile from China reported by Eric Buffetaut.
Fossil – Derodymus choristodera family, Early Cretaceous reptile from China reported by Eric Buffetaut.
Anon. 2007a. Ancient reptile had two heads. Iran Daily 21 January 2007 Science:1.
Congenital – Axial bifurcation in Cretaceous choristodere (plesiosaur morphotype, but without the flipper-like legs) *Hyphalosaurus lingyuanensis.*
Fossil – Axial bifurcation in Cretaceous choristodere (plesiosaur morphotype, but without the flipper-like legs) *Hyphalosaurus lingyuanensis.*
Anon. 2007b. River watcher. Orville Mercury Register 7 June 2007:1.
Congenital – Dicephalic gopher snake.
Anon. 2007c. 2-headed snake dies. Desert Morning News 21 June 2007:1.
Congenital – Dicephalic rat snake.
Anon. 2007d. City museum's 2-headed snake dies a celebrity. St. Louis Post-Dispatch 21 June 2007:C1.
Congenital – Dicephalic rat snake.

Anon. 2007e. Snake, and both of its heads, dies at aquarium. St. Petersburg Times 21 June 2007:6A.
Congenital – Dicephalic rat snake.
Anon. 2007f. Turtle with 2 heads. Youngstown Vindicator September 28, 2007:A4. (based on Associated Press: Two-headed turtle goes on display in Pa.)
Congenital – Conjoined red-eared slider. Duplication of anterior half of bodies including forelegs and heads. Also noted another "conjoined-twin turtle" 20 years ago, without any other information.
Anon. 2007g. Turtle makes do double take: Pet store's oddity may live 20 years. Grand Rapid Press 1 October 2007:A4.
Congenital – Derodymus red-eared slider with upper body duplication.
Anon. 2007h. 2 headed snake on a plane. Augusta Chronicle 21 November 2007:A01.
Congenital – Dicephalic snake.
Anon. 2007i. These carry-ons are just plane crazy. Daily News 21 November 2007:20.
Congenital – Dicephalic snake.
Anon. 2007j. Turtle racing to top of pet ownership list. Dayton Daily News 21 December 2007:D7.
Congenital – Dicephalic yellow-bellied turtle.
Anon. 2008a. Georgia museum on fiscal wish list. Atlanta Journal-World 21 January:1B.
Congenital – Dicephalic snake.
Anon. 2008b. Why field trips shouldn't become a thing of the past. Atlanta Journal-Constitution 24 February 2008:3J.
Congenital – Dicephalic snake.
Anon. 2009. 20 Amazing reptile and amphibian anomalies. http://webecoist.com/2009/07/31/20-amazing-reptile-and-amphibian-anomalies.
Congenital – States that first dicephalic reptile was *Hyphalosaurus*, dated at 120 million years. Derodymous turtle, lizards, and snakes (illustrated). Three-headed turtle – video! Frogs with supernumerary front limbs (illustrated). Lizard duplication of front two third of body. Conjoined (rear to rear) turtle, lizard, and Bangkok gator farm crocodile (illustrated).
Fossil – States that first dicephalic reptile was *Hyphalosaurus*, dated at 120 million years.
Anthony J, Serra RG. 1951. Novos casos de fraturas consolidadas em posição viciosa em serpentes sul americanas [New cases of fractures healed in vicious positions in South American serpents]. Revista Brasileira de Biologia 11(1):101–103 [Portuguese].
Trauma – South American snakes that suffered fractures in the hemimandible (*Lachesis muta, Bothrops atrox*) and quadrate (*Bothrops insularis*), and whose bones healed in a bad position.
Antinoff N. 1997. Osteomyelitis in reptiles. Proceedings of the Association of Reptilian and Amphibian Veterinarians 1997:149–152.
Infection – Digits and foot abscesses are common in reptiles. In monitor lizards, this may be related to obesity; in others, attributed to rough environmental surfaces, nail avulsions, or constricting lesions. Lytic lesions are present, but periosteal reaction is usually absent. When present, it is less prominent than in mammals (according to Silverman and Janssen 1996). However, snake vertebral bone proliferation, fusion, and deformity may be produced by infection.
Antunes A. 1963. Classificação dos monstros de Isidore Geoffroy Saint Hilaire, ampliada e adaptada. [Classification of monsters of Isidore Geoffroy Saint Hilaire amplified and adapted]. Boletim do Departamento de Anatomia Patológica, Faculdade de Veterinária da Universidade de São Paulo 1 (5):4–12; 1 (6):4–11 [Portuguese].
Congenital – Repetition of classification of Geoffroy Saint-Hilaire.
Anver MR, Pond CL. 1984. Biology and Diseases of Amphibians. In Laboratory Animal Medicine, JG Fox, BJ Cohen, FM Loew, eds. New York: Academic Press, pp. 427–447.
Congenital – Bone deformation in frogs and newts, citing Reichenbach-Klinke and Elkan (1965).
Appleby EC, Siller WG. 1960. Some cases of gout in reptiles. Journal of Pathology and Bacteriology 80:427–430.
Metabolic – Articular gout (recognized by presence of white or pale yellow crystals) occurs in *Varanus exanthematicus, Testudo sulcata, Kinixys belliana, Testudo radiata* (forelimb), Morish tortoise *Testudo graeca* and *Testudo hermanni, Alligator sclerops, Tupinambis*, and sharp-nosed crocodile *Crocodilus americanus*. Large fusiform urate kidney stones were found in a Bell's hinge-back tortoise *Kinixys belliana.*
Applegarth JS. 1991a. Care and feeding of a two-headed common garter snake, *Thamnophis sirtalis fitchi*. Oregon Herpetological Society News (28):10–11.
Congenital – Dicephalic common garter snake, *Thamnophis sirtalis fitchi* with short mandible.
Applegarth JS. 1991b. Notes on two-headed snakes. Oregon Herpetological Society News (29):3–4.
Congenital – Dicephalic common garter snake *Thamnophis sirtalis fitchi*, Northwest garter snakes *Thamnophis ordinoides* (2), and racer *Coluber constrictor.*
Araújo TH, Pavla de Faria F, Katchburian E, Freymüller Haapalainen E. 2010. Ultrastructural changes in skeletal muscle of the tail of the lizard *Hemidactylus mabouia* immediately following autotomy. Acta Zoologica (Stockholm) 91:440–446.
Trauma – Autotomy in *Hemidactylus mabouia.*

Arefare 1892. A double-headed gopher snake. Forest and Stream, p. 148.
Congenital – Derodymous gopher snake *Pituophis conifer*.

Arias E, Zavanella T. 1979. Teratogenic effects of manganese ethylenebisdithiocarbamate (maneb) on forelimb regeneration in the adult newt, *Triturus cristatus carnifex*. Bulletin of Environmental Contamination and Toxicology 22:297–304.
Toxicology – Manganese ethylenebisdithiocarbamate produces delayed regrowth after amputation in *Triturus cristatus carnifex*. Thiourea and phenylthiourea similarly delayed *Xenopus laevis* and *Rana temporaria* growth (Fox and Turner 1967).

Arias E, Zaffaroni NP, Zavanella T. 1983. A study on the genesis of Maneb-induced malformations of the regenerating limb of the adult crested newt. In Vago C, Matz G. eds. Comptes Rendus du Premier Colloque International de Pathologie des Reptiles et des Amphibiens. Angers, France, pp. 223–226.
Toxicology – Dithiocarbamate fungicide effect produced limb malformation in *Triturus cristatus*.

Arias E, Zavanella T, Pacces-Zaffaroni N. 1989. Teratogenic effects of 2,4-D on the regenerating limb of the crested newt. Herpetopathologia 1(2):1–4.
Toxicology – Regenerating crested newt *Triturus cristatus carnifex* exposed to 2,4-dichlorophenoxyacetic acid developed carpal fusions and reduction of metacarpals and phalanges.

Arias J, Martret G, Filella E, Subira I, Martinez-Silvestre A. 1999. Polimelia a *Triturus marmoratus* (Latreille, 1800) (Urodela, Salamandridae). [Polymelia in *Triturus marmoratus* (Latreille, 1800) (Urodela, Salamandridae)]. But Soc Cat d'Herp 14:98–99 [Italian].
Congenital – Polymelia in *Triturus marmoratus*.

Ariel E, Ladds PW, Buenviaje GN. 1997. Concurrent gout and suspected hypovitaminosis A in crocodile hatchlings. Australian Veterinary Journal 75:247–249.
Metabolic – Hypovitaminosis produced visceral, not articular gout in *Crocodylus johnstoni* and *porosus*.

Aristotle 1984. History of Animals. In The Complete Works of Aristotle, J Barnes, ed. The Revised Oxford Translation. Bollingen, Princeton University Series, pp. 774–993.
Trauma – Lizard and serpent tails regrow.

Aristotle. 1910. De generatione animalium. [On the generation of animals]. In Platt A: The works of Aristotle translated into English under the editorship of Smith JA and Ross WD. Oxford, 240 pp [Greek → English].
Congenital – Noted rarity of dicephalic snakes.

Arnold EN. 1984. Evolutionary aspects of tail shedding in lizards and their relatives. Journal of Natural History 18:127–169.
Trauma – Ability to shed and regenerate tail is widespread in lizards and amphisbaenians and also occurs in a few snakes and in the tuatara.
This ability appears to have been independently lost in members of many groups, but some agamids and snakes have regained the ability. Amphisbaenians do not regenerate.
Climbing lizards have higher incidences of broken tails than ground-dwelling species, perhaps because the tail is usually less important in locomotion. *Anolis carolinensis* tail loss reduces perching ability.
Sceloporus grow more slowly, when replacing tail. *Podarcis sicula*, *Cophosaurus texanus*, *Uma notata*, and *Dipsosaurus dorsalis* with broken tails are slower. Speed is reduced in 36% of *Cnemidophorus sexlineatus* that have lost tails. Loss of distal third of tail prevents Basiliscus from running on its hind legs (Snyder 1949). Additionally, 50% of total growth energy is used in tail regeneration in *Eumeces gilberti*, and regeneration of autotomized tails has future reproduction costs (Maiorana 1977).
Tail regeneration developed five times in agamids: *Physignathus, Amphibolurus Lophognathus, Agama s. str., Pseudotrapelus*, and *Stellio* (Moody 1980).
Noted frequencies of tail fracture/loss and regeneration are presented below:

Species	Percent with Broken tails	Percent with Regeneration
Agama aculeate	19	0
adramitana	73	23
agama	49	12
agilis	1	0
agrorensis	75	75
anchietae	7	0
annectans	44	16
atra	35	5
atricollis	28	3
batilifera	40	0
benueensis	71	3

(continued)

Species	Percent with Broken tails	Percent with Regeneration
bibronii	62	6
blanfordi	1	0
boueti	50	25
caucasica	56	10
caudospinosa	100	7
cyanogaster	39	15
distanti	10	0
doriae	50	8
flavimaculata	1	0
gracilimembris	50	0
hartmanni	75	0
himalayana	29	21
hispida	15	0
kirki	11	0
lehmanni	17	0
megalonyx	0	0
melanura	50	38
mossambica	43	0
mutabilis	1	0
mwanae	18	0
nupta	51	27
pallida	1	0
paragama	50	4
persimilis	25	0
phillipsii	52	23
planiceps	65	21
robecchii	0	0
rubrigularis	0	0
ruderata	1	0
rueppelli	63	3
sanguinolenta	0	0
sankaranika	68	4
sinaita	54	6
spinosa	75	13
stellio	57	9
sylvanus	11	11
tuberculata	47	25
weidholzi	93	0
yemenensis	53	20
Amphibolurus		
caudicinctus	16	3
cristatus	17	0
isolepis	21	0
maculatus	11	0
pictus	18	0
Diporophora bilineata	21	0
Lophognathus temporalis	44	15
Psammophilus dorsalis	67	0
Otocryptis wiegmanni	75	0
Sitana ponticeriana	40	0

The majority of lizards autotomize throughout the tail up to the basal, pygal series of vertebrae. Autotomy was not found in Platynota (Varanidae, Lanthanotidae, or Helodermatidae) or Chamaeleonidae. Fractures planes for autotomy occur in Dibamidae, Anelytropsidae, Lacertidae, Cordylidae, Xantusiidae, and Anniellidae. Autotomy planes are well developed in Scincidae, with exception of *Egernia, Tiliqua* [*Omolepida, Trachydosaurus* (such planes disappears in adult *T. rugosus*)], *Corucia*, and *Tribolonotus*. Among Iguanines, autotomy planes were absent in *Amblyrhynchus, Brachylophus, Conolophus*, and *Iguana delicatissima*; lost/reduced during ontogeny in *Iguana iguana, Ctenosaura acanthura, Sauromalus*, and in some *Cyclura* species. Teiidae have well-developed autotomy planes, which may be obliterated in adult *Tupinambis teguixin*. The very short-tailed diplodactyline Gekkonidae lacking fracture planes was *Nephrurus asper*. While most agamid lizards do not autotomize, site of autotomy occurrence is between, rather than within (fracturing), vertebrae. Intravertebral fracture planes and reduction of autotomy occur in *Pachydactylus, Stenodactylus, Anolis, Iguana, Ophryoessoides, Tiliqua, Ophisaurus*, and *Amphisbaena*.

Arnold SJ. 1988a. Quantitative genetics and selection in natural populations: Microevolution of vertebral numbers in the garter snake *Thamnophis elegans*. In proceedings of the Second International Conference of Quantitative Genetics, BS Weir, MM Goodman, EJ Eiser, G Namtung, eds. pp. 619–636. Sinauer Assoc Inc, Sunderland, Massachusetts.

Congenital – Body vertebral number corresponds to ventral scale number, while tail vertebra number corresponds to subcaudal scales in snakes.

Variation in snake vertebral numbers (Kerfoot 1969, often with a normal or log-normal distribution (Kerfoot and Kluge 1971). Coastal populations have fewer vertebrae than inland desert snake populations (Klauber 1941), while local populations manifest less than 1% divergence, contrasted with 10% between populations (Klauber 1941). This is eight times the variation that has been seen with temperature effects (Osgood 1978). Garter snake *Thamnophis elegans* males have five more body vertebrae and ten more tail vertebrae than females.

	Gender	California Coastal	California Inland
Body vertebrae	Male	156	172
	Female	151	167
Tail vertebrae	Male	79	90
	Female	70	81

F1 generation from mating coastal and inland resulted in intermediate numbers.

Trauma – Snake tails do not regenerate.

Arnold EN. 1988b. Caudal autotomy as a defense. In C Gans, RB Huey (eds.) Biology of the reptilian. New York: Alan R. Liss, pp. 235–273.

Trauma – Tail breakage in *Sceloporus magister*, and possibly, *Xantusia vigilis* was attributed to territorial disputes. It was more common in climbing geckos (Werner 1968) and *Liolaemus* (Jasiöand Fuentes 1980) and attributed to predator pressure in *Cnemidophorus* which is greater in northern than southern locations (Pianka 1970).

Arnold S, Bennet A. 1988. Behavioral variation in natural populations. V. Morphological correlates of locomotion in the garter snake (*Thamnophis radix*). Biological Journal of the Linnean Society 34:175–190.

Congenital – Eighteen percent of *Thamnophis radix* had abnormal number of ventral scutes, and 9% had undivided subcaudal scales. Body anomalies were associated with less tail anomalies, with more crawling speed. Constricting colubroids have a greater number of body vertebrae than non-constrictors, the latter (e.g., *Nerodia fasciata*) having faster crawling velocities than constrictors (e.g., *Elaphe obsoleta*).

Arnold SJ, Peterson CR. 2002. A model for optimal reaction norms: The case of the pregnant garter snake and her temperature-sensitive embryos. American Naturalist 160:306–316.

Environmental – Cited temperature effects on vertebral number (Fox 1948; Jockusch 1997; Lecyk 1965; Peabody and Brodie 1975; Yntema 1960), but disputed garter snake *Thamnophis elegans* variation in scale counts as false impression of thermile lability – as temperatures were outside ecologically relevant range.

Arntzen JW. 1994. Allometry and autotomy of the tail in the golden-striped salamander, *Chioglossa lusitanica*. Amphibia-Reptilia 15:267–274.

Trauma – 6–33% of golden-striped salamander, *Chioglossa Lusitanica* had lost their tails, which was associated with lower female fecundity.

Arntzen JW, Wallis GP. 1991. Restricted gene flow in a moving hybrid zone of the newts *Triturus cristatus* and *T. marmoratus* in western France. Evolution 45:805–826.

Congenital – Digital abnormalities were present in 17% of *Triturus cristatus* x *T. marmoratus* hybrids.

Arvy C, Fertard B. 2002. Pathologie des tortues: étude synthétique. [Pathology of turtles: summary study]. Bulletin de la Société Herpétologique de France 100:1–152 [French].

Congenital – Achondroplasia in *Trachemys scripta* (Frye and Carney 1974) was manifested as long bone shortening, cartilage hypertrophy with ossification spicules.

Trauma – Carapace fractures are noted (Bourdeau 1988; Devaux 1992; Jackson 1990; Jacobson 1994; Mautino and Page 1993; McArthur 1997).

Infection – Osteomyelitis in turtles can be either mycotic (especially *Mucor*) or bacterial and is typically humidity-related. Attributed organisms include *Pseudomonas, Citrobacter*, and *Klebsiella* (Bourdeau 1988b; Brogard 1990; Cooper 1992; McArthur 1997). "Septic ulcerative shell disease" was caused by Citrobacter (Bourdeau 1988; Cooper 1992) and digital/limb loss (Bourdeau 1988; Gabrisch and Zwart 1990; Mautino and Page 1993), although Garner et al. (1997) suggest that bacterial infection was secondary. Osteomyelitis was reported in association with aseptic (? actually septic) arthritis (Rhodin et al. 1990).

Metabolic – Vitamin D deficiency produces general skeletal and carapace deformity (Bourdeau 1988a). Gout is reported in turtles (Frye 1979; McArthur 1997). Dysplasia "en pyramide" in turtles is a metabolic osteodystrophy (Jacobson 1994). Carapace necrosis in *Gopherus agassizii* was reported as dyskeratosis with hyperuricemia (Jacobson et al. 1994).

Shell disease – Plaston erosion considered captivity-related (Bourdeau 1988b). Ulcerative shell disease occurs in *Trionyx, Sternotherus, Chrysemys, Chelonia, and Podonemis* (Bourdeau 1988b; Brogard 1990; Kaplan 1957). Mycotic (especially *Mucor*) and bacterial (*Pseudomonas, Citrobacter, Klebsiella*) infection is prominent under conditions of high humidity. (Bourdeau 1988b; Brogard 1990; Cooper 1992; McArthur 1997). Septic ulcerative shell disease was caused by *Citrobacter* (Bourdeau 1988; Cooper 1992) and with secondary bacterial infection (Garner et al. 1997). Carapace necrosis in *Gopherus agassizii* was related to dyskeratosis with hyperuricemia (Jacobson et al. 1994). Coquelet (1983) even suggested that the dyskeratosis represented psoriasis.

Neoplasms – Neoplasms are rare.

Astort ED. 1981. Bicefalia en *Bothrops alternatus duméril*, Bibron y Duméril (Ophidia-Crotalinae) [Bicephaly in *Bothrops alternatus duméril*, Bibron and Duméril (Ophidia-Crotalinae)]. Resúmenes VI Jornadas Argentinas de Zoología: 11–13 [Spanish].

Congenital – Derodymus *Bothrops alternatus*, also referred to as teratodymus cryptoderodymus iniodymus (joined laterally by temporal regions, but independent skulls and kyphotic vertebral columns). There was ventral scale fusion in the pre-cloacal region.

NOTE: A copy of this paper was obtained through colleagues for review, but the bibliographical citation could not be verified. *Resúmenes VI Jornadas Argentinas de Zoología 1981* appears to be a different (although associated) publication from *VI Jornadas Argentinas de Zoología Symposia 1981*.

Ataeva AA. 1986. [A case of supernumerary limb in *Rana ridibunda* in waters of Turkmenistan]. Izvestiya Akademii Nauk Turkmenskoi SSR, Seriya Biologicheskikh nauk 1986(4):78 [Russian].

Congenital – Supernumerary limb was present in *Rana ridibunda*.

Auer E. 1909. Über einige Krokodile der Juraformation. [About a Jurassic crocodile]. Palaeontographica 55:217–294 [German].

Other – Pathological changes (reduction or addition of bone substance) were noted on palatine, femoral, and on sacral vertebra in *Metriorhynchus cf. moreli*. The right femur had "corrosion" below caput femoris; the left femur, a deformed caput femoris. The trochanter femoris was thickened by bone deposits with a deformed distal end, and sacral vertebra was considerably thickened.

Fossil – Pathological changes (reduction or addition of bone substance) were noted on palatine, femoral, and on sacral vertebra in *Metriorhynchus cf. moreli*. The right femur had "corrosion" below caput femoris; the left femur, a deformed caput femoris. The trochanter femoris was thickened by bone deposits with a deformed distal end, and sacral vertebra was considerably thickened.

Auffenberg W. 1988. Gray's monitor lizard. University of Florida Press, Gainesville, 419 pp.

Trauma – Thirty two of 118 fugivorous bataans (Gray's monitor) had minor healed injuries. Eight percent lacked tail tip (contrasted with 16% of *Varanus salvator*, the latter being a carrion feeder). Mostly, female bataans injuries were attributed to intraspecific aggression. Seventy five percent of *Varanus olivaceus* (correct name of *Varanus grayi* after Auffenberg on priority) injuries were in females, but were not intraspecific aggression-related. One bataan had hip (proximal femur) arthritis, attributed to trauma.

Austwick PK, Keymer IF. 1981. Fungi and actinomyctes. In Cooper JE, Jackson OF. eds. Diseases of the Reptiles. Vol 1:193–234. New York, Academic Press.

Infection – *Coniothyrium fuckelianum* was associated with shell disease in *Testudo graeca* (Goodwin 1976) and *Fusarium solani* in *Caretta caretta* (Rebell et al. 1971). Plaston involvement by fungi was reported in *Chelus fimbriatus* (Hammerton 1934), *Testudo gigantea elephantiae* (Hammerton 1935), and *Chelodina longicollis* (Hammerton 1939).

Shell disease – *Coniothyrium fuckelianum* was associated with shell disease in *Testudo graeca* (Goodwin 1976) and *Fusarium solani* in *Caretta caretta* (Rebell et al. 1971). Plaston involvement by fungi was reported in *Chelus fimbriatus* (Hammerton 1934), *Testudo giantea elephantiae* (Hammerton 1935), and *Chelodina longicollis* (Hammerton 1939).

Avel M, Baer J-G. 1929. Un cas de duplication du bras droit n'obeissant pas a la loi de Bateson, observe chez *Discoglossus pictus* Otth. [A case of right arm duplication, not obeying Bateson's law in *Discoglossus pictus* Otth.] Revue Suisse de Zoologie 36:641–646 [French].

Congenital – Supernumerary forelimb noted in *Discoglossus pictus.*

Avery B. 1990. Surgery of reptiles. American Association of Zoo Veterinarians 1990:168–171.

Trauma – Pathologic fractures reported from nutritional secondary hyperparathyroidism.

Metabolic – Pathologic fractures reported from nutritional secondary hyperparathyroidism.

Avilla LS, Fernandes R, Ramos DF. 2004. Bite marks on a crocodylomorpha from the Upper Cretaceous of Brazil: Evidence of social behavior? Journal of Vertebrate Paleontology 24:971–973.

Trauma – Crocodylomorpha Mesoeucrocodylia *Baurusuchus pachecoi* from the Upper Cretaceous of Brazil had osteoderm scratches and linear, shallow, 22.1 mm long scratches and cracks. This compares with 20% of *Crocodylus niloticus*. Extensive cicatrization indicated survival.

Fossil – Crocodylomorpha Mesoeucrocodylia *Baurusuchus pachecoi* from the Upper Cretaceous of Brazil had osteoderm scratches and linear, shallow, 22.1 mm long scratches and cracks. This compares with 20% of *Crocodylus niloticus*. Extensive cicatrization indicated survival.

Babcock HL. 1930. Variations in the number of costal shields in *Caretta*. American Naturalist 64:95–96.

Shell disease – Enormous amount of variation in costal shields in *Caretta caretta* (Boulenger 1889). Atlantic Ocean *Caretta caretta* have 5 costal shields on either side, with 7 of 130 showing 1 unit deviation. Pacific and Indian Ocean forms have six to seven, often more and not always symmetrical. One Atlantic Ocean form had only three pairs of costals (Coker 1910).

Bachs Taberner M. 1998. Metabolic bone disease in common iguanas. Reptilia (GB) (2):12–13.

Metabolic – The article discusses metabolic bone disease (MBD) in iguanas, and states it is also known as secondary nutritional hypoparathyroidism, fibrous osteodystrophy, and osteoporosis. It also states that metabolic bone disease (MBD) is the most frequent disease found in captive iguanas and can also be found in other reptiles including chelonians. MBD occurs because the calcium in the bloodstream (required for the transmission of nervous impulses and muscle contractions) is insufficient, and the organism resorts to decalcification of the bones to raise the blood level of calcium. Among other symptoms, MBD causes hard swelling of the extremities, softening and deformation of the bones in general, deformation of the vertebral column, and multiple fractures of the bones. Causes of the disease include (1) improper diet, mainly the wrong calcium to phosphorus ratio; (2) inadequate ingestion or synthesis of vitamin D3, which is required for absorption of calcium from the intestine; and (3) inadequate exposure to UVB radiation, which is required for the synthesis of the biologically active form, vitamin D3.

Bacon J, Linzey D, Rogers R, Fort D. 2006. Deformities in cane toad (*Bufo marinus*) populations in Bermuda: Part I. Frequencies and distribution of abnormalities. Applied Herpetology 3:39–65.

Congenital – 19–30% of cane toad *Bufo marinus* had hind limb abnormalities, with lesser number with forelimb abnormalities. Abnormalities included brachydactyly and polymely. No *Ribeiroia* metacercariae were found in these Bermuda specimens.

Bagatto B, Guyer C, Hauge B, Henry RP. 1997. Bimodal respiration in two species of Central American Turtles. Copeia 1997:834–839.

Vascular – Mexican giant musk turtle *Staurotypus triporcatus* has a reduced plastron, with large cutaneous surface area for respiration, but had equal reliance on aquatic respiration as white-lipped mud turtle *Kinosternon leucostomum. Staurotypus triporcatua* has a thick, leathery skin, contrasted with that of *A. spinifera.* It also had twice as many breathing bouts as *Kinosternon leucostomum* but took only one sixth of the number of breaths per bout. Carbon dioxide excretion exceeds oxygen absorption, related to carbon dioxide solubility in water. They suggest the lungs are for oxygen absorption and the skin for CO2 excretion.

Bailey SW. 1987. Treatment of a terrapin for shell rot. Rephibearary 113:6–7.

Infection – Asiatic leaf or brown stream terrapin *Cyclemys dentata* with rugose, pitted shell from osteomyelitis/shell rot.

Shell disease – Asiatic leaf or brown stream terrapin *Cyclemys dentata* with rugose, pitted shell from osteomyelitis/shell rot.

Baker W. 1795. The two headed snake. Rural Magazine, Rutland, Vermont November 1795:563.

Congenital – Dicephalic water snake.

Bakken DJ, Bakken LA. 1987. Dicephalism in the Russian rat snake. *Elaphe schrencki schrencki.* Bulletin of the Chicago Herpetological Society 22:2.

Congenital – Dicephalic Russian rat snake *Elaphe schrencki schrencki.*

Balazs GH. 1977. See Wright and Balazs 1977.

Balazs GH. 1980. Synopsis of biological data on the green turtle in the Hawaiian Islands. 141 pp. National Oceanic and Atmospheric Administration Technical Memorandum National Marine Fisheries Service-SouthWest Fisheries Center-7 United States Department of Commerce, 141 pp.

Trauma – Carapace and plastron injuries in Chelonia from "violent contact with rocks and substrate in rough turf close to shore" (Wright and Lay).

Infection – *Mycobacetrium avium* infection in green turtle *Chelonia mydas*, but no comment on bone involvement (Brock et al. 1976).

Balinsky BI. 192. Über experimentelle Induktion der Extremitätenanlage bei Triton mit besonderer Berücksichtigung der Innervation und Symmetrieverhaltnisse derselben. [On experimental induction of extremity anlage in *Triton* with special regard to innervation and symmetrical relationships]. Roux Archiv EntwMech. Org. 110:71–88.

Trauma – Surgical induction of extremities in *Triton taeniatus.*

Baljet B, Heyke GCM. 1992. Zur Geschichte der Klassifikationssysteme der Doppelmißbildungen unter besonderer Berücksichtigung des Klassificationssystems von Louis Bolk (1866–1930) [On the history of the classification system of double malformations under special consideration of the classification system of Louis Bolk (1866–1930)]. Annals of Anatomy 174:361–368 [German].

Congenital – Terminology offered as (1) C. diplopagi simplex caudad, (2) diplopagi simplex craniad, (3) diplopagi simplex mesad, but herps were not mentioned.

Ball HA. 1946. Melanosarcoma and rhabdomyoma in two pine snakes (*Pituophis melanoleucus*). Cancer Research 6:134–138.

Neoplasm – Malignant melanoma in pine snake *Pituophis melanoleucus.*

Ball JC. 1995. Axial bifurcation. Case study: A two-headed yellow rat snake. Reptile and Amphibian Magazine Jan–Feb:36–43.

Congenital – Dicephalism in *Elaphe obsoleta lindheimeri, Vipera russellii, Crotaphopeltis hotamboeia, Bothrops moojeni, and Elaphe obsoleta quadrivittata.*

Craniodichotomy – In *Elaphe obsoleta lindheimeri.*

Prodichotomy – In *Thamnophis elegans vagrans.*

Opisthodichotomy – In *Lamprophis fulginosus* and in a *Lampropeltis alterna* x *mexicana* cross.

Ball RL, Dumonceaux G, MacDonald C. 1999a. Hypertrophic osteopathy associated with renal gout in a green iguana, *Iguana iguana*. Proceedings of the Association of Reptilian and Amphibian Veterinarians 1999:49–50.

Other – Diffuse periosteal proliferation in green iguana *Iguana iguana* diagnosed as hypertrophic osteoarthropathy. Pulmonary masses were present, and individual was cachectic from bite wound to rear foot, had a thorn in the left rostal mandible, and *Acinetobacter baumannii* and *Trichosporon beigelii* grown from coelomic fluid accumulation.

Ball RL, Dumonceaux G, MacDonald C. 1999b. Hypertrophic osteopathy associated with renal gout in a green iguana, *Iguana iguana*. Proceedings of the Annual Conference of the Association of Reptilian and Amphibian Veterinarians 6:49–50.

Infection – Cachectic *Iguana iguana* with bite wound on rear foot, a thorn in the left rostal mandible, and *Acinetobacter baumannii* and *Trichosporon beigelii* grown from coelomic fluid accumulation.

Metabolic – Femoral, tibial, and fibular periosteal proliferation (up to 1 cm in thickness) with "intertrabecular fibrosis" noted in a green iguana *Iguana iguana*, with 2–3 mm nodular lung masses and multifocal pinpoint kidney calcifications. Diagnosis is difficult to assess, but renal calcifications are highly suggestive of hyperparathyroidism. Renal osteodystrophy is an alternative diagnostic consideration, but cannot be further analyzed as no photographs were provided.

Neoplasia – Femoral, tibial, and fibular periosteal proliferation (up to 1 cm in thickness) with "intertrabecular fibrosis" noted in a green iguana *Iguana iguana*, with 2–3 mm nodular lung masses, and multifocal pinpoint kidney calcifications was attributed to hypertrophic osteoarthropathy. Diagnosis is difficult to assess, but renal calcifications are highly suggestive of hyperparathyroidism. Renal osteodystrophy is an alternative diagnostic consideration, but cannot be further analyzed as no photographs were provided.

Other – Femoral, tibial, and fibular periosteal proliferation (up to 1 cm in thickness) with "intertrabecular fibrosis" noted in a green iguana *Iguana iguana*, with 2–3 mm nodular lung masses and multifocal pinpoint kidney calcifications.

Ballengée B, Sessions SK. 2009. Explanation for missing limbs in deformed amphibians. Journal of Experimental Zoology (Molecular Development and Evolution) 312B:1–10.

Congenital – Three of 35 Hoyland Bank Pond, Barnsley, West Yorkshire, England *Bufo bufo* had missing eyes. The rest had limb or tail absence or reduction or tapered cartilage growth. Also reported, *Rana temporaria* with abnormal hind limb and smooth newt *Triturus vulgaris.*

Trauma – Selective predation by invertebrate predators, selectively snipping, or chewing off small pieces (Brodie et al. 1978; Johnson et al. 1975; Licht 1974; Manteifel and Reshetnikov 2002).

Frequency of deformed toads were 1.3% at Havercroft, 1.2% at Campsall Clay, and 9.8% at Upton Colliery (chi square = 22.139). Predators at sites included three-spined stickleback fish *Gasterosteus aculeatus*, newts *T. helveticus* and *T. vulgaris*, diving beetles *Dytiscus*, water scorpions *Nepa cinerea*, and dragonfly odonata nymphs *sympetrum*. Dragon fly nymphs *Aeshna mixta, Libellula* depressa and especially *sympetrum* (probably striolatum or sanguineum) were the only ones that generated deformities. *Dytiscus* larvae maimed and killed, but it did not produce deformities.

Environmental – Increased salinity in freshwater habitats increases tadpole susceptibility to predation (Cook 1971; Squires et al. 2008).

Ballinger RE, Tinkle DW. 1979. On the cost of tail regeneration to body growth in lizards (Reptilia: Lacertilia). Journal of Herpetology 13:374–375.
Trauma – *Anolis carolinensis* (which uses its tail for balance) and *Cnemidophorus sexlineatus* (which increases speed, using tail as counter balance) do not loose tails as frequently as *Lygosoma laterale* or *Coleonyx variegatus*. Body growth is reduced in tailless *Sceloporus undulatus*, *S. jarrovi*, and *S. scalaris*.
Ballinger RE, Nietfeldt JW, Krupa JJ. 1979. An experimental analysis of the role of the tail in attaining high running speed in *Cnemidophorus sexlineatus* (Reptilia: Squamata: Lacertilia). Herpetologica 35:114–116.
Trauma – Autotomy produces 36% reduction of speed in *Cnemidophorus sexlineatus*.
Balls M, Clothier RH. 1974. Spontaneous tumors in amphibia. Oncology 29:501–519.
Trauma – Alleged osteogenic sarcoma of femur in *Rana pipiens*, but more probably fracture with callus.
Neoplasia – Alleged osteogenic sarcoma of femur in *Rana pipiens*, but more probably fracture with callus.
Balsamo-Crivelli G. 1865. Sopra alcuni nuovi casi di Polimelia (membra soprannumerari) osservati in alcuni individui del genere *Rana*. [About some new cases of polymely (supernumerary members) observed in some individuals of the genus *Rana*]. Rendiconti Reale Istituto Lombardo di scienze e lettere. Classe di scienze matematiche e naturali 2:261–263 [Italian].
Congenital – Three cases of polymely in *Rana esculenta* (aka *Rana viridis*) in the collections of the Cabinet of Comparative Anatomy of the University of Pavia (Italy). The first case was a small specimen with three hind limbs, of which the supernumerary limb was attached to the left side, behind the articulation of the left leg, next to the pubic symphysis. It was shorter and more gracile than the normal legs. Balsamo-Crivelli believed the foot resulted from the fusion of two feet because the remaining three digits were also abnormal. The third was very small and the metatarsal was mostly fused with the metatarsal of the longest digit. The second case had a supernumerary hind limb articulated to the posterior part of the articulation for the left femur. It had only four metatarsals and four digits, of which the two in the middle were very long, and the two on the outsides were very short. Balsamo-Crivelli believed this foot also resulted from the fusion of two because in it he saw the external and medial digits of one foot, and the medial and external digits of another. The third case had a supernumerary limb articulated to the superior aspect of the left femur. The foot had only two digits, which Balsamo-Crivelli thought were the outermost digit and the following digit.
Bancroft E. 1769. An essay on the natural history of Guiana, in South America: containing a description of many curious productions in the animal and vegetable systems of that country. Together with an account of the religion, manners, and customs of several tribes of its Indian inhabitants. Interspersed with a variety of literary and medical observations. In several letters from a gentleman of the medical faculty during his residence in that country. VI + 402 pp.; London: T. Becket and P.A. de Hondt in the Strand. John Adams Library in the Boston Public Library: Adams 261.2. (pages 100–250: animals mainly snakes).
Congenital – Dicephalic snake (probably *Tropidonotus fasciata sipedon*).
Bangs O. 1896. An important addition to the fauna of Massachusetts. Proceedings of the Boston Society of Natural History 27:159–161.
Congenital – *Malaclemys terrapin* with variation in size and shape of skull.
Other – Makes statement related to spelling of author names: "A name is merely a combination of letters, and one is at liberty to use a wholly arbitrary combination."
Banta BH. 1963. An anomalous California red-backed alligator lizard. Turtox News 41:207.
Trauma – Red-backed alligator lizard *Gerrhonotus multicarinatus multicarinatus* Blainville, California Academy of Science 71342, with dorsoventrally bifurcated tail.
Banta BH. 1966. A six-legged anuran from California. Wasmann Journal of Biology 24:67–69.
Congenital – Supernumerary legs in yellow-legged frog *Rana boylii* Baird.
Banta BH. 1973. A supernumerary forelimb on a spring peeper, *Hyla crucifer crucifer* Wied (Amphibia: Salientia) from south central Michigan. Herpetgon 7(1):6–7.
Congenital – Supernumerary forelimb in spring peeper, *Hyla crucifer crucifer*.
Bantle JA, Dumont JN, Finch RA, Linder G. 1991. Atlas of abnormalities: a guide for the performance of FETAX. 68 pp.; Stillwater, Oklahoma: Oklahoma State Publications Department.
Congenital – *Xenopus* tail flexure, wavy tail, axial shortening, and dicephalism.
Trauma – *Xenopus* tail flexure, wavy, and bifid tail.
Barber LW. 1944. Correlation between wound healing and regeneration in forelimbs and tail of lizards. Anatomical Record 89:441–451.
Trauma – Tails of lizards regenerate, but limbs do not.
Barber P. 1991. Kyphosis with lung tissue expansion. Journal of Biological Photography 59:29–30.
Metabolic – Kyphotic hunchback *Trionyx ferox*.
Barbour EH. 1888. A young tortoise, *Chrysemys picta*, with two heads. American Journal of Science (3) 36:227–230.
Congenital – Bicephalic *Chrysemys picta*.
Barbour EH. 1896a. Eine zweiköpfige Schildkröte (*Chrysemys picta*). [A two-headed turtle]. Zoologischer Garten 30 – Jahrgang 2:61–62. [German]

Congenital – Dicephalic turtle, *Chrysemys picta*, observed in captivity for three and a half months. The heads alternated control of the body, depending on which was asleep.

Barbour EH. 1896b. (*Chrysemis picta*) bicèphale. [Dicephalic *Chrysemis picta*]. Revue Scientifique (4) 6 (9):281 or 348 (d'aprés Sciences). [French]

Congenital – Dicephalic turtle *Chrysemys picta* observed in captivity for three and a half months. The heads alternated control of the body, depending on which was asleep.

Barbour EH. 1896c. A two-headed tortoise, *Chrysemys picta*. Science (N.S.) 4:159–160.

Congenital – Derodymus *Chrysemys picta*.

Barbour T. 1912a. A Contribution to the Zoogeography of the East Indian Islands. Memoirs of the Museum of Comparative Zoology 44:1–203.

Congenital – Javanese cobra *Naja naja sputatrix* with 19 or 21 body scales and variable body scales163–183; 21–23 neck, 153–159 ventral, and 45–57 caudal scales in *Enhydris enhydris*; 175–186 ventral and 53–58 caudal scales in Ular pi-ter *Holarchus octolineatus*.

Trauma – Reproduced tails in *Ptychozoon kuhli*.

Barbour T. 1912b. A Contribution to the Zoogeography of the East Indian Islands. Memoirs of the Museum of Comparative Zoology 44:1–203.

Congenital – Javanese cobra *Naja naja sputatrix* with 19 or 21 body scales and variable body scales163–183; 21–23 neck, 153–159 ventral, and 45–57 caudal scales in *Enhydris enhydris*; 175–186 ventral and 53–58 caudal scales in Ular pi-ter *Holarchus octolineatus*.

Trauma – "Reproduced" tails in *Ptychozoon kuhli*.

Barbour T. 1926. Reptiles and amphibians: Their habits and adaptations. Houghton Mifflin Co, Boston, 125 pp.

Congenital – Derodymus Eastern king snake.

Barbour T, Stetson HC. 1929. The squamation of *Homoesaurus*. Bulletin of the Museum of Comparative Zoology 69:99–104.

Trauma – Change in caudal scaling patterns as evidence of tail regeneration in *Homeosaurus maximiliani* v. Meyer from Solenhofen. New scaling as evidence of tail regeneration in *Sphenodon* and *Pseudopus* (Barbour 1912; Boulenger 1888; Ginkos and Skinks).

Barfurth D. 1895a. Die experimentelle Herstellung der Cauda bifida bei Amphibienlarven. [The experimental production of cauda bifida (forked or bifid tail) in amphibian larvae]. Roux's Archiv für Entwicklungsmechanik 9:115 [German].

Congenital – Supernumerary digits regenerated in newt; double foot in Axolotl.

Barfurth D. 1895b. Die experimentelle Regeneration überschüssiger Gliedmaßentheile (Polydaktylie) bei den Amphibien. [The experimental regeneration of supernumerary parts of extremities (polydactyly) in ambibians]. Roux' Archiv für Entwicklungsmechanik 1:91–116 [German].

Congenital – Frog with three hind legs in Aschaffenburg, Germany, noting "monstrosities" reported by Duméril (1867), Fraisse (1885), Reuter (1875), Gervais (1864), Cisternas (1865), Giebel (1867), Leydig (1877), Goette (1879), and Wiedersheim (1875).

Barfurth D. 1895. Sind die Extremitäten der Frösche regenerationsfähig? [Do frog extremities regenerate?] Roux's Archiv für Entwicklungsmechanik 1: 117–123 [German].

Trauma – Frog legs can regenerate.

Barfurth D. 1899a. Sind die Extremitäten der Frösche regenerationsfähig? [Do frog extremities regenerate?] Roux's Archiv für Entwicklungsmechanik 9:2 [German].

Trauma – Tail duplicity in *Asyntaxia caudalis*.

Barfurth D. 1899b. Ein *Triton* mit einer überschüssigen fünfzehiger Vordergliedmaße (Atavistische Regeneration). [A *Triton* with supernumberary five toed forefoot]. Verhandlungen der Anatomischen Gesellschaft 13:131–132 [German].

Congenital – *Triton taeniatus* with additional front leg with five fingers.

Barfurth D. 1899c. Die experimentelle Herstellung der Cauda bifida bei Amphibienlarven. [The experimental production of cauda bifida (forked or bifid tail) in amphibian larvae]. Roux's Archiv für Entwicklungsmechanik 9:1–26 [German].

Trauma – Reports citation of bifurcated tails in lizards by Plinius, Porta and Marcgrav Marchant, Needham, Arnoult de Nobleville, Lacépède and Gachet (from Fraisse 1885), Aldrovandus (1596), and Leydig 1872, with four and two tails. Whereas Porta and Aldrovandus favor an ontogenetic origin of bifid tails, the regeneration of damaged tails has been cited as reason by Needham (1750), Lacépède, Bosc and Glückselig (from Fraisse 1885) and by Leydig (1872), and Tornier.

Congenital – Cauda bifida in *Pelobates fuscus, Triton cristatus, Bufo viridis*, and *Rana fusca* (Bruch 1864).

Barfurth D. 1903. Die Erscheinungen der Regeneration bei Wirbeltierembryonen. [The appearance of regeneration in vertebrate embryos]. In Hertwig O. ed. Handbuch der vergleichenden und experimentellen Entwicklungslehre der Wirbeltiere. [Handbook of Comparative and Experimental Vertebrate Ontogeny Series]. III (3) 1:1–129.; Jena: Gustav Fischer [German].

Congenital – Dicephalic snake with two heads, forearm duplication in *Rana esculenta, Triton taeniatus* (one with five toes), duplicated pelvic region and polydactyly, and loss of legs in *Rana fusca.*

Trauma – Bifurcated tails in *Lacerta agilis, Triton vulgaris*, and axolotl *Siredon pisciformis*; loss and regeneration of hind legs in *Rana fusca and Lacerta vivipara.*

Barker JB, Antonio FB. 1983. Life history notes. Testudines. *Gopherus polyphemus* (gopher tortoise). Scutellation. Herpetological Review 14:75–76.

Congenital – Abnormal scutes in *Gopherus polyphemus* (gopher tortoise).

Shell disease – Abnormal scutes in *Gopherus polyphemus* (gopher tortoise).

Barnes KM, Hiller N. 2010. The taphonomic attributes of a Late Cretaceous plesiosaur skeleton from New Zealand. Alcheringa 34:333–344.

Trauma – Elasmosaurid plesiosaur with gouges on distal humerus and distal femur (consistent with mosasaur *Prognathodon waiparaensis* teeth), coalescing subcircular pits (16 mm deep with concave floors) on femoral capitulum (possibly chemical taphonomy), and superficial scratches that gradually deepen, widen, and end abruptly (attributed to a *Cretoxyrhina mantelli*-sized shark.

Fossil – Elasmosaurid plesiosaur with gouges on distal humerus and distal femur (consistent with mosasaur *Prognathodon waiparaensis* teeth), coalescing subcircular pits (16 mm deep with concave floors) on femoral capitulum (possibly chemical taphonomy), and superficial scratches that gradually deepen, widen, and end abruptly (attributed to a *Cretoxyrhina mantelli*-sized shark.

Barnett S. 2003. Shell infections: When there are chinks in the armor. Newsletter Mid-Atlantic Turtle and tortoise Society: Terrapin Tales (MATTS) October:1–8.

Infection – Ulcerative shell disease traceable to bacteria, fungi, viruses, and possibly algae. Dry form usually attributed to fungus with wet form blamed on *Pseudomonas*, *Citrobacter*, and *Klebsiella.* Algae cause superficial pitting of keratin.

Shell disease – Ulcerative shell disease traceable to bacteria, fungi, viruses, possibly algae, and "internal disorders." Dry form usually attributed to fungus with wet form blamed on *Pseudomonas*, *Citrobacter*, and *Klebsiella.* Algae cause superficial pitting of keratin.

Barten SL. 1982a. Recovery from nutritional shell deformity in a softshell turtle, *Trionyx muticus.* Bulletin of the Chicago Herpetological Society 18:42.

Metabolic – Carapace curling in *Trionyx muticus* related to new born mouse diet, partially correcting with calcium supplementation and fluorescent light.

Shell disease – Carapace curling in *Trionyx muticus* related to new born mouse diet, partially correcting with calcium supplementation and fluorescent light.

Barten SL. 1982b. Fatal mineralisation in a red-footed tortoise. Veterinary Medicine Small Animal Clinician. April 1982:595–597.

Metabolic – Overdose of vitamin D (related to cat food ingestion in red-footed tortoise *Geochelone carbonaria* resulted in generalized mineralization, especially of oviducts, urinary bladder.

Barten SL. 1983. Nutritional bone disease in a softshell turtle, *Trionyx muticus*, fed pinky mice. Bulletin of the Chicago Herpetological Society 17:51–53.

Metabolic – Carapace curling in *Trionyx muticus.*

Shell disease – Carapace curling in *Trionyx muticus* related to new born mouse diet, partially correcting with calcium supplementation and fluorescent light.

Barten SL. 1991. Clinical problems of iguanas. Reptile and Amphibian Magazine 1991(Jan/Feb):40–45.

Metabolic – Metabolic bone disease in iguanas manifests as shortened or swollen/lumpy mandibles, firm long bone swelling, pathological fractures, and collapsed vertebrae, blamed on too much dietary protein. Vitamin D and calcium produce metastatic calcification in iguana.

Stones – Uroliths in iguana.

Barten SL. 1996a. Lizards. In Mader DR. ed. Reptile Medicine and Surgery. pp. 323–332; Philadelphia: Saunders.

Congenital – Axial bifurcation and conjoined twins in slow worm *Anguis fragilis*, sand lizard *Lacerta agilis*, and blue-tongued skink *Tiliqua scincoides*. Upper jaw shortening, cleft palate, and short tails, citing Bellairs (1981).

Infection – Ossifying spondylosis related to age, osteomyelitis, or "metabolic disease." Actually insufficiently illustrated for actual diagnosis.

Metabolic – Articular and periarticular gout in green iguana *Iguana iguana* and and pseudogout in Egyptian spiny-tailed lizard *Uromastyx.* Ossifying spondylosis related to age, osteomyelitis, or "metabolic disease." Actually insufficiently illustrated for actual diagnosis. Nutritional hyperparathyroidism, fibrous osteodystrophy, osteomalacia, osteoporsis, pathologic fractures, and folding fractures in lizards.

Vertebral – Ossifying spondylosis related to age, osteomyelitis, or "metabolic disease." Actually insufficiently illustrated for actual diagnosis.

Barten SL. 1996b. Shell damage. In Mader DR. ed. Reptile Medicine and Surgery. pp. 413–417; Philadelphia: Saunders.

Trauma – Carapace fracture in three-toed box turtle *Terrapene carolina triunguis.*

Infection – Shell abscess in diamondback terrapin *Malaclemys terrapin.*

Shell disease – Plastron erosion in red-eared slider *Trachemys scripta elegans.* Carapace fracture in three-toed box turtle *Terrapene carolina triunguis.* Shell abscess in diamondback terrapin *Malaclemys terrapin.*

Barten SL. 2000. Distal leg necrosis in a green iguana, *Iguana iguana.* Journal of Herpetological Medicine and Surgery 10(1):48–50.

Infection – Distal leg necrosis in green iguana *Iguana iguana* with diffuse long bone periosteal reaction and subcutaneous emphysema, diagnosed as hypertrophic osteoarthropathy.

Other – green iguana *Iguana iguana* with diffuse long bone periosteal reaction and subcutaneous emphysema, diagnosed as hypertrophic osteoarthropathy.

Barten SL. 2006a. Lizards (Biology and Husbandry section). In Mader DR. ed. Reptile Medicine and Surgery. pp. 59–77; Philadelphia: Saunders.

Trauma – Tail autotomy in five-lined skink *Eumeces fasciatus.*

Barten SL. 2006b. Lizards (Differential Diagnoses by Symptoms section). In Mader DR. ed. Reptile Medicine and Surgery. pp. 683–695; Philadelphia: Saunders.

Congenital – Axial bifurcation and other congenital abnormalities occur in lizards.

Conjoined slow worm *Anguis fragilis*, sand lizard *Lacerta agilis*, and blue-tongued skink *Tiliqua scincoides.*

Trauma – Autotomy occurs in lizards.

Infection – Osteomyelitis in horn-headed lizard *Acanthosaura crucigera.* Periodontal disease in bearded dragons, water dragon *Physignathus lesuerii*, frilled lizard *Chlamydosaurus kingii*, sailfin lizard *Hydrosaurus pustulatus*, and Jackson's chameleon *Chamaeoleo jacksonii.*

Metabolic – pseudogout, gout, and fibrous dystrophy occur in lizards.

Vascular – Avascular necrosis of tail in green iguana *Iguana iguana.*

Barten SL. 2006c. Shell damage. In Mader DR. ed. Reptile Medicine and Surgery. Pp. 893–899; Philadelphia: Saunders.

Infection – Shell abscess in *Chelydra serpentina.*

Shell disease – Shell necrosis in box turtle *Terrapene carolina*, abscess in *Chelydra serpentina*, erosions in red-eared slider *Trachemys scripta elegans.*

Bassarukin AM, Borkin LJ. 1984. Distribution, ecology and morphological variability of the Siberian salamander, *Hynobius keyserlingii* of the Sakhalin Island. Proceedings of the Zoological Institute Russian Academy of Science, Leningrad 124:12–54.

Congenital – Thirty-one percent phalangeal or digit brachydactyly (especially forelegs) in hynobiid salamander *Salamandrella (Hynobius) keyserlingii* in Sakhalin Island and Kamchatka Peninsula, attributed to aggressive behavior. Unclear how aggression would shorten a limb.

Trauma – Thirty-one percent phalangeal or digit brachydactyly (especially forelegs) in hynobiid salamander *Salamandrella (Hynobius) keyserlingii* in Sakhalin Island and Kamchatka Peninsula, attributed to aggressive behavior. Unclear how aggression would shorten a limb.

Bassi R. 1874. Di una Rana comune con un arto anteriore soprannumerario [Of a common frog with a supernumerary forelimb]. Il Medico Veterinario, Torino, Series 4, 8:120–123 [Italian].

Congenital – Polymely and polydactyly in a male *Rana esculenta*: An atrophied, insensitive, supernumerary forelimb was attached to the thorax just behind the normal right forelimb. It had five digits (normal count is four digits), of the middle digit (which was also the supernumerary digit) was rudimentary (one phalange was missing, and the other two were very small). The old fisherman that provided the specimen to Bassi stated he had seen others like it. Bassi obtained the specimen alive and was able to study it in that condition. Once dead, he desiccated and removed the skin over the supernumerary limb to study the musculature and osteology. A lithograph was provided illustrating the specimen, the dissected supernumerary limb in different aspects, and lastly, the muscular tissue that connected it to the thorax. The specimen was an adult, 63 mm long, and apparently, normal except for the atrophied and polydactylous supernumerary limb.

Bateman PW, Fleming PA. 2009. To cut a long tail short: A review of lizard caudal autotomy studies carried out over the last 20 years. Journal of Zoology 277:1–14.

Trauma – In-depth review of the autotomy literature, noting its occurrence in tuatara *Sphenodon* and 13 of 20 lizard families, citing Downes and Shine (2001). Repeated autotomies occur proximal to last one in *Niveoscincus metallicus*, citing Chapple and Swain (2004). Excellent review of effect of tail loss, loss of tail resources, cost of regeneration, and behavioral changes. Some reptiles eat their own or another individual's tails, citing Neill (1946), Clark (1971), Matuschka and Bannert (1987), Gillingham et al. (1995).

Autotomy occurs in 3% of *Callisaurus draconoides* (Bulova 1994), in contrast to 82% of *Mabuya frenata* (Van Sluys et al. 2002). Active ("albeit secretive") forager *M. frenata* had higher autotomy frequency than ambush forager *Tropidurus itambere* (Van Sluys et al. 2002).

Tail loss is generally gender independent, except for male predominance in *Mabuya heathi* (Vitt 1981).

Regenerated tail is a cartilaginous tube.

Incidence of tail loss by species	Average	Upper range
Gekkonidae	29	65
Pygopodidae	43	50
Scincidae	43	75
Lacertidae	45	70
Teiidae	35	42
Gymnophthalmidae	37	55
Phrynosomatidae	35	53
Polychrotidae	10	
Crotaphytidae	6	
Tropiduridae	35	50
Liolaemidae	60	73
Crotaphytidae	6	
Agamidae	40	62
Chamaeleonidae		
Anguidae	43	75
Xenosauridae		
Varanidae		
Lanthanotidae		

Bateman PW, Flemming PA. 2011. Frequency of tail loss reflects variation in predation levels, predator efficiency, and the behavior of three populations of brown anoles. Biological Journal of the Linnean Society

Trauma - Tail loss or regeneration in 61% of *Niveoscincus metallicus* in sites of low predation; 78%, in high predation sites, contrasted with 33 and 88% for *Podarcis muralis* and 50 and 83% for *Podarcis liffordi.* High rates of autotomy and regeneration were associated with viperid snake presence. Predators that preferentially attack cranially are associated with less regeneration, contrasted with caudal attacks (e.g, *Lampropeltis triangulum* on *Scincella lateralis, Thamnophis couchiion, Dicamptodon ensatus.*

Bateson W. 1894. Materials for the study of variation treated with especial regard to discontinuity in the origin of species. XVI + 598 pp.; London, New York: Macmillan a. Co.

Congenital – Snake dicephaly, citing Mitchill (1826):

Coluber constrictor, citing Wyman (1862).

Ophiobolus getulus, citing Yarrow (1878).

Pituophis.

Pelamis bicolor, citing Boettger (1890).

Partial or complete snake dicephaly, but species "not clear," citing Redi 1778 or 1684??, Lacépède 1790, Bancroft 1769, Lanzoni 1690 (Boston Society of Medicine Imp., Catalogue Museum #856, quoted from Wyman), Edwards 1751, Dorner 1873 and 1872, Geoffroy St. Hilaire 1838, Duméril & Bibron 1884.

Partial or complete lizard dicephaly, citing Geoffroy St Hilaire (1869).

Partial or complete turtle dicephaly, citing Edwards (1751), Mitchill and Barbour (1888).

Frog with supernumerary hind limb, citing Kingsly (1882).

Tiger boa *Python tigris* with partial duplication of vertebra 148 and vertebra 166 with duplication of vertebra and ribs on right, citing Baur (1891). Museum of the College of Surgeons #602.

African rock python *Python sebae* with abnormal vertebra 195, citing Albrecht (1883), Brussels museum #87.

Python sp. with abnormal vertebra 168.

Pelamis (*Hydrophis*) *bicolor* with abnormal vertebra 212 (Yale museum #763).

Cimoliasaurus plicatus with malformed cervical vertebra, citing Lydekker (1889) and Sutton (1889), British Museum #48001.

Polyodontophis subpunctatas with ventral shields varied in number from 151 to 240, citing Boulenger (1890).

Heloderma horridum with first caudal as either 36th or 37th vertebra.

Heloderma suspectum with first caudal as 38th or 39th vertebra, citing Bauer (1891).

Rana temporaria with partial vertebral bifurcation, citing Bourne (1884), and partial of ninth to trunk vertebra, citing Howes (1886).

Bombinator igneus with variation in pelvic girdle vertebrae.

Alytes obstetricans with enlarged proximal urostyle, citing Lataste (1879), and coccyx hypertrophy, citing Bedriaga (1879).

Bauchot R. 1994. Snakes: A Natural History. New York: Sterling Publ. Co. 220 pp.

Congenital – Derodymous smooth snake *Natrix natrix*, bull snake *Pituophis melanoleucus*, and cycloptic viper *Vipera xanthina*. Shortening or absence of mandible also reported.

Baur G. 1886. Osteologische Notizen über Reptilien. [Osteologic notes on reptiles]. Zoologischer Anzeiger 9:685–690 [German].

Congenital – *Gavialis gangeticus* with 25 instead of 24 presacral vertebrae (additional vertebra between ninth and tenth vertebra); also noted in *Aelodon priscus* (Meyer 1860) and in *Rhacheosaurus gracilis* (Quenstedt 1867). Three instead of two sacral vertebrae in *Alligator sclerops* (last presacral vertebra transformed, thus only 23 presacral vertebrae), *Crocodylus acutus*, one adult and one juvenile specimen (first caudal vertebra transformed to third sacral vertebra, after Reinhardt 1873), and in *Alligator mississippiensis* (last presacral vertebra in the process to become a sacral vertebra, with small sacral rib which does not reach the ilium).

Baur G. 1889. Revision meiner Mitteilungen im Zool. Anz., mit Nachträgen. [Revision of my Reports, with additions]. Zoologischer Anzeiger 12:238–243 [German].

Congenital – Different development of left and right sacral ribs on vertebrae of *Crocodylus acutus* and different vertebrae connected to right and left ilium. Two specimens of *Crocodylus biporcatus* with 23, instead of 24, presacral vertebrae.

Baur G. 1891. On intercalation of vertebrae. Journal of Morphology 4:329–336.

Congenital – African rock python *Python sebae* intercalated half vertebra. Another with ankylosis of vertebrae 195 and 196 with two ribs and one intervertebral foramen on the right, three ribs and two foramen on the left, citing similar case reported by Albrecht (1883) in *Tropidonotus* and by Owen 1853 of tiger boa *Python tigris* with ankylosis of vertebrae 148 and 149 with one rib on the left, two on the right.

Pelamis bicolor YPM 763 with 212th vertebrae "simple with one rib on the left, double with two ribs on the right."

Gavialis gangeticus with 25 presacral vertebrae – new vertebra inserted between ninth and tenth.

Heloderma with perforation of small first caudal rib. *H. horridum* with 36th or 37th as the first caudal. *H. suspectum* with 38th or 39th as the first caudal.

Bauwens D. 1981. Survivorship during hibernation in the European common lizard, *Lacerta vivipera*. Copeia 1981:741–744.

Trauma – Juvenile European common lizard *Lacerta vivipera* with partially regenerated (less than half original length) tails, survived winter less well than juveniles with intact or nearly regenerated tails.

Bean BA. 1889. Double-headed animals. Forest and Stream 33(9):164.

Congenital – Dicephalic Carolina king snake.

Beatty BL, Heckert AB. 2009. A large archosauriform tooth with multiple supernumerar carinae from the Upper Triassic of New Mexico (USA), with comments on carina development and anomalies in the Archosauria. Historical Biology 21:57–64.

Dental – Carina is an adaptive structure to enhance crack propagation in mechanically tough foods.

Late Triassic non-dinosaurian, possibly "rauisuchian" archosauriform tooth. New Mexico Museum of Natural History NMMNH P-18306 with two supernumerary cariane, one on posterior half of lingual size and the other one near and parallel to the anterior carina. Split carina has been reported in phytosaurs (Hungerbühler 2000; von Meyer 1861), *Nicrosaurus* (Hungerbühler 2000).

Fossil – Carina is an adaptive structure to enhance crack propagation in mechanically tough foods. Late Triassic non-dinosaurian, possibly "rauisuchian" archosauriform tooth. New Mexico Museum of Natural History NMMNH P-18306 with two supernumerary cariane, one on posterior half of lingual size and one near and parallel to the anterior carina. Split carina has been reported in phytosaurs (Hungerbühler 2000; von Meyer 1861), *Nicrosaurus* (Hungerbühler 2000).

Bechtel HB. 1995. Reptile and amphibian variants: Colors, patterns and scales. Krieger Publ Co, Malabar, 206 pp.

Congenital – Dicephalic Texas rat snake *Elaphe obsoleta lindheimeri* and corn snake *Elaphe g. guttata*.

Bedriaga J. 1879. Ueber *Bombinator pachypus* Fitz [On *Bombinator pachypus* Fitz]. Zoologischer Anzeiger 45: 664–668 [German].

Congenital – Clarifies Gené's 1939 report of sacrum and coccyx appearance in *Bombinator pachypus*, stating that they represented a developmental abnormality.

Beetschen JC. 1952. Extension et limites du pouvoir régénérateur des membres après la métamorphose chez *Xenopus laevis* Daudin [Extent and limits of the power of limb regeneration in post-metamorphic *Xenopus laevis* Daudin]. Bulletin Biologique de France et Belgique 86:88–108 [French].

Trauma – Post-metamorphic *Xenopus laevis* retain only limited ability for partial, atypical limb regeneration, usually with cartilaginous anlage.

Bélanger LF, Drouin P. 1966. Osteolysis in the frog: The effects of parathormone. Canadian Journal of Physiology and Pharmacology 44(6):919–922.

Metabolic – Parathyroid extract induced osteolysis in *Rana catesbeiana* (bullfrog), as manifested by enlarged osseous lacunae.

Bélanger LF, Domond MT, Colp DH. 1973. Histological observations on bone and cartilage of growing turtles treated with calcitonin. General and Comparative Endocrinology 20:297–304.

Metabolic – Calcitonin produces larger epiphyses (hypertrophic, though well-mineralized cartilage) in *Pseudemys scripta elegans*, *Graptemys kohni*, and *Chelydra serpentina serpentina.* Parathyroid hormone produces large lacunae.

Belfit S, Nienaber L. 1983. A dicephalic kingsnake (*Lampropeltis getula yumensis*) from Arizona. Herpetological Review 14(1):9.

Congenital – Dicephalic kingsnake with radiograph.

Belkin DA. 1968. Aquatic respiration and underwater survival of two freshwater turtle species. Respiration Physiology 4:1–14.

Metabolic – Musk *(Sternotherus odoratus)* turtle, soft shell with buccopharyngeal membrane respiration. Skin permeability of *S. minor* and *P. scripta.*

Bell GL Jr. in press. Skull wounds as evidence of competitive interaction among mosasaurs. in Rothschild BM, Shelton S. eds. Paleopathology. University of Texas Museum of Natural History, Austin (in press).

Trauma – Skull element healed and infected wounds in mosasaur support intraspecific fighting, based on size and tooth spacing. Broad and shallow grooves 6–15 mm wide, 10–140 mm long, occasionally 3–7 m round were found on external surface of dentaries, maxillae, and premaxillae. Important observations on Red Mountain Museum (Birmingham, Alabama) specimens not accessible for research at this time.

Infection – Infected skull wounds in mosasaur.

Fossil – Skull element healed and infected wounds in mosasaur support intraspecific fighting, based on size and tooth spacing. Broad and shallow grooves 6–15 mm wide, 10–140 mm long, occasionally 3–7 m round were found on external surface of dentaries, maxillae, and premaxillae. Important observations on Red Mountain Museum (Birmingham, Alabama) specimens not accessible for research at this time. Infected skull wounds in mosasaur.

Bell GL Jr, Barnes KR. 2007. First record of stomach contents in *Tylosaurus nepaeolicus* and comments on predation among mosasauridae. Second Mosasaur Meeting Abstract Booklet and Field Guide. Hays, Kansas: Sternberg Museum of Natural History, p. 9–10.

Trauma – *Tylosaurus nepaeolicus* MRM KB-M-3 gastric contents included digestive dissolution of cortical bone with trabecular exposure and dissolution of tooth crowns with open pulp cavities of immature *Platecarpus* af. *Platecarpus planifrons*; *Tylosaurus proriger* GSM 1 containing *Clidastes* sp.; *Hainosaurus* sp. RSM P 2588.1 containing *Plioplatecarpus* sp.; and an unknown mosasaur, *Mosasaurus* sp. TMP 83.126.1 containing a plioplatecarpine. Cites Bjork (1981) and Martin and Bjork (1987) reporting *Tylosaurus proriger* SDSM&T 10439 containing *Platecarpus tympaniticus.*

Fossil – *Tylosaurus nepaeolicus* MRM KB-M-3 gastric contents included digestive dissolution of cortical bone with trabecular exposure and dissolution of tooth crowns with open pulp cavities of immature *Platecarpus* af. *Platecarpus planifrons*; *Tylosaurus proriger* GSM 1 containing *Clidastes* sp.; *Hainosaurus* sp. RSM P 2588.1 contianing *Plioplatecarpus* sp.; and an unknown mosasaur, *Mosasaurus* sp. TMP 83.126.1 containing a plioplatecarpine. Cites Bjork (1981) and Martin and Bjork (1987) reporting *Tylosaurus proriger* SDSM&T 10439 containing *Platecarpus tympaniticus.*

Bell GL Jr, Martin JE. 1995. Direct evidence of aggressive intraspecific competition in *Mosasaurs conodon* (Mosasauridae: Squamata). Journal of Vertebrate Paleontology 15 (Supplement):18A.

Trauma – Conspecific tooth penetrating quadrate and splenial puncture in *Mosasaurus conodon.*

Fossil – Conspecific tooth penetrating quadrate and splenial puncture in *Mosasaurus conodon.*

Bellairs A. d'A. 1965. Cleft palate, microphthalmia and other malformations in embryos of lizards and snakes. Proceedings of the zoological Society London 144:239–251.

Congenital – Cleft palate in snake embryos, citing Bellairs and Boyd (1957). Common lizard (Bellairs and Gamble 1960) and monitor lizards *Varanus exanthematicus* (Shaw 1963) with cleft palate and short upper jaw. Premaxilla in 1 reduced to tiny nodule. *Lacerta lepida* with six eyes, short upper jaw, and cleft palate. *Lacerta vivipara* (bilateral with short jaw) and adder *Vipera berus* with cleft palate. Partial Siamese twinning (teratopagus) in sand lizard *Lacerta agilis.* Axial duplication in *Lacerta agilis*, citing Andersen (1930), and in *Lacerta saxicola*, citing Darewski and Kulikowa (1961).

Trauma – Short tail in *Lacerta lepida.*

Bellairs A. d'A. 1970. The life of Reptiles. 590 pp. New York: Universe Books, Weidenfeld and Nicolson: 590 pp.

Congenital – Eyed lizard *Lacerta lepida* with short upper jaw.

Anaconda *Eunectes murinus* with cleft lip/palate.

Siamese twins are fairly common in reptiles as is dicephalism in lizards, snakes, and turtles and are quite frequent.

Absence of a limb.

Trauma – Tail regrowth in lizards.

Bellairs R. 1971. Developmental Processes in Higher Vertebrates. Coral Gables Florida: University of Miami Press, 366 pp.

Toxicology – Lithium induces cyclopia in amphibians.

Bellairs A. d'A. 1981. Congenital and Developmental Diseases. In Cooper JE, Jackson OF. 1981. Diseases of the Reptilia. Vol 2, pp. 469–485. New York: Academic Press.

Congenital – General comments on Siamese twins, citing Bellairs (1971), Strohl (1925), Heasman (1933), Brongersma (1952); Virtual absence of head and cyclops (Ewert 1979); absence of heads and lower jaw in lizards; malformations in crocodilians but no known malformation in *Sphenodon.*

Cool conditions produce malformed young in garden snakes *Thamnophis elegans*, citing Fox et al. (1961).

Caucasian rock lizard *Lacerta saxicola* with monstrous embryos, citing Darevsky (1966).

Conjoined *Trionyx ferox* (Hildebrand 1930, 1938) and *T. sabnis.*

Fifteen percent of *Chelydra serpentina* had abnormal tail, hind limb, or carapace, citing Yntema (1960).

Dicephalic alligator and allegedly bisexual dicephalic *Coluber constrictor* (Dexter 1976).

Derodymus *Coluber florulentus* (Heasman 1933), *Pituophis melanoleucus annectens* (Fitzsimons 1932), *Chrysemys* sp. (Schmidt and Inger 1957), and *Pseudemys nelsoni* (with forelimb duplication).

Partial head duplication (or with body duplication) in parthenogenetic sand lizard *Lacerta agilis* (Bellairs 1965).

Shortening of upper jaw and cleft palate in chelonians (Ewert 1979) and *Lacerta* (Bellairs 1965; Bellairs and Gamble 1960).

Cleft palate in *Lacerta vivipara, Lacerta lepida*, anaconda *Eunectes murinus, Natrix natrix* and *Vipera berus* (Bellairs 1965; Bellairs and Boyd 1957; Bellairs and Gamble 1960), and alligator (Ferguson 1981).

Shortened, pug-like snout in crocodile (Kälin 1937).

Absence of limb in *Trionyx* and *Emydae* (Dutta 1931); of flipper in *Caretta* (Ewert 1979).

Polydactyly in alligator (Giles 1948) and supernumerary limb at dorsum of pelvis (with truncated tail).

Kyphosis in Trionychidae (Smith 1947). Williams (1957) suggested that excess yolk produced the kyphosis.

Short/kinked tails in *Lacerta lepida* (Bellairs 1965), geckos and crocodilians (Bustard 1969), and turtle (Ewert 1979).

Absence of tail in *Calotes versicolor* (Mathur and Goel 1974).

High temperature produces tail defects in lizards and crocodilians, citing Bustard (1969, 1971), and syndactyly, citing Ferguson (1981).

Tail duplication (Darevsky 1966).

Bellairs A d'A. 1983. Partial cyclopia and monorhina in turtles. 150–158. In Rhodin AGJ, Miyata K. eds. Advances in herpetology and evolutionary biology. Essays in honor of Ernest E. Williams. Cambridge, Mass.: Museum of Comparative Zoology.

Congenital – Cyclops in *Caretta caretta caretta*, citing Adelmann (1936) and Ewert (1979).

Bellairs A. d'A, Boyd JD. 1957. Anomalous cleft palate in snake embryos. Proceedings of the Zoological Society of London 129:525–539.

Congenital – Cleft palate in anaconda *Eunectes murinus*, grass snake *Natrix natrix*, and adder *Vipra berus.* Diminished premaxilla and vomer lateal process and narrowing of front part of palatine. Anterior process of septomaxilla is isolated by the cleft. Bilateral cleft palate in *Lacerta vivipara.*

Bellairs A d'A, Bryant SV. 1985. Autotomy and regeneration in reptiles. In C Gans and F Billett (eds.) Biology of the reptilian. Volume 15 Development, pp. 301–410. New York: Wiley.

Congenital – Tail arising from *Podarcis muralis* hind limb (Brindley 1898).

Trauma – Tail is not shed when functional (e.g., used in swimming). Tail breakage was more common in warm than in cold lizards (22–30°C).

Ophisaurus ventralis tails fly into pieces at slightest touch.

Possible regeneration in the Lower Permian *Araeoscelis* (Vaughn 1955). Tail loss in Upper Jurassic *Palaeolacerta* (Hoffstetter 1964).

Only partial tail present in *Melanosuchus niger.*

Partial limb regeneration in lizards – Increases stump in *Lacerta vivipara, Chalcides ocellatus* (Hellmich 1951; Marcucci 1930; Poyntz and Bellair 1965; Mather 1978).

Thirty seven percent of intact *Coleonyx variegatus*, escaped night snake *Hypsiglena torquata achrorhyncha,* while no tailless escaped.

Thirty five percent of *Anguis fragilis* and 59–69% of *Anniella* had tail loss.

Fossil – Possible regeneration in the Lower Permian *Araeoscelis* (Vaughn 1955). Tail loss in Upper Jurassic *Palaeolacerta* (Hoffstetter 1964).

Only partial tail present in *Melanosuchus niger.*

Bellairs A. d'A, Gamble HJ. 1960. Cleft palate, microphthalmia and other malformations in an embryo lizard (Lacerat vivipara Jacquin). British Journal of Herpetology 2:171–176.

Congenital – One of six embryos from a *Lacerta vivipara* had bilateral cleft palate (floor of each nasal cavity) with upper jaw shorter than lower.

Bellairs A, Miles E. 1960. Apparent failure of tooth replacement in monitor lizards, with remarks on loss of teeth in other reptiles. British Journal of Herpetology 2:189–194.

Dental – Failure of teeth to develop in crocodiles with skeletal deformities (Kälin 1937). Captive crocodiles sometimes fail to replace teeth.

Absence of all but the most anterior teeth in a large, apparently old *Varanus niloticus* (In 1961, they renamed it *Varanus exanthematicus*). Residual tooth at tip of dentaries, in front of right maxilla adjacent to suture; two on the right fused maxilla and three on the left. The most posterior tooth was out of alignment.

Australian *Varanus gouldii* with only one residual tooth at front of left maxilla and one near front of right dentary.

Absent "functional teeth" *Varanus albigularis* in premaxilla, front of maxillae, and dentary, though there "was evidence of some replacement in progress."

Alligator mississippiensis "George" at London zoo between 1912 and 1953 had only a few short teeth.

Bellairs A, Miles E. 1961. Apparent failure of tooth replacement in monitor lizards, addendum. British Journal of Herpetology 3:14–15.

Dental – *Varanus niloticus* in previous article has been reinterpreted as either *Varanus exanthematicus microstictus* or *Varanus exanthematicus albigularis.*

Suggests that explanation for absent teeth is that the "greater part of the dentition was in a simultaneous phase of replacement" and also relates that explanation to the the "Mertens specimen" of *Varanus spenceri.*

Belluomini HE. 1957/8. Bicefalia em *Xenodon merremii* (Wagler 1824) (Serpentes) [Dicephaly in *Xenodon merremii* (Wagler 1824) (Serpentes)]. Memórias do Instituto Butantan 28:85–90 [Portuguese].

Congenital – Derodymous *Xenodon merremii* snake with radiographs.

Belluomini HE. 1965. Serpenti bicefali. Revisione del materiale esistente nell' Istituto Butantan, Dipartimento di Zoologia e nell' Istituto Pinheiros, São Paulo, Brasile [Dicephalous serpents. Review of the existing material in the Butantan Institute, Zoology Department, and in the Pinheiros Institute, São Paulo, Brazil]. Archivio Zoologico Italiano 50:129–144 [Italian].

Congenital – Review and revision of the material used for the descriptions of all nine dicephalic snakes in the collections of the Butantan Institute, the Department of Zoology, and the Pinheiros Institute of São Paulo, Brazil, including three cases described by Do Amaral (1927), two by Prado (1942, 1946), one by Belluomini (1957–58), one by Vanzolini (1917), and two by Pereira (1950), estimating the occurrence at 1 in 100,000. Detailed descriptions and photographs (including radiographs in most cases) were provided. Belluomini presented a modified, expanded key (which included new sub-categories and terms) for the classification of two-headed snakes, based on Nakamura (1938). Snakes with two vertebral column segments that may bifurcate in different places in different individuals – from the 7th vertebra to the 29th. These individuals may externally show two necks or only one (even in some cases with two vertebral column segments).

Belluomini HE, Lancini AR. 1960. bicefali em *Leptodeira annulata ashmeadii* (Hallowell) 1845 (Serpentes). Descrição de um teratódimo deródimo [Dicephaly in *Leptodeira annulata ashmeadii* (Hallowell) 1845 (Serpentes). Description of a teratodymus derodymous]. Memórias do Instituto Butantan, São Paulo 29:175–180 (for 1959) [Portuguese].

Congenital – Derodymous *Leptodeira annulata ashmeadii* snake with radiographs in the collections of the Museum of Natural Sciences of Caracas, Venezuela, under number Herp. 109, 16/1/1957.

Belluomini HE, de Biasi P, Borelli V. 1974. Bicefalia em serpente *Crotalus durissus terrificus* (Laurenti)—"cascavel" (resumo) [Dicephaly in the snake *Crotalus durissus terrificus* (Laurenti)—"cascavel" (abstract)]. Atualidades Veterinarias, São Paulo, 3(18):37–38 [Portuguese].

Congenital – axial bifurcation in *Crotalus durissus terrificus* (Laurenti). This case was the third case reported in this snake; the previous two, reported by Belluomini (1965).

Belluomini HE, De Biasi P, Puorto G, Borelli V. 1976–7. Bicefalia em *Crotalus durissus terrificus* (Laurenti) (Serpentes: Viperidae, Crotalinae) [Dicephaly in *Crotalus durissus terrificus* (Laurenti) (Serpentes: Viperidae, Crotalinae)]. Memórias do Instituto Butantan 40–41:117–121 [Italian].

Congenital – Dicephalic *Crotalus durissus terrificus.*

Belz A. 1981. Die Lurche und Kriechtiere Wittgensteins. [Amphibians and reptiles of Wittgenstein]. Wittgenstein 45:143–162 [German].

Congenital – *Triturus vulgaris* (Teichmolch) with two tails.

Bemis DA, Owston MA, Lickey AL, Kania SA, Ebner P, Rohrbach BW, Ramsay EC. 2003.

Comparison of phenotypic traits and genetic relatedness of *Salmonella enterica* subspecies arizonae isolates from a colony of ridgenose rattlesnakes with osteomyelitis. Journal of Veterinary Diagnostic Investigation 15:382–387.

Infection – Osteomyelitis in ridgenose rattlesnakes.

Bender O. 1906a. Zur Kenntnis der Hypermelie beim Frosch. [Addition to my contribution. On the knowledge of hypermely in frogs]. Gegenbaurs Morphologisches Jahrbuch 35:395–412.

Congenital – *Rana fusca* with supernumerary hind limb.

Bender O. 1906b. Nachtrag zu meiner Abhandlung: Zur Kenntnis der Hypermelie beim Frosch. [Addition to my report: On knowledge of hypermely in frogs]. Gegenbaurs Morphologisches Jahrbuch 36:90–91 [German].

Congenital – Cites literature on polymelia.

Benest S. 1994. DOUBLE TAKE. Courier-Mail 16 June 1994:1.

Congenital – Dicephalic turtle.

Benham W. 1894a. Notes on a particularly Abnormal Vertebral Column of the Bidl-frog; and on Certain other Variations in the Anuran Column. Proceedings of the Zoological Society 1894:477.

Congenital – *Rana mugicus* with malformation of vertebrae (after Tornier 1901).

Benham W. 1894b. Notes on an abnormal vertebral column of the bullfrog and on certain other variations in the anuran column. Proceedings of the Zoological Society 1894:477–481.

Congenital – *Rana mugicus* with multiple fused vertebrae and components: C1–2 are fused, followed by a fusion consisting of 2–1/2 vertebrae, followed by 1–1/2 vertebrae; C1–2 fusion in *Buto agua*; and eighth and ninth fused in *Bufo pantherinus*. This is similar to extra ½ vertebra in a python reported by Bateson (1894). 1/212 toads had fusion by Aldolphi – Morph Jahrb.

Vertebral – C1–2 fusion in *Buto agua*, eighth and ninth fused in *Bufo pantherinus*.

Benick L. 1933. Über unsere Ringelnatter, *Tropidonotus n. natrix* L. [On our grass snake *Tropidonotus n. natrix* L.] Zoologischer Anzeiger 101:294–299 [German].

Congenital – *Tropidonotus n. natrix* specimen with arching of vertebral column in the anterior third of the body, shortened upper jaw, and fused eyes (cyclops).

Bennett R. 1989. Reptilian surgery. Part II. Management of surgical diseases. Compendium on Continuing Education for the Practicing Veterinarian 11(2):122–133.

Trauma – shell fracture repair in *Terrapine ornata*.

Healing of bone takes 6–8 months for complete healing, citing Frye (1981), but in which the statement is not substantiated).

Pathologic fractures from metabolic bone disease in *Iguana iguana*.

Metabolic – Pathologic fractures from metabolic bone disease in *Iguana iguana*.

Shell – Pathologic fractures from metabolic bone disease in *Iguana iguana*.

Bennett RA. 1994. Fracture management in reptiles. Seminars in Avian and Exotic Pet Medicine 3(2):108–112.

Trauma – Treatment of fractures, but no discussion of specific reptiles.

Metabolic – Mention of metabolic bone disease, but no discussion of specific reptiles.

Bennett RA. 1996a. Neurology. In Mader DR. ed. Reptile Medicine and Surgery. pp. 141–148; Philadelphia: Saunders.

Infection – *Salmonella*, *Proteus mirabilis*, and *Bacteroides* (possibly contaminants, rather than the infectious cause) were isolated from a California king snake with osteomyelitis.

Vertebral – Mistakingly rules out spondyloarthropathy (which he/she referred to as ankylosing spondylitis) in boids, *Crotalus*, and southern copperheads because of involvement of costovertebral joints.

Bennett RA. 1996b. Fracture management. In Mader DR. ed. Reptile Medicine and Surgery. pp. 258–264; Philadelphia: Saunders.

Metabolic – Metabolic bone disease in green iguanas.

Bennett SC. 2003. A survey of pathologies of large pterodactyloid pterosaurs. Palaeontology 46:185–198.

Trauma – Deformed *Pteranodon* bones and lumps from injury and necrosis. Several instances of *Pteranodon* mandibular ramus injury with infection were also reported. He also reviewed Tischlinger's 1993 *Pterodactylus* femur fracture, noting it was not "greenstick" because of the pseudoarthrosis (false joint).

Infection – *Pteranodon* mandibular ramus injury with infection.

Osteoarthritis – Osteoarthritis (grooving) of intersyncarpal, carpometacarpal, and metacarpophalangeal joints of pterodactyloid pterosaurs, but not *Pteranodon*.

Fossil – Deformed *Pteranodon* bones and lumps from injury and necrosis. Several instances of *Pteranodon* mandibular ramus injury with infection were also reported. He also reviewed Tischlinger's 1993 *Pterodactylus* femur fracture, noting it was not "greenstick" because of the pseudoarthrosis (false joint).

Osteoarthritis (grooving) of intersyncarpal, carpometacarpal, and metacarpophalangeal joints of pterodactyloid pterosaurs, but not Pteranodon.

Bennett J. 2004. Two headed wonder. Not Miss museum home to rare snake. The Clarion-Ledger, Jackson 20 April 2004:1.

Congenital – Dicephalic gray rat snake.

Bennett RA, Mehler SJ. 2006. Neurology. In Mader DR. ed. Reptile Medicine and Surgery. pp. 239–250; Philadelphia: Saunders.

Infection – Polymicrobial infection in California king snake *Lampropeltis getula*.

Vertebral – Proliferative spinal osteopathy in boids, crotalus, southern copperhead *Agkistrodon contortrix*, especially in those fed mice, citing Frye (1991), questioning possible role of virus, slow neoplasm, or hypovitaminosis D. They described sclerosis of end plates, bony proliferation, and costovertebral and "dorsolateral" articular facet involvement, citing Divers and Lawton (2000).

Neuropathic – Proprioceptive loss in green iguana *Iguana iguana* from a fourth cervical lesion (Divers and Lawton 2000), may be partially related to the neuropathic phenomenon recently recognized in cane toads.

Bergendal D. 1889. Über eine dritte vordere Extremität eines braunen Frosches. [On a third front (leg) extremity in the brown frog]. Bihang kungliga svenska Vetenskaps Akademiens Handlingar 14(IV, 8):1–35 [German].

Congenital – Extra forelimb, short with fairly normal humerus, and radio-ulna on the left side with only two digits in *Rana temporaria*, with review of literature on polymely in frogs.

Berger L. 1967. Embryonal and larval development of F1 generation of green different combinations. Acta Zoologica Cracoviensia 12(7):123–160.

Vertebral – Exposure to variable temperatures of *Rana lessonae, esculenta*, and *ridibunda* was associated with variable rates of metaphorphosis and occasional asymmetry and scoliosis.

Environmental – Exposure to variable temperatures of *Rana lessone, esculenta*, and *ridibunda* was associated with variable rates of metaphorphosis and occasional asymmetry and scoliosis.

Berger L. 1968. The effect of inhibitory agents in the development of green-frog tadpoles. Zoologica Poloniae 18:381–390.

Environmental – Abnormal or absent forelimbs in *Rana lessonae* tadpoles raised at 19°C.

Berger L. 1971. Viabilitiy, sex and morphology of F2 generation within forms of *Rana esculenta* complex. Zoologica Poloniae 21:345–393.

Congenital – Hybridization among *Rana lessonae, ridibundi*, and especially *esculente* produced abnormalities, including spine curvature and missing foreleg fingers. Those with *lessonae* had the only abnormal forelegs; *ridibundi*, hind leg. Bent leg was found in *esculenta* hybrids.

Berger H, Günther R. 1996. Bergmolch – *Triturus alpestris* (Laurenti, 1768). [Alpine newt – *Triturus alpestris* (Laurenti, 1768)]. In Günther R. ed. Die Amphibien und Reptilien Deutschlands. [The Amphibians and Reptiles of Germany]. pp. 104–120; Jena: Gustav Fischer.

Congenital – Finger duplication in *Triturus alpestris*.

Bergman J. 2004. Dazzling day geckos. Reptiles 12(2):44–59.

Congenital – Giant day gecko (*Phelsuma madagascariensis grandis*) with a double tail.

Bernard JB, Oftedal OT, Barboza PS, Citino SB, Ullrey DE, Montali RJ. 1991. The response of vitamin D-deficient green iguanas (*Iguana iguana*) to artificial ultraviolet light. Proceedings of the American Association of Zoo Veterinarians, Calgary, Alberta, pp. 147–150.

Environmental – Rickets and osteomalacia are common in certain sun-basking reptiles (e.g., green iguana), when housed indoors with inadequate UV radiation and dietary vitamin D^3 exposure. However, both bone demineralization and soft tissue mineralization have been noted.

Dermal and epidermal conversion of 7-dehydrocholesterol to cholecalciferol requires 285–315 wavelength ultraviolet light exposure.

Berrocal A, Morales JA, Werner D. 1995. Pathological changes in green iguanas (*Iguana iguana*) from Costa Rica. Herpetopathologia 2 (Proceedings of the 5th International Symposium on Pathology of Reptiles and Amphibians, Alphen a Rijn):187–189.

Metabolic – Two cases of articular gout in *Iguana iguana*.

Berry CT. 1935. A freak or evolution? The Scientific Monthly 40(6):566–569.

Congenital – Two box tortoise *Cistudo carolina* with extra costal plate.

Shell – Two box tortoise *Cistudo carolyna* with extra costal plate.

Bertacchini P. 1900. Morfogenesi e teratogenesi negli anfibi anuri. (IIIa Serie: Anomalie spontanee). [Morphogenesis and teratogenesis in the anuran amphibians (3rd series: Spontaneous abnormalities)]. Internationale Monatsschrift für Anatomie und Physiologie 17:26–87 [Italian].

Congenital – Microsomy, bifid spines, and anterior duplicity experimentally observed.

Beuchat CA. 1988. Temperature effects during gestation in a viviparous lizard. Journal of Thermal Biology 13:135–142.

Congenital – Viviparous *Sceloporus jarrovii* has more "abnormal" offspring when maintained at 26°C, 36°C, and 38°C, paradoxically contrasted to 28°C and 34°C and those that "behaviorally thermoregulated."

Environmental – Viviparous *Sceloporus jarrovii* has more "abnormal"offspring when maintained at 26°C, 36°C, and 38°C, paradoxically contrasted to 28°C and 34°C and those that "behaviorally thermoregulated."

Biddulph CH. 1936. A mugger (*Crocodylus porosus*) with a broken jaw. Journal of the Bombay Natural Historical Society 29:421.

Trauma – Mandible fracture in *Crocodylus porosus*.

Birge WJ, Black JA. 1977a. Sensitivity of Vertebrate Embryos to Boron Compounds. Washington DC: Office of Toxic Substances Environmental Protection Agency EPA-560/1–76–008, 66 pp.

Toxicology – *Rana pipiens, Bufo fowleri* teratogenesis 64–84% and 96–100% at 188 ppm boron, respectively. Sixteen percent were in vertebral column (kyphosis, vertebral coiling); 1%, in cranium.

Birge WJ, Black JA. 1977b. A continuous flow system using fish and amphibian eggs for bioassay determinations on embryonic mortality and teratogenesis. Washington, D.C., U.S. Environmental Protection Agency Office of Toxic Substances EPA-560/5–77–002, 59 pp.

Environmental – Reports amphibian mortality, but not deformity.

Birge WJ, Black JA, Westerman AG, Hudson JE. 1979. The effects of mercury on reproduction of fish and amphibians. In Nriagu JO, ed. Bioogeochemistry of Mercury in the Environment. Amsterdam, Holland, Elsevier/North Holland Biomedical Press, pp. 629–655.

Environmental – Amphibian (*Gastrophryne carolinensis, Hyla chrysoscelis, Hyla squirella, Hyla gratiosa, Hyla versicolor, Hyla crucifer, Rana pipiens, Acris crepitans blanchardi, Bufo punctatus, Bufo debilis debilis, Rana heckscheri, Bufo fowleri, Rana grylio, Ambystoma opacum*) mercury-related mortality, but teratogenesis was only assessed in fish.

Birge WJ, Black JA, Westerman AG, Ramey BA. 1983. Fish and amphibian embryos: A model system for evaluating teratogenicity. Fundamentals of Applied Toxicology 3:237–242.

Toxicology – Aquatic toxicant-induced head and vertebral column abnormalities, dwarfism, partially twinning, microcephaly, and amphiarthrodic jaws.

Birkett J, Mccracken H. 1990. Osteodystrophy in an Asian agamid (*Hydrosaurus pustulatus*). Herpetofauna 20(2):32–36.

Metabolic – Philippine sail-tailed water dragon agamid *Hydrosaurus pustulatus* with swollen thighs with very thin, fractured and angulated radius and ulna called osteodystrophy.

Bischler V. 1923. Rôle du squelette dans la régénération des membres du *Triton*. [Role of the skeleton in regeneration of *Triton* limbs]. Compte rendu des séances de la Societé de Physique et d'Histoire Naturelle de Geneve 40:158–160. [French]

Trauma – *Triton cristatus* and *Spelerpes ruber* limb regeneration after amputation.

Bishop M. 1908. Heart and anterior arteries in monsters of the dicephalic group; a comparative study of Cosmobia. American Journal of Anatomy 8:441–472.

Congenital – Dicephalic *Chrysemys picta*, but cephalad doubling of carapace, plaston, and shoulder girdle.

Bishop DW. 1947. Polydactyly in the tiger salamander. Journal of Heredity 38:290–293.

Congenital – Polydactyly in the tiger salamander *Ambystoma tigrinum.*

Bishop DW. 1948. Naturally occurring polydactyly and limb duplication in *Ambystoma*. Genetics 33:96.

Congenital – Fifteen of 19 *Ambystoma tigrinum* from 1946–1947. Muskee Lake Colorado had one to six extra digits, and four were polydactylous.

Bishop DW. 1949. Spontaneous occurrence of polydactylous salamanders. Proceedings of the 8th International Congress of Genetics 537–538.

Congenital – Polydactylous *Ambystoma tigrinum*.

Bishop DW, Hamilton R. 1947. Polydactyly and Duplication of Members. Science 106:641–642.

Congenital – First recorded naturally occurring anomalies in 17 of 19 tiger salamander *Ambystoma tigrinum* collected as larva and in 1/2 adults from Muskee Lake west of Boulder, Colorado. Hind limb polydactyly in 15, supranumerary digits in 3.

Bishop C, Carey J, Brooks R. 1989. Case studies on turtles and mammals. Proceedings of the Workshop on Cause-Effect Linkages. March 28–30, 1989, p. 14.

Congenital – Snapping turtles from Great Lakes and Hudson River had missing claws and deformed carapaces, fore and hind limbs, tails and jaws, possibly related to PCB exposure. Dwarfism was found in Hamilton Harbor. Greatest incidence of deformities was in Cranberry Marsh. Frequency in Lake Ontario was greater than in Laeke Erie or Algonquin Park, correlated with organochlorine levels. They reported that this was also noted in amphibians.

Toxicology – Snapping turtles from Great Lakes and Hudson River had missing claws and deformed carapaces, fore and hind limbs, tails and jaws, possibly related to PCB exposure. Dwarfism was found in Hamilton Harbor. Greatest incidence of deformities was in Cranberry Marsh. Frequency in Lake Ontario was greater than in Laeke Erie or Algonquin Park, correlated with organochlorine levels. They reported that this was also noted in amphibians.

Bjorndal KA, Bolten AB, Martins HR. 2003. Estimates of survival probabilities for oceanic-stage loggerhead sea turtles (*Caretta caretta*) in the North Atlantic. Fishery Bulletin 101:732–736.

Vascular – The pelagic phase in loggerhead sea turtles *Caretta caretta* is dominated by immature turtles.

Blahak S. 1994. Urolithiasis beim Grünen Leguab (*Iguana iguana*). Urolithiasis is green iguana *Iguana iguana*)]. Tierärztliche Praxis 22:187–190. [German]

Metabolic – Urate stone in *Iguana iguana*.

Stone – Urate stone in *Iguana iguana*.

Blair WF. 1960. The Rusty Lizard. A Population Study. XIV + 185 pp.; Austin, TX: University of Texas Press.

Trauma – Tail breakage by conspecific rusty lizard *Sceloporus olivaceus* males, as well as predators: 14 of 82 had two breaks; 2 males had three. Five males and six females had healed breaks.

Frequency of tail breaks:

Age	1 year	2 years	3 years +
Female	15%	22%	35%
Male	24%	35%	50%

Frequency of bifurcation by point of tail loss/fracture:
Proximal fourth – female 52% – male 37%
Proximal ½ – female 81% – male 79%
Third quarter – female 15% – male 16%
Distal quarter – female 4% – male 5%

Blair J, Wassersug RJ. 2000. Variation in the pattern of predator-induced damage to tadpole tails. Copeia 2000:390–401.

Trauma – Tail injury in % of larval amphibians and up to 75% of tail lost.

	Tail damage:		
Species	Any	>25%	> 50%
Ascaphus truei[a]	50.6%		
Bufo americanus	48.0%		
Phyllomedusa tomopterna[b]	28.9%		
Pseudacris crucifer	51%	4.0%	4.0%
Rana catesbeiana	87.7%		
Rana sylvatica[b] Maryland	70.7%	7.7%	1.7%
– Nova Scotia	43.0%	1.4%	0.2%
Rhinophrynus dorsalis[b]	25.0%		

[a] Only species, that exhibited increased tail damage as tadpoles aged.
[b] Pelagic.

Black JH. 1976. Kyphosis in the three-toed box turtle. Chelonia 2:2–3.

Vertebral – Kyphosis in *Terrapene carolina triunguis*.

Blanc CP. 1979. Notes sur les reptiles de Tunesie. VII. Observations sur un serpent bicéphale du g. *Natrix* (Reptilia: Colubridae). [Notes on Tunesian reptiles. VII. Observations on a dicephalic serpent *Natrix* (Reptilia: Colubridae)]. Archives d'Institute Pasteur Tunis 56:81–90 [French].

Congenital – Derodymous viperine snake *Natrix maura* (Linne 1758).

Blanco PA. 1997. Enfermedades degenerativas óseas y articulares en caimán del Orinoco (*Crocodylus intermedius*), caimán de la costa (*Crocodylus acutus*), presentación de tres (3) casos. [Degenerative osseous and articular diseases in the Orinoco caiman (*Crocodylus intermedius*), caiman of the coast (*Crocodylus acutus*), presentation of three (3) cases]. Proceedings of the 4th Regional Meeting, IUCN—Crocodile Specialist Group of South America and the Caribbean Region. Centro de Innovación Agroindustrial, SC, Villahermosa, Tabasco, Mexico, pp. 26–37 [Spanish].

Metabolic – Two specimens of Venezuelan *Crocodylus intermedius* and one of *Crocodylus acutus* were studied *postmortem*. One of the Crocodylus intermedius specimens had osteochondrosis, and secondary fibrous osteodystrophy linked to osteomalacy. The other individual had secondary fibrous osteodystrophy and osteochondrosis. The specimen of *Crocodylus acutus* had osteoporosis, osteochondrosis, and secondary fibrous osteodystrophy linked to rickets. Blanco concluded that the main causes of the osteodystrophies reported were nutritional: Calcium, Phosphorous and Vitamin D deficiencies.

Bland-Sutton – See Sutton.

Blatchley WS. 1906. On some reptilian freaks from Indiana. Philosophical Proceedings of the Academy of Natural Science of Philadephia 58:419–422.

Congenital – Dicephalic (Mitchill 1826) *Bascanium constrictor* (and a cranial Siamese), *Tropidonotus sipedon* (Wyman 1862), common chain or king snake *Ophibolus getulus* (Yarrow 1878), milk snake *Ophibolus doliatus triangulus* (Smith 1882), garter snake (Collett 1893), spreading adder *Heterodon platirhinus* (Mitchill 1926), turtle (Collett 1893), *Chrysemys marginata*, and *Testudines* (Mitchill 1826).

Derodymus garter snake, house or milk snake *Ophibolus doliatus triangulus*.

Rana palustris (Kingsley 1878; Smith 1882) and *Rana virescens* with supranumerary limbs.

Trauma – *Plethodon cinereus* with two tails.

Blaustein AR, Johnson PT. 2003a. Explaining frog deformities. Scientific American 288(2):60–65.
Congenital – Malformations have been reported in more than 60 species of frogs. Noted 5/22 leopard frogs with supranumerary limbs and occasionally with no limbs in a Henderson, Minn. Pond. In healthy populations, no more than 5% have missing or extra elements. Flatworm or fluke infection has been hypothesized in its occurrence.
Environmental – Malformations have been reported in more than 60 species of frogs. 5/22 leopard frogs had supranumerary limbs and, occasionally, with no limbs in a Henderson, Minn. Pond. In healthy populations, no more than 5% have missing or extra elements. Flatworm or fluke infection has been hypothesized in its occurrence.

Blaustein AR, Johnson PT. 2003b. The complexity of deformed amphibians. Frontiers in Ecology and the Environment 1(2):87–94.
Congenital – Summarizes deformity levels in 21,000 wood frogs (*Rana sylvatica*) from three ecoregions of Western Canada. Frequency of less than 2%was found of polymely, polyphalangy, ectomely, and amely, relating them to physical trauma, with "dramatic deformities … recorded more often than … deformed digits." Most severely affected areas were Midwestern and Western US and Southeastern Canada, with sixfold increase between 1963 and 1997. Curiously, abstract identical to that of Eaton et al. (2004).

Blaustein AR, Kiesecker JM, Chivers DP, Anthony RG. 1997. Ambient UV-B radiation causes deformities in amphibian embryos. Proceedings of the National Academy of Sciences 94:13735–13737.
Infection – Increased susceptibility of western toad *Bufo boreas* to fungus *Saprolegnia ferax.*
Vertebral – Concave spinal curvature in Western toad *Bufo boreas* (Worresyt and Kimeldorf 1975, 1976).
Environmental – UVB (280–315 nm) produced deformities (predominantly lateral flexure of tail) in 25 of 29 long-toed salamanders *Ambystoma macrodactylum*, but only 1 of 190, when mylar was used to reduce UV exposure.

Blaustein AR, Kiesecker JM, Chivers DP, Hokit DG, Marco A, Belden LK, Hatch A. 1998.
Effects of ultraviolet radiation on amphibians: field experiments. American Zoologist 38:799–812.
Environmental – Discusses UV mortality, without comment on actual pathology.

Blount IW. 1935. The anatomy of normal and reduplicated limbs in Amphibia, with special reference to musculature and vascularization. Journal of Experimental Zoology 69:407–457.
Trauma – Surgically produced limb duplication in *Ambylstoma punctatum*, examining vascular and nerve supply and muscles.

Blum HF, Butler EG, Schmidt SE. 1958. Regeneration of limb abnormalities after ultraviolet irradiation. Journal of Cellular and Comparative Physiology 52:177–186.
Environmental – UV light-induced supernumerary limb in *Ambystoma.*

Blundell PA, de la Pompa J-L, Meijers JH, Trumpp A, Zeller R. 1990. The limb deformity gene encodes evolutionarily highly conserved proteins. In Hinchliffe JR, Hurle JM, Summerbell D, eds Developmental Patterning of the Vertebrate Limb. pp. 25–30; New York: Plenum Press.
Congenital – Concept of limb deformity genes, but non-herpetological.

Boardman W, Blanchard B. 2006. Biology, captive management, and medical care of tuatara. In Mader DR. ed. Reptile Medicine and Surgery. pp. 1008–1012; phia: Saunders.
Metabolic – Secondary hyperparathyroidism with rubber jaw and osteopenia in tuatara.
Gout in tuatara.

Boardman WS, Sibley MD. 1991. The captive management, diseases and veterinary care of tuatara. Proceedings of the American Association of Zoo Veterinarians 1991:151–167.
Trauma – *Sphenodontia tuatara* with broken jaws and lost tails are not uncommon (Cree and Daugherty 1990), noting tails regenerate.

Bodemer CW. 1958. The development of nerve-induced supernumerary limbs in the adult newt (*Triturus viridescens*). Journal of Morphology 102:361–392.
Congenital – Supernumerary limbs in *Triturus viridescens.*

Boessenecker R. 2010. Barnacle colonization of Middle Pleistocene sea lion (Carnivora: Pinnipeda) boneselucidate the biostratinomy of a fossil marine mammal. Journal of Vertebrate Paleontology 30:61A.
Shell – Appearance of barnacle attachment residue is a circle with centripetal ridges.
Fossil – Appearance of barnacle attachment residue is a circle with centripetal ridges.

Boettger O. 1898. Katalog der Reptilien-Sammlungen im Museum der Senckenbergischen Naturforschenden Gesellschaft. II. Teil (Schlangen). [Catalogue of the reptile collection in the museum of the Senkenberg]. 160 pp.; Frankfurt: Gebrüder Knauer. [German]
Congenital – No dicephalic snake found (rhinodemus specimen should be there after Vanzolini (1947)).

Bohl E. 1997. Mißbildungen der Extremitäten von Amphibienlarven in Aufseß (Oberfranken). Versuche zur Ursachenermittlung. [Malformations of the extremities of amphibian larvae in Aufsess (Upper Franconia). Attempts at cause ascertainment]. Münchener Beiträge zur Abwasser-Fischerei und Flußbiologie 50:160–189 [German].

Congenital – Malformation include missing extremities (monopedia, apedia), missing phalanges (peromely, phocomely), and incomplete phalanges (micromely) in the lake of Aufseß (99% of damages in hind limbs), mostly one damage per animal; most likely cause: amputation by leech *Erpobdella* in early embryonic development.

Trauma – Malformations include missing extremities (monopedia, apedia), missing phalanges (peromely, phocomely), and incomplete phalanges (micromely) in the lake of Aufseß (99% of damages in hind limbs, mostly one damage per animal; most likely cause: amputation by leeche *Erpobdella* in early embryonic development).

Bohl E, Bohl E, Heise J, Fischer R, Popp M. 1996. Untersuchungen zu Mißbildungen an Amphibien larven im Natureich des Beispielsbetriebs in Aufseß (Oberfranken). [Studies on malformations in amphibian larvae in nature verification of the example operation in Aufseß (Oberfranken). Report to the Bayerisches Landesamt für Wasserwirtschaft, Institut für Wasserforschung, München, Germany.

Trauma – Leech predation produces ectromelia.

Böhme W. 1970. Extreme Wirbelsäulenverkrümmung bei einer Wüstenagame (*Agama mutabilis* Merrem, 1820). [Extreme spinal column curvature in a desert agama (*Agama mutabilis* Merrem, 1820)]. Aqua Terra 7:78–79 [German].

Congenital – Extreme spinal column curvature in *Lacerta agilis*.

Böhme G. 1982. Osteologische Anomalien bei anuren Amphibien (Salientia). [Osteological anomalies in anuran amphibians (Salientia).] Wissenschaftliche Zeitschrift der Humboldt Universität zu Berlin, Mathematisch-Naturwissenschaftliche Reihe 31:201–207 [German].

Congenital – Anomalies of the vertebral column of frogs described and figured from a Holocene locality: Symmetrical fusion of sacral vertebrae with pre- and postsacral vertebrae or with urostyle, reduced lateral apophyses, sacral diapophyses on urostyle, "sacroid" postsacral vertebrae, urostyloid, and expanded lamina horizontalis on urostyle. Asymmetrical lamina horizontalis, reduced lateral apophysis, reduced sacral diapophysis on vertebrae or urostyle, variable prezygapophyses, fusion of sacral vertebrae, and presacral vertebrae.

Similar evaluation of extant water frogs revealed 2.5% with fusion of sacral vertebrae with presacral vertebrae (3.5% in hybrid "*esculenta*") In *Bufo*, fusion sacrum with urostyle was common, in contrast to absence of fusion in *Rana*.

Bolkay StJ. 1919. Prinosi herpetologiji zapadnoga dijela balkanskog poluostrva [Additions to the herpetology of the western Balkan peninsula]. Glasnik Zem Muzeja Bosni i Hercegovini (Sarajevo) 31:1–38 [Slovak].

Congenital – Male *Lacerta muralis* with abnormally short head. *Rana ridibunda* with polydactyly.

Bolkay StJ. 1923. On the Origin of the double sacrum among the tailless batrachians. Glasnik Zem Muzeja Bosni i Hercegovini (Sarajevo) 35:75–80.

Congenital – *Bombinator pachypus* with sacral duplication.

Bolkay St J. 1924. Uloga regeneriranog repa guštera kod uzdržanja vrsta (Prilog biologiji *Lacerta taurica fiumana* Wern). [The rôle of the regenerated tail of the Lacertidae at the preservation of the species. A contribution to the biology of *Lacerta taurica fiumana* Wern)]. Glasnik Zemaljskog muzeja Bosne i Hercegovine u Sarajevu 36:91–94 [Slovak].

Trauma – Tail regeneration in *Lacerta taurica fiumana*.

Bolkay StJ. 1926. Über einen merkwürdigen Fall der Hyperdaktylie bei *Rana ridibunda* Pall. [On a curious case of hyperdactyly in *Rana ridibunda* Pall]. Novitates Musei Sarajevoensis 4:1–5 [German].

Congenital – *Rana ridibunda* with seven toes in the hind legs and a lateral outgrowth on the inside of both first fingers.

Bonaparte C. 1832–41. Iconografia della fauna italica per le quattro classi degli animali vertebrati [Iconography of the Italian fauna for the four classes of vertebrate animals]. Vol. II Anfibi [Amphibians]. Fasc. X. Roma. [Italian].

Pseudopathology – Remarked that caecilians are said to have two heads, but that is not the case.

Congenital – Presented a drawing of a dicephalous *Vipera aspis* Mas. According to Cantoni (1921), this small viper was dicephalic atloidic. It was captured in Ascoli by Orsini, a naturalist friend of Bonaparte. Bonaparte remarked that he felt no disdain in presenting it in his work because there were writers who apparently opposed the existence of such monsters.

Bonaparte C. (Actually Cuvier). 1930. Suite de l'extrait du compte-rendu des travaux de l'académie des sciences, pendant l'année 1829. [Extracts of report of work of the academy of sciences, during the year 1829]. Transactions Médicales. 1:409–419 [French].

Congenital – Related to the atlodimous (form of derodymous) viper of Dutrochet.

Bonin J, Ouellet M, Rodriguez J, DesGranges J-L. 1997. Measuring the health of frogs in agricultural habitats subjected to pesticides. In Green DM. ed. Amphibians in decline: Canadian studies of a global problem. Volume 1. Herpetological conservation. pp. 246–257; St. Louis, Missouri: Society for the Study of Amphibians and Reptiles.

Congenital – Hind limb deformities (agenesis, ectomelia, ectromely) and one forearm deformity were found in 7.4% of *Rana clamitans* from potato field ditches.

Bonin F, Devaux B, Dupré A. 2006. Turtles of the World. Baltimore: Johns Hopkins University Press, 416 pp.

Shell – *Pelusios adansoni* with carapace holes. *Pelusios marani, Lissemys scutata, Chitra vandijki, Kinosternon acutum, Kinosternon flavescens, Mauremys caspica, Rhinoclemmys nasuta, Sacalia bealei, Graptemys gibbonsi, Morena petersi,* and *Apsideretes gangeticus* with carapace damage.

Bonnet C. 1779. Sur la reproduction des members de la Salamandre aquatique. Observations sur la physique par Rozier. [On the reproduction of aquatic salamander limbs. Observations on the physical by Rozier]. 2. Mémoire vol. 13: Paris, 488 pp [French].

Trauma – Arm and tail reproduction in salamander.

Bonnet A, Rey M. 1935 Sur quelques monstruosités présentées par la grenouille. [On several monstrosities in the frog]. Bulletin de la Société zoologique de France 60:338–341 [French].

Congenital – Supernumerary small tibia and fibula.

Bonnet A, Rey M. 1937 Sur quelques cas de polydactylie et de schistodactylie observés en série chez la grenouille. [On several polydactylies and schistodactylies observed in frogs]. Bulletin de la Société zoologique de France 62:21–25 [French].

Congenital – Hind extremity with five to seven phalanges, rarely in front leg of *Rana esculenta.*

Borgert H. 1896. Über eine zweiköpfige Kreuzotter. [On a two-headed adder]. Verhandlungen des Naturwissenschaftlichen Vereins in Hamburg (3) 4:50–57 [German].

Congenital – Dicephalic snake *Vipera* (*Pelias*) *berus* figured + X-ray.

Boring AM, Chang LF, Chang WH. 1948. Autotomy and regeneration in the tails of lizards. Peking Natural History Bulletin 17:85–106.

Trauma – Caudal autotomy does not occur when tail use is for other than protection:

Prehension – Chameleons, *Gekko ceratolophis*

Swimming – Varanidae, *Amblyrhynchus, Physignathus*

Water steering – *Basiliscus*

Air steering – *Ptychozoon*

Air balancing – *Draco*

Very brittle tails in Iguanidae, Teiidae, Lacertidae, Scincidae, and Anguidae. Thirty three percent of Chinese skink *Eumeces chinensis* had regenerated tails as hollow cartilaginous tubes, starting from the centrum and neural arch of terminal vertebra half.

Fifty percent of *Takydromus septentrionalis* had regenerated tails.

Borkin LJ, Pikulik MM. 1986. The occurrence of polymely and polydactyly in natural populations of anurans of the USSR. Amphibia-Reptilia 7(3):205–216.

Congenital – *Rana arvalis, lessonae, esculenta,* and *Bufo bufo* polymely and polydactyly.

Born G. 1876. Die sechste Zehe der Anuren. [The sixth toe of anurans]. Morphologisches Jahrbuch 1: 435–453 [German].

Congenital – Additional sixth toe in tarsus of *Rana esculenta, R. temporaria, Bufo variabilis, B. calamita, Phryne vulgaris, Pelobates fuscus, Bombinator igneus,* and *Hyla arborea.*

Born G. 1880. Nachträge zu "carpus und tarsus". [Addition to carpus and tarsus"]. Morphologisches Jahrbuch 6:49–78 [German].

Congenital – Additional sixth toe in tarsus of *Rana fusca, R. esculenta, Hyla arborea, Phrygne vulgaris, Cystignathus ocellatus, Bufo variabilis, Bombinator igneus,* and *Phrynischus cruciger* (Argentinien).

Born G. 1887a. Eine Doppelbildung bei *Rana fusca.* [A double formation of *Rana fusca*]. Zoologischer Anzeiger 4:135–139 [German].

Congenital – Reports a double-faced (diprosopia) *Rana fusca* in early ontogeny (in egg).

Born G. 1882b. Ueber Doppelbildungen beim Frosch und deren Entstehung. [On double formation of frogs and their origin]. Breslauer aerztliche Zeitschrift 1882, No. 14:162–163 [German].

Congenital – 12 dicephalic *Rana esculenta* embryos in one spawn and reports dicephalic embryo of *Salamandra maculosa.*

Bosch GJ. 1984. Enige opmerkingen over phelsuma's. [Some remarks on phelsumas]. Lacerta (Schiperoord) 42(10/11):212–216 [Dutch].

Metabolic – calcium deficiency induced pathology in gecko *Phelsuma lineata chloroscelis.*

Botha J, Chinsamy A. 2000. Growth patterns deduced from the bone histology of the cynodonts *Diademodon* and *Cynognathus.* Journal of Vertebrate Paleontology 20:705–711.

Metabolic – Cynodont *Diademodon* zonal cortex with alternating fibro-lamellar and lamellar bone – cyclic growth contrasts with Cynodont *Cynognathus* with uninterrupted fibor-lamellar bone – continuous rapid growth and more vascularized bone.

Fossil – Cynodont *Diademodon* zonal cortex with alternating fibro-lamellar and lamellar bone – cyclic growth contrasts with Cynodont *Cynognathus* with uninterrupted fibro-lamellar bone – continuous rapid growth and more vascularized bone.

Bougon D. 1902. Les serpens à deux têtes. [The serpent with two heads.] Le naturaliste 24:205–206 [French].

Congenital – Symbolic discussion of two-headed serpents.
Bouin P. 1929. Eléments d'histologie. [Elements of Histology]. I:334 pp. Paris: Felix Alcan. [French]
Congenital – Experimental production of supernumerary partial hind leg in *Triton* and *Ambystoma punctatum*.
Boulenger GA. 1882. Catalogue of the Batrachia Salientias. Ecaudata in the collection of the British Museum. London: Taylor & Francis.
Congenital – Omosternum generally absent or reduced to a narrow cartilage in Bufonidae.
Boulenger GA. 1889. Catalogue of Chelonians, Rhynchocephalians and Crocodiles in the British Museum (Natural History). London: Taylor & Francis.
Shell disease – Enormous variation in *Caretta caretta* shields in the British Museum collection.
Boulenger GA. 1896. On a case of simous malformation in snakes. Annals & Magazine of Natural History (6)18:399.
Congenital – Aborted snout ("pug-nose") tropical American opisthoglyph snake *Stenorhina degenhardti*.
Boulenger GA. 1898. The Tailless Batrachians of Europe. 2 vol.:376 pp. London: Ray Society.
Vertebral – Vertebral abnormalities (citing Goette, Lateste, Camerano, Adolphi and Howes) including fusion and increased number of segments in *Bombinator pachypus* and cites same in *Alytes obstetricans* by Lataste.
Boulenger GA. 1908. A revision of the oriental Pelobatid Batrachians (Genus Megalophrys). Proceedings of the Zoological Society, London 1908:427–428.
Dental – *Megalophrys* (*Leptobrachium*) *carinense* with variable presence of vomerine teeth, noting that one female had vomerine teeth on the left and no trace on the right.
Boulenger GA. 1913. The Snakes of Europe. London: Methusen & Co., 151 pp.
Congenital – Dicephalic grass or ring snake *Tropidonotus natrix* and "even a" tricephalic snake.
Bourdeau P. 1988a. Pathologie des tortues. 1ère partie: Examen clinique et maladies générales. [Pathology of turtles. 1st part. Clinical manifestations and general maladies]. Point Vétérinaire 20(17):761–775 [French].
Metabolic – Osteomalacia in turtles.
Osteodystrophy with deformed, non-calcified extremities in *Chrysemys neurrie*.
Shell – Ulcerative shell disease in *Terrapene*.
Bourdeau P. 1988b. Pathologie des tortues. 2ème partie: Affections cutanées et digestives. [Pathology of turtles. 2nd part. Cutaneous and digestive afflictions]. Point Vétérinaire 20(18):871–884 [French].
Trauma – Carapace fractures in turtles.
Infection – Septicemic cutaneous ulcerative disease in turtles.
Metabolic – Nutritional osteodystrophy in *Gopherus*.
Shell – Carapace fractures and ulcerative disease in turtles.
Bourne AG. 1884. On certain abnormalities in the Common Frog (*Rana temporaria*). Quarterly Journal of Microscopical Science Series 2 24:86.
Congenital – A *Rana temporaria* with transverse process malalignment, duplication with fusion, bifurcation or single zygapophysis on various vertebrae.
Bouwhuis H. 1972. Een aangeboren afwijking bij *Pseudemys scripta elegans*. [A congenital abnormality of the carapace in *Pseudemys scripta elegans*.] Lacerta 30:143 [Dutch].
Congenital – Carapace anomaly in *Pseudemys scripta elegans*.
Shell – Carapace anomaly in *Pseudemys scripta elegans*.
Bowerman J. 2003. Timing of trematode-related malformations in Oregon spotted frogs and Pacific Tree frogs. Northwest Naturalist 84:142–145.
Congenital – Oregon spotted frog *Rana pretiosa* with *Ribeiroia ondatrae*-induced supernumerary limbs and polydactyly in 22%, but less in Pacific tree frog *Pseudacris regilla* because it breeds right after ice melts. Therefore, less time for parasites.
Bowerman J, Johnson PT, Bowerman T. 2010. Sublethal predators and their injured prey: Linking aquatic predators and severe limb abnormalities in amphibians. Ecology 91:242–251.
Trauma – Annual frequency of abnormalities in western toad *Bufo boreas* ranged from 1% to 35%. Of limb abnormalities, 53.1% were missing digits; 14.6%, missing foot; 23%, partially missing limb; 8.2% missing more than 2/3 of limb; 0.6%, micromelia; and 0.6% represented forelimb damage.
Cascade frog *Rana cascadae* had 5–21% with hind limb abnormalities attributed to sticklebacks *Gasterosteus aculeatus* at Lake Aspen and corduliid dragonfly larvae *Somatochlora albicincta*. Of limb abnormalities, 1.7% were missing digits; 8.6%, missing foot; 29.3%, partially missing limb; 53.4% missing more than 2/3 of limb; 5.2%, micromelia; and 1.7% represented forelimb damage.
Tail injury in larval amphibians was present in 30% and produced up to 75% loss (Blair and Wassersug, 2000; Feder 1983; Lawler et al. 1999).
Boy JA. 1971. Zur Problematik der Branchiosaurier (Amphibia, Karbon-Perm). [To the problem of Branchiosaurus]. Paläontologische Zeitschrift 45:107–119. [German]
Congenital – One specimen of *Micromelerpeton* with five (instead of four forefoot) phalanges. Few specimens of *Branchiosaurus* with three or two segments in phalanx IV. One *Branchiosaurus* with small median roof bone (? intertemporal). All features were explained as atavism.

Fossil – One specimen of *Micromelerpeton* with five (instead of four forefoot) phalanges. Few specimens of *Branchiosaurus* with three or two segments in phalanx IV. One *Branchiosaurus* with small median roof bone (? intertemporal). All features were explained as atavism.

Boy JA. 1972. Die Branchiosaurier (Amphibia) des saarpfälzischen Rotliegenden (Perm, SW-Deutschland). [The Branchiosaurians (Amphibia) of the Rotliegend of Saar-Palatine, Permian, southwest Germany]. Abhandlungen Hessisches Landesamt für Bodenforschung 65:3–137 [German].

Congenital – Right parietal grown together with right supratemporal in *Micromelerpeton*, and another specimen with separate second supratemporal (? intertempora), explained as variability.

Fossil – Right parietal grown together with right supratemporal in *Micromelerpeton*, and another specimen with separate second supratemporal (? intertempora), explained as variability.

Boyer TH. 1991. Common problems and treatment of green iguanas (*Iguana iguana*). Bulletin of the Association of Amphibian and Reptilian Veterinarians 1(1):8–11.

Infection – Avascular necrosis of toes and tail from septicemia, citing Barten (1991).

Metabolic – Metabolic bone disease in green iguanas *Iguana iguana* presenting as fibrous osteodystrophy, fractures.

Vascular – Avascular necrosis of toes and tail from septicemia, citing Barten (1991).

Stones – Cystic calculi.

Boyer TH. 1996a. Turtles, tortoises, and terrapins. In Mader DR. ed. Reptile Medicine and Surgery. Philadelphia: Saunders; pp. 332–336.

Shell disease – Attributed caseated crateriform ulcers on ventral surfaces of *Trionychidae* to *Citrobacter freundii.*

Boyer TH. 1996b. Metabolic bone disease. In Mader DR. ed. Reptile Medicine and Surgery. pp. 385–392; Philadelphia: Saunders.

Metabolic – Metabolic bone disease attributed to dietary calcium or vitamin D deficiency; negative Ca to PO4 ratio (e.g., lean beef 1:16, beef heart 1:38, insects, fruits, lettuce, as charted in article); lack of UV light exposure; altered vitamin D metabolism secondary to kidney, liver, intestinal, thyroid, or parathyroid disease; and hypocalcemia. Category also includes osteomalacia, fibrous osteodystrophy. Pliable mandible or maxilla; fractures of femur, humerus, radius, ulna, occasionally tibiotarsus; fibrodysplasia of long bones or jaw; kyphosis; lordosis; scoliosis; and widened radiolucent ends of long bone may be noted (*Iguana iguana*). Further examples include uneven scutes in common snapping turtle *Chelydra serpentina*; curling of carapate edges (parrot beak); overgrowth of beak and broken legs or net-like porosity of carapace in ornate box turtle *Terrapene ornata*; overgrowth of bridge between carapace and plaston in *Terrapene triunguis*; deformed jaw, spine, or legs; abnormal tooth angles in crocodilians; and failure of calcification in desert tortoise *Gopherus agassizii.*

Shell disease – Small shell in common snapping turtle *Chelydra s. serpentina.*

Boyer TH. 2006. Turtles, tortoises and terrapins. In Mader DR. ed. Reptile Medicine and Surgery. pp. 696–704; Philadelphia: Saunders.

Infection – *Beneckeia chitonovora* shell rot on *Apalone.*

Shell – Healed pits in box turtles. *Beneckeia chitonovora* shell rot on *Apalone*. Nutritional shell distortion.

Differential – Swollen bones in turtles from fracture, neoplasia, osteomyelitis, or fibrous dysplasia.

Boylan S. 2003. Prosthesis to protect traumatized hind limbs in box turtles. Exotic DVM 5(5):12.

Trauma – Eastern box turtle *Terrapene carolina carolina* with self-amputed hind limb and fractured (shattered) tibia and fibula.

Brady TJ. 1991. The scene in the nation and the World. The Philadelphia Inquirer 12 June 1994:A03.

Congenital – Dicephalic turtle.

Bragg R. 1991. Two-headed snake wonder of nature at San Diego Zoo. Toronto Star 27 July 1991:E2.

Congenital – Dicephalic corn snake.

Brain CK. 1958. Web-foot geckos of the Namib (*Palmatogecko rangei* Andersson). African Wild Life 12:67–70.

Trauma – Relatively low tail loss rate, comparable to *Chondrodactylus angulifer*, *Ptenopus garrulous*, and *Narudasia festiva.*

Braña F, Ji X. 2000. Influence of incubation temperature on morphology, locomotor performance, and early growth of hatchling wall lizards (*Podarcis muralis*). Journal of Experimental Zoology 286:422–433.

Environmental – Asymmetry of forelimb length increased with increased incubation temperature in wall lizard *Podarcis muralis.* Male tail length at 26°C was 38.67 ± 1.57 mm, 39.62 ± 3.99 mm at 29°C, and 36.69 ± 2.59 mm at 32°C; contrasted with female tail lengths at 38.4 ± 2.41 mm, 39.37 ± 2.00 mm, and 35.23 ± 4.07 mm, respectively.

Branch WR. 1979. Dicephalism in the African worm snake *Leptotyphlops bicolor*. Herpetological Review 10:7.

Congenital – Dicephalic African worm snake (*Leptotyphlops bicolor*).

Branch WR. 1982. Dicephalic *Chersina angulata*. Journal of the Herpetological Association of Africa 27:12–15.

Congenital – Dicephalic *Chersina angulata.*

Brandt W. 1931. Becken- und Gelenkbesonderheiten bei einem Naturfund von *Rana fusca* mit drei Hinterbeinen. [Peculiarities of pelvic and articulation in a find in nature of *Rana fusca* with three hind legs] Anatomischer Anzeiger 72:348–352 [German].

Congenital – Supernumerary hind leg attached to an additional median articulation on an additional ileum.

Brandt W. 1932. Ein fünfbeiniger Frosch. [A frog with five legs] Blätter zur Aquarien- und Terrarienkunde, Braunschweig, 43:62–64 [German].

Congential – *Rana fusca* with supernumerary legs.

Brandt W. 1933. Weitere Mitteilungen über Abnormitäten bei Amphibien. [Additional communications on abnormalities of amphibians] – Blätter zur Aquarien- und Terrarienkunde, Braunschweig, 44:395–401 [German].

Congenital – *Rana temporaria* with two left arms; one shortened with shortened hand and fused phalanges (ventral) and one long with one upper arm and duplication of the lower arm.

Trauma – *Rana esculenta* with shortened and stiffed left hind leg and enlarged right hind leg with wrong healed knee articulation.

Triton cristatus carnifex with duplication of tail.

Brandt W. 1934. Naturfunde von Doppel- und Dreifach-Bildungen bei Fischen und Amphibien. (3. Mitteilung) [Finds of double and tripel formations in fishes and amphibians in nature] Blätter zur Aquarien- und Terrarienkunde, Braunschweig, 45:217–222 [German].

Congenital – Duplication of right lower arm in axolotl (*Ambystoma mexicanum*), triplication of right anterior leg in *Rana fusca*.

Brandt W. 1935. Weitere Naturfunde von Doppel-Bildungen bei Amphibien. (4. Mitteilung) [Additional finds of double formations in amphibians. (4. communication)] Blätter zur Aquarien- und Terrarienkunde, Braunschweig, 46: 147–152 [German].

Congenital – Dicephalic *Salamandra maculosa* larva.

Rana temporaria with additional left front leg and *Triton marmoratus* with additional branch on right arm at elbow.

Brandt W. 1937. Weitere Naturfunde von Gliedmaßen- und Schwanzmißbildungen bei Amphibien. (5. Mitteilung) [Additional finds of malformations of extremities and tail in amphibians. (5. communication)] Blätter zur Aquarien- und Terrarienkunde, Braunschweig, 49:67–69 [German].

Congenital – Axolotl with duplication of right hind leg.

Rana temporaria with malformation of the phalanges of left hind leg.

Trauma – *Limnodynastes tasmaniensis* with tail duplication.

Brandt W. 1938b. Spiralig aufgerollte Larven vom Feuersalamander (6. Mitteilung über Naturfunde von Mißbildungen bei Amphibien). [Spiral rolled up larva of fire salamander (6. Reports of natural finds of malformations in amphibians]. Blätter zur Aquarien- und Terrarien-Kunde 49:81–83 [German].

Congenital – Spiral (laterally curved) vertebral column in *Salamandra maculosa*, noting report in *Triton cristatus carnifex* and of cyclops (Tornier 1919, p. 161).

Braun M. 1876. Notiz über Zwillingsbildungen bei Wirbeltieren. [Note on twin formations in vertebrates]. Verhandlungen der physikalisch-medizinischen Gesellschaft Würzburg Serie 2, 10:67–73 [German].

Congenital – Discussion of origin of double formation: two separate embryonic anlage which fuse, or division of anlage.

Early ontogenetic partial or total duplication in *Salamandra maculosa*.

Dicephalic *Tropidonotus natrix*.

Braus H. 1909. Gliedmaßenpropfung und Grundfragen der Skeletbildung. 1. Die Skeletanlage vor Auftreten des Vorknorpels und ihre Beziehung zu den späteren Differenzierungen. [Extremity graft and basic questions of the formation of the skeleton. First the analoga of the skeleton before the appearance of the pre-cartilage and its relation to later differentiation]. Gegenbaurs Morphologisches Jahrbuch 39:155–301 [German].

Congenital – Transplanting is the way to check the potency of developing regions (check the complex of cells and forces which form a new leg or shoulder girdle). Detailed anatomical description and figures of transplanted frontleg and shoulder girdle in *Bombinator*.

Brazaitis P. 1981. Maxillary regeneration in a Marsh crocodile, *Crocodylus palustris*. Journal of Herpetology 15:360–362.

Trauma – Closure of tooth sockets and replacement of third maxillary tooth after maxilla bitten off of Marsh crocodile, *Crocodylus palustris*.

Dental – Closure of tooth sockets and replacement of third maxillary tooth after maxilla bitten off of Marsh crocodile, *Crocodylus palustris*.

Breedis C. 1950. Induction of accessory limbs in salamanders with mixture containing carcinogens. Cancer Research 10:205–206.

Toxicology – Methylcholanthrene-induced accessory limb in *Triturus viridescens*.

Breedis C. 1951. Transplantable sarcoma of the salamander induced by methylcholanthrene. Cancer Research 11:239.

Toxicology – Methylcholanthrene-induced accessory limbs and "limb parts" in *Triturus viridescens*.

Breedis C. 1952. Induction of accessory limbs and of sarcoma in the newt (*Triturus viridescens*) with carcinogenic substances. Cancer Research 12:861–866.

Toxicology – Reduplication of injected limbs in newt *Triturus viridescens* produced by 4% methylcholanthrene, benzpyrene, and acetylaminofluorene and scarlet red in 40% contrasted with 26% with tar alone. Vaseline induced a single accessory digit-like growth; beryllium hydroxide in 6/18.

Bregulla D. 1987. Fund einer anencephalen und teilalbinotischen Feuresalamander-Larve *Salamandra salamandra* (Lnnnaeus, 1758) (Cauidata: Salamandridae). Case of an anencephalic and partially albino fire salamandar-Larva *Salamandra salamandra* (Linnaeus, 1758) (Caudata: Salamandridae)] Salamandra 23:65–67 [German].

Congenital – Anencephalic *Salamandra salamandra.*

Bridges CM. 1926. Could bite them coming and going: This snake has extra head. Fresno Morning Republican 12 January 1926:7.

Congenital – Derodymus rattlesnake.

Bridges W. 1974. Gathering of animals: an unconventional history of the New York Zoological Society. 518 pp.; New York: Harper & Row.

Congenital – Dicephalic snake.

Bridges CM. 2000. Long-term effects of pesticide exposure at various life stages of the southern leopard frog (*Rana sphenocephala*). Archives of Environmental Contamination and Toxicology 39:91–96.

Toxicology – Pesticide carbaryl produced visceral and limb malformations in 18% of southern leopard frog (*Rana sphenocephala*), compared with <1% of controls.

Bridges CM, Semlitsch RD. 2005. Xeonbiotics. In Amphibian Declines: The Conservation Status of United States Species, Lanoo M, ed. Berkeley, California: University of California Press, pp. 89–92.

Toxicology – Carbaryl-induced missing or supernumerary limbs and bent tails (Bridges 2000).

Environmental – Low pH–induced altered development (Pierce and Montgomery 1989).

UVB radiation-induced deformity (Ankley et al. 2000; Blaustein et al. 1998; Broomhall et al. 2000; Lizana and Pedraza 1998; Starnes et al. 2000).

Unionized ammonia-induced malformations (Jofre and Karasov 1999; Schuytema and Nebeker 1999).

Organophosphate-induced malformations (Fulton and Chambers 1985; Pawer et al. 1983; Sparling 2000).

Briggs R. 1973. Developmental genetics of the axolotl. In Genetic Mechanisms of Development. 31st Symposium of the Society for Developmental Biology. New York, Academic Press, pp. 169–199.

Congenital – R gene mutation in axolotl associated with phocomelia, delay in development of long bones, short toes, and absent ribs (citing Humphrey (1974); Martin and Signoret (1968)).

V gene mutation in axolotl associated with arrested forelimbs (citing Humphrey (1974)).

H gene mutation in axolotl associated with abnormal "hands" and curved trunk and tail (citing Humphrey (1974)).

U gene mutation in axolotl associated with arrested limb development (citing Humphrey (1974)).

Y gene mutation in axolotl associated with arrested limb development (citing Humphrey (1974)).

O gene mutation in axolotl associated with reduced capacity for limb regeneration.

Brimley CS. 1903. Notes on the reproduction of certain reptiles. The American Naturalist 37(436):261–266.

Congenital – 1/17 king snake egg embryos were dicephalic with duplication of anterior portion of bodies.

Brindley HH. 1894. On a specimen of *Hemidactylus gleadovii* with a bifid renewed tail. Journal of the Bombay Natural History Society 9:30–33.

Trauma – *Hemidactylus gleadovii* with bifid tail. Notes that renewed tail differ from congenital because of irregular cartilaginous tube, rather than actual vertebrae. *Trogonophis wiegmanni* (Amphisboenidae) with second tail, bifid tail in *Lygosoma telfairi* (Scincidae), and *Anolis grahami* with trifid tail.

Brindley HH. 1898. Some cases of caudal abnormality in *Mabuya carinata* and other lizards. Journal of the Bombay Natural History Society 11:680–689.

Trauma – Bifid tails with irregular cartilaginous tube in *Mabuya carinata, Lacerta agilis, Lygosoma telfairi, Trogonophis weigmanni* (Brindley 1894), *Hemidactylus gleadovii* (Brindley 1894), and *Lacerta agilis* (Giebel 1864).

Supernumerary tail in *Lacerta muralis* from dorsal surface of hind leg; two dwarf supernumerary tails in *Anolis grahami*; trifid in *Lacerta agilis*, *Calotes cristatellus*, and *Anolis grahami.*

Also noted bifid tails in "supernumerary tail growing out from an injured place" in *Tupinambis nigropunctatus* (Quelch or Timehri 1890).

Broadley DG. 1975. Dicephalism in African reptiles. Journal of the Herpetological Association of Africa (13):8–9.

Congenital – Dicephalic wolf snake *Lycophidion c. capense*; common house snake *Boaedon f. fulliginosus*, citing Fitzsimons (1932); dicephalic Herald snake *Crotaphopeltis hotamboeia*; and reported dicephalic common-striped skink *Mabuya striata striata* with duplication of forelimbs and anterior half of body.

Broadley DG. 1976. A review of the *Mabuya striata* complex in south-east Africa (Sauria: Scincidae). Occasional Papers of the National Museum Monum Rhodesia (natural sciences) 6:45–79.

Congenital – Two dicephalic *Mabuya striata punctatissimus*, dicephalic *M. striata striata*, and second, with twinning of anterior half of body from Pretoria.

Trauma – *Mabuya striata punctatissimus* with bifid tail.

Broadley DG. 1987. Caudal autotomy in African snakes of the genera *Natriciteres* Loveridge and *Psammophis* Boie. Journal of the Herpetological Association of Africa 33:18–19.

Trauma – 30–50% of *Natriciteres variegata sylvatica, Psammophis notostictus, P. sibilans sibilans, P. phillipsii* and *dromophis* had tail loss, with blunt cone regeneration.

Broadley DG, Findlay C. 1972. Dicephalism in the African striped skink *Mabuya striata striata*. Arnoldia (Rhodesia) 5(21):1–2.

Congenital – Dicephalic striped skink *Mabuya striata striata*.

Brockes JP. 1997. Amphibian limb regeneration: Rebuilding a complex structure. Science 276:81–87.

Trauma – Morphological analysis of amphibian limb regeneration, using Emperor newt *Tylotriton verrucosus* and red-spotted newt *Notophthalmus viridescens*.

Brockes JP, Kumar A. 2005. Appendage regeneration in adult vertebrates and implications for regenerative medicine. Science 310:1919–1923.

Trauma – Appendage regeneration in *Notophthalmus viridescens* as model for salamanders.

Brodie ED Jr, Formanowicz DR Jr, Brodie ED. 1978. The development of noxiousness of *Bufo americanus* tadpoles to aquatic insect predators. Herpetologica 34:302–306.

Trauma – Noxious defense against predators, but no comments on skeleton.

Brogard JJM. 1980. Les maladies bacteriennes et virales des reptiles. Étude bibliographique. [Bacterial and viral maladies of reptiles. Bibliographic study]. Thèse pour le Doctorat Vétérinaire Diplôme d'État. École Nationale Vétérinaire de Toulouse, 168 pp. [French]

Infection – *Salmonella typhimurium, regent,* and *amrina* in chelonians and saurophidians.

Beneckea chitinovora shell disease in *Trionyx, Chrysemys, Sternotherus,* and *Podocnemis.*

Shell – Ulcerative shell disease in *Trionyx, Chrysemys, Sternotherus,* and *Podocnemis* produced by *Beneckea chitinovora.*

Brogard J. 1987a. Quel est votre diagnostic?][What is your diagnosis?]. Point Vétérinaire 19(109):667–668 [French].

Infection – Infectious arthritis of toes of *Iguana iguana.*

Arthritis – Infectious arthritis of toes of *Iguana iguana.*

Brogard J. 198b. Les maladies des reptiles. [The Maladies of Reptiles]. 334 pp.; Maisons-Alfort: Point Vétérinaire [French].

Congenital – Anomalies in crocodilians, turtles, *Calotes versicolor.*

Thoracopagia in *Lacerta saxicola.*

Ododymus *Lacerta agilis* (Bellairs 1981).

Dicephalic turtles (Hildebrand 1930), *Malaclemmys* and *Alligator mississippiensis* (Bellairs 1981), derodymous *Malaclemmys.*

Cleft palate in *Lacerta, Natrix, Vipera,* and anaconda *Eunectes* (Bellairs 1981).

Achondroplastic *Pseudemys scripta* (Frye and Carney 1974).

Ectromelia in turtles (Hildebrand 1930).

Absence of forefoot in *Caretta caretta.*

Supernumerary in *Malaclemmys* and *Alligator mississippiensis* (hind leg with polydactyly from pelvis).

Trauma – *Malaclemmys* with bifid tail.

Infection – *Beneckea chitinovora* shell disease in *Trionyx, Chrysemys, Sternotherus, Chelonia,* and *Podocnemis.*

Pyogenic/septic polyarthritis in *Iguana iguana* (Chauvier 1981) and associated with osteomyelitis in leatherback turtle *Dermochelys coriacea* (Ogden et al. 1981).

Suppurative arthritis in teju *Tupinambis teguixin* (Ackerman et al. 1971).

Metabolic – Ramollisement as term for metabolic disease. Osteitis deformans suggested, but not confirmed.

Osteoporosis in crocodilians.

Calcium pyrophosphate depositon disease in *Uromastyx* and *Pseudemys.*

Osteofibrosis in *Iguana iguana* and *Lacerta viridis.*

Neoplasia – Osteochondroma in *Varanus bengalensis* (Frye and Dutra 1973).

Neurofibrosarcoma of vertebra in *Agkistrodon halys* (Effron 1977).

Arthritis – Arthritis in *Brachylophus fasciatus* (Ensley et al. 1981).

Vertebral – Vertebral periosteal reaction in *Iguana* (Chiodini and Nielsen 1983).

Shell – Carapace deformity in turtles.

Shell disease caused by Beneckea chitinovora in Trionyx, Chrysemys, Sternotherus, Chelonia, and Podocnemis.

Broili F. 1917. Unpaare Elemente im Schädel von Tetrapoden. [Unpaired elements in the skull of tetrapods.] Anatomischer Anzeiger 49:561–576 [German].

Congenital – Literature survey revealed unpaired, median bone, "Centroparietal," between parietals and postparietals in Triassic amphibian, *Aphaneramma rostratum* (see Wiman 1915). Unpaired, median bone, "Preparietal" (= suture or Wormian bone comparable to fontanelle bones in humans), between parietals and frontals in dicynodont therapsids (mammal-like reptiles), *Lystrosaurus, Diictodon, Galepus, Gordonia* and

Dicynodon, and in gorgonopsid therapsids, *Scylacops*. Unpaired, median bone between frontals and nasals, "Inter-Naso-Frontale" or "Naso-Frontale," in eryopoid temnospondyl, *Osteophorus, Eryops*, and microsaur, *Ricnodon*. The internasofrontal in eryopoid temnospondyls, *Eryops*, and *Trematosaurus*, compared with mesethmoid of fishes by Meyer (1859) and Wiman (1916). *Lonchorhynchus* with calcified and weakly ossified cartilage compared with ethmoidal cartilage. Unpaired, median bone, "ethmoidal," in caecilian (Gymnophiona), *Siphonops*. Unpaired, median bone, internasal between the internasal processes of the premaxillae in microsaur, *Micropholis*, comparable to the mesethmoid of teleosts.

Fossil – Literature survey revealed unpaired, median bone, "Centroparietal," between parietals and postparietals in Triassic amphib, *Aphaneramma rostratum* (see Wiman 1915). Unpaired, median bone, "Praeparietale" (= suture or Wormian bone comparable to fontanell bones in humans), between parietals and frontals in dicynodont therapsids (mammal-like reptiles), *Lystrosaurus, Diictodon, Galepus, Gordonia* and *Dicynodon*, and in gorgonopsid therapsids, *Scylacops*. Unpaired, median bone between frontals and nasals, "Inter-Naso-Frontale" or "Naso-Frontale," in eryopoid temnospondyl, *Osteophorus, Eryops*, and microsaur, *Ricnodon*. The internasofrontal in eryopoid temnospondyls, *Eryops*, and *Trematosaurus*, compared with mesethmoid of fishes by Meyer (1859) and Wiman (1916). *Lonchorhynchus* with calcified and weakly ossified cartilage compared with ethmoidal cartilage. Unpaired, median bone, "ethmoidal," in Caecilian (Gymnophiona), *Siphonops*. Unpaired, median bone, internasal between the internasal processes of the premaxillae in microsaur, *Micropholis*, comparable to the mesethmoid of teleosts.

Brongersma LD. 1952. On two cases of duplicitas anterior in snakes. Proceedings Koninklijke Nederlandse Akademie Wetenshap Series C 55(1):49–61.

Congenital – Dicephalic with anterior vertebral column duplication and 9th–10th vertebrae "in close contact" and fusion of 11th–19th as a compact mass, without sufficient details to determine diagnosis in *Homalopsis buccata*. Cites dicephalic individual reported by Flower (1899), Wall (1905), Smith (1917) and Cunningham (1937).

Broom R. 1904. On some points in the anatomy of the anomodont skull. Records of the Albany Museum, Grahamstown 1(2):75–82.

Congenital – Unpaired, median bone, "preparietal," between parietals and frontals in dicynodont therapsids (mammal-like reptiles), *Lystrosaurus*. P. 79.

Fossil – Unpaired, median bone, "preparietal," between parietals and frontals in dicynodont therapsids (mammal-like reptiles), *Lystrosaurus*. P. 79.

Broom R. 1913a. On the Gorgonopsia, a suborder of the mammal-like reptiles. Proceedings of the Zoological Society London 1913:225–230.

Congenital – Unpaired, median bone, "preparietal," between parietals and frontals in the gorgonopsian and dicynodont therapsids (mammal-like reptiles), *Galepus*.

Fossil – Unpaired, median bone, "preparietal," between parietals and frontals in the gorgonopsian and dicynodont therapsids (mammal-like reptiles), *Galepus*.

Broom R. 1913b. On some new genera and species of dicynodont reptiles with notes on a few others. Bulletin American Museum of Natural History 32:441–457.

Congenital – Unpaired, median bone, "preparietal," between parietals and frontals in dicynodont therapsids (mammal-like reptiles), *Diictodon*. *Emydorhynchus palustris* with unusually large postorbital and different in shape from known anomodonts.

Fossil – Unpaired, median bone, "preparietal," between parietals and frontals in dicynodont therapsids (mammal-like reptiles), *Diictodon*. *Emydorhynchus palustris* with unusually large postorbital and different in shape from known anomodonts.

Broom R. 1913c. Studies on the Permian temnospondylous stegocephalians of North America. Bulletin American Museum of Natural History 32:563–595.

Congenital – Unpaired, median bone between frontals and nasals, "interfrontal," in eryopoid temnospondyl, *Eryops*.

Fossil – Unpaired, median bone between frontals and nasals, "interfrontal," in eryopoid temnospondyl, *Eryops*.

Brown WS. 1971. Morphometrics of *Terrapene coahuila* (Chelonia, Emydidae), with comments on its evolutionary status. Southwestern Naturalist 16:171–184.

Congenital – Supernumerary scutes in *Terrapene coahuila*.

Brown BA. 1981. Extra limbs in the small mouthed salamander, *Ambystoma texanum*. Proceedings of the Indiana Academy of Science 90:443–445.

Congenital – Supernumerary limbs in small-mouthed salamander, *Ambystoma texanum*.

Brown PR. 1991. A two-headed snake! Herpetology (SW Herpetological Society) 21(2):42

Congenital – Dicephalic San Diego gopher snake *Pituophis catenifer annectens*.

Brown T. 2003. The whole damn hoopla. Herald Sun (Melbourne, Australia) 8 March 2003:24.

Congenital – Dicephalic turtle.

Brown CK, Ruby DE. 1977. Sex-associated variation in the frequencies of tail autotomy in *Sceloporus jarrovii* (Sauria: Iguanidae) at different elevations. Herpetologica 33:380–387.

Trauma – Tail autotomy in Yarrow's spiny lizard *Sceloporus jarrovii* was more common in southern latitudes and at higher altitudes.

Brown JD, Sleeman JM. 2002. Morbidity and mortality of reptiles admitted to the wildlife center of Virginia, 1991–2000. Journal of Wildlife Diseases 38:699–705.

Trauma – Seventy four of 694 reptiles presenting to Virginia wildlife center had evidence of trauma, but did not comment on osseous involvement.

Metabolic – One instance of nutritional disorder, but no comment on bony involvement.

Brown RM, Taylor DH, Gist DH. 1995. Effect of caudal autotomy on locomotor performance of wall lizards (*Podarcis muralis*). Journal of Herpetology 29:98–105.

Trauma – De novo tail loss in wall lizards *Podarcis muralis* increased terrestrial speed and distance traveled, but reduced arboreal speed and distance. However, individuals with regenerated tails were fastest arborially after their regenerated tails were removed.

Brown DR, Nogueira MF, Schoeb TR, Vliet KA, Bennett RA, Pye GW, Jacobson ER. 2001a. Pathology of experimental mycoplasmosis in American alligators. Journal of Wildlife Diseases 37:671–679.

Infection – Experimental *Mycoplasma alligatoris* infection in an elbow and knee of alligators.

Brown DR, Schumacher IM, Nogueira MF, Richey LJ, Zacher LA, Schoeb TR, Vliet KA, Bennett RA, Jacobson ER, Brown MB. 2001b. Detection of antibodies to a pathogenic mycoplasma in American alligators (*Alligator mississippiensis*), broad-nosed caimans (*Caiman latirostris*) and Siamese crocodiles (*Crocodylus siamensis*). Journal of Clinical Microbiology 39(1):285–292.

Infection – Multifocal infectious arthritis associated with *Mycoplasma alligatoris* pneumonia in captive American alligator *Alligator mississippiensis*, broad-nosed caiman *Caiman latirostris*, and Siamese crocodile *Crocodylus siamensis*.

Brown DR, Demcovitz D, Plourde D, Potter S, Rotstein DS. 2005. *Mycoplasma iguanae* proposed sp. Nov., from green iguanas (*Iguana iguana*) with vertebral disease. International Journal of Systematic Evolution and Microbiology. Abstracts of the American Society for Microbiology 105th General Meeting. Abstract G-001.

Infection – Green iguana *Iguana iguana* with multifocal vertebral abscesses attributed to *Mycoplasma iguanae* and osteomalacia.

Brown DR, Demcovitz D, Plourde D, Potter S, Rotstein DS. 2006. *Mycoplasma iguanae* proposed sp. Nov., from green iguanas (*Iguana iguana*) with vertebral disease. International Journal of Systematic Evolution and Microbiology. International Journal of systemic Evolution and Microbiology 56:761–764.

Infection – *Mycoplasma iguanae* vertebral abscesses in green iguanas *Iguana iguana*.

Brown GP, Shilton C, Phillips BL, Shine R. 2007. Invader under stress: Spinal arthritis in invasive cane toads. Proceedings of the National Academy of Science USA doi:10:1073/pnas.0705057104.

Neuropathic – Ten percent of cane toad *Bufo marinus* (introduced Australian invaders) have vertebral damage with fusion of one or more vertebrae. Frequency of vertebral damage increased with size. Seven of 9 cultures grew *Ochrobactrum anthropi*, a soil bacteria related to *Brucella*.

Vertebral – Cane toad *Bufo marinus* has vertebrae separated by synovial joints, not cartilaginous disks. Ten percent of these introduced Australian invaders have vertebral damage with fusion of one or more vertebrae. Frequency of vertebral damage increased with size. Seven of 9 cultures grew *Ochrobactrum anthropi*, a soil bacteria related to *Brucella*.

Brown-Séquard. 1849. 3° cas de polydactylie chez un triton [Third case of polydactyly in a triton]. Gazette Médicale (Paris) 1849:901 [French].

Congenital – *Rana esculenta* hindfoot with six digits.

Bruce HM, Parkes AS. 1950. Rickets and osteoporosis in *Xenopus laevis*. Journal of Endocrinology 7:64–81.

Metabolic – Decalcification, bowing and expanded irregular metaphyses from rickets in *Xenopus laevis*).

Bruch C. 1863. Neue Beobachtungen zur Naturgeschichte der einheimischen Batrachier. [New observations to the natural history of native batrachians]. Würzburger naturwissenschaftliche Zeitschrift 4:92–151 [German].

Congenital – Malformed larvae of *Bufo viridis* (arched vertebral column).

Bruch C. 1864. Über Missbildungen der Chorda dorsalis (Dichordus), nebst Bemerkungen über Doppelbildungen. [On malformations of the chorda dorsalism (Dichordus) and remarks on double]. Würzburger medicinische Zeitschrift 5:1–35 [German].

Trauma – Duplication of posterior tail in larval *Pelobates fuscus* and *Triton cristatus*.

Vertebral – Arching of tail in *Rana esculenta*, larval *Pelobates fuscus*, and *Hyla arborea*.

Bruch C. 1866. Ueber die Entstehung der Doppelbildungen. [On the origin of double formation]. Würzburger medicinische Zeitschrift 7:257–320 [German].

Congenital – Referred in the text to dicephalic snakes published by Lacépède (1790), Edwards (1751) and Mitchill (1826), dicephalic turtles published by Catesby (1731–1743) and Peale, and doubled posterior body of a lizard described by Tiedemann.

Trauma – Eight *Pelobates fuscus* larvae with double tails originate by doubling of the anlage.

Bruch C. 1873. Ueber Dreifachbildungen. [On tripple formation]. Jenaische Zeitschrift 7:142–175 [German].

Trauma – Larva of *Pelobates fuscus* with twice split chorda dorsalis in tail resulting in deep tail.

Brunst V. 1937. [Anomalies in amphibians] Priroda no. 7:105–110 [Russian].

Congenital – Supernumerary forelimbs and polydactyly in *Rana graeca, Rana esculenta, Alytes obstetricans,* and *Salamandra maculosa*, lateral body wall supernumerary in *Triton cristatus* and anterior duplication in unnamed.

Supernumerary limbs in frogs – Two individuals with forelimb and four individuals with three, four, and six hind limbs and one with polymelia and a newt with a supernumerary forelimb, summarized from Przibram (1921).

Polydactyly in a frog, citing Korschelt (1927) and Tornier (1898).

Cites Hellmich (1929) for supernumerary hind limb in *Rana graeca* and *esculanta* and *Salamandra maculosa* and Stockard (1910), for *Alytes obstetrican.*

Brunst V. 1952. The problem of roentgen stimulation. American Journal of Roentgenology and Radium Therapy 68:281–289.

Toxicology – Supranormal X-ray dose of 1,000–6,000 roentgens inhibited axolotls *Siredon mexicanum* tail development in the field, but stimulated growth in areas peripheral to the X-ray field.

Brunst V. 1955. The axolotl (*Siredon mexicanum*) II. Morphology and pathology. Laboratory Investigation 4:429–449.

Congenital – States "the Mexican axolotl (*Siredon mexicanum*) is the neotenic larval form of *Ambystoma mexicanum*."??

Three axolotl *Siredon* with no right forelimb, three with no left, and three lacking both; occasionally extra fingers, toes, limbs or tails.

Fusion of tarsal or carpal bones was blamed by Schmalhausen (1925) on malnutrition or abnormally high temperature.

Environmental – Fusion of tarsal or carpal bones was blamed by Schmalhausen (1925) on malnutrition or abnormally high temperature.

Brunst V. 1961. Some Problems of Regeneration. The Quarterly Review of Biology 36(3):178–206.

Congenital – Polydactyly in *Pelobates fuscus, Rana esculenta, Triton taeniatus, Siredon pisciformis, and Triton cristatus*, and cites Hellmich (1929) for *Alytes obstetricans* with 12 supernumerary limbs.

Bryant SV, Bellairs A d'A. 1967. Tail regeneration in the lizards *Anguis fragilis* and *Lacerta dugesii.* Journal of the Linnean Society (Zoology) 46:297–305.

Trauma – Slow worm *Anguis fragilis* and Madeira wall lizard *Lacerta dugesii* regeneration rate varied with amount lost. *Lacerta dugesii* who lost more than 50% of tails regenerated more rapidly than those with less loss.

Bryant SV, Bellairs Ad'A. 1970. Development of regenerative ability in the lizard, *Lacerta vivipara.* American Zoologist 10:167–173.

Trauma – Embryonic *Lacerta vivipara* do not regenerate as well as older ones, but *Lacerta* regenerate tails rapidly and completely (Bryant and Bellairs 1967; Moffat and Bellairs 1964; Slotopolsky 1922; White 1925). Ability is linked to tail autotomy.

Bryant SV, Gardiner DM. 1992. Retinoic acid, local cell-cell interactions, and pattern formation in vertebrate limbs. Developmental Biology 152(1):1–25.

Trauma – Regeneration of amputated distal parts in urodele amphibians at any time of their life cycle.

Toxicology – Retinoid beads implanted at amputation cites produce supernumerary digits, citing Maden (1982); Niazi and Saxena (1978).

Amputated tails of marbled balloon frog *Uperodon systoma* exposed to retinoic acid instead regenerate legs (Mohanty-Hejmady et al. 1992), while Pietsch (1987) observed growth inhibition, and Niazi and Saxena (1979), truncation.

Bryner J. 2011. World's first known toothache revealed in ancient reptile. Life Science April 18 www.livescience.com/13759-oldest-toothache-reptile-fossils.html.

Dental – Cites Robert Reisz' report of 275-million-year-old-*Labidosaurus hamatus* with damaged jaw and missing teeth with abscess.

Bube A, Burkhardt E, Weiss R. 1992. Spontaneous chromomycosis in the marine toad (*Bufo marinus*) Journal of Comparative Pathology 106(1):73–77.

Infection – Chromycotic granulomas produced by *Cladosporium cladosporioides* in the bone marrow cavities of the skull of marine toad *Bufo marinus*, but no comment on gross or radiologic appearance of bones.

Buenviaje GN, Ladds PW, Melville L, Manolis SC. 1994. Disease-husbandry associations in farmed crocodiles in Queensland and the Northern Territory. Australian Veterinary Journal 71:165–173.

Metabolic – Five cases of articular gout in *Crocodylus johnstoni*, but not *C. porosus* in an crocodile farm.

Buffetaut E. 1983. Wounds on the jaw of an Eocene mesosuchian crocodilian as possible evidence for the antiquity of crocodilian intraspecific fighting behavior. Paläontologische Zeitschrift 57(1–2):143–145.

Trauma – Toothmark wounds on lower jaw of Eocene crocodilian *Telemsisuchus lavocati.*
Fossil – Toothmark wounds on lower jaw of Eocene crocodilian *Telemsisuchus lavocati.*

Buffetaut E, Li J, Tong H, Zhang H. 2007. A two-headed reptile from the Cretaceous of China. Biology Letters 3:80–81.
Congenital – Lepidosaur *Sinohydrosaurus lingyuanensis* or *Hyphalosaurus lingyuanensis*, embryo or neonate from the Yixian Formation, Lower Cretaceous of Liaoning Province, northeastern China.
Fossil – Lepidosaur *Sinohydrosaurus lingyuanensis* or *Hyphalosaurus lingyuanensis*, embryo or neonate from the Yixian Formation, Lower Cretaceous of Liaoning Province, northeastern China.

Buffon, GL. 1802. Histoire naturelle des quadrupedes ovipares et des serpents. [Natural history of oviparous quadrapedal and serpents]. Edinburg: Smellie. [French]
Congenital – Dicephalic snake.
Pseudopathology – Fable of two-headed *Amphisbaena.*

Buguet A. 1899. Régénérations osseuses, suivies à l'aide de la radiographie. [Bone regeneration followed by radiography]. Comptes rendus de l'Académie des Sciences Paris 129:174–175 [French].
Trauma – Axolotl with five regenerated digits.

Bumpus HC. 1897. A contribution to the study of variation. Journal of Morphology 12:455–484.
Congenital – Skeletal variations of *Necturus maculosus.*

Burger J, Zappalorti RT, Gochfeld M. 1987. Developmental effects of incubation temperature on hatchling pine snakes *Pituophis melanoleucus.* Comparative Biochemistry and Physiology 87A:727–732.
Congenital – Kinked tails in 8% and kinked bodies in 5% of pine snakes *Pituophis melanoleucus* raised at 21–23°C.
Environmental – Cites temperature-related skeletal abnormalities in Burmese, Indian, or Ceylon python *Python molurus* at 27.5°C (Vinegar 1973, 1974) and increased vertebral number in *Natrix fasciata* incubated at both high and low temperature (Osgood 1978).

Burghardt GM. 1991. Cognitive ethology and critical anthropomorphism: A snake with two heads and hognose snakes that play dead. pp. 53–90 in Ristau CA. ed. Cognitive ethology. The minds of other animals. 90 pp.; Hillsdale, NJ: Lawrence Erlbaum.
Congenital – Prey competition of each head of a dicephalic black rat snake *Elaphe obsoleta* from Oak Ridge.

Burghardt GM, Batts BG, Bock BC. 1982. An analysis of feeding behavior in a bicephalic black rat snake. Abstract Society for the Study of Amphibians and Reptiles/Herpetologists League Meeting, Raleigh, North Carolina:60.
Congenital – Dicephalic *Elaphe obsoleta obsoleta.*
Rana alticola lacking hind limbs.

Burke RL. 1994. *Apalone spinifera* (spiny softshell). Extreme kyphosis. Herpetological Review 25(1):23.
Congenital – Kyphosis in *Apalone spinifera* (spiny softshell).

Burkett RD, Huggins JA. 1985. Life history notes: *Nerodia rhombifer* (diamondback water snake). Twinning. Herpetological Review 16(2):56.
Congenital – Siamese twin diamondback water snake *Nerodia rhombifer* joined from heart to ovaries, with smaller one dead at birth.

Burkhart JG, Helgen JC, Fort DJ, Gallagher K, Bowers D, Propst TL, Gernes M, Magner J, Shelby MD, Lucier G. 1998. Induction of mortality and malformation in *Xenopus laevis* embryos by water sources associated with field frog deformities. Environmental Health Perspectives 106 (12):841–848.
Congenital – Background frequency of *Xenopus laevis* supernumerary limbs and ectomelia in Minnesota was 2.25% in 1997.
Trauma – *Uperodon systoma* amputated tails may be replaced by hind limbs, prevented by blocking thyroid hormone function.
Toxicology – Retinyl palmitate exposure (Luo et al. 1993; Mohant-Hejmadi et al. 1992) in *Xenopus laevis* produced abnormalities.
Environmental – Increase in ectromelia and ectrodactyly in frog, *Rana pipiens*, green frog *Rana clamitans*, bullfrog *Rana catesbeiana*, and mink frog *Rana septentrionalis*, possibly related to UV radiation.
Xenopus abnormalities blamed on "groundwater in some sites."

Burkhart JG. Ankley G, Bell H, Carpenter H, Fort D, Gardner D, Gardner H, Hale R, Helgen JC, Jepson P, Johnson D, Lannoo M, Lee D, Lary J, Levey R, Magner J, Meteyer C, Shelby MD, Lucier G. 2000. Strategies for assessing the implications of malformed frogs for environmental health. Environmental Health Perspectives 108:83–90.
Environmental – Abnormalities in green frog *Rana clamitans*, northern leopard frog *Rana pipiens* (5–23% missing and partial hind legs), bull frog *Rana catesbeiana*, and mink frog *Rana septentrionalis* in Vermont.
Buschendorf J, Günther R. 1996. Teichmolch – *Triturus vulgaris* (Linnaeus, 1758) [Smooth newt *Triturus vulgaris* (Linnaeus, 1758)]. In Günther R. ed. Die Amphibien und Reptilien Deutschlands. [The Amphibians and Reptiles of Germany]. pp. 174–195; Jena: Gustav Fischer [German].
Congenital – Frequent *Triturus vulgaris* dicephalism and abnormalities in extremities (one specimen with right hind leg with two feet [H. Wawrzyniak].
Trauma – Frequent *Triturus vulgaris* with bifid tails.

Bustard HR. 1964. Defensive behavior shown by Australian geckos, genus *Diplodactylus*. Herpetologica 20:198–200.
Trauma – 7/80 (all but 1 male) *Diplodactylus* had lost tails.
Bustard HR. 1968. Temperature dependent tail autotomy mechanism in gekkonid lizards. Herpetologica 24:127–130.
Trauma – *Anolis carolinensis* perching is reduced with tail loss.
Bustard HR. 1969. Tail abnormalities in reptiles resulting from high temperature egg incubation. British Journal of Herpetology 4:121–123.
Environmental – *Oedura ocellata* with a short tail attributed to high incubation temperatures.
Bustard HR. 1971. Temperature and water tolerances of incubating crocodile eggs. British Journal of Herpetology 4:198–200.
Environmental – *Crocodylus novaeguineae* eggs exposed to 38°C produce deformed tails.
Bustard HR, Hughes RD. 1966. Gekkonid lizards: Average ages derived from tail-loss data. Science 153:1670–1671.
Trauma – Tail loss was a function of age in geckos *Gehyra variegata* and *Heteronotia binoei*.
Butler EG. 1933. The effects of x-radiation on the regeneration of the forelimbs of *Ambystoma* larvae. Journal of experimental Zoology 65:271–303.
Environmental – 60 kV, 720 mas X-ray exposure inhibits regeneration in *Ambystoma*.
Butler EG, Blum HF. 1955. Regenerative growth in the urodele forelimb following ultraviolet radiation. Journal of the National Cancer Institute 15:877–889.
Environmental – UV light produces accessory limbs in *Ambystoma*.
Butler EG, Blum HF. 1963. Supernumerary limbs of urodele larvae resulting from localized ultraviolet light. Developmental Biology 7:218–233.
Environmental – *Ambystoma maculatum* and *Ambystoma opacum* exposed to UV light develop supernumerary limbs.
Byerly TC. 1925. Note on the partial regeneration of the caudal region of *Sphenodon punctatum*. Anatomical Record 30:61–66.
Trauma – Replacement of 12 or 14 regenerated segments in 4 of 5 *Sphenodon punctatum* by unsegmented rod of cartilage, continuous with the last remaining caudal vertebra. Notochord was not regenerated.
Byrd EE. 1939. Certain aspects of the anatomy of a two headed turtle. Journal of the Tennessee Academy of Science 14(1):102–106.
Congenital – Dicephalic *Pseudemys troostii*.
Bystrow AP, Efremov JA. 1940. [*Benthosuchus sushkini* EFR. a labyrinthodont from the Eotriassic of Sharzhenga River.] Akademija Nauk SSSR, Trudy Paleontologicheskii Institut 10(1):1–152 [Russian].
Congenital – Infraspecific variation in the labyrinthodont *Benthosuchus sushkini* (long English summary; details in Russian on p. 137).
Fossil – Infraspecific variation in the labyrinthodont *Benthosuchus sushkini* (long English summary; details in Russian on p. 137).
Caetano MH. 1991. Anomalies et regeneration des membres chez *Triturus marmoratus* (Latreille, 1800). [Limb anomalies and regeneration in *Triturus marmoratus* (Latreille, 1800)]. Bulletin de la Sociétè Herpétologie de France 57:53–58 [French].
Congenital – Southern Portugal *Triturus mammoratus* had supernumerary digis in 0.7%, bifurcation in 1.6%, absent digits in 2%, atrophied digits in 1.6%, and supernumerary limbs in 0.2%, but none were found in Northern Portugal.
Anurans have more teratologic abnormalities than Urodeles.
Supernumerary limbs in *Salamandra maculosa* (Hellmich 1929) and *Cynops pyrrhogaster* (Meyer-Rochow and Asashima 1988).
Cagle FR. 1946. Tail loss and regeneration in a Pacific island gecko. Copeia 1946(1):45.
Trauma – Twenty nine percent of female, 33% of male, and 10% of juvenile *Hemidactylus garnotii* from Tinian in the Mariannas, had partially or completely regenerated.
Tails; 8%, for the second time.
Cagle FR. 1950. The life history of the slider turtle, *Pseudemys scripta troostii* (Holbrook). Ecological Monographs 20:31–54.
Congenital – 4.8% of Illinois slider turtle *Pseudemys scripta troostii* plaston had anomalies: 15 with extra plates, 6 divided, 33 irregular sutures, and 12 of 860 adults with both inserted plates and irregular sutures, compared to 0.8% of hatchlings with irregular sutures. 5.7% of carapaces had anomalies: 11 inserted, 20 divided plates, 18 irregular sutures, compared to 20.2% of hatchlings: 73 with inserted or divided plates, 2 with irregular sutures. Many with "extreme anomalies." Costals and neurals were most affected, with eight individuals with "extremely malformed neurals."
Trauma – *Pseudemys* sp. carapaces gnawed on by muskrats (Stophlet 1947).
Cahn A.R. 1937. The turtles of Illinois. Illinois Biological Monographs 16:1–218.
Congenital – Highly domed, abnormally wide costal scutes of *Chelydra serpentina*.
Calori L. 1858. Sullo scheletro della *Lacerta viridis* Linn., sulla riproduzione della coda nelle lucertole, e sulle ossa cutanee del teschio de' Sauri. [On the skeleton of *Lacerta viridis* Linn., on the reproduction of the tail in the lizards,

and on the dermal bones of the skull of saurians]. Rendiconto delle sessioni dell'Accademia delle scienze dell'Istituto di Bologna 9:46–50 [Italian].

Lizards with regenerated and bifid tails. One of the two tails has vertebrae; the other has an ossified, originally cartilaginous rod. Noted Buffon as first to make that observation.

Calori L. 1859. Sulla riproduzione di una doppia coda nelle lucertole [On the reproduction of a double tail in lizards]. Rendiconto delle sessioni dell'Accademia delle scienze dell'Istituto di Bologna 10:64–65 [Italian].

Trauma – Bifid tail in gecko *Platydactylus muralis*.

Other – Skeleton of *Stellio vulgaris* Daudin at variance with report by Eichwald in *Fauna Caspio-Caucasica*, and by Blanchard in *Organisation du Regne Animal*.

Calmonte A. 1968. Zwei bemerkenswerte Schildanomalien bei *Testudo hermanni hermanni* Gmelin 1798, der griechischen Landschildkröte. [Two signs of plate abnormalities in *Testudo hermanni hermanni* Gmelin 1798, the Greek Tortoise]. Aquaterra 5:34–36 [German].

Congenital – Segmentation of posterior marginal or vertebral plates in *Testudo hermanni hermanni.*

Calmonte T. 1983. Eine doppelköpfige Rattenschlange (*Elaphe obsoleta lindheimeri*, Baird and Girard 1853) aus Texas. [A double-headed rat snake (*Elaphe obsoleta lindheimeri*, Baird and Girard 1853) of Texas]. Herpetofauna 5(26): 12–13 [German].

Congenital – Dicephalic *Elaphe obsoleta* described and figured.

Camerano L. 1879 (not 1880). Nota intorno allo scheletro del *Bombinator igneus* (Laur.) [Note about the skeleton of *Bombinator igneus* (Laur.)]. Atti della Reale Accademia delle Scienze di Torino 15:445–450 [Italian].

Congenital – In a study of all the *Bombinator igneus* in the collections of the Zoological Museum of Turin, Camerano encountered and described four individuals with abnormal development of the sacral vertebra and the coccyx as it regards the transverse processes and lateral extensions. He provided illustrations for each, and case by case descriptions of each abnormality. He also cited Bedriaga (1879), Genè (1839), and Gotte (1875), reproducing line illustrations from Gotte (1875) and Genè. He agreed with Bedriaga, who stated that the drawing provided by Genè of the sacral vertebra and coccyx of *B. igneous* does not represent the normal condition but a developmental anomaly. He concluded that in *B. igneous*, the sacral vertebra and the coccyx frequently develop abnormally as it regards the transverse processes and lateral extensions.

Camerano L. 1882. Di un caso di polimelia in un *Triton taeniatus* (Schneid.) [On a case of polymely in a *Triton taeniatus* (Schneid.)]. Atti della Società Italiana Scienze naturali e del Museo Civico di Storia Naturale in Milano 25:113–116 [Italian].

Congenital – Polymely reported in a salamander, *Triton taeniatus* (Schneid.). The foot articulated with the bifurcated femur through a cylinder of bone with a thin cartilaginous structure, which Camerano suggested may represent the supernumerary femur. However, he also presented alternative hypotheses to explain the bifurcation of the regular femur, all related to locomotive pressure from use of the supernumerary and normal limbs during growth as a juvenile.

Camerano L. 1888. Descrizione di un girino anomalo di *Rana esculenta* Linneo [Description of an anomalous tadpole of *Rana esculenta* Linné]. Bolletino dei Musei di Zoologia ed Anatomia comparata della Reale Università di Torino 3(36):1–2 [Italian].

Congenital – *Rana esculenta* tadpole with extraordinarily small hind limbs. Hypotheses offered included arrest of growth in the affected area or violent extraction of the legs by another animal, which resulted in regenerated small limbs.

Trauma – *Rana esculenta* tadpole with extraordinarily small hind limbs. Hypotheses offered included arrest of growth in the affected area or violent extraction of the legs by another animal, which resulted in regenerated small limbs.

Camerano L. 1896. Note zoologiche, VIII. Di una *Molge vulgaris* polimelica [Zoological note VIII. On a polymelus *Molge vulgaris*]. Bolletino dei Musei di Zoologia ed Anatomia comparata della Reale Universita di Torino 11 (No. 251):4 [Italian].

Congenital – Stated that urodele amphibians captured in the wild very rarely exhibit polymely. Reported a *Molge vulgaris* in the Zoologic Museum of Torino with extra left leg, which possessed two bifurcated feet.

Camp CL. 1930. A study of the phytosaurs with description of new material from western North America. Memoirs of the University of California 10:1–161.

Trauma – Pits and swelling on phytosaur snout dorsum related to bites in combat (Abel 1922, Mooie 1922, Nopcsa 1926).

Missing teeth with irregular grooves in distal premaxilla of *Machaeroprosopus gregorii* Univ Cal Mus Paleontol A270/27200.

Infection – Multiple variably sized and shaped pits in the parietal region of phytosaur *Machaeroprosopus zunii* Univ Cal Mus Paleontol 7307/27048, interpreted by Abel (1922) as abscesses from bites. Abel similarly attributes snout swelling.

Fossil – Pits and swelling on phytosaur snout dorsum related to bites in combat (Abel 1922; Mooie 1922; Nopcsa 1926).

Missing teeth with irregular grooves in distal premaxilla of *Machaeroprosopus gregorii* Univ Cal Mus Paleontol A270/27200.

Multiple variably sized and shaped pits in the parietal region of phytosaur *Machaeroprosopus zunii* Univ Cal Mus Paleontol 7307/27048, interpreted by Abel (1922) as abscesses from bites. Abel similarly attributes snout swelling.

Campbell H. 1967. Two-headed snapping turtle *Chelydra serpentina.* International Turtle and Tortoise Society Journal 1(4):24–29.

Congenital – Dicephalic common snapping turtle.

Campbell JW. 1970. Excretory nitrogen metabolism in reptiles and birds. In: Nitrogen Metabolism and Excretion, Walsh PJ, Wright P, eds, pp. 147–178.

Metabolic – Cites Coulson and Hernandez on tendency of crocodilians to develop gout because of uric acid synthesis, especially when over-eat in winter, the result of a high protein diet.

Campione NE, Reisz RR. 2010. *Varanops brevirostris* (Eupelycosauria: Varanopidae) from the lower permian of Texas, with discussion of Varanopid morphology and interrelationships. Journal of Vertebrate Paleontology 30:724–746.

Vertebral – *Varanops brevirostris* vertebrae (Fig. 8) with apparent syndesmophyte characteristic of spondyloarthropathy.

Fossil – *Varanops brevirostris* vertebrae (Fig. 8) with apparent syndesmophyte characteristic of spondyloarthropathy.

Canella MF. 1932. Transformismo e teratologia [Transformism and teratology]. Rivista di psicologia: organo della Societá italiana di psicologia e degli istituti universitari di psicologia sperimentale 28:291–313 [Italian].

Congenital – Canella discussed teratology vs. pathological anatomy, theories of heredity, and cited the report of a dicephalic turtle by Cederstrom (1931).

Canella MF. 1935a. Problemi generali e particolari di Embriologia teratologica [General and particular problems in teratological embryology]. Archivio Zoologico Italiano 21:69–160 [Italian].

Congenital – Stated that anterior duplicity was obtained only experimentally in amphibians, and that it was extremely rare in sauropsids. Cited Terni's notation of the frequency of occurrence of anterior duplicity in Sauropsids: 19 total, 16 in birds and 3 in reptiles (no year given).

Double-faced tadpole of *Bufo vulgaris* figured.

Canella M.F. 1935b. Osservazioni teratologiche in Anfibi anuri [Teratological observations on anuran amphibians]. Communicazione al XX Convegno di Zoologia (Bologna, Settembre 1934). Bolletino di Zoologia 6 (1–2):195–201 [Italian].

Congenital – Doubled face in a tadpole of *Bufo vulgaris*.

Supernumerary forelimbs in two *Rana esculenta*, noting majority of the known cases in *Rana* belong to the *esculenta* species.

Canella M.F. 1937. Contributo allo studio degli arti doppi speculari. Due casi di polimelia toracica in *Rana esculenta*, L. [Contribution to the study of double limbs. Two cases of thoracic polymely in *Rana esculenta*, L.] Scritti Biologici, raccolti da Luigi Castaldi 12:3–78 [Italian].

Congenital – *Rana esculenta* with small supernumerary limbs situated ventrally and medially in relation to the anterior right normal limb, the one on the left had four stubs of various lengths instead. A second specimen had a lateral right arm and a medial left arm. Discussed causes and experimental induction of supernumerary limbs.

Cantoni AF. 1921. Casi di dicefalia in *Tropidonotus natrix* [A case of dicephaly in *Tropidonotus natrix*]. Atti della Società ligustica di Scienze Naturali e geografiche 32:131–142 [Italian].

Congenital – Derodymous *Tropidonotus natrix* snake, male specimen, probably captured a few weeks after birth. Right head was dominant over the left. Cantoni provides a description of the internal organs, and later reviews the literature on dicephalic snakes. This article also provides brief mention of some unpublished, preserved in alcohol, dicephalic *Tropidonotus natrix* that reside in Italian museums. Among these are: Two specimens in the Museo Civico di Milano (one of which had already been cited by De Betta in 1865), two other individuals in the Comparative Anatomy museum of the University of Pavia, and one neonate specimen in the Museo Civico di Savona, which was captured in 1916.

Capocaccia L. 1966. Variabilita della popolazione mediterranea di *Caretta caretta* (L.) (Testudines). [Population variability of Mediterranean *Caretta caretta* (L.) (Testudines)]. Annali del Museo Civico di Storia Naturale di Genova 75:1–22 [Italian].

Congenital – Individual variability of Mediterranean *Caretta caretta* number of costal plates, vertebral plates, principal ventral plates and number of nails, inframarginal plate.

Cardeilhac PT. 1981. Nutritional disease of alligators. Proceedings of the First Alligator Production Conference, Gainesville, Florida, pp. 58–64.

Metabolic – Nutritional bone disease–induced skeletal demineralization, rubber snout, and loss of teeth.

Gout develops when food intake reaches 35% of body weight.

Arthritis – Erroneously claimed that chronic recurrent arthritis is not seen in reptiles.

Stones – Erroneously claimed urinary tract stones are not seen in reptiles.

Carpenter L. 1879. A two-headed snake. The Weekly Inter Ocean 30 October 1879:7.

Congenital – Alleged dicephalic snake.

Carpenter L. 1879. A two-headed snake. The Daily Inter Ocean 27 October 1879:6.

Pseudopathology – Southwest Ohio water snake with alleged head on each end.

Carpenter CC. 1956. Carapace pits in the three-toed box turtle, *Terrapene carolina triunguis* (Chelonia-Emydidae). Southwest Naturalist 1:83–86.

Shell disease – 6.5% of three-toed box turtle *Terrapene carolina triunguis* with carapace pits.

Carpenter CC. 1958. An unusual Ouachita map turtle. Herpetologica 14:116.

Congenital – Kyphosis in *Graptemys (Malaclemys) pseudogeographica ouachitensis.*

Carpenter K, Lindsey D. 1980. The dentary of *Brachychampsa montana* Gilmore (Alligatorinae; Crocodylidae), a Late Cretaceous turtle-eating alligator. Journal of Paleontology 54:1213–1217.

Dental – Ninth, tenth, and eleventh teeth are strongly compressed laterally in a Late Cretaceous turtle-eating alligator *Brachychampsa.*

Shell disease – Heavy etching and pitting in turtle shell were attributed to gastric acid, similar to that found in the stomach of *Alligator mississippiensis* USNM 14736.

Fossil – Ninth, tenth, and eleventh teeth are strongly compressed laterally in a Late Cretaceous turtle-eating alligator *Brachychampsa.*

Heavy etching and pitting in turtle shell were attributed to gastric acid, similar to that found in the stomach of *Alligator mississippiensis* USNM 14736.

Carpenter CC, Yoshida JK. 1967. One-egg twins in *Agama agama.* Herpetologica 23:57–59.

Congenital – *Agama agama* eggs with Siamese twins joined at thorax.

Cappio A-S. 1985. La Conservation des Tortues dans le Monde: Trois exemples. [The conservation of turtles in the world: Three examples]. Chambray-les-Tours, France: Universite Claude-Bernard-Lyon I Doctoral Thesis, 193 pp [French].

Trauma – Olive turtle *Lepidochelys olivacea* with hindfoot loss secondary to boat accident.

Carr AF Jr. 1940. A Contribution to the Herpetology of Florida. Univesity of Florida Publication, Biological Science Series 3(1):1–118.

Trauma – Tail loss in ground skink *Leiolopisma unicolor.*

Carr A. 1952. Handbook of Turtles. 542 pp.; Ithaca, New York: Comstock Publ Co.

Congenital – Kyphosis in *Pseudemys* (*Chrysemys*) *scripta scripta.*

Carretero MA, Llorente GA, Santos X, Montori A. 1995. Un caso de polidactilia en lacértidos. [A case of polydactyly in Lacertidae.] Boletín de la Asociación Herpetológica Española 6:11–13 [Spanish].

Congenital – Polydactylous adult female specimen of *Podarcis pityusensis* from an introduced population in Barcelona, Spain in the collection of the Department of Animal Biology (vertebrates) of the Faculty of Biology of the University of Barcelona, under number DZBH-2597. The left forelimb had six fingers, all fully formed and of normal aspect in external view. Radiograph revealed an unequal bifurcation of metacarpal IV, which identified finger IV as the double digit. One had five phalanges (IVa) and the other (IVb) had only four.

Carrick T, Reddacliff G. 1980. Repair of traumatic shell fractures in Australian side-necked turtles. International Zoo Yearbook 20:241–243.

Trauma – Apparent automobile-related traumatic shell fractures in Australian side-necked turtles *Chelodina longicollis* and *Emydura krefftii.*

Carroll RL. 1968. The postcranial skeleton of the Permian microsaur *Pantylus.* Canadian Journal of Zoology 46:1175–1192.

Congenital – Structural modification with alternating height vertebral neural spines for efficient vertebral column dorsiflexion and lateral flexion in microsaur *Pantylus.*

Fossil – Structural modification with alternating height vertebral neural spines for efficient vertebral column dorsiflexion and lateral flexion in microsaur *Pantylus.*

Carswell LP, Lewis TE. 2003. Embryo and hatchling abnormalities in loggerhead sea turtles on St. Vincent Island, Florida. NOAA Technical Memorandum NMFS-SEFSC 503, August: 185–186.

Congenital – 0.32% abnormalities in 14,361 loggerhead turtle eggs in 1992 and 2001, manifest as small size, carapace deformities, folded or foreshortened flippers, misshapened mouthparts and twinning.

Case EC. 1896. Abnormal sacrum in alligator. American Naturalist 30:232–234.

Congenital – Vertebral number anomalies.

Three instead of two sacral vertebrae in *Alligator mississippiensis.*

Last lumbar converted to sacral in *A. sclerops.*

Last caudal converted to sacral in *Crocodylus acutus.*

Extra presacral in *Gavialis gangeticum.*

Extra presacral in *A. mississippiensis.*

Lumbar with sacral rib (sacralization) in *A. mississippiensis* and *Crocodylus acutus.*

Case EC. 1924. Some new specimens of Triassic vertebrates in the museum of geology of the University of Michigan. Papers of the Michigan Academy of Sciences etc. 4:419–423.

Infection – *Leptosuchus imperfecta* with "heavy rugose swelling of median bone… at the angle wither snout begins its descent. The top of the swollen area is occupied by a nearly hemispherical cavity… Surface is porous or spongy…" diagnosed as "abscess or deep-seated tumor." Probably an abscess.

Fossil – *Leptosuchus imperfecta* with "heavy rugose swelling of median bone… at the angle wither snout begins its descent. The top of the swollen area is occupied by a nearly hemispherical cavity… Surface is porous or spongy…" diagnosed as "abscess or deep-seated tumor." Probably an abscess.

Case EC. 1930. On the lower jaw of *Brachysuchus megalodon*. Contributions from the Museum of Paleontology, University of Michigan 3:155–161.

Trauma – Displaced teeth and fragments with wrinkled roots, distorted pulp, and partially occluded distal right jaw sockets in phytosaur *Brachysuchus megalodon.*

Fossil – Displaced teeth and fragments with wrinkled roots, distorted pulp and partially occluded distal right jaw sockets in phytosaur *Brachysuchus megalodon.*

Casimire-Etzioni AL, Wellehan JF, Embury JE, Terrell SP, Raskin RAE. 2004. Synovial fluid from an African spur-thighed tortoise (*Geochelone sulcata*). Veterinary Clinical Pathology 33:43–46.

Metabolic – African spur-thighed tortoise (*Geochelone sulcata*) with synovial fluid confirmed gout, but no comment on skeletal changes.

Catesby M. 1754. The natural history of Carolina, Florida, and the Bahama Islands: Containing the figures of birds, beasts, fishes, serpents, insects, and plants: Particularly, the forest trees, shrubs, and other plants, not hitherto described, or very incorrectly figured by authors, with their descriptions in English and French. To which are added observations on the air, soil, and waters, with remarks upon agriculture, grain, pulse, roots, etc. To the whole is prefixed, a new and correct map of the countries; with observations on their natural state, inhabitants, and productions. Histoire Naturelle de la Caroline, de la Floride, et des isles de Bahama. London: Marsh, Wilcox and Stichall. 2 v. 220 col. plates, double map. 53 cm.

Pseudopathology – Horn snake "armed with death at both ends." (II, p. 43). Actually, *Amphisbaena.*

Caullery M. 1931. Remarques sur des cas de bicéphalie. [Remarks on cases of dicephaly]. Bulletin de la Société zoologique de France 56:362–363 [French].

Congenital – Dicephalism and two pairs of arms in *Testudo ibera.*

Caullery M. 1932. Addition à la bibliografie relative à la bicephalic chez les serpents. [Additional biography related to dicephaly in serpents]. Bulletin de la Société zoologique de France 56:398 [French].

Congenital – Quotes Strohl (1931) for dicephaly.

Cavanna G. 1877a. Decrizione di alcuni batraci anuri polimeliani e considerazioni intorno alla polimelia. [Description of some polymelus batrachian anurans and considerations about polymely]. Pubblicazioni del Reale Istituto di Studi superiori pratici e di perfezionamento di Firenze, Sezione di scienze fisiche e naturali 1:1–38 [Italian].

Congenital – Review of 20 cases of polymelia in anurans, adding two new cases in *Rana esculenta* Linn. in the zoological collections of the Real Museo di Fisica e Storia Naturale of the Institute of Higher Studies of Firenze (supernumerary underdeveloped limbs) and the museum of the Illustre Accademia dei Fisiocritici of Siena (hind limbs with abnormal phalangeal counts).

Cavanna stated that Fabretti (1866) mistakenly thought one of his specimens (a big, female *R. esculenta*) was a *R. temporaria*.

His assessment of polymelia was as follows:

1. Supernumerary limbs are always homologous to normal limbs.
2. Pelvic polymely is more common than thoracic polymely.
3. Rarely, supernumerary limbs have no osteological attachment to the normal limb in the pelvic or thoracic belt.
4. Dorsal and ventral attachments of extra limbs have no apparent difference in frequency, but when the two extra limbs develop, ventral attachment appears more common.
5. Simple polymelia is more common than double polimelia.
6. Left-handed polimelia is much more frequent than right.
7. The genus *Rana* has the most polymelia among anurans.
8. Supernumerary limbs often have all the osseous segments of the normal homologous limb, but with varying degrees of development.
9. Absence, deformity, or reduction of segments of the supernumerary limbs is more common in the short bones, central or peripheral (sternum, scapula, carpal, tarsus, metatarsus, etc.), than in the long bones (humerus, radius-ulna, femur, tibia-fibula).

Table: Polymelia

	Author	Taxon	Attachment	Remarks
Thoracic Polymelia				
1	de Supperville (1744)	*Rana* sp.?	1. Right, dorsal	
2	Gervais (1864/5)	*Pelobates cultripes* Tsch.	1. Left	Little different than normal, imperfectly developed. The normal limb has only three digits, the two laterals being rudimentary.
3	Strobel (1876)	*Rana esculenta* auct	1. Left, ventral	All osseous segments are represented, imperfectly. Digits of the abnormal limb are in normal numbers, but very divergent.
4	Lunel (1868)	*Rana esculenta* auct	2. Left side	Well formed. Digits divergent but in normal number.
Pelvic Polymelia				
5	Guettard (1783)	*Rana* sp.?	1. Right, ventral	
6	Otto (1816)	*Rana* sp.?	1. Right	Well formed.
7	Duméril (1865)	*Rana esculenta* auct.	1. Right, dorsal	Short, lacks femur. One supernumerary finger in the abnormal arm, which was thus the result of the fusion of two feet according to Duméril.
8	Duméril (1865)	*Rana temporaria* auct.	1. Right, dorsal	All osseous segments are represented. No supernumerary digits, but according to Duméril the abnormal foot was the result of the fusion of two feet.
9	Cisternas (1865)	*Alytes obstetricans* Wagl.	1. Right, dorsal	All osseous segments are represented. Only four fingers, irregular, in the abnormal limb.
10	Thomas (in Duméril)	*Bufo vulgaris* Linn.	1. Left? Dorsal?	All osseous segments are represented. Only three fingers in the abnormal limb.
11	Strobel (1876)	*Rana temporaria* auct	1. Left, dorsal	All osseous segments are represented. Only two fingers in the abnormal limb.
12	Strobel (1876)	*Rana esculenta* auct.	1. Right, ventral	Lacks part of the tarsus, metatarsi and phalanges. Complete ectodactily in the abnormal limb.
13	Fabretti (1866)	*Rana esculenta* auct.	1. Left, ventral	Well formed.
14	Fabretti (1866)	*Rana esculenta* auct.	1 or 2 fused? Left, ventral	
15	Van Deen (1835)	*Rana esculenta* auct.	2. Left?	Smaller than the normal one more or less well developed, the other very brief and lacking metatarsi and phalanges. Complete ectodactily in one of the abnormal limbs.
16	Duméril (1865)	*Rana clamata* Daud	2. Middle, ventral	Well formed. In the right thoracic normal limb exists only one digit, in the left only three.
17	Lunel (1868)	*Rana esculenta* auct.	2. Left, ventral	Well formed.
18	Lunel (1868) (indirectly mentioned by Ducret)	*Rana* sp.?	2. Left	Rudimentary.
19	Cavanna (1877) (taken indirectly from his table)	*Rana esculenta* auct.	2. Dorsal	Well formed. Fingers in normal numbers.
20	Cavanna (1877)	*Rana esculenta* auct.	2. Ventral	Incomplete. Ectodactilous.

Cavanna G. 1877b. Description de quelques Batraciens anoures polymèles et considérations sur la polymélie. [Description of polymely in several Batrachian Anurans and consideration of polymely]. Journal de Zoologie 6:417–418. [French]

Congenital – Supernumerary limbs inserted at sacro-iliac articulation in *Rana viridis.*

Cavanna G. 1879. Ancora sulla Polimelia nei Batraci anuri [Once more about polymely in batrachian anurans]. Pubblicazioni del Reale Istituto di Studi superiori pratici e di perfezionamento di Firenze, Sezione di scienze fisiche e naturali 1:1–14 [Italian].

Congenital – Thoracic polymely in *Rana esculenta* in the Anatomical Museum of the Liberal University of Ferrara (Italy). Extra left forelimb (with coracoid duplication) attached just posterior to the normal left forelimb with fusion of two primitive forearms (laterally composed of two ulnae and a rudimentary bone that must have represented the two radii) and lacking some phalanges.

Reviewed reports by Balsamo-Crivelli (1865), Strobel (1877), and Sordelli (1877, 1878).

Cederstrom JA. 1931. A two-headed turtle. Journal of Heredity 22:137–138.
Congenital – Dicephalic hard shell snapping turtle *Chelydra serpentina.*
Chabaraud P. 1923. Description d'un *Chamaeleon* nouveau d'Indochineet d'un explaire monstreaux d'*Enhydris hardwicki* Gray. [Description of a new *Chamaeleon* from Indochina and explaining a monstrous *Enhydris hardwicki* Gray]. Bulletin de la Museum de Paris 29:209–210 [French].
Congenital – *Enhydris hardwicki* with large scale replacing three or four smaller ones and absence of symphysial.
Chaffin K. 2005. Two-headed snake finds home at Dan Nichols. Salisbury Post, Salisbury, NC (31 August 2005).
Congenital – Dicephalic kingsnake and cornsnake.
Derodymus copperhead.
Cites 2002 National Geographic, but no related article found that year.
Chalaux J. 1952. Un cas de duplication du bras gauche et de sa ceinture observé chez un femelle de *Rama exciemta* L. [Left arm duplication and its banding observed in a female *Rama exciemta* L]. Compte Rendu Congrès Société des Savantes Pris Départements (Science) 76:85–92 [French].
Congenital – *Rana esculenta* with supernumerary (the only one seen among 5,000 *Rana*) arm with incomplete radio-cubitus and no digits on metacarpals 2 and 4 and only one digit on metacarpals 1 and 3.
Cites Tornier (1898) with three right arms, de Falvard (1931) with supernumerary leg in *Rana esculenta*, Avel and Baer (1929) with arm duplication in *Discoglossus pictus* and dicephalic *Rana* (Born 1881; Jourdain 1877; Loyez 1897).
Chan AR. 1937. The turtles of Illinois. Illinois Biological Monographs 16:1–218.
Congenital – Kyphosis in *Chelydra serpentina* and *Amyda (Trionyx) spiniferus.*
Chan JG, Young LL. 1985. *Bufo marinus* (marine toad): anomaly. Herpetological Review 16:23–24.
Congenital – Hindfoot Amelia in *Bufo marinus.*
Chan JG, Young LL, Chang PR, Shero CM, Watts C. 1984. Morphological anomalies of two geckos, *Hemidactylus frenatus* and *Lepidodactylus lugubris*, and the toad, *Bufo marinus*, on the island of Hawaii. Proceedings Fifth Conference in Natural Sciences, Hawaii Volcanoes National Park June 5–7 1984:41–50.
Congenital – Hindfoot amelia in *Bufo marinus*; bifurcated tails in two house geckos *Hemidactylus frenatus.* Zigzag tail with kinking.
Zigzag garter snake *Thamnophis radix haydeni.*
Trauma – Multilobed, bulbous multi-tipped tails in two mourning gecko *Lepidodactylus lugubris.*
Chandra M, Mukherjee R. 1980. On the occurrence of bifurcated tail in Agama lizard from Simla hills, Himachal Pradesh. Journal of the Bombay Natural history Society 77(2):343.
Trauma – Tail bifurcation in *Agama tuberculata.*
Chang T-K, Boring AM. 1935. Studies in variation among the Chinese Amphibia. I. Salamandridae. Peking Natural History Bulletin 9:327–360.
Congenital – *Triturus orientalis, Triturus sinensis, Pachytriton brevipes* with premaxilla suture as individual variation. 31–38 caudal vertebrae in *Triturus orientalis*, 31–35 in *Triturus sinensis* and *Pachytriton brevipes.*
Digits are shortened because of loss or fusion of phalanges. One *Pachytriton brevipes* with 5 tarsi and intermediale and centrale fused with tibiale and tarsale 1 with tarsale 2. *Triturus orientalis* with fused third and fourth metatarsals, partially united fourth phalanges. *Triturus sinensis* with fused phalanges and cartilaginous tarsals, another with third and fourth digits fused and fifth fused with an extra one. Another has enlarged, cylindrical femur, tibia, and fibula with fused fourth and fifth digits. A *Pachytriton breviceps* had fusion of fourth and fifth metatarsals. A *Triturus orientalis* had only three fingers. Polydactyly was noted in one *Pachytriton brevipes.* Variation in phalangeal formula was also noted.
Chang CG, Witschi E, Ponseti IV. 1954. Teratologic development in *Xenopus* larvae caused by sweet pea seeds (*Lathyrus odoratus*) and their extracts. Anatomical Record 120:816.
Toxicology – Sweet pea diet causing scoliosis in *Xenopus* with dislocation of hind limbs.
Chang CG, Witschi E, Ponseti IV. 1955. Teratogenic effects of *Lathyrus oderatus* seeds on development and regeneration of vertebrate limbs. Proceedings of the Society of experimental Biology and Medicine 90:45
Toxicology – *Lathyrus odoratus* seeds produced dislocation of knee and ankle joints in *Xenopus.*
Chapman H. 1960. So long, Dudley Duplex! San Diego Union/Evening Tribune 15 June 1960: 1.
Congenital – Dicephalic California king snake.
Chapple C. 1999. Snake's alive, and it has 2 heads! Collector shows off 'flukes of nature'. The Times-Picayune, New Orleans, LA 30 September 1999:B3.
Congenital Two Siamese red-eared slider turtles.
Dicephalic canebrake rattlesnake and red-eared slider turtles and two derodymous red-eared sliders with upper limb duplication.
Chapple DG, Swain R. 2004. Inter-populational variation in the cost of autotomy in the metallic skink (*Niveoscincus metallicus*). Journal of Zoology (London) 264:411–418.
Trauma – Previous tail loss was present in 72% of 368 metallic skink *Niveoscincus metallicus*, in a gender independent manner (61–78%, dependent on site, but not altitude).

Charles H. 1944. Abnormal frog. Turtox News 22:179.
Congenital – Supranumerary limb in a frog.
Chauvier G. 1981. Pathologie des reptiles. 1) Polyarthrite purulent de l'iguane (*Iguana Iguana* L). [Reptile pathology. 1. Purulent polyarthritis in the iguana (*Iguana iguana* L). Revue française Aquarioliste 8 (26 June 1981):23 [French].
Infection – *Iguana iguana* loss of a phalanx from septic necrosis.
Chauvin E. 1920. Précis de Tératologie. [Summary of Teratology]. Paris: Masson & Cie. 160 pp. [French].
Congenital – Multiple supernumerary forelimbs in a frog and trauma induced supernumerary limbs in batrachians and polydactyly in a triton.
Cherepanov GO. 1994. Anomalies of bony carapace in turtles. Zoologicheskii Zhurnal 73(6):68–78.
Congenital – Anomalies in 121 of 510 *Mauremys caspica, Emys orbicularis, Testudo graeca, Agrionemys horsfieldii*, related to number of neurals and costals, peripherals and agenesis.
Chevemetieva (or Scheremetjewa) EA, Brunst VV. 1938. [Preservation of the capacity for regeneration in the middle part of the limb of the newt and its simultaneous loss in the proximal and distal parts of the same limb.] Biulleten' eksperimental'noi biologii I meditsiny. 6:723–724 [Russian].
Trauma – Experimental vascular destruction producing amelia.
Chiodini RJ, Nielsen SW. 1983. Vertebral osteophytes in an iguanid lizard. Veterinary Pathology 20:372–375.
Vertebral – Reported as exostoses in *Iguana iguana*, but actually dense enlargements around coccygeal vertebrae. Actually, Spondyloarthropathy.
Cholodkovsky N. 1896. Sur quelques exemples de polydactylie. [On several examples of polydactyly]. Comptes rendus de séances de la Société Impérial des naturalistes de Saint-Pétersbourg 27:74–80 [Russian with German summary].
Congenital – Two examples of *Rana esculenta* with eight toes at the hind legs and five toes on the front legs.
Chou LM. 1979. Shell deformity in a box turtle. Herpetological Review 10(2):56.
Shell – Shell deformity in box turtle *Cuora amboinensis*.
Christaller J. 1983. Vorkommen, Phänologie und Ökologie der Amphibien des Enzkreises. [Occurrence, phenology and ecology of amphibians of Eur County], Jahreshefte der Gesellschaft für Naturkunde Baden-Württembergs 138:153–182 [German].
Congenital – Abnormalities in *Bufo viridis* mentioned (1980 50%, 1981 1–2%).
Christiansen JL, Feltman H. 2000. A relationship between trematode metacercariae and bullfrog limb abnormalities. Journal of the Iowa Academy of Science 15:529–540.
Congenital – More trematode metacercariae infection in *Rana catesbeiana* (only one of six amphibian species to be afflicted) with abnormal limbs (Amelia, supernumerary, withered, shortened predominantly hind legs, or abnormal toes). Cites Sessions and Ruth (1990) of metacercariae at point of abnormal limb bifurcation in Pacific tree frog *Hyla regilla* and long-toed salamanders *Ambystoma macrodactylum*.
Infection – More trematode metacercariae infection in *Rana catesbeisana* (only one of six amphibian species to be afflicted) with abnormal limbs. Cites Sessions and Ruth (1990) of metacercariae at point of abnormal limb bifurcation in Pacific tree frog *Hyla regilla* and long-toed salamanders *Ambystoma macrodactylum*.
Christopher M, Berry KH, Henen BT, Nagy KA. 2003. Clinical disease and laboratory abnormalities in free-ranging desert tortoises in California (1990–1995). Journal Wildlife Diseases 39:35–56.
Trauma – Traumatic lesions in 42% of desert tortoise *Gopherus agassizii* from Colorado desert.
Shell disease – Shell lesions "consistent with cutaneous dyskeratoses" in 86% of desert tortoise *Gopherus agassizii* from Colorado desert.
Chuen-Im T, Areekijseree M, Chongthammakun S, Graham SV. 2010. Aerobic bacterial infections in captive juvenile green turtles (*Chelonia mydas*) and hawksbill turtles (*Eretmochdelys imbricate*) from Thailand. Chelonian Conservation and Biology 9:135–142.
Shell disease – Stomatitis and ulcerative shell disease in 2/17 green (*Chelonia mydas*) and hawksbill turtles (*Eretmochdelys imbricate*) in 2006 in Thailand, and in another table, ulcerative shell and traumatic ulcerative dermatitis in 29% in 2006 and 11%, in 2007. They attributed ulcerative damage in farmed turtles to biting, citing Glazebrook and Campbell (1990).
Cisneros M. 1997. Museum becomes ghoulish success. Press Enterprise (Riverside, CA) 28 February 1997:1.
Congenital – Parasitic twin turtle (with two extra legs).
Dicephalic turtle
Trauma – Sudan armor-plated lizard with two tails.
Cisneros JC, Cabral UG, de Beer F, Damiani R, Fortier DC. 2010. Spondarthritis in the Triassic. PLoS One 5:1–5 (e13425).
Arthritis – Early Triassic (Olenekian) basal archosaurian reptile BP-1–5796 at Bernard Price Institute for Palaeontological Research, Johannesburg, with fused vertebral bodies. Anulus fibrosus and nucleus pulposus were completely ossified. Neutron images suggested destructive changes in one of three vertebrae. Interpretation was spondyloarthropathy, but infection needs also be considered.

Vertebral – Early Triassic (Olenekian) basal archosaurian reptile BP-1–5796 at Bernard Price Institute for Palaeontological Research, Johannesburg, with fused vertebral bodies. Anulus fibrosus and nucleus pulposus were completely ossified. Neutron images suggested destructive changes in one of three vertebrae. Interpretation was spondyloarthropathy, but infection needs also be considered.

Fossil – Early Triassic (Olenekian) basal archosaurian reptile BP-1–5796 at Bernard Price Institute for Palaeontological Research, Johannesburg, with fused vertebral bodies. Anulus fibrosus and nucleus pulposus were completely ossified. Neutron images suggested destructive changes in one of three vertebrae. Interpretation was spondyloarthropathy, but infection needs also be considered.

Cisternas R. 1865. Polymélie dans un *Alytes obstetricans* Wagl. [Polymelia in a *Alytes obstretricans* Wagl]. Revue et Magasin de Zoologie (2) 17:287–288 [French].

Congenital – Supernumerary limb in *Alytes obstetricans.*

Clabaugh K, Haag KM, Hanley CS, Latimer IKS, Hernandez-Divers SJ. 2005. Undifferentiated sarcoma resolved by forelimb amputation and prosthesis in a radiated tortoise (*Geochelone radiata*). Journal of Zoo and Wildlife Medicine 36:117–120.

Other – Radiated tortoise *Geochelone radiata* with subluxed elbow, related to sarcoma.

Clark DR. 1971. The strategy of tail autotomy in the ground skink *Lygosoma laterale*. Journal of Experimental Zoology 176:295–302.

Trauma – Tail autotomy in ground skink *Lygosoma laterale, Eumeces septentrionalis, Eumeces brevilineatus, Eumeces laticeps, Eumeces fasciatus* (Judd 1955), *Eumeces skiltonianus* (Grant 1957), *Anolis carolinensis,* and *Sceloporus undulatus.*

Clark WH. 1973. A San Joaquin lizard with a bifid tail. HISS News-Journal 1(5):158.

Trauma – San Jaoquin fence lizard *Sceloporus occidentalis biseriatus* with bifid tail and traumatic amputation of right hindfoot fourth and fifth toes.

Cites tail duplication in *Plethodon* by Lynn (1950).

Clark DR Jr, Callison GL. 1967. Vertebral and scute anomalies in a racer, *Coluber constrictor.* Copeia 1967:862–864.

Congenital – Racer, *Coluber constrictor* KU 105460 with failure of left ventral portion of fifth lymphapophysis and postzygapophysis and prezygapophysis of sixth vertebrae. Left sides of vertebrae 11 and 12 are fused, with right-sided doubling of next vertebrae. Each extra half vertebra had a fused rib and hemal process.

Clarke J, Etches S. 1991. Predation among Jurassic marine reptiles. Proceedings of the Dorset Natural History and Archaeological Society 113:202–205.

Trauma – Three Jurassic plesiosaur propodials (SEKC.K92, SEKC.K81, and SEKC.220) with tooth marks and grooves. Holes were compatible with pliosaur attack from front. There were no signs of healing.

Fossil – Three Jurassic plesiosaur propodials (SEKC.K92, SEKC.K81, and SEKC.220) with tooth marks and grooves. Holes were compatible with pliosaur attack from front. There were no signs of healing.

Clark B, Tytle T. 1983. Life history notes: Serpents: *Python molurus bivittatus* (Burmese python). Morphology. Herpetological Review 14(4):121.

Congenital – Derodymous Burmese python *Python molurus bivittatus.*

Clark NJ, Gordos MA, Franklin CE. 2008. Diving behavior, aquatic respiration and blood respiratory properties: A comparison of hatchling and juvenile Australian Turtles. Journal of Zoology 275:399–406.

Vascular – Fresh water turtles respire through the skin; Fitzroy River turtle *Rheodytes leukops* (63–73%, which is approximately four times greater than other turtles) and have high oxygen affinity.

Cloaca bursa in Fitzroy River turtle *Rheodytes leukops* has more highly vascularized and multibranching cloacal papillae than *Emydura signata, Elseya latisternum,* and *Elusor macrurus.*

Ecology as the evolutionary driving force for reliance on aquatic respiration.

Claudio C. 1999. The Komodo dragon. Scientific American 280(3):84–91.

Trauma – Infected bites by conspecific *Varanus komodoensis* [buaja darat (land crocodile) or ora or biawak raksasa (giant monitor)].

Claus KF. 1876. Beiträge zur vergleichenden Osteologie der Vertebraten. I. Rippen und unteres Bogensystem. 2. Verschiebungen des Darmbeines und der Sacralregion der Wirbelsäule von Amphibien. [Contributions to the comparative osteology of vertebrates. I. Ribs and lower arch system. 2. Shifts of the iliac bone and the spine of amphibians sacral region]. Sitzungsberichte der mathematisch-naturwissenschaftlichen Klasse der Kaiserlichen Akademie der Wien 74:785–818 [German].

Congenital – Sacral variation in *Triton cristatus* with 16 left and 17 right sacral components and 14 right and 15 left sacral components in *Triton taeniatus*, 21 right and 22 left in *Cryptobranchus japonicus* and asymmetry at the 16–17 level in *Salamandra maculosa.*

Asymmetrical vertebrae in *Triton taeniatus.*

Dorsal-lumbar vertebral number variation, 14–17 in *Triton cristatus*, 12–14 in *Triton taeniatus*, 12–13 in *Triton helveticus*, 11–12 in *Triton noachicus*, and 20 left and 21 right in *Menopoma alleghaniense.*

Clayton L, Mylniczenko N, Greenwell M. 2003. Review of etiology and treatment options for nontraumatic deep shell lesions in freshwater turtles. Proceedings of the Association of Reptilian and Amphibian Veterinarians 2003:103–108.

Shell disease – Multiple pits in *Podocnemis unifilis* and *Emydura subglobosa* and isolated pit in *Clemmys guttata* with multichambered cavity beneath pits, with histologic evidence of necrosis, bacteria and fungi.

Clement H. 1967. Two heads better than one? International Turtle & Tortoise Society Journal 1(2):4–5.

Congenital – Dicephalic *Graptemys pseudogeographica* and *Pseudomys scripta elegans.*

Shell disease – *Graptemys pseudogeographic* with flared, recurved shell margins and three parallel keels in carapace.

Cligny A. 1897. Un cas de gémellité chez la couleuvre. [Twinning in the smooth snake]. Comtes Rendus de la Société Biologique Paris série 10, 4(49):630 [French].

Congenital – Couleuvre lisse (smooth snake) *Coronella austriaca* monster with arrested development. Although not stated, probably Siamese twin.

Clinton DW. 1814. An introductory discourse delivered before the Literary and Philosophical Society of New York, on the fourth of May, 1814. Note FF. Transactions of the New York Literary and Philosophical Society 1:160–162.

Congenital – Dicephalic *Coluber bicep.*

Pseudopathology – Mistaken *Amphisbaena* as having a tail on either end of the body in "Essay on the Natural History of Guiana," but erroneously lists Bancroft as the author.

Cloudsley-Thompson JL. 1970. On the biology of the Desert tortoise *Testudo sulcata* in Sudan. Journal of Zoology, London 160:17–33.

Congenital – Carapace asymmetries in *Testudo sulcata.*

Shell disease – Carapace asymmetries in *Testudo sulcata.*

Coates MI, Clack JA. 1990. Polydactyly in the earliest known tetrapod limbs. Nature 347:66–69.

Congenital – Polydactyly as the normal condition in earliest tetrapod limbs. Eight digits in forelimb of *Acanthostega*, seven in hind limb of *Ichthyostega*, both from Upper Devonian of East Greenland. Cites six digits in Russian *Tulerpeton*, by OA Lebedev 1984.

Fossil – Polydactyly as the normal condition in earliest tetrapod limbs. Eight digits in forelimb of *Acanthostega*, seven in hind limb of *Ichthyostega*, both from Upper Devonian of East Greenland. Cites six digits in Russian *Tulerpeton*, by OA Lebedev 1984.

Cocteau T. 1835. Sur un genre peu connu et imparfaitment décrit de Batraciens anoures à carapace dorsale osseuse, et sur une nouvelle espèce de ce genre. [On a little known and imperfectly described genera of batrachian anurans with dorsal osseous carapace, and a new species of this genus]. Magasin de Zoologie Classe 3, Pls. 7,8 (note this is a location designation for an article which contained no page numbers [French].

Congenital – *Bufo ephippium* with fracture or incomplete ossification and one with carapace (shell fusion). *Bufo ephippium* with three fingers on each foot, citing Spix (1824).*Ceratophrys boiei* with no trace of dorsal carapace (a shell allegedly present in this species) in Museum National de Histoire Naturalle de Paris.

Trauma – *Bufo ephippium* with fracture or incomplete ossification.

Cohen E. 1937. A two-headed snake. Bulletin of the Natural History Society of Maryland 8(2):9–10.

Congenital – Dicephalic garter snake *Thamnophis sirtalis sirtalis.*

Cohen H. 1986. *Terrepine carolina*: Twinning. Herpetological Review 17:25.

Congenital – Eastern box turtle *Terrepine carolina* twins, but not conjoined.

Cohen T. 1996a. Two heads are not always better than one; South African snake herald of bad fortune, superstitions say. The Ottawa Citizen, Ottawa, Canada 4 March 1996:A2.

Congenital – Dicephalic herald snake.

Cohen T. 1996b. Some for the two-headed snake. The Ottawa Citizen 5 March 2006:1.

Congenital – Dicephalic herald snake.

Cohen MM. Jr. 2001. Frog decline, frog malformations, and a comparison of frog and human health. American Journal of Medical Genetics 104(2):101–9.

Environmental – Malformations in frogs related to environmental pollution by *Ribeiroia*. Reduced pelvic elements were noted on 2–3% background hind limb genetic abnormalities. *Ribeiroia* (16–48 cercariae) exposure of *Hyla regilla* produced 11.8–25.5% hemimely and ectomely; 2.1% ectodactyly, at highest concentration. Also noted were nine 3–17.6% syndactyly, 2.9–6.9% polydactyly, 32.3–55.3% polymely.

Coker RE. 1905. Orthogenetic variation? Science New Series 22(574):873–875.

Shell disease – Disputes Gadow's 1905 numbers, as to whether frequency of scute abnormalities increase in size.

Coker RE. 1910. Diversity in scutes of chelonia. Journal of Morphology 21:1–75.

Congenital – Kyphosis in *Caretta caretta.*

Shell disease – Variation in number of costal plates in Atlantic Ocean *Caretta caretta*, where most have 5 costal shields on either side, with 7 of 130 showing 1 unit deviation. One Atlantic Ocean form had only three pairs of costals.

Collett J. 1893. Indianapolis News, July 19, 1893.
Congenital – Dicephalic garter snake and turtle (after Blatchley 1906).

Collins HH. 1932. Morphological anomalies in the amphibian, *Triturus viridescens*. Proceedings of the Pennsylvania Academy of Science 6:165–167.
Congenital – Deformities in vermilion-spotted newt *Triturus viridescens*:
Malformed hand, abnormalities include fused third and fourth digits, shortened second manual digit, blunt third digit, partial fusion of pedal digits 3 and 4, fusion of second and third manual digits, bifid third pedal digit, only first joints of digits 3 and 4, bifid third manual digit, forearm double size of other side, supernumerary manual digit, foot duplication, and shortened twisted foot.
Trauma – Partial tail amputation and regenerated tails.
Bifid tail in vermilion-spotted newt *Triturus viridescens.*

Colton HG. 1923. The anatomy of a five-legged frog. Anatomical Record 34:247.
Congenital – Supernumerary arm coming off a shortened coracoid in *Rana pipiens.*

Colwell GJ. 1992. *Microlophus bilineatus* (Galapagos lava lizard). Morphology. Herpetological Review 23:118.
Congenital – Tail bifurcation in *Microlophus bilineatus* (Galapagos lava lizard).

Congdon JD, Vitt LJ, King WW. 1974. Geckos adaptive significance and energetics of tail autotomy. Science 184:1379–1380.
Trauma – High frequency (74%) of regenerated tails in *Coleonyx variegatus.*
Regenerated tails in eublepharid geckos may be larger than original (Günther 1864).

Converse KA, Mattsson J, Eaton-Poole L. 2000. Field surveys of Midwestern and Northeastern Fish and Wildlife Service lands for the presence of abnormal frogs and toads. Journal of the Iowa Academy of Science 197:160–167.
Congenital – US Fish and Wildlife Service regions:

1. Minnesota, Wisconsin, Michigan, Iowa, Missouri, Illinois, Indiana, Ohio: External malformations were present in 110 of 6,632 frogs and toads from 61% of these sites. Rate was 0.4–9.2% at each of the sites. Toes and hind limb deformities represented 33% and 21%, respectively. The rest were front limb, foot, mandible, and chest deformities.

Species affected	Captured/malformed
Northern leopard frog *Rana pipiens*	3,598/73 – missing/deformed toes, missing/malformed hind limb, retained fused tail, supernumerary forelimb, missing or extra hindfoot.
Southern leopard frog *Rana utricularia*	969/18 – missing/deformed toes, missing hind/front limb.
Green frog *Rana clamitans melanota*	485/5 – missing/deformed toes, malformed hind limb, retained fused tail, supernumerary forelimb, missing or supernumerary hindfoot, deformed mandible.
Gray treefrog *Hyla versicolor*	174/9 – malformed/missing hind limb, missing/malformed frontlimb.
Wood frog *Rana sylvatica*	253/2 – missing hindfoot.
Dakota toad *Bufo hemiophrys*	225/1
Mink frog *Rana septentrionalis*	159/1
Bullfrog *Rana catesbeiana*	1/187 – malformed mandible.
Cricket frog *Acris crepitans*	153/0
American toad *Bufo americanus*	139/0
Fowler's toad *Bufo woodhousii fowleri*	147/0
Spring peeper *Pseudacris crucifer*	66/0
Chorus frog *Pseudacris triseriata*	45/0
Green tree frog *Hyla cinerea*	28/0
Plains leopard frog *Rana blairi*	4/0
Pickerel frog *Rana palustris*	0/0

2. New York, New Jersey, Pennsylvania, Maryland, Delaware, West Virginia and Virginia: External malformation in 58 of 2,267 frogs and toads, ranging from 0.9–9.9% at each site. Mink frogs *Rana septentrionalis* had especially high rates at 15.6%.

Species affected	Captured/malformed
Northern leopard frog *Rana pipiens*	404/18 – partial/missing/malformed hind limb, missing/fused toes, malformed foot, partial front limb, missing front foot.
Southern leopard frog *Rana utricularia*	107/0
Green frog *Rana clamitans melanota*	895/16 – partial/missing/malformed hind limb, missing toes, malformed foot, malformed lower mandible.
Gray treefrog *Hyla versicolor*	55/3 – partial/missing hind limb, partial front limb.
Wood frog *Rana sylvatica*	1/0
Dakota toad *Bufo hemiophrys*	0
Mink frog *Rana septentrionalis*	32/5 – extra toe, hind limb webbing.
Bullfrog *Rana catesbeiana*	370/8 – partial/missing/malformed hind limb, missing toes, malformed foot.
Cricket frog *Acris crepitans*	2/0
American toad *Bufo americanus*	98/3 – partial/missing hind limb, malformed hindfoot.
Fowler's toad *Bufo woodhousii fowleri*	44/0
Spring peeper *Pseudacris crucifer*	42/0
Chorus frog *Pseudacris triseriata*	5/0
Green tree frog *Hula cinerea*	
Plains leopard frog *Rana blairi*	
Pickerel frog *Rana palustris*	212/5 – missing hindfoot, malformed hind limb, supernumerary hind limb.

Cook AS. 1971. Selective predation by newts on frog tadpoles treated with DDT. Nature 229:275–276.

Toxicology – Hyperactivity induced by DDT increases predation of warty newts *Triturus cristatus* on *Rana temporaria* 400–500%.

Cook AS. 1972. The effects of DDT, dieldrin, and 2,4-D on amphibian spawn and tadpoles. Environmental Pollution 3:51–68.

Toxicology – DDT and Dieldrin-induced small snouts in *Rana temporaria.*

Contejean C. 1890. Sur l'autotomie chez la sauterelle et le lizard. [On autotomy in the grasshopper and the lizard]. Comptes Rendus Hebdomadaires des séances de l'Académie des Sciences Paris 61:611–614.

Trauma – Autotomy in lizard.

Cooper JE. 1958. Some albino reptiles and polydactylous frogs. Herpetologica 11:149.

Shell disease – Scrapings of *Emys orbicularis* revealed *Naricula, Seminuluma* algae, *Chlamydomones, Scenedesmus, Sigeolonium, Oscillatoria and Phormidin.*

Cooper JE. 1981. Investigation of abnormalities in captive-bred reptiles, with particular reference to the possible role of inbreeding. International Herpetological Congress Proceedings, Oxford, 1981:11–17.

Congenital – Cephalic abnormalities in slow-worms *Anguis fragilis* (Raynaud 1960).

Kinked tails in Round island or Telfair's skink *Leiolopisma telfairii.*

Environmental – Noted high temperature producing tail lesions in lizards (Bustard 1969), while cool temperatures produced malformed garter snakes (Fox et al. 1961).

Cooper JE. 1983. Investigation of abnormalities in captive-bred reptiles, with particular reference to the possible role of inbreeding. ASRA Journal 2(2):11–17.

Congenital – Dicephalic snake in Pathology museum of the Royal College of Surgeons of England (reference T13 B1).

Round Island skink *Leiolopisma telfairii* develop tail kinks and "vertebral malformations."

Cooper JE. 1984. Physical influences. In Hoff HL, Frye FL, Jacobson ER. Diseases of amphibians and reptiles. pp. 607–634; New York: Plenum.

Environmental – Temperature as a factor in malformations.

Cooper JE. 1992. Post-mortem examination. In: Manual of Reptiles, PH Beynon, MP Lawton, JE Cooper, eds, pp. 40–48. Cheltenham: BSAVA.

Metabolic – Nutritional osteodystrophy with calcium deposits in Rick Island gecko *Phelsuma guentheri.*

Cooper JE, Cooper ME. 2007. Introduction to Veterinary and Comparative Forensic Medicine. Ames, Iowa: Blackwell, 415 pp.

Congenital – Scoliosis in a skink.

Cooper JE, Jackson OF. 1981. Diseases of the Reptilia. Academic Press, London, New York, Tokio … Vol. 1: XI + 383 + XXXII pp.: Background and Infectious Diseases. Vol. 2: X + 385–584 + XXXII pp.: Non-infectious Diseases: Chapter 11: Traumatic and Physical Diseases by Frye, F. L. (387–407).

Trauma – Rostral abrasions, bites and miscellaneous lacerations, bites by cage mates, crashing injuries, damage to keratin shields by invertebrates.

Shell disease – Rostral abrasions, bites and miscellaneous lacerations, bites by cage mates, crushing injuries, damage to keratin shields by invertebrates.

Cooper JE, West CD. 1988. Radiological studies on endangered Mascarene fauna. Oryx 22(1):18–24.

Congenital – Scoliosis or kinking in Round island or Telfair's skink *Leiolopisma telfairi.*

Trauma – In 10/123 *Phelsuma guentheri.* Healed femoral fracture, regeneration of missing tail tip and missing digits.

Infection – Infection in 3/123 *Phelsuma guentheri.*

Metabolic – Poor mineralization, soft jaw in 5/11 Round Island gecko *Phelsuma guentheri.* Increased long bone diaphyseal diameter, growth plate distortion, and osteodystrophy (5%). The latter is reduced or eliminated by ultraviolet exposure. Nutritional/developmental abnormalities in 63/120.

Cooper JE, Arnold L, Henderson GM. 1982. A developmental abnormality in the Round Island skink *Leiolopisma telfairi.* Dodo, Journal of the Jersey Wildlife Preservation Trust 19:78–81.

Congenital – Round Island skink *Leiolopisma telfaii* with epignathia (35–71% with undershot jaw), 10–50% with thoracolumbar, 10–36% with pelvic, 90–100% with caudal vertebrae and 90–100% with tail abnormalities. Radiology revealed on "unusual degrees of curvature or deviation in the spine," but "no evidence of osteoporosis… or exostoses." Some with thoracolumbar or pelvic pathology dragged their hind limbs, suggesting neuropathic disease.

Cooper JE, Jackson OF, Harshbarger JC. 1983. A neurilemmal sarcoma in a tortoise (*Testudo hermanni*). Journal of Comparative Pathology 93:541.

Neoplasia – Neurilemmal sarcoma in *Testudo hermanni.*

Cooper WE. 2003. Shifted balance of risk and cost after autotomy affects use of cover, escape, activity, and foraging in the keeled earless lizard (*Holbrookia propinqua*). Behavioral Ecology and Sociobiology 54:179–187.

Trauma – Tail loss is associated with decreased running speed (Arnold 1984; Ballinger et al. 1979; Formanowicz et al. 1990; Pond 1981; Punzo 1982), reduced ability to escape (Dial and Fitzpatrick 1984), but in some species, increases speed (Brown et al. 1995; Daniels 1983). Tail loss may increase mortality rates (Congdon et al. 1974, Daniels 1984, Dial and Fitzpatrick 1984), decrease social status (Fox and Rostker 1982; Fox et al. 1990), reduce male access to females (Salvador et al. 1995), and reduce mating success (Martin and Salvador 1995). Autotomy in some snakes (Cooper and Alfieri 1993).

Tail loss in ground skink *Sincella lateralis* decreases activity, but allows predators closer (Formanowicz et al. 1990); in Iberian rock lizard *Lacerta monticola,* decreases activity and range; and in *Psammodromus algirus,* reduces range (Martin and Salvador 1995, 1997; Salvador et al. 1996). This contrasts with tailless female *Lacerta monticola,* which are more active during embryogenesis (Martin and Salvador 1995).

Tailless keeled earless lizard *Holbrookia propinqua* had reduced range, staying closer to cover, but predator approach distance was not altered. This contrasted with increased approach distance in *Sincella lateralis* (Formanowicz et al. 1990).

Cooper WE, Alfieri KJ. 1993. Caudal autotomy in the eastern garter snake, *Thamnophis s. sirtalis.* Amphibia-Reptilia 14:86–89.

Trauma – Eighteen percent tail breakage in Eastern garter snake *Thamnophis s. sirtalis,* and also found in *Nerodia s. sipedon, Thamnophis cyrtopsis* and *Elaphe obsoleta quadrivittata.*

Coquelet JP. 1983. Contribution à l'étude de la pathologie des tortues marines. Observations personnelles dans un élevage de *Chelonia mydas* (Linné) [Contribution to the study of marine turtle pathology. Personal observations on the raising of *Chelonia mydas* (Linné)]. Thèse Doctoral Médicine Vétérinaire, Nantes. 177 pp [French].

Congenital – Reported green turtle with second, atrophic head.

Infection – Cites Brogard's (1980) report of *Beneckea chitinivora* carapace osteomyelitis in *Chelonia mydas* and Chauvier's (1981) report of purulent polyarthritis in *Iguana iguana.*

Metabolic – Cites Glazebrook's (1980) report of nutritional osteodystrophy (but extremely rare in marine turtles), noting Keymer (1978) report of osteodystrophy in 8.3% of terrestial and 9.8% of aquatic turtles.

Avascular necrosis – Cites aseptic necrosis in proximal femur of green turtle (Balazs 1980).

Arthritis – Cites Ogden et al. (1981) report of arthritis in a marine turtle and in a lizard by *Serratia marcescens.*

Shell – Noted plaques in *Chelonia mydas* and cites Brogard (1980) report of carapace osteomyelitis by *Beneckea chitinivora.*

Cory CB. 1896. Hunting and Fishing in Florida. Boston: Estes & Lauriat, 304 pp.

Trauma – Alligator with broken jaw from conspecific fight.

Cosgrove GE. 1965. The radiosensitivity of snakes and box turtles. Radiation Research 25:706–712.

Environmental – Snakes are more sensitive to X-ray-induced teratogenesis (LD50 of 300–400 rads) than other vertebrates (1,000 rad). Turtles are intermediate (850 rad) to other reptiles. These exposures are still far above background or clinical exposures.

Cosgrove GE. 1971. Reptilian radiobiology. Journal of the American Veterinary Medical Association 159:1628–1684.

Environmental – Snakes LD^{50} was 300–400 rad.

Cites Pieau (1966) 780–2,400 rad produced malformed *Anguis.*

Cosgrove GE, Satterfield LC. 1982. Amyloidosis. In Hoff GL, Davis JW. Noninfectious Diseases of Wildlife. pp. 95–99. Ames: Iowa University Press.

Metabolic – Amyloidosis in *Constrictor constrictor*, Cook's or Amazon tree boa *Boa* (*Corallus*) *enhydris* and *Bufo peltocephalus*, without comment on osseous involvement.

Cott HB. 1961. Scientific results of an inquiry into the ecology and economic status of the Nile Crocodile (*Crocodylus niloticus*) in Uganda and Northern Rhodesia. Transactions of the Zoological Society of London 29:211–337.

Trauma – Fractures and amputations (especially tail) in 20% of Nile crocodile *Crocodylus niloticus* including loss of distal rami of jaws and broken or amputated limbs. 63% of male *Crocodylus niloticus* > 4 m in length had scars from prior injuries.

Synodontis pectoral or dorsal fin spines in palate of *Crocodylus niloticus.*

Infection – "Festering wound" in jaw of *Crocodylus niloticus.*

Cotton HS. 1922. The anatomy of a five legged frog. Anatomical Record 24:247–253.

Congenital – Supernumerary leg in *Rana pipiens*, citing literature of other cases in *Pelobates cultripes, Rana viridis, Rana esamelica, Rana temporaria, Rana esculenta, Bufo columbiensis, Rana palmipes, Rana halcicinum, Rana pipiens, Hyla aurea* by De Superville (1740), Gervais (1864), Lunel (1868), Mazza (1888), Sutton (1889), Bergendal (1889), Tornier (1898), Washburn (1899), Johnson (1901), Ergerman and Cox (1901), O'Donoghue (1910), and Wagner (1913), an imperfect extra pectoral girdle in a *Rana temporaria* (Hamilton 1933) and thoracic polymely in anurans.

Coulson RA, Hernandez T. 1970. Nitrogen metabolism and excretion in the living reptile. In JW Campbell, ed. Comparative Biochemistry of Nitrogen Metabolism. Volume 2. The Vertebrates. pp. 639–710; New York: Academic Press.

Metabolic – Tendency of crocodilians to develop gout because of uric acid synthesis, especially when over-eat in winter, the result of a high protein diet.

Cox PG. 1969. Some aspects of tail regeneration in the lizard *Anolis carolinensis*. I. A description based on histology and autoradiography. Journal of Experimental Zoology 171:127–150.

Trauma – Autotomy in *Anolis carolinensis.*

Cowan DF. 1968. Diseases of captive reptiles. Journal of the American Veterinary Medicine Association 153:848–859.

General – Pathology in captive reptiles: 92 chelonians, 208 saurians, 63 loricates, and 886 snakes, but no comment on bony involvement except as listed specifically.

Infection – *Endameba invadiens, Mycobacteria thamnopheos in garter snakes.*

Metabolic – Amyloidosis in *Constrictor constrictor* and Amazon tree boa *Boa* (*Corallus*) *enydris enydris.*

Neoplasms – lymphosarcomas, sarcomas, carcinomas, metastases, and adenomas (seven of ten were in snakes).

Vertebral – "Bone-periostitis" in *Elaphe taenura* resulting in vertebral exostoses and fusion.

Other – Fusiform swelling of long bones in *Iguana.*

Cowie AF. 1976. A Manual of the Care and Treatment of Children's and Exotic Pets. London: British Small Animal Veterinary Association, 116 pp.

Trauma – Tail fragility in lizards.

Metabolic – Nutritional osteodystrophy producing shell, maxillary, and mandibular deformities in *Chelonia.*

Crane SW, Curtis M, Jacobson ER, Webb A. 1980. Neutralization bone plating repair of a fractured humerus in an Aldabra tortoise. Journal of the American Veterinary Medical Association 177(9):945–948.

Trauma – Fractured humerus in an Aldabra tortoise.

Crawshaw GJ. 1989. Medical care of amphibians. Proceedings of the American Association of Zoo Veterinarians Meeting, Greensboro, North Carolina, October 14–19, 1989:155–169.

Congenital – High frequency of developmental defects in Puerto Rican crested toad *Peltophryne lemur.*

Metabolic – Scoliosis in Puerto Rican crested toad *Peltophryne lemur* from inadequate calcium supplementation.

Large percentage of amphibians dying at Toronto zoo had histologic evidence of renal disease.

Gout was noted in tips of digits of burrowing bullfrog *Pixicephalus*, similar to Frye's 1989 report in *P. delalandii.*

Vertebral – Epidemic of posterior paralysis with lower motor neuron dysfunction in Puerto Rican crested toad *Peltophryne lemur* producing collapsing vertebrae, as also noted in juvenile *Ceratophrys*. It responded to amino acid and vitamin B (but not calcium) supplementation.

Crawshaw GJ. 1993. Amphibian medicine. In Fowler ME ed. Zoo and wild animal medicine. Pp. 131–139; Philadelphia: Saunders.

Congenital – Dendrobatid spindly leg syndrome producing rigid forelimb.

Metabolic – Gout in African bullfrog *Pixicephalus*.

Cressey D. 2007. Invading cane toads suffer from arthritis. Nature doi:10:1038/news.2007.165

Neuropathic – Ten percent of *Bufo marinus* invading Australia have a "spinal arthritis" manifest as bony growths fusing joints between vertebrae, with cultures positive for *Ochrobactrum anthropi*. Cited from Brown et al. (2007).

Crosswhite E, Wyman M. 1920. Journal of Entomology and Zoology (Pomona College Claremont, California) 12:78. (Note that Wyman should probably be listed as first author, although cited as here by some authors).

Congenital – Figure and notes on abnormal toad, evidently *Bufo boreas halophilus* with fused leg bones and two normal feet derived from posterior sacral region.

Crouch CR. 1969. Odd Fellow. Wildlife in Australia 1969(6):101.

Congenital – Marbled gecko *Phyllodactylus marmoratus* with three tails.

Crowe S. 2000. Twinning in green tree pythons (*Morelia viridis*)? Notes from NOAH 27 (10):12.

Congenital – Twin green tree python *Morelia viridis* in egg. One was "deformed."

Crumly CR, Sánchez-Villagra MR. 2004. Patterns of variation in the phalangeal formula of land tortoises (Testudinidae): Developmental constraint, size, and phylogenetic history. Journal of Experimental Zoology (Molecular Development and Evolution) 302B:134–146.

Congenital – Phalangeal loss in testudininds occurs phylogenetically starting with first and then one at a time.

Crump ML. 1992. Cannibalism in amphibians. In: Cannibalism: Ecology and Evolution Among Diverse Taxa, MA Elgar, BJ Crespi, eds. New York: Oxford University Press, pp. 256–276.

Trauma – Disproportionately large head, wide mouth, slit-like eyes, elongate vomerine teeth, and slender body form characterize members of *Ambystoma tigrinum* (*A. t. mavortium, A. t. nebulosum, A. t. tigrinum*) that are cannibalistic, citing Rose and Armentrout (1976).

Cuadrado M. 1996. Tasa de polidactilia en el camaleón común (*Chamaeleo chamaeleo*) [Rate of polydactyly in the common chameleon (*Chamaeleo chamaeleo*)]. Boletín de la Asociación Herpetológica Española 7:23–24 [Spanish].

Congenital – Common chameleon *Chamaeleo chamaeleo* with six fingers in its left hind limb, observed in the gardens of the Real Observatorio de Marina in San Fernando (Cádiz, Spain), in 1995 and reported personal communication from Juan Carlos Neva of the same observation in Puerto de Santa María (Cádiz, Spain). Also reported that Octavio Paulo found variable incidences of polydactyly in diverse localities in Portugal.

Cullen LJ, Harshbarger JC. 1989. Dose dependent teratogenesis in spotted salamander (*Ambistoma maculatum*) embryos exposed to diethylnitrosamine. Herpetopathologia 1(1):41–48.

Toxicology – Reports no osseous alterations in spotted salamander *Ambystoma maculatum*, but cites *N*-methyl-*N'*-nitro-*N*-nitrosoguanine-induced "abnormalities of bony structures" in *Triturus* (*Cynops*) *pyrrhogaster* by Tsonis and Eguchi, 1981.

Cummins CP. 1987. Factors influencing the occurrence of limb deformities in common frog larvae raised at low pH. Annales de la Société Royale Zoologique Belgique 117 Suppl. 1:353–364 [French].

Metabolic – Common frog larvae raised at pH 4 have high frequencies of deformities.

Cummins CP. 1989. Interaction between the effect of pH and density on growth and development in *Rana temporaria* L. tadpoles. Functional Ecology FECOE5 3:45–52.

Metabolic – *Rana temporaria* limb deformities at low pH.

Cummins CP, Greenslade PD, Mcleod AR. 1999. A test of the effect of supplemental UV-B radiation on the common frog, *Rana temporaria* L., during embryonic development. Global Change Biology 5:471–479.

Environmental – UVB light did not affect malformation rate in *Rana temporaria*.

Cunningham B. 1927. Two-Headed Snakes. The Scientific Monthly 25(6):559–561.

Congenital – Dicephalic spreading adder.

"Ribs from one vertebra fusing with those of vertebra lying opposite" in *Amphisbaena alba* and *Pituophis sayi*.

Cunningham B. 1934. Axial bifurcation in reptiles. Journal Elisha Mitchell Science Society 50(1–2):36–38.

Congenital – Axial bifurcation was first described by Aristotle, also Pliny and Aelianus. Figured by Liceto (1643). Strohl (1925) presents a good bibliography and reports a snake with a single head by two vertebral columns. Heasman (1933) adds another bicephalic case. No citation list is provided.

The authenticity of the three-headed serpent of Cube 1536 is questioned.

Pseudopathology – The authenicity of the three-headed serpent of Cube 1536 is questioned.

Cunningham B. 1937. Axial Bifurcation in Serpents. An Historical Survey of Serpent Monsters Having Part of the Axial Skeleton Duplicated. 118 pp.; Durham, North Carolina: Duke University Press.

Congenital – Describes cephalic (head only), anterior (cephalo), posterior (body or tail), and amphi (head and tail, but not central portion of body) duplication, disallowing many of the early reports. Almost all are anterior, and well illustrated in this text:

Genus	Common name	Number
Agkistrodon blomhoffii[a]		1
mokasen	Copperhead	3
Bascanium constrictor	Black	4
Bothrops atrox	Fer-de-lance	3
Coluber constrictor		1
florulentus		2
natrix		5
Crotalus terrificus		1
Elaphe longissima		1
obsoleta		1
vulpina		1
Epicrates angulifer	Maja	2
Eutainia sirtalis		2
Heterodon platirhinus	Hognose	7
simus		1
Homalopsis buccata	Fish snake	1
schneiderlli		1
Hydrophis sirtalis		1
sublaevis		1
Lampropeltis calligaster		1
triangulum	Milk snake	2
Leptodira hotamboela		1
Liophis almadensis		1
Lycodon aulicus		2
Naja tripudians		1
Natrix fasciatus		2
natrix		1
sipedon		5
Notechis scutatus	Tiger snake	1
Opheodrys aestivus		1
Ophibolus	King	1
doliatus		1
getulus	common king	3
Pituophis		1
catenifer		2
sayi	bull	1
Psammophis		1
Regina leberis		1
Storeria dekayi	Dekay's snake	1
occipitomaculata		1
Thamnophis elegans		1
ordinoides		1
radix	plains garter	1
marcianus	spotted garter	2
sirtalis	garter	3
Tropidonotus fasciatis		7
natrix		1
quincunciatus		1
sipedon		1
tigrinus		1
Viper aspis		2
Zamenis florulentus – see *Coluber*		
	Common milk	1
	Common viper	2
	Timber rattler	1

[a] With double vertebral column for 40 vertebrae.

Posterior dichotomous lesions were seen only in a *Vipera berus* and amphi-dichotomous only in a *Bascanium constrictor, Natrix sipedon*, and a *coluber*.

Cunningham JD. 1955. Notes on abnormal *Rana aurora draytonii*. Herpetologica 11:149.
Congenital – One amelia and 1 supernumerary hind limb in *Rana aurora draytonii*.
Curran CH, Kauffeld C. 1937. Snakes and their ways. Harper & Brothers, New York, 285 pp.
Congenital – Derodymous milk snake and *Natrix sipedon* skull joined at braincase.
Currie PJ, Jacobsen AR. 1995. An azhdarchid pterosaur eaten by a velociraptorine theropod. Canadian Journal of Earth Science 32:922–925.
Trauma – Azhdarchid pterosaur bone from Upper Cretaceous strata of Dinosaur Provincial Park with theropod *Saurornitholestes langstoni* with tooth marks and imbedded broken tooth tip.
Fossil – Azhdarchid pterosaur bone from Upper Cretaceous strata of Dinosaur Provincial Park with theropod *Saurornitholestes langstoni* with tooth marks and imbedded broken tooth tip.
Curry-Lindahl K. 1963. Dicephalism in the adder (*Vipera berus*). British Journal of Herpetology 3(4):81.
Congenital – Two dicephalic and two derodymous adder (*Vipera berus*), one previously published in Swedish in Fauna och Flora 2 (Lönnberg 1907) and cited in previous reports by Dorner (1873) and Borgert (1896).
Curtis L. 1950. A case of twin hatching in the rough green snake. Copeia 1950(3):232.
Congenital – Twin *Opheodrys aestivus*, each with one eye.
Cuthbertson R. 2003. Morphological anomalies in a mosasaur, *Platecarpus*, from Manitoba. Journal of Vertebrate Paleontology 23 (supplement to 3):44A.
Congenital – Presence of extremely large skulls suggests possibility of progressive increase in body size of mosasaurs.
Fossil – Presence of extremely large skulls suggests possibility of progressive increase in body size of mosasaurs.
Cutler M. 1868. Extracts from an unpublished journal. Hours at Home 7(5):455–465.
Congenital – Reports Ben Franklin's dicephalic snake.
Cutler WP, Cutler JP. 1888. Life, journals and correspondence of Rev. Manasseh Cutler, LL.D. by his grandchildren. Cincinnati: Robert Clarke & Co., Vol. 1:524 pp.
Congenital – Reported Benjamin Franklin showing him a dicephalic dark brown-black snake with red/white checkered belly that had been found at the confluence of the Schuylkill and Delaware Rivers and a similar finding near Lake Champlain.
Cuvier G. 1812. Sur les ossemens fossiles de Crocodiles, et particulierement sur ceux des environs du Havre et de Honfleur, avec des remarques sur le squelettes de Sauriens de la Thuringe. [On fossil crocodile bones, particularly those from Haver and Honfleur, with remarks on saurina skeletons of Thringe (central Germany)]. Recherches sur les Ossemens Fossiles de Quadrupedes. Paris. Vol IV, part V, no III [French].
Congenital – Crocodile image in Fig. 10 appears to be congenital fusion, whereas Fig. 8 could be simple adherence of adjacent bones.
Vertebral – Crocodiles fossils on Pl I and Pl II. Figure 8 in Plate I and Fig. 10 in Plate II appear to represent vertebral fusion of atlas and axis.
Fossil – Crocodiles fossils on Pl I and Pl II. Figure 8 in Plate I and Fig. 10 in Plate II appear to represent vertebral fusion of atlas and axis.
Cuvier G. 1829. Le Règne Animal: Distribué d'après son Organisation pour servir de base a l'Histoire Naturelle des Animaux et Introduction a l'Anatomie Comparée. The Animal Kingdom: Distribution after its Organization as a Basis for Natural History of Animals and Introduction to Comparative Anatomy]. Paris: Déterville, 406 pp. [French].
Trauma – Tail damage in *Lacerta caeruleocephala*.
Cuvier M. 1830. Suite de l'extrait du compte-rendu des travaux de l'académie des sciences, pendant l'année 1829. [Extracts of report of work of the academy of sciences, during the year 1829]. Transactions Médicales. 1:409–419 [French].
Congenital – Related to the atlodimous (form of derodymous) viper of Dutrochet.

Annotated Bibliography D-G

Da Cunha OR. (Actually Rodrigues O). 1968. Um teratódimo deródimo em Jibóia (*Constrictor constrictor constrictor*, Linn., 1766) (Ophidia, Boidae) [A teratodymous derodymous of Jibóia (*Constrictor constrictor constrictor*, Linnaeus, 1766) (Ophidia, Boidae)]. Boletim do Museo Parense Emílio Goeldi; Nova Série, Zoologica 67:1–17 [Portuguese].

Congenital – Derodymous *Constrictor constrictor*. Left head could not digest food because it led to an intestine that had no excretory exit, and all organs in that side of the body were reduced or malformed. Osvaldo Rodrigues Da Cunha also provided a summary of reptile dicephalic studies in Brazil, and discussed embryogenic interpretations. He suggested the fusion of two simultaneously developing embryonic areas in the same egg as a cause of this teratology, as previously suggested by Vanzolini (1947) for a similar case.

In his summary of reptile dicephalic studies in Brazil, Osvaldo Rodrigues Da Cunha stated accounts of teratology in Brazilian reptiles were quite rare, with snakes appearing most often in the record. Rarer were instances in lizards and turtles, especially regarding cases of dicephaly. He referred the reader to Rosa (1966) for a review of cases in lizards, and stated that he was not aware of any cases reported in turtles in the Brazilian literature. However, he also reported an account from a credible person (which he did not name) of a small dicephalic turtle having lived in two aquarium tanks from 1910 to 1912 in the Museum Emilio Goeldi of Para, Brazil. According to him, this turtle was probably *Geoemyda punctularia punctularia* (Daudin 1802), but he was unable to find any records in the museum that could confirm the veracity of this account.

Rodrigues Da Cunha also wrote that the majority of cases of dicephaly were reported from snakes, and that cases had been reported sporadically in all of the Americas. The first reports for Brazil were by Do Amaral (1927), Prado (1942, 1943, 1945, 1946), and Vanzolini (1947). Pereira (1944 and 1950), Lema (1957, 1958, 1961), and Belluomini (1957/8) were also reviewed. Belluomini and Lancini (1960) were reported to have presented a case for Venezuela.

Dahme E, Reinacher M, Knochen S. 1981. Calcium regulation. In Sub-mammalian vertebrates, CG Packe (ed.). London: Academic Press, pp. 147–155.

Metabolic – Female Philippine house lizard *Cosmybotus platyurus* has calcareous material filling endolymphatic sacs in neck.

D'Alton, E. 1853. De monstris, quibus extremitates superfluae suspensae sunt [Of monsters whose superflulous extremities are suspended]. Halis: Formis Ploetzionis, pp. 1–72 [Latin].

Congenital – Supernumerary limb was composed of three sections: the first attached to the left humerus, the second was at an acute angle with the first, and had a convex curvature that touched the body. The third part was an almost normal hand, but the second digit was very short. It is unclear why Taruffi (1886) attributed findings to a *Rana esculenta*, as no obvious mention of species in text.

Cites Saint-Hiliare melomelia (1836) and cephalomelia (in 1829 dissertation) and Tiedemann (1831) reporting another example of monster.

Dämmrich K. 1967. Pathomorphologische Befunde am Panzer der Schildkröten. [Pathomorphological findings in the carapace of tortoises.] IX Internationales Symposium zu Erkrankungen der Zootiere, Prag: 271–274 [German].

Congenital – Kyphosis in turtles.

Trauma – Fractures with callus formation in turtles.

Infection – Necrosis caused by fungal infection in turtles.

Multiple sequestra in Caspic water turtle *Clemmys caspica caspica*.

Metabolic-Rickets in Mauric water turtle *Clemmys leprosa*, osteodystrophy fibrosa generalisata, and osteoporosis in Greek tortoise *Testudo hermanni*.

DOI 10.1007/978-1-4614-0824-6_6,

Dämmrich K. 1979. Zur Pathogenese der Skeletterkrankungen bei Zootieren. [To the pathogenesis of the sickness of the skeletons of zoo animals]. Verhandlungsbericht des Internationalen Symposiums über die Erkrankungen der Zootiere 21:65–71 [German].

Metabolic – Osteoporosis, osteodystrophia fibrosa, and rickets in tortoises.

Dämmrich K. 1985. Bewegungsorgane. Knochen, Gelenke und Skelett. [Organs of movement. Bones, articularions, and skeleton]. In: Ippen R., Zwart P., Schröder, HD. (eds.) 1985. Handbuch der Zootierkrankheiten. [Handbook of sickness in zoo animals] Band 1. Reptilien:215–240 [German, well illustrated].

Trauma – of leg bones, lower jaws, and ribs in lizards and crocodiles, broken ribs and fractures of vertebral column in snakes, fractures of the carapace of turtles.

Infection – Osteomyelitis of the jaw and skull bones.

Metabolic – Gout and pseudogout in lizards, tortoises, and snakes.

Rickets, osteodystrophia fibrosa generalisata in tortoises and turtles, and lizards.

Arthritis – Inflammatory, ankylosis.

Vertebral – Spondylopathia deformans, spondyloarthropathia deformans.

Shell disease – Necrosis of turtle carapace.

Other – Periostitis ossificans, panostitis.

Daneri GA. 1991. Teratología en *Dermatonotus muelleri* (Boettger 1885) (Microhylidae) [Teratology in *Dermatonotus muelleri* (Boettger, 1885)(Microhylidae)]. Boletín de la Asociación Herpetológica Argentina 6(1):6. [Spanish]

Congenital – *Dermatonotus muelleri* frog (Herpetological Division of the Argentine Museum of Natural Sciences "Bernardino Rivadavia" #33253) with agenesis of tibia-fibula, tarsals, and metatarsals. Two "pieces" were apparently ossified at the level of the two "digits," and the right femur markedly widened distally.

Daniel H. 1941. Una serpiente bicéfala [A dicephalous serpent]. Revista Facultad Nacional de Agronomía (Colombia) 3(12):1182–1185 [Spanish].

Congenital – Teratological studies in animals and plants in South America, particularly Chile, Bolivia and Brazil. Dicephalic *Bothrops atrox* snake, joined at maxillae.

Daniel H. 1955. Algunos aspectos de la lucha biológica. IV. Los reptiles y la agricultura. Una serpiente bicéfala [Some aspects of the biological fight. IV. Reptiles and agriculture. A bicephalous serpent]. Revista de la Facultad Nacional de Agronomía (Medellín, Colombia) 17(48):48–52 [Spanish].

Congenital – Dicephalic *Bothrops atrox* snake, reported in Daniel's 1941 paper, preserved in the museum of the school of San José. Photograph included.

Daniels CB. 1983. Running: An escape strategy enhanced by autotomy. Herpetologica 39:162–165.

Trauma – *Phyllodactylus marmoratus* increases running speed with tail (which had dragged on the ground) loss.

Danielyan FD. 1970. Embryonic abnormalities in rock lizards of Armenia. Zoologiski Zhurnal 49(7):1064–1068.

Congenital – 3–6.8% of parthenogenetic and 1.1–1.8% of bisexual Armenian rock lizards had embryonic abnormalities including body and tail duplication, amelia, polydactyly, and failure of maxilla development.

Danielyan FD. 1987. [Artificial hybridization of two bisexual species of the rock lizards in natural conditions.] Trudy Zoologicheskogo Instituta, Akademija Nauk SSSR 158:179–183, 9–10 [Russian].

Congenital – Eighteen percent of *Lacerta mixta x valentini* reported as "ungliness" without photos or further description.

Danini ES. 1946. Histological processes as observed in the regeneration of the carpace of the tortoise *Emys orbicularis*. L. Bull Acad Sci USSR Biol 5:581–594.

Trauma – Carapace regeneration after trauma in European pond turtle *Emys orbicularis*.

Dareste C. 1874. Sur l'origine et la mode de formation des monstres doubles. [On the origin of a mode of formation of double monsters]. Archives de Zoologie Expérimental Paris 3:74–118 [French].

Congenital – Body duplication in a salamander.

Dareste C. 1877. Recherches sur la production artificielle des monstrosités, ou Essai de tératogénie expérimentale. [Research on the artificial production of monsters. Or essay on experimental teratology] 320 pp.; Paris: Reinwald & Cie. [French].

Teratology – Experimental teratology: predominantly theory and avian.

Dareste C. 1891. Recherches sur la production artificielle des monstrosités, ou Essai de tératogénie expérimentale. [Research on the artificial production of monsters. Or essay on experimental teratology] XVI+587pp.; Paris:Reinwald & Cie. [French].

Teratology – Experimental teratology: predominantly theory and avian.

Darevsky IS. 1960. Urodstva pri parthenogeneticheskom razmnojenii skalnykh iasherits *Lacerta saxicola* Eversmann. [Monsters of the rock-lizard *Lacerta saxicola* Eversmann which develop parthenogenetically.] Doklady Akademija Nauk SSSR 132(1):234–237 [Russian].

Congenital – Partial dicephaly, supernumerary limbs, polymelia, ectomelia and teratopagus in rock lizard *Lacerta saxicola* developing parthenogenetically.

Darevsky IS. 1966. Natural parthenogenesis in a polymorphic group of Caucasian rock lizards related to *Lacerta saxicola* Eversmann. Journal of the Ohio Herpetological Society 5:115–152.

Congenital – Frequency of "monstrosities" in Caucasian rock lizards related to *Lacerta saxicola* includes:

	# Curvature	Jaw	Bicephalic	Absent lower jaw	Disproportionate trunk
L. armeniaca	276	4	8	6	0
L. dahli	111	1	0	0	7
L. rostombekovi	98	1	0	0	4
L. saxicola defilippii	152	2	0	0	0
L. saxicola valentini	63	1	0	0	0

Darevsky IS, Kupiyanova LA, Uzzell TM. 1985. Parthenogenesis in reptiles. pp. 412–526. in C Gans, F. Billett (eds.) Biology of the Reptilia. 15B. John Wiley and Sons, New York.

Congenital – Partial dicephaly and partial body duplication and absent lower jaw in parthenogenetic *Lacerta armeniaca*, tail duplication, and abnormal head in *Lacerta dahli*. Frequency of "monsters" in pathogenic species *Lacerta armeniaca* (5%), *dahlia* (8%), *rostombekovi* (5%), *unisexualis* (3%), contrasted with bisexual *Lacerta radii nairensis* (1%) and *valentine* (2%) (Danielyan 1970; Darevsky 1966).

Darewski IS, Kulikowa WN. 1961. Natürliche Parthenogenese in der polymorphen Gruppe der kaukasischen Feldseidechse (*Lacerta saxicola* Eversmann). [Natural parthenogenesis in the polymorphic group of the kaukasian rock lizard (*Lacerta saxicola* Eversmann)]. Zoologische Jahrbücher (Systematik) 89:119–176 [German].

Congenital – Embryos and young Kaukasian rock lizard with shortening and arching of lower jaw, partial duplication of head, duplication of trunk (fused in anterior body and head) and loss of lower jaw, which were more common in parthenogenetic forms than in bisexual reproducing forms.

Das GM. 1932. Observations on the trifid tails in two specimens of *Hemidactylus flaviridis*, Rüppel, with a note on the artificial regeneration of double and triple tails of the "Tokhak" lizard, *Gekko verticillatus*, Laurenti. Journal of the Bombay Natural History Society 35(3):657–662.

Trauma – Trifid tail in *Hemidactylus flaviridis* and produced in laboratory by injuring *Gekko verticillatus*.

Das P, Mohanty-Hejmadi P. 1999. Histological effects of vitamin A on the tail-amputated tadpoles of *Polypedates maculatus* with special reference to homeotic transformation. Cell Tissues Organs 164:90–101.

Toxicology – Vitamin A-induced anteroposterior, proximodistal and dorsoventral duplications and single, paired and supernumerary limbs in *Polypedates maculatus*.

Das P, Mohanty-Hejmadi P. 2000. Vitamin A mediated limb deformities in the common Indian toad, *Bufo melanostictus*. Indian Journal of Experimental Biology 38:258–264.

Toxicology – Ectomely, meromely, phocomely, and polymely from excess vitamin A in *Bufo melanostictus*.

Dasgupta S, Grewal MS. 1970. Inheritance of vertebral fusion in the skipper frog. The journal of heredity 61:174–176.

Vertebral – Vertebral fusions in 8–12% of skipper frog *Rana cyanophylictis* as a dominant trait in vicinity of Delhi.

Dathe H. 1959. Erstaunliche Lebensfähigkeit einer Riesenschlange. [Surprising survival of gigant snake]. Bijdragen tot de Dierkunde 29:71–72 [German].

Trauma – Radiographs of reticulated or regal *Python reticulatus* with broken and healed ribs (callus formation) and broken vertebral column (final cause of death).

Dathe H. 1960. Schwanz-regeneration beim Brillenkaiman. [Tail regeneration in Yacare caiman]. Natur und Volk 90:289–292 [German].

Trauma – Tail regeneration in *Caiman crocodilus* and *Alligator mississipiensis*, citing similar regeneration in *Melanosuchus niger* (Kälin 1937).

Fossil – Tail regeneration in *Melanosuchus niger* (Kälin 1937).

David Th. 1976. Mikrochirurgie in der Kleintierpraxis: Osteosynthese des Humerus beim Leguan. [Micro-surgery in the practice of small animals: Osteosynthesis in lizard humerus]. Tierärztliche Praxis 4:543–546 [German].

Trauma – Procedures to repair fractured distal epiphysis of right humerus in lizard.

Davies M. 1973. A double-headed grass snake, *Natrix natrix helveticus*, found at Ladock, Truro, Cornwall. Journal of the Cambrone-Redruth Natural History Society 2(5):1–2.

Congenital – Dicephalic grass snake *Natrix natrix helveticus* and adder *Vipera berus*.

Davies M. 1974. A double-headed grass snake, *Natrix natrix helveticus*, found at Ladock, Truro, Cornwall. British Journal of Herpetology 5:452–453.

Congenital – Dicephalic grass snake *Natrix natrix helveticus* and adder *Vipera* berus.

Davies M. 1976. A complete record of incidents of dicephalism occurring in the British snakes. British Herpetological Society, Newsletter 15:18–19.

Congenital – Dicephalic grass snake *Natrix natrix* (Doe 1915) and Summarizes literature of four cases of *Natrix natrix* and one of *Vipera berus*.

Davies B. 1992. More on spindly leg… British Dendrobatid group Newsletter #15 as reprinted in American Dendrobatid Society Newsletter 12:3–4.

Metabolic – Vitamin "dusting" prevents spindly leg in *Hyla arborea* and *Epipedobates tricolor.*

Davila PF. 1757. Cataloque systématique et raisonné des curiosités de la nature et de l'art, qui composent le cabinet de M. Davila. [Systematic and rational catalogue of natural curiosities and art composing the collect of M. Davila]. [57pp.; Paris: Briasson. 571 pp. [French]

Congenital – Dicephalic serpent.

Davis KR, Schultz TW, Dumont J. 1981. Toxic and teratogenic effects of selected aromatic amines on embryos of the amphibian *Xenopus laevis*. Archives of Environmental Contamination and Toxicology 10:371–391.

Toxicology – Aromatic amines induce abnormalities in *Xenopus laevis*, but no description. One image revealed a bent tail.

Dawe CJ, Banfield WG, Small JD, Woronecki DE. 1981. Chondrosarcoma of a corn snake and nephroblastoma of a rainbow trout in cell culture. In RJ Montali, G Migaki, eds. The Comparative Pathology of Zoo Animals. Washington DC: Smithsonian Institution Press; pp. 603–612.

Neoplasia – Chondrosarcoma in corn snake *Elaphe guttata* vertebrae.

Dawbin WH. 1962. The tuatara in its natural habitat. – Endeavour 21(81):16–24.

Trauma – Forked tail regeneration in tuatara.

Tail bifurcation in *Agama tuberculata*, citing Chandra and Mukherjee

Dawson AB. 1932. A ventral accessory tail in *Triturus viridescens* and its duplication experimentally. The Anatomical Record 52:139–149.

Congenital – Accessory (extra) tail in *Triturus viridescens.*

Dawydoff K. 1898. Zur Frage über die Autotomie der Eidechsen. [On the question of autotomy in reptiles]. Trudy Imperatorskago S. Petersbourg [Travaux de la Societé Imperiale des Naturalistes de St. Pétersbourg] 29:325–327 [German].

Trauma – Autotomy in gecko occurs during the day, rarely at night.

De Albuquerque NR, Arruda WS, Costa AS, Galharte RC, Vargas LG, Moreno IH. 2010. A dicephalic yellow anaconda snake, *Eunectes notaeus* (Serpentes: Boidae), from Southern Pantanal, Brazil. Journal of Natural History 44:31/32:1989–1994.

Congenital – Derodymous yellow anaconda snake *Eunectes notaeus* Coleção Zoológica de Referência of the Campus do Pantanal/UFMS (CEUCH 6024). This is 28th Boidae dicephalism, 2nd from South America, and 28th in Brazil.

Other dicephalic Boidae include *Acrantophis dumerilii, Boa constrictor, Epicrates angulifer, Epicrates cenchria, Epicrates maurus, Epicrates striatus, Lichanura roseofusca*, and *Gongylophis conicus* (Cunha 1968; Hoser and Harris 2005; Smith and Pérez-Higareda 1988; Vinegar 1973; Wallach 2007).

Cites dicephalism in lizards (Gekkonidae, Scincidae, and Lacertidae), snakes (Leptotyphlopidae, Pythonidae, Boideae, Tropidophiidae, Viperidae, Elapidae, Hydrophiidae, and Colubridae) (Mcallister and Wallach 2006; Spadola and Insacco 2009) and that 1,087 cases have been found to date, with 171 species in 95 genera.

Dean JN, Glenn JL, Straight RC. 1980. Bilateral cleft labia and palate in the progeny of a *Crotalus viridis viridis* Rafinesque. Herpetological Review 11:91–92.

Congenital – Bilateral cleft palate in prairie rattlesnake *Crotalus viridis viridis.*

De Andrade DV, Abe AB. Or AS. 1993. Malformações em ninhadas de caiçaca, *Bothrops moojeni* (Serpentes: Viviperidae) [Malformations in hatchlings of caiçaca, *Bothrops moojeni* (Serpentes: Viviperidae)]. Memórias do Instituto Butantan 54(1992):61–67 [Portuguese].

Congenital – Kinks and dicephalism in 7.7% of *Bothrops moojeni.*

Dearlove GE, Dresden MH. 1976. Regenerative abnormalities in *Notophthalmus viridescens* induced by repeated amputations. Journal of Experimental Zoology 196:251–261.

Trauma – Repeated amputations in *Notophthalmus viridescens*, normal at first, become increasing abnormal, leaving a simple regenerative spike in 14%.

Regeneration is affected by temperature (27% at 25°C and 33% at 30°C) (Schmidt 1968; Schaube and Nentwig 1974), body size and age (Goodwin 1946; Manner et al. 1960; Pritchett and Dent 1972; Zamaraev 1974), season of year (Schauble 1972), starvation (Twitty and Delanney 1939), and level of amputation (Iten and Bryant 1973; Smith et al. 1974).

De Betta E. 1857. Erpetologia delle provincie Venete e del Tirolo meridionale [Herpetology of the Venetian provinces and south Tyrol]. Atti dall' Accademia di Agricoltura, Arti e Commercio di Verona 35:XVI + 365 pp. [Italian]

Congenital – Monstrosities and conjoined twin lizards occur less commonly than multiple-tailed lizards, citing dicephalic snake reports by Aldrovandi, Lanzoni, and Redi, and stating similar cases were later reported on lizards. Duméril preserved a derodymous lizard in the museum of Paris, and another was presented at the 1831 Science Academy of Paris by Beltrami and Rigol.

Trauma – Lizards with two or three tails are frequently encountered, and according to Bonaparte (1832–41), some have been observed to have four or seven tails (all of which are internally cartilaginous).

De Betta E. 1863. Materiali per una Fauna Veronese [Materials for a fauna of Verona]. Atti e memorie della Accademia di agricoltura, scienze e lettere di Verona 42:92–113 [Italian].

Congenital – Dicephalic viper from Vicenza province, preserved in De Betta's collection of European reptiles.

De Betta E. 1864. Monografia degli amfibi urodeli italiani [Monograph of Italian urodel amphibians]. Memorie del Reale istituto veneto di scienze, lettere ed arti 11:497–508 [Italian].

Congenital – Refers the reader to a future paper (De Betta, 1865) on a dicephalic *Vipera aspis*, captured in 1861, which was and also reported in Gazzetta di Fiume on December 2nd of 1861.

De Betta E. 1865. Nota sopra un caso di dicefalia atloidica in una giovane vipera (*Vipera aspis* Merr.) raccolta nel Vicentino [Note about a case of atloidic bicephaly in a young viper (*Vipera aspis* Merr.) collected in Vicenza]. Atti del Reale Istituto Veneto di Scienze, Lettere ed Arti, Serie 3, 10 (7):916–925 [Italian].

Congenital – Dicephalic snake *Vipera aspis*. Additionally, review of the history of serpent dicephaly.

De Betta E. 1874. Fauna d'Italia. Parte Quarta: Rettili ed Anfibi [Fauna of Italy. Part Four: Reptiles and Amphibians]. 104 pp.; in L'Italia sotto l'aspetto fisico, storico, artistico e statistico. Milano: Dottor Francesco Vallardi [Italian].

Congenital – Two cases of dicephalic *Vipera aspis* (specimen in De Betta 1865 and one additional specimen).

De Betta E. 1877–1878. Alcune Note Erpetologiche per servire allo studio dei Rettili ed Anfibi d'Italia [Some herpetological notes to serve the study of Reptiles and Amphibians of Italy]. Atti del Reale Istituto Veneto di Scienze (5) 4:963–981 [Italian].

Congenital – Three dicephalic *Tropidonotus natrix* (*Natrix natrix*) Wagl., one in the Civic Museum of Milan, Italy).

De Betta E. 1883. Terza serie di note erpetologiche per servire allo studio dei Rettili ed Anfibi d'Italia [Third series of herpetological notes to serve the study of Reptiles and Amphibians of Italy]. Atti del Reale Istituto Veneto di Scienze, Lettere ed Arti, Serie 6, 1:919–951 [Italian].

Congenital – De Betta reviewed various herpetological species found in Italy, and remarked in which cases he knew of anomalies. He cited himself (1874 and 1878, respectively) for examples of dicephalic *Vipera aspis* and *Tropidonotus natrix*. He also mentioned Geoffroy Saint-Hilaire reported other cases in snakes (1827) and one dicephalic lizard (Histoire des anomalies, Tom. III, pg. 132). De Betta remarked that he knew of only three cases of dicephalic lizards. One was the case published by St.-Hilaire, another was a young lizard with two heads and two distinct necks found in the collections of the museum of Paris (France), and the third was the very young two-headed *Anguis fragilis* reported by Lessona (1877). He provided an in-depth review of the latter. De Betta also cited the work of Strobel (1876) on polimelic *Rana*, and discussed his ideas on the frequency of the anomaly in batracians. De Betta mentioned that Strobel had reported two cases of polimely in *Rana esculenta*, and one in *R. temporaria*, and later two more cases in *R. esculenta*. He also mentioned Sordelli (1876), but refered to the species reported therein as *Rana esculenta* (not *R. viridis*). Additonally, De Betta cited Cavanna (1877), and his two cases of polimely in *R. esculenta*. Most importantly, De Betta reported a new case of polimely in *R. esculenta*. The specimen was captured in the fall of 1880 in the province of Mantova, and given to De Betta in September of 1881. It was not fully grown, and its normal limbs were normally developed. It had one right supernumerary hindlimb that attached in the pubic region and was atrophied in all its parts, and ended in a stump without a foot or toes. Lastly, he reviews Camerano (1882) and polimely in *Triton taeniatus*.

de Buffrenil V, Mazin J-M. 1990. Bone histology of the ichthyosaurs: Comparative data and functional interpretation. Paleobiology 16:435–447.

Metabolic – *Omphalosaurus, Stenopterygius*, and *Ichthyosaurus* limb bone cortex composed of woven fibers, suggesting rapid postnatal growth with endotherm-like metabolic rate.

Fossil – *Omphalosaurus, Stenopterygius*, and *Ichthyosaurus* limb bone cortex composed of woven fibers, suggesting rapid postnatal growth with endotherm-like metabolic rate.

de Chaignon H. 1905. Notes zoologiques. Bulletin Sociétè d'histoire naturelle d'Autun 18:150–157 [French].

Trauma – *Lacerta* with bifid tail.

Deeming DC. 2004. Post-hatching phenotypic effects of incubation in reptiles. pp. 229–251. In: DC Deeming, ed. Reptilian Incubation: Environment, Evolution and Behavior, Nottingham, Nottingham, University Press.

Environmental – Grass lizard *Takydromus wolteri* hatched at 33°C had shorter, narrower heads than from lower temperature incubation (Pan and Ji 2001).

Sceloporus virgatus, Bassiana duperreyi, Oligosoma suteri, and Chinese skink *Eumeces chinensis* had shorter tails when incubated in the cold (15–25°C) (Elphick and Shine 1998; Hare et al. 2002; Ji and Zhang 2001; Qualls and Andrews 1999). Wall lizard *Podarcis muralis* head, tail, and femur were longest at 26–29°C (versus 32°C, where head was widest) (Braña and Ji 2000; Ji and Braña 1999). *Oedura ocellata* had tail defects at 30°C (Bustard 1969). Pine snake *Pituophis melanoleucus* incubated at low 21–23°C were shorter, with shorter tails (Burger et al. 1987). Burmese, Indian, or Ceylon *Python molorus* incubated at low temperature of 27.5°C had tail defects (Vinegar 1974).

De Falvard G. 1931. Un cas d'Hyperrégénération chez *Rana esculenta*. [A case of hyperregeneration in *Rana esculenta*]. Bulletin Biologique de la France et de la Belgique 65:267–290 [French].

Congenital – Supernumerary limb in *Rana* esculenta, citing Avel and Baer's (1929) report in *Discoglossus pictus*.

De Jesus SM, Jiménez Fuentes E, Mulas Alonso E. 1989. Un pelomedusido (Chelonia) con malformaciones patológicas, del Eoceno de Zamora [A pelomedusid (Chelonia) with pathologic malformities from the Eocene of Zamora]. Studia Geologica Salmanticensia 26:355–364 [Spanish].

Trauma – Eocene Pelomedusidae *Neochelys* aff. *salmanticensis* with trauma-induced asymmetrical xifiplastron and pelvis.

Fossil – Eocene Pelomedusidae *Neochelys* aff. *salmanticensis* with trauma-induced asymmetrical xifiplastron and pelvis.

De Jong JK. 1928. Uber eine abnorme Bezahnung bei *Crocodylus porosus* (Schn.). [On the abnormal dentition of *Crocodylus porosus* (Schn.]. Miscellanea Zoologica Sumatrana 31:1–5 [German].

Dental – One female of *Crocodylus porosus* with abnormal teeth: Many together in one alveoli (normal number of alveoli in jaw), malformed teeth (different sizes and shapes).

De Kerville G. 1909. Note sur les radiographies de huite vertébrés monstrueux. (Chatons déradelphe et opodyme, faisandeau pelvadelphe, poussin déradelphe, canetons métopage et dérodyme, et saurian à queue bifurquée). [Notes on radiography of 8 vertebral monsters. (Kittens with posterior body duplication and partial skull duplication, young pheasant with defective pelvis, chick with, chicken defective pelvis, young duck twins conjointed at top of skull and with body duplication and saurien has bifurcated tail). Bulletin de la Société des Amis des Sciences Naturelles de Rouen 1909:285–292 [French].

Trauma – Bifurcated tail in *Rhacodactylus trachyrhynchus*, *Lacerta muralis* and *Hemidactylus mabouia*, noting previous reports by Pline, Gesner, and Aldrovande.

Delfino M, Scheyer TM, Fritz U, Sánchez-Villagra MR. 2010. An integrative approach to examining a homology question: Shell structures in soft-shell turtles. Biological Journal of the Linnean Society 99:462–476.

Shell disease – Multiple isolated and grouped round (occasionally irregular shaped) pits in *Lissemys punctata* MTD 42711, noting on p. 466 that "pathological condition of surface frequently occurring in soft-shelled turtles."

Della Valle P. 1913. Studii sui rapporti fra differenziazione e rigenerazione. 1. La doppia rigenerazione inversa nelle fratture delle zampe di *Triton*. Analisi della legge Bateson in relazione ai fenomeni di polarità e di differenziazione [Studies on the relation of differentiation and regeneration. 1. The inverse double regeneration in the fractures of the legs of *Triton*. Analyses of Bateson's law in relation to the phenomenon of polarity and differentiation]. Bollettino della Società dei Naturalisti in Napoli 25:95–161 [Italian].

Congenital – Review of hypermely in wild urodels, both caused by natural trauma and that purposefully caused in laboratory experiments on not only urodels, but also larval anurans. Results on the inverse double regeneration of the leg in *Triton* experimentally verified Bateson's law. The hypermely seen in wild urodels, that obtained from trauma, and that obtained from experiments in larvae were only special cases of the same order of phenomena.

Trauma – Review of hypermely in wild urodels, both caused by natural trauma and that purposefully caused in laboratory experiments on not only urodels, but also larval anurans. Results on the inverse double regeneration of the leg in *Triton* experimentally verified Bateson's law. The hypermely seen in wild urodels, that obtained from trauma, and that obtained from experiments in larvae were only special cases of the same order of phenomena.

Dely OG. 1960. Une grenouille verte (*Rana esculenta* L.) à cinq extrémités. [A green frog verte (*Rana esculenta* L.) with 5 extremities. Vertebrata Hungarica 2:41–47 [French].

Congenital – Supernumerary limb in *Rana esculenta.*

De Mortillet G. 1865. Matériaux pour l'histoire positive et philosophique de l'homme. [Material for history and philosophy of humans]. 574 pp., Paris: Édouard Blot.

Congenital – Supernumerary hind limbs in *Rana esculenta* and *Rana temporaria*, citing Strobel and in bactarins, citing Gervais.

Dennert C. 2005. Ernährung von Landschildkröten. [Diet of tortoises]. 3rd ed. Münster, Germany: Natur und Tier – Verlag GmbH, 143 pp [German].

Metabolic – Osteomalacia/rickets from vitamin D, phosphate or UV deficiency in *Geochelone pardilis* and Greek turtle.

Osteodystrophia fibroa from excess phosphate.

Gout in Greek turtle.

Deraniyagala PE. 1939. The Tetrapod reptiles of Ceylon. Volume I. Testudinates and crocodilians. 412pp.; London: Dulau & Co.

Congenital – Kyphosis in *Melanochelys trijuga thermalis.*

Deraniyagala PEP. 1944. A teratological frog from Ceylon. Journal of the Ceylon Branch of the Royal Asiatic Society 36:224–225.

Congenital – *Rana cyanophylictis* with two pelvic girdles and supernumerary limbs.

Deraniyagala PEP. 1958. A double headed Russell's viper. Spolia Zeylanica Geology Zoology 28:167–168.

Congenital – Derodymous Tit polonga *Vipera russelli* and *Ptyas mucosus.*

Derickson SH. 1927. A twin turtle. The Scientific Monthly 25(6):562–565.

Congenital – Derodymous painted turtle *Chrysemys picta*, noting similar report by Barbour (1888).

Shell disease – *Chrysemys picta* fifth neural plate on right carapace reduced in size and triangular in shape.

De Saint Vincent B. 1829. Dictionnaire class. D'histoire naturelle. [Classical Dictionary of Natural History]. Vol. 9:338.

Trauma – Tail duplication in lizards.

Desai RN. 1984. A report on the rare occurrence of two-headed Russell's earth-snake or red-earth boa *Eryx conicus* (Ophidia: Boidea). Journal of the Bombay Natural History Society 1984:483.

Congenital – Dicephalic Russell's earth-snake or red-earth boa *Eryx conicus*.

De Superville D. 1744 or 1740 or 1739. Some reflections on generation, and on monsters, with a description of some particular monsters. Philosophical Transactions of the Royal Society of London 41:294–307 or 302–320 (Am. J. Med. Genet. 80:74–89).

Congenital – *Rana* sp.? with a supernumerary arm to the right of its back (but Gervais claimed left instead (after Mazza 1888).

Detwiler SR, Copenhaver WM. 1941. Further Experiments upon the production of developmental abnormalities in *Ambystoma*. Journal of Experimental Zoology 88:399–411.

Metabolic – Oxygen deprivation producing teratogenesis in *Ambystoma*.

De Vosjoli P, Fast F, Rephoshy A. 2003. *Rhacodactylus*: The Complete Guide to their Selection and Care. Vista, California: Advanced Vison, 281 pp.

Congenital – Spine kink attributed to low calcium in *Rhacodactylus*.

Metabolic – Metabolic bone disease and spine kink attributed to low calcium in *Rhacodactylus*.

De Vosjoli P, Temper R, Klingenberg R. 2005. The Herpetocoulture of Leopard Geckos. Twenty-seven generations of living art. China: Advanced Vision, 259 pp.

Trauma – Autotomy in leopard gecko.

Metabolic – Metabolic bone disease with leg deviation and rubber jaw in leopard gecko.

Dexter RW. 1976. F.W. Putnam's original description (1862) of a double-headed bisexual black racer snake (*Coluber constrictor* Linn.) – a historical record. Herpetological Review 7:158–159.

Congenital – Derodymous bisexual black racer snake (*Coluber constrictor*).

Dial BE, Fitzpatrick LC. 1981. The energetic cost of tail autotomy to reproduction in the lizard *Coleonyx brevis* (Sauria: Gekkonidae). Oecologia 51:310–317.

Trauma – Tail autotomy in 2 of 8 salamander families and 13 of 20 lizard families (Shaffer 1978; Wake and Dresner 1967).

Species that autotomize the tail use it as a major site of energy storage.

Total energy reserves accounted for 53% of reproductive energy investment in tailed *Coleonyx brevis* contrasted with 29% in tailless.

Dial BE, Fitzpatrick LC. 1984. Predator escape success in tailed versus tailless *Scincella lateralis* (Sauria: Scincidae). Animal Behavior 32:301–302.

Trauma – Tailless *Scincella lateralis* have less locomotor efficiency and did not escape snake *Lampropeltis triangulum* predation, in contast to escape by 73% of those with tails. The latter used autotomy 50% of the time to escape.

Diaz-Figueroa O, Mitchell MA. 2006. In Mader DR. ed. Reptile Medicine and Surgery. Pp.145–162; Philadelphia: Saunders.

Neoplasia – Cornsnake *Elaphe gutata guttata* with rib osteolysis from colon cancer, citing Latimer and Rich (1998).

Diethelm G. 2006. Digit abnormalities. In Mader DR. ed. Reptile Medicine and Surgery. pp. 774–777; Philadelphia: Saunders.

Infection – Mycobacterial foot infection in shingle-backed skink *Trachydosaurus rugosa*.

Metabolic – Fibrous dystrophy.

Metatarsal phalangeal joint typically affected in gout.

Vascular – Tail constriction producing avascular necrosis.

Neoplasia – Toe sarcoma in green iguana *Iguana iguana*.

Diner R, Geyer M, Wietelmann A, Schreiyäck C, Fiehn C, Müller C, Müller-Ladner U, Borchardt T. 2008. Towards an animal model for joint regeneration in osteoarthritis. Arthritis and Rheumatism 58:S654.

Arthritis – Treatment of *Notophthalmus viridescens viridescens* with collagenase leads to osteoarthritis.

Dinsmore CE. 1977. Tail regeneration in the Plethodontid salamander, *Plethodon cinereus*: Induced autotomy versus surgical amputation. Journal of Experimental Zoology 199:163–176.

Trauma – Tails of most plethodontid salamanders have an autotomy-"designed" basal constriction and internal weak point modification, but the terrestrial *Plethodon cinereus* has each myosepta as a potential autotomy plane.

Mufti and Simpson (1972) discussed tail regeneration in the plethodontid salamander *Desmognathus fuscus* and noted that the newt *Notophthalmus viridescens* does not undergo tail autotomy.

Ditmars RL. 1903. Observations on the development of reptiles. 7th Annual Report New York Zoological Society 1902:145–153.

Congenital – Dicephalic *Thamnophis sirtalis.*

Ditmars RL. 1907. The Reptile Book. New York: Doubleday, Page & Company, plate XXIV.

Pseudopathology – *Charina bottae* is known as the two-headed, silver and worm snake because the tail is almost as blunt as the head.

Ditmars RL. 1936. The reptiles of North America. XVI+476pp.; New York: Doubleday Doram.

Pseudopathology – *Charina bottae* is known as the two-headed, silver and worm snake because the tail is almost as blunt as the head.

Ditmars RL. 1949. The Reptiles of North America. A Review of the Crocodilians, Lizards, Snakes, Turtles and Tortoises Inhabiting the United States and Northern Mexico. 476pp., Garden City, New York: Doubleday & Company.

Trauma – Supernumerary tail in western swift *Sceloropus biseriatus.*

Divers SJ. 1996. The structure and diseases of the chelonian shell. Reptilia 4(6):51–55.

Environmental – Chronic calcium or UV light (290–320 nm) and vitamin D or improper calcium/phosphorus dietary ratio produces osteopenia (general lack of mineralization with thin cortices and deformed bones) in Argentine horned frog *Ceratophrys ornata.*

Divers SJ. 1997. The diagnosis of nutritional metabolic bone disease and the treatment of hypocalcemic tetany in an Argentine horned frog *(Ceratophrys ornata)*. Proceedings of the 1996 Annual Conference of the Association of Reptilian & Amphibian Veterinary 3:7–9.

Metabolic – Generalized lack of bone mineralization in Argentine horned frog *Ceratophrys ornata* as manifestation of nutritional metabolic bone disease.

Divers SJ, Lawton MP. 2000. Spinal osteomyelitis in a green iguana, *Iguana iguana*: Cerebrospinal fluid and myelogram diagnosis. Proceedings of the Association of Reptilian and Amphibian Veterinarians Annual Conference 2000:77.

Infection – Green iguana *Iguana iguana* with loss of limb proprioceptive deficits secondary to fourth cervical vertebra osteomyelitis.

Neuropathic – Green iguana *Iguana iguana* with loss of limb proprioceptive deficits secondary to 4th cervical vertebra osteomyelitis.

Do Amaral A. 1927. Collectanea ophiologica [Ophiological anthology]. – 13 – Bicephalia em ophidios [Bicephaly in ophidians]. Revista do Museu Paulista, São Paulo 15:93–101 [Portuguese].

Congenital – Nine cases (in total) of dicephalic specimens of the snakes *Elaphe vulpina* (NMNH50.003), *Lampropeltis getula* (NMNH 21.164), *Natrix sipedon* (NMNH 38.045, MCZ 7.043), *Bothrops atrox* [MCZ 14.103, Instituto de Butantan (Brazil) 3.109, 3.110], and *Crotalus terrificus* [Instituto de Butantan (Brazil)3.101].

Dobson J. 1972. Descriptive Catalogue of the Pathological Series in the Hunterian Museum of the Royal College of Surgeons of England. Part II. A Selection of surviving specimens illustrating Hunter's opinions on the nature of diseases, experiments and observations on cases in surgery. 242 pp.; London: E & S Livingstone.

Trauma – Chameleon with spheroidal osseous swelling on ribs interpreted as probable fracture, although tuberculosis also considered.

Infection – Chameleon with spheroidal osseous swelling on ribs interpreted as probable fracture, although tuberculosis also considered.

Dobson GE. 1873. On a double-headed snake presented to the Indian Museum by Dr. R. F. Thompson, civil surgeon. Proceedings of the Asiatic Society of Bengal:23–24.

Congenital – Derodymous *Lycodon aulicus* and dicephalic *Naja tripudians.*

Dodd C Jr. 2001. North American Box Turtles: A Natural History. Norman, OK: University of Oklahoma Press.

Congenital – Extrauterine eggs in a North American box turtle and supernumerary scutes.

Shell disease – Pits in North American box turtle.

Dodd CK Jr, Franz R, Johnson SA. 1997. Shell injuries and anomalies in an insular population of Florida box turtles (*Terrapene carolina bauri*). Herpetological Natural History 5:66–72.

Congenital Supernumerary scutes in Florida box turtles *Terrapene carolina bauri.*

Shell disease – 6.5% of Florida box turtles *Terrapene carolina bauri* with pits.

Doe GM. 1915. Two-headed snake. Reports and Transactions of the Devonshire Association for the Advancement of Science, Literature and Art 47:85.

Congenital – Dicephalic common ringed or grass snake.

Dollinger P, Pagan O, Jermann T, Baumgartner R, Honegger RE. 1997. Husbandry and pathology of land tortoises (Testudinidae) in Swiss zoos. Erkrankungen der Zootiere: Verhandlungsbericht des Internationalen Symposiums über die Erkrankungen der Zoo- und Wildtiere 38:7–16 [German].

Congenital – Galapagos tortoise from Zurich zoo with "congenital carapax anomaly."

Metabolic – Galapagos tortoise from Zurich zoo with metabolic bone disease.

Dollo L. 1882. Note sur l'ostéologie des Mosasauridae [Note on the osteology of Mosasauridae]. Bulletin Musée Royal d'Histoire Naturelle Belgique 1:55–75 [French].

Vertebral – Dorsal vertebral fusion in *Plioplatecarpus marshi* compatible with congenital or spondyloarthropathy etiology.

Fossil – Dorsal vertebral fusion in *Plioplatecarpus marshi* compatible with congenital or spondyloarthropathy etiology.

Dollo L. 1885. I^{e} note sur le Hainosaure [Primary note on *Hainosaurus*]. Bulletin Musée Royal d'Histoire Naturelle Belgique 4:25–33 [French].

Vertebral – Ankylosis of 10th and 11th vertebrae in *Hainosaurus*.

Fossil – Ankylosis of 10th and 11th vertebrae in *Hainosaurus*.

Dollo L. 1892. Nouvelle note sur l'ostéologie des Mosasauridae [New note on the osteology of the Mosasauridae]. Mémoire de la Société Belgique de Géologie 6:219–259 [French].

Vertebral – Vertebral fusion in *Plioplatecarpus*-types.

Fossil – Vertebral fusion in *Plioplatecarpus*-types.

Done LB. 1996a. Neoplasia. In Mader DR. ed. Reptile Medicine and Surgery. pp. 125–141; Philadelphia: Saunders.

Neoplasia – Enchondroma in Indian monitor *Varanus dracaena* (Jacobson 1981; Machotka 1984).
Osteochondroma in *Varanus bengalensis* (Frye 1991 in Frye)
Osteosarcoma in emerald lizard *Lacerta viridis* (Machotka 1984)
Chondro-osteofibroma in rhinoceros iguana *Cyclura cornuta* (Jacobson 1981; Machotka 1984).
Osteosarcoma in Burmese python *Python molurus bivittatus*
(Frye 1991 in Frye) and in rufous-beaded snake *Rhamphiophis rostratus* (Machotka 1984).
Osteochondrosarcoma in black cobra *Naja melanoleuca* (Machotka and Whitney 1980).
Chondrosarcoma in corn snake *Elaphe guttata* (Harshbarger 1974).

Done LB. 1996b. Postural abnormalities. In Mader DR. ed. Reptile Medicine and Surgery. pp. 406–411; Philadelphia: Saunders.

Metabolic – *Ameiva* spine fracture from hyperparathyroidism.
Red-tailed boa *Eryx conicus* with osteitis deformans – Actually a proliferative spine calcification.
Other – Red-tailed boa *Eryx conicus* with proliferative spine calcification.

Done LB, Willard-Mack CL, Ruble G, Cranfield M. 1993. Diagnostic exercise: Ulcerative dermatitis and cellulitis in American toads. Laboratory Animal Science 43:619–621.

Infection – American toad *Bufo americanus* with *Mycobacterium marinum* destruction of tibiotarsal joint and cervical vertebrae.

Donoghue S. 2006. Nutrition. In Mader DR. ed. Reptile Medicine and Surgery. pp. 251–298; Philadelphia: Saunders.

Congenital – Box turtle and *Geochelone sulcata* with deformed carapace.
Trauma – *Chamaeleo deremensis* with fractures radius and ulna.
Iguana iguana with limb fracture.
Standings day gecko *Phelsuma standingii* with tail loss.
Metabolic – Bearded dragon *Pagona vitticeps* with deformed carpus and scoliosis secondary to nutritional hyperparathyroidism.
Stone – Urinary calculi related to excess vitamin D, vitamin C, calcium, sodium, phosphorus, or magnesium, low moisture in food, urine retention, abnormal excretion of calcium, sodium, phosphorus, and to pH abnormalities: acidic producing calcium oxalate stones and alkaline producing calcium phosphate stones.

Dorner H. 1873. Eine Kreuzotter mit zwei Köpfen. [A *Pelias berus* with two heads] Zoologischer Garten 14:407–410 [German].

Congenital – Dicephalic snake *Vipera* (*Pelias*) *berus*.

Douglas T. 1941. Chinese settled early in state; Oriental band welcomed Nye in Carson City. Nevada State Journal, Reno, NV (22 June 1941):6.

Congenital – Dicephalic snake.

Douglas TC, Pennino M, Dierenfeld ES. 1994. Vitamins E and A, and proximate composition of whole mice and rats used as feed. Comparative Biochemistry and Physiology 107:419–424.

Metabolic – Vitamin A content of mice and rats – suggested as pertinent to argument as to induction of hypervitaminosis A in amphibians and reptiles.

Downes SJ, Shine R. 2001. Why does tail loss increase a lizard's later vulnerability to snake predators? Ecology 82:1293–1303.

Trauma – Autotomy as a major predator escape mechanism in 13 of 20 lizard families. Fourteen percent of 66 consumed *Lampropholis guichenoti* lost tails prior to being eaten, contrasted with 56% of 9 uneaten tailless lizards.

Doyle TS. 1971. Two-headed snake. Conservationist 25(6):47.

Congenital – Dicephalic common garter snake.

Drew ML, Phalen DN, Berridge BR, Johnson TL, Bouley D, Weeks BR, Miller LA, Walker MA. 1999. Partial tracheal obstruction due to chondromas in ball pythons (*Python regius*). Journal of Zoo and Wildlife Medicine 30(1):151–157.
Neoplasia – Tracheal obstruction by chondromas in ball python *Python regius*, but no bone involvement.
Droin A, Fischberg M. 1980. Abnormal limbs (ABL), a recessive mutation affecting the tadpoles of *Xenopus l. laevis*. Experientia 36:1286–1288.
Congenital – Bradymely, syndactyly, polydactyly, brachydactyly, clinodactyly and supernumerary claws in *Xenopus laevis* offspring, but not from heterozygous crosses.
Droin D, Uehling V, Reynaud J. 1970. Une mutation letale, recessive, "bt" (bent tail) chez *Xenopus laevis*. [A lethal recessive mutation, "bt" (bent tail) in *Xenopus laevis*]. Revue Suisse de Zoologie 77:596–602 [French].
Congenital – Bent tail as lethal recessive mutation in *Xenopus laevis*.
Dubois A. 1968. Sur deux anomalies de la grenouille verte (*Rana esculenta*). [On two anomalies in the green frog (*Rana esculenta*)]. Bulletin Mensuel de la Société Linnéenne de Lyon 37:316–310 [French].
Congenital – Suggests that anomalies are not hereditary. In Champdieu, he noted *Rana esculenta* 6–8 polydactyly frequencies as follows:
1950 – 1.4% of 70 specimens
1951 – 1.7% of 960 specimens
1952 – 4.6% of 238 specimens
1964 – 0.2% of 835 specimens
1965 – 0% of 120 specimens
1966 – 3.2% of 597 specimens
Dubois A. 1974. Polydactylie massive, associée à la clinodactylie, dans une population de *Rana graeca*. Remarques sur la polydactylie faible et la clinodactylie chez *Bufo bufo* (amphibians, anoures). [Massive polydactyly linked with clinodactyly in a population of *Rana graeca*. Remarks on slight polydactyly and clinodactyly in *Bufo bufo* (Amphibia, Anura).] Bulletin de la Société Zoologique de France 99:505–521 [French].
Congenital – A benign mass polydactyly, occasionally linked with a clinodactyly due to hyperphalangy, is described in a population of young *Rana graeca* (9% of 553 specimens with polydactyly, 2.9% bilateral; clinodactyly predominantly unilateral and affected posterior elements) in former Yugoslavia (Montenegro) and in a population of *Bufo bufo* (24.9% of 185 specimens with polydactyly; 25% of 32 specimens with clinodactyly) in the Paris area.
Dubois A. 1977. Une mutation dominante determinant l'apparition de diverses anomalies digitales chez *Rana temporaria* (Amphibiens, Anoures). [A dominant mutation determining the appearance of diverse digital anomalies in *Rana temporaria* (Amphibia, Anura)]. Bulletin de la Société Zoologique de France 102:197–213 [French].
Congenital – Cross of female *Rana temporaria* with normal produced 50% hypophalanges, symphalanges, prachyphalanges, syndactylies, ectodactylies and clinodactylies as dominant mutation.
Dubois A. 1979a. Anomalies and mutations in natural populations of the *Rana "esculenta"* complex (Amphibia, Anura). Mitteilung aus dem Zoologischen Museum Berlin 55(1):59–87.
Congenital – Allegedly non-congenital polymely, polydactyly, clindactyly and syndactyly in *Rana "esculenta."*
Dubois A. 1979b. Néoténie et pédogenèse. A propos d'une anomalie du développement chez *Bombina variegate* (Amphibiens, Anoures). [Neotony and pedomorphism. Anomaly of development in *Bombina variegate* (Amphibia, Anura)]. Bulletin de Muséum national d'Histoire naturelle Paris 4 (1), section A(2):537–546 [French].
Congenital – Retention of *Bombina variegate* forearm under skin, similar to *Litoria aurea* (Richardson and Barwick 1957) and *Bufo bufo* (actually, recurved, saber-like forelimb; animal also retained its tail – a partial neoteny). (Olivier 1893) and cited ectodactyly in amphibians (Dubois 1974, 1977; Rostand 1958).
Dubois A. 1983. L'anomalie P des grenouilles vertes (complexe de *Rana* kl. *esculenta* Linné, 1758) et les anomalies voisines chez les amphibians. [Anomaly P in green frogs (*Rana esculenta*) and anomalies seen in amphibians]. In Vago C, Matz G. eds. Comptes Rendus du Premier Colloque International de Pathologie des Reptiles et des Amphibiens, Angers, France, pp. 215–221 [French].
Congenital – General review of anomalies:
Trevignon from 1949–1967 had 0–80% larval anomalies and 2.6–14.5 in adult *Rana esculenta;* Champdieu, 1950–1976 – 0–3.2% and 0–6.3%;
Linge 1961–1970 – 14–70%.
Polydactyly in Champdieu, Loire France, Switzerland, Linge, Indre France, Amsterdam Low country, Saint-Philbert-de-Grand-Lieu, Loire Atlantic France, Kenitra Marocco, Voronezh USSR, Greece, Poland, north of Low country, East Germany.
Luxeuil-les-Bains Haute-Saone France – 1/296
Faverois, Territoire-de-Belfort France – 1/76
Fenetrange Moselle, France – 7/109
Tassenierees Jura France – 3/80
Other examples in *Rana lessonae, Rana perezi, Rana ribidunda, Hyla arborea, Bufo bufo, Triturus helveticus, Triturus cristatus, Chioglossa lusitanica, Ambystoma tigrinum, Rana pipens, Ambystoma maculatum, Rana ridibunda, Salamandra salamandra, Hyla regilla.*

Dubois A, Fischer J-L. 1975. Un leptodactyle pentadactyle ectrodactyle (amphibiens, anoures). [A 5 fingered Leptodactyl ectrodactyly (Amphibia, Anura)]. Bulletin mensuel de la Société Linnéenne de Lyon 44(4):111 [French].

Congenital – Ectodactyly in 1 of 121 *Leptodactylus pentadactylus labyrinthicus*.

Dubois A, Thireau M. 1972. Polydactylie chez *Rana iberica* Boulenger, 1879 (amphibiens, anoures). [Polydactyly among *Rana iberica* Boulenger, 1879 (Amphibia, Anura)] Bulletin du Muséum national d'Histoire naturelle, Paris, section A, Zoologie 22:157–161 [French].

Congenital – Polydactyly in *Rana iberica*.

Dubois A, Vachard D. 1969. Sur trois anomalies digitales de la Grenouille rousse (*Rana temporaria*). [On 3 digital anomalies in the red frog (*Rana temporaria*)]. Comptes Rendus de la Société Biologique 163:2255–2257 [French].

Congenital – Ectodactyly in 1–2% of *Rana temporaria* from Hennezel and polydactyly in one of 65 from Vesinet (Yvelines).

Dubois A, Vachard D. 1971a. Sur l'élevage et la reproduction de quelques anoures indigènes, et particulièrement *Rana temporaria* et *Bufo bufo*. [On increased reproduction of several indigenous anourans, particluarly *Rana temporaria* and *Bufo bufo*]. Comptes Rendus de la Société d'Herpétologie de France, Angers, 26 et 27 mars 1971 [French].

Congenital – *Rana temporaria* – ectodactyly in 14/21 crosses of an ectodactylic with normal. Thus apparently autosomal dominant.

Dubois A, Vachard D. 1971b. Sur la descendance d'une Grenouille rousse (*Rana temporaria*) ectrodactyle. [On the decendents of an ectrodactylic red frog (*Rana temporaria*)]. Comptes Rendus de la Société Biologique 165:26–29 [French].

Congenital – Ectodactyly (14%), syndactyly, and clinodactyly in *Rana temporaria*.

Posterior ectodactyly in 15% of offspring of affected *Rana temporaria*.

Dubouy – actually Dupouy.

Ducey PK, Brodie ED. 1983. Salamanders respond selectively to contacts with snakes: Survival advantage of alternative anti-predator strategies. Copeia 1983:1036–1041.

Trauma – Presence of tails facilitated *Eurycea bislineata* survival of attacks by garter snake *Thamnophis sirtalis*, as permitted autotomy. 25/109 tailed had autotomy, with 17 escaping (15.6% of tailed, 68% of those autotomizing).

Ducey PK, Brodie ED Jr, Baness EA. 1993. Salamander tail autotomy and snake predation: Role of antipredator behavior and toxicity for three neotropical *Bolitoglossa* (Caudata: Plethodontidae). Biotropica 25:344–349.

Trauma – *Bolitoglossa rostrata* and *rufescens* autotomize more frequently (85–90%) than *Bolitoglossa subpalmata* (3%). Hubbard (2003) suggested that propensity for salamander tail autotomy was inversely related to effectiveness of their toxin.

Duda PL, Gupta VK. 1977. A kyphotic individual of *Lissemys punctata punctata* (Bonnaterre). Herpetologica 33:427–433.

Congenital – Kyphotic *Lissemys punctata punctata*.

Duellman WE. 1979. The Biology of and equatorial herpetofauna in Amazonian Ecuador. Publications of the University of Kansas Museum of Natural History 65:1–352.

Trauma – Tail readily breaks in *Dendrophidion dendrophis*.

Duerden JE, Essex R. 1922. Degeneration of limbs in serpentiform lizards (Chamaesaura). South African Journal of Science 19:269–275.

Congenital – Reduction of number of digits in fore- and hind limbs in *Chamaesaura anguina* from three to two or one digit. Retrogressive evolution in the genus from *Chamaesaura aenea* (five digits) to *Chamaesaura macrolepis* (no forelimb, one digit in hind limb).

Dugès MA. 1829. Mèmoire sur les espèces indigenes du genera *Lacerta*. [Report on indigenous species of genera Lacerta]. Annales des Sciences Naturelles 16:337–402 [French].

Trauma – Reported short tail in lizards (e.g., *Lacerta*), noting regeneration produces single cartilage section, not bone. Reported several lizards with two or three tails.

Dugès A. 1834. Recherches sur l'ostéologie et la myologie des Batraciens à leurs différents ages. [Research on the osteology and myology of batrachians at different ages]. 216 pp.; Paris: Ballière [French].

Congenital – Occasional failure of intervertebral spheres to unite with the center, resulting in the amphicoelous (ancestral) condition persistence to maturity in *Pelobates cultripes*.

Duguy R, Moriniere P, Le Millinaire C. 1998. Facteurs de mortalité observes chez les tortues marines dans le Golfe de Gascogne. [Mortality factors among Golf of Gascogne marine turtles]. Oceanological Acta 21:383–388 [French].

Shell disease – Fifty percent of loggerhead and 25% of Kemp's Ridley turtles had dermatomycotic scutes, sometimes ulcerating underlying bone.

Duméril A. 1865a. Observations sur la Monstruosité dite Polymélie on augmentation du nombre des Membres chez les Batraciens anoures. [Observations on the mostrosity of polymelia in a number of Bactrachian anurans]. Nouvelles Archives du Muséum d'Histoire Naturelle Paris 1:309–319 [French].

Congenital – *Pelobates cultripes* with polymely, and *Rana temporaria, Rana clamata, Rana viridis* and *Rana esculenta* with supernumerary limbs, noting Thomas' 1861 observation of supernumerary limb in *Rana esculenta* (which Dumeril published in 1865).

Duméril A. 1865b. Trois cas de polymélie (members surnuméraires) observés sur les batraciens du genre *Rana*. [Three cases of polymelia (supernumerary limbs) in *Rana* batrachians]. Comptes Rendues hébdomadairesdes Séances de l'Académie des Sciences de Paris 9:911–913 [French].

Congenital – Supernumary limb in *Pelobates cultripes, Rana temporaria, Rana viridis, Rana clamata* and Polymelia in *Pygomeles*.

Duméril A. 1865c. Trois nouveaux cas de polymélie. [Three cases of polymelia]. Revue et Magasin de Zoologie (2) 17:170 [French].

Congenital – Polymelia in *Pelobates cultripes, Rana temporaria* and *Rana viridis*.

Duméril A. 1866. V. Observations sur la reproduction, dans la Ménagerie des Reptiles du Muséum d'histoire naturelle, des Axolotls, batraciens urodèles à branchies extérieures, du Mexique, sur leur dévelopement et sur leurs métamorphoses. [Observations in the reptile collection of the Museum of Natural History of Axolotls, urodele batrachians with external armes from Mexico, on their development and metamorphoses]. Nouvelles Archives du Museum d'Histoire Naturelle 2:265–292 [French].

Congenital – "Mutilated" Axolotl. Right forefoot is short and left tarsals are "defective."

Duméril A. 1867a. Description de diverses monstruosités, observées à la menagerie des reptiles du muséum d'histoire naturelles sur les batraciens urodèles à branchies extérieures dits Axolotls. [Description of diverse monstrosities observaed in the reptile collection of the Museum of Natural History of Axolotl urodele batrachians with external armes]. Nouvelles Archives du Muséum d'Histoire Naturelle Paris 3:119–130 [French].

Congenital – Polydactyly and melomelia in *Ambystoma argus*.

Duméril A. 1867b. Métamorphoses des batraciens urodéles a branchies extérieures du Mexique dits axolotls. Observéés a la menagerie des reptiles du museum d'histoire naturelle. [Metamorphosis of bactracian urodeles exterior referred to as Mexican axolotls]. Annales des Sciences Naturelles. Zoologie et biologie animale série 5, 7:229–254 [French].

Congenital – *Pelobates cultripes* with polymely, and *Rana temporaria, Rana clamata, Rana viridis*, and *Rana esculenta* with supernumerary limbs.

Duméril A., Bibron G. 1844. Erpetologie Générale ou Histoire Naturelle complete des Reptiles [General herpetology or complete natural history of reptiles]. 600 pp. Paris: Libraire Encyclopédique de Roret 6. [French].

Congenital – Supernumerary limbs or doubling of trunk portions.

Cites dicephalic serpent reports by Aritstole, d'Aldrovandi, de Rédi, Geoffroy, and Mitchill (1925).

Dumpert K, Zietz E. 1984. Platanna (*Xenopus laevis*) as a test organism for determining the embryotoxic effects of environmental chemicals. Ecotoxicology and Environmental Safety 8:55–74.

Toxicology – 0.0004 ppm Methylmercury chloride-induced bent tails, supernumerary forelimb and femoral-tibial joint abnormality in plantana *Xenopus laevis*.

Dunayev EA. 1997. A record of the green toad (*Bufo viridis*) with five legs in Moscow Province. Advances in Amphibian Research in the former Soviet Union 2:169–171.

Congenital – Supernumerary leg in green toad *Bufo viridis*.

Dunson WA. 1960. Aquatic respiration in *Trionyx spinifer asper*. Herpetologica 16:277–283.

Vascular – The ventral skin of soft shelled turtle has a rich blood vessel network, allowing respiratory function. Pharyngeal respiration was almost as effective as respiratory in Gulf Coast soft-shelled turtle *Trionyx spinifer asper*, which had less effective dermal respiration.

Dupouy W. 1958. Un caso de bicefalia ofídica en Venezuela [A case of ophidian dicephaly in Venezuela]. Boletín del Museo de Ciencias Naturales (Caracas) 2/3(1–4):55–61 [Spanish].

Pseuodpathology – The folk of Venezuela refer to the *Amphisbaena* as two-headed snakes, but these are in fact in the Sauria and are not snakes at all, and are known as two-headed because their tail resembles their head, not because of dicephalism.

Congenital – Report of dicephalic Mapanare *Bothrops atrox* (L.) (Science Museum of Caracas, Venezuela No. 365, with photograph and review of dicephalic snake cases reported by Do Amaral (1927).

Durban M. 1909. An analysis of the rate of regeneration throughout the regenerative process. Journal of Experimental Zoology 7:397–420.

Trauma – Regenerative process in *Rana clamitans* replaces ½ to 2/3 of tail.

Dürigen B. 1897. Deutschlands Amphibien und Reptilien. Eine Beschreibung und Schilderung sämmtlicher in Deutschland und den angrenzenden Gebieten vorkommenden Lurche und Kriechthiere. [Germany's amphibians and reptiles. A description of all amphibians and reptiles in Germany and surrounding regions]. VIII+676 pp.; Magdeburg: Creutz'sche Verlagsbuchhandlung [German].

Congenital – Dicephalic snakes, lizards, *Salamandra salamandra* (one larva with only one eye) and *Rana fusca*.

Supernumerary legs in urodeles and frogs. "Anomalies" of legs and shell in *Emys europaea*.

Trauma – Double tails in lizards.

Dusseau. 1865. Catalogue de la collection d'anatomy humaine, comparee, et pathologique de MM Ger et W Vrolik. [Catalog of the human, comparative, and pathologic anatomy collection of MM Ger and W Vrolik]. Amsterdam: WJ de Roever Krober, 464 pp [French].

Congenital – Derodymous *Tropidonotus natrix* and tortoise.

Frog with extremity loss, probably from trauma.

Trauma – Frog with extremity loss, probably from trauma.

Dutrochet H. 1830. Monstruosités: une vipère à deux têtes. [Monstrosities: A viper with two heads]. Transactions Médicales Paris 1:415–417 [French].

Congenital – Often cited in herpetology literature, but actually human Siamese twins, with no report of herpetologic cases.

Dutta SK. 1931. Congenital absence of limbs in tortoises of the genera Trionyx and Emyda. Allahabad University Studies 8:1–8.

Congenital – *Trionyx gangeticus* with imperfect acetabulum (convex, rather than concave) lacking right hind limb.

Comments on Reichenow (1908) *Rana esculenta* with missing leg and another with three legs and a third with four legs. Row (1916) describes absent first digit in *Rana temporaria.*

Trauma – *Emyda granosa* lacking right foreleg (with proximal portion manifest as small nodule).

Dwyer CM, Hanken J. 1990. Limb skeletal variation in the Jemez Mountains salamander, *Plethodon neomexicanus*. Canadian Journal of Zoology 68:1281–1287.

Congenital – Seventy five percent of Jemez Mountains salamander *Plethodon neomexicanus* had phalangeal formula atypical for what was "once considered characteristic of this species." Two thirds had 1–2–3–3–2, ¼ had 1–2–3–3–1.The following were also found: 1–2–3–2–2, 1–2–2–2–2, 1–2–3–2, and 1–2–2–1. Eighteen percent had fusion of ulnare and intermedium. One had fusion of ulnare, intermedium, and centrale. Distal tarsals 4 and 5; distals 1,2, and 3; and tibial and centrale 1 fusion were reported.

Earle DE. 2001. Pathology of amphibian. In: Amphibian Medicine and Captive Husbandry. KM Wright, BR Whitaker, eds. Malabar, Florida: Krieger Publishing Company; pp. 401–485.

Congenital – Pelvic duplication (Lopez, Maxson 1990, Reynold, Stephens 1984, Sessions, Ruth 1990).

Ninety percent of supernumerary limbs are hind limbs in North America (Meyer-Rochow, Modzelewski, Koebke 1986), while extra forelimbs are more common in Europe (Flindt et al. 1968) and in long-toed salamander *Ambystoma macrodactylum* in California, equally distributed (Sessions, Ruth 1990).

Recessive and dominant alterations in African clawed frog *Xenopus laevis* and axolotl *Ambystoma mexicanumn* (Graveson, Armstrong 1994, Gurdon, Woodland 1975, Humphrey 1967, Lipsett 1941, Malacinski, Brothers 1974, Uelhinger 1966, Washabaugh et al. 2005).

35.5% of male and 20.5% of female *Triturus cristatus carnifex* had missing digits (less than half due to trauma), supernumerary phalanges and digits, fused carpals, atrophied and distorted phalanges, and a bent radius and ulna in a newt.

2.26% of Japanese firebelly newt *Cyunops pyrrhogaster* had supernumerary libs, digits or tails, or missing limbs or digits (Meyer-Rochow, Asashima 1988) 60% of one population of Western slimy salamander *Plethodon albagula* in Texas had syndactyly, forked digit or carpal, and digit duplication (Lazell 1995).

Spindly leg, aka match(-stick) legs (from Steichholzbeinchen in German) and luciferpootjes in Dutch in dendrobatids, oriental firebelly toad *Bombina orientalis*, orange-legged leaf frog *Phyllomedusa hypocondrialis* and painted frog *Discoglossus pictus*, manifest as absence of bones distal to humeri and maldeveloped humeri and carpal-metacarpal joints (Hakvoort et al. 1995).

Trauma – 35.5% of male and 20.5% of female *Triturus cristatus carnifex had* missing digits (less than half due to trauma).

2.26% of Japanese firebelly newt *Cyunops pyrrhogaster* had supernumerary tails or missing limbs or digits (Meyer-Rochow, Asashima 1988).

Metabolic – Periarticular gout in pixie frog *Pixicephalus delalandii*, manifest as swollen forelimb digit.

Hypervitaminosis A may produce osteoporosis, multiple fractures and mandibular protrusion (Bruce, Parkes 1950).

Feeding African clawed frog *Xenopus laevis* low calcium or high phosphorus diets produces mandibular protrusion and humpback deformity.

Environmental – Yeast feeding or deionized water environment produced forelimb deformities in African clawed frog *Xenopus laevis* (Pollack, Leibig 1989).

Ammonium nitrates in fertilizer produce bent tails in American toad *Bufo americanus*, Western chorus frog *Pseudacris triseriata* and northern leopard frog *Rana pipiens*.

Eaton BR, Eaves S, Stevens C, Puchniak A, Paszkowski CA. 2004. Deformity levels in wild populations of the wood frog (*Rana sylvatica*) in three ecoregions of western Canada. Journal of Herpetology 38:283–287.

Congenital – Summarizes deformity levels in 21,000 wood frogs (*Rana sylvatica*) from three ecoregions of Western Canada. Frequency of less than 2% were found of polymely, polyphalangy, ectomely, and amely, relating them to physical trauma, with "dramatic deformities… recorded more often than… deformed digits."

Trauma – Summarizes deformity levels in 21,000 wood frogs (*Rana sylvatica*) from three ecoregions of Western Canada. Frequency of less than 2% was found of polymely, polyphalangy, ectomely, and amely, relating them to physical trauma, with "dramatic deformities… recorded more often than… deformed digits." Curiously, abstract identical to that of Blaustein and Johnson (2003).

Eaton-Poole L, Pinkney AE, Green DE, Sutherland DR, Babbitt KJ. 2003. Investigaton of frog abnormalities on National Wildlife Refuges in the Northeast US pp. 63–78. In: Multiple Stressor effects in Relation to Declining Amphibian Populations. West Conshochocken, Pennsylvania: ASTM International.

Environmental – 1997 Lake Champlain Basin of Vermont sampling revealed 2–45.4% northern leopard frog abnormalities (Fort et al. 1999).

Malformations were noted in 22/89 abnormal frogs, of which 14 had calcium deficiency. Malformed included 10 from Great Bay, New Hampshire, 8 from Great Dismal Swamp, Virginia, 8 from Great Swamp, New Jersey, 3 from Iroquois, New York, 24 from Missisquoi, Vermont, 14 from Patuxent, Maryland, 19 from Rachel Carson, Maine and 7 from Wallkill River, New Jersey.

Species	National Wildlife Refuges(1991–2001)	Percent abnormal
Green frog	Bombay Hook, Delaware	0
	Great Bay, New Hampshire	0–15.4
	Great Swamp, New Jersey	5.7–9.0
	Iroquois, New York	4.0
	Missisquoi, Vermont	2.8–10.6
	Moosehorn, Maine	1.9–5.8
	Rachel Carson, Maine	1.2–6.4
	Wallkill River, New Jersey	0–1.8
American toad	Erie, Pennsylvania	0
Fowler's toad	Back Bay, Virginia	0
	Barnegat, New Jersey	5.1
	Blackwater, Maryland	0–1.8
	Cape May, New Jersey	5.3–7.2
	Eastern Neck, Maryland	1.6–6.0
	Eastern Shore of Virginia	3.6–6.7
	Great Dismal Swamp	0–2.0
	Patuxent, Maryland	0–6.9
	Prime Hook, Delaware	0–4.8
Southern leopard frog	Blackwater, Maryland	0–1.8
	Patuxent, Maryland	0–6.9
	Prime Hook, Delaware	0–2.2
Northern leopard frog	Aroostook, Maine	0
	Iroquois, New York	1.5–3.1
	Missisquoi, Vermont	0.9–10.6
	Montezauma, New York	0
	Sunkhaze Meadows, Maine	0–1.9
Northern cricket frog	Great Dismal Swamp, Virginia	0–2.0
	Patuxent, Maryland	0–6.9
Mink frog	Aroostock, Maine	0
Bullfrog	Great Meadows, Maine	0
	Moosehorn, Maine	1.9–5.8
Pickerel frog	Rachel Carson, Maine	1.2–9.3
WF (without identification)	Erie, Pennsylvania	0
	Walkill River, New Jersey	0–1.8

Edmund AG. 1969. Dentition. In: AC Gans, TS Parsons, eds. Biology of the Reptile. Vol 1, pp. 115–200. Academic Press, London.

Dental – Cites Bellairs and Miles (1960) report of possible failure of varanid tooth replacement in an aged individual and offers possibility of just failure of ankylosis, since this phenomena was seen only after maceration.

Edwards G. 1751. A natural history of uncommon birds, and of some other rare and undescribed animals, quadrupedes, reptiles, fishes, insects, &c. exhibited in two hundred and ten copper-plates,... colored after life. In four parts. 205 pp.; London: printed for the author.

Congenital – Dicephalic common English snake; only mouth separate in yellow snake of Barbados. Carolina tortoise with partial duplication of skull (laterally fused, but with separate eyes).

Trauma – Two great spotted Jamaican lizards *Lacerta major and Lacerta cinereus maculatus* with bifid tails.

Edwards G, Catesby. 1759. Vögel, Folio *Serpens dilute fuscus Busdanensis biceps.* Norimberg (Nürnberg): J Mich. Seligmann. Tab. CII. [German].

Congenital – Note that the *Busdanensis biceps* means dicephalic snake.

Effron M, Griner L, Benirschke K. 1977. Nature and rate of neoplasia found in captive wild mammals, birds and reptiles at necropsy. Journal of the National Cancer Institut 59:185–198.

Neoplasia – Neurofibrosarcoma of spine in Korean viper *Agkristrodon halys brevicaudus.*

Eggenschwiler U. 2000. Die Schildkröte in der tierärztlichen Praxis. Vom Praktiker für den Praktiker. [The turtle in veterinary practice. By practitioners for practitioners]. Schöneck Verlag, 150 pp. [German].

Metabolic – General comments on vitamin D, calcium metabolism, rickets and accelerated growth, noting formation of bumps in *Geochelone sulcata,* and attributed irregular growth in *Testudo hermanni* to rich food.

Shell disease – Carapace pits in Egyptian tortoise *Testudo kleinmanni.*

Carapace shortening. Carapace defects were illustrated but not explained.

Egger E. 1887. Ein Fall von Regeneration einer Extremität bei Reptilien. [A case of regeneration of an extremity in reptiles] Arbeiten aus dem Zoologisch-Zootomischen Institut in Würzburg 8(2):201–211 [German].

Trauma – Regenerated left hind leg of *Lacerta vivipara.*

Ehrenberg K. 1931. Neue Untersuchungen über Krankheitserscheinungen bei vorzeitlichen Tieren und ihre Ergebnisse. [New investigations of cases of sickness in prehistoric animals and their results]. Medizinische Klinik 1931(5): 172–180 [German].

Pseudopathology = Paleopathology: author discusses examples of alleged pachyostosis in *Mesosaurus, Stereosternum* [Mesosauria], *Pachypleura* [= nothsaur *Pachypleurosaurus*], *Proneusticosaurus* [nothosaurid], *Eidolosaurus* [varanoid], and *Pachyophis* [snake].

Fossil – Paleopathology: author discusses examples of alleged pachyostosis in *Mesosaurus, Stereosternum* [Mesosauria], *Pachypleura* [= nothsaur *Pachypleurosaurus*], *Proneusticosaurus* [nothosaurid], *Eidolosaurus* [varanoid], and *Pachyophis* [snake].

Ehret DA. 2005. The roof is on fire: Fire-scarring in Pleistocene box turtles (Terrapene) from the southeastern U.S. Journal of Vertebrate Paleontology 25 Suppl 3:53A.

Trauma – Fire scarred Irvingtonian and Rancholabrean (Pleistocene) box turtles with loss of sulci and moderate shell remodelling.

Shell disease – Fire scarred Irvingtonian and Rancholabrean (Pleistocene) box turtles with loss of sulci and moderate shell remodelling.

Fossil – Fire scarred Irvingtonian and Rancholabrean (Pleistocene) box turtles with loss of sulci and moderate shell remodelling.

Ehret DA, Bourque JR. 2011. An extinct map turtle Graptemys (Testudines, Emydidae) from the Late Pleistocene of Florida. Journal of Vertebrate Paleontology 31:575–587.

Shell disease – Elliptical defect in nuchal and surrounding carpace of Late Pleistocene map turtle Graptemys kerneri UF 10672.

Eigenmann CH., Cox UO. 1901. Some cases of saltatory variation. American Naturalist 35(409):33–38.

Congenital – *Rana pipiens* with supernumerary arm on the right side.

Eigner JB. 1994. Public Eye. The San Diego Union-Tribune 13 June 1994:E2.

Congenital – Dicephalic turtle.

Eikamp H. 1981. Pathologische Veränderungen und osteologische Anomalien am Skelett des Moorfrosches, *Rana arvalis* NILSSON 1842 (Amphibia: Salientia: Ranidae). [Pathological changes and osteological anomalies of the skeleton of the mud frog,, *Rana arvalis* NILSSON 1842 (Amphibia: Salientia: Ranidae)]. Salamandra 17(3–4):189–193 [German].

Congenital – Swamp frog *Rana arvalis* with abnormal sacral vertebra diapophyses #8 and 9, ilio-ischium and urostyle of unclear etiology.

Trauma – Swamp frog *Rana arvalis* with femur thickening and cystoid changes, suggesting fracture and osteochondromatous metaplasia.

Neoplasia – Swamp frog *Rana arvalis* with femur thickening and cystoid changes, suggesting fracture and osteochondromatous metaplasia.

Einarsson E, Lindgren J, Kear BP, Siverson M. 2010. Mosasaur bite marks on a plesiosaur propodials from the Campanian (Late cretaceous) of southern Sweden. GFF 132:123–128.

Trauma – Two deep diagonal gouges with well-defined edges on plesiosaurian propodial in Department of Earth and Ecosystem Sciences, Lund University, Lund Sweden LO10602, attributed to mosasaur *Dollosaurus* (which has latero-posterior marginal tooth crowns with strong posterior carina and a blade-like lateral profile). Cited polycotylid digested by *Tylosaurus* (Everhart 2004; Sternberg 1922).

Fossil – Two deep diagonal gouges with well-defined edges on plesiosaurian propodial in Department of Earth and Ecosystem Sciences, Lund University, Lund Sweden LO10602, attributed to mosasaur *Dollosaurus* (which has latero-posterior marginal tooth crowns with strong posterior carina and a blade-like lateral profile). Cited polycotylid digested by *Tylosaurus* (Everhart 2004; Sternberg 1922)

Eisler R, Belisle AA. 1996. Planar PCB hazards to fish, wildlife, and vertebrates: A synoptic review. National Biological Service Biological Report 31:1–75.

Environmental – PCB-induced mortality, without discussing skeletal changes.

Elkan E. 1976. Pathology in the amphibia. In Lofts B. Physiology of the Amphibia. New York: Academic Press; pp. 273–312.

Trauma – Healing pelvis and elbow fracture in *Ceratophrys ornata.*

Elkan E. 1983. Random samples from herpetopathology. In Vago C, Matz G. eds. Proceedings of the First International Colloquium on the Pathology of Reptiles and Amphibians 1982 Angers:1–8.

Trauma – Ankylosis of pelvis and right elbow after comminuted fracture in *Ceratophrys ornata.*

Ellis MM. 1909. The relation of the amount of tail regenerated to the amount removed in tadpoles of *Rana clamitans.* Journal of Experimental Zoology 7:421–455.

Trauma – Tail regeneration was proportional to amount removed in *Rana clamitans.*

Elphick MJ, Shine R. 1998. Longterm effects of incubation temperatures on the morphology and locomotor performance of hatchling lizards (*Bassiana duperreyi*, Scincidae). Biological Journal of the Linnean Society 63:429–447.

Environmental – Hot-incubated scincid lizard *Bassiana duperreyi* had longer tails than those cold-incubated, with difference manifest for at least the first 6 weeks of life.

Engelmann W-E, Obst FJ. 1981a. Mit gespaltener Zunge. Aus der Biologie und Kulturgeschichte der Schlangen. [With forked tongue: On the histology and cultural history of snakes]. Herder, Freiburg, Basel and Wien 1981: 1–217 [German].

Congenital – Dicephalic ?Texas rattle snake ?*Crotalus atrox* observing that; most malformed snakes do not live long.

Engelmann W-E, Obst FJ. 1981b. Snakes. 221 pp.; London: Croom Helm.

Congenital – Dicephalic western diamondback *Crotalus atrox.*

Ensley PK. 1981. Hypothermic surgery for Krinkles. Zoonooz 54(3):14–15.

Neoplasia – Bones in fifth pedal digit is replaced by a circular-shaped growth in *Alligator mississippiensis*. X-ray image is not sufficiently clear for further comment.

Ensley PK, Reichard T, Liston S. 1981. What is your diagnosis? Case 2. Journal of the American Veterinary Medical Association 179, 11:1312–1314.

Infection – Septic arthritis of knee in Fiji Island iguana *Brachylophus fasciatus* with associated osteomyelitis.

Epure E, Pogorevici N. 1940. Über einen seltenen Fall von Opodymus bei *Emys orbicularis* L. [On a rare case of opodymus in *Emys orbicularis* L.]. Zoologischer Anzeiger 131:39–44 [German].

Congenital – Dicephalic snakes (*Tropidonotus, Vipera, Zamenis, Coluber, Coronella,* etc.), lizards, and turtles. *Emys orbicularis* specimen with double nuchal plate and duplication of anterior head (two nasal openings on left side, none on right, separated mouth slits.

Erber 1876. Die Lebensweise von *Siren lacertina* in der Gefangenschaft [The way of life of *Siren lacertina* in husbandery]. Verhandlungen der zoologisch-botanischen Gesellschaft Wien 6. Dezember:114–116 [German].

Trauma – Regeneration of legs and tail in Lissamphibia.

Ercolani G. 1882. Della polidactylia, e della polimelia nei Rettili e di alcune forme di mostruosita con queste corrispondenti, osservate negli organi della locomozione nei Pesci [On polydactyly, and on polymely in Reptiles and on some forms of monstrosity corresponding to this, observed in the locomotor organs of fish]. Memorie della Accademia delle scienze dell'Istituto di Bologna (serie 4) 3:810–824 [Italian].

Congenital – Review of literature on polydactyly, melomelia, and polymelia in reptiles and amphibians, suggesting that fusion of two developing embryos and abnormally excessive regeneration produce the anomalies.

Reported the following:

1) *Rana esculenta* with two feet in the left leg, and with ten digits. It was missing the right leg when purchased at a market, and hence it was unknown if the latter was abnormal.
2) Two specimens of *Rana esculenta* with three left forelimbs each.
3) *Rana esculenta* with three right forelimbs.
4) *Triton taeniatus* with an extra left leg. It possessed a very thick thigh, and from its medial side to the knee region, protruded an accessory limb that ended in two feet.
5) *Rana esculenta* with extra, gracile leg that hangs freely from the pubic region.

Erickson BR. 1982. The Wannagan Creek Quarry and its reptilian fauna (Bullion Creek Formation, Paleocene) in Billings County, North Dakota. Report of Investigations No. 72, North Dakota Geological Survey: 15 pp.

Trauma – Fractures, punctures, and plastic deformities, most with healing, in *Leidosuchus.*

Shell – Perforation and scars in posterior carapace (especially of snapping turtle *Protochelydra* from the Paleocene of North Dakota associated with *Leidosuchus* teeth).

Fossil – Fractures, punctures, and plastic deformities, most with healing, in *Leidosuchus.*

Perforation and scars in posterior carapace (especially of snapping turtle *Protochelydra* from the Paleocene of North Dakota associated with *Leidosuchus* teeth).

Erickson BR. 1984. Chelonivorous habits of the Paleocene crocodile *Leidyosuchus formidabilis.* Scientific Publications of the Science Museum of Minnesota, New Series 5(4):3–9.

Trauma – Paleocene crocodile *Leidyosuchus formidabilis* suggested as responsible for wounds with puncture wounds on posterior and posterodorsal carapace surfaces with thickened bone in *Protochelydra,* a trionychid and "at least two other forms."

Shell – Reported Carpenter and Lindsey (1980) suggestion for Cretaceous alligator *Brachychampsa* of responsible for etched/pitted turtle shell, although suggested that these may have been ingested as stomach stones.

Fossil – Paleocene crocodile *Leidyosuchus formidabilis* suggested as responsible for wounds with puncture wounds on posterior and posterodorsal carapace surfaces with thickened bone in *Protochelydra,* a trionychid and "at least two other forms."

Reported Carpenter and Lindsey (1980) suggestion for Cretaceous alligator *Brachychampsa* of responsible for etched/pitted turtle shell, although suggested that these may have been ingested as stomach stones.

Erickson GM. 1996. Toothlessness in American alligators, *Alligator mississippiensis.* Copeia 1996:739–743.

Trauma – Webb and Manolis (1989) reported shattered teeth (also noted by Erickson in 98% of *Alligator mississippiensis* also noting toothless alveoli in 70% of wild-caught, as well as puncture marks. Pooley (1962) reported teeth ripped out of sockets. While trauma is suspected in toothless alveoli, Erickson found that only 10 of 234 (4.3% could be directly attributed to cranial injuries.

Dental – Localized toothlessness occurs when replacement teeth formation ceases (Auffenberg 1988; Bellairs and Miles 1960; Mertens 1959). This is common in crocodilians (Bellairs and Miles 1960; Neill 1971; Poole 1961) rare in lepidosaurs (Bellairs and Miles 1960, 1961; Edmund 1969). Nearly toothless crocodilians reported by Neill (1971). Tornier (1911), Bellairs and Miles (1960), Bellairs (1970) and Guggisberg (1972) suggested this limited to captives, but Hall (1985) documented toothlessness in wild *Crocodylus novaeguinae* and Webb and Manolis (1989), Poole (1961), and Neill (1971) in *Crocodylus porosus.* Kälin (1937) suggested this is a phenomenon of old age, while Hall (1985) reported edentulous alveolar in subadults. Webb and Manolis (1989) reported shattered teeth (also noted by Erickson in 98% of *Alligator mississippiensis* also noting toothless alveoli in 70% of wild-caught, as well as puncture marks… Pooley (1962) reported teeth ripped out of sockets. While trauma is suspected in toothless alveoli, Erickson found that only 10 of 234 (4.3% could be directly attributed to cranial injuries.

Erickson BR, Sawyer GT. 1996. The estuarine crocodile *Gavialosuchus carolinensis* n. sp. (Crocodylia: Eusuchia) from the late Oligocene of South Carolina, North America. Monograph of the Science Museum of Minnesota 3:1–47.

Trauma – Late Oligocene estuarine crocodile *Gavialosuchus carolinensis* SCSM (South Carolina State Museum) 90.93.1 with 4 tibial punctures (with midshaft fistula)) and 2 fibular gouges. Both were angulated, suggesting healed fracture, apparently from a conspecific. Possible bite marks on right distal humerus, right distal metatarsal 1, 2 osteoscutes and ribs. Lateral spines of two dorsal vertebrae had healed fractures, as did metatarsal V. Type specimen ChM (Charleston Museum) PV 4279 had coracoid fracture causing thickening probably related to overlap.

Fossil – Late Oligocene estuarine crocodile *Gavialosuchus carolinensis* SCSM (South Carolina State Museum) 90.93.1 with 4 tibial punctures (with midshaft fistula)) and 2 fibular gouges. Both were angulated, suggesting healed fracture, apparently from a conspecific. Possible bite markes on right distal humerus, right distal metatarsal 1, 2 osteoscutes and ribs. Lateral spines of two dorsal vertebrae had healed fractures, as did metatarsal V. Type specimen ChM (Charleston Museum) PV 4279 had coracoid fracture causing thickening probably related to overlap.

Ernst CH. 1960a. Dichocephaly in a northern water snake. Bulletin of the Philadelphia Herpetology Society 8(5):17.

Congenital – Dicephalic northern water snake *Natrix s. sipedon* and eastern garter snake *Thamnophis s. sirtalis* attached at gular region.

Ernst CH. 1960b. No title. Bulletin of the Philadelphia Herpetological Society 8(6):6.

Congenital – Dicephalic *Thamnophis sirtalis.*

Ernst CH. 1965. A case of dichocephaly in *Lampropeltis doliata triangulum.* Transactions of the Kentucky Academy of Science 26:67–68.

Congenital – Dichocephalic *Lampropeltis doliata triangulum* and in *Natrix s. sipedon* (Ernst 1960), citing 6 cases in 700,000 snakes in Institute Butantan records (Belloumini 1949), Meyer (1958), Triplehorn (1955) – *Storeria dekayi* with two bodies, Steward (1961), Curry-Lindahl (1963), Klauber (1956), Neill (1960).

Ernst CH. 1971. Observations of the painted turtle, *Chrysemys picta.* Journal of Herpetology 5:216–220.

Congenital – Kyphosis in *Chrysemys picta* and *Clemmys guttata.*

Ernst CH, Lovich JE. 2009. Turtles of the United States and Canada. 2nd Ed. Baltimore: Johns Hopkins University Press, 827.

Trauma – 40% of nesting female loggerhead turtle *Caretta caretta* had wounds, possibly from sharks.

Olive Ridley turtle *Lepidochelys olivacea* with wounded shells and amputated limbs.

11.9% of male and 15% of female spotted turtle *Clemmys guttata* with amputated limbs, tails, and damaged shells.

31% of Blanding's turtle *Emydoidea blandingii* had injuries or missing body parts.

60% of adults and 28.6% of juvenile wood turtle *Glyptemys insculpts* had wounds in Ontrario, contrasting with 9% in Pennsylvania

Shell injury from alligator in ringed map turtle *Graptemys nigrinoda.*

12% of female and 8% of male diamond-backed terrapin *Macaclemys terrapin* had missing limbs from predator attacks

Essex R. 1927. Studies in reptilian degeneration. Proceedings of the general meetings for scientific business of the Zoological Society of London 2:879–945.

Congenital – Variation on number of phalanges, carpals and tarsals, with fusions of the latter in Chamaesaura anguina.

Etheridge R. 1967. Lizard caudal vertebrae. Copeia 1967(4):699–721.

Trauma – Functional autotomy may disappear from progressive replacement of cartilaginous septum with bone, starting in posterior region of tail, from top of neural arch, extending through septum. This may be so extensive in *Iguana iguana* as to leave tail totally non-autotomic.

Production of a cartilage rod may occur in lizards without fracture planes, e.g., in *Brachylophus fasciatus* and *Agama agama* and *Caiman crocodilus.*

Autotomic (fracture plane) vertebrae in Gekkonidae and Pygopodidae (*Lucasius, Cnemaspis, Uroplatus, Nephrurus laevis, Stenodactylus sthenodactylus*, and *Ceramodactylus doriae* are restricted to the 2 or 3 vertebrae without processes. *Nephrurus asper* had no septa.). Dibamidae and Anelytropsidae septae are in front of process bases. The first caudal vertebrae of *Dibamus argenteus* and the first five of *Anelytropsis* are non-autotomic. The first five or six of Xantusiidae are non-autotomic. Autotomic septa in Iguanidae are behind the bases in *Callisaurus, Chalarodon, Ctenoblepharis, Crotaphytus, Gambelia, Cupriguanus, Halbrookia, Leiocephalus, Liolaemus, Morunasaurus, Ophryoessoides, Oplurus, Petrosaurus, Phrynosaura, Phymaturus, Platynota, Plica, Prototretus, Sator, Sceloporus, Stenocercus, Strobilurus, Tropidurus, Uma, Urosaurus*, and *Uta.* Non-autotomic was noted in *Crotaphytus, Hoplocercus, Leiosaurus, Phrysoma*, and *Uracentron.* Non-autotomic never more than one-fourth of caudal vertebrae. Autotomic in some *Anolis* (septum in front of processes) and all *Corythophanes, Phenacosaurus, Chamaeleolis, Basiliscus, Laemanctus, Enyalius, Enyalioides, Polychrus, Polychroides, Urostrophus*, and *Uranoscodon.* Septum in *Basiliscus, Enyalioides*, and *Uranoscodon* is posterior. There is no autotomy in *Phenacosaurus, Chamaeleolis, Corythophanes, Laemanctus, Enyalius, Polychrus, Polychroides, Urostrophus*, and somne *Anolis.* Autotomy is lacking in *Chamaelinorops, Anisolepis*, and *Aptycholaemus.*

Fractures in *Iguana iguana, Ctenosaura, Cyclura, Enyaliosaurus, Sauromalus*, and *Dipsosaurus* occurs between processes. *Iguana delicatissimja, Conolophus, Amblyrhynchus*, and *Brachylophus* are non-autotomic. Autotomy does not occur in agamids, although the vertebrae are fragile. Autotomy septa were present in all Teiidae vertebrae. In Lacertidae, Cordylidae, and Gerrhosauridae, septum is between the processes. In Scincidae, the fracture plane is through the transverse processes. *Brachylophus* and some *Acontias, Mabuya*, and *Emoia* have septum anterior to the processes; *Chalcides, Ophiomorus*, and *Egernia depressa*, posterior. Anguids *Diploglossus monotropus*, and *Sauresia sepsoides* differ by having skink type autotomy. Autotomy occurs in all Anguidae, Anniellidae, and Feyliniidae, except *Ophisaurus apodus* and *compressus.* Autotomy does not occur in Helodermatidae, Xenosauridae, Lanthanotidae, and Varanidae. The compressed tail of *Dracaena*, prehensile tail of *Abronia*, and *Ophisaurus* are autotomous.

Fracture planes are present in *Sphenodon* and Jurassic rhynchocephalians *Homoeosaurus* and *Sapheosaurus* (Price 1940), Gekkonidae, Pygopodidae, Dibamidae, Anelytropsidae, Xantusiidae, Feyliniidae, Cordylidae, Gerrhosauridae, Lacdertidae, Teiidae, and Anniellidae, almost all Anguidae and Scincidae. They are absent in Agamidae, Chamaeleontidae, Xenosauridae, Helodermatidae, Varanidae, and Lanthanotidae.

Tail fragility was found in *Scaphiodontophis* and *Sibynophis*, and the unrelated *Pliocercus.*

Frequency of lost tails in nature were:

50% of male and 35% of female *Sceloporus olivaceous* (Blair 1960)

50% of *Takydromus septentrionalis* (Boring et al. 1948)

33% of *Gekko swinhonis* (Boring et al, 1948)

72% of *Uta stansburiana hesperis*

45% of *Sceloporus occidentalis*

31% of *Dipsosaurus dorsalis*

30% of *Sator grandavus*

Mutilated tails were found in colubrid subfamily Sibynophbinae (Taylor 1954; Taylor and Elbel 1958).

Fossil – Fracture planes are present in Jurassic rhynchocephalians *Homoeosaurus* and *Sapheosaurus* (Price 1940).

Evans SE. 1983. Mandibular fracture and inferred behavior in a fossil reptile. Copeia 1983:845–847.

Trauma – Seven percent of 3,000 Lower Jurassic eosuchian *Gephyrosaurus bridensis* jaw fragments have "in vivo damage." Fracture with callus and tooth row/meckelian fossa malignment was illustrated.

Dental – Lower Jurassic eosuchian *Gephyrosaurus bridensis* jaw with tooth row/meckelian fossa malignment were illustrated.

Fossil – Seven percent of 3,000 Lower Jurassic eosuchian *Gephyrosaurus bridensis* jaw fragments have "in vivo damage." Fracture with callus and tooth row/meckelian fossa malignment was illustrated.

Lower Jurassic eosuchian *Gephyrosaurus bridensis* jaw with tooth row/meckelian fossa malignment were illustrated.

Everhart M. 1999. Evidence of feeding on mosasaurs by the Late Cretaceous lamniform shark, *Cretoxyrhina mantelli*. Journal of Vertebrate Paleontology 19(supplement to 3):43A–44A.

Trauma – *Cretoxyrhina mantelli* tooth marks on mosasaur bones with partially digested bones and preserved stomach contents.

Fossil – *Cretoxyrhina mantelli* tooth marks on mosasaur bones with partially digested bones and preserved stomach contents.

Everhart M. 2004. Conchoidal fractures preserved on elasmosaur gastroliths are evidence for processing food. Journal of Vertebrate Paleontology 24(supplement to 3):56A.

Trauma – Gastroliths in elasmosaurs with arc-shaped markings caused by 2–5 mm conchoidal fractures of the chest with varying degrees of wear.

Fossil – Gastroliths in elasmosaurs with arc-shaped markings caused by 2–5 mm conchoidal fractures of the chert with varying degrees of wear.

Everhart MJ. 2005. Oceans of Kansas: A Natural History of the Western Interior Sea. Bloomington: Indiana University Press, 322 pp.

Trauma – Serrated teeth marks on marine reptiles (Bardet et al. 1998; Schwimmer et al. 1997).

Protostega shell and limbs with bite marks and embedded teeth attributed to *Cretoxyrhina mantelli* Fort Hays State University Museum FHSM VP-2158 (Shamida and Hooks 2004).

Bite marks attributed to *Squalicorax* sp. on *Toxychelys*? LA County LACM 50974 humerus, *Toxychelys* Fort Hays FHSM VP 13449 with posterior aspect missing, *Dermatochelys lowi* University of Kansas KUVP 32401 and 32405 humeri, *Protostega* University of Wisconsin-Madison 1503.57 humerus and a *Toxychelys* FHSM VP 13449 with posterior aspect missing.

Mosasaurus conodon skull with bites marks and a conspecific tooth embedded in left quadrate (Bell and Martin 1995).

Tylosaurus sp. FHSM VP-2295 with bite marks across top of skull and lower jaw.

Mosasaur FHSM VP-13283 articulated vertebrate series with two embedded shark teeth (Everhart 1999; Shimada 1997, p. 928 Figure 4).

Partially digested mosasaur bones FHSM VP-13283 with embedded shark teeth (Everhart 1999), suggesting they were severed by a shark bite.

Vascular – Plesiosaur *Elasmosaurus platyurus* Philadelphia Academy of Sciences ANSP 10081 with apparent proximal humeral and femoral irregularities suggestive of avascular necrosis.

Vertebral – Fusion of C1 and C2 in Philadelphia Academy of Sciences plesiosaur *Elasmosaurus platyurus* ANSP 10081, suggestive of spondyloarthropathy.

Other – Bumps on inferior aspect of mandible of polycotylid plesiosaur *Trinacromerum* (*bentonianum*) *willistoni* KUVP 5070.

Fossil – Serrated teeth marks on marine reptiles (Bardet et al. 1998; Schwimmer et al. 1997).

Protostega shell and limbs with bite marks and embedded teeth attributed to *Cretoxyrhina mantelli* Fort Hays State University Museum FHSM VP- 2158 (Shamida and Hooks, 2004).

Bite marks attributed to *Squalicorax* sp. on *Toxychelys*? LA County LACM 50974 humerus, *Toxychelys* Fort Hays FHSM VP 13449 with posterior aspect missing, *Dermatochelys lowi* University of Kansas KUVP 32401 and 32405 humeri, *Protostega* University of Wisconsin-Madison 1503.57 humerus and a *Toxychelys* FHSM VP 13449 with posterior aspect missing.

Mosasaurus conodon skull with bites marks and a conspecific tooth embedded in left quadrate (Bell and Martin 1995).

Tylosaurus sp. FHSM VP-2295 with bite marks across top of skull and lower jaw.

Mosasaur FHSM VP-13283 articulated vertebrate series with two embedded shark teeth (Everhart 1999; Shimada 1997, p. 928 Figure 4).

Partially digested mosasaur bones FHSM VP-13283 with embedded shark teeth (Everhart 1999), suggesting they were severed by a shark bite.

Plesiosaur *Elasmosaurus platyurus* Philadelphia Academy of Sciences ANSP 10081 with apparent proximal humeral and femoral irregularities suggestive of avascular necrosis.

Fusion of C1 and C2 in Philadelphia Academy of Sciences plesiosaur *Elasmosaurus platyurus* ANSP 10081, suggestive of spondyloarthropathy.

Bumps on inferior aspect of mandible of polycotylid plesiosaur *Trinacromerum* (*bentonianum*) *willistoni* KUVP 5070.

Everhart MJ. 2008. A bitten skull of *Tylosaurus kansasensis* (Squamata: Mosasauridae) and a review of mosasaur-on-mosasaur pathology in the fossil record. Transactions of the Kansas Academy of Science 111:251–262.

Trauma – Partially healed frontal, parietal, and articular puncture wounds attributed to mosasaur attack.

Rib fractures in mosasaurs (Fort Hayes State Musuem (FHSM) VP- 3366 and FHSM VP-4516.

Tylosaurus tooth wedged between cervical vertebrae of *Tylosaurus kansasensis* FHSM VP-15631. Partially healed premaxilla puncture in *Tylosaurus nepaeolicus* FHSM VP-2209. *Tylosaurus kansasensis* FHSM VP-2295 with frontal, prefrontal, and jaw (with lower jaw gouges) bite marks and damaged right sclerotic ring. There was an acute deviation at the atlas–axis junction, suggesting possible broken neck from attack my bite of a large mosasaur. Attack was from slightly below.

Platecarpus tympaniticus FHSM VP-322 with fused tail vertebrae and exostotic chevron bridging, partially healed puncture wound on surangular and left front paddle with fusion of two phalanges.

Clidastes propython PR495 with frontal and prefrontal haled punctures.

Cites Mudge (1877) that healed saurian rib fractures are not uncommon in the Cretaceous of Kansas, a specimen with dorsal spine "fracture" with loss and also ankylosis of spinous processes.

Cites Williston (1898) account of exostoses on mosasaur lower jaws, vertebrae, and digits, with coalesced forearm, carpus, and metacarpals in a *Platecarpus.*

Cites Bell and Martin (1995) report of *Mosasaurus conodon* with embedded conspecific tooth in quadrate.

Cites Lingham-Soliar (1998) attributing skull fractures in *Mosasaurus hoffmanni* to ramming by *Hainosaurus.*

Cites Lingham-Soliar (2004) fractured dentary in *Platecarpus ictericus.*

Cites Bell and Barnes (2007) *Tylosaurus nepaeolicus* containing three *Platecarpus* and a *Tylosaurus proriger* with partially digested *Clidastes* vertebrae.

Vertebral – Cites Sternberg (1909) reporting "ankylosed" mosasaur bones.

Fossil – Partially healed frontal, parietal, and articular puncture wounds attributed to mosasaur attack.

Rib fractures in mosasaurs (Fort Hayes State Musuem (FHSM) VP-
3366 and FHSM VP-4516.

Tylosaurus tooth wedged between cervical vertebrae of *Tylosaurus kansasensis* FHSM VP-15631. Partially healed premaxilla puncture in *Tylosaurus nepaeolicus* FHSM VP-2209. *Tylosaurus kansasensis* FHSM VP-2295 with frontal, prefrontal, and jaw (with lower jaw gouges) bite marks and damaged right sclerotic ring. There was an acute deviation at the atlas–axis junction, suggesting possible broken neck from attack my bite of a large mosasaur. Attack was from slightly below.

Platecarpus tympaniticus FHSM VP-322 with fused tail vertebrae and exostotic chevron bridging, partially healed puncture wound on surangular and left front paddle with fusion of two phalanges.

Clidastes propython PR495 with frontal and prefrontal haled punctures.

Cites Mudge (1877) that healed saurian rib fractures are not uncommon in the Cretaceous of Kansas, a specimen with dorsal spine "fracture" with loss and also ankylosis of spinous processes.

Cites Williston (1898) account of exostoses on mosasaur lower jaws, vertebrae, and digits, with coalesced forearm, carpus, and metacarpals in a *Platecarpus.*

Cites Bell and Martin (1995) report of *Mosasaurus conodon* with embedded conspecific tooth in quadrate.

Cites Lingham-Soliar (1998) attributing skull fractures in *Mosasaurus hoffmanni* to ramming by *Hainosaurus.*

Cites Lingham-Soliar (2004) fractured dentary in *Platecarpus ictericus.*

Cites Bell and Barnes (2007) *Tylosaurus nepaeolicus* containing three *Platecarpus* and a *Tylosaurus proriger* with partially digested *Clidastes* vertebrae.

Cites Sternberg (1909) reporting "ankylosed" mosasaur bones.

Everhart M. 2009. http://www.oceansofKansas.com/mosapath.html.

Trauma – Fractured, healed mosasaur ribs in FHSM VP-3366 and 4516.

Cites FB Mudge (1878) as one of first paleontologist to discuss mosasaur pathology, including fractured ribs, fracture of five contiguous vertebrae forming a "firm mass" (likely infected) and fusion of caudal vertebrae.

Crookedly healed humerus, radius, and ulna from infection allegedly from bite of large mosasaur.

Platecarpus tympaniticus FHSM VP-322 with fused tail vertebrae, fused or misshapened chevrons (attributed to infection, but healed fractures definite), puncture wounds in skull (surangular) and left front paddle attributed to bite, associated with region of expanded ribs.

Cited Williston (1898, p. 214) report of "exostosial" growth in lower jaws, vertebrae (especially tail), and paddles (especially digits), false joint from distal spinous process fractures in *Platecarpus* and fusion of forearm, carpus, and metacarpals by "exostosis." and partial tail loss attributed to a bite.

Infection – Crookedly healed humerus, radius, and ulna from infection allegedly from bite of large mosasaur.

Mosasaur FHSM VP-13750 with three to four vertebrae fused by infection, in association with shark bite.

Tylosaurus proriger in Hobetsu, Japan museum with broken bones on right side of skull (premaxilla, with rearward displacement of first maxillary tooth.

Jaw fragment apparently from *Platecarpus tympaniticus* with three holes with underlying cavity, associated with scratches, suggesting infection after shark bite.

Avascular necrosis – Cites Rothschild and Martin (1987) on decompression syndrome in mosasaurs.

Other – Region of expanded ribs in *Platecarpus tympaniticus* FHSM VP-322

Fossil – Fractured, healed mosasaur ribs in FHSM VP-3366 and 4516.

Cites FB Mudge (1878) as one of first paleontologist to discuss mosasaur pathology, including fractured ribs, fracture of five contiguous vertebrae forming a "firm mass" (likely infected) and fusion of caudal vertebrae.

Crookedly healed humerus, radius, and ulna from infection allegedly from bite of large mosasaur.

Platecarpus tympaniticus FHSM VP-322 with fused tail vertebrae, fused or misshapened chevrons (attributed to infection, but healed fractures definite), puncture wounds in skull (surangular) and left front paddle attributed to bite, associated with region of expanded ribs.

Cited Williston (1898, p. 214) report of "exostosial" growth in lower jaws, vertebrae (especially tail), and paddles (especially digits), false joint from distal spinous process fractures in *Platecarpus* and fusion of forearm, carpus, and metacarpals by "exostosis." and partial tail loss attributed to a bite.

Crookedly healed humerus, radius, and ulna from infection allegedly from bite of large mosasaur.

Mosasaur FHSM VP-13750 with three to four vertebrae fused by infection, in association with shark bite.

Tylosaurus proriger in Hobetsu, Japan museum with broken bones on right side of skull (premaxilla, with rearward displacement of first maxillary tooth.

Jaw fragment apparently from *Platecarpus tympaniticus* with three holes with underlying cavity, associated with scratches, suggesting infection after shark bite.

Cites Rothschild and Martin (1987) on decompression syndrome in mosasaurs.

Region of expanded ribs in *Platecarpus tympaniticus* FHSM VP-322

Everhart MJ, Everhart PA, Shimada K. 1995. A new specimen of shark bitten mosasaur vertebrae from the Smoky Hill Chalk (Upper Cretaceous) in western Kansas. Presented at 127th Annual Meeting of the Kansas Academy of Sciences. Kansas Academy of Sciences 14:19.

Trauma – Shark-bitten mosasaur vertebrae, but no signs of healing.

Fossil – Shark-bitten mosasaur vertebrae, but no signs of healing.

Eversmann E. 1858. Erinnerungen aus einer Reise in's Ausland 1857–1858. [Memories of a travel into foreign country]. Bulletin de la Société impériale des Naturalistes de Moscou 31 (3):265–305 [German].

Trauma – Double and triple tail in lizards in Algeria, North Africa.

Ewert MA. 1979. The embryo and its egg. Development and natural history. In Harless M, Morlock H. eds. Turtles: Perspectives and Research. New York: John Wiley & Sons; pp. 333–413.

Congenital – Hunchback in alligator snapping turtle *Macroclemys temminckii.*

Absent mouth and jaw in a box turtle. Cyclops in *P. scripta elegans.*

Faarke A. 2007. Reexamination of paleopathology in plesiosaurs and implications for behavioral interpretations. Journal of Vertebrate Paleontology 27:724–726.

Pseudopathology – Attempts to criticize use of statistics in an article by Rothschild and Storrs (2003). Faarke's major contention is that the paleontologic record is not amenable to Chi Square statistics, because it is a retrospective analysis of an incomplete record and therefore biased. If that were true, the paleontologic record would not be amenable to scientific analysis. Since all available specimens were analyzed and there was no bias in their collection, statistical analysis is actually valid. His suggestion that "the G-test is more appropriate" belies a fundamental misunderstanding of statistics. The same challenge is true for any "retrospective" experiment, including the G-test. However, the G-test is an alternative to the Chi square test, not as a replacement for the Fisher exact test! The sample size is inadequate for the G-test to be applied, and the analysis in question was with the Fisher exact test.

Fossil – Attempts to criticize use of statistics in an article by Rothschild and Storrs (2003). Faarke's major contention is that the paleontologic record is not amenable to Chi Square statistics, because it is a retrospective analysis of an incomplete record and therefore biased. If that were true, the paleontologic record would not be amenable to scientific analysis. Since all available specimens were analyzed and there was no bias in their collection, statistical analysis is actually valid. His suggestion that "the G-test is more appropriate" belies a fundamental misunderstanding of statistics. The same challenge is true for any "retrospective" experiment, including the G-test. However, the G-test is an alternative to the Chi square test, not as a replacement for the Fisher exact test! The sample size is inadequate for the G-test to be applied, and the analysis in question was with the Fisher exact test.

Fabretti F. 1875 (not 1866). Cenni en due casi di Polìmelia nei Batraci [Hints on two cases of polymely in Batrachians]. Rivista Scientifico-Industriale delle principali scoperte ed invenzioni fatte nelle scienze e nelle industrie [Rivista Scientifico-Industriale di G.Vimercati] 7:214–218. (extr. dalle pubbl. Accademia Medico-Chirurgica di Perugia) [Italian].

Congenital – Polymely in two specimens of *Rana esculenta*: one specimen had a left supernumerary limb inserted ventrally. Another had a left supernumerary limb inserted ventrally that had five digits.

Fabrezi M. 1999. Duplicación de la extremidad anterior en *Lepidobatrachus llaenensis* (Anura: Leptodactylidae). [Duplication of the anterior extremity of *Lepidobatrachus llaenensis* (Anura: Leptodactylidae).] Cuadernos de Herpetologia 13:99–100 [Spanish].

Congenital – Supernumerary anterior limb in *Lepidobatrachus llaenensis*.

Fahrenholz C. 1928. Hohe Zahl von Jungen beim Feuersalamander. [High number of offspring in the fire salamander]. Blätter zur Aquarien- und Terrarienkunde 39:366 [German].

Congenital – Three abnormal *Salamandra maculosa* larvae, among 63.

Farke AA. 2004. Paleopathology in archosaurs from the Upper Cretaceous Maevarano Formation of Madagascar. Abstracts with programs. Meeting of the Geological Society of America 36 (5):61.

Trauma – Thirty six of 420 elements with pathology in archosaurs from Upper Cretaceous Maevarano Formation of Madagascar. Thirty one of these were in phalanges and caudal vertebrae. Titanosauran sauropods (*Rapetosaurus* and unnamed) had "bone scars, osteophytes," and vertebral asymmetry. Abelisaurid theropod *Majungatholus* had pedal "osteophytes," a fractured healed pedal phalanx,and caudal vertebrae with "bone spurs, intervertebral fusions, and vertebral spine malformation." "Osteophytes" were present on multiple phalanges of crocodyliform *Mahajangasuchus* and an unnamed crocodiliform had a healed humeral fracture. Future publication of photos of the images from this abstract will should allow assessment if these are truly osteophytes.

Fossil – Thirty six of 420 elements with pathology in archosaurs from Upper Cretaceous Maevarano Formation of Madagascar. Thirty one of these were in phalanges and caudal vertebrae. Titanosauran sauropods (*Rapetosaurus* and unnamed) had "bone scars, osteophytes," and vertebral asymmetry. Abelisaurid theropod *Majungatholus* had pedal "osteophytes," a fractured healed pedal phalanx, and caudal vertebrae with "bone spurs, intervertebral fusions, and vertebral spine malformation." "Osteophytes" were present on multiple phalanges of crocodyliform *Mahajangasuchus* and an unnamed crocodiliform had a healed humeral fracture. Future publication of photos of the images from this abstract will should allow assessment if these are truly osteophytes.

Farlow JO, Pianka ER. 2000. Body form and trackway pattern in Australian desert monitors (Squamata: Varanidae): comparing zoological and ichnological diversity. Palaios 15(3):235–247.

Trauma – Monitors regularly lose distal portions of tails.

Fauvelle M. 1886. Des doigts supernuméraries développés chez l'adulte, leur mode de développement et leur disposition. [Supernumerary digit development in adults, their embryology and disposition]. Bulletins de la Société d'anthropologie de Paris 9:38,44. [French]

Polydactyly in an axolotl.

Feder, M. E. 1983. The relation of air breathing and locomotion to predation on tadpoles, *Rana berlandieri*, by turtles. Physiological Zoology 56:522–531.

Trauma – Experimental turtle predation on *Rana berlandieri tadpoles*.

Federsoni, PA. Jr. 1981. Casos teratogênicos em *Bothrops atrox* (Serpentes: Viperidae: Crotalinae) [Teratological cases in *Bothrops atrox* (Serpentes: Viperidae: Crotalinae)]. Memórias do Instituto de Butantan, São Paulo (1978–1979) 42/43: 49–64 [Portuguese].

Congenital – A specimen of the snake *Bothrops atrox* collected pregnant in Iquitos, Peru, gave birth to 33 healthy offspring in Brazil in 1974, some of which were bred in captivity with their siblings. The inbreeding produced several teratological specimens (although Federsoni also considered the possibility of changes in humidity, atmospheric pressure, and temperature as factors giving rise to the teratologies). The first case was born in 1977 and had a mandibular malformation. The lower jaws were much smaller than the upper jaws. It died 51 days after birth. The second case was born in 1979, and was the most severe case. While its anterior third was normal, the posterior two thirds were rolled up in an abnormal position, with fused scales as ties in each roll. This portion resembled a ball of thread. There, malformations in the vertebral column were easily visible and appeared as a series of hunch backs. Empty spaces and an absence of vertebrae were found in between the bends in the vertebral column. Federsoni proposed the term TERATELESSOMATURO to indicate a teratology involving the distal portion of the body and tail because this type of teratology had not been described in the past, and did not fit the existing classifications. The specimen died 45 days after birth. Two radiographs of it were published in this paper. The third case was also born in 1979, and had malformations in its head – it lacked eyes and orbits. It had slightly larger lower jaws than its upper jaws. It lived 24 h. The fourth case was born dead, and had a head malformed in the left side: it lacked the eye and orbit. The fifth case lacked the right eye, but had an orbit. The sixth case lacked the eye and orbit in the left side of the head.

Fei L., Liang Y-S. 1987. Two cases of abnormality on supernumerary limbs in pelobatid toads (*Vibrissophora boringii*). Chinese Journal of Zoology 22(3):7–11.

Congenital – Supernumerary limbs in *Vibrissophora boringii*.

Fejérváry GJ v. 1916. Beiträge zur Kenntnis von *Rana mehelyi* BOLKAY. [Contribution to the knowledge of *Rana mehelyi* BOLKAY]. Mitteilungen Jahrbuch der königlich-ungarischen geologischen Reichsanstalt 23(3):133–155 [German].

Trauma – Fracture of os ilium in *Rana mehelyi*.

Feldman ML. 1983. Life history notes Testudines. *Chelydra serpentina* (common snapping turtle). Morphology. Herpetological Review 14:47–48.

Congenital – Dicephalic *Chelydra serpentina* (common snapping turtle) with kinked tail.

Feldman ML, Feldman MB. 2011. A tale of two shell diseases: Observations of two shell diseases present in breeding operations of the red-eared slider turtle (*Trachemys scripta elegans*). Turtle and Tortoise Newsletter 15:25–29.

Trauma – Posterior plaston lesions in *Trachemys scripta elegans* attributed to injury climbing a roughly surfaced ramp and abrading against rock surfaces; those on anterior portion suggest collisions with bottom of pool from shallow water diving.

Infection – *Trachemys scripta elegans* plaston with discolored area surrounded by hyperemic blood vessels ulcerating to bone, referred to as septicemic cutaneous ulcerative disease with gram negative bacteria seen.

Environmental – Posterior plaston lesions in *Trachemys scripta elegans* attributed to injury climbing a roughly surfaced ramp and abrading against rock surfaces; those on anterior portion suggest collisions with bottom of pool from shallow water diving.

Shell disease – *Trachemys scripta elegans* plaston with discolored area surrounded by hyperemic blood vessels ulcerating to bone, referred to as septicemic cutaneous ulcerative disease with gram negative bacteria seen. Posterior plaston lesions attributed to injury climbing a roughly surfaced ramp and abrading against rock surfaces; those on anterior portion suggest collisions with bottom of pool from shallow water diving. Also reported white, rough, elevated plaston, and carapace with erosion, attributed to continuous moisture exposure.

Feldmann R, Klewen R. 1981. Feuersalamander – *Salamandra salamandra terrestris* Lacépède. [Feuersalamander – *Salamandra salamandra terrestris* Lacépède]. [Fire salamander – *Salamandra salamandra terrestris* Lacépède]. In Feldmann R. ed. Die Amphibien und Reptilien Westfalens. Abhandlungen aus dem Landesmuseum für Naturkunde Münster 43(4):30–44 [German].

Congenital – *Salamandra salamandra terrestris* with three hands on one front leg.

Felisbino SL, Carvalho HF. 2002. Ectopic mineralization of articular cartilage in Bullfrog *Rana catesbeiana* and its possible Cell Tissue Research 307(3):357–365.

Pseudopathology – Normal cartilage mineralization in ontogeny mimics periosteal reaction in bullfrog *Rana catesbeiana*.

Ferguson MJW. 1981a. Review: The value of the American alligator (*Alligator mississippiensis*) as a model for research in craniofacial development. Journal of Craniofacial Genetics and Developmental Biology 1:123–144.

Congenital – Offspring of young females have high frequency of spontaneous malformations. Spina bifida and anencephaly comprise 70–80% of spontaneous malformations in eggs of young and old alligators.

Ferguson MJW. 1981b. Developmental mechanisms in normal and abnormal palate formation with particular reference to the aetiology, pathogenesis, and prevention of cleft palate. British Journal of Orthodontology 8:115–137.

Congenital – *Alligator mississippensis* is a model for cleft palate research as it is the only external egg derived animal with a true mammal-like secondary palate.

Ferguson MJW. 1984. Craniofacial development in *Alligator mississippiensis*. In Ferguson MWJ. ed. The Structure, Development and Evolution of Reptiles. A Festschrift in honour of Professor A. d'A. Bellairs on the occasion of his retirement. Symposia of the Zoological Society of London 52:223–273. London: Academic Press.

Congenital – Malformations in *Alligator mississippiensis* related to maternal age: Young females have 10% malformation rates (including spina bifida, anencephaly, and cleft palate); old individuals, 5%; and middle aged, < 0.5%.

Cites Ferguson (1982c) and Ferguson (in press): cleft palate in *Lacerta vivipara, Lacerta lepida*, anaconda *Eunectes murinus, Natrix natrix, Vipera berus, Crotalus viridis*, turtles.

Toxicology – 5-fluoro-2-desoxyuridine induced left palate, cyclopia, and absent lower jaw in *Crocodylus johnstoni.*

Ferguson MJW. 1985. Reproductive biology and embryology of the crocodilians. In Gans C, Billett FS, Maderson PFA. Eds. Biology of the Reptilia, vol. 14:329–491.

Congenital Pathology	Phylogeny
Cleft palate	*Alligator mississippiensis, Crocodylus niloticus*
Microcephaly	*Alligator mississippiensis, Crocodylus niloticus*
Hydrocephaly	*Alligator mississippiensis*
Cyclops	*Alligator mississippiensis, O. tetrospis* (Tryon 1980)
Reduced upper or lower jaw	*Alligator mississippiensis, Crocodylus porosus Crocodylus johnsoni, Crocodylus niloticus, Crocodylus palustris, Crocodylus tetrapis*
Skewed lower jaw	*Alligator mississippiensis, Crocodylus porosus, Crocodylus johnsoni, Gavia gangeticus* (Singh and Bustard 1982)
Duplicate teeth in Socket	*Crocodylus porosus* (De Jong 1928)
Teeth erupting into nose	*Alligator mississippiensis*
Post-head duplication	*Alligator mississippiensis*
Dicephalic	*Alligator mississippiensis*
Dicephalic with hind-quarter duplication	*Crocodylus niloticus*
Spina bifida	*Alligator mississippiensis*
Supernumerary limb	*Alligator mississippiensis*
Extra toes	*Alligator mississippiensis, Crocodylus porosus*
Absent digits	*Alligator mississippiensis, Crocodylus niloticus*
Syndactyly	*Alligator mississippiensis*
Extra vertebrae	*Alligator mississippiensis* (Case 1896), *Crocodylus porosus* (Baure 1886, 1889; Rhenhardt 1874), *Crocodylus niloticus* (Kälin 1936a,b)
Missing vertebrae	*Crocodylus porosus* (Webb et al. 1983).
Kinked tail	*Alligator mississippiensis, Gavia gangeticus* (Singh and Bustard 1982), *Crocodylus novaeguineae* (Bustard 1969, 1971), *Crocodylus niloticus, Crocodylus palustris, Crocodylus porosus.*
Reduced or absent tail	*Alligator mississippiensis, Crocodylus porosus, Crocodylus niloticus.*
Twisted back	*Alligator mississippiensis, Crocodylus niloticus, Gavia gangeticus* (Singh and Bustard 1982)

Ferguson MJW. 1989. Birth defects in American alligators. In CA Ross CA, Garnett S. eds. Crocodiles and Alligators, London: Meerhurst Press. p. 98

Environmental – Spina bifida, cyclopia, hydrocephalus, reduced jaw, opisthodichotomy in *Alligator mississippiensis*, especially with egg incubation at viability limits – below 29–34°C or with dehydration.

Ferguson GW, Jones JR, Gehrmann WH, Hammack SH, Hudson R, Talent LG, Frye FL, Dierenfeld ES, Fitzpatrick MP, Holick MF, Chen TC, L u Z, Vogel JJ. 1995. Effects of vitamins A and D and ultraviolet irradiation on the life history of captive panther chameleons (*Chamaeleo pardalis*). In Holick MF, Jung EG. eds. Biologic Effects of Light. pp. 77–79; Berlin: Walter de Gruyter & Co.

Metabolic – High doses of vitamin A produces renal gout in panther chameleon (*Chamaeleo pardalis*).

Ferguson GW, Jones JR, Gehrmann WH, Hammack SH, Talent LG, Hudson DR, Dierenfeld ES, Fitzpatrick MP, Frye FL, Holick MF, Chen TC, Lu Z, Gross TS, Vogel JJ. 1996. Indoor husbandry of the panther chameleon (*Furicifer pardalis*): Effects of dietary vitamins A and D and ultraviolet radiation on pathology and life-history traits. Zoo Biology 15:279–299.

Metabolic – High doses of vitamin A resulted in mild soft-tissue mineralization in panther chameleon *Furicifer pardalis*, in contrast to soft, easily cut, poorly formed bone, vertebral kinking, and irregular cement lines with low doses (deficiency).

Ferguson GW, Gehrmann WH, Karsten KB, Hammnack SH, McRae M, Chen TC, Lung NP, Holick MF. 2003. Do panther chameleons bask to regulate endogenous vitamin D_3 production? Physiological and Biochemical Zoology 76:52–59.

Metabolic – Panther chameleon *Furcifer pardalis* basks to regulate endogenous vitamin D^3 production.

Ferigolo J. 1993. Fratura com osteomielite em femur de crocodiliano do Neogeno do Estado do Amazonas, Brasil. [Fracture with osteomyelitis in a femur of a crocodilian from the Neogene of the State of Amazonas, Brazil.] In IX Jornadas Argentinas de paleontología de vertebrados. Ameghiniana 30:105–106 [Portuguese].

Trauma – Left femur of a crocodilian from the Neogene of Amazonas, Brazil with misaligned healed fracture. It also has a hole of 1.9 cm at its greatest diameter on its anterior face, due to a cloaca that in life drained a purulent secretion.

Infection – Brazilian Amazonian fossil crocodilian femur from the Neogene with a misaligned femoral fracture with a hole of 1.9 cm at its greatest diameter on its anterior face, due to a cloaca that in life drained a purulent secretion. The distal portion of the bone, in its anterior aspect, between the cloaca and the condyle, shows a cortical thickening because of an extensive periosteal reaction due to a minor fracture and/or osteomyelitis. In between the cloaca and the latter lesion, there is a less marked periosteal reaction, which suggests involucrum.

Fossil – Left femur of a crocodilian from the Neogene of Amazonas, Brazil with misaligned healed fracture. It also has a hole of 1.9 cm at its greatest diameter on its anterior face, due to a cloaca that in life drained a purulent secretion.

Brazilian Amazonian crocodilian femur from the Neogene with a misaligned femoral fracture with a hole of 1.9 cm at its greatest diameter on its anterior face, due to a cloaca that in life drained a purulent secretion. The distal portion of the bone, in its anterior aspect, between the cloaca and the condyle, shows a cortical thickening because of an extensive periosteal reaction due to a minor fracture and/or osteomyelitis. In between the cloaca and the latter lesion, there is a less marked periosteal reaction, which suggests involucrum.

Ferreira J. B. 1923. Trabalhos de Erpetologia do Museu Bocage. I. Emydosáurios da Colecção Antiga, provenientes da exploração do Dr. Alexandre Rodrigues Ferreira (1783–1793) [Works on Herpetology of the Museum of Bocage. I. Emydosaurids of the old collection, obtained by the exploration expedition of Dr. Alexandre Rodrigues Ferreira (1783–1793)]. Jornal de Academia de Ciências de Lisboa, Ciências Mathematicas, Physicas e Naturaes, Terceira Série, (14): 77–89 [Portuguese].

Congenital – Ferreira reported the Emydosaurids in the Lisbon Museum collections that resulted from the expedition of Dr. Alexandre Rodrigues Ferreira to northern Brazil (1783–1793). Among these, he remarked that in volume I of the illustrations of the expedition was figured a small, juvenile dicephalic turtle of the species *Podocnemis unifilis* Troschel, a species from the superior course of the Amazon, also found in Pará and Guyana. Figure 3 of Ferreira's paper provided three illustrations of the teratodymous specimen, copied from a watercolor in the illustrations volume of Viagem Scientifica of Dr. Rodrigues Ferreira.

Feuer A. 2005. Pickled Piglets and Other Curiosities, in Exile. The New York Times 4 June 2005:B1.

Congenital – Dicephalic turtle.

Fife JD. 2007. Star tortoise: The Natural History, Captive Care and Breeding of *Geochelone elegans* and *Geochelone platynota*. Serpents Tail, 116 pp.

Stones – Urate bladder stones in Burmese star tortoise.

Figureres JM. 1997. Treatment of articular gout in a Mediterranean pond turtle, *Mauremys leprosa*. Proceedings of the Association of Reptilian and Amphibian Veterinary 7(4):5–7.

Infection – Mediterranean pond turtle *Mauremys leprosa*, with ulcerated abscesses at tarsal joints, but with clinical tarsal and femorotibial joint involvement, without comment on X-ray changes.

Filatow D. 1927. Aktivierung des Mesenchyms durch eine Ohrblase und einen Fremdkorper bei Amphibien. [Activation of the mesenchyme by an ear bladder and a foreign body in amphibians]. Roux Archiv für Entwicklungsmechanik der Organismen Organ für d. gesamte kausald Morphologie 110:1–32 [German].

Trauma – Surgically induced extra forelimb in *Triton taeniatus*.

Finkler MS, Claussen DL. 1997. Use of the tail in terrestrial locomotor activities of juvenile *Chelydra serpentina*. Copeia 1997:884–887.

Congenital – Acaudal condition in common snapping turtle *Chelydra serpentina* was associated with reduced righting from overturned position.

Finlayson R, Woods SJ. 1977. Arterial disease of reptiles. Journal of Zoology, London 183:397–410.

Metabolic – Subperiosteal erosion and generalized skeletal rarefaction with multiple fractures, especially of the appendicular skeleton in General Hardwicke's dab lizards *Uromastyx hardwickii* and Bell's dab-lizards *Uromastyx acanthinurus*.

Vascular – Vascular calcification in 6/12 General Hardwicke's dab-lizards *Uromastyx hardwickii* and 1 of 2 Bell's dab-lizards *Uromastyx acanthinurus*.

Finnell TC. 1864. Reports of Societies: New York Pathological Society. American Medical Times of New York; March 21, vol 6:138.

Congenital – Dicephalic snake.

Fisher GJ. 1868. Diploteratology. An essay on compound human monsters, comparing the history, literature, classification, description and embryology of double and triple formation; including parasitic monsters, foetus in foetu and supernumerary development. 311 pp.; Albany, New York: Van Benthuysen's Steam Printing House.

Congenital – Terminology:

Diprosopus – Two faces.

Cephalopagus – Two bodies

Pygopagus – Butt to butt fusion.

Dicephalic snake *Tropidonotus sipedon*, snake (Conant 1862, Bancroft 1769) and viper (Geoffroy St Hilaire 1836)

Dicephalic tortoise with anterior shell fusion,

Fischer EC. 1896. A two-headed snake. Scientific American 75(23):413.

Congenital – Derodymous hog nose snake *Heterodon simus*.

Fischer CE. 1907. Aberration in scales of regrown tail of *Agama tuberculata*, Gray. Journal of the Bombay Natural History Society 18:208.

Trauma – Regeneration of tail of *Agama tuberculata*.

Fisher EM. 1928. Note on a Two-Headed Specimen of the Pacific Garter Snake. Copeia 167:40–42.

Congenital – Dicephalic Pacific garter snake *Thamnophis sirtalis infernalis*.

Fischer J-L. 1971. Luxation et raideur permanente des members postérieurs chez la grenouille rousse. [Luxation and permanent loss of posterior limbs of red frog]. Bulletin mensuel de la Société Linnéenne de Lyon 10:292–294 [French].

Congenital – One percent micromelia and 1% syndactyly in *Rana temporaria* from the Vincennes woods of Nogent-sur-Marne.

Fischer J-L. 1973. Polydactylie faible chez la grenouille rousse. [Mild polydactyly in red frog]. Bulletin mensuel de la Société Linnéenne de Lyon 42:1–4 [French].

Congenital – Polydactyly in *Rana temporaria*. 10.5% of 229 with bifid distal digits, 3 with microdactyly, 2 with ectodactyly and 1 with clindactyly, noting report by Dubois and Varchard (1971) of ectodactyly.

Fischer J-L. 1977. Un mime morphologique de la polydactylia faible: la fissuration de phalanges distales chez *Rana temporaria*(Amphibiens, Anoures). [Morphologic mimic of mild distal phalanx polydactyly in *Rana temporaria* (Amphibia, Anura)]. Bulletin mensuel de la Société Linnéenne de Lyon 46:143–146 [French].

Congenital – Polydactyly presenting as cleft distal phalanges in 20% of *Rana temporaria*, noting that Dubois (1974) – reported such polydactyly and clindactyly in *Rana graeca*.

Fischer J-L. 1983. Notice sur l'histoire de la teratologie des amphibiens. Note on the history of teratology in amphibians]. In Vago C, Matz G. eds. Comptes Rendus du Premier Colloque International de Pathologie des Reptiles et des Amphibiens, Angers, France, pp. 191–196 [French].

Congenital – General article on origin of monster and teratology appellation, with excellent early references. Drawings of dicephalic "aquatic salamander" illustrated.

Fischer J-L. 1986. De la genèse fabuleuse à la morphogenèse des monsters. [Fabulous generation of monster morphology]. Cahiers d'histoire et de philosophie des sciences no. 13: 150pp.; Paris: Belin edition. [French].

Congenital – General review of teratology, but not specifically of osseous phenomena in herps.

Fitzgerald KT, Vera R. 2006. Spinal osteopathy. In Mader DR. ed. Reptile Medicine and Surgery. Pp. 906–912; Philadelphia: Saunders.

Congenital – Rhoecosis is combination of kyphosis, lordosis, and scoliosis in green iguana *Iguana iguana*.

Infection – Osteopathy in Burmese python *Python molarus bivitattus* from rat bite. Rat bite derivation of infectious phenomenon in Burmese python raises question of its involvement in other animals because of prolonged incubation period for rat bite-related organisms – perhaps longer than the observation period for cultures and therefore missed?

Neoplasia – Osteosarcoma, fibrosarcoma, chondrosarcoma in green iguana *Iguana iguana*.

Vertebral – Proliferative spinal osteopathy in turtles, lizards, and snakes, citing White and Murphy (1972), Hadeen (1976), Grogan (1976), Plymale et al. (1978), Smith and Fitzgerald (1983), Chiodini and Nielsen (1983), Antinoff (1997) and Ogden et al. (1981).
Osteopathy in Ball python *Python regius*, with endplate sclerosis, body lysis, and bone proliferation with fusion of facet joints and vertebral bridging and pathologic fractures in snakes, suggesting possible etiologies including viral, hypovitaminosis D, hypervitaminosis A, inactivity, neoplasia, and septicemia.
Fitzgerald SD, deMaar TW, Thomas JS, Berry DE. 2004. Pathologic limb fracture attributed to mycobacterial infection in a marine toad, *Bufo marinus*, with systemic mycobacteriosis and chromomycosis. Journal of Herpetological Medicine and Surgery 14(3):19–23.
Trauma – Marine toad *Bufo marinus* with comminuted fracture of dital tibia.
Infection – Marine toad *Bufo marinus* with mixed infection (granulomas) of tibial fracture with long thin acid-fast bacteria (*Mycobactertium marinum* and brown pigmented yeast, referred to chromomycosis. The latter is caused by *Cladosporium, Fonsecaea, Phialophora*, or *Scolecobasidium*, citing Done et al. (1993) report of *Bufo americanus* tibiotarsal and cervical vertebral involvement by *Mycobacterium marinum. Mycobacterium terrae*, isolated from the *Bufo marinus* liver, has unclear significance related to the joint infection.
Fitzsimons FW. 1932. Snakes. London: Hutchinson & Co., XIV + 286 pp.
Pseudopathology – *Psammophis* (mistaken dicephalism because of blunt tail) and burrowing snake *Typhlops mucroso*.
Congenital – Derodymous unnamed ? *Crotaphopeltis hotamboeia*
Imaged dicephalic herald or red-lipped snake (*Leptodira hotamboeia*).
Flax NL, Borkin LJ. 1997. High incidence of abnormalities in anurans in contaminated industrial areas (eastern Ukraine). In Böhme W, Bischoff W, Ziegler T. eds. Herpetologia bonnensis Federal Republic of Germany: SHE:119–123.
Congenital – Cites microcephaly, reduction of feet and leg bone size, and chondrodysplastic tail lesions in *Rana chensinensis, R. amurensis* and *Bufo bufo gargarizans* by Mizgireuv et al. (1984).
Environmental – Assymmetry of limbs, hind-limb segment reduction and polydactyly in 28.6% of *Bombina bombina*, 0.3% of *Bufo viridis*, 42% of *Rana ridibunda*, correlated with environmental pollution, with contaminated industrial areas of Nikopol and Dnepropetrovsk Districts 51.6% and 45.8% contrasted with 15.6% from the less polluted Novomoskovsk District.
Fletcher KC, Kardon A. 1979. Treatment of traumatic perforative fractures of nasal cavities and paranasal sinus in an American alligator. Veterinary Medicine 74:1537–1541.
Trauma – Traumatic nasal cavities and paranasal sinus fractures in *Alligator mississippensis*.
Flindt R. 1985. Untersuchungen zum Auftreten von mißgebildeten Wechselkröten (*Bufo viridis*) in einem Steinbruch in Vaihingen-Roßwag. [Investigation of the occurrence of malformed toads (*Bufo viridis*) in a quarry in Vaihingen-Roßwag]. Jahreshefte der Gesellschaft für Naturkunde Württembergs 140:212–233 [German].
Congenital – Supernumerary front limbs in *Bufo viridis*: examples with 2 left, 2–3 right or without front limbs in 0.3–3.2% of the population. Rejected consideration of recessive gene or multiple allele (similar in bastardization between *Bufo viridis* and *Bufo bufo* [Erdkröte], as serum electrophoresis does not give positive indication, making herbids more likely, and suggested loss of limbs by mechanical cause (cannibalism or attack by dragonfly larvae).
Flindt VR, Hemmer H, Shipp R. 1968. Zur Morphogenese von Missbildungen bei Bastardlarven *Bufo calamito* X *Bufo viridis*: Störungen in der Ausbildung des Axialskeletts. [The morphogenesis of malformation in bastard larvae *Bufo calamito* X *Bufo viridis*: Irregularities in the formation of the axial skeleton]. Zoologisches Jahrbuch, Abteilung für Anatomie und Ontogenie der Tiere 85:51–71 [German].
Neoplasia – Shortly after leaving the egg strong arching of larvae caused by primary changes in notochord: local accumulation of chordoblasts, additional growth of chordal sheath tissue, irregularities in size, and formation of chordocytes; secondary changes with start of second week: atypical growth of chordal tissue, chorda sheath, and perichord (with side effects in somatic musculature, neural cord, aorta, and mesonephros). This results in extreme lateral and dorsal enlargement of diameter of notochord and nearly complete displacement of spinal tube, rupture of chordal sheath, and proliferation of chordocytes and chordoblasts into neural tube. Comparison with chordoms in amphibians.
Flower SS. 1899. Notes on a second collection of reptiles made in the Malay Peninsula and Siam. Proceedings of the Zoological Society London 1899:677.
Congenital – Dicephalic *Homalopsis buccata*.
Flyaks NL, Borkin LJ. 2003. Morphological abnormalities and heavy metal concentrations in anurans of contaminated areas, eastern Ukraine. Applied Herpetology 1:229–264.
Environmental – High incidences of morphological abnormalities (mostly of limbs) and tumor-like dysplasias were recorded in three species of anurans [fire-bellied toad *Bombina bombina* – 48%, green toad *Bufo viridis* – 6% (nonaqueous habitat), and marsh frog *Rana ridibunda* – 63%] over 5 years in three areas of Dnepropetrovsk Province, eastern Ukraine, with significant correlations between frequencies of deformities and levels of environmental contamination. These included brachydactyly and polydactyly. They noted

increased (17%) digital abnormalities in *Triturus cristatus* x *T. marmoratus* hybrids. 31% phalangeal or digit brachydactyly especially of forelegs in hynobiid salamander *Salamandrella (Hynobius) keyserlingii* in Sakhalin Island and Kamchatka Peninsula.

Foggin CM. 1987. Diseases and disease control on crocodile farms in Zimbabwe. In Webb GJW, Manolis SC, Whitehead PJ (eds.) Wildlife management: crocodiles and alligators. Surrey Beatty and Sons/The Conservation Commission of the Northern Territory. Chipping Norton. Pp. 351–362.

Metabolic – Lack of Calcium and vitamin D causes rickets which is reflected in uneven growth of jaws, soft pliable bones (particularly the mandible and maxilla) and fractures of the spine with permanent posterior paralysis in crocodiles.

Congenital deformities like coiled and curled tails and abnormalities of the head, limbs, and spine were present in crocodiles (some caused by extreme incubation temperature).

Foggin CM. 1992. Diseases of farmed crocodiles. In: Conservation and Utilization of the Nile Crocodile in South Africa. In Smith GA, Marais J (eds.). Handbook on Crocodile Farming, South Africa: The Crocodilian Study Group of Southern Africa. pp. 197–140.

Infection – Swollen limbs in crocodilians from infections.

Metabolic – Rubber jaw in crocodilian from osteomalacia which he related to calcium deficiency.

Gout in alligators due to overfeeding of red meat at suboptimal temperatures (especially nutria) (Cardeilhac 1981).

Fohrmann U, Hintze-Podufal C. 1987. Anatomische Untersuchungen an ausgewählten Mißbildungen von *Xenopus laevis* (Daudin). [Anatomical investigation of selected malformation in *Xenopus laevis* (Daudin)]. Zoologische Jahrbücher, Abteilung für Anatomie und Ontogenie der Tiere 115:127–146 [German].

Congenital – Deformations of hind legs, pelvic girdle, forelegs, and pectoral girdle in *Xenopus laevis*.

Fojtl J. 1989. Oligodaktýlie u želvy. [Oligodactyly in tortoises.] Akvárium Terárium 32(4):32–33 [Czech].

Congenital – *Testudo hermanni* with oligodactyly (only two phalanges) in right front leg.

Fojtl J. 1994. A double tail in the lizard *Lacerta agilis*. Akvárium Terárium 37(5):46.

Congenital – Tail duplication in *Lacerta agilis*.

Formanowicz DR, Brodie ED, Bradley PJ. 1990. Behavioral compensation for tail loss in the ground skink *Scincella lateralis*. Animal behavior 40:782–784.

Trauma – Reduction in ground skink *Scincella lateralis* activity level, escape speed and flight distance after tail loss.

Fong A, Hero JM, Vina R, Bignotte-Gir?I. 2010. Population ecology of the riparian frog *Eleutherodactylus cuneatus* in Cuba. Biotropica 42:348–354.

Trauma – Rare toe regeneration in riparian frog *Eleutherodactylus cuneatus*.

Forrest R. 2003. Evidence for scavenging by the marie crocodile *Metriorhynchus* on the carcass of a plesiosaur. Proceedings of the Geologists' Association 114:363–366.

Trauma – Bite marks on right neural arch and several centra of Peterborough Museum and Art Gallery R54(B) *Cryptoclidus* correspond in spacing to marine crocodilian *Metriorhynchus brachyrhynchus*.

Fossil – Bite marks on right neural arch and several centra of Peterborough Museum and Art Gallery R54(B) *Cryptoclidus* correspond in spacing to marine crocodialian *Metriorhynchus brachyrhynchus*.

Forsman A, Merilä J, Lindel EL. 1994. Do scale anomalies cause differential survival in *Vivera berus*? Journal of Herpetology 28:435–440.

Congenital – Higher mortality among deviant snake phenotypes (Travis 1989 – but citation is wrong).

Ventral scale and skeletal anomalies (Bernström 1943; King 1959; Peters 1969; Plummer 1980; Arnold 1988; Arnold and Bennett 1988; Schwaner 1990; Merilä et al. 1992).

Ventral scale anomalies (40% according to Merilä et al 1992) in snakes are often associated with fused vertebrae or ribs or vertebral duplications (King 1959; Clark and Callison 1967; Plummer 1980; Merilä et al. 1992).

Ventral scale anomalies are negatively relted to speed in garter snakes *Thamnophis radix* (Arnold 1988).

Anomalies increased with litter size.

Environmental – Morphological anomalies from extreme temperatures (Fox 1948; Osgood 1978) and are often associated with high degree of homozygosity (Leary et al. 1984; Soulé 1979; Schwaner 1990).

Fort DJ, Stover EL, Rogers RL, Copley HF, Mortan LA, Foster ER. 2000a. Chronic boron or copper deficiency induces limb teratogenesis in *Xenopus*. Biological Trace Element Research 77:173–187.

Metabolic – Copper deficiency produces abnormal cartilaginous development with atrophied hind limbs in *Xenopus laevis*.

Boron deficiency produces abnormal (shortened) femur development and forelimb digit malformation in *Xenopus laevis*, especially in F2 generation. Craniofacial malformation in *Xenopus laevis* induced by boron deficiency (Fort et al. 2000, pp. 159–172).

Microcephaly induced by copper deficiency in *Xenopus laevis* (Fort et al. 2000, pp. 159–172).

Fort DJ, Stover EL, Lee CM, Adams WJ. 2000b. Adverse developmental and reproductive effects of copper deficiency in *Xenopus*. Biological Trace Element Research 77:159–172.

Metabolic – Hydrocephaly and microcephaly by copper deficiency in *Xenopus laevis*.

Fort DJ, McLaughlin DW, Burkhart JG. 2003. The FETAX of today – and tomorrow. pp. 23–45. In Linder GS, Krest D, Sparling D, Little E, eds. Multiple Stressor effects in Relation to Declining Amphibian Populations. American Society for Testing Materials International, West Conshoshocken, Pennsylvania.

Metabolic – Boron deficiency produces fore and hind-limb malformation in *Xenopus laevis* (Fort et al. 2000).

Fort DJ, Rogers, RL, Bacon JP. 2006a. Deformities in cane toad (*Bufo marinus*) populations in Bermuda. Part II. Progress towards characterization of chemical stressors.

Environmental – Surface water blamed for malformations, dependent on site:

Species	Microphthalmia	Mouth abnormal	Craniofacial abnormal	Ectrodactyly	Hemimelia
Bufo marinus	10–35%	50–100%	75–100%	25–40%	30–35%
Xenopus laevis	10–25%	75–100%	100%	50–75%	25–50%

Fused digits in 25–30% of *B. marinus.*

Notochord flexion lesions were present in 90% of *Xenopus laevis.*

As toxicity was reduced by iron exchange, diesel-range petroleum hydrocarbons, metals (Al, As, Sn, Cr, Cu, Pb, Mn, Ni, Zn), and ammonia were blamed.

Fort DJ, Rogers RL, Buzzard BO, Anderson GD, Bacon JP. 2006b. Deformities in cane toad (*Bufo marinus*) populations in Bermuda. Part III. Microcosm-based exposure pathway assessment. Applied Herpetology 3:257–277.

Environmental – Frequencies of deformities was locale (in Bermuda)-dependent, with affliction rate varying from 3% to 48%. Mouth deformities were present in 0.04–0.6% of free roaming animals; eye deformities, in 0.1–0.7%; forelimb deformities, in 0.9–1.3% (polyphalangy in 0–19%, brachydactyly in 32–77%) and hind-limb deformities in 13–23%. These included hemimelia, phocomelia, ectrodactyly, and brachydactyly in 87–97% of *Bufo marinus.*

Fowler HW. 1922. Additions to the museum: 1921. Annual Report of the Academy of Natural Science Philadelphia (1921) 11:48–54.

Congenital – Dicephalic water snake *Natrix sipedon.*

Fowler ME. 1978. Metabolic bone disease. In Fowler ME. ed. Zoo and Wild Animal Medicine. Saunders, Philadelphia. pp 53–76.

Metabolic – Defines metabolic bone disease as incorporating osteoporosis, osteomalacia, rickets, simian bone disease, osteogenesis imperfecta, cage paralysis, nutritional secondary hyperparathyroidism, bone atrophy, juvenile osteoporosis, paper bone disease, groundou, Paget's disease, Ricklinghausen's disease, fibrous osteodystrophy, and osteodystrophia fibrosa cystica. Ingestion of high fat diet results in insoluble calcium salts in gut and hypocalcemia. Snakes are rarely affected because of whole animal ingestion, in contrast to lizards, crocodilians, and chelonians.

Admonishes that normal lizard, chelonian, and crocodilian long bones are curved.

Illustrates caiman with skeletal demineralization, common iguana with thin cortices and femoral "cortical alterations" in a Figi Island iguana and a second with "periosteal and subperiosteal membrane."

Fox H. 1941. Kyphoscoliosis in a painted tortoise. Report of the Penrose Research Laboratory 1941:16.

Congenital – Kyphoscoliosis in *Terrapene ornata.*

Fox W. 1948. Effect of temperature on development of scutellation in the garter snake, *Thamnophis elegans atratus.* Copeia 1948(4):252–262.

Environmental – Cooler temperature (65–85°C versus 75–95°C) produced fewer scale rows in garter snake *Thamnophis elegans atratus.*

Fox SF, Rostker MA. 1982. Social cost of tail loss in *Uta stansburiana*, Science 218:692–693.

Trauma – Loss of two third of tail in *Uta stansburiana* resulted in more loss of dominance than one third and reduced habitat quality.

Fox W, Gordon C, Fox MH. 1961. Morphological effects of low temperatures during embryonic development of the garter snake, *Thamnophis elegans*. Zoologica 46(5):57–71.

Environmental – Night chilling of garter snake, *Thamnophis elegans* is teratogenic.

Fox SF, Heger NA, Delay LS. 1990a. Social cost of tail loss in *Uta stransburiana*: Lizard tails as status-signalling badges. Animal Behavior 39:549–554.

Trauma – Tail loss in side-blotched lizard *Uta stansburiana* result in reduced social status. This is more common if interactions are frequent, but risk of agonistic interaction injuries are low, in absence of individual recognition.

Fox SF, Heger NA, Delay LS. 1990b. Social cost of tail loss in *Uta stansburiana*: Lizard tails as status-signalling. Animal Behavior 39:549–554.

Trauma – Tail loss in side-blotched lizard *Uta stansburiana* lowered social status, which was restored in females by artificial tail replacement. Subordinate females became aggressive toward prior dominant opponents when they lost their tails.

Fraisse B. 1885. Die Regeneration von Geweben und Organen bei Wirbelthieren, besonders Amphibien und Reptilien. [The regeneration of tissue and organs in vertebrae, especially in amphibians and reptiles]. VIII+164pp.; Cassel, Berlin: Theodor Fischer [German].

Congenital – Additional legs or toes in urodeles (Duméril 1865; Reuter 1875; Siebold 1828) and anurans (Cisternas1865; Duméril 1865; Gervais 1864; Giebel 1867; Leydig 1877) and own observations.

Trauma – Cites Aristotle in his historical treatment of regeneration; early mentioning of lizards with two or three tails (Plinius the younger, Porta 1650; Aldrovandus 1645, 1647; Gesner 1686; Marcgrav 1648; Marchant 1718; Nédham 1750; de Nobleville 1756; de Bomare 1776; Lacépède 1790; Bosc 1817; de Saint Vincent 1829; Dugès 1829; Gachet 1834; Eversmann 1858; Glückselig 1863; Müller 1864; Giebel 1865).

Regeneration of the tail in lizards in two (*Platydactylus guttatus*) or three (*Podinema teguixin*) tails or seven tail-like processes (*Hemidactylus* sp.).

Frank W. 1965. "Gelenk- und Viszeralgicht" bei Panzerechsen (*Tomistoma schlegelii* und *Gavialis gangeticus*) (Reptilia, Crocodilia). ["Articulation and visceral gout" in armoured lizard (*Tomistoma schlegelii* und *Gavialis gangeticus*) (Reptilia, Crocodilia)]. Tropicana 22:217–234 [German].

Metabolic – Gout (uratic deposition in joints) in two crocodiles, *Tomistoma schlegelii* and *Gavialis gangeticus*.

Frank W. 1976a. Amphibien-Reptilien. In H-G Klös, EM Lang, eds. Zootier Krankheiten. Krankheiten von Wildtieren im Zoo, Wildpark, Zirkus und in Privathand sowie ihre Therapie. [Handbook of Zoo Medicine: Diseases and Treatment of Wild Animals in Zoos, Game Parks, Circuses and Private Collections]. pp. 290–305; Berlin, Hamburg: Paul Parey [German].

Trauma – Rib fractures in chameleon.

Metabolic – Ostitis deformans of vertebrae in *Boa constrictor* (without subspecies, common name cannot be determined).

Neoplasia – osteochondromas in reptiles, carapace defects in tortoises.

Frank W. 1976b. Amphibia-Reptiles. In H-G Klös, EM Lang, eds. Handbook of Zoo Medicine: Diseases and Treatment of Wild Animals in Zoos, Game Parks, Circuses and Private Collections. pp. 355–374; New York: Van Nostrand Reinold Company.

Metabolic – Rickets producing "insufficient mineralization" and osteodystrophia fibrosa in turtles, *Varanus* and iguana.

Frank W. 1976c. Amphibien-Reptilien [Amphibians-Reptiles]. In H-G Klös, EM Lang, eds. Zootier Krankheiten: Krankheiten von Wildtieren im Zoo, Wildpark, Zirkus und in Privathand sowie ihre Therapie. [Zoo Animal Diseases: Diseases of Wild Animals in the Zoo, Wildlife park, Circus, and in Private Hands and its Treatment] Berlin: Paul Parey, pp. 290–305 [German].

Infection – Tail abscess in lizard, but no comment on bone alteration.

Metabolic – Vitamin D deficiency producing osteodystrophia fibrosa in iguana.

Visceral gout I alligator.

Neoplasia – Osteochondroma in reptiles.

Frauscher K. 1903. Eine Ringelnatter mit zwei Köpfen. [Grass or water snake with two heads]. Carinthia, Jahrgang 93:203–204 [German].

Congenital – Dicephalic grass or water snake (?*Tropidonotus*) with two heads.

Frédéricq L. 1883. Sur l'Autotomie. Ou mutilation par voie reflexe comme moyen de defense chez les animaux. [On autotomy. Or mutilation by reflex as a means of defence among animals]. Archives of Zoology Exp Gen 1:413–426 [French].

Trauma – Tail autotomy in *Anguis fragilis*, lizard and *« Serpent de verre »*.

Frédéricq L. 1886. Les Mutations spontanées ou l'autotomie. [Spontaneous mutations or autotomy]. Revue Scientifique (Ser 3) 18:613–620 [French].

Trauma – Tail autotomy in *Anguis fragilis*, lizard and *« Serpent de verre »*.

Frédéricq L. 1887. L'autotomie chez les etoiles de mer [Autotomy in starfish]. Revue Scientifique (Ser 3) 19:589–592 (7 May 1987) [French].

Trauma – Tail loss in *Anguis fragilis* and *Lacerta vivipara, Lacerta muralis*, and "many other lizards."

Frédéricq L. 1888. L'Autotomie ou la mutilation active [Autotomy or active mutilation]. Travaux du Laboratoire de Léon Frédéricq 2:201–221 [French].

Trauma – Tail autotomy in *Anguis fragilis*, lizard and *« Serpent de verre »*.

Frey E, Buchy M-C, Stinnesbeck W, Guadalupe López-Oliva J. 2002. *Geosaurus vignaudi* n.sp. (Crocodyliformes: Thalattosuchia), first evidence of metriorhynchid crocodilians in the Late Jurassic (Tithonian) of central-east Mexico (State of Puebla). Canadian Journal of Earth Science 39:1467–1483.

Trauma – Metriorhynchid crocodilian *Geosaurus vignaudi* with skull surface alterations effaced by subsequent abrasion. Author was unclear if they might represent bites?

Fossil – Metriorhynchid crocodilian *Geosaurus vignaudi* with skull surface alterations effaced by subsequent abrasion. Author was unclear if they might represent bites?
Freye H-A. 1958. Wirbelsäulenverkrümmung beim Krallenfrosch (*Xenopus laevin* Daudin). [Arching of vertebral column in clawed frog (*Xenopus laevin* Daudin)]. Zoologischer Garten Leipzig 23:45–49 [German].
Congenital – Kyphoscoliosis in larva of the African clawed frog *Xenopis laevis* and in the adult stage of the same animal after metamorphosis.
Freytag D. 1941. Eine sechsbeinige Knoblauchkröte. [A common spade foot toad]. Wochenschrift für Aquarien- und Terrarienkunde 38:325–326 [German].
Congenital – *Pelobates fuscus* with two large (as large as the specimen) additional hind legs.
Freytag GE. 1945. Eine sechsbeinige Knoblauchkröte. [A common spade foot toad]. Blätter für Terrarienkunde (52. Jahrgang der "Blätter für Aquarien- und Terrarienkunde") 33:325–326 [German].
Congenital – *Pelobates fuscus* with two large additional posterior extremities.
Freytag D. 1952. Über einen jungen Wasserfrosch mit fünf überzähligen Gliedmaßenanlagen an drei Beinen. [A young water frog with five additional arms and three legs]. Zoologischer Garten 19:154–157 [German].
Congenital – Water frog *Rana esculenta* with five additional anlage of extremities (on both front legs and on left hind leg).
Freytag GE. 1954a. Der Teichmolch. [Common newt]. Die Neue Brehm-Bücherei 117:71pp.; Wittenberg Lutherstadt: A. Ziemsen Verlag [German].
Congenital – Dicephalic *Triturus vulgaris* with tail duplication.
Freytag GE. 1954b. Molche mit Wirbelsäulenverkrümmungen. [Newts with pinal curvature]. Vom Praktiker für den Praktiker. [By practitioners for practitioners]. Aqu u Terr 1:184–185 [German].
Congenital – Japanese fire belly newt *Cynops pyrrhogaster*, Tiger salamander *Ambystoma tigrinum mavortium* and *Triturus cristatus danubialis* with spinal irregularity producing a hump.
Triturus marmoratus marmoratus with kinked tail.
Freytag GE. 1955. Feuersalamander und Alpensalamander. [Fire salamander and Alpen salamander]. Die Neue Brehm-Bücherei 142:79 pp.; Wittenberg Lutherstadt: A. Ziemsen Verlag [German].
Congenital – *Salamandra salamandra quadric-virgata* (Feuersalamander) with double anterior body and only one tail, mentions duplication of legs.
Freytag GE, Hübener HE. 1958. Drei Anomalien in einer Bergmolchzucht. [Three anomalies in an alpine newt]. Deutsche Aquarien und Terrarien-Zeitschrift 11:82–83 [German].
Congenital – *Triturus alpestris apuanus* with arched vertebral column, reduced legs.
Freytag GE, Petzold H-G. 1978. Ein weiterer Beitrag zur Kenntnis der Gattung *Paramesotriton*, insbesondere des nordvietnamesischen Wassermolches *Paramesotriton deloustali* (Bourret 1934) (Amphibia:Caudata:Salamandridae) from North Viet Nam. [Another contribution to the knowledge of the genus *Paramesotriton deloustali* (Bourret 1934) (Amphibia:Caudata:Salamandridae)]. Salamandra 14:117–125 [German].
Congenital – Polydactyly hind legs with 8 and 11 toes.
Fritsch C. 1911. Experimentelle Studien über Regenerationsvorgänge des Gliedmaßenskelets der Amphibien. [Experimental studies of regeneration process of the skeleton of the extremity of amphibians]. Zoologische Jahrbücher, Abteilung für Allgemeine Zoologie und Physiologie der Tiere 30:377–472 [German].
Trauma – Internal structure (bones) of regenerated extremities of *Triton taeniatus*.
T. alpestris. Distinction between true polydactyly and false/rudimentary polydactyly (see Barfurth).
Fry, A. E. 1966. Hind-limb regeneration in *Rana pipiens* larvae. Copeia 1966:530–534.
Trauma – Hind-limb regeneration in *Rana pipiens*.
Frye FL. 1972. Surgical removal of a cystic calculus from a desert tortoise. Journal of the American Veterinary Medical Association 161(6):600–602.
Stone – Urate calculus in desert tortoise *Gopherus agassizii*.
Frye FL. 1973. Husbandry, Medicine and Surgery in Captive Reptiles. 140pp; Bonner Springs, Kansas: VM Publishing, Inc.
Congenital – Caiman with fore-shortened mandible (nutritional osteoporosis); *Iguana* with misshapen mandible. South American red-footed tortoise, *Pseudemys scripta elegans*, and gopher tortoise with asymmetrical shells.
Trauma – Gopher tortoise with carpace and plastron fracture.
Metabolic – Osteoporotic fracture and misshapened carapace (osteomalacy) in *Pseudemys*, *Iguana* with moth-eaten spongy bones attributed to nutritional fibrous osteodystrophy.
Visceral and left axilla gout in *Iguana*.
Neoplasia – Osteochondroma in a monitor lizard.
Vertebral – Boa with scoliosis and reactive bone.
Shell disease – Concave plaston in red-footed tortoise, misshapened carapace (osteomalacia) in *Pseudemys*; *Trionyx* with shell curvature.
Frye FL 1981. Biomedical and surgical aspects of captive reptile husbandry. Volume 1. 325 pp.; Malabar, Florida: Krieger Publishing Company.
Infection – *Boa constrictor* (without subspecies, common name cannot be determined) with vertebral osteomyelitis.

Greek tortoise *Testudo graeca* with septic elbow.
Metabolic – European chameleon *Chamaeleo chamaeleo* with alteration labeled osteolysis – actually osteoporosis.
Hyperparathyoridism in tortoises and iguana.
Fibrous osteodystrophy: Lizard and crocodilian mandibles bow outward or become foreshortened because of tongue musculature tension.
Enlargements are more like fibrous dystrophy, ossifying fibromas.
Iguana with mandibular fracture related to osteodystrophy fibrosa.
Hyperuricemia from water deprivation, excess protein intake and renal disease from improper use of nephrotoxic drugs.
Articular gout in green iguanas and tortoises.
Basiliscus lizard *Basiliscus plumjifrons* with elbow chondrocalcinosis (Calcium pyrophosphate deposition disease).
Toxicology – Ergot induced avascular necrosis of digit or tail.
Neoplasia – *Testudo hermanni* elbow osteolysis with osteochondromatosis-probable neuropathic joint.
Neuropathic – *Testudo hermanni* elbow osteolysis with osteochondromatosis-probable neuropathic joint.
Shell disease – Tortoise deformation from downward tension on carapace underside at girdle attachments. Shallow concavities are found in bone and shell overlying limb attachments.
Ulcerative shell disease.

Frye FL. 1983. Urinary calculosis and cystotomy in a lizard. Veterinary Medicine & Small Animal Clinician 78:431–433.
Stone – Urate calculus in San Estaban Island chuckwalla lizard *Sauromalus varius.*

Frye FL 1984. Nutritional disorders in reptiles 633–660. In Hoff GL, Frye FL, Jacobson ER. eds. Diseases of amphibians and reptiles. Plenum Press, New York. pp. 633–660.
Metabolic – Gout in *Gopherus agassizi* and metabolic bone disease variously called secondary nutritional hyperparathyroidism, osteomalacia, renal rickets, osteogenesis imperfecta, cage parlaysis, osteodystrophia fibrous osteodystrophy.

Frye FL. 1989a. Developmental Anomalies in Reptiles. Proceeding Eastern States Conference, Orlando, Florida, p. 254.
Congenital – Anomalies due to improper egg incubation or genetics include achondroplastic dwarfism, multiple osteocartilaginous exostoses, anophthalmia, cyclopia, microphthalmia, cleft palate, kyphosis, scoliosis, coccygeal agenesis, tail aplasia, rattle agenesis, shell irregularities, and deformities and situs inversus.
Environmental – Anomalies due to improper egg incubation or genetics.
Shell – Shell irregularities and deformities.

Frye FL. 1989b. Periarticular gout in a Pixie frog *Pyxicephalus delalandi.* 3rd International Colloquium on Pathology of Amphibians and Reptiles Orlando, Abstracts: 108.
Metabolic – Periarticular gout in thumb of a Pixie frog *Pyxicephalus delalandi.*

Frye FL. 1989c. Temperature related epigenetic developmental anomalies in the tortoise, *Xenobates agassizi.* First World Congress of Herpetology; University of Kent at Canterbury, United Kingdom, Abstracts: S4.
Environmental – Tortoise *Xenobates agassizi*, incubated at 33.5°C developed anophthalmy, maxillofacial clefts, harelips, forelimb, and partial hind-limb agenesis, coccygeal hypoplasia and scute, and shell plate duplication.

Frye FL. 1991a. Biomedical and surgery aspects of captive reptile husbandry. 2nd enlarged edition. 2 vol.: XV + 637 +16 pp. Malabar, Florida: Krieger Publishing Co.
See separate titles below.

Frye FL. 1991b. Developmental abnormalities. In Frye FL. Biomedical and surgery aspects of captive reptile husbandry. 2nd enlarged edition. pp. 393–419; Malabar, Florida: Krieger Publishing Co.
Congenital – Notes no anomalies have been recorded in *Sphenodon punctatus*, but the animal is extremely rare.
Parasitic twin of California desert tortoise *Xerobates agassizi* attached to caudal shell, conjoined slider turtles, citing Koch's day gecko *Phelsuma madagarascarensis* joined at dorsum of cranium (Tytle et al. 1984) and conjoined twin Australian blue-tongued skink *Tiliqua scincoides* (Willis 1932) – but review of actual article revealed no bone involvement.
Cites previous reports of double monster snakes or chelonians and occasional lizards: Belfit and Nienaber (1983); Campbell (1967); Cohen (1986); Dexter (1976); Feldman (1983); Nakamura (1938); Newman (1917, 1923).
Dicephalic Burmese pytheon *Python molurus bivittatus* (Clark and Tytle 1983), terrapin (Hildebrand 1930), garter snake (Nelson and Slavens 1975); *Kinosternon flavescens* (Porras and Baeraducci 1980); Amphidichotomous black racer *Coluber c.* constrictor (Tuck and Hardy 1970).
Derodymus California king snake *Lampropeltis getula californiae*, Pacific gopher snake *Petuophis melanoleucus catenifer*, gopher snake, northern red-bellied snake *Storeria o. occipitomaculata*, black rat snake

Elaphe o. obsoleta; Great Basin gopher snake *Pituophis deserticola*, Northern black racer *Coluber c. constrictor*, fox snake *Elaphe vulpina*, Derodymous diamondback terrapin *Malaclemys terrapin*, California king snake *Lampropeltis getula californiae*, gopher snake, and red-eared slider turtle *Trachemys scripta elegana.*

Anakatamesodidymus black racer *Coluber constrictor.*

Ododymus queen snake *Regina septemvittata* (Herman and Johnson 1986).

Labial and palatine clefts in Prarie rattlesnakes *Crotalus viridis viridis* (Dean et al. 1980).

Shell and limb agenesis in painted turtles.

Achondroplastic dwarf red-eared slider *Trachemys scripta elegans* (Frye 1981; Frye and Carney 1974) and inbred California desert tortoises *Xerobates agassizi.*

California desert tortoise *Xerobates* (*Gopherus*) *agassizii* with absent forearms, cleft superior mouth parts, upper jaw foreshortening, caudal agenesis and carapacial, and plastal plate and scute abnormalities (Frye 1989a).

Inbred California desert tortoise with increased marginal carapacial and asymmetrical plastral plates. Wedge-shaped vertebral plates in red-footed tortoise, *Geochelone carbonaria.* Anomalies of shell, scute, etc in painted turtle *Chrysemys picta picta* and common snapping turtle *Chelydra serpentina serpentina* observed by Lynn and Ullrich (1950).

Supernumerary vertebral and marginal or reduced carapacial plates in desert tortoises. Anomalous vertebral plates in red-eared slider turtle *Trachemys scripts elegans.*

Plastal axial deviation in red slider turtle and desert tortoise. 180° torsion in *Sternotherus odoratus.* Massive carapacial and plastal asymmetry in box turtles *Terepene* sp.

Raised conical plates in star *Geochelone elegans* and leopard tortoises *Geochelone pardalis.*

Absent tail in *Calotes versicolar* (Mathur and Goel 1974) and California desert tortoise *Xerobates agassizzii.*

Blacktail rattlesnake *Crotalus molossus* lacking a rattle.

Neoplasia – Multiple osteocartilaginous exostoses in varanids (Frye 1981; Frye and Dutra 1973).

Frye FL. 1991c. Common pathologic lesions & disease processes. In Frye FL. Biomedical and surgery aspects of captive reptile husbandry. 2nd enlarged edition. pp. 530–617; Malabar, Florida: Krieger Publishing Co.

Infection – Granulomatous osteitis producing pathologic fractures – Frye and Kiel (1985), Kiel (1977), Marcus 1981.

Metabolic – Osteomalacia in common iguana.

Fibrous dysplasia-like disease in iguanids, similar to ossifying fibroma.

Arthritis – Osteoarthritis in aged lizards, especially *Varanus*, affecting hip, elbow and shoulder and occasionally temporomandibular joints.

Vertebral – Ossifying spondylosis in an iguana – ill-defined image – unclear diagnosis. Also in lizards and crocodilians.

Spine deformation and spinal canal narrowing in corn snakes *Elaphe guttata.*

Osteochondritis – Osteochondritis dessicans in proximal humerus of alligator *Alligator mississippiensis.*

Environmental – Subcutaneous calcification from excess vitamin D and calcium in diet or chronic overheating.

Stones – Uroliths in pseudobladder (produced by pressure from a tumor) in spitting cobra *Naja nigricollis.*

Other – Osteitis deformans in boa constrictor *Boa constrictor*, rattlesnake – calcific overgrowths of unclear diagnosis, but not osteitis deformans.

Subcutaneous calcification from excess vitamin D and calcium in diet or chronic overheating.

Ossification of African leopard tortoise *Geochelone pardalis babcocki* liver and other internal organs.

Frye FL. 1991d. Common pathological lesions and disease processes: Neoplasia. In Frye FL. ed. Reptile Care: An Atlas of diseases and Treatments. Vol 2:576–609. TFH Publishing, Neptune City, New Jersey.

Congenital:

Terminology:

Teratopagus – Conjoined twins with axial skeletal fusion.

Craniophagus – Fusion by cranium.

Cephaloderopagus – Cranial and cervical vertebral fusion.

Anakatamesodidymus – Fused at trunk only.

Pygopagus – Fused at pelvis.

Teratodidymus – Duplications.

Rhinodidymus – Double nosed.

Opodidymus – Cranial bifurcation.

Derodidymus – Cervical or thoracic vertebral bifurcation.

Theracopagus – Thoracic vertebral bifurcation.

Psodidymus – parasacral bifurcation.

California desert tortoise *Xerobates agassizi* with parasitic twin attached to caudal shell. Back to back fusion of Australian blue-tongued skink *Tiliqua scincoides*.

Aristotle described a two-headed snake.

Dicephalism in California king snake *Lampropeltis getula californiae*, Pacific gopher snake *Pituophis melanoleucus catenifer*, northern red-bellied snake *Storeria o. occipitomaculata*, black rat snake *Elaphe o. obsoleta*, *Kinosternon flavescens*, red eared slider *Trachemys scripta elegans*, California desert tortoise *Xerobates agassizi*, queen snake *Regina septemvittata*, Burmese python *Python molurus bivittatus*.

Axial bifurcation was described in black racer *Coluber c constrictor*; cranial and thoracic duplication in Great Basin gopher snake *Coluber constrictor* and fox snake *Elaphe vulpina*.

Common or red tail boa constrictor *Boa c constrictor* with duplication and kyphoscoliosis.

Kyphosis and scoliosis in *Crocodylus siamensis* and red-spotted garter snake *Thamnophis sirtalis concinnus*.

Coccygeal agenesis in *Calotes versicolor, xerobates agassizii, Crotalus molossus* and *viridis oreganus* (no rattle).

Achondroplastic dwarfism in *Trachemys scripta elegans* and *Xerobates agassizi*.

Plastral and carapacial asymmetry in box turtle *Terrapene*.

Diamondback terrapin *Malaclemys centrata* with incomplete scutation. 180° twist of a musk turtle *Sternotherus odoratus* shell.

Arthritis – Spondyloarthropathy-type changes in iguana, alligator, and boa constrictor *Boa constrictor*. Additionally, discusses osteoarthritis in *Varanus* sp – which actually appears to be spondyloarthropathy.

Environmental – Abnormalities in shell plates in painted turtle *chrysemys picta picta* and common snapping turtle *Chelydra serpentina serpentina* if eggs allowed to dry between the 35th and 50th day of the 60–75-day incubation period. 32°C incubation of *Xerobates* (*Gopherus*) *agassizii* with lack of forelimbs, upper jaw foreshortening, and supernumerary or anomalous carapacial or plastal plates.

Osteochondritis – Osteochondritis was reported in *Alligator mississippiensis*.

Neoplasia – Osteochondroma in *Varanus bengalensis;* osteosarcoma, in Burmese python *P. molurus bivittatus* and rofous-beaded snake *R. rostratus*.

Frye FL. 1991e. Reptile Care: An Atlas of diseases and Treatments. Vol 1: 324 pp. TFH Publishing, Neptune City, New Jersey.

Trauma – Healed fractures of carapace and plaston in *Terrapene* sp., limb fracture in soft shell turtle from rock falling on it.

Infection – Septic elbow in Greek tortoise *Testudo graeca*. Infected vertebrae in *Elaphe guttata*. Iguana with vertebral osteomyelitis and fracture.

Mycobacterial osteitis of leg of desert tortoise *Gopherus agassizi*.

Miliary tuberculosis affecting bone in Children's python *Liasis childreni*.

Metabolic – Fibrous osteodystrophy producing folding fractures, outward bowing and foreshortening of mandibles (e.g., *Iguana iguana*), and deformed carapaces.

Rickets producing rachitic rosettes and thickened, ballooned bones.

Renal rickets in terrestrial chelonian with accentuated "interplates" and growth lines.

Primary hyperparathyroidism in tortoises and iguana.

Pseudogout in red-eared slider turtle *Trachemys scripta elegans* and in shoulder and elbow of *Uromastix*.

Hypervitaminosis D produces soft tissue calcification.

Suggestion of Paget's disease in (without subspecies, common name cannot be determined) *Boa constrictor*, but not confirmable.

Arthritis – Figures 7–39 illustrate what he called rheumatoid arthritis, the X-ray image illustrated is not sufficiently clear to recognize the osseous pathology.

Neuropathic – Elbow fragmentation in *Testudo hermanni*.

Vascular – Digit necrosis in red-eared slider turtle *Trachemys scripta elegans*.

Gangrene of tail caused by thromboembolisms from intravascular helminths, trauma, ergot or fescue induced digital avascular necrosis.

Other – Osteitis in corn snake.

Massive osteolysis in *Chamaeleo chameleo*.

Frye FL. 1994a. Diagnosis and Surgical Treatment of Reptilian Neoplasma with a compilation of Cases 1966–1993. In Vivo 8:885–892.

Metabolic – Parathyroid adenomas in three *Iguana iguana, Geochelone carbonaria, Testudo gracca, Xerobates agassizi*.

Neoplasia – Captive live longer than wild and often develop multiple neoplasias.

No comment on presence of bony involvement:

Ossifying fibroma in two lizards: *Iguana iguana*

Osseous dysplasia in two snakes and two lizards

Osteoma/osteosarcoma in 1 snake: Burmese python *Python m. bivittatus*

Ostochondrosarcoma in 1 lizard: *Varanus bengalensis*
Stone – Urocystolithiasis from testicular tumor in a cobra which produced pseudobladder by pressure on ureter, with eight urinary calculi.

Frye FL. 1994b. Reptile Clinician's Handbook: A Compact Clinical and Surgical Reference. Pp. 171–172; Malabar, Florida: Krieger Pub Co P.
Stone – Urocystolithiasis from testicular tumor in a cobra, which produced pseudobladder by pressure on ureter, with eight urinary calculi.

Frye FL. 1995. Self-assessment color review of reptiles and amphibians. Self-assessment color review of reptiles and amphibians 1995:93–94.
Metabolic – Green iguana *Iguana iguana* with swollen limbs and slight raised ridges on rib cage "particularly when the iguana took in a deep breath." Massive pulmonary silicosis and severe hepatic fibrosis were noted, supporting diagnosis of hypertrophic osteoarthropathy, although osteomalacia must also be considered (given the rib pathology).
Other – Green iguana *Iguana iguana* with swollen limbs and slightly raised ridges on rib cage "particularly when the iguana took in a deep breath." Massive pulmonary silicosis and severe hepatic fibrosis were noted, supporting diagnosis of hypertrophic osteoarthropathy, although osteomalacia must also be considered (given the rib pathology).

Frye FF. 1997. The importance of calcium in relation to phosphorus, especially in folivorous reptiles. Proceedings of the Nutrition Society 56:1105–1117. (listed as FF. instead of FL.).
Metabolic – Diets with too much phosphorus, relative to calcium produce "fibrous osteodystrophy." Mandibles (especially of green iguanas) become diffusely swollen and bowed outward.

Frye FL. 2000. Polyostotic appendicular osteitis deformans (Paget's disease) in an ornate uromastyx lizard, *Uromastyx ornata*. Proceedings of the Association of Reptilian and Amphibian Veterinarians 2000:19.
Other – Uromastyx lizard, *Uromastyx ornata*, with swollen coxofemoral, radiohumeral, radiocarpal, tarsal, and metatarsal joints with "multiple rounded firm excrescences of yellowish tissue arising from the joints." Bony excrescences were composed of "variably thick compact bony trabeculae… formed into lamellae with multiple and markedly irregular cement lines. Intertrabecular spaces were filled with normal appearing bone marrow and small amounts of fat." This actually differs from Paget's disease because of lack of marrow fibrosis and absence of the classic giant osteoclast-type/like cells.

Frye FL, Carney J. 1974a. Osteitis deformans (Paget's disease) in a boa constrictor. Veterinary Medicine, Small Animal Clinician 69:186–188.
Vertebral – Boa constrictor (without subspecies, common name cannot be determined) *C. constrictor* boa constrictor with thickened calcification in the form of excrescences and coalescing masses on vertebrae with irregular cement lines.
Other – Boa constrictor (without subspecies, common name cannot be determined) *C. constrictor* boa constrictor with thickened calcification in the form of excrescences and coalescing masses on vertebrae with irregular cement lines.

Frye FL, Carney J. 1974b. Achondroplastic dwarfism in a turtle. Veterinary Medicine, Small Animal Clinician 69:299–301.
Congenital – *Chrysemys (Pseudemys) scripta elegans*, red-eared turtle with enlarged brachycephalic head, foreshortened rostrum, and long bones.

Frye FL, Carney JD. 1975. Parathyroid adenoma in a tortoise. Veterinary Medicine, Small Animal Clinician 70:582–584.
Metabolic – South American red-legged tortoise *Geochelone carbonaria* with soft, deformed carpace and plaston, caused by hyperparathyroidism.

Frye FL, Dutra F. 1973. Multiple osteocartilaginous exostoses in a monitor lizard. Veterinary Medicine, Small Animal Clinician 68:1414–1416.
Neoplasia – Long-necked monitor lizard, *Varanus benegalensis nebulosus*, with "extensions from a portion of the osseous cortex" of ribs, carpometacarpus, and tarsometatarsus with cortical replacement. While called osteocartilaginous exostoses, this neoplasia is of unclear derivation and identity.

Frye FL, Dutra FR. 1974. Mycotic granulomata involving the forefeet of a turtle. Veterinary Medicine, Small Animal Clinician 69:1554–1556.
Infection – Musk turtle *Sternotherus odoratus* digit and metatarsal "involvement" but does not comment if bone affected.

Frye FL, Dutra FR. 1976. Articular pseudogout in a turtle (*Chrysemys [Trachemys] p. elegans*). Veterinary Medicine, Small Animal Clinician 71:655–659.
Metabolic – Articular pseudogout (calcium pyrophosphate deposition disease) in elbow of a turtle, *Chrysemys elegans*.

Frye FL, Kaiser HE. 1995. Characteristic of some notable monostotic mandibular lesions in green iguanas (*Iguana iguana*) Proceedings of the Fifth International Conference on Anticancer Research Corfu Greece, 17–22 October, 1995:1.

Other – Mandibular lesions in green iguanas *Iguana iguana.*

Frye FL, Kiel JL. 1985. Multifocal granulomatous spondylitis in a rattlesnake. Annual Proceedings of the American Association of Zoo Veterinarians 1985:60–66.

Infection – Hybrid speckled rattlesnake, *Crotalus mitchellii*, x ridge-nose rattlesnake, *Crotalus willardi*, with osteolytic and blastic lesions with granulocytic infiltrate and rod-shaped bacteria with a solitary polar flagellum. Osteomyelitis.

Frye FL, Williams DL. 1995 Self-Assessment Colour Review of Reptiles and Amphibians. 192 pp.; London: Manson Publishing/The Veterinary Press.

Other – Massive bony overgrowth of long bones and "bilateral slightly raised longitudinal ridges on rib cage" in *Iguana iguana* with massive pulmonary silicosis and fibrosis, severe hepatic fibrosis. Cross-section appears to show expanding bony reaction, but unclear if represents periosteal reaction. Hepatic disease can cause hypertrophic osteoarthropathy, but uncertainty remains in this case.

Frye FL, Centofanti BV, Harris JM. 1991. Successful treatment of iatrogenic (diet-related) hypervitaminosis-D and hypercalcemia in four iguanas, *Iguana iguana.* 4th International Colloquium on Pathology and Medicine of Reptiles and Amphibians; Bad Nauheim:244–249.

Metabolic – *Iguana iguana* with swollen femora and healed humeral fracture, osteopenic and deformed mandibles.

Frye FL, Claxton RF, Bailey JF, Vasser JR, McNeely HE. 1994. Hypertrophic osteoarthropathy in an iguana, *Iguana iguana.* Second World Congress of Herpetology, Adelaide, South Australia, Abstracts:91–92.

Other – "Massive and extremely sclerotic neo-osseous tissue which eventually involved the diaphyses of most… leg bones" of *Iguana iguana,* referred to as hypertrophic osteoarthropathy.

Fujinami A. 1901. Ueber die Gewebsveränderungen bei der Heilung von Knochenfrakturen. Eine vergleichende pathologische-anatomische Studie. [On the tissue changes during the healing of bone fractures]. Beiträge zur pathologischen Anatomie und allgemeinen Pathologie 29:432–485 [German].

Trauma – Healing of artificially broken bones (callus formation).

Fukada H. 1978. A case of twin hatching in the snake, *Elaphe climacophora.* Japanese Journal of Herpetology 7(4): 88–91.

Congenital – Fraternal twin Japanese rat snake *Elaphe climacophora.*

Fulton MH, Chambers JE. 1985. The toxic and teratogenic effects of selected organophosphorus compounds on the embryos of three species of amphibians. Toxicology Letters 26:175– 180.

Toxicology – Organophosphate induced spine deformities in "less than 10%" (p. 178) of gray treefrog *Hyla chrysoscelis*, narrow-mouthed toad *Gastrophryne carolinensis* and 9% of leopard frog *Rana sphenocephala.* Cites bent tails produced by di(2-ethyhexyl) phthalate, methylmercury chloride, and a thalidomide analog (Dumpert and Zietz 1984) and malformations of the head and spine induced by sodium selenite, acridine, aniline, pyridine, and quinolone (Browne and Dumont 1980; Davis et al. 1981; Schultz et al. 1982).

Funk RS. 1996. Tail damage. In Mader DR. ed. Reptile Medicine and Surgery. pp. 417–418; Philadelphia: Saunders.

Trauma – Lizard with second tail. Tail amputation produced by avascular necrosis from trematode *Alaria.*

Funk RS. 2006a. Snakes. In Mader DR. ed. Reptile Medicine and Surgery. pp. 675–682; Philadelphia: Saunders.

Neoplasia – Chondroma in milk snake *Lampropeltis triangulum multistriate.*

Funk RS. 2006b. Tail damage. In Mader DR. ed. Reptile Medicine and Surgery. pp. 913–915; Philadelphia: Saunders.

Trauma – Forked tails.

Vascular – Tematode *Alaria*-induced avascular necrosis of tail in garter snake *Thamnophis sirtalis.* Thromboembolic avascular necrosis of the tail in Burmese python *Python molurus bivitattus.*

Ascending avascular necrosis in green iguana *Iguana iguana.*

Gabor M, Puţariu F, Deac C. 1973. Action tératogène des aflatoxines sur les têtards [Teratogenic action of aflatoxins on tadpoles]. Archives Roumaines de Pathologie Experimentales et de Microbiologie 32:269–275 [French].

Toxicology – Aflatoxin induced hemimely and ectomely in *Rana temporaria.*

Gabrisch K, Zwart P. 1992. Tortues. [Turtles]. In Gabrisch K, Zwart P. eds. La Consultation des Nouveaux Animaux de Compagnie [Consultation on new country animals], Maisons-Alfort: Editions du Point Veterinaire, pp. 243–286 [French].

Metabolic – Secondary hyperparathyroidism producing osteodystrophy fibrosa with residual lacy calcification in *Chrysemys s. elegans.*

Infection – Carapace mycosis in Hermann's tortoise *Testudo hermanni.*

Irregular lytic carapace lesions attributed to granulomatous disease in *Chelodina longicollis.*

Shell – Carapace necrosis forming coalescing holes in *Emydoidea blandingii.*

Gachet H. 1834. Mémoire sur la reproduction de la queue des reptiles sauriens. [Memoir on regeneration of the tail in saurian reptiles]. Actes de la Société Linnéenne de Bordeaux 36:213–259 [French].

Trauma – Double tail in *Lacerta ocellata, L. viridis, L. muralis, Anolis, Iguana, Ameiva* (after Fraisse 1885).

Lizards with two or three tails (Aldrovande. Jo Baptistoe Portoe Neapolitani, magioe naturalis libri viginti. Rhotomagi MDCL lib II. De variis animalibus gignendis. Chap XVIII serpentes pluribus capitibus caudibusque). Also sites de Bomare (1776), p. 96: bifurcation. Cites Cuvier - Regn anim t II (1829), p. 29 for tail duplication in *Ameiva.* Tail duplication in lézard ocellé, lézard vert piqueté, lézard des murailles *(Lacerta muralis), Anolis a crête, Iguana ardiosé, Lac. teguixin, Lac. coeruleocephala, Lac. bipes.*

Gadd JP. 1982. Observations on the sexual behavior of the boa constrictor, *Constrictor constrictor*, in captivity, with notes on an unsuccessful parturition. Bulletin of the British Herpetological Society 6:39–41.
Congenital – Scoliotic vertebral column in boa constrictor (without subspecies, common name cannot be determined), *Constrictor (Boa) constrictor*.

Gadow, H. 1901 (2. ed.1909, 3. ed. 1920, 4. ed. 1923). Amphibia and Reptiles. London: McMillan and Co. 668 pp.
Congenital – Right ilium in a *Bombinator* and a *Alytes* articulated with tenth, while left articulated with ninth vertebra.

Gadow H. 1905. Orthogenetic variation. Science New Series 22(568):637–640.
Congenital – Corrects Coker's 1905 comments on loggerhead turtle *Thallassochelys caretta* scute variation. 9/28 with increased scutes, 2 with decreased costal scutes. This was found in 7% of newborns, 33% of 3–8 in., 22% of 8–24 in., 24% of larger, and 1 of 7 large turtles.

Gaffney ES, Tong H. 2008. Redescription of the skull of *Ummulisani rutgersensis* Gaffney, Tong, and Meylan, 2006, a bothremydid side-necked turtle from the Eocene of Morocco. American Museum Novitates 3615:1–20.
Congenital – Bothremydid side-necked turtle from the Eocene of Morocco *Ummulisani rutgersensis* AMNH 30569 lacking frontal bones, a unique pathology in turtles. Complete absence of septum orbitotemporal was attributed to absence of frontals (bones involved in the pleurodire septum orbitotemporale.
Fossil – Bothremydid side-necked turtle from the Eocene of Morocco *Ummulisani rutgersensis* AMNH 30569 lacking frontal bones, a unique pathology in turtles. Complete absence of septum orbitotemporale was attributed to absence of frontals (bones involved in the pleurodire septum orbitotemporale.

Gage SH, Gage SP. 1886. Aquatic respiration in soft-shelled turtles; A contribution to the physiology of respiration in vertebrates. American Naturalist 20:233–236.
Vascular – Aquatic respiration in *Amyda mutica* and *Aspidonectes spirifer.*
A submerged 1 kg turtle extracted 71 mL of free oxygen and added 318 mg of carbon dioxide in 10 h.

Gaggero P. 1960. Caso de monstruosidad en el sapo *Bufo arenarum* Hensel [A case of monstrosity in the frog *Bufo arenarum* Hensel]. Actas y Trabajos del Primer Congreso Sudamericano de Zoología 4:69–72 [Spanish].
Congenital – Supernumerary hind limb in common Argentine frog *Bufo arenarum* Hensel, in the teratologic collection of the division of Herpetology of the Museum of La Plata.

Gál J, Jakab C, Balogh B, Tóth T, Farkas B. 2007. First occurrence of periosteal (juxtacortical chondroma) in *Uromastyx maliensis* (Reptilia: Sauria: Agamidae). Acta Veterinaria Hungarica 55(3):327–331.
Neoplasia – Periosteal (juxtacortical) chondroma of the "shoulder" in *Uromastyx maliensis.*
Cites osteochondrosarcoma in black-lipped forest cobra *Naja melanoleuca* and in corn snake *Pantherophis guttatus* (Mader 1996), osteochondroma in Bengal monitor *Varanus benegalensis* (Frye 1991; Mader 2006), enchondroma in Indian monitor *Varanus dracaena (indicus)*, osteosarcoma in green lizard *Lacerta viridis*, chondro-osteofibroma in rhinoceros iguana *Cyclura cornuta* (Mader 2006), and osteochondroma and osteosarcoma in green iguana *Iguana iguana* (Jacobson 2003), chondroma in trachea of ball pythons *Python regius* (Diethelm et al. 1996, Drew et al 1999) and chondrosarcoma in *Gerrhosaurus* (Garner et al 2004).

Galbraith DA. 2008. Population biology and population genetics. In Biology of the Snapping Turtle (*Chelydra serpentina*). Steyermark AC, Finkler MS, Brooks J, eds. Baltimore: John Hopkins University Press, pp. 168–180.
Shell – Vertebral scute perforations in snapping turtle *Chelydra serpentina.*

Galois P, Ouellet M. 2001. 2001 Health and disease in Canadian reptile populations. In C. N. L. Seburn and C. A. Bishop (eds.), Ecology, Conservation and Status of Reptiles in Canada. Herpetological Conservation. Society for the Study of Amphibians and Reptiles, Bethesda, MD. Vol. 2. Salt Lake City, Utah, pp. 131–168.
Congenital – Note only 75 published reports of abnormalities in Canadian reptiles.
They express the urgent need for baseline information.
This article could be considered the "clarion" for this annotated bibliography.

Galois P, Ouellet M. 2007. Traumatic injuries in Eastern Spiny Softshell turtles (*Apalone spinifera*) due to recreational activities in the Northern Lake Champlain basin. Chelonian Conservation and Biology 2007 6(2):288–293.
Trauma – Traumatic injury in eastern spiny softshell turtle *Apalone spinifera.*

Gamble KC, Garner MM, West G, Didier ES, Cali A, Alvarado TP. 2005. Kyphosis associated with microsporidia in San Marcos salamanders *Eurycea nana*. Journal of Herpetological Medicine and Surgery 15:14–18.
Environmental – Kyphosis from myositis associated with microsporidia in San Marcos salamanders *Eurycea ana.*

Gans C. 1961. Notes on amphisbaenids (Amphisbaena: Reptilia). 2. *Amphisbaena occidentalis* Cope from the Coastal Plain of Northern Peru. Postilla 56:1–17.
Trauma – *Amphisbaena occidentalis townsendi* has a fracture plane and autotomy

Gans C. 196a. Notes on amphisbaenids (Amphisbaena: Reptilia). 5. A redefinition and a bibliography of *Amphisbaena alba* Linné. American Museum Novitiates 2105:1–31.
Trauma – Autotomy in *Amphisbaena* species at a single level 4–8 segments posterior (22–26) to cloaca, but no fracture planes or autotomy in *Amphisbaena alba*

Gans C. 1962b. Notes on amphisbaenids (Amphisbaena: Reptilia). 6. Redescription and range extension of *Amphisbaena spurrelli* Boulenger. Breviora 171:1–11.
Trauma – In *Amphisbaena spurrelli*, autotomy occurs after the second annulus.

Gans C. 1964. Notes on amphisbaenids (Amphisbaena: Reptilia). 12. Redescription of *Amphisbaena dubia* Müller (Amphisbaenia: Reptilia). Breviora 205:1–11.

Trauma – No fracture planes or autotomy in *Amphisbaena alba*.

Gans C. 1971. Studies on amphisbaenians (Amphisbaenia, Reptilia). 4. A review of the amphisbaenid genus *Leposternon*. Bulletin of the American Museum of Natural History 144:379–464.

Congenital – Lack of regular correlation between dermal annular scales and vertebral number in *Leposternon*. Variation in vertebral number in *Leposternon microcephalum* was 93–105; *Leposternon scutigerum*, 119–123; *Leposternon polystegum*, 122–135; *wucherei*, 96–99.

Gans C. 1978. The characteristics and affinities of the Amphisbaenia. Transactions of the Zoological Society of London 34:347–416.

Pseudopathology – Limbless lizard whose tail is sometimes mistaken for a second head.

Gans C, Alexander AA. 1962. Studies on amphisbaenids (Amphisbaenia, Reptilia). 2. On the amphisbaenids of the Antilles. Bulletin of the Museum of Comparative Zoology 128:65–158.

Trauma – Autotomy in *Amphisbaena caeca* but not *Amphisbaena antillensis* or *punctata*.

Garza-Garcia AA, Driscoll PC, Brockes JP. 2010. Evidence for the Local Evolution of Mechanisms Underlying Limb Regeneration in Salamanders. Integrative and Comparative Biology 50:528–535

Trauma – Evolution of limb regeneration in salamanders. Developmental models discussed included *Ambystoma mexicanum, maculatum*, and *trigrinum, Xenopus tropicalis* and *laevis, Rana chensinensis*, red-legged salamander, Japanese common newt, Wehrle's salamander and Eastern newt

Gardiner DM, Hoppe DM. 1999. Environmentally induced limb malformations in Mink Frogs (*Rana septentriolis*). Journal of Experimental Zoology 284:207–216.

Environmental – All Mink frog (*Rana septentrionalis*) from Minnesota had deformed hind limbs, with environmental retinoids suggested as etiology.

Garner MM. 1995. Vertebral chondrosarcoma in a corn snake (*Elaphe gutatta gutatta*). In: Proceedings of Joint Conference of the American Association of Zoo Veterinarians, Wildlife Disease Association, and American Association of Wildlife Veterinarians, Houston, pp. 332–334.

Neoplasia – Reported corn snake *Elaphe guttata* with vertebral chondrosarcoma and pathologic fracture. Cites multiple vertebral chondromas in emerald lizard *Lacerta viridis* (Stolk 1958), vertebra osteochondrosarcoma in black cobra *Naja melanoleuca* (Wadsworth 1954), spinal column osteosarcoma in rufous-beaked snake *Ramphiophis rostratus* (Hill and Osman 1954) and vertebral cohondrosarcoma in (Dawe et al. 1981).

Garner MM. 2004. Trends in reptilian neoplasia: A diagnostician's perspective. Ithaca, New York, American College of Veterinary Pathologists & American Society of Veterinary Clinical Pathology: Middleton Internet Publisher: International Veterinary Information Service, P1226. http://www.ivis.org/proceedings/ACVP/2004/Garner/chapter_frm.asp?LA=1.

Neoplasia – Neoplasia is common in captive reptile, frequency in snakes > turtles, lizards, and crocodilians. Chondrosarcomas in colubrids and arising from lizard "cartilage of legs."

Garner MM. 2006. Overview of biopsy and necropsy techniques. In Mader DR. ed. Reptile Medicine and Surgery. Pp. 569–630; Philadelphia: Saunders.

Trauma – Fracture nonunion in Chinese water dragon *Physignathus cocincinus*.

Infection – Elbow osteomyelitis in green iguana *Iguana iguana*.

Garner MG, Johnson C, Funk R. 1994. Liposarcoma in a shingle-back skink (*Trachydosaurus rugosus*). Journal of Zoo and Wildlife Medicine 25(1):150.

Neoplasia – Liposarcoma in a shingle-back skink *Trachydosaurus rugosus*, but no comment on osseous involvement, despite suggestion by Mauldin and Done LB (2006).

Garner MM, Collins D, Joslin J. 1995. Vertebral chondrosarcoma in a corn snake. In Jung RE, ed. Proceedings Joint Conference American Association of Zoo Veterinarians, Wildlife Disease Association, American Association of Wildlife Veterinarians, August 12–17, 1995:332–333.

Neoplasia – Vertebral chondrosarcoma in a corn snake.

Garner MM, Herrington R, Howerth EW, Homer BL, Nettles VF, Isaza R, Shotts EB Jr, Jacobson ER. 1997. Shell disease in river cooters (*Pseudemys concinna*) and yellow-bellied turtles (*Trachemys scripta*) in a Georgia (USA) lake. Journal of Wildlife Diseases 33:78–86.

Shell – Shell rot in river cooter *Pseudemys concinna* and yellow-bellied turtle *Trachemys scripta* has been attributed to *Mucorales, Rusarium, Geotrichum, Trichosporon, Caniothyrium*, algae, and *Baneckea chitinovora*. Shell necrosis with ulcer and hemosiderosis in all; pancreatitis, in 2/4. Serpiginous lesions with local thickening and osteoclasts lining. Nodules also present with cortical bone replaced by woven or lamellar bone.

Garner MM, Hernandez-Divers SM, Raymond JT. 2004. Reptile neoplasia: A retrospective study of case submissions to a specialty diagnostic service. Veterinary Clinics: Exotic Animal Practice 7:653–671.

Neoplasia – General commentary on referred neoplasia, without indicating presence or absence of bone involvement in Northwest ZooPath records, indicating prevalence: Snakes > lizards > chelonians > crocodilians, noting that metastasis was uncommon.

Snakes (15–21.8% colubrids, 15.9% crotalids, 12.7% vipers, 8.6% boids).

Lizards (8.5–9.9% monitors, 9.2% skinks, 8.6% agamids, 7.7% geckos, 4.8% iguanas, 3.4% chameleons, 3.9% miscellaneous). Chelonians (2.7–3.2% turtles/terrapins, 1.4% tortoises) Crocodilians (2.2.

Chondrosarcoma in 9/325 (2.7%) snakes (0.8% of colubrids, 0.8% of crotalids and 0.4% of vipers), 1/162 (0.6%) lizards, but not in 4 crocodilians or 29 chelonians.

Fibrosarcoma in 19 snakes, 5 lizards, and 2 chelonians.

Hemangiosarcoma in five snakes and one lizard.

Leukemia in five snakes and five lizards.

Lymphoma in 33 snakes (8 colubrids, 15 lizards, and 1 chelonian

Myeloproliferative disease in two snakes.

Osteosarcoma in 3 (1%) snakes (0.1% of colubrids, 0.4% of crotalids, and 0.2% of vipers).

Reports without indicating presence or absence of bone involvement in:

Osteosarcoma in snakes (Frye 1994; 1 Colubrid, 1 Crotalid, and 1 Viper) and a lizard (Machotka 1984) and chondrosarcoma in corn snakes (Dawe et al. 1980; Garner 1995).

Chondrosarcoma in five colubrids, two crotalids, and two vipers and of the proximal humerus in a plated lizard *Gerrhosaurus* sp. They arise from vertebral articulations (Dawe et al. 1980; Garner 1995).

Fibrosarcoma in 5 Colubrids, 1 Crotalid, 10 Boids, and 6 Vipers;

1/455 Iguanids,1/370 Agamids, 2/306 Geckos, 2/203 Monitors, 1/174 Skinks, and 0/149 Chameleons.

Hemangioma was present in 1 snake (0.3%) and 4 lizards [2.5%, representing 2 of 370 agamids (0.6%), 1 of 304 geckos (0.3%) and 1 of 250 miscellaneous.

Hemangiosarcoma in 5 snakes [1.5%, representing 1 (0.1% of 651 colubrids, 2 (0.2%) of 842 boids and 0.4% of 442 vipers)] and 1 unnamed lizard (0.6%).

Myeloproliferative disease in 1 Crotalid and 1 Boid.

Leukemia in 3 Crotalids and 2 Boids; 2 Geckos and 1 Chameleon.

Lymphoma in 8/651 Colubrids, 4/251 Crotalids, 10/842 Boids, and 13/442 Vipers; 3/455 Iguanids, 6/370 Agamids, 1/304 Geckos, 4/203 Monitors, 1/174 Skinks, but not in 149 Chameleons.

Gassel R. 1988. Wiederholtes Auftreten von Chondromen in einer thüringer Bergmolchpopulation, *Triturus alpestris* (Laurenti 1768). [Repeated occurrence of chondromes in a population of *Triturus alpestris* (Laurenti 1768) in Thuringia]. 47–49. In H-J Herman (ed) Tagungsmaterial Amphibien: Haltung und Vermehrung von Amphibien in Labor und Terrarium. Naturhistorisches Museum Schloss Bertholdsburg, Schleusingen. [German]

Neoplasia – Cartilaginous outgrowth (referred as Chondromen in text, but does not fulfill criteria for Chondroma) on the tip of the upper jaw in a localized *Triturus alpestris* population in Thuringia.

Gates WH. 1929. A two-headed snake. Journal of Heredity 20:555–556.

Congenital – Derodymous common grass snake *Opheodrys aestivas.*

Gautschi B, Widmer A, Joshi J, Koella JC. 2002. Increased frequency of scale anomalies and loss of genetic variation in serially bottlenecked populations of the dice snake, *Natrix tessellata.* Conservation Genetics 3:235–245

Environmental – Scale anomalies are more common in introduced (1–9, average = 3), compared with naturally occurring (one abnormal scale) populations, also citing Mebert 1993. Three types of scale anomalies – cleft, incompletely formed and unpaired (Mebert 1993).

Extreme temperatures (Fox 1948), increased population homozygosity (Soulé 1979) and genetics (Forsman et al. 1994) have been blamed.

Gay G. 1909. Su la coda mostruosa delle lucertole [On the monstruous tail of the lizards]. Monitore Zoologico Italiano (Firenze) 20(2–3):84 [Italian].

Trauma – Presented at the eighth ordinary assembly and convention of the Italian zoological union in Bormio in 1908 on a lizard *Podarcis muralis* with a double tail.

Geen GH, McKeown BA, Watson TA, Parker DB. 1984. Effects of acephate (orthene) on development and survival of the salamander, *Ambystoma gracile* (Baird). Journal of Environmental Sciences and Health B19:157–170.

Congenital – Bent vertebral column in *Ambystoma gracile*, but no discussion of bone changes.

Gehlbach FR. 1972. Coral snake mimicry reconsidered: The strategy of self-mimicry. Forma Functio 5:311–320.

Trauma – Tail injuries in coral snakes 3- /8% of *Micrurus fulvius* and *Micruroides euryxanthus*, contrasted with 6.2% in sympatric colubrids (e.g., *Sonora episcopa)*, with difference attributed to tail displays in coral snakes.

Geisel O, Kriegleder H. 1978. Uratsteine bei Reptilien. [Urate stones in reptiles]. Berliner und Münchner Tierärztliche Wochenschrift 91:267–268.

Stone *Iguana iguana* with 75 g urate stone with 5.50 cm diameter in urinary bladder, and in unidentified turtle.

Geist N, Carpenter S. 2002. Chemical and morphological analysis of soft tissue preservation in a mosasaur. Journal of Vertebrate Paleontology 22 (supplement to 3):57A.

Congenital – *Platecarpus* three-dimensional preservation of tracheal rings and apparently of visceral organs, liver and possible kidney.

Fossil – *Platecarpus* three-dimensional preservation of tracheal rings and apparently of visceral organs, liver and possible kidney.

Gemmill JF. 1906. Supernumerary limb in a frog. Journal of Anatomy and Physiology 40:387–395.

Congenital – Supernumerary limb in *Rana temporaria*, with slightly distorted pelvis, elongated acetabular cavity and double proximal and distal femoral ends, fused astragalus and calcaneum with only three digits and two distorted phalanges.

Cites Sutton (Evolution and Disease, p. 111) supernumerary forelimb from sternum, right shoulder (De Superville), right scapula (Otto) in *Pelobatyes fuscus*, adherent to left limb (D'Alton), with accessory coracoid and scapula (Gervais), thoraco-abdominal boundary (Strobel), all cited from Taruffi, accessory coracoid (Sutton), bilateral scapula (Johnson), attached to outer coracoid (Bergendal).

Supernumerary posterior limb to symphysis pubis (Balsamo Crivelli; Fabretti), with acetabular cavity (Ercolani) inserted to posterior inferior tibial tuberosity, superior border of symphysis, to inside of right limb (Parona), right pelvis (Otto), right coccyx (associated with seven toes) (Vallisneri), accessory ilium from symphysis to transverse process of ninth vertebra (Fabretti), internal ilium, left of anus, left pelvis in *Alytes obstetricans*, between left limb and anus (Sordelli), between sacrum in hip with syndactyly in *Bufo vulgaris*, left of posterior aspect of urostyle (Tuckerman) (from Taruffi), accessory acetabulum (Kingsly Proceedings of the Boston Society of Natural History 21:169, 173), misshapen from midline abdomen (Alessandrini), from abdomen (Guittard).

Supernumerary forearm and partial metacarpals (Cavana), anterior with supernumerary toe (Bassi) with five toes (Eigenman and Cox 1901).

Supernumerary posterior leg with accessory bifid femur (Ercolani; Strobel), attached to ilium with enlargement of distal tibia fibula with three bones replacing astragalus and calcaneum and eight metatarsals (Johnson), fusion of inner surfaces (Ryder), fused between right flank and anus without femur but with seven toes (Strobel), femur attached to posterior inferior iliac tuberosity with nine toes (Parona in Taruffi), slender without femur but with six metatarsals and six toes (Dumeril) and with seven toes (Sutton).

Double supernumerary anterior limbs with unilaterally two fused accessory scapulae (Lunel; Sordelli, Ercolani), from a pedicle (Johnson).

Double supernumerary posterior limbs unilateral with rudimentary pelvis (Van Deen, Lunel, Cavanna), attached to symphysis (Van der Hoeven), with two cotyloid cavities and accessory pubis (Dumeril cited by Taruffi, Cavanna).

Partial duplication of limb as second leg in humerus with three toes in *Axolotl* (*Sordelli*), leg from inner aspect of normal (Otto), femur with distal femoral bifurcation and leg duplication in *Triton taeniatus* (Camerano, duplication of hind foot (Ercolani), foot duplication after experimental amputation (Barfurth).

Gemmill JF. 1912. The Teratology of Fishes. Glasgow: James Maclehose and Sons. 71 pp.

Congenital – Three-headed snakes, citing Bruch (1873) and Aldrovandi (1642).

Trauma – Tripartite tail in reptiles and in *Pelobates*, citing Bruch (1873).

Genè J. 1839. Synopsis reptilium Sardiniae indigenorum. [Synopsis of the creeping things of the natives of Sardinia]. Memorie della Reale Accademia delle Scienze di Torino. 1839. 31:257–285 [Latin].

Congenital – Illustraton of *Bombinator ignei* with very wide and long sacral alae.

Geoffroy Saint-Hilaire I. 1827?. Histoire générale et particulière des anomalies de l'organisation chez l'homme et des animaux: Ouvrage comprenant des recherches sur les charactères, la classification, l'influence physiologique et pathologique, les reppots généraux, les lois et le… [General and special history of organizational anomalies among humans and animals. Work understanding research on the characters, classification, physiologic and pathologic influences, general reports, the laws, and causes of monstrositites, the varieties of abnormal conformation or traits of teratology]. Bruxelles. Tom. III, page 132 [French].

Congenital – Dicephalic lizard, captured in the small French province of Roussilon in October of 1829 (after De Betta 1883).

Geoffroy St.-Hilaire I. 1832. Histoire générale et particulière des anomalies de l'organisation chez l'homme et les animau. Ouvrage comprenant des recherches sur les caractères, la classifaction, l'influence physiologique et pathologique, les rapports généraux, les lois et les causes des monstrousités, des variétés et vices de conformation, ou traité de teratology. [General and special history of organizational anomalies among humans and animals. Work understanding research on the characters, classification, physiologic and pathologic influences, general reports, the laws, and causes of monstrositites, the varieties of abnormal conformation or traits of teratology]. Volume I. 746 pp.; Paris: Ballière. Librarie de l'Academie Royale de Medicine [French].

Congenital – Dicephalic viper.

Triton with six toes on right hindfoot; seven, on left.

Geoffroy St.-Hilaire I. 1836a. Histoire générale et particulière des anomalies de l'organisation chez l'homme et les animaux ou, Traité de teratology, ouvrage comprenant des recherches sur les caractères, la classifaction, l'influence physiologique et pathologique, les rapports généraux, les lois et les causes des monstrousités, des variétés et vices de conformation. [General and special history of organizational anomalies among humans and animals. Work understanding research on the characters, classification, physiologic and pathologic influences, general reports, the laws, and causes of monstrositites, the varieties of abnormal conformation or traits of teratology]. Volume III:618 pp. Paris: Ballière [French].

Congenital – Derodymous lizard, dicephalic viper, and trigonocephalic snake of Lacépède and viper.

Geoffroy St. Hilaire I. 1836b (not 1838). Histoire Générale et Particulière des Anomalies de l'organisation chez l'homme et les Animaux, ouvrage comprenant des Recherches sur les Caractères, la Classification, l'influence Physiologique et Pathologique, les Rapports Généraux, les lois and les causes des Monstruosités, des variétés et des Vices de Conformation, ou Traité de Tératologie. [General and particular history of anomalies of organization in humans and animals, work examinge research on character, classification, physiological and pathologic influences, general reports, laws, and causes of monstrosities, of varieties and defects of conformation, or treatise on teratology]. Volume III. Paris Cosson, p. 197 [French].

Congenital – Dicephalic serpent.

Georgi JA, Krause DW. 2010. Postcranial axial skeleton of *Simosuchus clarki* (Crocodyliformes: Notosuchia) from the Late Cretaceous of Madagascar. Journal of Vertebrate Paleontology 30 Supplement to Number 6:99–121.

Vertebral – Late Cretaceous crocodyliform *Simosuchus clarki* FMNH PR 2597 with fused neural spines of the first and second dorsal vertebrae.

Other – Late Cretaceous crocodyliform *Simosuchus clarki* Université d'Antananarivo, Antananarivo, Madagascar UA 8679 with rugous dorsal ribs.

Fossil – Late Cretaceous crocodyliform *Simosuchus clarki* FMNH PR 2597 with fused neural spines of the first and second dorsal vertebrae and *Simosuchus clarki* Université d'Antananarivo, Antananarivo, Madagascar UA 8679 with rugous dorsal ribs.

Gerecht A. 1929. Über die Richtung der phyletischen Wanderung der Sacral region bei *Triton cristatus* und *Triton taeniatus*. [A comparative patholgical-anatomical study on the direction of phylogenetic migration of the sacral region in *Triton cristatus* und *Triton taeniatus*]. Latvijas Biologijas Biedribas *Raksti* (Bulletin de la Société Biologique de Lettonie) 1:9–12 [German].

Congenital – Change in vertebra number (no. 16–18 in *Triton cristatus* and no. 13–15 in *Triton taeniatus*), which forms the sacrum = variability interpretated as phylogenetic migration caudad or craniad.

Supernumerary ribs in one specimen of *Triton cristatus* at the 13. vertebra and in one specimen of *Triton taeniatus* on different vertebrae + fusion of vertebrae.

Gerlach J. 2004. Effects of diet on the systematic utility of the tortoise carapace. African Journal of Herpetology 53:77–85.

Metabolic – Distortion of carpace (major character used in Chelonia identification) caused by calcium/phosphate dietary deviations (metabolic bone disease producing thick spongy carapace with abnormally porous bones). They cause hyperparathyroidism and scute pyramiding, but did not compromise "taxonomically useful characters" in *Testudo ibera* and *Dipsochelys dussumieri*.

Gervais P. 1864a. Cas de polymélie observé sur un batrachian (*Pelobates cultripes*) et sur une espèce du genre Raie (*Raja clavata*). [Case of polymelia observed in a batrachian (*Pelobates cultripes*) and on a species of genus *Raja (Raja* clavata)]. Comptes rendus hebdomadaires de l'Académie de Science Paris 59:800–802 [French].

Congenital – Polymelia in *Pelobates cultripes* and *Raja clavata*.

Gervais P. 1864b Cas de polymélie (membres surnuméraires) observés sur un Batracien du Genre *Pelobates* et sur un éspèce du genre *Raja*). [Case of polymelia (supernumerary limbs) observed in a batrachian of genus *Pelobates* and on a species of genus *Raja*]. Comptes Rendus de l'Académie des Sciences de Paris 59:800–802 [French].

Congenital – In *Pelobates cultripes*.

Gervais P. 1864c. Description du *Mesosaurus tenuidens*. Reptile fossile de l'Afrique australe. [Description of *Mesosaurus tenuidens*. Fossil reptile of southern plain of Africa]. Mémoire de l'Académie de Montpelier Section de Science 6:167–175 [French].

Pseudopathology – Pachyostosis of ribs in *Mesosaurus tenuidens*.

Fossil – Pachyostosis of ribs in *Mesosaurus tenuidens*.

Gervais P. 1864d. No title. Revue et Magazin de Zoologie 16:356 [French].

Congenital – Supernumerary extremities in anurans (nach Fraisse 1885).

Gessner C. 1586 (also listed as 1686). Historiae animalium. Lib. II, qui et de quadrupedis oviparis [History of Animals. Book II, that are quadrupedal egg-layers]. 119 pp.; Francofurdi [Latin].

Trauma – Lizards with double tail [p. 34] (after Fraisse 1885; Leydig 1872).

Geus A. 1966. Wirbelsäulenverkrümmung und Gabelschwanz bei Wildfängen von *Lacerta agilis*. [Vertebral column curvature and forked tail in wild-caught *Lacerta agilis*]. Deutsche Aquarien- und Terrarien-Zeitschrift 19:281–282 [German].

Congenital – Spiral vertebral column in *Lacerta agilis*.

Trauma – Bifid tail in *Lacerta agilis*.

Ghate HV, Dodakundi GB, Mulherkar L. 1979. Effect of dye factory effluent on the developing embryos of *Microhyla ornata*. Indian Journal of Environmental Health 20:359–365.

Toxicology – Dye factory (rhodamine, auramine, methylene blue, methyl violet) effluent produces bent tail abnormalities in *Microhyla ornata.*

Ghorab HM. 1959. On the occurrence of an accessory fore-limb in the Egyptian maculated *regularis* Reuss. A'In Shams Science Bulletin (4):109–113.

Congenital – Supernumerary forelimb with extra coracoid in Egyptian maculated toad *Bufo regularis.* Cites Wagner 1913 *Rana tigrina*, Havass 1942 *Rana esculenta, Rana temporaria* and Pachori 1950 in *Rana tigrina.*

Giazzon MD. 2001. Extra hind limb in a small mouth salamander (*Ambystoma texanum*) from central Illinois. Bulletin of the Maryland Herpetological Society 37:143–150.

Congenital – Supernumerary limb in smallmouth salamander *Ambystoma texanum.*

Gibbons W, Green J. 2009. Turtles: The Animal Answer Guide. Baltimore: Johns Hopkins University Press, 163 pp.

Shell – Barnacle on diamondback terrapin carapace, noting oceangoing and diamondback terrapin that live in brackish water frequently acquire barnacles.

Giebel CG. 1864. Eine doppelschwänzige Eidechse. [A double-tailed lizard]. Zeitschrift für die gesamten Naturwissenschaften 24:248–49 [German].

Trauma – Lizard (*Lacerta agilis)* with two tails.

Giebel CG. 1867. Monstrositäten von *Cervus capreolus, Rana temporaria* und *Bombinator igneus.* [Monsters of *Cervus capreolus, Rana temporaria* and *Bombinator igneus*]. Zeitschrift für die gesamten Naturwissenschaften 29: 503–504 [German].

Congenital – *Rana temporaria:* 3. left front leg with four phalanges; and *Bombinator igneus*: lacking left hind leg.

Giles LW. 1948. Polydactylism in an Alligator. Copeia 1948(3):214.

Congenital – *Alligator mississippiensis* with three supernumerary manual toes (middle four with claws); with eight toes on each forefoot; four on left and three on right hind feet.

Gillespie D. 1994. Reptiles. In Saunders Manual of Small Animal Practice. SJ Birchard, RG Sherding, eds. Philadelphia: Saunders; pp. 1390–1411

Congenital – Achondroplastic dwarfism in turtles.

Metabolic – Rickets, osteomalacia and soft shells from vitamin D deficiency in reptiles. Fibrous osteodystrophy and cystic calculi from secondary hyperparathyroidism.

Osteopenia and metastatic calcification (including joints) from hypervitaminosis D.

Neoplasia – Multiple cartilaginous exostoses in lizards.

Vertebral – Spinal osteoperiostitis with hard rib/spine swellings, especially in boas.

Gillies C.D., Taylor B.B. 1918. An abnormal vertebral column in *Hyla caerulea* White. Proceedings of the Royal Society of Queensland 30:166–169.

Congenital – *Hyla caerulea* with multiple malformations. Atlas had steeper lamina with rudimentary left postzygapophysis, unequal C2 prezygapophysis, C3 oblique anteriolateral fissure, C4 with reduced right diapophysis, C5 with dilated diapophysis, C8 absent right posterior zygapophysis and fusion of sacral vertebra with urostyle.

Gillilland MG, Muzzal PM. 2002. Amphibians, trematodes, and deformities: An overview from Southern Michigan. Comparative Parasitology 69:81–85.

Congenital – Parasites (*Fibricola, Clinostomum, Mesocestoides, Spiroxys*) found in hind limbs were not responsible for the deformities.

Infection – Parasites (*Fibricola, Clinostomum, Mesocestoides, Spiroxys*) found in hind limbs were not responsible for the deformities.

Gillingham JC, Carmichael C, Miller T.. 1995. Social behavior of the tuatara, *Sphenodon punctatus.* Herpetological Monograph 9:5–16.

Congenital – Autotomy in tuatara *Sphenodon punctatus.*

Gilmore CW. 1919. A mounted skeleton of *Dimetrodon gigas* in the United States National Museum with notes on the skeletal anatomy. Proceedings of the United States National Museum 56:525–539.

Trauma – *Dimetrodon gigas* NMNH 8635 with healed fractures (with "exostoses") of vertebral spinous processes.

Fossil – *Dimetrodon gigas* NMNH 8635 with healed fractures (with "exostoses") of vertebral spinous processes.

Gilmore CW. 1946. Notes on recently mounted reptile fossil skeletons in the United States National Museum. Proceeding of the United States National Museum 96:195–203.

Trauma – Eocene crocodile *Crocodylus clavis* with extensive cranial and postcranial pathologies – fractured scapula healed in overlapped position, second metatarsal shaft expanded to twice normal size (probably infected) and injuries to coracoid, anterior thoracic ribs, caudal vertebrae, chevrons, and skull.

Infection – Second metatarsal shaft of Eocene crocodile *Crocodylus clavis* expanded to twice normal size (probably infected).

Second specimen USNM 129900 Bridger, Eocene crocodile with spongy growth surrounding interventricular articular surfaces (probable infection).
Fossil – Eocene crocodile *Crocodylus clavis* with extensive cranial and postcranial pathologies – fractured scapula healed in overlapped position, second metatarsal shaft expanded to twice normal size (probably infected) and injuries to coracoid, anterior thoracic ribs, caudal vertebrae, chevrons, and skull.
Second metatarsal shaft of Eocene crocodile *Crocodylus clavis* expanded to twice normal size (probably infected)
Second specimen USNM 129900 Bridger, Eocene crocodile with spongy growth surrounding interventricular articular surfaces (probable infection).

Girard C. 1891–1892a. Une tortue (*Crisemys picta*) bicèphale – Avec 1 fig. [A dicephalic turtle (*Crisemys picta*) with 1 figure]. Le naturaliste 13 Annèe 93:19 [French].
Congenital – Dicephalic turtle *Chrysemys picta*. The heads alternated control of the body depending on which was asleep.

Girard C. 1891–1892b. Une tortue (*Crisemys picta*) bicèphale. Note additionelle. [A dicephalic turtle (*Crisemys picta*). Additional note]. Le naturaliste 14 Annèe 130:174. [French]
Congenital – Dicephalic turtle *Chrysemys picta*. The heads alternated control of the body depending on which was asleep.

Girard C. 1895. Polydactylie provoquée chez *Pleurodeles waltii* [Induced polydactyly in *Pleurodeles waltii*]. Comptes Rendues de la Societe de Biologie Paris 10(2):789–792 [French].
Congenital – Triton *Pleurodeles waltii* forefoot with six digits. Observed that it is not always easy to distinguish between inherited and acquired polydactyly. Cites Leydig (1879) as to a *Pleurodeles waltii* with six toes on hindfoot and four (instead of normal five) toes on forefeet.

Glauert L. 1947. A two-headed bob-tailed lizard. Proceedings of the Royal Zoological Society of New South Wales for 1946–1947:34.
Congenital – Conjoined twin Stump-tailed or shingle-back lizard *Trachysaurus rugosus* with two heads, three forelimbs, and four hind limbs.

Glaw F. 1987. Probleme mit Streichholzbeinen bei der Nachzucht von *Discoglossus pictus*. [Problems with matchlegs in the progeny of *Discoglossus pictus*]. Herpetofauna 9:30–33 [German].
Congenital – Thin front legs in *Discoglossus pictus*.

Glazebrook JS. 1980. Diseases in farmed sea turtles. Pp 42–55. Management of Turtle Resources, Glazebrook, ed. James Cook University, North Queensland.
Metabolic – Two farmed green turtle *Chelonia mydas* with osteodystrophy, presenting as osteopenia, curled shells, misshapened mandible or lower jaw or "lateral prominences in the occipital regions" (p. 51), shorter long bones, widened calcific rings near epiphyseal junctions.

Glazebrook JS, Campbell RS. 1990a. A survey of the diseases of marine turtles in northern Australia. I. Farmed turtles. Diseases of Aquatic Organisms 9:83–95.
Infection – Osteomyelitis in 0.9% of 66 farmed *Chelonia mydas* and *Eretmochelys imbricata*. *Vibrio alginolyticus*, *Pseudomonas* and *Cytophaga-Flavobacterium* were isolated.
Parasite load-induced plastron shrinkage.
Metabolic – Fibrous osteodystrophy (nutritional secondary hyperparathyroidism in 1.9% of farmed *Chelonia*.
Stones – Cystic calculus in 0.9% of farmed *Chelonia mydas*.
Shell disease – Ulcerative shell disease in 1.9% of farmed *Chelonia mydas* and *Eretmochelys imbricate*. *Vibrio alginolyticus* and *Pseudomonas* were isolated.
Plastron shrinkage from parasite load.

Glazebrook JS, Campbell RS. 1990b. A survey of the diseases of marine turtles in northern Australia. II. Oceanarium-reared and wild turtles. Diseases of Aquatic Organisms 9:97–104.
Infection – Fifty percent of *Chelonia mydas*, *Eretmochelys imbricate* and *Caretta caretta* had carapace-plastron barnacles.
Barnacle *Cheloniba* on carapace and plastron of *Chelonia mydas*.
Barnacle *Platylepas* on *Eretmochelys imbricate* carapace and plaston.
Heavy infestation by cardiovascular flukes (Digenea: Spirochiidae) produces plastron shrinkage.
Shell disease – Fifty percent of *Chelonia mydas*, *Eretmochelys imbricate* and *Caretta caretta* had carapace-plastron barnacles.
Barnacle *Cheloniba* on carapace and plastron of *Chelonia mydas*.
Barnacle *Platylepas* on *Eretmochelys imbricate* carapace and plaston.
Heavy infestation by cardiovascular flukes (Digenea: Spirochiidae) produces plastron shrinkage.

Glückselig MC. 1863. Einige Beobachtungen über das Leben der Eidechsen. [Some observations on the life of lizards]. Verhandlungen des Zoologisch-Botanischen Verhandlungen der Zoologische-Botanischen Gesellschaft Wien 13:1133–1136 [German].

Trauma – Formation of double tail in lizards by split of wound before regeneration of tail (after Fraisse 1885 and Leydig 1872).
Gnaedinger LM, Reed CA. 1948. Contribution to the natural history of the plethodont salamander *Ensatina eschscholtzii*. Copeia 1948(3):187–196.
Congenital – One aberrant male *Ensatina eschscholtzii* had tail 1–1/2 times the length of its body.
Ensatina recent tail loss in 4.9% of 169 and 14 others with partially regenerating tails.
Götte A. 1875. Die Entwicklungsgeschichte der Unke (*Bombinator igneus*) als Grundlage einer vergleichenden Morphologie der Wirbelthiere. [The anatomy of *Bombinator igneus* as base of comparative morphology of vertebrates]. 965 pp.; Leipzig: Leopold Voss [German].
Congenital – Deformation of *Bombinator igneus* sacrum: right lateral process of 10. vertebra broad compared to the left rudimental (= normal) process (Pl. 19, Fig. 346).
Götte A. 1879. Ueber Entwickelung und Regeneration des Gliedmassenskelets der Molche. [On ontogeny and regeneration of the extremity skeleton of urodeles]. 47pp.; Leipzig: Leopold Voss [German].
Congenital – Supernumerary fingers and toes in salamanders caused by dichotomous division during regeneration (Siebold 1828). Reported observation on one foot, where the third toe is divided and the third metatarsus has a lateral toe fused with the second and referred to similar case in Wiedersheim (1875).
Goh Lilian. 2005. Two-headed turtle sets off alarm bells. South China Morning Post 12 December 2005:5.
Congenital – Dicephalic turtle.
Goin, CJ. 1971. Introduction to Herpetology. 2nd ed.: 353 pp.; San Francisco: WH Freeman,
Trauma – Cuban anole *Anolis sagrei* with bifid tail.
Goldberg SR. 1967. Prize snake for Dale: It has two heads. The News, Frederick, Maryland, Sept. 14:2. (photograph).
Congenital – Dicephalic unborn garter snake.
Goldberg SR, Holshuh HJ. 1991. A case of leukemia in the desert spiny lizard (*Sceloporus magister*). Journal of Wildlife Diseases 27:521–525.
Neoplasia – Bone marrow affected by leukemia in desert spiny lizard *Sceloporus magister*.
Golder F. 1985. Ein gemeinsamer Masseneiablageplatz von *Natrix natrix helvetica* (LACÉPÈDE, 1789) und *Elaphe longissima longissima* (LAURENTI, 1768) mit Daten über Eizeitigung und Schlupf. [A joint place of deposition of large number of eggs of *Natrix natrix helvetica* (LACÉPÈDE, 1789) and *Elaphe longissima longissima* (LAURENTI, 1768) with data on injury of eggs and hatching]. Salamandra 21:10–16 [German].
Congenital – New born snakes with arching of vertebral column in region of tail of *Coronella austriaca* and so strong that grown together ventrally in *Natrix natrix*.
Golding A. 1955. The Excellent and Pleasant Worke Collectanea Rerum Memorabilium of Caius Julius Solinus. 225 pp.; Gainesville, Fl: Scholars' Facsimiles & Reprints..
Pseudopathology – *Amphisbaena* as having a head on each end.
Congenital – Dicephalic *Amphisbaena*.
Also reported the mouth of a river as two heads.
Gollman G, Hödl W, Ohler A. 1984. A tadpole from a *Bombina* hybrid population - a hopeless monster. Amphibia-Reptilia 5:411–413.
Congenital – Puffed up *Bombina*, probably from lymphatics.
Gollmann G. 1991. Osteological variation in *Geocrinia laevis*, *Geocrinia victoriana*, and their hybrid populations (Amphibia, Anura, Myobatrachinae). Zeitschrift für zoologische Systematics und Evolution-forschrift 29:289–303.
Congenital –Five percent frequencies of malformed vertebral columns in *Geocrinia*, including fusions and gross asymmetries, fusion of first and second vertebrae, fusion of presacral and sacral vertebrae (sacralization of lumbar vertebrae) and fused sacral vertebrae.
Three *Geocrinia laevis* and *Geocrinia victoriana* with asymmetry of first digit phalangeal number, extra phalanges, reduced phalanges or fused phalanges.
Twelve had abnormal carpal pattern. 22 of 211 tarsi were abnormal. Foot with fused metatarsals. Hand with Y-shaped metacarpal. Supernumerary finger.
Trauma - 35% of *Geocrinia laevis* and *Geocrinia victoriana* had lost phalanges, especially terminal ones. Isolated or missing feet were noted in
Good HM. 1987. Shell anomalies in the desert tortoises (*Gopherus agassizii*) populations of the Beaver Dam Slope, Utah, and Desert Tortoise Natural Area, California. Desert tortoise Council, Proceedings of 1984 Symposium:95–104.
Congenital – *Gopherus agassizii* from Desert Tortoise Natural Area California had 11.22% carpace scute anomalies, compared with 20.40% for Beaver Dam Slope, Utah/Arizona. Reduced marginals and supernumerary of all areas, but Desert Tortoise Natural Area had more supernumerary vertebrals than Beaver Dam Slope. Abnormal scute shape was rarely noted. One Desert Tortoise Natural Area tortoise had a plaston abnormality. One individual had two fused toes. Suggested environmental factors, including soil moisture, temperature variation, oxygen deprivation and handling/rotation of eggs or natural (uranium/thorium) or plutonium exposure.

Goodwin PA. 1946. A comparison of regeneration rates and metamorphosis in *Triturus* and *Ambystoma.* Growth 10:75–87.
Trauma – Young *Ambystoma punctatum, Ambystoma jeffersonianum,* and *Triturus viridescens* regenerate more rapidly than old ones.
Goodwin LG. 1976. Nuffield Institute of Comparative Medicine. Journal of Zoology London 178:534.
Shell disease – *Coniothyrium fuckelianum* shell disease in *Testudo graeca.*
Gordos M, Franklin CE. 2002. Diving behavior of two Australian bimodally respiring turtles, *Rheodytes leukops* nad *Emydura macquarii,* in a natural setting. Journal of Zoology 258:335–342.
Vascular – Dive duration correlated with aquatic respiration in *Apalone spinifera, Apalone ferox, Trachemys scripta. Rheodytes leukotops* obtained 41.5% of oxygen supply aquatically, contrasted with *Emydrura macquarii,* which derived only 10.3%.
Goss RJ. 1961. Regeneration of vertebrate appendages. – In Abercrombie M, Brachet J. eds. Advances in Morphogenesis. Vol. 1:103–152; New York: Academic Press.
Trauma – Regeneration of limbs, tails, jaws, and snouts in urodeles is greater than in anurans. Tail cartilaginous replacement and jaw replacement claimed in some lizards and turtles. Limited leg regeneration.
Postmetamorphic *Xenopus laevis* produced malformed, branched cartilaginous outgrowths (Beetschen 1952; Skowen and Komala 1957).
Goss RJ. 1974. Regeneration, Probleme - Experimente – Ergebnisse [Regeneration, problematic experiments - results]. VI+288 pp.; Stuttgart: Georg Thieme Verlag [German].
Trauma – Processes of regeneration of limbs in urodele amphibians depending on kind of transplanted skin, innervation, hormonal influence, left over bone fragment, and UV-light.
Regeneration of jaws and tails described, lizard, *Lacerta agilis.*
Grosse W-R, Bauch S. 1988. Fehlentwicklung beim Laubfrosch *Hyla arborea* L. [Abnormal development in the European tree frog *Hyla arborea*]. Feldherpetologie, Erfurt 1988:25–28 [German].
Trauma – *Hyla arborea* specimens missing one hind leg or part of a hind leg; supposed causes: Hormonal disturbance (possible of thyroid) through introduction of pollutants into the environment. Author discards injuries by dragonfly larvae because only hind legs damaged, never front legs.
Metabolic – *Hyla arborea* specimens missing one hind leg or part of a hind leg;
supposed causes: Hormonal disturbance (possible of thyroid) through introduction of pollutants into the environment. Author discards injuries by dragonfly larvae because only hind legs damaged, never front legs.
Goswami B. 1994. Report on an additional forelimb in a common Indian toad, *Bufo melanostictus* Schenider. Journal of the Bengal Natural History Society 13:2,53.
Congenital – Supernumerary forelimb in common Indian toad, *Bufo melanostictus.*
Gotte A. 1875. Entwicklungsgeschichte der Unke (*Bombinator igneus*) als Grundlage einer vergleichenden Morphologie der Wirbelthiere. [Evolution of the snake (*Bombinator igneus*) as the basis of a comparative morphology of vertebrates]. Leipzig. Figure 346, Table XIX [German].
Congenital – Unrecognized sacrum variability in *Bombinator igneus* – caudization of a sacral vertebra.
Grant C. 1936a. An extraordinary tortoise shell, Copeia 1936:231–232.
Congenital – *Testudo tabulate* with extra neurals, but only two right and one left costal.
Chrysemys picta marginata with fourth neural separating costals.
Grant C. 1936b. The southwestern desert tortoise, Gopherus agassizi. Zoologiea 21:225–229.
Congenital – Plastron asymmetry in southwestern desert tortoise *Gopherus agassizii.* 90% had inter-gular suture on right; 6%, median; and 3%, left. Among 366, 19 female, and 4 males had major acute abnormalities.
Cites asymmetry by Ditmar (1907) plate XXIV and Hay (1908).
Trauma – Occasional old female with cut (semicircular nick) or caudal worn flat caused by male anal plate rubbing.
Grant C. 1957. Autophagy in *Eumeces.* Herpetologica 13:156.
Trauma – Tail loss in *Eumeces skiltonianus.*
Grant F. 2001. Tests on two-headed lizard. Daily Telegraph 17 January 2001:13.
Congenital – Ododymus blue tongue lizard.
Gräper L. 1909. Über eine dreischwänzige Eidechse mit sieben Schwanzskeletten. [On a three-tailed lizard with seven tail skeletons]. Roux's Archiv für Entwicklungsmechanik 27:640–652 [German].
Trauma – *Lacerta agilis* with three tails (after Volante 1923).
Gray RH. 2000. Historical occurrence of malformations in the cricket frog, *Acris crepitans,* in Illinois. Transactions of the Illinois State Academy of Science 93:279–284.
Congenital – Thirty years ago, 0.39% of cricket frog *Acris crepitans* had missing, deformed or extra limbs, digits and mouthparts, and no increase when Mackinaw River resampled in 1998.
Gray BS. 2006. The reptiles and amphibians of the Asbury Woods Greenway, Erie County, Pennsylvania. Bulletin of the Maryland Herpetological Society 42(2):115–126.
Trauma – Shell abrasions in a captive midland painted turtle.
Metabolic – Abnormal shell growth from poor "maintenance" in a midland painted turtle.

Shell – Shell abrasions in a captive midland painted turtle.

Gray BS, Lethaby M. 2010. Observations of limb abnormalities in amphibians from Erie County, Pennsylvania. Journal of Kansas Herpetology 35:14–16.

Congenital – But noted trauma-induced (see below) supernumerary limbs.

Noted that Bridges and Semlitsch (2005) reported low dose pesticide (e.g., Carbaryl) induced supernumerary limbs.

Trauma – Selective predation as explanation for both missing and supranumerary limbs (Ballengée and Sessions 2009).

2.3% of *Ambystoma maculatum* at Asbury Woods, Errier Pennsylvania had legs truncated at the femur.

4% of *Lithobates sylvaticus* had unilateral missing distal hand limb, with truncated tibiale and fibulare, as also noted by Kiesecker (2002) in Centre County, Pennsylvania. This contrasts with UV-related truncations, which are usually bilateral (Blaustein et al. 1997; Cohen 2001).

Northern leopard frog *Lithobates pipiens* lacking most of left hind limb.

Ambystoma jeffersonianum with supernumerary forelimb – between second and third costal grooves.

Ambystoma jeffersonianum observed nipping sibling's tail an partially severing left hindfoot. Healing produced an extra limb at site of injury (AMNH A163453).

Cites Giazzon (2001) report of small mouthed salamander *Ambystoma texanum* with supernumerary hind limb.

Toxicology – Low dose pesticide (e.g., Carbaryl) produce missing or supernumerary limbs (Bridges and Semlitsch 2005).

Gray HM, Green DM, Peters MJ. 1999. *Physalaemus pustulosus* (Túngara frog). Predation. Herpetological Review 30:93.

Trauma – Predation on Túngara frog *Physalaemus pustulosus* by tarantula *Sericopelma rubronitens*.

Gray B, Smith HM, Woodling J, Chiszar, D. 2001. Some bizarre effects on snakes, supposedly from pollution, at a site in Pennsylvania. Bulletin of the Chicago Herpetological Society 36(7):144–148.

Environmental – Mandibular agenesis in eastern garter snake *Thamnophis s. sirtalis* Millcreek Hazardous Waste Site (heavy metals, phenols, aromatic).

Gray B, Smith HM, Chiszar, D. 2003. Further anomalies in the litters of a garter snake from a hazardous waste site. Bulletin of the Chicago Herpetological Society 38(1):4–6.

Congenital – Garter snake *Thamnophis s. sirtalis* with foreshortened snout and prognathous lower jaw.

Greekhamer A. 1995. Schwanzanomalien bei Geckos der Gattung *Phelsuma* GRAY, 1825 (Squamata: Sauria: Gekkonidae) [Tail anomalies in geckos of the genus *Phelsuma* GRAY, 1825 (Squamata: Sauria: Gekkonidae)]. Herpetozoa 8:35–42.

Trauma – "Crumpled tail" (*Phelsuma abbotti chekei*) and "buckled tail" (*Phelsuma madagascariensis grandis, madagascariensis madagascariensis, dubia, comorensis, laticauda laticauda, abbotti sumptio, cepediana, pusilla pusilla*, and *P. sundbergi sundbergi*) connected with lordosis and kyphosis caused by inadequate holding.

Greek TJ. 2000. Cystic calculi in the frog *Phyllomedusa sauvagei*. Proceedings of the Association of Amphibian and Reptile Veterinarians Annual Conference 2000:3–5.

Stone – Urate calculus in nine Chacoan monkey tree frogs *Phyllomedusa sauvagei*.

Greek TJ. 2001. Osteomyelitis in a green iguana, *Iguana iguana*. Proceedings of the Eighth Association of Reptilian and Amphibian Veterinarians Annual Conference 2001:163–164.

Infection – Severe "proliferation of vertebrae of entire tail up to the midlumbar region with a fracture at tail base" attributed to osteomyelitis (with negative culture) in green iguana *Iguana iguana*.

Vertebral – Severe "proliferation of vertebrae of entire tail up to the midlumbar region with a fracture at tail base" attributed to osteomyelitis (with negative culture) in green iguana *Iguana iguana*.

Green DE, Harshbarger JC. 2001. Spontaneous neoplasia in amphibia. In Wright KM, Whitaker BR. eds. Amphibian Medicine and Captive Husbandry. pp. 335–400; Malabar: Krieger Publisher.

Congenital – Thirty five percent of male and 20.5% of female *Tristurus cristatus carnifex* in Italy had supernumerary phalanges, polydactyly, carpal fusion, phalangeal "distortion," syndactyly, forked digit, carapla, and digit duplication and a bent radius and ulnar in a newt. Two percent of Japanese firebelly newt *Cynops pyrrhogaster* had supernumerary tails, polydactyly and missing limbs or digits. 6/10 terrestrial Western slimy salamander *Plethodon albagula* had limb abnormalities.

Ninety percent of supernumary limbs in North American anurans affect the hind limbs, with forelimb involvement rare (contrasted with European anurans. California long-toed salamander *Ambystoma macrodactylum* had equal polydactyly and supernumary fore as hind limbs. Pelvic duplication has also been noted. Spindly leg in oriental firebelly toad *Bombina orientalis*, orange-legged leaf frog *Phyllomedusa hypocondrialis* and painted frog *Discoglossus pictus*.

Trauma – What Ohlmacher (1898) called an osteosarcoma was actually a fracture callus in the femur of "*Rana virescens*." It appeared as a spindle-shaped diaphyseal swelling with bony trabeculae surrounding by bone marrow.

Metabolic – Hypervitaminosis A produces thin cortices, subject to fracture and projection of mandible beyond maxilla.

Gout in pixie frog *Pixicephalus delalandii*.

Tumor – Tumor-like osteochondrous dysplasia in 11.5% of Inkiapo frog *Rana chensinensis* exposed to effluents from a paper factory and 5.5% from sewage, contrasted with none in other site.
Chondrodysplastic lesions are 1–7 mm nodules referred to as "osteochondromas" (actually enchondromas) or bone cysts. Other lesions are larger, with atypical pleomorphic chondroblasts, possibly representing chondrosarcoma.
Chondromyxoma was described in Pacific chorus frog *Pseudacris regilla* and *Rana fusca*.
What Ohlmacher (1898) called an osteosarcoma was actually a fracture callus in the femur of "Rana virescens," It appeared as a spindle-shaped diaphyseal swelling with bony trabeculae surrounding by bone marrow. Chordoma were reported in hybrid toads, as a result of experimental lathyrism – from sweet pea *Lathyrus odorata*.
Environmental – *Xenopus laevis* exposed as tail-stage embryos to microgravity developed lordosis and deformed notocord, but not those fertilized in space.

Greene HW. 1973. Defensive tail display by snakes and amphisbaenians. Journal of Herpetology 7:143–161.
Trauma – Malaysian pipe snake *Cylindrophis rufus* RMNH 8264 with missing terminal spine, 6 of 15 Calabar ground python *Calabaria reinhardtii* and 19 of 27 Indian sand boa *Eryx johnii* with missing or damaged tail tips, contrasted with 0 of 28 Russell's sand boa *Eryx conicus* and none in *Anilius scytale* and only rarely in *Cylindrophis rufus*.
Myers (1967) reported 29% tail loss in *Rhadinaea flavilata*.
Neill (1960) thought tail rot caused the missing tails.
Indian and Pakistani snake charmers sometimes mutilate sand boa tails to produce a two-headed appearance.

Greene HW. 1988. Antipredator mechanisms in reptiles. In C Gans, RB Huey (eds.). Biology of the Reptilia. New York: Alan R Liss, 16B:1–152.
Trauma – Anti-predator mechanisms in reptiles.

Greene HW. 1989. Natural death associated with skeletal injury in the terciopelo, *Bothrops asper* (Viperidae). Copeia 1989(4):1036–1037.
Trauma – Fracture with lateral rotation of dentary in terciopelo *Bothrops asper* preventing engagement of dentary teeth, producing starvation. Food could not be ingested.

Greenhouse G. 1976. Evaluation of the teratogenic effects of hydrazine, methylhydrazine, and dimethylhydrazine on embryos of *Xenopus laevis*, the South African clawed toad. Teratology 13:167–178.
Toxicology – Tail kinks, microcephaly from hydrazine in *Xenopus laevis*.

Greer AE. 1997. Developmental and evolutionary implications of some limb abnormalities in a sample of green and golden bell frogs, *Litoria aurea* (Hylidae). Journal of Herpetology 31:596–599.
Congenital – Fifty four percent of green and golden bell frogs *Litoria aurea* from Sydney Australia with supernumerary metatarsals and phalanges.

Greer AE, Byrne M. 1995. Sex ratio and frequency of osteological abnormalities in the Australian hylid frog *Litoria aurea* from two apparently unpolluted localities in Sydney, New South Whales. Australian Zoologist 30:43–47.
Congenital – 1/41 Australian hylid frog *Litoria aurea* from Homebush Bay had absence of terminal digits two to four of manus.

Greer LL, Strandberg JD, Whitaker BR. 2003. *Mycobacterium chelonae* osteoarthritis in a Kemp's Ridley sea turtle (*Lepidochelys kempii*). Journal of Wildlife Diseases 39:736–741.
Infection – *Mycobacterium chelonae* plastron lesions and swollen left elbow with osteolytic radial and ulnar lesions with sclerotic margins in a Kemp's Ridley sea turtle (*Lepidochelys kempii*).

Gressitt JL. 1936a. Camel-back turtle. Zoologica 39:246.
Congenital – Kyphosis in *Amyda (Trionyx) sinensis* and *Trionyx steindachneri*.

Gressitt JL. 1936b [1937]. Soft-shelled turtle monstrosities from Hainan Island. Peking Natural History Bulletin 11:413–415.
Shell – *Trionyx steindachneri* with carapace hump from Island of Hainan, south of southern tip of China.

Griffith AS. 1941. The susceptibility of the water (or grass) snake (*Trepidonotus natrix*) to the avian tubercule bacillus and to reptilian strains of acid-fast bacilli. The Journal of Hygiene 41:248–288.
Vertebral – "Superficial necrosis of vertebral bones by *Mycobacterium marinum* in grass snake *Trepidonotus natrix*.

Griffiths R. 1981. Physical abnormalities and accessory limb growth in the smooth newt, *Triturus vulgaris*. British Journal of Herpetology 65:180–182.
Congenital – Four percent frequency of supernumerary or bifurcate limbs in smooth newt, *Triturus vulgaris*.

Grimme A. 1907. Eine Missbildung von *Rana temporaria* Ant. [Abnormality of *Rana temporaria* Ant.] Abhandlungen und Berichte des Vereins für Naturkunde zu Kassel 51:126 [German].
Congenital – *Rana temporaria* with supernumerary hind leg between the two normal legs.

Grochmalicki J. 1909. Über Missbildungen von Salamanderlarven im Mutterleib. [About abnormalities in the womb of salamander larvae]. Archiv für Entwicklungsmechanik 28:181–209.

Congenital – Posterior duplication in *Salamandra maculosa.*

Grogan WL Jr. 1976. Scoliosis in the African lizard, *Agama a. anchietae* (Bocage) (Reptilia, Lacertilia, Agamidae). Journal of Herpetology 10:262–263.

Congenital – Scoliosis in *Agama a. anchietae* and *mutabilis* and *Lacerta agilis* (the latter with a forked tail).

Grosse W-R, Bauch S. 1988. Fehlentwicklung beim Laubfrosch *Hyla arborea* L. [Wrong development of European tree frog]. Feldherpetologie, Erfurt 1988:25–28.

Trauma – *Hyla arborea* specimens missing one hind leg or part of a hind leg; supposed causes: hormonal disturbance (possible of thyroid) through introduction of Schadstoffe into the environment. Author discards injuries by dragonfly larvae because only hind legs damaged never front legs.

Metabolic – *Hyla arborea* specimens missing one hind leg or part of a hind leg; supposed causes: hormonal disturbance (possible of thyroid) through introduction of Schadstoffe into the environment. Author discards injuries by dragonfly larvae because only hind legs damaged never front legs.

Grosse W-R, Günther R. 199a. Triton – *Triturus cristatus* (Laurenti, 1768). [Crested newt – *Triturus cristatus* (Laurenti, 1768)]. In Günther R. ed. Die Amphibien und Reptilien Deutschlands. pp. 120–141; Jena: Gustav Fischer.

Congenital – Double tail and duplication of extremities in *Triturus cristatus.*

Grosse W-R, Günther R. 1996b. Laubfrosch – *Hyla arborea* (Linnaeus, 1758). [Tree frog – *Hyla arborea* (linnaeus, 1758)]. In Günther R. ed. Die Amphibien und Reptilien Deutschlands. pp. 343–364; Jena: Gustav Fischer.

Trauma – Missing extremities in *Hyla arborea* caused by dragonfly larvae.

Grünberg H. 1963. The Pathology of Development: a Study of Inherited Skeletal Disorders in Animals. Oxford: Blackwell Scientific Publications, 309 pp.

Trauma – Notochord is not required for salamander tail regeneration (Holtzer 1956; Holtzer et al., 1955). Weaker skeletal elements regress or fuse with stronger neighbor in *Xenopus* (Bretscher 1949).

Grünberg W. 1963/64. Harnsteine bei Tieren. 1. Mitteilung: Vorkommen, Untersuchungsmaterial und angewandte Methoden. [Kidney stones in animals. 1. Communication.Research material and used methods] Zentralblatt für allgemeine Pathologie und pathologische Anatomie 105:256–271 [German].

Stone – Oldest fossil kidney stone from the Upper Cretaceous (Voigt 1960).

Affected species	Type of stone	Author
Marine turtle	Ca-carbonate, phosphate, sulfate, Mg-phosphate	Virchow (1878)
Bufo calamita	Ca-phosphate	Ebstein (1899)
Testudo graeca	Urate	Ebstein (1899)
Testudo gigantean	Urate	Hamerton (1934, 1938)
Iguana tuberculata	Urate	Plimmer (1910)
Sauromalus showi	Urate, Ca-phosphate	Cornelius and Bishop (1961)
Tupinambis teguixin		Hamerton (1939), Osman Hill (1952)
Tupinambis nigropunctatus	Urate	Hamerton (1933)
Gerrhosaurus vallidus		Osman Hill (1952)
Gavialis gangeticus	Urate	Hamerton (1933)
Crocodylus americanus	Urate	Hamerton (1933)
Sepedon haemachates	Urate	Hamerton (1939)
Naja melanoleuca	Urate	Hamerton (1939)
Coluber jugularis	Urate	Hamerton (1947)
Naja hannah	Urate	Hamerton (1937)
Megalobatrachus japonicus		Grünberg (1963/64)
Xenopus laevis		Grünberg (1963/64)
Testudo graeca		Grünberg (1963/64)
Testudo elephantopus		Grünberg (1963/64)
Boa c. occidentalis		Grünberg (1963/64)
Phyton sebae		Grünberg (1963/64)

Grünberg W. 1964. Vergleichende Untersuchungen zur Biokristallographie tierischer Harnsteine. [Comparative analyses of the crystallography of kidney stone sof animals] Pathologia veterinaria 1:258–268 [German].

Stones – Composition (formulas based on X-ray analyses) of kidney stones.

Guardabassi A. 1955. Su alcune anomalie dello scheletro assile di *Xenopus laevis*. [On some anomalies of the axial skeleton of *Xenopus laevis*]. Archivio Zoologico Italiano (Napoli) 40:321–346 [Italian].

Congenital – Axial skeleton anomalies of tadpoles in their last larval stage, and in metamorphosed specimens of *Xenopus laevis*:

1. Fusion of vertebrae I and II or of others, leading to reduction of the total number of vertebrae as an individual variation.
2. Presence of one or two supplementary small ribs at the atlas level, in addition to the three pairs of normal ribs (corresponding to vertebrae II, III, and IV) which are present as separate formations in fully developed larvae of *Xenopus*.
3. Anteriolateral asymmetry concerning the ribs.
4. Fusion of the distal parts of the first and second pairs of the normal ribs in some cases and tendency to fusion of the distal parts of the second and third pairs.
5. Increased development of one or both processes of vertebrae VIII and X, which may take part, along with the processes of vertebrae IX (the true sacral) in the articulation with the Ilium.

Gubin Y, Novikov IV. 1999. Aberrant skull roof morphology of some temnospondyl amphibians. Journal of Vertebrate Paleontology 19 (supplement to 3):48A–48B.

Congenital – Anomalous skull roof structures affected *Apateon, Aphaneramma, Batrachosuchoides, Benthosuchus, Branchierpeton, Chelyderpeton, Gonioglyptus, Kestrosaurus, Konzhukovia, Lydekkerina, Melanerpeton, Platyoposaurus, Schoenfelderpeton, Sclerocephalus, Thoosuchus, Trematosuchus, and Wetlugasaurus.* 4–12% (especially younger individuals) with additional longitudinal midline ossifications and atypical sutures among nasals, frontals, parietals and postparietals. The question if this is due to lack of fusion or developmental/ontologic.

Fossil – Anomalous skull roof structures affected *Apateon, Aphaneramma, Batrachosuchoides, Benthosuchus, Branchierpeton, Chelyderpeton, Gonioglyptus, Kestrosaurus, Konzhukovia, Lydekkerina, Melanerpeton, Platyoposaurus, Schoenfelderpeton, Sclerocephalus, Thoosuchus, Trematosuchus, and Wetlugasaurus.* 4–12% (especially younger individuals) with additional longitudinal midline ossifications and atypical sutures among nasals, frontals, parietals, and postparietals. The question if this is due to lack of fusion or developmental/ontologic.

Gubin YM, Novikov IV, Morales M. 2000. ???? ?????? ? ?????? ????? ???? ?????????? ??????? ??? ?????. [A review of anomalies in the structure of the skull roof of temnospondyloous labyrinthodonts]. Palaeontological Journal 34:S154–S164. [Russian]

Congenital – Preorbital anomalies in zone between paired frontal and nasal bones were internasal in Early Triassic dissorophid *Micropholis*, between the nasals in *Benthosuchus korobkovi*, internasofrontal in *Osteophorus roemeri, Eryops megacephalus, Mordex calliprepes, Sclerocephalus haeuseri haeuseri, Trematosuchus sobeyi, Kestrosaurus dreyeri, Benthosuchus korobkovi, Lydekkerina huxleyi*, and *Batrachosuchoides impressus*, and interfrontal in *Onchiodon, Branchierpetron*, and *Apateon flagrifera.* Postorbital anomalies were intertemporal in *Cochleosaurus, Edops, Sclerocephalus haeuseri haeuseri, Chelyderpeton latirostre, Melanerpeton eisfeldi, Schoenfelderpetron prescheri, Gonioglyptus kokeni, Platyoposaurus watsoni, Benthosuchus sushkini*, and *Apateon flagrifera*, interfrontoparietal in *Benthosuchus sushkini, Thoosuchus yakovlevi*, and *Sclerocephalus haueseri jeckenbachensis*, interparietal in *Gonioglyptus kokeni*, lateroparietal in *Apateon flagrifera, Chelyderpeton latirostre*, and *Platyoposaurus watsoni*, centroparietal in *Apateon flagrifera, Aphaneramma rostratum, Wetlugasaurus rostratum, Volgasaurus kalajevi*, and *Lyddekkerina huxleyi*, interpostparietal in *Apateon flagrifera*, laterosquamosal in *Wetlugasaurus angustifrons, Thoosuchus yakovlevi*, and *Benthosuchus sushkini*, laterotabular in *Benthosuchus korobkovi*, and posttabular in *Thoosuchus yakovlevi.* Other anomalies include inclusion of frontals in border of orbits in *Platyoposaurus watsoni, Konzhukovia vetusta, Benthosuchus korobkovi,* and *Melosaurus platyrhinus*, absence of pineal foramen in *Thoosuchus yakovlevi*, laterally expanded frontal in *Thoosuchus yakovlevi* and ear notch formation in *Micropholis.*

Fossil – Preorbital anomalies in zone between paired frontal and nasal bones were internasal in Early Triassic dissorophid *Micropholis*, between the nasals in *Benthosuchus korobkovi*; internasofrontal in *Osteophorus roemeri, Eryops megacephalus, Mordex calliprepes, Sclerocephalus haeuseri haeuseri, Trematosuchus sobeyi, Kestrosaurus dreyeri, Benthosuchus korobkovi, Lydekkerina huxleyi*, and *Batrachosuchoides impressus*; and interfrontal in *Onchiodon, Branchierpetron* and *Apateon flagrifera.* Postorbital anomalies were intertemporal in *Cochleosaurus, Edops, Sclerocephalus haeuseri haeuseri, Chelyderpeton latirostre, Melanerpeton eisfeldi, Schoenfelderpetron prescheri, Gonioglyptus kokeni, Platyoposaurus watsoni, Benthosuchus sushkini*, and *Apateon flagrifera*; interfrontoparietal in *Benthosuchus sushkini, Thoosuchus yakovlevi*, and *Sclerocephalus haueseri jeckenbachensis*; interparietal in *Gonioglyptus kokeni*, lateroparietal in *Apateon flagrifera, Chelyderpeton latirostre*, and *Platyoposaurus watsoni*; centroparietal in *Apateon flagrifera, Aphaneramma rostratum, Wetlugasaurus rostratum, Volgasaurus kalajevi*, and *Lyddekkerina huxleyi*;

interpostparietal in *Apateon flagrifera*; laterosquamosal in *Wetlugasaurus angustifrons, Thoosuchus yakovlevi*, and *Benthosuchus sushkini*; laterotabular in *Benthosuchus korobkovi*; and posttabular in *Thoosuchus yakovlevi*. Other anomalies include inclusion of frontals in border of orbits in *Platyoposaurus watsoni, Konzhukovia vetusta, Benthosuchus korobkovi*, and *Melosaurus platyrhinus*, absence of pineal foramen in *Thoosuchus yakovlevi*, laterally expanded frontal in *Thoosuchus yakovlevi* and ear notch formation in *Micropholis*.

Gubin YM, Petrovichev NN, Solovyev YN, Kochergina NV, Lukyanchenko AB, Markov SM. 2001. Новообразование в черепной кости раннего триаса амфибия. [Neoplasm in cranial bone of Early Triassic amphibian]. Voprosy onkologii 47 (4):449–455 [Russian].

Neoplasia – Neoplasm in cranial bone of *Parotosuchus* sp. identified by comparison of flat bone with that of extant animals. Skeletal tumor interpreted as a parostotic osteosarcoma.

Fossil – Neoplasm in cranial bone of *Parotosuchus* sp. identified by comparison of flat bone with that of extant animals. Skeletal tumor interpreted as a parostotic osteosarcoma.

Guderyahn L. 2006. Nationwide Assessment of Morphological Abnormalities Observed in Amphibians Collected from United States National Wildlife Refuges. US Fish and Wildlife Service CBFO-C0601Ball State University: Muncie, Indiana.

Congenital – Terminology:

Amelia – Missing limb.

Taumelia – Long bone bent back on itself forming >90° angle.

Brachydactyly – Normal number of metatarsals but abnormal number of phalanges.

Hemimelia – Shortened bone.

Micromelia – Entire limb present but all elements shortened.

US Region 1 West – 1 brachydactyly, 1 ectodactyly, 3 ectomelia, 11 syndactyly of forelimbs; 1 with amelia, 8 with bony expansions, 13 with brachydactyly, 8 with ectodactyly, 30 with ectomelia, 2 with hemimelia, 2 with syndactyly, and 5 with taumelia of hind limbs among:

Rana aurora 71

Rana catesbeiana 10

US region 6 – Mountain and plains – 1 with forelimb ectomelia and 3 with Hind-limb bony expansion, 26 with brachydactyly, 11 with ectodactyly, 53 with ectomelia, 22 with hemimelia, 23 with micromelia, 1 with polymelia, 2 with syndactyly, and 2 with taumelia among:

Bufo boreas 10

Pseudacris triseriata 1

Rana pipiens 6

US region 3 – upper midwest – 1 with brachydactyly, 2 ectomelia, 1 micromelia, 1 syndactyly of forelimb, 2 with amelia, 7 with bony expansion, 40 with brachydactyly, 10 with ectodactyly, 33 with ectomelia, 10 with hemimelia, 3 with micromelia, 1 with polydactyly, 8 with syndactyly, and 3 with taumelia of hind limbs among:

Rana blairi 8

Rana clamitans 9

Rana pipiens 49

Rana sphenocephala 29 and 33

US region 4 – South and Virginia – 2 with amelia, 16 brachydactyly, 3 ectodactyly and 1 with ectomelia of forelimb; 4 with amelia, 3 bony expansion, 132 brachydactyly, 23 ectodactyly, 37 ectomelia, 3 hemimelia, 2 syndactyly of hind limb among:

Acris crepitans 21

Acris gryllus 3

Bufo fowleri 53

Gastrophryne carolinensis 2

Rana clamitans 1

Rana sphenocephala 111

Hyla chrysoscelis 12

Hyla femoralis 2

Hyla squirella 2

Pseudacris crucifer 2

Pseudacris feriarum 16

Rana virgatipes 1

US region 7 – Alaska – 2 with amelia, 1 brachydactyly, 2 ectodactyly, 6 ectomelia, 10 micromelia, 2 syndactyly, and 6 taumelia of forelimb; 11 with amelia, 3 with bony expansion, 26 brachydactyly, 11 ectodactyly,

53 ectomelia, 22 hemimelia, 23 micromelia, 1 polymelia, 2 syndactyly, and 2 taumelia of hind limbs among:
Rana sylvatica 164
US region 5 – New England and mid-Atlantic – 1 each with amelia, brachydactyly, ectodactyly, and ectomelia of forelimb; 6 with amelia, 17 bony expansion, 17 brachydactyly, 9 ectodactyly, 32 ectomelia, 4 hemimelia, 1 polydactyly, 3 polymelia, 2 syndactyly, and 4 taumelia of hind limb among:
Rana clamitans 29
Rana palustris 5
Rana pipiens 5
No significant variation in location of anomaly by region, but region 4 less commonly had multiple anomalies in a given animal.
Rana sylvatica with shortened long bones is illustrated.

Gudynas E, Gambarotta JC. 1981. Two *Philodryas patagoniensis* from one egg. Herpetological Review 12(2):54.
Congenital – Twin *Philodryas patagoniensis*

Guirley R. 2003. Keeping and Breeding Freshwater Turtles. Ada, Oklahoma: Living Art, 297 pp.
Metabolic – Metabolic bone disease in freshwater turtles.
Shell – Shell rot from *Beneckia chitinovova* in yellow-headed temple turtle
Hieremys annandalii.

Guettard JE. 1783. Première Mémoire. Sur différentes monstruosités des Plantes et d'Animaux. [On different monstrosities in plants and animals]. Mémoires sur différentes parties des sciences et arts. Tome 5:1–49. Paris (Philippe-Denys Pierres) [French].
Congenital – Supernumerary right leg in a *Rana temporalis*, figured on pl. XVIII, fig. 3.

Guggisberg CA. 1972. Crocodiles: Their natural history, folklore and conservation. 195 pp. Stackpole Books, Harrisburgh, PA.
Trauma – *Alligator mississippiensis* with ½ of upper jaw bitten off.

Gumbel A. 1999. American Times: Hollywood, California: The Doyenne of death heads for Tinseltown. The Independent (London) 8 December 1999: 15.
Congenital – Dicephalic turtle.

Gunckel Lüer H. 1944. Un caso teratológico en un ofidio chileno [A teratological case in a Chilean ophidian]. Boletín de la Sociedad Biológica de Concepción, Chile 19:83–85 [Spanish].
Congenital – Dicephalic snake *Tachymenis chilensis* from Cherquenco, province of Cautín, in southern Chile in 1943 in the collections of the Museo Araucano of Temuco, Chile.

Gund W. 2001. Bericht über dreifache Zwillingsgeburt bei *Python regius*. [Report on a triple birth in a *Python regius*]. Elaphe 9(1):24 [German].
Congenital – Twins of ball or royal *Python regius*, poor pictures: one case with two anterior bodies.

Gundlach J. 1880. Contribución a la Erpetología Cubana (Habana). [Contribution to Cuban Herpetology (Habana)] 99 pp.; Habana: G. Montiel y Cia.. [Spanish]
Pseudopathology – In page 72, Gundlach observes that during the Cuban boa *Epicrates angulifer* mating season, sometimes some males wrap themselves around the female, which may be the origin of the belief in the existence of snakes with many heads.
Congenital – Monstrous dicephalic Cuban boa *Epicrates angulifer* in the collections of the Museum of the Royal Academy of Sciences of la Habana (Page 73).

Günther A. 1867. Contribution to the anatomy of *Hatteria*. Philosophical Transactions 157:595–629
Trauma – *Hatteria* tail less easily broken than Lacertidae, but regenerates.

Günther A. 1897. Note on some reptiles and a frog from Argentina. Annals and Magazine of Natural History, Series 6, 20:365–366.
Congenital – *Lepidosternum affine* with parietal and occipital shield asymmetry.

Günther R. 1990. Die Wasserfrösche Europas: (Anura-Froschlurche). [The water frogs of Europe]. Neue Brehm-Bücher 600:1–228. Wittenberg: A. Ziemsen [German].
Congenital – Anomalies in larvae like bowed vertebral column, no eyes or only One.
Polymelia and Polydactyly in *Rana esculenta*. Anomaly P caused by teratogene virus.
Trauma – 2.8% of *Rana esculenta* of France with fractures, 7.9% of *Rana esculenta* in the Museum national d'histoire naturelle (Paris) with fractures.

Günther R. 1996a. *Bufo bufo* Linnaeus, 1758. In Günther R. ed. Die Amphibien und Reptilien Deutschlands. pp. 274–302; Jena: Gustav Fischer [German].
Congenital – Polydactyly in *Bufo bufo*.

Günther R. 1996b. Teichfrosch – *Rana* kl. *esculenta* Linnaeus, 1758. [Edible frog – *Rana* kl. *esculenta* Linnaeus, 1758]. In Günther R. ed. Die Amphibien und Reptilien Deutschlands. [Amphibians and Reptiles of Germany]. pp. 455–475; Jena: Gustav Fischer [German].

Congenital – *Rana* kl. *esculenta* with two additional hind legs figured, shortened and deformed toes and legs, scoliosis in larvae and other malformations; Polymelia and polydactyly as consequence of regeneration in animals damaged by teratogene virus or larvae of cercariae.

Günther R, Lehnert M. 1996. Aspis viper – *Vipera aspis* (Linnaeus 1758). In Günther R. ed. Die Amphibien und Reptilien Deutschlands. Pp. 699–710; Jena: Gustav Fischer [German].

Congential – Dicephalic *Vipera aspis*.

Günther R, Meyer F. 1996. Kreuzkröte – *Bufo calamita* Laurenti, 1768. [Natterjack toad – *Bufo calamita* Laurenti, 1768]. In Günther R. ed. Die Amphibien und Reptilien Deutschlands. pp. 302–321; Jena: Gustav Fischer [German].

Congenital – Anatomical deformities (missing toes, or extremities) rare in *Bufo calamita*.

Günther R, Nabrowsky H. 1996. Moorfrosch – *Rana arvalis* Nilsson, 1894. [Moor frog – *Rana arvalis* Nilsson, 1894]. In Günther R. ed. Die Amphibien und Reptilien Deutschlands. [The Amphibians and Reptiles of Germany]. pp. 364–388; Jena: Gustav Fischer [German].

Congenital – Polydactyly and osteological anomalies in posterior vertebral column and pelvic in *Rana arvalis*.

Günther R, Podloncky R. 1996. Wechselkröte – *Bufo viridis* Laurenti, 1768. [Green frog – *Bufo viridis* Laurenti, 1768]. In Günther R. ed. Die Amphibien und Reptilien Deutschlands. pp. 322–343; Jena: Gustav Fischer [German].

Congenital – Supernumerary extremities in *Bufo viridis*.

Günther R, Scheidt U. 1996. Geburtshelferkröte – *Alytes obstretricans* (Laurenti, 1768). [Midwife frog – *Alytes obstretricans* (Laurenti, 1768)]. In Günther R. ed. Die Amphibien und Reptilien Deutschlands. pp. 195–214; Jena: Gustav Fischer [German].

Congenital – Superregeneration of left hind leg in larvae; 12 legged regeneration in *Alytes obstretricans*.

Günther R, Schneeweiss N. 1996. Rotbauchunke – *Bombina bombina* (Linnaeus, 1761). [Fire bellied toad – *Bombina bombina* (Linnaeus, 1761)]. In Günther R. ed. Die Amphibien und Reptilien Deutschlands. pp. 215–232; Jena: Gustav Fischer [German].

Congenital – Malformation of toes in *Bombina bombina*.

Günther R, Völkl W. 1996a. Waldeidechse – *Lacerta vivipara* Jacquin, 1787. [Common or Viviparous lizard – *Lacerta vivipara* Jacquin, 1787]. In Günther R. ed. Die Amphibien und Reptilien Deutschlands. pp. 588–600; Jena: Gustav Fischer [German].

Congenital – Missing toes and double tail in *Lacerta vivipara*.

Günther R, Völkl W. 1996b. Blindschleiche – *Anguis fragilis* Linnaeus, 1758. [Slow worm – *Anguis fragilis* Linnaeus, 1758]. In Günther R. ed. Die Amphibien und Reptilien Deutschlands. pp. 617–631; Jena: Gustav Fischer [German].

Congenital – *Anguis fragilis* with two trunks and one head, and siamesese twins in a second.

Günther R, Völkl W. 1996c. Aspis viper – *Vipera aspis* (Linnaeus 1758). In Günther R. ed. Die Amphibien und Reptilien Deutschlands. [The Amphibians and Reptiles of Germany]. Jena: Gustav Fischer, pp. 699–710 [German].

Congenital – Dicephalic *Vipera aspis*.

Günther R, Völkl W. 1996d Schlingnatter – *Coronella austrica* Laurenti, 1768. [Smooth snake – *Coronella austrica* Laurenti, 1768]. In Günther R. ed. Die Amphibien und Reptilien Deutschlands. [The Amphibians and Reptiles of Germany]. Jena: Gustav Fischer, pp. 631–647 [German].

Congenital – Dicephalic *Coronella austrica*.

Günther R, Völkl W. 1996e. Ringelnatter – *Natrix natrix* (Linnaeus, 1758). [Grass snake – *Natrix natrix* (Linnaeus 1758)]. In Günther R. ed. Die Amphibien und Reptilien Deutschlands. [The Amphibians and Reptiles of Germany]. Jena: Gustav Fischer, pp. 666–684 [German].

Congenital – Dicephalic *Natrix natrix* often found, in association with arched vertebral column.

Günther R, Waitzmann M. 1996. Äskulapnatter – *Elaphe longissima* (Laurenti 1768). [Aesculapian snake – *Elaphe longissima* (Laurenti 1768)]. In Günther R. ed. Die Amphibien und Reptilien Deutschlands. [The Amphibians and Reptiles of Germany]. Jena: Gustav Fischer, pp. 647–666 [German].

Congenital – Deformed head and arched vertebral column (scoliosis) in *Elaphe longissima*.

Gupta BB, Goyal RK, Sharma CS. 1982. A case of reduplication of hind limb in an Indian frog *Rana tigrina* David (Anura: Amphibia). Uttar Pradesh Journal of Zoology 2:106–108.

Congenital – Supernumerary hind limb in Indian frog *Rana tigrina*.

Gurdon JB, Woodland HR. 1975. *Xenopus*. In Handbook of Genetics. Volume 4. Vertebrates of Genetic Interest, RC King, ed. New York: Plenum, pp. 35–50.

Congenital – Named recessive anomalies in *Xenopus laevis*:

Kinky tail (kt) are microcephalic (Uehlinger and Reynaud 1965).

Yolky rectum (yr) are microcephalic (Uehlinger and Reynaud 1965).

Polydactyly – Uehlinger (1969)

Named dominant anomaly in *Xenopus laevis*:

Screwy (S) have twisted backbone and tail (Uehlinger 1966).

Metabolic – Exclusive liver diet produces severe skeletal abnormalities because of calcium and vitamin D deficiency (Bruce and Parkes 1950).

Gurlt EF. 1877a. Ueber thierische Missgeburten. Ein Beitrag zur pathologischen Anatomie und Entwicklungs-Geschichte. [On animal deformed fetuses. A contribution of pathological activity and ontogeny]. 97 pp., Berlin: Hirschwald [German].

Congenital – Abnormal embryos/newborns of domestic mammals mainly (all the figures).

"*Peromelus monoscelus*" frog lacking hind leg.

"*Emprostomelophorus tetrascelus*" frog two supernumerary legs extending ventrally from the pelvic.

Gurlt EF. 1877b. Die neuere Literatur über menschliche und thierische Missgeburten. [The new literature on human and animal malformed fetuses]. Virchow's Archiv für pathologische Anatomie und Physiologie und für klinische Medizin 74:504–527 [German].

Congenital – Compilation of literature on deformities (mainly human and mammals).

Frog with supernumerary hind legs.

Gurushankara HP, Krishnamurthy SV, Vasudev V. 2007. Morphological abnormalities in natural populations of common frogs inhabiting agroecosystems of central Western Ghats. Applied Herpetology 4:39–45.

Environmental – Abnormality rates of 3.92% in water bodies, contrasted with

3.98 in agricultural habitats and 4.64 in coffee plantations, contrasted with 0–2% in the wild (Ouellet 2000).

Species	Environment	Abnormalities (%)
Limnonectes limnocharis		3.26
Limnonectes keralensis		3.35
Limnonectes brevipalmata	Semiaquatic	4.09
Tomopterna (Spherotheca) rufescens	Litter-dwelling	7.51

Frequency (%) of malformations by habitat and species:

Species	Habitat	An/microophthalmia	Hemimelia	Brachydactyly	Ectrodactyly	Ectromelia
L. limnocharis	Water	0.847%/0	0.424%	1.695%	0.424%	0.847%
	Paddy	0.101/0.152	0.404	1.011	0.506	1.264
	Coffee	0.273/0	0	0.273	0.273	1.366
	Forest	0	0	0	0	0
L. keralensis	Water	0.494/0	0.247	0.247	1.235	0.494
	Paddy	0.719/0	0.359	0.539	0.539	1.377
	Coffee	0/0	0	0	1.370	3.196
	Forest	0	0	0	0	0
L. brevipalmata	Water	1.351/1.351	0	1.351	1.351	0
	Paddy	0.398/0	0.398	0.398	1.992	0
	Coffee	1.613/0	0	1.613	4.032	0
	Forest	0	0	0	0	0
T. rufescence	Water	4.000/0	0	4.000	4.000	0
	Paddy	0.803/0	0	4.418	1.205	2.008
	Coffee	0/0	0	4.444	0	4.444
	Forest	0	0	0	0	0

Neoplasia – Femoral tumor in 3 (0.152%) of *L. limnocharis*, 2 (0.12%) of *L. keralensis* and 1 (0.398) of *L. brevipalmata* from paddy fields and 2 *L. limnocharis* and 1 (2.222%) *T. rufescence* from coffee plantations of Western Ghats.

Guthrie JE. 1929. Snake notes. Proceedings of the Iowa Academy of Science 36:349–359.

Congenital – Derodymous bull snake *Pituophis sayi*, dicephalic copperhead, garter snake *Thamnophis radix*, *Thamnophis sirtalis*, hog-nosed snake *Heterodon platorhinus*, king snake *Ophibolus getulus*, and milk snake *Coluber doliatus triangulus*.

Gutzke WH, Packard GC. 1987. Influence of the hydric and thermal environments on eggs and hatchlings of bull snakes *Pituophis melanoleucus*. Physiologic Zoology 60(34):9–17.

Environmental – Abnormalities of "tail or cranial deformity" in 10%, 4%, and 32% of bull snake *Pituophis melanoleucus* raised at 22°, 27°, and 32°.

Guyénot E. 1921. Mutations et monstruosités. [Mutations and monsters]. Revue scientifique 21(12 November 1921):611–617 [French].

Congenital – Ectromelia in a lizard.

Amelia in snakes.

Apodia in salamanders.

Guyénot E. 1928. Territoire de regeneration chez les lezards: *Lacerta muralis*. [Regeneration landscape in lizard: *Lacerta muralis* C. R. Soc. Biol. Paris 99:27–28 [French].

Trauma – Regeneraton in *Lacerta muralis.*

Guyénot E, Matthey R. 1928. Les processus régénératifs dans la patte postérieure du Lézard. – [The regenerative process in the posterior lizard foot]. Roux's Archiv für Entwicklungsmechanik 113:520–529.

Trauma – *Lacerta muralis* hind-limb regeneration as a featureless extension.

Annotated Bibliography H-L

Haagner GV, 1993. *Lycophidion capense capense* Cape wolf snake: Dicephalism and reproduction. Journal of the Herpetological Association of Africa 42:39.

Congenital – Dicephalic *Lycophidion capense capense* Cape wolf snake.

Haagner GV. 1994. Life history notes: Reptilia: Elaphidae: *Hemachatus haemachatus*, Rinkhals: Dicephalism. African Herpatology News 21:22.

Congenital – 1/19 *Hemachatus haemachatus* embryos was dicephalic, with labial fusion, a common jaw and two tails.

Haagner G, van Rhyn J. 1991. Dicephalism in two southern African colubrid snakes. Naturalist (Port Elizabeth) 35(3):1–3.

Congenital – Dicephalism in *Lamprophis fuliginosus* and *Crotaphopeltis hotamboeia*.

Hackbarth R. 1990. Reptile Diseases. [Translated by U E Friese from Krankheiten der Reptilien. Franckh'sche Verlagshandlung. First Edition 1985.] 127 pp.; Neptune City, New Jersey: TFH Publications.

Other – Only discusses bone disease in generalities.

Hackenbroich Ch, Floeck A, Hetzel U, Koenig A, Kramer M, Berthold LD, Alzen G, Bonath KH. 2001. Discospondylitis in a Cuban crocodile (*Crocodylus rhombifer*): Diagnosis and Therapy. Erkrankungen der Zootiere, Verhandlungsbericht des 40. Internationalen Symposiums über die Erkrankungen der Zoo- und Wildtiere:41–49.

Infection – Infectious discospondylitis in a Cuban crocodile *Crocodylus rhombifer*.

Haddad CFB, Cardoso AJ, Castanho LM. 1990. Hibridação natural entre *Bufo ictericus* e *Bufo crucifer* (Amphibia: Anura). [Natural hybridization between *Bufo ictericus* and *Bufo crucifer* (Amphibia: Anura)]. Revista Brasileira de Biologia 50(3):739–744. [Portuguese].

Congenital – Natural hybridization of *Bufo ictericus* and *Bufo crucifer* produced fertilized eggs, half of which were taken to a lab and reared in two aquaria. Egg abnormalities were observed from the beginning of development. Irregular shapes, contour irregularities, and conical growths were observed during cleavage, and 60 h after fertilization, only 6% of the embryos appeared normal. Among the abnormalities of the embryos before metamorphosis were the tail twisted at the distal end, the tail twisted at the proximal end, and the vertebral column twisted in the central region of the body. These abnormalities caused mobility problems to the larvae, which swam in circles if they had a twisted vertebral column, or swam unidirectionally very slowly if they had a proximally twisted tail. Very few lived to metamorphosis. The post-metamorphosis abnormalities included a missing eye (in most instances the left) and a twisted vertebral column in the central region of the body. Those individuals missing an eye were most numerous after metamorphosis and would likely have been inefficient capturing food in the wild. All of the individuals (including the only normal-looking one) died a few days after metamorphosis.

Hadeen SE. 1976. Scoliosis and hydrops in a *Rana catesbeiana* (Amphibia, Anura, Ranidae) tadpole population. Journal of Herpetology 10:261–262.

Congenital – Scoliosis in *Rana catesbeiana*, without comment on bony alteration.

Häfeli W, Zwart P. 2000. Panzerweiche bei jungen Landschildkröten und deren mögliche Ursachen. [Weakening of the carapace in a young tortoise and causes of this disease.] Der Praktische Tierarzt 81:129–132. [German]

Metabolic – Hypoplastic osteoporosis in *Testudo marginata*.

Haft J. 1994. Bemerkungen zu den Suppenschildkröten bei Xcacel, Halbinsel Yucatan, Mexiko. [Remarks on green turtle of Xcacel, Yucatan Peninsula, Mexico]. Salamandra 30(4):254–259. [German]

Congenital – Siamese twins (two heads, four forelimbs) of *Chelonia mydas*.

B.M. Rothschild et al., *Herpetological Osteopathology: Annotated Bibliography of Amphibians and Reptiles*,
DOI 10.1007/978-1-4614-0824-6_7, © Springer Science+Business Media, LLC 2012

Haines R.W. 1939. The structure of the epiphyses in *Sphenodon* and the primitive form of secondary centers. Journal of Anatomy London 74:80–90.
Congenital – Documented epiphyseal centers in *Sphenodon* and large centers of calcified cartilage in anuran amphibians. Those in *Phyllodactylus* persist throughout life.
Hajkova P, Knotek Z. 1998. Osteitis deformans in black snake (*Elaphe obsoleta*). Proceedings EAZWA 2:469–472.
Infection – Mistakenly claimed that "multifocal proliferative changes of vertebral bodies and ankylosis are characteristic for" osteitis deformans. They described proliferative changes and ankylosis with radiolucent reaction, actually characteristic of osteomyelitis.
Metabolic – Mistakenly claimed that "multifocal proliferative changes of vertebral bodies and ankylosis are characteristic for" osteitis deformans. They described proliferative changes and ankylosis with radiolucent reaction, actually characteristic of osteomyelitis.
Hakvoort H, Gouda E. 1990. Voorpootmisvorming bij de driekleurige gifkikker (*Epipedobates tricolor*). [Foreleg malformation in the three colored poison frog (*Epipedobates tricolor*)]. Lacerta 48:117–125 [Dutch].
Congenital – Elbow fusion in *Epipedobates tricolor*
Dendrobates leucomelas with bilaterally fused elbows.
Arthritis – Elbow fusion in *Epipedobates tricolor*
Dendrobates leucomelas with bilaterally fused elbows.
Hakvoort JH, Gouda EJ, Zwart P. 1995. Skeletal and muscular underdevelopment (SMUD) in young tree frogs (Dendrobatidae). Proceedings 5th International Symposium of Pathology of Reptiles and Amphibians, Alphen a Rijn 271–275.
Congenital – Amelia and phocomelia in frogs and toads.
Haldane JB, Huxley J. 1927. Animal Biology. London: Oxford University Press, 344 pp.
Metabolic – Failure of leg development in frogs with thyroid removed.
Halfpenny S. 1992. Spindly-leg syndrome. British Dendrobatid group Newsletter #13 as reprinted in American Dendrobatid Society Newsletter 12:1–2
Metabolic – Suggestion that iodine supplementation prevents dendrobatid spindly leg (presenting as small, "ineffective" forelimbs).
Hall BK. 1984. Developmental processes underlying the evolution of cartilage and bone. Symposium of the Zoological Society of London 52:155–176. In: The Structure, Development and Evolution of Reptiles: A festschrift in honour of Professor A. d'A. Bellairs on the occasion of his retirement, WJ Ferguson, ed. London: Academic Press.
Congenital – "What changed from reptiles to mammals was not bone structure but associations" (pp. 162–163).
Sphenodon's secondary center of calcified cartilage is from diaphysis, not from independent center (Haines 1939). This contrasts with most lizards, which have epiphyses.
Intratendinous calcification.
Hall PM. 1985. Brachycephalic growth and dental anomalies in the New Guinea crocodile *Crocodylus novaeguineae*. Journal of Herpetology 19:300–303.
Congenital – two of 333 New Guinea crocodyle *Crocodylus novaeguineae* skulls had short snouts with expanded mandibles.
Trauma – Scars, punctures, and growths of skull, mandibular asymmetry, supratemporal fenestrae ossification.
Hall L. 2003. The magic of television comes south. Richmond Times Dispatch (Virginia) 23 April 2003:1.
Congenital – Dicephalic turtle.
Hall BK. 2005. Bones and Cartilage. Developmental and Evolutionary Skeletal Biology. XXVIII + 760 pp. Elsevier Academic Press,
Congenital – Extra cartilages, known as paraphalangeal bones, illustrated in gecko lizards.
Sesamoids are more common in reptiles than amphibians.
Sesamoids in hind limb of South African dwarf frog, *Hymenochirus boettgeri*, some of which sesamoids depend on normal hind-limb function. Proximal os sesamoides tarsale, tibialis anticus, and cartilages sesamoides failed to form in the right hind limb of individuals.
Ankylosaurus has combination of exostoses of dermal cranial bones, sesamoids, and ectopic extracranial bones.
The many sesamoids in the garden lizard *Calotes versicolor* are examples of metaplastic mineralization.
Fossil – *Ankylosaurus* has combination of exostoses of dermal cranial bones, sesamoids, and ectopic extracranial bones.
Halliday T, Adler K. 2002. Firefly Encyclopedia of Reptiles and Amphibians. Firefly Books, Buffalo, New York.
Environmental – Minnesota as deformed frog hotspots. Illustrated supernumerary limbs in Pacific tree frog *Hyla regilla*.
Hallmann G, Krüger J, Trautmann G. 1997. Faszinierende Taggeckos – Die Gattung *Phelsuma*. [Fascinating day gecko – the genus *Phelsuma*]. 229pp.; Münster: Natur und Tier-Verlag. [German]
Metabolic – Rickets/osteomalacia (Fig. 33) in *Phelsuma astriata astriata* Vascular – Dying of tail tip (Fig. 34) in *Phelsuma abbotti sumptio*)

Halloy M. 1990 (1991). Nota sobre dos malformaciones de una puesta del colúbrido *Hydrodynastes gigas* [Note about two malformed hatchlings of the snake *Hydrodynastes gigas*]. Boletín de la Asociación Herpetológica Argentina 6(1):8–9. [Spanish]

Congenital – Two malformed *Hydrodynastes gigas* hatchlings, one with deformed vertebral column (mostly along the saggital plane) with fusion of many neural arches. Halloy described it as rhoecotic, as defined by Smith and Fitzgerald (1983). The other hatchling lacked the left eye and possessed a small left orbit, resulting in a slightly distorted skull.

Hamilton AG. 1950. Polymely in a frog. Nature 166:611–612.

Congenital – Supernumerary *Rana temporaria* left forefoot articulated with sternum by means of a glenoid. Notes Wagner (1913) described extra imperfect limbs attached to "extra basal piece overlying the sternum" and Cotton's 1922 description of an immobile forelimb attached to an extra half pectoral girdle.

Hamilton G. 1995. '94 wasn't without its oddities in Citrus. St. Petersburg Times (Florida) 2 January 1995:1.

Congenital – Dicephalic turtle.

Hamley T. 1990. Functions of the tail in bipedal locomotion of lizards, dinosaurs and pterosaurs. Memoirs of the Queensland Museum 28:153–158.

Trauma – Tail loss was noted in 40% of Eastern water dragon *Physignathus lesueurii*. Loss of 80% led to tripedal trackways, but did not reduce speed. Geckos with "impaired" tails have impaired bipedal ability (Snyder 1949).

Hammerton AE. 1933. Report on the deaths occurring in the Society's Gardens during the year 1932. Proceedings of the Zoological Society London 1933:451–482.

Metabolic – *Crocodylus niloticus* with osteomalacia.

Neoplasia – Hardwick's lizard *Uromastix hardwickii* with vertebral osteoma.

Hammerton AE. 1934. Report on deaths occurring in the Society's Garden for the year 1933. Proc Zool Soc London 104:389–422.

Infection – Matamata terrapin *Chelus fimbriatus* with plaston fungal infection. Numskull frog *Rana adspersa* with necrosis of cranial bones and cellulitis.

Shell disease – Snapping turtle *Chelydra serpentina* with perforating plaston ulcer. Fungal infection of Matamata terrapin *Chelus fimbriatus* plaston.

Hammerton AE. 1935. Report on deaths occurring in the Society's Garden for the year 1934. Proc Zool Soc London 105:443–474.

Infection – Austwick and Keymer (1981) claimed fungal infection of *Testudo gigantea elephantidae* plaston, but was not in the cited 1981 report.

Shell disease – Austwick and Keymer (1981) claimed *Testudo gigantea elephantidae* plaston fungal infection, but was not in the cited 1981 report.

Hammerton AE. 1939. Report on deaths occurring in the Society's Garden for the year 1938. Proc Zool Soc London 109B:281–327.

Infection – Long-necked terrapin *Chelodina longicollis* with fungal infection of plaston. Coachwhip snake *Coluber flagellum* with pyogenic tail gangrene.

Metabolic – *Iguana iguana* with osteomalatic fractures and broad fronted crocodile *Osteolaemus tetraspis* with generalized osteomalacia. Great Teguexin *Tupinambis teguixin* with renal and articular gout.

Shell – Fungal infection of long-necked terrapin *Chelodina longicollis* plaston.

Hanken J. 1982. Appendicular skeletal morphology in minute salamanders, genus *Thorius* (Amphibia: Plethodontidae): Growth regulation, adult size determination, and natural variation. Journal of Morphology 174:57–77.

Congenital – Alternative carpal arrangements were noted in 53.9% of *Thorius pennatulus*; 15% of *Thorius macdougalli*, 25.1% of *Thorius minutissimus*, 16.7% of *Thorius schmidti*, 10.3% of *Thorius narisovalis*, 5.4% of *Thorius troglodytes* and 9.1% of *Thorius pulmonaris*.

Alternative tarsal arrangements were noted in 43.6% of *Thorius pennatulus*; 28.9% of *Thorius macdougalli*, 14.2% of *Thorius minutissimus*, 47.2% of *Thorius schmidti*, 20% of *Thorius narisovalis*, 7.7% of *Thorius troglodytes* and 18.2% of *Thorius pulmonaris*.

Hanken J. 1983. High incidence of limb skeletal variants in a peripheral population of the red-backed salamander, *Plethodon cinereus* (Amphibia: Plethodontidae), from Nova Scotia. Canadian Journal of Zoology 61:1925–1931.

Congenital – Increased phalangeal number in *Plethodon cinerus* from Nova Scotia, with variable pattern of fusion of intermedium and centrale, tibiale and centrale, distal tarsals one, two, and three, distal tarsals three and four, distal tarsals one, two, three, and four, and distal tarsals four and five.

Hanken J, Dinsmore CE. 1986. Geographic variation in the limb skeleton of the red-backed salamander, *Plethodon cinereus*. Journal of Herpetology 20(1):97–101.

Congenital – Variation in frequency of anomalies in red-backed salamander, *Plethodon cinereus*, with Glasgow Mountain, Nova Scotia, having much higher frequencies (eight patterns = intermedium-ulnare, distal carpals three and four, distal carpal four-centrale, intermedium-centrale, distal carpals one, two, and three, distal carpals one, two, and three and four, radiale-centrale one, distal tarsals four, five, distal tarsals three, four, and five) than Dismal Swamp, Suffolk, Virginia, and Mount Desert Island, Maine (three and four patterns). Carpal and tarsal fusions, mesopodial patterns, phalangeal formulae – manual 1231, 1121, and 1032 versus 1222 and 1132, pedal 12331, 1231 znc 12232 versus 12232, 12322, 02332, 12331, with asymmetry equally represented. The four patterns seen in Virginia and Maine were also the most common patterns seen in Nova Scotia. All populations showed the primitive plethodontid phalangeal formula (1–2–3–2 in the hand; 1–2–3–3–2 in the foot), with variants represented by phalangeal loss.

Hansemann V von. 1903. Ueber säurefeste Bacillen bei *Python reticularis*. [On granulomatous bacterial infection in *Python reticularis*]. Centralblatt für Bakteriologie, Parasitenkunde und Infektionskrankheiten 34:212–213. [German]

Infection – Tubercular bacterial infection in *Python reticulatus*, but no comment on osseous involvement.

Hardy JD Jr. 1964. The spontaneous occurrence of scoliosis in tadpoles of the leopard frog, *Rana pipiens*. Chesapeake Science 5:101–102.

Congenital – Caudal scoliosis in *Rana pipiens*.

Hare KM, Miller KA. 2010. Frequency of tail loss does not reflect innate predisposition in temperate New Zealand lizards. Naturwissenschaftern 97:197–203.

Trauma - Field rate and latency of autotomy was similar among geckos *Hoplodactylus chrysosireticus* and *maculatus* and skinks *Oligosoma* species except for the large nocturnal skink *Oligosoma macgregori*, which was less likely to autotomise. Failure of latency and field rates of autotomy to correlate suggests that autotomy rates relate to efficiency of predation attacks, social interactions and as yet unidentified environmental factors.

Harmon R, Wayne B. 2000. Ossifying fibroma in a green iguana, *Iguana iguana*. Proceedings of the 7th Annual Conference of the Association of Reptilian and Amphibian Veterinarians: 37–38.

Metabolic – Swollen appendages and pliable jaw and dorsoventral flattening in iguana with secondary nutritional hyperparathyroidism.

Harms CA, Lewbart GA, Beasley J. 2002. Medical management of mixed nocardial and unidentified fungal osteomyelitis in a Kemp's Ridley sea turtle, *Lepidochelys kempii*. Journal of Herpetological Medicine and Surgery 12(3):21–26.

Infection – Carpal swelling with proximal first metacarpal radiolucency, obliteration of distal carpal row, ulnar, pissiforms, and metacarpals II and III in Kemp's ridley sea turtle, *Lepidochelys kempii* by *Nocardia* and an unidentified fungus

Harris VA. 1964. The Life of the Rainbow Lizard. 174 pp.; Hutchinson Tropical Monographs, London.

Dental – Tooth row fracture and Meckelian fossa malignment in rainbow lizard.

Harris HS. Jr. 1968. Letter on two-headed *Crotalus h. horridus*. Bulletin of the Maryland Herpetological Society 4(1):28–29.

Congenital – Derodymous *Crotalus h. horridus*.

Harris ML, Bishop CA, McDaniel TV. 2001. Assessment of rates of deformity in wild frog populations using in situ cages: A study of leopard frogs (*Rana pipiens*) in Ontario, Canada. Biomarkers 6:52–63.

Congenital – Limb deformities in mud puppies (*Necturus maculosus*) and leopard frogs (*Rana pipiens*) in the St. Lawrence River basin since 1990s. 2.2% in high risk wetlands, especially (3–10%) in the "young-of-the-year frogs," manifest as fused, missing or extra digits, or disproportionate hind limb length.

Harrison RG. 1898. The Growth and Regeneration of the Tail of the Frog Larva Studied with the Aid of Born's method of Grafting. Roux's Archiv für Entwicklungsmechanik 7:430–485.

Trauma – Tail regeneration from grafted bud, which later atrophies.

Harshbarger JC. 1974. Activities report registry of tumors in lower animals. 1965–1973: pp. 53,61–65,69,72,80,82. Washington DC: Smithsonian Institute Press.

Neoplasia – Chondrosarcoma in corn snake *Elaphe guttata*.

Hasel S. 1992a. The tale of a tiny turtle with 2 heads, 2 spines. St. Petersburg Times (Florida) 7 October 1992:2.

Congenital – Derodymous elegant slider turtle.

Hasel S. 1992b. 2-headed, 2 spined turtle's tale. St. Petersburg Times (Florida) 6 October 1992:1.

Congenital – Derodymous elegant slider turtle.

Hatt JM. 2008. Raising giant tortoises. In Fowler ME, Miller RE, eds. Zoo and Wild Animal Medicine Current Therapy 6. St. Louis Missouri: Saunders Elsevier, p. 144–153.

Metabolic – Metabolic "developmental disease" has been observed only in captives among giant tortoises. Aldabra tortoises develop it within 10–12 years with softening of carapace, pyramiding and fibrous osteodystrophy, in contrast to Galápagos tortoises. Alfalfa pellets are unacceptable litter for giant tortoises, as their protein content is too high. It stimulates fast growth resulting in pyramiding.

Shell – Aldabra tortoises within 10–12 years develop softening of carapace and pyramiding in captivity, in contrast to Galápagos tortoises. Alfalfa pellets are unacceptable litter for giant tortoises, their high protein content is stimulating fast growth and pyramiding, Häfeli and Zwart (2000) suggested endoparasites, kidney disease, and hepatitis as possible causes for pyramidal shell changes.

Haughton SH. 1915. Investigations in South Africa fossil reptiles and Amphibia. 1. On a new species of *Trematosaurus* (*T. sobeyi*). Annals of South African Museum 12:47–51.

Congenital – Presence of internasofrontal (his interfrontal) in eryopoid temnospondyls, *Trematosaurus*.

Fossil – Presence of internasofrontal (his interfrontal) in eryopoid temnospondyls, *Trematosaurus*.

Hauser B, Mettler F, Honegger RE. 1977. Knochenstoffwechselstörungen bei Seychellen-Riesenschildkröten (*Testudo* [*Geochelone*] *gigantea*). [Dysfunction of bone metabolism in giant tortoises of the Seychelles (*Testudo* [*Geochelone*] *gigantea*)]. Verhandlungsbericht des Internationalen Symposiums über die Erkrankungen der Zootiere 19:121–125 [German].

Congenital – Osteodystrophia fibrosa and rickets in *Geochelone gigantea* from island Aldabra in the Indian Ocean.

Hauver RC. 1958. Studies on natural anomalies of the hind limb of *Rana catesbeiana*. V+30pp.; Masters Thesis. Miami Univ, Oxford, Ohio.

Congenital – 10% of *Rana catesbeiana* from Ripley Ohio with supernumerary limbs and one with pelvic duplication and poorly developed limbs.

Hawkins WR Jr. 1990. Taphonomy of an Upper Cretaceous (Maastrichtian) mosasaur, Braggs, Alabama. Journal of Vertebrate Paleontology 10(Supplement 3):26A.

Trauma – Grooves on mosasaur internarial bar mosasaur to mosasaur or shark, but no evidence of healing.

Fossil – Grooves on mosasaur internarial bar mosasaur to mosasaur or shark, but no evidence of healing.

Hay WP. 1904. A revision of *Malaclemmys*, a genus of turtles. Bulletin of the Bureau of Fisheries. 24:3–19.

Congenital – Great variation in size and shape of *Malaclemmys* skull and mouth, citing Bangs (1896). Supernumerary plates in *Malaclemmys* from longitudinal division of nuchal or vertebral plates. Dwarfed *Malaclemmys* with broad short shell, twisted to one side.

Trauma – *Malaclemys centrata* with missing feet, irregular stumps, or irregular growths. One individual had lost the posterior half of its body, including one leg.

Shell disease – Pitted *Malaclemys centrata* carapace and plaston. "Disease similar to necrosis of bone" in *Malaclemmys* starting as white plaston spot, which enlarges producing a cheesy nodule which falls out, leaving a cavity in the bone. It may be covered by scar tissue, but may grow until "a large area of the shell is eaten away." Pitted *Emys concentrica* shell citing Sowerby and Lear (1872).

Hay OP. 1908. The Fossil Turtles of North America. Carnegie Institution Publication 75:1–556, 113 plates.

Congenital – *Echmatemys wyomingensis* at Yale University Museum with extra neural, two extra costal plates and one extra peripheral.

Shell disease – Carapace pits in *Baena arenosa* AMNH 5973, *Osteopytis chelydrinis* AMNH 1131 with multiple pits, *Osteopytis g. bicarape* at Yale, *Baptemys wyomingensis* AMNH 1494, *Baptemys fluviatilis* with multiple pits, *Echmatemys naomi* AMNH 5975, and *Hadrianus corsoni* AMNH 6027. Plastron pits in *Baena callosa* Carnegie Museum 330, *Hadrianus corsoni* AMNH 6027.

Fossil – *Echmatemys wyomingensis* at Yale University Museum with extra neural, two extra costal platns, and one extra peripheral. Carapace pits in *Baena arenosa* AMNH 5973, *Osteopytis chelydrinis* AMNH 1131 with multiple pits, *Osteopytis g. bicarape* at Yale, *Baptemys wyomingensis* AMNH 1494, *Baptemys fluviatilis* with multiple pits, *Echmatemys naomi* AMNH 5975, and *Hadrianus corsoni* AMNH 6027. Plastron pits in *Baena callosa* Carnegie Museum 330, *Hadrianus corsoni* AMNH 6027.

Hayden F.V. 1878. Notes on the Tertiary and Cretaceous Period of Kansas. U.S. Geological and Geographical Survey of the Territories 1878:279–294.

Trauma – Healed rib fractures in saurians. Bite marks in saurian bones from *Galeocerdo* shark. Fractured saurian vertebrae which have fused and become confluent.

Fossil – Healed rib fractures in saurians. Bite marks in saurian bones from *Galeocerdo* shark. Fractured saurian vertebrae which have fused and become confluent

Hayes FE, Beaman KR. 1985. Life history notes. *Geochelone elephantopus vanderburghi* (Galapagos giant tortoise: Volcan Alcedo race). Morphology. Herpetological Review 16:81–82.

Congenital – Shell anomaly in Galapagos giant tortoise *Geochelone elephantopus vanderburghi*

Hayes-Odum L, Dixon JR. 1997. Abnormalities in embryos from a wild American alligator (*Alligator mississippiensis*) nest. Herpetological Review 28:73–75.

Congenital – Short lower jaw, scoliosis, and maxillary hyperplasia in American alligator (*Alligator mississippiensis*) attributed to heat during incubation.

Heard DJ, Jacobson ER, Clemmons RE, Campbell GA. 1998. Bacteremia and septic arthritis in a West African dwarf crocodile. Journal of the American Veterinary Medicine Association 192:1453–1456.

Infection – Radius and ulna *Morganella morganii*, *Seratia marcescens*, and *Staphylococcus sp.* osteomyelitis and in *Osteolaemus tetraspis* West African dwarf crocodile.

Heasman WJ. 1933. The anatomy of a double-headed snake. Journal of Anatomy 67:331–345

Congenital – Dicephalic *Coluber florulentus, Elaphe vulpina, Lampropeltis getula getula, Natrix sipedon fasciata, Bothrops atrox, Crotalus terrificus, Pituophis sayi* (= *Coluber melanoleucus*), *Thamnophis sirtalis*, as previously cited by Do Amaral 1927, Pole 1927 and Fisher 1928.

Heath HD. 1953. Regeneration and growth of chimaeric amphibian limbs. Journal of Experimental Zoology 122:339–366.

Congenital – *Triturus torosus, Ambystoma punctatum, tigrinum* and *punctatum* chimera produced by interspecies interchange of parts.

Heatley JJ, Mitchell MA, Williams J, Smith JA, Jully TN Jr. 2001. Fungal periodontal osteomyelitis in a chameleon, *Furcifer pardalis*. Journal of Herpetological Medicine and Surgery 11(4):7–12.

Infection – Proliferative swelling of rostal and left hemimandibles of *Furcifer pardalis* by *Nocardia*.

Heatwole H, Suárez-Lazú PJ. 1965. Supernumerary legs in *Bufo marinus* and abnormal regeneration of the tail in *Ameiva exsul*. Journal of the Ohio Herpetological Society 5:30–31.

Congenital – three Puerto Rico *Bufo marinus* with five legs, one with five fingers. Another with two supernumerary front legs; a third, with rudimentary leg with an amputated stump.

Trauma – Bifurcate and occasionally trifurcate tails were noted in *Anolis*. Second tail apparently grew from a wound in *Ameiva exsul*.

Hebard WB, Brunson RB. 1963. Hind limb anomalies of a western Montana population of the Pacific tree frog, *Hyla regilla* Baird and Girard. Copeia 1963:570–572.

Congenital – 14/73 with supernumerary hind limbs in Pacific tree frog *Hyla regilla* from Salish Mountain pond near Polson, Montana. Elongate appendage with no digits, bifurcated appendage, or lobster-claw appearance was noted. *Ambystoma* and *Rana* from same pond had no lesions.

Hedeen SE. 1976. Scoliosis and hydrops in a *Rana catesbeiana* (Amphibia, Anura, Ranidae) tadpole population. Journal of Herpetology 10:261–262.

Congenital – Caudal scoliosis or lateral tail curvature in *Rana sylvatica* and *Rana pipiens*.

Hedstrom I, Bolaños F. 1986. *Dendrobates auratus* (poison arrow frog). Predation. Herpetological Review 17:88–89.

Trauma – Predation on poison arrow frog *Dendrobates auratus* by "machaca" *Brycon guatemalensis*.

Heimann E. 1992. Sind Panzeranomalien bei Schildkröten vererbbar? [Are anomalies of the carapace hereditary?] Sauria 14(3):27–29. [German]

Congenital – 5% to 8% of captive bred tortoise *Testudo marginata* with anomalities in the scutellation: different number of centralia or/and lateralia, fusion or division of marginalia and postcentralia, lack or doubling of precentralia, presence of small supramarginalia. Hereditary cause.

Heimes P. 1994. Morphologische Anomalien bei Äskulapnattern (*Elaphe longissima*) im Rheingau-Taunus. [Morphological anomalies in *Elaphe longissima* in the Rheingau-Taunus]. Salamandra 30(4):268–271. [German]

Congenital – Keratin abnormalities: Pholidose aberrations like undivided anal, 1. ventral shield in front of anal divided, sporadic appearance of fused postocularia, different number of supralabials, microphthalmy of right eye (half of left eye), one hydrocephalus within 200 juveniles, and pholidosis like irregularities in supralabials.

Heiss E, Natchev N, Beisser C, Lemell P, Weisgram J. 2010. The fish in the turtle: On the functionality of the oropharynx in the common musk turtle *Sternotherus odoratus* (Chelonia, Kinosternidae) concerning feeding and underwater respiration. Anatomical Record 293:1416–1424.

Vascular – Papillous skin of submerged common musk turtles *Sternotherus odoratus* used mainly for oxygen uptake (Root 1949; Stone et al. 1992), but not in kinosternids (Bagatto et al. 1997).

Highly vascularized oropharyngeal mucosa in sister group of kinosternids, the soft-shell turtles Trionychidae (Dunson 1960; Gage and Gage 1886), and common musk turtle *Sternotherus odoratus*, less so in *Emys orbicularis*.

Cloacal gas exchange in pleurodiran turtles (Clark et al. 2008; Gordos and Franklin 2002; King and Heatwole 1994), but not kinosternids (Dunson 1960).

Helgen J, McKinnell RG, Gernes MC. 1998. Investigation of malformed northern leopard frogs in Minnesota. In Lannoo MJ. ed. Status and Conservation of Midwestern Amphibians. Pp. 288–297; Iowa City: University of Iowa Press.

Congenital – Malformed leopard frogs *Rana pipiens* in 1995: one third of Le Sueur County Minnesota specimens were abnormal; 20 of 137 with amelia, 20 with ectomelia, 12 with supernumerary leg, and one supernumerary arm derived from pectoral girdle. Fifty with "spike on posterior end." Parasitic cysts were noted in these Sibley County.

Meeker County – 91/93 had hind limb abnormalities. Seventy three had only stumps of legs. No parasitic cysts were present. One malformed Elk River gray tree frog *Hyla versicolor/chrysoscelis*. One malformed *Bufo* sp. from St. Paul.

Hellmich W. 1929a. Mehrfachbildungen von Extremitäten bei Amphibien (Naturfunde) [Multiple formation of extremities in amphibians (finds in nature)]. Wilhelm Roux' Archiv für Entwicklungsmechanik 115:409–414. [German]

Congenital – *Rana graeca* with three hind legs (north of Saloniki), *Rana esculenta* with four hind legs (southern Germany), *Salamandra maculosa* with fifth leg (= shortened right front leg) on back (near Triest), and *Alytes obstetricans* with left hind leg "abnormal" legs.

Hellmich W. 1929b. Überzählige Gliedmaßen bei Amphibien (Naturfunde). [Supernumerary extremities in amphibians (natural finds)]. Blätter zur Aquarien- und Terrarienkunde 40:302–304. [German]

Congenital – *Rana esculenta* with four hind legs, the two additional ones turned by 90°; *Salamandra maculosa* with leg (= front leg with three toes) on the back above the middle of the abdomen. *Alytes obstelricans* larva with outgrowth with 12 additional legs of different size besides left hind leg.

Hellmich W. 1930. Untersuchungen über Herkunft und Determination des regenerativen Materials bei Amphibien. [Investigations of origin and determination of the regenerative materials of amphibians]. Roux' Archiv für Entwicklungsmechanik 121:135–203 [German].

Congenital – Refigures *Alytes obstetricans* larva with outgrowth with 12 additional legs of different size besides left hind leg, and *Salamandra maculosa* with leg (= front leg with three toes) on the back above the middle of the abdomen (Hellmich 1929).

Hellmich W. 1941. Abnorme Regeneration einer Hintergliedmaße bei einer Larve von *Alytes obstetricans* [Geburtshelferkröte]. [Abnormal regeneration of a hind leg in a larva of *Alytes obstetricans*]. Blätter für Aquarien- und Terrarienkunde, 52:326–327. [German]

Congenital – One larva of *Alytes obstetricans* with multiple hind legs on the left side.

Hellmich G. 1951. A case of limb regeneration in the Chilean iguanid *Liolaemus*. Copeia 1951:241–242.

Trauma – One among 2000 *Liolaemus* with limb regeneration, although "there is an extraordinary proportion of specimens with regenerated tails." Cylindrical regeneration of a bitten off hind limb.

Henkel F-W, Schmidt W. 1991. Geckos. Biologie, Haltung und Zucht. [Gecko Biology: Keeping and raising]. 224 pp.; Stuttgart: Eugen Ulmer Verlag. [German]

Trauma – Double tail related to incomplete loss of autotomed tail.

Metabolic – Rickets in Gekkos caused by deficiency of vitamin D3 resulting in soft bones (e.g., jaws), arching of vertebral column and tail, and shortened extremities and jaws.

Henle K. 1981. A unique case of malformation in a natural population of the green toad (*Bufo viridis*) and its meaning for environmental politics. Bulletin of the British Herpetological Society 1981(4):48–49.

Congenital – Giant and underdeveloped skeletons, phocomelia, and absent jaw in green toad *Bufo viridis*.

Hennig C. 1870. Eine zweiköpfige Eidechse [Double headed lizard]. Zoologischer Garten, 11. Jahrgang: 196. [German]

Congenital – Describes behavior of a dicephalic lizard.

Herklotz O. 1871. Über den Wiederersatz verloren gegangener Gliedmaßen bei *Triton cristatus*.[On the new formation of lost extremities in *Triton cristatus*]. Verhandlungen der zoologisch-botanischen Gesellschaft Wien 21, 1. Sitzungsbericht: 54–55. [German]

Trauma – Report about repeated loss of toes (eaten by fish) and four times regeneration in *Triton cristatus*.

Herman DW, Johnson N. 1986. Life history notes. Serpents. *Regina septemvittata* (queen snake). Herpetological Review 17:22.

Congenital – Dicephalic queen snake *Regina septemvittata*.

Hermanns JF, Hermans-Le T. 1990. Ectromelie par dysplasie du tibia chez *Rana arvalis*. [Ectomelia by tibial dysplasia in *Rana arvalis*]. Revue Verviétoise d'Histoire Naturelle 1990:44–45. [French]

Congenital – Ectomelia and tibial dysplasia in *Rana arvalis*.

Hernandez-Divers SJ. 2002. The Thai water dragon, *Physignathus cocincinus*. Journal of Herpetological Medicine and Surgery 12:41–44.

Metabolic – Secondary nutritional hyperparathyroidism or nutritional metabolic bone disease with pathologic fractures in Thai water dragon *Physignathus cocincinus*.

Hernandez-Divers SJ. 2006. Diagnostic techniques. In Mader DR. ed. Reptile Medicine and Surgery. Pp. 490–532; Philadelphia: Saunders.

Trauma – Autotomy in Caledonian lizard *Rhacodactylus* sp. Limb fracture from proprioceptive loss from spinal compression in green iguana *Iguana iguana*.

Metabolic – Submandibular swelling from nutritional hyperparathyroidism in green iguana *Iguana iguana*.

Hernandez-Divers SJ, Knott CD, MacDonald J. 2001. Diagnosis and surgical treatment of thyroid adenoma-induced hyperthyroidism in a green iguana (*Iguana iguana*). Journal of Zoo and Wildlife Medicine 32(4):465–475.

Metabolic – Missing dorsal spines in green iguana *Iguana iguana* with hyperparathyroidism were actually dermal appendages, not vertebral spinous processes, not osseous!

Hernandez-Divers SM, Garner MM. 2003. Neoplasia of reptiles with an emphasis on lizards. Veterinary Clinic of Exotic Animals 6:251–273.

Neoplasia – Multiple enchondroma in *Varanus dracoena*, osteochondroma in *Varanus bengalensis*, and osteosarcoma in emerald lizard *Lacerta viridis*.

Hernandez-Divers SJ, Hensel P, Gladden J, Hernandez-Divers SM, Buhlmann KA, Hagen C, Sanchez S, Latimer KS, Ard M, Camus AC. 2009. Investigation of shell disease in map turtles (*Graptemys* spp.). Journal of Wildlife Diseases 45:637–652.

Shell disease – From trauma, malnutrition, and bacterial and fungal infections (Barton 2006; Frye 1991; Homer et al. 1998; Keymer 1978; LaFortune et al. 2005; Mader 2006). Damage most severe on plaston of *Gopherus pseudogeographica kohnii* and *Graptemys nigrinoda*, spreading from seams between scutes, with osteoclastic resorption into dermal bone with extensive remodeling (Garner et al. 1997; Lovich et al. 1996). Clusters of round-oval green-yellow 2–4 micron organisms were seen. Additionally, there were small clefts and pitting lesions with "no evidence of erosion, ulceration, or inflammation" (page 11 of html version). Filamentous organisms, presumed algal, were also noted. *Fusarium semitectum* was isolated from infected scutes. Biopsies revealed *Chryseobacterium indologenes and Morganella morganni* in *Graptemys pseudogeographica kohnii*, *Aeromonas hydrophila* in a *Graptemys geographica*. Mucoid colonies of unclear nature were isolated from *Graptemys pseudogeographica kohnii*, *Graptemys geographica*, and *Graptemys ouachitensis*; flat gray colonies from *Graptemys ouachitensis*; white colonies from *Graptemys pseudogeographica kohnii*, *Graptemys geographica*, and *Graptemys ouachitensis*; and alpha-hemolytic colonies from *Graptemys pseudogeographica kohnii*. Organism identified included *Clostridium butyricum*.

Héron-Royer 1884. Cas tératologiques observes chez quelques tetrads de batraciens anoures et de la possibilité de prolonger méthodiquement l'état larvaire chez les batraciens. [Teratology cases observed among several batrachian anurans and the possibility of prolonged modification of the larval state in batrachians. Bulletin de la Société Zoologique de France 9:162–168. [French]

Congenital – *Alytes obstetricans* with a grossly enlarged head. Supernumerary hind limb in *Pleurodeles waltl* noting Cámerano's (1882) report of supernumerary hind limb in *Triton taeniatus*. Cites irregular sacral vertebrae in *Bombinator igneus* (Cámerano 1880).

Herre W. 1934. Der Schädel eines ungewöhnlichen *Triturus alpestris apuana* Bon. [The skull of an unsual *Triturus alpestris apuana* Bon]. Bonner Zoologischer Anzeiger 106:39–46. [German]

Congenital – Especially wide skull of reproduced *Triturus alpestris apuana* with extreme lateral reaching quadrates.

Herrera y Tordesillas A. 1725–1726. The general history of the vast continent and islands of America, commonly called the West Indies, from the first discovery thereof: with the best accounts the people could give of their antiquities. Vol. IV: 422pp. Translated into English by Captain John Stevens, Batley, London.

Pseudopathology – Account on page 97 of vol. 4, of very venomous snakes in Brazil that "have two mouths, the one in the head, and the other in the tail…" This appears not to be a description of a dicephalic snake, but of a snake with a tail that resembles its head. It is puzzling, however, that the account continues to relate that these snakes bite with "both" heads. The snakes are further described as white and very short, and it is written that the people that are bitten by them "live only a few hours."

Herrmann H-J, Uteshov VK. 1991. Abnorme Extremitätenausbildung bei *Litoria caerulea* (Shaw 1790). [Abnormal formation of extremities *Litoria caerulea* (Shaw 1790).] in Veröffentlichung Naturhistorisches Museum Schloss Bertholdsburg Schleusingen 6:105–107. [German]

Metabolic – Hormonal stimulation of artificially raised hylid *Litoria caerulea* produced four individuals with malformations among several thousand individuals (= normal rate of mutation). Three were missing one front extremity; one, both.

Hertwig G. 1953. Das Auftreten einer Chordoms bei einer Kröten-bastardlarve und die möglichen Ursachen seiner Entstehung. [The occurrence of chordomas in a toad bastard larva and the possible causes of its origin]. Zentralblatt für Allgemeine Pathologie und Pathologische Anatomie 91:56–64. [German]

Congenital – Hybrid European tadpoles with notochord hypertrophy or hyperplasia.

Hes O, Honsa V, Čada F, Vrabec V. 2001. Benirni osteom u hada-popis dvou případů. [Two cases of benign osteomas in snakes]. Veterinářstvi 51:366–367 [Czech].

Neoplasia – Vertebral osteomas in yellow rat snake *Elaphe obsoleta quadrivittata* and garden tree boa *Corallus enydris cooki*. Suggested previous cases in *Lacerta viridis*, *Uromastyx hardwicki*, and *Varanus salvator* (Stolk 1958).

Hes O, Honsa V, Vrabec V. 2002. Nádory páteře u hadů – popis tří případů. [Spinal tumors in snakes – a description of three cases.] Akvárium Terárium 45:74–75. [Czech]

Neoplasia – Osteoma in Elaphe obsoleta quadrivittata and Corallus enydris cooki and malignant sarcoma in Corallus endris cooki.

Heselhaus R. 1983. Zum Problem der Streichholzvorderbeine bei Dendrobatiden-Nachzuchten. [The problem of match legs in propagated Dendrobatids.] Herpetofauna 5(26):22–234. [German]

Congenital – Thin front legs in *Discoglossus tricolor*.

Heusner JC. 1824. Descriptio Monstrorum avium, amphibiorum, piscium quae extant in Museo Univ. Lit. Berol. Eorumque cum monstris mammalium comparatio [Description of avium, amphibium and fish monsters in the Museum of the University of Berolini compared with mammalian monsters]. Dissertation in Auguralis Medica, Universitat Litteraria Berolinensi. Berlin: Bruschckianis. 46 pp. [Latin]

Congenital – Dicephalism is rare in amphibians and noted occasional dicephalic turtle and noted additional examples in the Philadelphia Peal Museum and the Paris Museum and cited Edwards' (1751) report of dicephalic snake from Barbados and Bancroft's (1769) report of dicephalic snake, *Lacerta* and *Rana* with supernumerary legs.

Hibberd EM. 1996. Abnormalities in eggs, embryos and juvenile farmed estuarine crocodiles *Crocodylus porosus*. In Crocodiles. Proceedings of the 13th Working Meeting of the Crocodile Specialist Group. IUCN – The World Conservation Union, Gland Switzerland, pp. 290–295.

Congenital – Farmed estuarine crocodile *Crocodylus porosus* with malformed/absent tail, kyphoscoliosis, reduced upper/lower jaw.

Higginbottom J. 1847. Researches to determine the number of species and the mode of development of the British *Triton*. Proceedings of the Royal Society of London 5:669.

Trauma – Regeneration of limbs in *Triton*, if temperature between 58° and 75° Farenheit.

Higginbottom J. 1853. Researches to determine the number of species and the mode of development of the British *Triton*. Annals of Natural History 12:369–384.

Trauma – Regeneration of limbs in *Triton*, if temperature between 58° and 75° Farenheit.

Highfield AC. 1990a. Notes on skeletal and carapace deformity in captive-bred tortoises (genus Testudo) related to diet with observations on the use of vitamins and mineral supplements. Herptile 15:110–113.

Metabolic – Large turtle with deformed, raised scute, allegedly due to excess dietary protein. Renal osteodystrophy produced thick, porous, raised scutes.

Highfield AC 1990b. Keeping and Breeding Tortoises in Captivity. VI+149pp.; R&A Publishing Ltd, London.

Infection – Septic arthritis in *Testudo graeca* and *Testodo hermanni*.

Metabolic – Humpbacked turtle secondary to diet deficient in vitamin D and calcium, but with elevated protein. Elongation of body shape, deformed, raised scutes.

Shell disease – Ulcerative shell disease/shell-rot, with *Pseudomonas,Citrobacter*, and *Klebsiella* isolated.

Arthritis – *Testudo graeca* and *Testudo hermanni* septic arthritis.

Highfield AC 1990c. The causes of "pyramiding" deformity in tortoises: A summary of a lecture given to the Soiciedad Herpetologica Velenciana Congreso Tortugas on October 30 2010.

Environmental – Pyramiding in tortoises is produced by dehydration. Scute growth accelerates and thickens, exerting an amplified force on the skeleton.

Highton R. 1960. Heritability of geographic variation in trunk segmentation in the red-backed salamander *Plethodon cinereus*. Evolution 14:351–360.

Congenital – Red-backed salamander *Plethodon cinereus* with hereditary trunk vertebral counts: 20 trunk vertebrae in those from New Brunswick, Indiana, New Jersey, Virginia, North Carolina; 21–23 from Georgia. Using costal groove counts (94% accuracy) to count vertebrae, 21 or more were found in eastern Long Island, southern Delmarva Peninsula, southern coastal plain of Maryland, Piedmont and coastal Virginia, and North Carolina. Western New York, Pennsylvania and Maryland, northeast Ohio and east/central Virginia had 19. Additionally, 2–6% had asymmetrical vertebral articulations of sacral ribs with ilia.

Hildebrand SF. 1930. Duplicity and other abnormalities in diamond-back terrapins. Journal of the Elisha Mitchell Science Society 46:41–53.

Congenital – Scoliosis in diamond-back terrapin *Malaclemys centrata*.

Hildebrand SF. 1938. Twinning in turtles. Journal of Heredity 29:243–253.

Congenital – Literature contained 21 cases of conjoined twins, to which this article adds 6. Two diamond-back terrapins (of 100,000 hatched over 25 years) were dicephalic. A dicephalic Florida terrapin *Pseudemys floridana* and a Florida terrapin *Pseudemys floridana* with Siamese plaston to plaston posterior to the abdominal plates, the larger of which had an incomplete left foot and a Siamese river terrapin was attached end to end at lower portion of abdomen. Notes report of a dicephalic sea turtle in a 1751 report by George Edwards in "A Natural History of Birds." EH Barbour in 1888 reported dicephalic Eastern painted turtle *Chrysemys picta* with abnormally broad carpace with extra plates and slight scoliosis. One painted turtle *Chrysemys picta* was Siamese, attached at the posterior ends. There were three other examples of partial twinning. A dicephalic snapping turtle *Cheyldra serpentina* was also reported, as was a dicephalic yellow-bellied terrapin *Chrysemys scaber* with abnormally broad carapace and a diamond-back terrapin *Malaclemys centrata* with two heads, two tails, and rudimentary fifth leg. Another specimen with duplicated head and front feet and double platon to abdominal shields. Cyclopic Florida terrapin *Pseudemys floridana* with twisted skull and upper jaw.

Hill WC. 1954. Report of the Society's Prosector for the year 1953. Proceedings of the Zoological Society London 124:303–311.

Congenital – No congenital malformations among 250 reptiles and amphibians for 1953 at Regent's Park zoo.

Metabolic – Eight with "diseases of metabolism" among 250 reptiles and amphibians for 1953 at Regent's Park zoo.

Neoplasia – Multiple spinal column osteosarcoma in rufous-beaked snake *Ramphiophis rostratus*.

Hill WC. 1955. Report of the society's prosector for the year 1954. Proceedings of the Zoological Society London 125:533–539.

Neoplasia – Osteosarcoma in spine of rufous-beaked snake *Ramphiophis rostratus*, noting that this is apparently not uncommon in serpents according to Wadsworth (1954).

Hillenius D. 1959. Ein zweiter Fall der "Anomalie P" (Rostand) bei *Rana esculenta* Linné. [A second case of "anomaly P" (Rostand) in *Rana esculenta* Linné]. Medizinische Klinik 54:99–100. [German]

Congenital – Brachymelia, shortened tibia, and polydactyly (six to nine toes) in *Rana esculenta.*

Hiller A. 1962. Sondert die Knoblauchkröte bei Gefahr einen Knoblauchgeruch ab? [Does garlic toad *Pelobates fuscus* secret a smell of garlic in case of danger?] Aquarien Terrarien 9:59 [German]

Congenital – *Pelobates fuscus* with supernumerary hind legs.

Hiller A. 1986. Über die Entwicklung einer Larve der Knoblauchkröte (*Pelobates fuscus*) mit Mehrfachbildung an der Schwanzwirbelsäule. [On the development of a larva of the toad (*Pelobates fuscus*) with repeated formation of the vertebral column of the toad]. Salamandra 22:93–94 [German].

Congenital – *Pelobates fuscus* tadpole with duplication of tail end.

Hiller A. 1990. Eine Europäische Sumpfschildkröte *Emys orbicularis* (Linnaeus, 1758) mit angeborener letaler Missbildung. Salamandra 26:215–217.

Congenital – Arhinencephalia, anophthalmia, and brachycauda in *Emys orbicularis.*

Hilts PJ, Sawyer K. 1989. Fiery 2-headed snake upsets behavior theory. Washington Post 6 February 1989:A2.

Trauma – Dicephalic black snake.

Hinchliffe JR, Hurle JM, Summerbell D. 1991. Developmental Patterning of the Vertebrate Limb. NATO-ASI-Advanced Science Institute Series. Series A-Life Sciences 205:1–452; New York: Plenum Press.

Congenital – General review of limb development, ontogeny, and phylogeny.

Metabolic – Review of retinoid effects.

Hingley K, Hemmings M. 1991. A couple of double-headers! Herpetile 16(4):175.

Congenital – Dicephalic African house snake and cites J Cole (Snake Breeders magazine) August 1991) report of a dicephalic California kingsnake.

Hisoka KK, List JC. 1957. The spontaneous occurrence of scoliosis in larvae of *Rana sylvatica.* Transactions of the American Microscopical Society 76:381–387.

Congenital – Caudal scoliosis in *Rana sylvatica.*

Hlasek L. 1987. Chrupavčitý Nádor u Leguána Zeleného. [Double head in *Elaphe dione*]. Akvárium Terárium 30(4):26–27. [Czech].

Congenital – Derodymous *Elaphe dione.*

Hnizdo J. 2000. Neue Erkenntnisse zur Fortpflanzungsbiologie von *Ahaetulla prasina* (Boie, 1827). [New findings on the reproductive biology of *Ahaetulla prasina*]. Sauria 22(2):41–44. [German]

Toxicology – Birth of two deformed Malayan whipsnake *Ahaetulla prasina* (Boie 1827), blamed on treatment of purulent cloacitis with Enrofloxacin.

Hnizdo J, Velebny H. 2003. Parcialni mandibulektomie v chirurgii malych zvirat a plazu. [Partial mandibulectomy in small animal and reptile surgery]. *Veterinarstvi* 53(11):479–488. [Czech]

Neoplasia – Mandibular osteoma in *Iguana iguana.*

Hobson BM. 1958. Polymely in *Xenopus laevis.* Nature 181:862.

Congenital – Additional forearm in *Xenopus laevis.*

Hoby S, Wenker C, Robert N, Jermann T, Hartnack S, Segner H, Aebischer C-P, Liesegang A. 2010. Nutritional metabolic bone disease in juvenile veiled chameleons (*Chamaeleo calyptratus*) and its prevention. Journal of Nutrition 140:1923–1931.

Metabolic – Veiled chameleons *Chamaeleo calyptratus* with bending at proximal metaphyses of radius and ulnar and distal metaphyses of tibia and fibula. Histology revealed endosteal scalloping with thinned cortices and layers of unmineralized osteoid on the periosteal surface, prevented by calcium and vitamin A supplementation. Metastatic calcification in green iguana related to low 25OH vitamin D levels (Richman et al. 1995). Dystrophic calcification in *Uromastyx* sp. related to low 25OH vitamin D levels (Raphael et al. 1999). Vitamin A deficiency associated with inadequate endosteal bone resorption (Thompson 2007).

Hoda S, Alim A, Razi S. 1977. A report on the occurrence of five limbed water skipping frog, *Rana cyanophlyctis* (Schneider). Geobios 4:60–61.

Congenital – Supernumerary limb in water skipping frog *Rana cyanophlyctis.*

Hödl W, Amézquita A. 2001. Visual signaling in anuran amphibians. In Anuran Communication, MJ Ryan (Ed.). Washington DC: Smithsonian Institution Press, pp. 121–141.

Other – Displays – e.g., pedal luring:

Twitching/curling three middle toes – horned toad *Ceratophrys ornata.*

Toe trembling – in male Leptodactylidae *Leptodactylus mealnotus*, Brazilian torrent frog *Hylodes asper*, *Hylodes dactylocinus*, Hylidae *Hyla parviceps* and independent of gender in Dendrobatidae *Colostethus trinitatis, collaris* and *palmatus, Epipedobates parvulus/bilinguis, pictus, pulchripectus, tricolor, anthonyi/tricolor, boulengeri, silverstonei, bassleri, trivittatus, femoralis, Phyllobates terribilis, lugubris, vittatus, Dendrobates auratus, truncatus, leucomelas, tinctorius, azureus, fantasticus, reticulatus, variabilis, imitator, quinquevittatus complex, granuliferus, pumilio, speciosus, lehmanni*, and *histrionicus.*

Hind foot lifting – in male *Crossodactylus gaudichaudii.*

Arm waving – in male *Brachycephalus ephippium*, Bufonidae *Atelopus limosus, zeteki, chiriquiensis*, Leptodactylidae *Crossodactylus gaudichaudii*, and independent of gender in Bufonidae *Atelopus varius* and Hylidae *Hyla parviceps.*

Limb shaking – in male Leptodactylidae *Hylodes asper, dactylocinus*, Dendrobatidae *Colostethus palmatus, Dendrobates granuliferus*, female Dendrobatidae *Epipedobates tricolor, Phyllobates terribilis, Dendrobates histrionicus*, and unknown gender in *Epipedobates pictus, anthonyi/tricolor, boulengeri, bassleri, Phyllobates lugubris, vittatus, Dendrobates auratus, truncatus, leucomelas, tinctorius, reticulatus, variabilis, imitator, quinquevittatus complex, pumilio, speciosus*, and *lehmanni.*

Wiping – in male Dendrobatidae *Dendrobates*, female Dendrobatidae *Epipedobates tricolor*, independent of gender in Dendrobatidae *Phyllobates terribilis* and *Dendrobates histronicus*, and unknown gender in Dendrobates *Epipedobates bassleri, Phyllobates vittatus, Dendrobates leucomelas, tinctorius, azureus, reticulatus, variabilis, imitator, quinquevittatus* complex, *granuliferus, speciosus*, and *lehmanni.*

Leg stretching – in male Leptodactylidae *Crossodactylus gaudichaudii, Hylodes asper, Hylodes dactylocinus*, Myobatrachidae *Taudactylus eungellensis*, Hylidae *Litoria rheocola*, and Dendrobatidae *Epipedobates parvulus/bilinguis.*

Foot flagging – in male Leptodactylidae *Hylodes asper, Hylodes dactylocinus*, Myobatrachidae *Taudactylus eungellensis*, Hylidae *Litoria genimaculata, nannotis, fallax, Hyla parviceps, Phyllomedusa sauvagii*, and in Ranidae rock skipper *Staurois parvus, Staurois latopalmatus.*

Two-legged pushups – in male Myobatrachidae *Taudactylus eungellensis*, female Dendrobatidae *Colostethus collaris, Dendrobates histrionicus*, and unknown gender in Dendrobatidae *Epipedobates bassleri, Phyllobates terribilis, Dendrobates auratus, truncatus, tinctorius, reticulates, variabilis, imitator, quinquevittatus complex, lehmanni.*

Hoff GL, Frye FL, Jacobson ER. 1984. Diseases of Amphibians and Reptiles. 784 pp.; New York: Plenum Press.

Neoplasia – *Elaphe guttata* snake with chondrosarcoma *Ramphiophis rostratus* rufous-beaked snake with spinal osteosarcoma *Naja melanoleuca* with spinal osteochondrosarcoma.

Hoffmann 1873–78 Sechster Kreis. Amphibien. [Sixth Circle. Amphibian]. In Dr. H. G. Bronn's Klassen und Ordnungen der Amphibien wisenschaftlich dargestellt in Wort und Bild. [Orders of amphibians presented scientifically in description and picture]. 724 pp.; Leipzig, Heidelberg: C.F. Winter'sche Verlagshandlung. [German]

Other – P. 56: fusions, but these are "phylogenetic fusions."

Hoffstetter R. 1964. Les sauria du Jurassique supérieur et spécialement les Gekkota de Bavière et de Mandchourie. [Saurians of the Upper Jurassic and especially the Gekkota of Bavaria and Manchuria]. Senckenbergiana Biologica 45:281–316. (also cited with same title as J Amin Morph Physiol 28:117–127. [French]

Trauma – Jurassic gecko *Bavarisaurus macrodactylus* with proximal autotomous caudal vertebrae. Late Jurassic *Palaeolacerta bavarica* with distal tail.

Other – Jurassic gecko *Bavarisaurus macrodactylus* with apparent surface holes of unclear etiology in distal humerus. Jurassic gecko *Bavarisaurus macrodactylus* with ulnar overgrowth of unclear etiology illustrated. Late Jurassic *Palaeolacerta bavarica* with apparent extra coracoid holes of unclear etiology.

Fossil – Jurassic gecko *Bavarisaurus macrodactylus* with proximal autotomous caudal vertebrae. Late Jurassic *Palaeolacerta bavarica* with distal tail. Jurassic gecko *Bavarisaurus macrodactylus* with apparent surface holes of unclear etiology in distal humerus. Jurassic gecko *Bavarisaurus macrodactylus* with ulnar overgrowth of unclear etiology illustrated. Late Jurassic *Palaeolacerta bavarica* with apparent extra coracoid holes of unclear etiology.

Holahan J. 2003. 50 years of spiders, snakes and stars; in 1953 the North Museum opened its doors. For generations of kids of all ages, it's revealed the wonders of nature. Lancaster New Era (Pennsylvania) 9 October 2003:1.

Congenital – Dicephalic turtle.

Holder LA. 1960. The comparative morphology of the axial skeleton in the Australian Gekkonidae. Journal of the Linnean Society Zoology 44:300–335.

Trauma – Arboreal Australian leaf-tailed geckos *Phyllurus cornutus* and rupicolouos *Phyllurus platurus* have only two, three fracture planes at base of tail. The short-tailed *Nephrurus asper* lacks autotomy splits.

Holfert T. 1999. Doppelköpfigkeit (Bicephali) bei *Rhacodactylus auriculatus* (Bavay 1869). [Dicephaly in *Rhacodactylus auriculatus* (Bavay 1869)]. Elaphe, Rheinbach 7(2):13 [German].

Congenital – Dicephalic embryo of *Rhacodactylus auriculatus.*

Holland E. 2006. Two-headed snake gets no bite, will stay here. Online auction found no one willing to pay $150,000. St. Louis Post-Dispatch, St. Louis, MO (9 March 2006):D1.

Congenital – Dicephalic snake.

Holliday TR, Adler K. 1986. The Encyclopedia of Reptiles and Amphibians. New York: Facts on File, 143 pp.

Trauma – Texas banded gecko with tail loss. Australian marbled gecko lives longer with tail intact, but illustrated one with three tails. Autotomy is absent in chameleons, beard lizards, Bornean earless lizard, and xenosaurs.

Holman JA. 1984. Life history notes. Testudines. *Terrapene carolina carolina* (Eastern box turtle). Morphology and behavior. Herpetological Review 15:114.

Congenital – Rigid plaston without hinge in *Terrapene carolina carolina* (Eastern box turtle).

Holman JA. 2003. Fossil Frogs and Toads of North America. Indiana University Press, 246 pp.

Trauma – Late Miocene (Hemphillian) *Bufo campi* LACM 276/5120 tibiofibulare with pathologic ridge, typical in appearance to that seen with "typical anuran healing after fracture." Ridges may be swollen or thin (Holman 1992).

Fossil – Late Miocene (Hemphillian) *Bufo campi* LACM 276/5120 tibiofibulare with pathologic ridge, typical in appearance to that seen with "typical anuran healing after fracture." Ridges may be swollen or thin (Holman 1992).

Holtzer SW. 1956. The inductive activity of the spinal cord in urodele tail regeneration. Journal of Morphology 99:1–39.

Congenital – Spinal cord position determined tail regeneration in *Ambystoma punctatum* and *opacum*.

Holtzer H, Holtzer S, Avery G. 1955. An experimental analysis of the development of the spinal column. IV. Morphogenesis of the tail vertebrae during regeneration. Journal of Morphology 96:145–172.

Trauma – Regeneration of tails in *Ambystoma opacum* occurs by replacement with a segmented cartilaginous rod.

Holz PH, Slocombe R. 2000. Systemic *Fusarium* infection in two snakes, a carpet python, *Morelia spilota variegata*, and a red-bellied snake, *Pseudacris porphyriacus*. Journal of Herpetological Medicine and Surgery 10:18–20.

Infection – Red-bellied snake *Pseudacris porphyriacus* with *Fusarium oxysporum* fungal vertebral osteomyelitis.

Homer BL, Berry KH, Brown MB, Ellis G, Jacobson ER. 1998. Pathology of diseases in wild desert tortoises from California. Journal of Wildlife Disease 34(3):508–523.

Metabolic – Among 24 ill or dead desert tortoises *Gopherus agassizii* were three with urolithiasis, one of which also had articular gout involving shoulder and hip.

Stone – Among 24 ill or dead desert tortoises *Gopherus agassizii* was one with articular gout.

Shell disease – Among 24 ill or dead desert tortoises *Gopherus agassizii* were two with shell necrosis. Osteoclastic resorption and osteopenia characterized the shell necrosis.

Homma GI. 1979. [Three cases of *Hynobius nigrescens* with malformed legs.] Japanese Journal of Herpetology 8(2):41–46. [in Japanese with English abstract]

Congenital – *Hynobius nigrescens* with segmentation disorder and brachydactylia.

Honig LS. 1984. Pattern formation during development of the amniote limb. Symposium of the Zoological Society of London 52:197–221. In: The Structure, Development and Evolution of Reptiles: A festschrift in honour of Professor A. d'A. Bellairs on the occasion of his retirement, WJ Ferguson, ed. London: Academic Press.

Toxicology – Vitamin A-induced limb duplication in *Bufo* and *Rana* (Maden 1982; Niazi and Saxena 1978).

Hooijer DA. 1972. *Varanus* (Reptilia, Sauria) from the Pleistocene of Timor. Zoölogische Mededelingen Rijksmuseum van Natuurlijke Histoire te Leiden 47:445–448.

Neoplasia – Vertebral osteochondroma in Pleistocene *Varanus* cf. *komodoensis* from Raebia, Timor.

Fossil – Vertebral osteochondroma in Pleistocene *Varanus* cf. *komodoensis* from Raebia, Timor.

Hopkins WA, Congdon J, Ray JK. 2000. Incidence and impact of axial malformations in larval bullfrogs (*Rana catesbeiana*) developing in sites polluted by a coal-burning power plant. Environmental and Toxicological Chemistry 19:862–868.

Environmental – 37% of bullfrogs *Rana catesbeiana* larvae had lateral curvatures of the spine from coal combustion waste sites, contrasted with 0% and 4% from non-exposed sites.

Hopley PJ. 2001. Plesiosaur spinal pathology: The first fossil occurrence of Schmorl's nodes. Journal of Vertebrate Paleontology 21:253–260.

Vertebral – Schmorl's nodes in 24 of 27 long-necked plesiosaurs in City of Bristol Museum and Art Gallery (BRSMG) 26539) with two wedged vertebrae. But, now intervertebral disk hypothesis replaced by recognition that the vertebral bodies are connected by synovial lined spaces – so not Schmorl's nodes.

Fossil – Schmorl's nodes in 24 of 27 long-necked plesiosaurs in City of Bristol Museum and Art Gallery (BRSMG) 26539) with two wedged vertebrae. But, now intervertebral disk hypothesis replaced by recognition that the vertebral bodies are connected by synovial lined spaces – so not Schmorl's nodes.

Hoppe D, Mottl E. 1997. Anuran species differences in malformation frequency and severity in Minnesota. In: Anon. Proceedings of the Society of Environmental Toxicology and Chemistry 18th Annual Meeting, 16–20 November 1997, Pensacola, Florida, p. 90.

Congenital – Amelia, ectomelia, supernumerary limbs, and malformed limbs in northern leopard frog *Rana pipiens*, mink frog *Rana septentrionalis*, wood frog *Rana sylvatica*, spring peeper *Pseudacris crucifer*, gray tree frog *Hyla versicolor*, and American toad *Bufo americanus*, especially the most aquatic species *Rana septentrionalis* (50%), with 10% in *Rana pipiens* and 3% in other species.

Horne EA. 2010. Cricket frogs talk… with their feet. Kansas Herpetological Society 37th Annual Meeting Program 6–7 November 2010, Topeka Zoo, Topeka, Kansas.

Congenital – Foot shuffles and long leg or forelimb extensions during breeding season, especially during hottest part of day. Noting stereotypic extension in tropical species.

Hoser R, Gibbons D. 2003. A bicephalic costal Queensland carpet snake, *Morelia spilota macdowelli* (Serpentes: Pythonidae). Herpetofauna 33:111.

Congenital – Dicephalic costal Queensland carpet snake *Morelia spilota macdowelli*.

Hoser RT, Harris P. 2005. A second case of bicephalism in Queensland carpet snakes (*Morelia spilota macdowelli*) (Serpentes: Pythonidae). Herpetofauna 35(1):61.

Congenital – Dicephalic *Acanthophis wellsi x Acanthophis pyrrhus/Acanthophis wellsi* hybrid, *Pseudonaja affinis* (Maryan 2001) and Queensland carpet snake *Morelia spilota macdowelli* (Hoser and Gibbons 2003). Derodymous Queensland carpet snake *Morelia spilota macdowelli.*

Houck WJ, Henderson C. 1953. A multiple appendage anomaly in the tadpole of *Rana catesbeiana.* Herpetologica 9:76–77.

Congenital – Supernumerary hind limb in *Rana catebeiana.*

Hovelacque A. 1920. Anatomie et morphogenie d'une anomalie hereditaire des membres abdominaus (Absenced congenitale du tibia). [Anatomy and morphology of a hereditary abdominal limb anomaly (congenital absence of tibia)]. Bulletin Biologique de la France et de la Belgique 1 Supplement 3:1–156. [French]

Congenital – Absence of tibia in *Triton taeniatus.*

Howes E. 1631. A two-headed snake. In: I Stow: Annales, or, a general chronicle of England. Continued and Augmented with Matters Foraigne and Domestique, Ancient and Moderne, vnte the End of this Present Yeere, 1631. London: Richardi Meighen.

Congenital – Dicephalic snake.

Howes GB. 1886. On some abnormalities of the frog's vertebral column, *Rana temporaria.* Anatomischer Anzeiger 1:277–281.

Congenital – *Rana temporaria* supernumerary tenth vertebra. Ninth was concave in front, instead of being convex and had secondary facets. Similar report by Bourne (1886) in a frog.

Howes GB. 1890. Vertebral skeleton of a fire toad. Journal of Anatomy and Physiology 24:xvi-xix.

Congenital – Fire toad *Bombinator* sp. with sacralization of last pre-sacral vertebra, as also noted by Walterstorff (1885–1886) in extinct *Palaeobatrachus*, which had three sacral vertebrae. It also had an enlarged ninth (normal) right anterior zygapophysis. Walterstorff (1886) noted extreme sacral variation in *Bombinator.* Abnormal sacra in anura (Bourne 1884; Götte 1875; Howes 1886; Morgan 1886; Sasserno), and in tailed amphibian *Menopoma* (Huxley 1875; Lucas 1886).

Howes GB. 1893. Notes on variation and development of the vertebral and limb-skeleton of the amphibian. Proceedings of the Zoological Society London 1893:14.

Congenital – majority of variations affected head of urostyle and sacrum. 1/212 *Rana esculenta* with united 8th and 9th vertebrae and Rod-like body in a possible *Pelobates* at urostyle. Tridactyle in a spotted salamander *Salamandra maculosa.*

Howes GB, Davies AM. 1888. Observations upon the morphology and genesis of supernumerary phalanges, with especial reference to those of the amphibian. Proceedings of the Zooological Society 1888:495–511.

Congenital – Supernumerary phalanges as normal state in amphibians.

Howes GB and Swinnerton HH. 1901. On the development of the tuatara, *Sphenodon punctatus*: With remarks on the egg, on the hatching, and on the hatched young. Transactions of the Zoological Society of London 16:1–86.

Trauma – Autotomy in *Sphenodon punctatus* is at caudal vertebrae 6–8.

Hruban Z, Meehan T, Frye FL, Harshbarger JC. 1989. Neoplasia in reptiles and amphibians in the Lincoln Park Zoological Garden. Third International Colloquium on the Pathology of Reptiles and Amphibians, Abstracts:76–77.

Metabolic – Cook's tree boa *Boa cookie*, with hemochromatosis.

Hua S. 1995. À propos d'un crane de *Dyrosaurus phosphaticus* (THOMAS, 1893), (Crocodylia, Mesosuchia): Contribution à l'interpretation paléoécologique et biogéographique des Dyrosauridae. [Skull of *Dyrosaurus phosphaticus* (THOMAS, 1893), (Crocodylia, Mesosuchia): Contribution to the paleoecological interpretation and biogeography of Dryosauridae]. Bulletin de la Société belge de Géologie 104:109–118. [French]

Congenital – *Dyrosaurus phosphaticus* skull with an osseous anomaly near the external naris (irregular and left nasal aperture).

Fossil – *Dyrosaurus phosphaticus* skull with an osseous anomaly near the external naris (irregular and left nasal aperture).

Hua S. 1999. Le Crocodilien *Machimosaurus mosae* (Thalattosuchia, Teleosauridae) du Kimmeridgien du Boulonnais (Pas de Calais, France). [The crocodilian *Machimosaurus mosae* (Thalattosuchia, Teleosauridae) from the Kimmeridgian of the Boulonnais (Pas de Calais, France).] Pp.141–170; Stuttgart: E. Schweizerbart'sche Verlagsbuchhandlung (Nägele u. Obermiller). [French]

Trauma – Crocodilian *Machimosaurus mosae* (Thalattosuchia, Teleosauridae) with exostoses on femur and right pubis possibly related to infection.

Fossil – Crocodilian *Machimosaurus mosae* (Thalattosuchia, Teleosauridae) with exostoses on femur and right pubis possibly related to infection.

Hubbard ME. 1903. Correlated protective devices in some California salamanders. University of California Publications in Zoology 1:157–170.

Trauma – *Batrachoseps* drops tail readily and at any position on tail and regenerates. *Plethodon* drops tail only under major stress, but only at constriction behind anus, and regenerates.

Hübner H. 1911a. Die Doppelbildungen des Menschen und der Tiere [Double formations of man and animals]. In: Lubarsch O, Ostertag R. (eds.). Ergebnisse der allgemeinen Pathologie und Pathologischen Anatomie des Menschen und der Tiere. [Results of general pathology and pathological anatomy of man and animals]. I. Abteilung: 650–796. Wiesbaden (J. F. Bergmann). [German]

Congenital – Literature review and generalities. Figured artificially produced hind legs in *Pelobates fuscus* (Tornier 1905), snake with doubled anterior body (Johnson 1902), doubled lizard *Mabuya multifasciata* (Tur 1904), and grass or water snake (Braun 1893; *Klauss*ner 1890). Stated that double formation in amphibians is rare, noting *Salamandra maculata* (Braun 1893; Klaussner 1890; Born 1881; Grochmalicki 1909) and frog (Ragnotti, Loyez 1897).

Hübner H. 1911b. Die Doppelbildungen des Menschen und der Tiere. [Double formations of man and animals]. In: Lubarsch O, Ostertag R. (eds.). Ergebnisse der allgemeinen Pathologie und Pathologischen Anatomie des Menschen und der Tiere. [Results of general pathology and pathological anatomy of man and animals]. II. Abteilung:1–348. Wiesbaden (J. F. Bergmann). [German]

Congenital – Duplications predominantly noted in mammals and man, sometimes birds, rarely in reptiles and amphibians, noting dicephalic snakes (Johnson 1902), turtles (Wanser 1896; Bishop 1908; Barbour 1896), and frog, *Rana esculenta* (Loyez 1897).

Huchzermeyer FW. 1986. Osteomalacia in young captive crocodiles (*Crocodylus niloticus*). Journal of the South African Veterinary Association 57(3):167–168.

Metabolic – Osteomalacia in *Crocodylus niloticus*.

Huchzermeyer FW. 2003. Crocodiles: Biology, Husbandry and Diseases. 337pp.; Oxon, UK: CABI Publishing.

Congenital – Brachycephaly and dental abnormalities in New Guinea crocodiles (Hall 1985). Brachycephaly in *Alligator mississippiensis, Crocodylus niloticus*, and *C. porosus* (Kälin 1936). Asymmetrical skull (Youngprapakorn et al. 1994). Incisors of short lower jaw protruding through maxilla, producing false nostrils in *C. johnsonii*. Upward curvature of snout in gharial (Singh and Bustard 1982). Cleft lip, cleft palate, and cleft chin (Youngprapakorn et al. 1994). Skewed jaws in *C. porosus*. Nile crocodile with polydactyly; additional toes in *Alligator mississippiensis*(Giles 1948). Additional limb (with single claw) attached to Nile Crocodile navel (Youngprapakorn et al. 1994). Ectromelia in *Crocodylus moreletii* (Rainwater et al. 1999) and caimans (Troiano and Román 1996). Tail shortening, or absence (e.g., Nile crocodile and *Osteolaemus tetraspis*) (Kar and Bustard 1982; Ferguson 1989; Singh and Bustard 1982; Youngprapakorn et al. 1994; Hibberd 1996; Troiano and Román 1996; Webb and Manolis 1998).

Metabolic – Nutritional bone disease producing rubber jaws and glassy teeth with osteomalacia (Huchzermeyer 1986; Foggin 1992). Combined fibrous osteodystrophy, osteochondrosis, and osteoporosis in *Crocodylus intermedius* and *C. acutus* (Blanco 1997). Gout is more common in winter months (Pooley 1986) related to inability to metabolize digested food and decreased urate solubility at low temperatures (McNease and Joanen 1981) or vitamin A induced (Foggin 1987; Ariel et al. 1997). *C. porosus* was spared, while *C. novaeguineae* and *C. johnstoni* have gout (Buenviaje et al. 1994; Ladds et al. 1995), as do *Alligator mississippiensis* (Cardeilhac 1981), spectacles caiman (Jacobson 1984), Nile crocodiles (Pooley 1986; Foggin 1987), Indo-Pacific crocodile (Bueviaje et al. 1994), false gharial (Frank 1965), and gharial (Frank 1965). Congenital articular gout (Foggin 1992)?

Neoplasia – Bone tumors on *Caiman crocodilus* museum skeletons (Kälin 1937).

Environmental – High temperatures (35°C) produce spine abnormalities, tail curvature, premature ossification with skull deformities and protruding mandibles in *Crocodylus porosus* and *C. johnstoni* (Ferguson 1989; Webb and Manolis 1998).

Other – Erosion on diaphysis (not articular head) of juvenile Nile Crocodile attributed to osteoporosis. Etiology is unclear.

Huchzermeyer FW, Cooper JE. 2000. Fibriscess, not abscess, resulting from a localized inflammatory response to infection in birds and reptiles. TheVeterinary Record 147:515–517.

Infection – Systemic spread of bacterial infections in birds and reptiles is allegedly constrained by fibrin deposition, rather than development of pus. Therefore, they favored the term "fibriscess" rather than "abscess," although they acknowledge that abscess is the term used in the veterinary literature. They demur use of the term "abscess," claiming that encapsulation does not occur. They also referred to bumblefoot as related to fibriscesses.

Huene F. v. 1911. Beiträge zur Kenntnis und Beurteilung der Parasuchier. [Contributions to our knowledge and understanding of parasuchians]. Geologische und Palaeontologische Abhandlungen 10:65–121. [German]

Trauma – *Mystriosuchus plieningeri* with healed damage on snout.

Fossil – *Mystriosuchus plieningeri* with healed damage on snout.

Huene F. v. 1915. On reptiles of the New Mexican Trias in the Cope Collection. Bulletin American Museum of Natural History 34(15):485–507.

Congenital – *Typothorax* with "aberrant type of ribs and peculiar skeletal bones with extraordinarily thickened articular ends."

Fossil – *Typothorax* with "aberrant type of ribs and peculiar skeletal bones with extraordinarily thickened articular ends."

Huene F. v. 1930. Über zwei Fälle von Nearthrose bei fossilen Ichthyosauriern. [Two cases of neoarthrosis in fossil Ichthyosauridae]. Anatomischer Anzeiger 70:108–109. [German]

Trauma – Two cases of healed ribs (pseudarthrosis and neoathrosis) in ichthyosaur *Leptopterygius trigonodon.*

Fossil – Two cases of healed ribs (pseudarthrosis and neoathrosis) in ichthyosaur *Leptopterygius trigonodon.*

Huey RB, Dunham AE, Overall KL, Newman RA. 1990. Variation in locomotor performance in demographically known populations of the lizard *Sceloporus merriami*. Physiological Zoology 63:845–872.

Trauma – Speed unaffected by tail loss in *Sceloporus merriami.*

Humphrey RR. 1967. Genetic and experimental studies on a lethal trait ("short toes") in the Mexican axolotl (*Ambystoma mexicanum*). Journal of Experimental Zoologie 164:281–296.

Congenital – Short toes in Mexican axolotl *Ambystoma mexicanum* – lacking or shortened phalanges. Crossbreeding produced intermediate pattern.

Humphrey RR. 1973. Experimental studies on a new lethal trait in Mexican axolotls of the Wistar Institute white strain. Journal of Experimental Zoology 183:201–208.

Congenital – Gene s is for short toes in Mexican axolotls of the Wistar Institute white strain. Gene y produces short forelimb stump, usually with notched or forked ends, occasional with two elongated digits with one or more phalanges in Mexican axolotls of the Wistar Institute white strain.

Humphrey RR 1975. The axolotl *Ambystoma mexicanum*. Handbook of Genetics. In Vertebrates of Genetic Interest, RC King, ed. New York: Plenum, Volume 4, pp. 3–17.

Congenital – Short toe "s" gene, phocomelia ph gene, and hand "h" gene (with first digit at right angle, resembling a thumb) in Mexican axolotl (water dog, water twin, water sprite, water slave) *Ambystoma mexicanum.*

Hungerbühler A. 2000. Heterodonty in the European phytosaur *Nicrosaurus kapffi* and its implications for the taxonomic utility and functional morphology of phytosaur dentitions. Journal of Vertebrate Paleontology 20:31–48.

Dental – Alveolus closure in right maxilla SMNS 4379 from injury or failed replacement. Anomalous tooth orientaton in SMNS 13078. Double anterior carina in SMNS 13078.

Trauma – Phytosaur jaw and teeth injuries reported by Camp 1930, Case 1930, Mehlo 1915.

Fossil – alveolus closure in right maxilla SMNS 4379 from injury or failed replacement. Anomalous tooth orientaton in SMNS 13078. Double anterior carina in SMNS 13078. Phytosaur jaw and teeth injuries reported by Camp 1930, Case 1930, Mehlo 1915.

Hunt TJ. 1957. Notes on diseases and mortality in testudines. Herpetologica 13:19–23.

Trauma – Occurs in turtles.

Infection – General comment that mycosis invades plaston more frequently than carapace.

Metabolic – Rickets is found in captive turtles.

Hutchinson J, Simmonds M. 1991. A review of the effects of pollution on marine turtles. Greenpeace Ecotoxicology Project 1:1–28.

Toxicology – Organochlorines produce missing claws, deformed carapaces, tails, limbs, crania, deformed nostrils and upper jaws and deformed lower jaws, and dwarfism in snapping turtles, citing Bishop et al. (1989).

Hutchison JH, Frye FL. 1989. Pathologies of the shell in Eocene turtles. Journal of Vertebrate Paleontology 9 (Supplement to 3):26A.

Shell disease – Circular to irregular shaped pits with well-defined margins in Wyoming Eocene *Echmatemys* and *Baptemys*, probable scratch marks in *Baptemys* and *Chisternon* and subscale rot in *Baptemys* and modern *Dermatemys*.

Fossil – Circular to irregular shaped pits with well-defined margins in Wyoming Eocene *Echmatemys* and *Baptemys*, probable scratch marks in *Baptemys* and *Chisternon* and subscale rot in *Baptemys* and modern *Dermatemys*.

Hutchison JH, Frye FL. 2001. Evidence of pathology in early Cenozoic turtles. Paleobios 21(3):12–19.

Trauma – Linear excavations in *Baptemys garmanii* attributed to scratches. *Echmatemys* UCMO 128283 with 9 x 9 mm shallow pits and full-thickness partially remodeled pits, attributed to crocodilian bite. *Hadrianus corsoni* with conical perforations from predation.

Infection – Emydidae UCMP 128418 with near-half-thickness crossing sutures associated with sclerotic bone interpreted as osteomyelitis.

Metabolic – Possibility that high frequency of pitting in *Chrysemys picta* females might be related to calcium/phosphate withdrawal for egg laying.

Shell disease – Parasite or predator injury? Linear excavations in *Baptemys garmanii* attributed to scratches. Emydidae UCMP 128418 with near-half-thickness crossing sutures associated with sclerotic bone interpreted as osteomyelitis. *Echmatemys euthenta* with 40+ 1–3 mm punctate pitting lesions. *Echmatemys* UCMO 128283 with 9 x 9 mm shallow pits and full-thickness partially remodeled pits, attributed to crocodilian bite. *Hadrianus corsoni* with conical perforations from predation. Divided pits into circular to ovoid with flat bottoms, versus rounded bottoms, irregular pits with discrete margins, track (linear) and rot, represented by large areas of irregular depression or dead lamellar bone. Possibility that high frequency of pitting in *Chrysemys picta* females might be related to calcium/phosphate withdrawal for egg laying.

Fossil – Linear excavations in *Baptemys garmanii* attributed to scratches. *Echmatemys* UCMO 128283 with 9 × 9 mm shallow pits and full-thickness partially remodeled pits, attributed to crocodilian bite. *Hadrianus corsoni* with conical perforations from predation. Emydidae UCMP 128418 with near-half-thickness crossing sutures associated with sclerotic bone interpreted as osteomyelitis. Possibility that high frequency of pitting in *Chrysemys picta* females might be related to calcium/phosphate withdrawal for egg laying. Parasite or predator injury? Linear excavations in *Baptemys garmanii* attributed to scratches. Emydidae UCMP 128418 with near-half-thickness crossing sutures associated with sclerotic bone interpreted as osteomyelitis. *Echmatemys euthenta* with 40+ 1–3 mm punctate pitting lesions. *Echmatemys* UCMO 128283 with 9 × 9 mm shallow pits and full-thickness partially remodeled pits, attributed to crocodilian bite. *Hadrianus corsoni* with conical perforations from predation. Divided pits into circular to ovoid with flat bottoms, versus rounded bottoms, irregular pits with discrete margins, track (linear) and rot, represented by large areas of irregular depression or dead lamellar bone. Possibility that high frequency of pitting in *Chrysemys picta* females might be related to calcium/phosphate withdrawal for egg laying.

Huxley TH. 1864. Indian pre-Tertiary Vertebrata. On a collection of vertebrate fossils from the Panchet rocks near Ranigunj, Bengal. Memoirs of the Geological Survey of India. Palaeontologica Indica, series 4, 1(1):1–24.

Congenital – Unpaired, median bone, "intercalary" or "Wormian bone," between parietals and frontals in dicynodont therapsids (mammal-like reptiles).

Fossil – Unpaired, median bone, "intercalary" or "Wormian bone," between parietals and frontals in dicynodont therapsids (mammal-like reptiles).

Huxley TH. 1875. Amphibia. In Enclyclopaedia Britanica, 9th edition, p. 750.

Congenital – Extra sacral rib in amphibian.

Hvass H. 1943. Ein Fall von Polymelie beim Wasserfrosch, *Rana esculenta* L. [A case of polymely of water frog, *Rana esculenta* L.]. Videnskabelige Meddelser Dansk Naturhistorisk Foreningen I København 106:233–240. [German]

Congenital – Supernumerary limb in *Rana esculenta.*

Hyde BTB. 1925. Two-headed snakes. A not uncommon phenomenon of the reptile world. Journal of the American Museum of Natural History 25:184–187.

Congenital – Dicephalic hog-nose snake *Heterodon contortrix*, black snake *Coluber constrictor*, king snake *Lampropeltis getula*, milk snake *Lampropeltis triangulum*, garter snake *Thamnophis radix*, copperhead *Agkisrodon mokasen*, and rattlesnake *Crotalus* sp.

Hyrtl J. 1853. Über normale Quertheilung der Saurierwirbel. [On normal transverse division of dinosaur vertebrae]. Sitzungsberichte der Akademie der Wissenschaften in Wien 4:185–192 [German].

Congenital – Vertebral septum in *Pygopus lepidopodus*, but not in *Chamaesaura anguina*. Many lizards (*Podinema teguixin, Crocodilurus amazonicus*) have fusion lines extending into the transverse process. He discusses *Cyclodus scincoides, Gongylus ocellatus, Sphenops capistratus, Eutropis multifasciata, Trachydosaurus rugosus, Seps chalcides, Ophiodes striatus, Anguis fragilis, Acontias meleagris, Pygodactylus gronovii, Gerrhonotus taeniatus, Chirocolus imbricatus, Ophiosaurus ventralis, Chamaesaurus anguinea, Bipes pallasii, Platydactylus, Ptychozoon homalocephalum, Scelotretus, Rhacoëssa, Phyllurus, Ascalabotes, Prototretus pectinatus, Ophryoëssa superciliosa, Enyalius catenatus, Hypsilophus tuberculatus, Cyclura pectinata, Tropidoliepis undulatus, Urostrophys vautieri, Ctenonotus cuvieri, Chrysolamprus ocellatus, Lacerta chloronotus, Tachydromus sexlineatus, Ctenodon nigropunctatus*, and *Cnemidophorus lemniscatus.*

Hyrtl J. 1864. Über Wirbelassimilation bei Amphibien. [On vertebrate asymmetry in amphibians]. Sitzungsberichte der Kaiserlichen Akademie der Wissenschaften. Mathematisch-Naturwissenschaftliche Classe. Abt. 1, Mineralogie, Botanik, Zoologie, Anatomie, Geologie und Paläontologie 49:264–273. [German]

Congenital – Vertebral asymmetry in batrachian *Menobranchus lateralis, Chelonia Testudo graeca, Chersina angulata, Clemmys cumberlandensis, Vipera aspis, Trigonocephalus, Trimeresurus, Lachesis, Crotalus*, Saurian *Ctenodon nigropunctatus.*

Oblique pelvic girdle in *Tupinambis teguixin, Oplurus cuvier, Oplurus torquatus, Hydrosaurus, Lophura amboinensis, Lacerta, Grammatophora barbata, Chrysolamprus ocellatus, Bipes pallasii, Chersina angulata* (4 vertebrae connected with sacra form S shape), *Trachyemes scripta, Clemmys cumberlandensis, Pyxis arachnoids* (right transverse process does not reach ilium), *Ophidia* (forked transverse process of last trunk vertebra) and *Coronella austriaca* (4 hemivertebrae between last trunk and first tail vertebrae. Unilateral sacralization of *Grammatophora barbata* second caudal vertebra.

Transverse processes are offset side to side in their attachment to *Ctenodon nigropunctatus* ilum.

Asymmetrical transverse processes in *Monopoma alleghanense* pelvic area with asymmetrical distal fusion of those processes.

In den Bosch HA, Musters CJ. 1987. Scalation and skull morphology of a cyclopian *Natrix maura*. Journal of Herpetology 21:107–114.

Congenital – First case of cyclops and reduced lower jaw in snake *Natrix maura* in the RMNH #21647, in Leiden.

Innis CJ, Garner MM. 2006. Spinal osteoarthropathy in green iguanas (*Iguana iguana*). Proceedings of the Annual Conference of the Association of Reptilian and Amphibian Veterinarians 13:40–42.

Vertebral – Spinal osteoarthropathy in *Iguana iguana* manifest by chondro-osseous metaplasia, reactive new bone formation with gangrene and "active" osteomyelitis (*Bacillus, Streptococcus* and *Pseudomonas aeruginosa* isolated), producing a proliferative, ankylosing osteoarthropathy with pathologic coccygeal fractures. The authors could not distinguish nutritional, traumatic, metabolic or even vascular explanation.

Innis CJ, Kincaid AL. 1999. Bilateral calcium phosphate ureteroliths and spirorchid trematode infection in a red-eared slider turtle, *Trachemys scripta elegans*, with a review of the pathology of spirochiasis. Proceedings of the Association of Reptilian and Amphibian Veterinarians 9(3):32–35.

Metabolic – Red-eared slider *Trachemys scripta elegans* with calcium phosphate kidney stones.

Stones – Red-eared slider *Trachemys scripta elegans* with calcium phosphate kidney stones.

Inukai T. 1929. Japanische Vipern mit zwei Koepfen. [Japanese viper with two heads]. Transactions of the Sapporo Natural History Society 11:21–24. [German]

Congenital – Two dicephalic Japanese vipers *Agkistrodon blomhoffii.*

Iordansky NN. 1973. The skull of the Crocodilia. In Gans C, Parsons TS. (eds.) Biology of the Reptilia. Vol. 4:201–262.

Congenital – *Crocodylus acutus* with mandibular rami asymmetry and first dentary tooth fitting into pit of opposite premaxillary, also noted by Kälin 1932. *Crocodylus porosus* and *niloticus* and *Alligator mississippiensis* with shortened snout with enlarged base, noting lateral mandibular margin projects outward. First dentary fits opposite premaxillary "fossa" in *C. porosus* (Klein 1863). Brachycephaly in a crocodilian.

Ippen R von. 1962. Die spontane Tuberkulose bei Kaltblütern [Spontaneous tuberculosis in cold-blooded animals]. Nordisk Veterin Medicin 14 (Suppl 1):185–192 [German].

Infection – *Mycobacterium fortuitum, M. Marinum, M. Thamnopheos, M. Platypoecilus, M. Chelonei* in Frogs (Rupprecht 1904, Küster 1906, Lichtenstein 1921), and specifically in *Pseudis paradoxa* (Griffith 1929), *Rana tigrina* and *Ceratophrys americana* (Zwart 1959), *Tropidontus natrix* (Sibley 1889), boa (Gibbes and Shurly 1890), python (Gibbes and Shurly 1890, Zwart 1960), *Python molurus* (Shattock 1902), *Python reticulatus* (van Hansemann 1903), African python *Python sebae* (Griffith 1941, Scott 1930), *Python spilotes* (Tempel 1959), *Tarbophobis fallax* (Griffith 1928, Scott 1930), *Thamnophis sirtalis* (Aronson 1929), red rattlesnake *Crotalus exsul* (Scott 1930), Aeskulapsal *Coluber longissimus* (Scott 1930), puff adder *Bitis arietans* (Scott 1930), *Natrix piscator* (Griffith 1941), *Lampropeltis getula holbrooki* (Griffith 1941), *Agkistrodon piscivorus* Iippen 1960), *Chelone corticata* (Friedmann 1903), *Trionyx gangeticus* (Scott 1930), *Trionyx triungus* (Besse 1949), *Lacerta viridis* (Bertarelli and Bocchia 1910), *Ctenosaura multispinis* (Scott 1930), and Caiman (Zwart 1959), including *Caiman sclerops* (Griffith 1928, Scott 1930), but no comment on osseous involvement. .

Ippen R. 1965a. Vergleichend pathologische Betrachtungen über einige Knochenerkrankungen bei Reptilien. [Comparative pathological considerations of some bone sickness in reptiles]. Zentralblatt für Allgemeine Pathologie und pathologische Anatomie 108:424–434. [German]

Trauma – Chameleon *Chamaeleo zeylanicus* with callus formation on fractured ribs.

Infection – *Boa constrictor* (without subspecies, common name cannot be determined) with destruction of vertebrae starting with necrotic granuloma formation caused by infection of *Pseudomonas aeruginosa.*

Metabolic – *Varanus niloticus* with hyperostotic osteodystrophia fibrosa.

Neoplasia – *Cyclura macleayi* with osteochondroma on left humerus, right femur, and other bones – ? multiple hereditary osteochondroma.

Ippen R. 1965b. Über Sektionsbefunde bei Reptilien. [On dissection results in reptiles]. Zentralblatt für allgemeine Pathologie 107:520–529. [German]

Metabolic – Osteodystrophy in snakes and lizards.

Vertebral – Necrosis of vertebrae.

Shell – Necrotic changes of carapace.

Ippen R. 1966. Vergleichend pathologische Betrachtungen über einige Knochenerkrankungen bei Reptilien. [Comparative pathological considerations of some bone sickness in reptiles]. Zentralblatt der allgemeinen Pathologie 108:424–434. [German]

Trauma – Rib fracture with callus formation in chameleon *Chamaeleo zeylanicus.*

Infection – *Pseudomonas aeruginosa* induced granuloma producing resorption and apposition in mountain viper *Vipera xanthina* and boomslang *Dispholidus typus.*

Metabolic – Rickets = Schlumberger (1958) and Reichenbach-Klinke (1963) Osteodystrophia fibrosa (hyperostotic kind) in *Varanus niloticus.*

Neoplasia – Osteochondroma in front and hind legs of Cuban lizard *Cyclura macleayi* and green lizard *Iguana iguana,* and in vertebrae and ribs of *Uromastyx hardwickii* from India.

Vertebral – Separation of vertebrae in *Boa constrictor* (without subspecies, common name cannot be determined).

Ippen R. 1975. Ein Beitrag zu einigen stoffwechselbedingten Veränderungen bei Zoo- und Wildtieren. [A contribution to some changes in metabolic condition of zoo and wild animals]. Verhandlungsbericht des Internationalen Symposiums über die Erkrankungen der Zootiere 17:187–196. [German]

Metabolic – Calcifications (sclerosis) of different organs in 43 of 596 reptiles. Gout and formation of concretions in reptiles.

Ippen, R. 1982. Ein Beitrag zu den angeborenen Mißbildungen der Reptilien. [A contribution to congenital malformation of reptiles]. Verhandlungsbericht des Internationalen Symposiums über die Erkrankungen der Zootiere 24:463–473. [German]

Congenital – Literature survey of malformations in reptiles (one each, unless otherwise indicated in parentheses:

Duplications (225)	snakes	Cunningham 1937
Dicephalic (17)	snakes	Brongersma 1952
Dicephalic (5)	*Lampropeltis getula*	Shaw 1971
Dicephalic (2)	*Lampropeltis getula*	Griner 1981
	Lampropeltis getula	Schmidt and Inger 1957
	Alligator mississippiensis	Ferguson 1981
	Crotalus horridus	Lasher 1980
	Elaphe obsolete	Cosgrove 1981
	Enhydris sp.	Ippen 1965
	Eryx conicus	Wall 1905
	Heterodon simus	Wall 1905
	Homalopsis buccata	Wall 1905
	Leptotyphlops sp.	Branch 1979
	Lycodon aulicus	Wall 1905
	Mabuya striata	Broadley, Findley 1972
	Natrix natrix	Davies 1974
	Natrix piscator	Wall 1905
	Pituophis melanoleucus	Cosgrove 1981
	Ptyas sp.	Cosgrove 1981
	Vipera berus	Dorner 1873
	Vipera russelli	Deranyiagala 1958
Derodymous	*Pseudemys nelsoni*	Bellairs 1981
Partial Cyclops	*Caretta caretta*	Ewert 1979
Two bodies fused at neck with three heads	*Lacerta agilis*	Bellairs 1965
Two bodies with one head	*Anguis fragilis*	Reichenbach-Klinke and Elkan 1965
Double trunk + tail	*Tiliqua scinoides*	Willes 1932
Double "formations"	*Chelonia mydas*	Glaesner 1924
Cephalothoracopagus	*Anguis fragilis*	Stemmler 1977
Two animals with Fused plastron	*Trionyx ferox*	Hildebrandt 1938
Three additional toes	*Alligator mississippiensis*	Giles 1948
Supernumerary leg	*Alligator* sp.	Ferguson 1981
Cleft palate (6)	*Crotalus viridis*	Dean et al. 1980
	Eunectes murinus	Bellairs 1965
	Lacerta lepida	Bellairs, Boyd 1957
	Lacerta vivipara	Bellairs, Gamble 1960
	Pseudomys scripta	Ippen 1982
	Thamnophis sp.	Fox et al. 1961
Cleft palate and short upper jaw	*Chelonia mydas*	Ewert 1979, Bellairs 1965
Shortened lower jaw	*Iguana iguana*	Griner 1981
Shortened snout	crocodile	Kälin 1937
Shortened body	*Natrix sipedon*	Rubin et al. 1967
Shortened tail	crocodiles	Bustard 1969
	Geckos	Bustard 1969
Arching of body	*Natrix maura*	Riches 1967
Kyphosis	*Chrysemys picta marginata*	Anderson 1981
Kyphoscoliosis	*Agkistrodon piscivorus*	Petzold 1963
Carapace kyphosis	Trionychidae	Smith 1947
Vertebral agenesis	*Physignathus concincinus*	Montali 1981
Deformations	*Crotalus durissus*	Montali 1981
Missing leg	Emydidae	Dutta 1931
	Caretta caretta	Ewert 1979
	Trionyx sp.	Dutta 1931

Ippen, R. 1985a. Geschwülste. [Tumors]. In Ippen R., Zwart P., Schröder, HD. eds. Handbuch der Zootierkrankheiten. [Handbook of Sickness in zoo animals]. Band 1. Reptilien. Pp.270–301; Berlin: Akademie Verlag. [German]

Neoplasia – Osteoma in varanid Varanus salvator, lizard Lacerta viridis and Uromastyx hardwicki.

Osteosarcoma in extremities of lizard *Iguana iguana* and vertebrae of snake *Rhamphiophis rostratus*.

Osteochondroma on extremities of Bengalian varanid *Varanus bengalensis*, Cuban lizard *Cyclura macleayi* femur, green lizard *Iguana iguana*, Indian spiny tail lizard *Uromastyx hardwicki* and Nile varanid, *Varanus niloticus* and on vertebrae of snake *Naja nigricollis*.

Osteochondrofibroma of green lizard *Lacerta viridis*.

Enchondroma in Pacific varanid *Varanus indicus*.

Osteochondrosarcoma on vertebrae of cobra *Naja nigricollis*; Chondrosarcoma in snake, *Elaphe guttata*.

Fibroma in jaw of tiger python *Python sebae*.

Ippen, R. 1985b. Entwicklungsbedingte Anomalien. [Developmental anomalies]. In Ippen R., Zwart P., Schröder, HD. eds. Handbuch der Zootierkrankheiten. Band 1. Reptilien. [Handbook of Animal Illness. Volume 1. Reptiles]. Pp. 302–316; Berlin: Akademie Verlag. [German]

Congenital – Survey of literature on anomalies in reptiles:

Malformation	species
	Turtles
Dicephalic	*Alligator mississippiensis*
	Crotalus horridus
	Elaphe obsoleta
	Enhydris sp.
	Eryx conicus
	Heterodon simus
	Homalopsis buccata
	Lampropeltis getula
	Leptotyphlops sp.
	Lycodon aulicus
	Mabuya striata
	Natrix natrix
	Natrix piscator
	Pituophis melanoleucus
	Pseudemys nelsoni
	Ptyas sp.
	Vipera berus
	Vipera russelli
Unspecified duplication	*Chelonia mydas*
Siamese twins	*Trionyx ferox*
Two bodies + three heads	*Lacerta agilis*
Two bodies	*Anguis fragilis*
	Tiliqua scincoides
Head and thorax fused	*Anguis fragilis*
Hydrocephalus	*Thamnophis sirtalis*
Brachygnathia	*Thamnophis sirtalis*
Shortened lower jaw	*Iguana iguana*
Cleft palate	*Chelonia mydas*
	Crotalus viridis
	Eunectes murinus
	Lacerta lepida
	Lacerta vivipara
	Pseudemys scripta elegans
Kyphosis	Trionychidae
	Chrysemys picta marginata
Scoliosis	*Agama anchietae*
	Trionyx triunguis
Kyphoscoliosis	*Agkistrodon piscivorus*
Arched vertebral column	*Thamnophis sirtalis*
	Natrix maura
Shortened body	*Natrix sipedon*
Shield anomalies	*Gopherus polyphemus*
	Pseudemys scripta
	Testudo carbonaria
Malformed scales	*Natrix faciata*
	Natrix natrix
	Pituophis melanoleucus
	Python molurus – Burmese, Indian, or Ceylon python
Agenesis	*Physignathus concincinus*
Missing extremity	snakes
	Emydidae
	Caretta caretta
	Trionyx sp.
Polydactyly	*Alligator mississippiensis* (Mississippi alligator)
Polymelia *Alligator* sp.	
Missing tail	*Calotes versicolor*
	Crocodiles

(continued)

Ippen R, Heinrich W-D. 1977. Pathologische Veränderungen an fossilen Extremitäten-Knochen von Anuren aus dem fossilen Tierbautensystem von Pisede bei Malchin.[Pathological changes in bones of extremities of fossil anurans from fossil animal burrow system in Pilsede near Malchin] Wissenschaftliche Zeitschrift der Humboldt-Universität Berlin 26:301–305. [German]

Trauma – Different fractures of femur and tibia in fossil *Rana* sp. and *Bufo* cf. *bufo* (Holocene of Pilsede, Mecklenburg, Germany) with callus formation. Infection – Infected fractures with chronic panostitis.

Fossil – Different fractures of femur and tibia in fossil *Rana* sp. and *Bufo* cf. *bufo* (Holocene of Pilsede, Mecklenburg, Germany) with callus formation. Infected fractures with chronic panostitis.

Ippen R, Schröder H-D. 1977. Zu den Erkrankungen der Reptilien. [On diseases of reptiles]. Verhandlungsberichte der Erkrankungen der Zootiere 19:15–30. [German]

Other – 13 osteopathologies in reptiles (one turtle, 10 lizards, two snakes).

Irwin S. 1996. Survival of a large *Crocodylus porosus*, despite significant lower jaw loss. Memoirs of the Queensland Museum 39(2):338.

Trauma – *Crocodylus porosus* with amputation of first and second right metatarsals. States that head injuries are the least commonly reported in large crocodiles.

Dental – *Crocodylus porosus* with dentary loss from symphysis to 11th tooth alveoli and right dentary 7.5 cm shorter than left.

Isaac M. 1982. Developmental abnormality in the eastern garter snake (*Thamnophis sirtalis*). Notes from Northern Ohio Asoociation of Herpetologists 9(5):13–14.

Congenital – Eastern garter snake *Thamnophis sirtalis* with duplication of skull of upper portion of skull and kinked back. He also reported a second individual with "body… separated in the middle, with the only connection being the esophagus."

Isaza E, Jacobson ER. 1995. Non-nutritional bone diseases in reptiles. Current Veterinary Therapy 12:1357–1361.

Trauma – Carapace fractures that cross the dorsal midline of chelonian, may be associated with spinal fractures.

Forest fire-induced carpace and plaston burns.

Infection – Osteomyelitis of fifth digit in *Crocodylus siamensis*.

Osteitis deformans-like changes from disseminated vertebral osteomyelitis.

Isaza R, Garner M, Jacobson E. 2000. Proliferative osteoarthritis and osteoarthrosis in 15 snakes. Journal of Zoo and Wildlife Diseases 31:20–27.

Infection – Infection causing segmental fusion of 2–14 vertebral bodies with foci of irregular proliferative bone from infection.

Salmonella osteomyelitis causing vertebral fusion in eastern hognose snake *Heterodon platirhinos*, Western diamondback rattlesnake *Crotalus atrox*, Taylor's cantil *Agkistrodon bilineatus*, Uracoan rattlesnake *Crotalus vergrandis*, and bush viper *Atheris nitschei*.

Proliferative bone changes with thickening of trabeculae with irregular cement lines in *Boa constrictor* (without subspecies, common name cannot be determined) – originally called Paget's disease.

Metabolic – Chinese stripe-tailed snake *Elaphe taeniura*, possibly related to vitamin D deficiency.

Vertebral – Ankylosis, but bones resembled Paget's disease histologically in a canebrake rattlesnake *Crotalus horridus* and in five *Boa constrictor* (without subspecies, common name cannot be determined).

Article intriguingly makes diagnosis of osteomyelitis when bacteria found and osteoarthritis, when none found. This is at variance with customary usage of term osteoarthritis.

Isenbügel E, Frank W. 1985. Heimtierkrankheiten. [Sickness of house animals]. Verlag Eugen Ulmer, Stuttgart. 420pp. [German]

Trauma – *Hydrosaurus pustulatus* with fracture.

Chamaeleo dilepis and *Varanus bengalensis* with multiple rib fractures. *Iguana iguana* with regenerated tail rod.

Infection – *Iguana iguana* with jaw described as osteoporosis (but more actinomycotic in appearance).

Metabolic – *Testudo radiata* again labeled as osteoporosis, but cystic changes, *Iguana iguana* with rachitic skull.

Uromastix sp. with possible pseudogout (manifest as calcific deposits).

Neoplasia – *Amphibolurus* with probable limb tumor.

Vertebral – Madagascar tree boa *Sanzinia madagascariensis* with scoliotic spinal bump.

Boa constrictor (without subspecies, common name cannot be determined) labeled as Paget's disease (but appears more like a spondyloarthropathy) and *Bitis nasicornis* with similar, but unilateral changes.

Other – *Alligator mississippiensis* with reactive bone and adherent calcification.

Iwamuro S, Sakakibara M, Terao M, Ozawa A, Kurobe C, Shigeura T, Kato M, Kikuyama S. 2003. Teratogenic and anti-metamorphic effects of bisphenol A on embryonic and larval *Xenopus laevis*. General and Comparative Endocrinology 133:189–198.

Toxicology – Bisphenol A-induced scoliosis and skull abnormality (reduction of distance between the eyes in *Xenopus laevis*.

Iwasawa H. 1968. Occurrence rate of skeletal abnormalities in *Bufo bufo japonicus* and *Rana catesbeiana*. Hachu ryoseiruigaku zasshi [Acta Herpetologica Japonica] 2:46 [Japanese].

Trauma – Fractures in female *Rana catesbeiana* was 11.1%, contrasted with 6.5% in overall group. This was greater than that found in *Bufo bufo japonicus*.

Neoplasia – "Bone tumor" was seen in 12.9% of 85 *Bufo bufo japonicus* in 1965.

Environmental – Variation in *Bufo bufo japonicus* size from 111 mm in 1962 to 78 mm in 1963 to 144 mm in 1965 in the Niiguta Prefectorate in Japan.

Skeletal anomalies in *Bufo bufo japonicus* and *Rana catesbeiana* included deformed or fused spine, absence or deformity of transverse processes, abnormal articular processes, glenoid or finger bones, and "mismatched" iliac bones. The prevalence in *Bufo bufo japonicus* ranged from 15% of 80 in 1962 to 22.5% of 80 in 1963 and 39.2% of 85 in 1965. The prevalence in *Rana catesbeiana* was 9.1%. Zeroth vertebral articular process was abnormal was "front/back" in 85.9% of 77 *Rana catesbeiana* and 14.1%, "posterior only."Third transverse process tended to have small protrusion in *Rana catesbeiana* (especially left side)," compared with 40–80% of female (right side in 73.6%) in *Bufo bufo japonicus*.

Iwasawa H, Takasu T. 1985. Study of thumb pad regions developed by administration in a young female of *Rana nigromaculata* with a supernumary forelimb. Japanese Journal of Herpetology 11:5–10.

Congenital – Supernumerary forelimb in *Rana nigromaculata*.

Izecksohn E. 1971. Variação no padrão vertebral de *Dendrophryniscus brevipollicatus* Espada (Amphibia, Anura) [Variation in the vertebral pattern of *Dendrophryniscus brevipollicatus* Espada (Amphibia, Anura)]. Arquivos do Museu Nacional (Rio de Janeiro) 54:129–138. [Portuguese]

Congenital – 92 specimens of the frog *Dendrophryniscus brevipollicatus* from four different localities in Brazil with vertebral variation: Specimens from Paranapiacaba and Serra da Bocaina had seven free presacral vertebrae and a sacrum formed by vertebra VIII + sacral vertebra + fused urostyle. Specimens from Santa Teresa had six free presacral vertebrae and a sacrum formed by vertebra VII + vertebra VIII + sacral vertebra + fused urostyle. In many specimens from Guanabara (86%), vertebra VIII was not fused with the sacral vertebra and some had several types of fusion among presacral vertebrae. A few instances of fusion of vertebrae I and II existed in three populations (Serra da Bocaina, Santa Teresa, and Tijuca).

Jäckel 1881. Ein fünfbeiniger *Triton cristatus*. [A five-legged *Triton cristatus*]. Der Zoologische Garten 22:156. [German]

Congenital – Female of *Triton cristatus* with additional short front leg attached before the base of the right thoracic limb.

Jackson CG. 1974. An unusual pattern of cervical central articulation in *Deirochelys reticularia*. Copeia 3:788.

Congenital – *Deirochelys reticularia* with biconvex (instead of opisthocoelous) third cervical vertebrae.

Vertebral – *Deirochelys reticularia* with biconvex (instead of opisthocoelous) third cervical vertebrae.

Jackson OF. 1980. The sick chelonian. Proceedings of the European Herpetological Symposium, Oxford. Buford, England: Cotswald Wild Life Pa; 1980:1–4.

Congenital – Deformed beaks in four of 100; deformed shell, in three; leg weakness and deformity, in eight.

Trauma – Foreleg was gnawed off turtle during hibernation by rat.

Metabolic – Nutritional osteodystrophy in softshell turtles and almost exclusively in terrapins, citing Keymer's 1978 report of 15% frequency of nutritional osteodystrophy in 122 freshwater chelonians on postmortem examination. Nutritional osteodystrophy diagnosed in 31 turtles, mostly related to low calcium, high phorphorus diet.

Jackson OF. 1987. Carapace and other bone injuries in chelonians. Testudo 2(5):18–21.

Trauma – Fractures in turtles.

Metabolic – Nutritional osteodystrophy in turtles presents as carapace lumps which become conical. Shell is undersized and beak, overgrown.

Jackson OF, Cooper JE. 1981. Nutritional diseases. Pp. 409–428 In Cooper JE, Jackson OF. Eds. Diseases of the Reptilia. Vol. 2. London: Academic Press.

Metabolic – Nutritional osteodystrophy in lizards, chelonians, and crocodilians, usually related to diet, presenting with softening, deformity and shortening of carapace, short mandibles, greenstick fractures, nonunion, swollen deformed legs, and kyphoscoliosis. Hermann's tortoise *Testudo hermanni* with short mandible, misshapened limb bones and humped carapace shields from nutritional osteodystrophy. Hydroxyapatite in *Uromastix* hip, wrist, spinal column, and stifle joints.

Stone – Cloacal (bladder) calculi.

Jackson OF, Duff Fasal M. 1981. Radiology in tortoises, terrapins and turtles as an aid to diagnosis. Journal of Small Animal Practice 22:707–716.

Metabolic – X-rays of nutritional osteodystrophy in *Testudo graeca*.

Jackson OF, Sainsbury AW. 1992. Radiological and related investigations. Pp. 63–72. In Beynon P, Lawton MP, Cooper JE. eds, Manual of Reptiles. Cheltenham: British Small Animal Veterinary Association.

Congenital – Kinked spine in reticulated python *Python reticulatus*.

Infection – Infectious osteolysis of *Testudo* distal tibia, fibula, and tarsals.

Metabolic – Nutritional osteodystrophy in *Iguana iguana* and red-eared terrapin *Trachemys scripta elegans*.

Jacobson ER. 1980a. Mycotic diseases of reptiles. In Montali RJ, Migaki G. eds. The Comparative Pathology of Zoo Animals. Smithsonian Institution Press, Washington DC. Pp. 283–290.

Infection – Shell necrosis in a turtle caused by *Mucorales.*
Shell disease – Shell necrosis in a turtle caused by *Mucorales.*

Jacobson ER. 1980b. Reptile neoplasm. In Murphy JB, Collins JT. eds. Reproductive Biology and Diseases of Captive Reptiles. Society for the Study of Amphibians and Reptiles. Contributions to Herpetology No. 1, Oxford, Ohio. Pp. 255–265.
Neoplasia – Chondro-osteofibroma in *Cyclura cornuta.*
Squamous cell cancer destroying fifth metatarsal and phalanges in *Tupinambis teguixin.*
Multiple enchondromas in *Varanus dracoena.*

Jacobson ER. 1981. Neoplastic diseases. In: Cooper JE, Jackson OF. eds. Diseases of the Reptilia. Vol 2. Academic Press, San Diego. Pp. 429–468.
Neoplasia – Enchondroma in distal metaphyses of humerus, metacarpals, hyoid, and cervical vertebrae in Indian monitor, *Varanus dracoena.*

Jacobson ER. 1984a. Imobilization, blood sampling, necropsy techniques and diseases of crocodiles: A review. Journal of Zoo Animal Medicine 15:38–45.
Metabolic – Curvature of spine, jaw growth abnormalities, and teeth at abnormal oblique angles from metabolic bone disease in crocodilians.

Jacobson ER. 1984b. Biology and diseases of reptiles. In Laboratory Animal Medicine, JG Fox, BJ Cohen, FM Loew, eds. New York: Academic Press, pp. 449–476.
Congenital – Anterior duplication of flat-bellied turtle *Chrysemys* nelsoni, also citing Appleby and Siller 1960 and Wallach 1969.
Infection – Florida softshell turtle *Trionyx ferox* with mucomycosis producing carapace lesion, citing Jacobson et al. 1980. *Moralus* fungus produced plaston lesions in Florida softshell turtle *Trionyx ferox*, citing Hunt 1957. *Fusarian* shell infection in radiated tortoise *Testudo radiata.*
Metabolic – Nutritional bone disease producing vertebral curvature and shell deformitiy in turtle. Gout in chelonian, citing Frye 1981.
Neoplasia occurrence in reptiles – Citing Jacobson 1981
Shell disease – Florida softshell turtle *Trionyx ferox* with mucomycosis producing carapace lesion, citing Jacobson et al. 1980. *Moralus* fungus produced plaston lesions in Florida softshell turtle *Trionyx ferox*, citing Hunt 1957. *Fusarian* shell infection in radiated tortoise *Testudo radiata.*

Jacobson ER. 1993. Diseases and medical problems of Iguanid lizards. In Proceedings of American Association of Zoological Parks and Aquariums (AAZPA) 1993:260–267.
Infection – Digital, long bone and pelvic girdle osteomyelitis is common in captive green iguanas, with *Pseudomonas* and *Escherichia coli* cultured. Neoplasia – Chondro-osteofibroma in a rhinoceros iguana.

Jacobson ER. 1994. Causes of mortality and diseases in tortoises – A review. Journal of Zoo and Wildlife Medicine 25:2–17.
Infection – Burmese brown tortoise *Manouria emys* with large abscess of forelimb and another with focal ulcerative lesion on plaston – linked to a coelomic cavity bacterial abscess. Chromomycosis of lower jaw in radiated tortoise (Frank 1970).
Metabolic – Osteopenia, and thinning of peripheral and lateral hypoplastron region trabeculae producing spongy appearance in Beaver Dam Slope Arizona and Utah tortoises, noting osteodystrophy in 12 of 144 chelonians reported by the Zoological Society of London (Keymer 1978). Parathyroid adenoma in red-footed tortoise Frye and Carney 1975).
Shell disease – Cites Rosskopf's 1986 claim that shell disease derives from infectious organisms.

Jacobson, ER. 2007a. Overview of reptile biology, anatomy, and histology. Infectious Diseases and Pathology of Reptiles: Color Atlas and Text, ER Jacobson, ed. Boca Raton, Fl: Taylor & Francis; pp. 1–130.
Metabolic – Growth arrest lines produced by hibernation in reptiles.

Jacobson ER. 2007b. Bacterial diseases of reptiles. Infectious Diseases and Pathology of Reptiles: Color Atlas and Text. ER Jacobson, ed. Boca Raton, Fl: Taylor & Francis, pp. 461–526.
Infection – Alligator with wound-induced shoulder abscess (Novak and Seigel 1986).
"Multifocal osteoarthritis" in green iguanas with mixed gram negatives (actually osteomyelitis); foot abscess-induced osteomyelitis in *Iguana iguana*; maxillary osteomyelitis from stomatitis in meadow viper *Vipera ursinii.*
"Osteoarthritis" caseous material in distal hind foot joint space with osteomyelitis in *Iguana iguana* – actually osteomyelitis.
Osteomyelitis of vertebrae and ribs with pitted proliferation crossing joint space in Russian rat snake *Elaphe schrenckii* and *Boa constrictor* (without subspecies, common name cannot be determined), vertebral bone lysis from streptococcal infection in eastern hognose (Isaza et al. 2000.) and in *Boa constrictor* (without

subspecies, common name cannot be determined) from *Edwardsiella*, *Bitis nasicornis* with *Enterococcus* osteomyelitis; *Crotalus willardi* osteomyelitis from *Salmonella enterica-arizonae* (Schroetter et al. 2005).

Neisseria iguanae induced proliferation of hyoid in green iguana dewlap abscess (Barrett et al. 1994; Plowman et al. 1987).

Serratia lipolysis facilitation of *Citrobacter* shell disease (Jackson and Fulton 1970).

Mycoplasma and *Chlamydia*-induced polyarthritis in Nile crocodiles in Israel (Levisohn, cited in Mohen et al. 1995). *Mycoplasma crocodyli* polyarthritis in *Crocodylus niloticus* (Mohan et al. 1995; Kirchoff et al. 1997). *Mycoplasma iguanae* granuloma with osteomyelitis in green iguana (Brown et al. 2005, 2006).

Nocardia-induced metacarpal osteolysis (Harmes et al. 2002).

Erysipothrix in *Crocodylus acutus* (Jasmin and Baucom 1967) and snapping turtle (Keymer 1978b)

Acid fast organisms in plaston of African soft-shelled *Trionyx triunguis* (Friedman 1903). *Mycobacterium chelonae* of elbow of Kemp's ridley sea turtle (Greer et al. 2003). *M. chelonae* ulcerative proliferative mouth lesion in *Boa.*

Arthritis – Multifocal arthritis in *Alligator mississippiensis* (Brown et al. 2001b).

Vertebral – Lytic and proliferative vertebral disease in snakes, citing Kiel 1977; Isaza 2000, noting this is not Paget's disease.

Boa constrictor (without subspecies, common name cannot be determined) with "degenerative vertebral osteoarthrosis and histological ankylosis with irregular cartilage and bone deposits" but no inflammation. Clearly not osteoarthritis, but etiology unclear – consider neuropathic?

Vascular – Thromboembolism-induced distal tail necrosis in Burmese python *Python molurus bivittatus.*

Neuropathic – *Boa constrictor* (without subspecies, common name cannot be determined) with "degenerative vertebral osteoarthrosis and histological ankylosis with irregular cartilage and bone deposits" but no inflammation. Clearly not osteoarthritis, but etiology unclear – consider neuropathic?

Shell disease – Disfiguring shell disease with dermal bone remodeling in river cooters *Pseudemys concinna* and yellow-bellied turtles (Garner et al. 1997; Lovich et al. 1996). *Constrictor*, river cooter *Pseudemys concinna* with disfigured shell carapace and plaston called segmental necrosis and remodeling, appearing as bulky proliferative lesions.

Ulcerative shell with scute shelling from *Beneckia chitinosora* in sliders, musk, side necked and painted turtles (Wallach 1975).

Serratia lipolysis facilitates *Citrobacter* shell disease (Jackson and Fulton 1970).

Jacobson ER. 2007c. Parasites and parasitic diseases of reptiles. Infectious Diseases and Pathology of Reptiles: Color Atlas and Text. ER Jacobson, ed. Boca Raton, Fl: Taylor & Francis, pp. 571–605.

Shell disease – Shell lesions from *Spirorchis* in emydine turtles.

Barnacles without information on alterations in turtle bone/shell.

Illustrates chicken turtle *Deirochelys reticularia* with *Spirochidiasis* ulcerations into dermal bone and barnacles on head and shell of *Caretta caretta.*

Jacobson ER, Calderwood MB, Clubb SL. 1980. Muromycosis in hatchling Florida softshell turtles. Journal of the American Veterinary Medical Association 17:835–837.

Shell disease – Florida softshell turtle *Trionyx ferox* with multifocal carapace surface disruptions attributed to *Mucor.*

Jacobson ER, Millichamp NJ, Gaskin JM. 1985. Use of a polyvalent autogenous bacterin for treatment of mixed gram-negative bacterial osteomyelitis in a rhinoceros viper. Journal of the American Veterinary Medical Association 187:1224–1225.

Infection – Osteomyelitis in rhinoceros viper.

Jacobson ER, Wronski TJ, Schumacher J, Reggiardo C, Berry KH. 1994. Cutaneous dyskeratosis in free-ranging desert tortoises, *Gopherus agassizii*, in the Colorado Desert of Southern California. Journal of Zoo and Wildlife Meducube 25(1):68–81.

Shell disease – Shell necrosis in desert tortoise *Gopherus agassizii*, especially affecting the plaston. Possible relationship to the spring food *Astragalus, Stanleya*, and *Xylorrhiza*, which produces aliphatic nitro compounds.

Jahan Q, Ovais M. 1979. On the presence of twin heads in a snake *Eryx conicus*. Journal of Scientific Research 1(2):65–66.

Congenital – Derodymous sand boa Sri Lanka or rough-scaled sand boa *Eryx conicus.*

Jahn K. 1998. A case of abnormal extremities in *Pelobates fuscus*. Ein Fall von Extremitätenanomalie bei *Pelobates fuscus*. Zeitschrift für Feldherpetologie 5:65–69. [German]

Congenital – Observations of growth of *Pelobates fuscus* male with two additional hind legs (X-rays): the third pair of legs grows together, no handicap in movements.

Jaksić FM, Busack SD. 1984. Apparent inadequacy of tail-loss figures as estimates of predation upon lizards. Amphibia-Reptilia 5:177–179.

Trauma – 19.5% of *Acanthodactylus erythrurus* (27% of 447 tail loss), 18% of *Lacerta lepida* (24.6% of 65), 20.3% of *Podarcis hispanica* (59.3% of 81), 23.5% of *Psammodromus algirus* (32.5% of 169), and 0.8% of *Tarentola mauritanica* (54.5% of 121) were prey. Tail loss was not proportional to amount of known predation.

Jaksić FM, Fuentes ER. 1980. Correlates of tail losses in twelve species of *Liolaemus* lizards. Journal of Herpetology 14:137–141.

Trauma – Frequency of tail regeneration in *Liolaemus altissimus* (53%), *chiliensis* (24%), *fuscus* (76%), *lemniscatus* (49%), *leopardinus* (43%), *monticola* (68%), *nigromaculatus* (58%), *nigroviridis* (77%), *nitidus* (63%), *platei* (41%), *schroederi* (38%), and *tenuis* (67%) was equal in juveniles and adults and gender and body size independent, but correlated with frequency of utilization of elevated perches.

Jaksić FM, Greene HW. 1984. Empirical evidence of non-correlation between tail loss frequency and predation intensity on lizards. Oikos 42:407–411.

Trauma – Tail break frequencies do not correlate with predation density, despite their having been used as such an index for snakes (Gehlbach 1972; Greene 1973; Zug et al. 1979), salamanders (Shaffer 1978), amphisbaenians (Papenfuss 1982), and in lizards, the relationship may actually be negative.

Eumeces gilberti had more tail loss than *Gerrhonotus multicarinatus, Sceloporus occidentalis, and Uta stansburiana.*

Confounding factors include activity periods and survivorship (Tinkle and Ballinger 1972; Vinegar 1975), intraspecific aggression (Blair 1960; Bustard and Hughes 1966; Norris 1953; Tinkle 1967; Vitt and Ohmart 1974), microhabitat preferences (Ballinger 1973; Jaksić and Fuentes 1980; Pianka and Huey 1978; Pianka and Pianka 1976; Werner 1968), and variable ease of autotomy (Congdon et al. 1974; Vitt et al. 1977). Schoener (1979) and with Schoener (1980) also suggested that injury with tail regeneration reflected predator inefficiency.

Jakstien K-P, Petzold H-G. 1960. Über eine Fraktur bei einer Wassermokassinschlange (*Agkistrodon piscivorus* [Lacép.]). [On a fracture in a water moccasin (*Agkistrodon piscivorus* [Lacép.])]. Tierärztliche Umschau 15:407–408. [German]

Congenital – Kyphoscoliosis from fracture in #19 thoracic vertebra and in #28 vertebra of *Agkistrodon piscivorus.*

Jan G. 1863. Iconographie des ophidiens. Elenco Sist. degli ofidi descritti e disegnati per l'icon. gener Milano [Iconography of ophidians. Systematic listing of ophidians described and drawn for the general iconography of Milano]. Tip. Lombardi. p. 71. [Italian]

Congenital – Atloidic dicephalic *Tropidonotus fasciatus* from North America, in the collections of the Museo Civico of Milano, although Cantoni (1921) was unable to locate it. [Italian]

Janensch W. 1934. Eine halbseitige überzählige Wirbelbildung bei einem Dinosaurier. [The formation of supernumerary vertebrae on one side in a dinosaur]. Sitzungsberichte der Gesellschaft Naturforschender Freunde zu Berlin 1933:458–462. [German]

Congenital – *Dysalotosaurus* with hemivertebra.

Veterbral – *Dysalotosaurus* with hemivertebra.

Fossil – *Dysalotosaurus* with hemivertebra.

Jarofke D, Lange J. 1993. Reptilien, Krankheiten und Haltung. [Reptiles, sickness and keeping]. 188pp.; Berlin, Hamburg: Parey. [German]

Congenital – Turtle kyphosis.

Trauma – jaw, extremity and shell fractures in turtles, fractures in lizards and snakes.

Metabolic – Osteoporosis, rickets, osteomalacia and osteodystrophia fibrosa in turtles and lizards and osteodystrophy in snakes.

Jayasinghe A. 1997a. Two-headed python pulls crowds in Sri Lanka. Agence France-Presse 18 November 1997:1.

Congenital – Dicephalic python, actually Russell's viper.

Jayasinghe A. 1997b. Two-headed 'python' turns out to be Russell's viper. Agence France- Presse 19 November 1997:1.

Congenital – Dicephalic Russell's viper.

Jayne BC, Bennett AF. 1989. The effect of tail morphology on locomotor performance of snakes: A comparison of experimental and correlative methods. Journal of Experimental Zoology 252:126–133.

Trauma – 52 of 445 *Thamnophis sirtalis fitchi* lost 3–80% of their tails, without significant effect on locomotion. Loss of two third of tail resulted in 4.5% reduction in speed.

Jayne BC, Bennett AF. 1990. Selection on locomotor performance capacity in a natural population of garter snakes. Evolution 44:1204–1229

Trauma – Negative correlation between tail length residual and running speed capability.

Jenkins JR. 1996. Digit abnormalities. In Mader DR, ed. Reptile medicine and Surgery. Pp. 365–67; Philadelphia: Saunders.

Infection – Proximal interphalangeal joint infection in unnamed reptile species.
Metabolic – Unnamed reptile species with foot gout.

Jennemann G. 2003. Panzergangrän bei Europäischen Landschildkröten: Ein Fallbeispiel zur Diagnose, Ursachenanalyse und Behandlung. [Carapace gangrene in European tortoises: an example for diagnosis, analysis of the causes and treatment]. Radiata 12(2):23–30 [German].
Shell disease – Gangrene as hole on the right side of the carapace caused by bacteria.

Jenssen TM, Marcellini DL. 1986. *Leiocephalus schreibersi* (curly-tailed lizard): Tail autotomy. Herpetological Review 17:89.
Trauma – Tail autotomy related to conspecific fights, as well as failed predation. Autotomy in curly tailed lizard *Leiocephalus schreibersi* was documented in a female during mating with a struggling female. Incidently, the male ate the severed tail subsequent to completion of mating.

Ji X, Zhang C-H. 2001. Effects of thermal and hydric environments on incubating eggs, hatching success, and hatchling traits in the Chinese skink (*Eumeces chinensis*). Acta Zoologica Sinica 47:256–265.
Environmental – Chinese skink *Eumeces chinensis* incubated at 30°C had longest tails and those at 32°C, the smallest heads (all versus 24°C and 26°C).

Ji X, Qui QB, Diong CH. 2002. Influence of incubation temperature on hatching success, energy expenditure for embryonic development, and size and morphology of hatchlings in the oriental garden lizard *Calotes versicolor* (Agamidae). Journal of Experimental Zoology 292:649–659.
Environmental – More deformed oriental garden lizard *Calotes versicolor* embryos when incubated at 33°C.

Jockusch EL. 1997. Geographic variation and phenotypic plasticity of number of trunk vertebrae in slender salamanders, *Batrachoseps* (Caudata: Plethodontidae). Evolution 51:196- 6–1982.
Congenital – Slider salamander *Batrachoseps* trunk vertebrae vary from 16 to 23, even within populations – a phenotypically plastic character. Costal *Batrachoseps attenuatus* had 19, while inland had 22. Females tended to have more than males, especially in *Batrachoseps major* and *Batrachoseps relictus*. Females with more than 21 trunk vertrebrae more often had more without ribs and fewer caudosacral vertebrae. Absent ribs correlated with more than two caudosacral vertebrae. "5.8% had different numbers of trunk vertebrae on the left and right" (page 1972). Jockusch determined the number of trunk vertebrae based on the position of the pelvic articulation. There are animals in which pelvic articulation is asymmetric, articulating with the 21st vertebra on one side and the 22nd on the other. Asymmetry ranged from 0% to 16.7%. Negative correlation between number of presacral vertebrae and number of phalanges (index of limb reduction occurs in skink *Lerista* and *Chalcides*. Reduced limbs are associated with increased elongation. Number of vertebrae correlated inversely with length of species. Performance differences correlate with numbers of vertebrae in some snakes.
Environmental – More trunk vertebrae were present in higher temperature environments, except *Batrachoseps nigriventris*, wherein trunk vertebral number was decreased between 7 and 13°C.

Jofre MB, Karasov WH. 1999. Direct effect of ammonia on three species of North American anuran amphibians. Environmental Toxicology and Chemistry 18:1806–1812.
Toxicology – Ammonia concentration of 0.6 mg/l (increasing to 100% at 0.9 mg/l) in green frogs *Rana clamitans* and 1.5 mg/l (increasing to 60% at 2.25 mg/l) in leopard frogs *Rana pipiens* induced deformities, but not in American toads *Bufo americanus* (as high as 0.9 mg/l). Asymmetric bodies, curled spine, short and/or deformed tails were noted.

Johnson RH. 1901a. Three polymelous frogs. American Naturalist 35:25.
Congenital – Supernumerary limbs in *Rana palmipes* and *Rana halecina*.

Johnson RH. 1901b. Axial bifurcation in snakes. Transactions of the Wisconsin Academy of Sciences, Arts and Letters, Madison 13:223–236, 523.
Congenital – Dicephalic probable *Tropidonotus fasciata fasciata*, 2 *Tropidonotus fasciata sipedon*, 2 *Bascanium constrictor*, *Ancistrodon piscivorus*, *Ophiobolus getulus*, *Ophiobolus getulus getulus*, *Pityophis*, *Pituophis catenifer* (Anonymous 1878), *Eutania sirtalis*, *Eutania sirtalis sirtalis*, *Thamnophis elegans lineolata*, a milk snake and a rat snake. He also reports an abnormal frog, previously described by Kingsley (American Naturalist XII:694–695).

Johnson RH. 1902. Axial bifurcation in snakes. (Part 2). Transactions of the Wisconsin Academy of Sciences, Arts and Letters 13:523–538.
Congenital – Dicephalic: Hog-nose snake *Heterodon simus*, black snake *Bascanium constrictor*, Western bull snake *Pituophis catenifer*, bull snake *Pituophis sayi*, king snake *Ophiobolus getulus*, milk snake *Lampropeltis triangulum*, water snake *Tropidonotus fasciata sipedon*, *Tropidonotus fasciata fasciata*, garter snake *Eutainia sirtalis sirtalis*, *Eutainia elegans lineolata* and cotton-mouth moccasin *Ancistrodon piscivorus*.

Johnson GD. 1988. An abnormal captorhinomorph vertebra from the Lower Permian of north-central Texas. Journal of Vertebrate Paleontology 8(3):19A.

Congenital – Captorhinomorph vertebra from the Lower Permian with two rib attachments on right; three, on left, noting incomplete/missing right posterior and left anterior processes and right anterior parapophysis.

Fossil – Captorhinomorph vertebra from the Lower Permian with two rib attachments on right; three, on left, noting incomplete/missing right posterior and left anterior processes and right anterior parapophysis.

Johnson JH. 2004. Husbandry and medicine of aquatic reptiles. Seminars in Avian and Exotic Pet Medicine 13:223–228.

Metabolic – Metabolic bone disease manifest as fracture or deformity from dietary/husbandry mismanagement.

Shell disease – Multifocal, circular gray areas, with *Aspergillus, Beauvaria,Cladosporium*, and *Paecilomyces* isolated.

Ulcerative shell disease attributed to *Baneckea chitonovora.*

Johnson PT, Chase JM. 2004. Parasites in the food web: Linking amphibian malformations and aquatic eutrophication. Ecology Letters 7:521–526.

Infection – *Ribeiroia* infection is predictor of malformation frequencies.

Johnson PJ, Sutherland DR. 2003. Amphibian deformities and *Ribeiroia* infection: An emerging helminthiasis. Trends in Parasitology 19:332–335.

Congenital – Widespread appearance of malformed amphibians in the mid-1990s. Laboratory studies implicated infection by a digenetic trematode – *Ribeiroia ondatrae*, suggesting exogenous agents (e.g., pesticides, nutrient runoff, introduced fishes) might be interacting with *Ribeiroia.*

Infection – Widespread appearance of malformed amphibians in the mid-1990s. Laboratory studies implicated infection by a digenetic trematode – *Ribeiroia ondatrae*, suggesting exogenous agents (e.g., pesticides, nutrient runoff, introduced fishes) might be interacting with *Ribeiroia.*

Johnson PTJ, Lunde KB, Ritchie EG, Launer AE. 1999. The effect of trematode infection on amphibian limb development and survivorship. Science 284:802–804.

Infection – Trematode *Ribeiroia ondatrae* cercariae cyst infestation-induced deformities in Pacific tree frogs *Hyla regilla.*

Johnson PTJ, Lunde KB, Haight RW, Bowerman J, Blaustein AR. 2001. *Ribeiroia ondatrae* (Trematoda: Digenea) infection induces severe limb malformations in western toads (*Bufo boreas*). Canadian Journal of Zoology 79:370–379.

Infection – Trematode *Ribeiroia ondatrae* cyst infestation induced deformities in western toads *Bufo boreas.*

Johnson PT, Lunde KB, thurman EM, Ritchie EG, Wray SN, Sutherland DR, Kapfer JM, Frest TJ, Bowerman J, Blaustein AR. 2002. Parasite (*Ribeiroia ondatrae*) infection linked to amphibian malformations in the western United States. Ecological Monographs 72:151–168.

Congenital – *Ribeiroia ondatrae* preferentitally infects limb buds. Pacific tree frogs have more abnormalities than western toads and less than California newts. More than 5% ectomelia, ectodactyly, supernumerary limbs, polymelia, polydactyly, and jaw malformations at sites where *Ribeiroia* is found. Pesticides and "metabolite compounds" were not differentially present. Affected animals include long-toed salamander *Ambystoma macrodactylum*, rough-skinned newt *Taricha granulosa*, California newt *Tarocja tprpsa*, western toad *Bufo boreas* (predominantly ectodactyly and ectomelia), Pacific tree frog *Hyla regilla*, Northern red-legged frog *Rana aurora*, Cascades frog *Rana cascadae*, American bullfrog *Rana catesbeiana*, and Columbia spotted frog *Rana luteiventris* especially at sites with *Ribeiroia* and fish present. 94% of involvement among anurans was hind limb; 56.9% in urodeles.

Infection – *Ribeiroia ondatrae* preferentitally infects limb buds. Pacific tree frogs have more abnormalities than western toads and less than California newts. More than 5% ectomelia, ectodactyly, supernumerary limbs, polymelia, polydactyly, and jaw malformations at sites where *Ribeiroia* is found. Affected animals include long-toed salamander *Ambystoma macrodactylum*, rough-skinned newt *Taricha granulosa*, California newt *Tarocja tprpsa*, western toad *Bufo boreas* (predominantly ectodactyly and ectomelia), Pacific tree frog *Hyla regilla*, Northern red-legged frog *Rana aurora*, Cascades frog *Rana cascadae*, American bullfrog *Rana catesbeiana*, and Columbia spotted frog *Rana luteiventris* especially at sites with *Ribeiroia* and fish present.

Johnson PTJ, Lunde KB, Zelmer DA, Werner JK. 2003. Limb deformities as an emerging parasitic disease in amphibians: Evidence from museum specimens and resurvey data. Conservation Biology 17:1724.

Congenital – Review of available information for nine historical accounts from California, Colorado, Idaho, Mississippi, Montana, Ohio, and Texas reported between 1946 and 1988 revealed malformations at six of eight sites were associated with infection by *Ribeiroia*, dating back as far as 1946. Malformations recorded historically at these sites were consistent with the documented effects of *Ribeiroia* infection, including extra limbs, cutaneous fusion, and bony triangles. Of the six sites that still supported amphibians upon resurvey, three continued to support severe limb malformations (polydactyly, polymelia of hind limbs) at frequencies of 75% in one or more species [*Ambystoma tigrinum, Ambystoma macrodactylum croceum* (with forelimb polymelia also), *Rana pipiens, Rana catesbeiana, Hyla regilla*]. Although no pesticides were detected, amphibians from each of these sites were infected with *Ribeiroia* metacercariae. Specifically reported in *Hyla regilla* were ectodactyly, polydactyly, ectomelia, hemimelia, polymelia, polypodia, taumelia, and micromelia.

Infection – Review of available information for nine historical accounts from California, Colorado, Idaho, Mississippi, Montana, Ohio, and Texas reported between 1946 and 1988 revealed malformations at six of eight sites were associated with infection by *Ribeiroia*, dating back as far as 1946. Malformations recorded historically at these sites were consistent with the documented effects of *Ribeiroia* infection, including extra limbs, cutaneous fusion, and bony triangles. Of the six sites that still supported amphibians upon resurvey, three continued to support severe limb malformations (polydactyly, polymelia of hind limbs) at frequencies of 75% in one or more species [*Ambystoma tigrinum, Ambystoma macrodactylum croceum* (with forelimb polymelia also), *Rana pipiens, Rana catesbeiana, Hyla regilla*]. Although no pesticides were detected, amphibians from each of these sites were infected with *Ribeiroia* metacercariae. Specifically reported in *Hyla regilla* were ectodactyly, polydactyly, ectomelia, hemimelia, polymelia, polypodia, taumelia, and micromelia.

Johnson PT, Chase JM, Dorsch KL, Hartson RB, Gross JA, KLarson DJ, Sutherland DR, Carpenter SR. 2007. Aquatic eutrophication promotes pathogenic infection in amphibians. Proceedings of the National Academy of Science 104:15781–15786

Environmental – Eutrophication (increased nitrogen and phosphorus content) of environments increases *Ribeiroia ondatrae* emergence, because of promotion of algal production, increasing density of snail hosts and of infection.

Joly P. 1966. Polydactylie bilateral naturelle chez des larves de *Salamandra salamandra* (L.). [Natural bilateral polydactyly in larval *Salamandra salamandra* (L.)]. Bulletin de la Société zoologique de France 91:336. [French]

Congenital – Polydactyly in *Salamandra salamandra*, noting that Bishop and Hamilton (1947) reported 17 instances of polydactyly among *Ambystoma tigrinum* from Colorado lake.

Jonsson G. 2004. Two-headed albino black rat snake is safe after cross-riper caper. St. Louis Post-Dispatch, St. Louis, MO (24 August 2004):B-1.

Congenital – Derodymous black rat snake.

Jonstonus J. 1657. Historiae naturalis de quadrupedibus libri cum Aeneis figures. [Natural History of Quadrupeds with figures]/ Amstelodami or Francofurti Libri II pp. 1–40. [Latin]

Trauma – Lizard with double tail as malformation [vol.1, book 4c. II art. I a. II, p. 134].

Jonstonus J. 1665. Historiae Naturalis de Serpentibus. [Natural History of Serpents]. Book 2. Haeredum Merianaeorum, Francofurti, 40 pp. [Latin]

Congenital – Derodymous *Serpens biceps.* Illustrated with elevated collar around neck that was twice as long as head. Membrane on body suggests ? sea snake?

Jones ME. 2006. Skull evolution and functional morphology in *Sphenodon* and other Rhynchocephalia (Diapsida: Lepidosauria). Unpublished PhD thesis. London: University London.

Neoplasia – *Sphenodon* BMB 101688 with circular mass on external surface of jugal.

Jones TD, Ruben JA, Martin LD, Kurochkin EN, Feduccia A, Maderson PF, Hillenius WJ, Geist NR, Alifanof V. 2000. Nonavian feathers in a Late Triassic archosaur. Science 288:2202–2205.

Congenital – *Longisquama insignis*, a Late Triassic central Asian archosaur has paired integumentary appendages. The basal region is similar to the calamus of modern feathers, and these appendages are interpreted as nonavian feathers, homologous with avian feathers, antedating the feathers of the Later Jurassic *Archaeopteryx.*

Fossil – *Longisquama insignis*, a Late Triassic central Asian archosaur has paired integumentary appendages. The basal region is similar to the calamus of modern feathers, and these appendages are interpreted as nonavian feathers, homologous with avian feathers, antedating the feathers of the Later Jurassic *Archaeopteryx.*

Jourdain M. 1877. Monstre ectromèle rencontré sur une *Rana temporia.* [Ectromelia in a Rana tempororia monster] Bulletin de la Société scientifique de Nancy 2(3):10. [French]

Congenital – Ectomelia in *Rana temporaria.*

Joyce WG. 2000. The first complete skeleton of *Solnhofia parsonsi* (Cryptodira, Eurysternidae) from the Upper Jurassic of Germany and its taxonomic implications. Journal of Paleontology 74:684–700.

Trauma – Upper Jurassic Cryptodiran turtle *Solnhofia parsonsi* with symmetrical posterior carapace rim holes with new bone formation.

Shell disease – Upper Jurassic Cryptodiran turtle *Solnhofia parsonsi* with symmetrical posterior carapace rim holes with new bone formation.

Fossil – Upper Jurassic Cryptodiran turtle *Solnhofia parsonsi* with symmetrical posterior carapace rim holes with new bone formation.

Ju B-G, Kim W-S. 1994. Pattern duplication by retinoic acid treatment in the regenerating limbs of Korean salamander larvae, *Hynobius leechii*, correlates well with extent of dedifferentiation. Developmental Dynamics 199:253–267.

Toxicology – Partial duplication of structures in Korean salamander *Hynobius leechii* induced by retinoic acid.

Judd WW. 1955. Observations on the blue-tailed skink, *Eumeces fasciatus*, captures in Rondeau Park, Ontario and kept in captivity over winter. Copeia 155:135–136.

Trauma – Blue-tailed skink *Eumeces fasciatus* with tail loss and regrowth.

Metabolic – Blue-tailed skink *Eumeces fasciatus* ejects pellet with white uric acid crystals after ingesting insects.

Jungfer KH, Weygoldt P. 1999. Biparental care in the tadpole-feeding Amazonian tree frog *Osteocephalus oophagus*. Amphibia Reptilia 20:235–249.

Trauma – Rare toe regeneration in Amazonian tree frog *Osteocephalus oophagus*.

Jungnickel J. 1986. Bemerkungen zur Problematik der Höckerbildung bei der Aufzucht von Schildkröten. [Remarks to the problem of bulge formation in raised turtles]. Sauria 8(1):19–22. [German]

Environmental – Prohibiting hump-like deformation of the carapace (centralia, lateralia) of tortoises by raising in open air, adding vitamin D3 and lime to the food; UV light required if raising indoor.

Kabisch K. 1974. Die Ringelnatter. [The grass snake]. Die Neue Brehm Bücherei 483:88pp.Wittenberg Lutherstadt: A. Ziemsen. [German]

Congenital – 100 cases of dicephalism known, leading to [Linné: *Coluber bicephalus*].

Trauma – *Natrix natrix*: fractures of ribs (nine cases, all well healed), fractures of vertebrae, loss of tail vertebrae.

Vertebral – Bowing of vertebral column in *Natrix natrix*.

Kabisch K. 1975. Ein fünfbeiniger Teichfrosch (*Rana esculenta* Linnaeus). [A five-legged edible or common water frog (*Rana esculenta* Linnaeus)]. Abhandlungen und Berichte des Naturkundlichen Museums 'Mauritianum' 9:1. [German]

Trauma – One specimen of *Rana esculenta* with two well-developed right forelimbs caused probably by injury of extremity bud in early larval stage.

Kabisch K. 1989. Carapaxanomalien bei vier Jungtieren von *Testudo graeca ibera* Pallas 1814. [Anomalies of the carapace in four young *Testudo graeca ibera*]. Sauria 11:23–25. [German]

Shell disease – Carapace anomalies in *Testudo graeca ibera*.

Kaestner S. 1898. Doppelbildungen bei Wirbeltieren. [Vertebrate double monsters]. Archiv für Anatomie und Physiologie 1898:81–94. [German]

Congenital – Duplications in *Amphioxus, Triton* and defines various types of duplication, but does not name specific reptiles.

Kaestner S. 1912. Die Entstehung der Doppelbildungen des Menschen und der höheren Wirbeltiere. [The formation of double formation in man and higher vertebrates]. Sammlung Anatomischer und Physiologischer Vorträge und Aufsätze Heft 18:54 pp.; Jena. [German]

Congenital – Early ontogeny; on p. 49 mentioning double tails in amphibians in a general manner, with no specifics.

Kahn RH. 1926. Anatomische und physiologische Untersuchung eines Falles von Hypermelie beim Frosche. [Anatomical and physiological study of a case of frog hypermelia]. Prager Archiv für Tiermedizin 6(1):7–24. [German]

Congenital – Supernumerary limb coming off accessory scapula of a frog.

Kaiser BW, Mushinsky HR. 1994. Tail loss and dominance in captive adult male *Anolis sagrei*. Journal of Herpetology 28:342–346.

Trauma – No effect of tail loss on dominance, aggressiveness, or territorial size in captive adult males of the short-lived species *Anolis sagrei*.

Kälin JA. 1934. Über Skelettanomalien bei Crocodiliden. [Skeletal anomalies in crocodilians]. Verhandlungen der Schweizerischen Naturforschenden Gesellschaft 115:380–381. [German]

Congenital – Broadening of skull in *Alligator mississippiensis*, *Crocodylus porosus*, and *C. vulgaris* ("Mops" skull) as in domestication in mammals.

Anomalies of jaw and snout in crocodilians.

Trauma – High number (25%) of injured and sick animals.

Exostoses are common crocodile skeletons (20% of specimens in the zoological collection of Munich).

Neoplasia – Osteomas of vertebral column in crocodilians.

Arthritis – Noted in crocodilians.

Vertebral – Spondylitis deformans in crocodilians.

Other – Necrosis on head of crocodilians.

Kälin JA. 1937a. Sur les anomalies du squelette chez les crocodilians. [Skeletal anomalies in crocodilians]. Bulletin du Muséum Histoire Naturale Paris 8 (series 2):385–387. [French]

Congenital – 20% of crocodilian skeletons had deformity (Kälin 1934)

Kälin JA. 1937b. Über Skeletanomalien der Crocodiliden. [Skeletal anomalies in crocodilians]. Zeitschrift für Morphologie und Oekologie der Tiere 32:327–347. [German]

Congenital – *Crocodylus porosus*: broad skull (pug-shaped head). Two *Alligator mississippiensis* and *Crocodylus porosus* specimens with broad short skull. *Caiman crocodilus* with asymmetrical snout. *Crocodylus palustris* with horizontal and vertical snout deformations; other snout anomalies in *Crocodylus americanus* and *Caiman crocodilus*.

Trauma – *Crocodylus porosus* with fractured left hind leg with callus formation and synostosis; second specimen with exostosis and pseudarthrosis; third with two fractures, one healed, the other a pseudarthrosis.

Crocodylus vulgaris with healed fractures of tibia and fibula.

Caiman crocodilus with exostoses on thoracal vertebrae 8, 9, 14, and 15; healed fractures on caudal vertebra 9, 20; exostosic rugosities on caudal vertebrae 22, 23, and 24; healed fracture in middle of right ulna; exostoses on left humerus; callus formation and exostosis on sternum; necrotic changes, exostoses and synostosis on skull. Another specimen had healed fracture of left humerus; a third had shortened tail.

Caiman latirostris with deformation of snout, exostoses of thoracal vertebrae and shoulder girdle, healed fractures in legs and vertebrae, necrotic changes in skull, and tail regeneration.

Melanosuchus niger with healed fracture of gastralia.

Paleosuchus niloticus with exostosis of 1 sacral vertebra, spongious bone on thoracal vertebrae 8–12 and left humerus with healed fracture (preserved vertebra) in *Caiman crocodilus*, and *Melanosuchus niger*.

Neoplasia – *Caiman crocodilus* with deformed caudal vertebra (irregular bone mass) and thickened left tibia.

Crocodylus porosus with bone tumor (labeled osteoma) between last lumbar and first caudal vertebrae.

Caiman latirostris with osteoma.

Dental – *Caiman crocodilus* with reduction of premaxillary teeth to three.

Crocodylus palustris and *C. porosus* lacked maxillary teeth four, five, and six (filled alveolar cavity).

Missing premaxillary tooth five in *Tomistoma schlegeli, Crocodylus porosus, C. siamensis*, and *C. cataphractus. Crocodylus vulgaris* with partly many teeth in one alveolar place.

Kamrin RP, Singer M. 1955. The influence of the spinal cord in regeneration of the tail of the lizard, *Anolis carolinensis*. Journal of Experimental Zoology 128:611–628.

Trauma – Regeneration is dependent on presence of nerve supply, especially sensory ganglia, in limbs and tail of urodeles, tadpole extremities (Schotté and Harland 1943), and *Anolis* tails.

Kaplan HM. 1957. Septicemic cutaneous ulcerative disease of turtles. Proceedings of the Animal Care Panel 7:272–277.

Infection – *Escherichia freundii* apparently responsible for ulcerative shell disease.

Shell disease – *Escherichia freundii* apparently responsible for ulcerative shell disease.

Kar SK. 1979. Malformation at birth in the saltwater crocodile (*Crocodylus porosus* Schneider) in Orissa, India. Journal of the Bombay Natural History Society 76:166–167.

Environmental – *Crocodylus porosus* with bent neck and twisted tail from increased incubation temperatures. *C. novaeguineae* with stunted tail.

Kar SK, Bustard HR. 1982. Embryonic tail deformation in the saltwater crocodile (*Crocodylus porosus*) in Orissa, India. British Journal of Herpetology 6:220–221.

Environmental – Curly tail in salt water crocodile *Crocodylus porosus* was attributed to high incubation temperatures.

Katsura Y. 2004. Paleopathology of *Toyotamaphimeia machikanensis* (Diapsida, Crocodylia) from the middle Pleistocene of central Japan. Historical Biology 16:93–97.

Trauma – Middle Pleistocene (370,000 ybp) *Toyotamaphimeia* (originally *Tomistoma) machikanensis* with distal mandible loss, irregular bumps and wrinkle-like ridges on anterior mandibular surface and malaligned, healed fracture of right tibia and fibula with shortening and holes in dermal scutes, apparently related to bites. The internal surfaces are smooth, with radial cracks surround the holes. Comments on American alligator *Alligator mississippiensis* with bite mark – although it appears to simply represent a healed, malaligned overlapping femoral fracture. States that healing in crocodilian cranial bones is often incomplete and that fracture "setting" takes at least a month.

Cites injuries in Australian crocodiles *Crocodylus johnstoni* (Webb and Manolis 1977), *Crocodylus porosus* (Webb and Messel 1977) and in the New Guinea freshwater crocodile *Crocodylus novaeguineae* (Montague 1984), Erickson (1996) for loss of large portions of jaws and snouts, Cott (1961) reporting loss of distal rami of *Crocodylus niloticus* jaws and broken or amputated limbs, Irwin (1996) reporting *Crocodylus porosus* missing anterior third of mandible and Sawyer and Erickson's (1987) American alligator with healed right foot amputation.

Williamson (1996) reported a Late Cretaceous alligatorid NMMNH P-25050 ?*Brachychampsa sealeyi* with puncture wounds on mandible.

Erickson and Sawyer (1996) reported fractures with punctures of tibia and fibula of Oligocene estuarine crocodile *Gavialosuchus carolinensis.*

Sawyer and Erickson (1998) reported Paleocene freshwater crocodile *Leidyosuchus* (= *Borealosuchus*) *formidabilis* with "extensive injuries and diseases." Buffetaut (1983) reported an Eocene mesosuchian crocodilian *Tilemsisuchus lavocati* with wounds on fragmentatry mandibular synthesis, blaming it on intraspecific conflict.

Fossil – Middle Pleistocene (370,000 ybp) *Toyotamaphimeia* (originally *Tomistoma*) *machikanensis* with distal mandibule loss, irregular bumps and wrinkle-like ridges on anterior mandibular surface and malaligned, healed fracture of right tibia and fibula with shortening and holes in dermal scutes, apparently related to bites. The internal surfaces are smooth, with radial cracks surround the holes.

Erickson and Sawyer (1996) reported fractures with punctures of tibia and fibula of Oligocene estuarine crocodile *Gavialosuchus carolinensis.*

Sawyer and Erickson (1998) reported Paleocene freshwater crocodile *Leidyosuchus* (= *Borealosuchus*) *formidabilis* with "extensive injuries and diseases."

Buffetaut (1983) reported an Eocene mesosuchian crocodilian *Tilemsisuchus lavocati* with wounds on fragmentatry mandibular synthesis, blaming it on intraspecific conflict.

Kavlock RJ. 1998. What's Happening to Our Frogs? Environmental Health Perspectives 106(12):773–774.

Environmental – Diluting pond water 25% eliminated malformations in *Xenopus laevis.*

Kawakami I. 1953. Studies of the effects of lithium chloride on *Triturus pyrrhogaster* embryos. II. Simultaneous exposure to lithium and hydrocyanic acid. Memoirs of the Faculty of Science, Kyushu University, Series E (Biology) 1(3):125–131.

Toxicology – Hydrocyanic acid prevents lithium-induced head deformity in *Triturus pyrrhogaster.*

Kawakami I, Kawakami I. 1951. Studies of the effects of lithium chloride on *Triturus pyrrhogaster* embryos. I. Suceptibility of embryos in every developmental stage. Dobutsugaku Zasshi [Zoological Magazine (Japan)] 60:59–62 (Japanese).

Toxicology – Lithium-induced head deformity in *Triturus pyrrhogaster.*

Kawamura T. 1939. Artificial parthenogenesis in the frog. II. The sex of parthogenetic frogs. Journal of Science of the Hiroshima University Series B Division 1 Vol 7:39–86.

Congenital – "Bent bodies" in parthogenetic female *Rana nigromaculata* and *Rana japonica.*

Kawamura T, Nishioka M. 1972. Viability and abnormalities of the offspring of nucleocytoplasmic hybrids between *Rana japonica* and *Rana ornativentris*. Scientific Report of the Laboratory for Amphibian Biology 1:95–209.

Congenital – *Rana japonica-Rana ornataventis* progeny had abnormal hind limbs. F-F1 mating resulted in abnormal fore and hind limbs in 17/77, associated with delayed metamorphosis. Half of F1-F2 hybrids had "ill-developed" (abnormally slender) forelimbs.

Kear BP. 2002. Dental caries in an Early Cretaceous ichthyosaur. Alcheringa 25:387–390.

Dental – Early Cretaceous (Late Albian) Queensland ichthyosaur *Platypterygius longmani* Australian Museum, Sidney Australia AM F98273 with anterior dentary tooth with prominent U-shaped cavity on medial-lingual surface, measuring 1.83 by 3.97 mm.

Cited dental abscess in Late Cretaceous Belgium mosasaur mandible (Moodie 1917) and crown breakage attributed to feeding by Massare (1987).

Fossil – Early Cretaceous (Late Albian) Queensland ichthyosaur *Platypterygius longmani* Australian Museum, Sidney Australia AM F98273 with anterior dentary tooth with prominent U-shaped cavity on medial-lingual surgace, measuring 1.83 by 3.97 mm.

Cited dental abscess in Late Cretaceous Belgium mosasaur mandible (Moodie 1917) and crown breakage attributed to feeding by Massare (1987).

Kearney M, Shine R. 2004. Developmental success, stability and plasticity in closely related parthenogenetic and sexual lizards (Heteronotia, Gekkonidae). Evolution 58:1560–1572.

Congenital – Parthenogenetic *Heteronotia binoei* was more likely (21% versus 6%) to have damaged legs and kinked or absent tails. There were two cases of conjoined twins. The same is true of *Darevskia* (Darevsky et al. 1985).

Keasey MS. III. 1969. Some records of reptiles at the Arizona-Sonora Desert Museum. International Zoo Yearbook 9:16–17.

Congenital – Dicephalic gopher snake *Pituophis melanoleucus affinis.*

Kelly A. 1909. A Yuma rattler. Forest and Stream 73(18):691.

Congenital – Dicephalic desert rattlesnake.

Kerfoot WC. 1969. Selection of an appropriate index for the study of variability of lizard and snake body scale counts. Systematic Zoology 18:53–62.

Congenital – Ventral scale counts of *Diadophis s. similis, Phyllorhynchus d. perkinsi, Thamnophis hammondii, Thamnophis o. ordinoides, Lampropeltis g. californae*, and *Geophis nasalis* manifest minimally variability, contrasted with *Crotalus v. viridis, Crotalus v. oreganus, Crotalus lucasensis*. Dorsal scale counts were highly variable in *Cnemidophorus tigris* and especially in *Cnemidophorus motaguae* in contrast to *Sceloporus undulatus, Sceloporus graciosus*, and *Uta stansburiana*. 59% of *Cnemidophorus* midbody scale count variability was attributable to bisexuality.

Kerfoot WC, Kluge AG. 1971. Impact of the lognormal distribution on studies of phenotypic variation and evolutionary rates. Systematic Zoology 20:459–464.

Congenital – Standard deviations of vertebral counts were proportion to the mean, a lognormal result, not a normal distribution in lizards and snakes.
Environmental – Cites temperature-shock-induced variation in vertebral number (Fox et al. 1991).

Kerr R. 1978. Translation The Natural History of Oviparous Quadrupeds and Serpents. Arranged and Published from the Papers and Collections of the Count de Buffon by the Count de la Cepede. Volume 4, 382 pp.
Congenital – Dicephalic snake, simply listed as "monstrous serpent with two heads" (facing p. 323 and illustrated on page with normal "worm-like snake").
Pseudopathology – Fable of two-headed *Amphisbaena*.

Keymer IF. 1978a. Disease of chelonians: (1) Necropsy survey of tortoises. The Veterinary Record December 23 and 30, 1978:577–582.
Infection – Notes salmonella infections in tortoises.
Osteopathy – "Miscellaneous osteopathies" were reported in 4.2% of turtles autopsied at Zoologic Society of London between 1965 and 1975, including one each of Greek or spur-thighed Mediterannean tortoise, *Testudo graeca*; starred tortoise, *Testudo elegans*; marginated tortoise, *Testudo marginata*; radiated tortoise, *Testudo radiata*; Home's hinged tortoise, *Kinixys homeana*; and Horsefield's tortoise, *Testudo horsfieldi*.
Metabolic – Nutritional osteodystrophies were found in 3%, including four Jaboty or rain forest tortoises *Testudo denticulata*, three Hermann's tortoises *Testudo hermanni*, two marginated tortoises *Testudo marginata*, one Bell's hinged tortoise *Kinixys belliana*, one Horsefield's tortoise *Testudo horsfieldi*, and one stared tortoise *Testudo elegans*.
Stones – "Cystic calculi were reported 4.2%, including two Greek or spur-thighed Mediterannean tortoises, *Testudo graeca*; one giant tortoise, *Geochelone gigantea*; one Hermann's tortoise, *Testudo hermanni*; one Kuhl's tortoise, *Kinixys homeana*; and one starred tortoise, *Testudo elegans*.

Keymer IF. 1978b. Disease of chelonians: (2) Necropsy survey of terrapins and turtles. The Veterinary Record December 16, 1978:548–552.
Infection – Fungus was reported as "common" in turtle, but without comment on whether carapace/plaston was affected.
Infection/Metabolic – 23 (19%) of 122 terrapins autopsied at Zoologic Society of London between 1965 and 1975 had "lesions of the skeletal system." Miscellaneous osteopathies (including nutritional and infectious) included two red-eared or elegant terrapins *Pseudemys scripta elegans*, two Spanish terrapins *Clemmys caspica leprosa*, one yellow-bellied or serrated terrapin *Pseudemys scripta scripta*, one Caspian turtle *Clemmys caspica rivulata*, one painted terrapin *Chrysemys picta picta*, one snapping turtle or alligator terrapin *Chelydra serpentina*, one three-keeled terrapin *Staurotypus triporcatus*, one blackish terrapin *Pelusios subniger*, and one Ouachita map terrapin *Graptemys pseudogeographica ouachitensis*.
Metabolic – Nutritional and metabolic disorders were reported in 19.7% of turtles, with actual nutritional osteodystrophies were listed in 9.8%, including six red-eared or elegant terrapins, *Pseudemys scripta elegans*; four Spanish terrapins, *Clemmys caspica leprosa*, one ornate terrapin *Pseudemys ornata callirostris*; and one Western Caspian turtle *Clemmys caspica rivulata*.
Suggested gout does not occur in terrapins, because of urinary excretion as urea and ammonia.
Stones – Cystic calculi reported as occurring in terrapins, without specifics.

Khaire A, Khaire N. 1984. Birth of a bicephalic snake. Hamadryad 9(3):7.
Congenital – Dicephalic *Vipera russelli*.

Khaire A, Khaire N. 1988. Occurrence of a house gecko (*Hemidactylus flaviridis*) with a bifid tail. Hamadryad 13:17–18.
Trauma – Bifid tail in house gecko *Hemidactylus flaviridis*.

Khosatzky LI. 1991. Двойной монстров рептилий. [A double monsters of reptiles.] Gerpetologicheskie Issledovaniya 1:164–170, 174. [Russian]
Congenital – Partial anterior duplications in *Pseudechis porphyriacus* and *Agkistrodon blomhoffi* (snakes) and *Chinemys reevesii* (turtle).

Kiel JL. 1977. Spinal osteoarthropathy in two southern copperheads. Journal of Zoo Animal Medicine 8:21–24.
Vertebral – Southern copperhead *Ancistrodon contortrix* with hard enlargements of cervical vertebrae and nodules on rib shafts. Exostoses, vertebral and costovertebral ankylosis. Etiology unclear.

Kiel JL. 1983. Paget's disease in snakes. Annual Proceedings of the American Association of Zoo Veterinarians 1983:201–207.
Infection – Southern copperhead with *Staphylococcus epidermidis* and *Pseudomonas fluorescens* (normal snake flora) with sclerotic vertebrae, called Paget's but probably infectious.
Vertebral – Southern copperhead with *Staphylococcus epidermidis* and *Pseudomonas fluorescens* (normal snake flora) with sclerotic vertebrae, called Paget's but probably infectious.

Kiesecker JM. 2002. Synergism between trematode infection and pesticide exposure: A link to amphibian limb deformities in nature? Proceedings of the National Academy of Science 99:9900–9904.
Environmental – Deformities (ectomelia, polymelia, and apodia most common, but ectodactyly and micromelia also present) more common in wood frogs *Rana sylvatica* exposed to trematode infections at sites adjacent to agricultural runoff (20–32% versus 5–8%), possibly by impairing immune system.

Killebrew FC. 1976. An unusual basisphenoid in one specimen of *Graptemys flavimaculata*. Herpetological Review 7(4):67–68.
Congenital – Unfused basisphenoid parasphenoid in one specimen of *Graptemys flavimaculata*.

Kim W-S, Stocum DL. 1986. Retinoic acid modifies positional memory in the anteroposterior axis of regenerating axolotl limbs. Developmental Biology 114:170–179.
Toxicology – Retinoic acid results in supernumerary regeneration in anurans but only rarely in mature urodeles. In *Ambystoma mexicanum*, metatarsal fusions often occur. Retinoic acid modifies memory of blastema cells in the AP axis in only one direction, posterior.

Kincel F. 1969. Ein Fall von Duplicitas anterior bei *Natrix natrix* L. [A case of duplicitas anterior in *Natrix natrix* L]. Jahrbuch der naturwissenschaftlichen Abteilung des Steiermärkischen Landesmuseums Joanneum, Graz 1969: 16–18 [German].
Congenital – Partial duplication of anterior head: third eye surrounded by orally open elevation, two frontals, two nasals, two prefrontals, two ethmoid bones, two septomaxillaria, two intermaxillaria; with other part of skull normal.
Referred to "Knaur's Tierreich in Farben" with picture of anterior duplication anterior in *Lampropeltis getula*.

King W. 1959. Vertebra duplication, an osteological anomaly widespread in snakes. Herpetologica 15:87–88.
Congenital – Hemi-duplication of vertebrae, with minimal length disturbance, unilateral duplicate accessory processes and two hypapophyses (but two ribs on one side; one, on the other) Two hypapophyses were present anomalously in the following species: *Coluber, Diadophis punctatus, Natrix natrix, Elaphe, Lampropeltis, Rhadinacea*, and *Thamnophis*, in contrast with extinct Paleophiidae where they are natural, although bilateral occurrences.
Vertebral – Hemi-duplication of vertebrae, with minimal length disturbance, unilateral duplicate accessory processes and two hypapophyses (but two ribs on one side; one, on the other) Two hypapophyses were present anomalously in the following species: *Coluber, Diadophis punctatus, Natrix natrix, Elaphe, Lampropeltis, Rhadinacea*, and *Thamnophis*, in contrast with extinct Paleophiidae where they are natural, although bilateral occurrences.
Fossil – Hypapophyses in extinct Paleophiidae are natural, although bilateral occurrences.

King OM. 1960. Observation on Oklahoma toads. The Southwestern Naturalist 5:102–103.
Congenital – Supernumerary front leg in *Bufo w. woodhousii*.

King REB. 1987. Color pattern polymorphism in the Lake Erie water snake, *Nerodia sipedon insularum*, Evolution 41:241–255.
Trauma – 17% of Lake Erie water snake *Nerodia sipedon insularum* had tail injury or loss.

King P, Heatwole H. 1994. Non-pulmonary respiratory surfaces of the chelid turtle *Elseya latisternum*. Herpetologica 50:262–265.
Vascular – The skin anterior to the rear leg of *Elseya latisternum* is heavily vascularized (sometimes even red). Buccopharyngeal breathing accounts for almost half of the non-pulmonary uptake by *Elseya latisternum*.
Cloacal bursae are responsible for a third of aquatic respiratory effect in *Elseya latisternum*.

Kingsley JS. 1878. Two-headed snakes. American Naturalist 12:694–695.
Congenital – Supernumerary limb in a *Rana*.
Pseudopathology – Two-necked "serpent," citing this as Amphisbaena from Guiana by "An essay on the Natural History of Guiana" by T. Becket and PA DeHondt in 1769.

Kingsley JS. 1880. An abnormal foot in *Ambystoma*. American Naturalist 14:594.
Congenital – Bifid foot in *Ambystoma punctatum*.

Kingsley JS. 1882. A case of polymely in the Batrachia. Proceedings of the Boston Society of Natural History 29:169–175.
Congential – No batrachian with supernumerary limbs in the Museum of Natural History in Paris. Gervais (1864) reported polymelia in *Pelobates cultripes*, Duméril (1865) in *Rana viridis, Rana temporaria*, and *Rana clumata*. Cisternas (1865) reported supernumerary limb in *Alytes obstetricans*, Balsamo-Crivelli (1865) in *Rana esculenta* and Lunel, in the common green frog of Europe *Rana halecina* at the Lyceum of Natural History of Williams College, Williamstown, Mass.

Kirchoff H, Mohan K, Schmidt R, Rung M, BrownDR, Brown MB, Foggin CM, Muvavariwa P, Lehman H, Flossdorf J. 1997. *Mycoplasma crocodyli* sp. nov., a new species from crocodiles. International Journal of Systemic Bacteriology 47:742–746.
Infection – *Mycoplasma crocodyli* exudative polyarthritis in *Crocodylus niloticus*.

Kirkland W. 1871. Double-headed snakes. (*Regina leberis, Tropidonotus septemvittatus*) American Naturalist 4:375.
Congenital – Dicephalic Regina leberis and Tropidonotus septemvittatus.

Kirsche, W. 1972. Über Panzeranomalien bei Landschildkröten. [On anomalies of the shell in tortoises]. Aquarien Terrarien 19:259–261. [German]
Congenital – Additional horny plates on tortoise shells.

Kirsche, W. 1982. Über einen besonderen Typus von Carapax-Mißbildungen aus drei Gelegen (1979–1981) einer *Testudo hermanni hermanni* GMELIN. [On a special type of carapace malforation of three in a *Testudo hermanni*

hermanni GMELIN]. In Vago C., Matz G. eds. Comptes Rendus du Premier Colloque international de Pathologie des Reptiles et des Amphibiens 29.9.-2.10.1982 – Angers. Pp. 239–244. [German]

Congenital – Horny scute anomalies in *Testudo hermanni hermanni.*

Klaphake E. 2010. A fresh look at metabolic bone diseases in reptiles and amphibians. Veterinary Clinics of North America: Exotic Animal Practice 13:375–392.

Metabolic – Metabolic bone disease in green iguanas *Iguana iguana*, bearded dragons *Pogona vitticeps*, Asian water dragons *Physignathus cocincinus*, chameleons, leopard geckos *Eublepharis macularius*, Uromastyx *Uromastyx, Testudo, Geochelone*, manifested as fractures, swelling, rubber jaw, stunted growth, scoliosis, abnormal shell growth.

Blood parameters related to bone were preserved across orders and classes.

Klauber LM. 1941. The correlation between scalation and life zones in San Diego County snakes. Bulletin of the Zoological Society of San Diego 17:73–80.

Environmental – Higher number of ventral scales *Salvadora grahamiae hexalepis, Salvadora grahamiae virgultea, Arizona elegans occidentalis, Pituophis catenifer deserticola, Pituophis catenifer annectens, Lampropeltis getula californiae, Rhinocheilus lecontei, Hypsiglena ochrorhynchus, Tantilla eiseni, Trimorphodon vandenburghi,* and *Crotalus ruber* in desert than in more humid "cismontane" region. Desert rosy boa *Lichanura roseofusca (trivirgata) roseofusca, Coluber flagellum frenatus,* and *Crotalus mitchellii pyrrhus* did not show that pattern. *Salvadora grahamiae hexalepis, Salvadora grahamiae virgultea, Lampropeltis getulus californiae, Rhinocheilus lecontei* and *Hypsiglena ochrorhynchus* had more caudals in desert than in the Cismontane region of San Diego County. Labials were constant, except for more scales in coastal Trimorphodon vandenburghi.

Klauber LM. 1972 Rattlesnakes. Their Habits, Life Histories and Influence on Mankind. 1536pp., Berkeley a. Los Angeles: University California Press.

Congenital – *Crotalus adamanteus*, citing similar cases in The two headed snake by Lovering 1978, p 194, Denamalibus ch 60; Johnstonus 1657, Nakamura 1938, Cunninghan 1937; Johnson 1902), reports of two headed rattlesnake (Anon. 1877, Baird 1856; Michler 1857), South American rattler *C. d. terrificus* (Do Amaral 1927, Vanzolin 1947), *C. b. basiliscus* from Mexican west coast (Wiley 1930), *C. h. horridus*: Rimkus 1947) and *Lampropeltis doliata*: Hyde 1925

Klauber LM 1982 Rattlesnakes. Their Habits, Life Histories and Influence on Mankind. Berkeley, University of California Press, 339 pp

Congenital – Dicephalic *C. h. horridus* and king snakes.

Klausewitz W. 1952. Doppelseitige Polydactylie bei einem Männchen von *Rana esculenta*. [Polydactyly of a male *Rana esculenta*]. Senckenbergiana 33(4/6):247–251. [German]

Congenital – (additional toe) formation on 1st toe of both hind limbs in two *Rana esculenta*

Klaussner F. 1890. Mehrfachbildungen bei Wirbeltieren. Eine teratologische Studie. [More formations of vertebrates: A teratological study]. 71 pp.; München: M. Rieger'sche Universitäts-Buchhandlung. [German]

Congenital – Anterior and posterior double formations in one embryo and double body and head with single tail in another embryo of *Salamandra maculata, Tropidonotus natrix,* and *Lacerta viridis.*

Cites Braun (1876) for dicephalic *Salamandra maculata* and of *Tropidonotus natrix* and Born (1882) for double embryos of *Rana esculenta* (double head, double head and body with single tail, fusion of notochord in middle region), and

Trauma – Cites Bruch (1866) with double to triple tail in *Pelobates fuscus.*

Kleinberg E. 1994. House of refuge last of its kind. Palm Beach Post (Florida) 12 June 1994; TRAVEL: 1.

Congenital – Dicephalic turtle.

Klembara J. 1995. Some cases of fused and concrescent exocracranial bones in the Lower Permian Seymouriamorph tetrapod *Discosauriscus* Kuhn, 1933. Geobios 19:263–267.

Congenital – Fused exocranial bones in Lower Permian Seymouriamorph tetrapod *Discosauriscus.*

Fossil – Fused exocranial bones in Lower Permian Seymouriamorph tetrapod *Discosauriscus.*

Klewen R. 1985. Untersuchungen zur Ökologie und Populationsbiologie des Feuersalamanders (*Salamandra salamandra terrestris* Lacépède 1788) an einer isolierten Population im Kreise Paderborn. [Investigations of the ecology and population biology of fire salamanders (*Salamandra salamandra terrestris* Lacépède 1788) in an isolated population in the country]. Abhandlungen des westfälischen Museums für Naturkunde 47:1–51. [German]

Trauma – Author describes healing process of tail and leg of *Salamandra salamandra* after damage by car, figures additional hands (three hands on left front leg).

Klös HG, Lang, EM 1976. Zootierkrankheiten. 365pp.; Berlin, Hamburg: Verlag Paul Parey. [German]

See Frank, 1976

Knoll CM. 1935. Shield variation, reduction, and age in a box terrapin, *Terrapene carolina.* Copeia 1935:100.

Congenital – Supernumerary scutes in box terrapin, *Terrapene carolina.*

Knox DW. 1980. Gout in reptiles and birds, with observations on a comparable syndrome in man. In: Montali RJ, Migaki G. eds. The Comparative Pathology of Zoo Animals. Smithsonian Institution Press, Washington DC. Pp. 137–141.

Metabolic – Uses human model for gout and notes induction of gout by hypovitaminosis A and renal disease, especially in crocodilians.
Kobrin M. 1945. The regenerated tail of *Anolis carolinensis* (Cuvier). Masters Thesis, Department of Zoology, Tulane University, 45 pp.
Trauma – *Anolis carolinensis* tail regeneration completed in 2–1/4 – 3 months.
Autotomy in *Hemidactylus* (Woodland 1920), *Lacerta* (White 1925), *Mabaya* (Sibtain 1938) and *Sphenodon* (Ali 1941).
Koenig R. 2008. Sanctuaries aim to preserve a model organism's wild type. Science 322:1456–1457.
Trauma – Tail regeneration in axolotl *Ambystoma mexicanum.*
Köhler G. 1992. Asymetrische zusammenhängende Doppelmissbildung (Duplicitas incompleta) bei *Sceloporus cyanogenys*. [Unsymmetric related congenital defects (Duplicatas incompleta) in *Sceloporus cyanogenys.*] Iguana Rundschreiben 1992(2):35–41. [German]
Congenital – Incomplete Siamese twins of *Sceloporus cyanogenys*, one specimen without head and tail; born, but not vital.
Köhler G. 2006. Diseases of Amphibians and Reptiles. Krieger Publishing Company; Malabar, Florida, 171 pp.
Congenital – List of publications and illustration of Siamese twinning in *Sceloporus cyanogenys.*
Trauma – Fractured carapace and plastron in *Gopherus polyphemus.*
Infection – Vertebral osteomyelitis in *Ctenosaura similis.*
Metabolic disease – Fibrous osteodystrophy in *Ctenosaura similis, Cyclura, Iguana*, and turtles.
Rickets in *Gekko gekko.*
Rickets and hypovitaminosis A with soft-shell in *Trachyemys scripta.*
Gout in *Uromastyx acanthinura.*
Vertebral fusion in *Boa constrictor* attributed to Paget's disease. Infection or even spondyloarthropathy more likely.
Arthritis – Arthrogryposis in frogs, but not salamaders.
Vascular – Tail necrosis in *Iguana, Ctenosaura, Cyclura* and *Tupinambis*.
Vertebral – "Spine deformity" in *Iguana iguana.*
Vertebral osteomyelitis in *Ctenosaura similis.*
Vertebral fusion in *Boa constrictor* attributed to Paget's disease. Infection or even spondyloarthropathy more likely.
Stones – Stones in anal sacs of *Iguana iguana* and *Trionyx triunguis*, noting contribution of oxalates, carbonates, and phosphates.
4.5 cm bladder stone in *Sauromalus varius* (Frye 1983).
65 gram, 8 by 5.8 by 3 cm stone in *Ctenosaura acanthura*, composed of struvite, xanthine, weddellite, and carbonate apatite, in animal with snout-vent length of 60 cm (Isenbügel and Frank 1985).
Ctenosaura similis with 3.4 by 2.6 by 1.6 cm urate stone (Kohler 1992b).
Shell disease – Plastron necrosis in *Chrysemys picta* and *Agrionemys horsfieldii.*
Carapace pyramiding in European tortoise.
Kolbe JJ, Losos JB. 2005. Hind-limb plasticity in *Anolis carolinensis*. Journal of Herpetology 39:674–678.
Environmental – Plasticity of hind-limb length of *Anolis sagrei* and *Anolis carolinensis* determined by size of perches in cages.
Kolbow H. 1928. Experimentel verursachte Bildung von Armen aus ursprünglichem Beinmaterial bei *Triton.* [Experimental true formation of front legs of original leg material in *Triton*]. Roux's Archiv für Entwicklungsmechanik 113:12–38. [German]
Trauma – Artificially introduced duplication and triplification of right hind leg in *Triton vulgaris.*
Kölle P, Hoffmann R. 2002. Incidence of nephropathies in European tortoises. Proceedings of the Association of Amphibian and Reptile Veterinarians Annual Conference 2002:33–35.
Metabolic – 64% of European tortoises (especially spur-thighed tortoises – as more sensitive to high protein diet than Afghan tortoises) had "renal alterations" on 1977 claimed that uric acid/urate calculi only occur in carnivorous species.
Stones – 16% of European tortoises had gouty tophi in the kidney.
Kölle P, Hoffman R, Wolters M, Hesse A. 2001. Cystic calculi in reptiles. Proceedings of the Association of Reptile and Amphibian Veterinarians 2001:190–192.
Stones – Present in 4% of necropsied tortoises and lizards. *Testudo, Geochelone*, and *Iguana iguana* are most often affected. All except one calculus were in a herbivorous reptile. But, Grünberg et al. D (Wallach 1969, 1971) or A deficiency (Mebs 1976) and to protein (Frye 1991) or oxalic acid rich diet (Spinach/wharf) (Eggenschwiler 2000), with necrotic cells as core (Keymer 1978).
Half contained two or more substances:
Uric acid dehydrate 18%
Ammonium urate 14%
Sodium urea 12%

Calcium urate 6%
Potassium urate 2%
Combination of urates 16%
Struvite, carbon apatite, and calcium phosphate rarely found.
Confused literature recomposition:
Marcus 1983 – 98% calcium phosphate
Eggenschwiler 2000 – urate and/or oxalic acid
Frye 1991 – uric acid and urates

Korschelt E. 1927a. Über geheilte Knochenbrüche bei einigen Wirbeltieren. [On healed bone fractures in some vertebrae]. Sitzungsberichte der Gesellschaft für Naturwissenschaften in Marburg 62:1–28. [German]

Congenital – Left costal plates of carapace of European pond turtle *Emys orbicularis* with 2.5 cm long opening with bony callus on inner side; healed break of femur (strong callus formation) in *Iguana*; healed broken pelvis of bullfrog *Rana mugiens*; healed (thick callus formation) broken right lower hind leg of *Rana temporaria*; and healed femur of Surinam toad *Pipa americana*.

Korschelt E. 1927b. Regeneration und Transplantation. I. Band: Regeneration. [Regeneration and Transplantation. I. Tome: Regeneration]. 818pp.; Berlin:Gebrüder Bornträger. [German]

Trauma – New formation of bone in salamanders and triton. X-rays of *Triton cristatus* show partial regeneration of bones in forearm. *Triton taeniatus* with regeneration of orbit. Polydactyly in urodele (external and internal views) and in *Ambystoma mexicanum*, polymelia in *Pelobates fuscus*. Duplication of left hind leg in *Rana* temporary, of front legs in *Triton taeniatus, T. cristatus*, and *Rana esculenta* and of tail in *Lacerta muralis*, triplication of tail in *Lacerta agilis*.

Korschelt E. 1931. Weitere Beobachtungen an geheilten Knochenbrüchen wildlebender Tiere. [Additional observations on healed bone fractures of wild living animals]. Sitzungsberichte der naturforschenden Gesellschaft Marburg 66 (Heft 5): 111–129. [German]

Trauma – Fractures and callus formation of extremities in *Caiman crocodilus* (humerus and left tibia), lizard *Calotes versicolor* (femur), *Iguana* sp. (femur), and *Crocodylus niloticus* (right tibia and fibula). Fractures and callus formation of ribs in *Varanus ocellatus*.

Fractures were frequent in snakes (*Boa constrictor* (without subspecies, common name cannot be determined)), anaconda *Eunectes murinus*, Burmese, Indian, or Ceylon python *Python molurus*, reticulated or regal python *Python reticulatus*).

Fractures and callus formation in frogs (humerus, femur, tibia, fibula), include *Rana mugiens* with double broken pelvis and *Rana temporaria* with broken and healed ischium.

Korschelt E. 1932. Über Frakturen und Skelettanomalien der Wirbeltiere. Zweiter Teil: Vögel, Reptilien, Amphibien und Fische. [On fractures and skeletal anomalies of vertebrates. Second part: Birds, reptiles, amphibians and fish]. Beiträge zur pathologischen Anatomie und zur allgemeinen Pathologie 89:668–717. [German]

Trauma – Anguis fragilis, *Hydrosaurus varanus, Leptopterygius trigonodon* (Huene 1930), *Dimetrodon* sp. (Moodie 1923), *Varanus niloticus, V. ocellatus* and V. sp. with healed broken ribs, callus formation. Rib fractures in *Cylindrophis rufus, Sinocephalus poensis, Tretanorhinus variabilis, Stenorhina degenhardti, Tropidonotus natrix, Python molurus, Python tigris, Python reticulatus, Eunectes murinus, Boa constrictor, Dipsadoboa unicolor, Zamenis mucosus, Herpetodryas carinatus, Elaps frontalis, Acanthophis antarcticus, Enhydris hardwickii, Naja bungarus, Causus rhombeatus, Crotalus horridus*.

Fractured humerus in *Rana temporaria, Caiman crocodilus* with callus formation.

Fractured radius *Dimetrodon* sp. radius (Moodie 1923).

Fractured femur in Ceylon lizard *Calotes versicolor* and *Rana temporaria* with callus formation (Westenhöfer (1926) and in *Iguana* sp. which healed as "dislocated" with large callus.

Fractured tibia in *Caiman crocodilus* with callus formation and *Calyptocephalus gayi* (Chile).

Fractured fibula in *Edaphosaurus cruciger* (Moodie 1923).

Fractured tibia and fibula in *Rana temporaria* and *Crocodylus niloticus* with callus formation.

Fractured carapace in turtles: *Clemmys leprosa, Emys orbicularis, Testudo graeca*, and *Chelone mydas*.

Multiple fractures – *Metriorhynchus* sp. reduction of articulation head of left femur, enlargement of trochanter with articulation areas of tibia and fibula torn and distorted (Auer 1909).

Fractured ischium and vertebrae (with fusion) in *Rana temporaria*.

Fossil – Fractured radius *Dimetrodon* sp. radius (Moodie 1923).

Fractured fibula in *Edaphosaurus cruciger* (Moodie 1923).

Multiple fractures – *Metriorhynchus* sp. reduction of articulation head of left femur, enlargement of trochanter with articulation areas of tibia and fibula "torn" and distorted (Auer 1909).

Korschelt E, Stock H. 1928. Geheilte Knochenbrüche bei wildlebenden und in der Gefangenschaft gehaltenen Tieren. [Healed bone fractures in wild living and animals in cages]. 176pp., Berlin:Gebrüder Bornträger. [German]

Trauma – *Caiman* sp. with many biting injuries healed by cartilaginous fillings and periostal callus formation.

Wide split in costal plates of *Emys orbicularis* L. filled with bone tissue.

Femur fracture in *Iguana* sp. and *Pipa americana* with callus.

Iliac fracture with callus formation in *Rana mugiens* and *Rana temporaria*.

Koskela P. 1974. Combination of partial adactylism and syndactylism in *Rana temporaria* L. Aquilo, series Zoologie 15:37–38.

Congenital – Absent (partial adactylism) digit five with conjoined (syndactylism) digits III and IV in *Rana temporaria*. One *Rana temporaria* among 10,000 from Northern Finland with partial adactylism and syndactylism.

Koussoulakos S, Sharma KK, Anton HJ. 1988. Vitamin A induced bilateral asymmetries in triturus forelimb regenerates. Biological Structures and Morphogenesis 1:43–48.

Toxicology – Asynchronous regeneration of amputated forelimbs in vitamin A treated *Triturus alpestris*.

Kovalenko EE. 1995. On some sacral anomalies in laboratory common platanna (*Xenopus laevis*). Russian Journal of Herpetology 1:170–173.

Congenital – Platanna *Xenopus laevis* with segmentation disorders. Variable presacral vertebrae and fusion with urostyle, scoliosis, unilateral sacral diapophysis, and non-widened diapophytes.

Kovalenko EE. 2000. !!!!!!! !!!!!!! !!!!!!!!!! ! !!!!!! ! !!!!. [Mass anomalies of extremities in Anura.] Zhurnal Obshchei Biologii 61:412–427. [Russian.]

Congenital – Mass anomalies in frog *Xenopus laevis*, anomalies involve hind limbs and/or forelimbs, and their girdles, present in up to 50 and 100 percent of offspring). The first year when anomalies where registered, it was the syndrome of unmotile forelimbs (SUF); two years later in some of the subjects who also had SUF another anomaly of hind limbs was registered (syndrome of pelvis and hind limbs anomalies SPHA), and later most of the offspring had both SUF and SPHA. Both syndromes are polymorphic and interconnected, and form morphological lines so that between norm and full-blown deformity there is a number of intermediate variants. Also it noted that SPHA exists only in subjects with SUF. Full-blown SPHA may be described in conventional terms as "ectrodactilia and phocomelia," but those terms do not describe intermediate variants that have great number of characteristics in different combinations. For example, one of the first registered variants of SPHA consisted of absent or so underdeveloped that hind limbs did not attached to it. Hind limbs were pointing not to the sides, but straight back and had limited motility (nonflexible limbs).

Cited "right-sided polymelia" (extra light hind leg) in *Rana ridibunda* (Voitkevich 1955); inherited polydactyly in *Bufo bufo* (Rostand 1949); and non-inherited? polydactyly in *Rana esculenta* (Dubois 1979)

Kovalenko EE, Danilevskaya OE. 1991. [Experimental receive of abnormal vertebral column of Anura.] Vestnik Leningradskogo Universiteta Biologiya 1991(2):11–23, 116. [Russian]

Environmental – Frequency of *Rana temporaria* embryo abnormalities increases with density of population.

Kovalenko EE, Danilevskaya SE. 1994. On unique forms of anomalous sacral structure in tailless amphibians. Russian Journal of Herpetology 1:30–36.

Congenital – *Xenopus laevis* and *Rana temporaria* with vertebral asymmetries from homeobox disorders. *Rana temporaria* with amelia, scoliosis, short stump with only humerus, rudimentary scapula, and hemi-development of vertebrae. *Xenopus laevis* missing leg and half of pelvis.

Kovalenko EE, Kovalenko YI. 1996. Certain pelvic and sacral anomalies in anura. Russian Journal of Herpetology 3:172–177.

Congenital – Segmentation disturbances in ground toad *Bufo bufo*. Asymmetry, acetabular or sacral absence with or without limb absence or two sacral diapophysis with supernumerary limbs.

Kramer T. 1980. Duplo. [Duplication]. De Schildpad 6(4):10–15. [Dutch]

Congenital – Dicephalic Terekay turtle *Podocnemis unifilis*.

Kramer MH. 2006. Granulomatous osteomyelitis associated with atypical mycobacteriosis in a bearded dragon (*Pogona vitticeps*). Veterinary Clinics of Exotic Animals 9:563–568.

Infection – Atypical (non-tubercular) mycobacterial infection in a bearded dragon *Pogona vitticeps* produced stifle (knee) destruction with sclerosis and new bone formation and periostisis. Histology revealed granulomatous osteomyelitis and periostitis.

Krintler K. 1988. Beobachtungen zum problem der 'Steicholzbein' bei Dendrobatiden und Hyliden. [Observation to the problem of spindley leg in dendrobatids and hylids]. Herpetofauna 10(32):30–31. [German]

Metabolic – Spindley legs in captivity rised *Dendrobates leucomelas, D. auratus*, and *D. azureus* related to food supply.

Kritzer J. 2002. Summer fun. News & Record (Greensboro, NC) 6 June 2002:1.

Congenital – Dicephalic yellow belly turtle.

Kroenlein KR, Zimmerman KL,, Saunders G, Holladay SD. 2011. Serum vitamin D levels and skeletal and general development of young bearded dragon lizards (*Pogona vitticeps*), under different conditions of UV-B radiation exposure. Journal of Animal and Veterinary Advances 10:229–234.

Environmental – Bone disorder in *Pogona vitticeps* that "does not fall squarely into previously described conditions" (page 233). They suggested a nutritional deficiency or abnormal calcium absorption, rather than vitamin D deficiency, blaming inadequate UV-B exposure – as all animals had some degree of osteodystrophy.

Krysko KL, Eady BE, Abdelfattah KR. 2001. The occurrence of a malformed smallmouth salamander (*Ambystoma texanum*) from Indiana. Bulletin of the Maryland Herpetological Society 37:22–24.

Congenital – Smallmouth salamander *Ambystoma texanum* with supernumerary forelimb.

Malformations in *Bufo* sp., southern leopard frog *Rana sphenocephala*, American toad *Bufo americanus*, Blanchard's cricket frog *Acris crepitans blanchardi*, chorus frog *Pseudacris triseriata*, northern leopard frog *Rana pipiens*, Fowler's toad *Bufo woodhousii fowleri*, green frog *Rana clamitans melanota*, cave salamander *Eurycea lucifuga*, Jefferson salamander *Ambystoma jeffersonianum*, longtail salamander *Eurycea longicauda*, bullfrog *Rana catesbeiana*, and smallmouth salamander *Ambystoma texanum* from Indiana.

Krysko KL, Sheehy CM, Decker JN. 2002. Cephalic dichotomy in the mangrove salt marsh snake, *Nerodia clarkii compressicauda* (Colubridae: Natricinae). Bulletin of the Maryland Herpetological Society 38:86–87.

Congenital – Dicephalic marsh snake *Nerodia clarkii compressicauda*.

Kuch U, Schnitzler R. 1988. Ein Fall von Mißbildung bei einer *Boiga dendrophila melanota* (Boulenger 1898). [A case of malformation in a *Boiga dendrophila melanota* (Boulenger 1898)]. Sauria 10(4):25–26. [German]

Trauma – *Boiga dendrophila melanota* with missing tail.

Kuhn G, Gruber P, Müller R, Rühli F. 2008. Three-dimensional evaluation of structures in small bones by micro-CT: Tail fracture planes of autotomizing lizards (Scincidae and Gecconidae families). Internet Journal of Biological Anthropology 1(2): Published online.

Trauma – Autotomy = self amputation related to fracture planes between vertebrae (intervertebral) in some salamanders (Wake and Dresner 1967); autotomy through verebral body and neural arch (transvertebral) in others (Bellairs and Bryant 1965), noting positions vary with lizard family:

Bassiana duperreyi (Scincidae) manifests autotomy through transverse processes; *Christinus marmoratus* (Gekkonidae) manifests autotomy distal to transverse processes.

Kühne K. 1913. Über die Variationen der Wirbelsäule und der Extremitätenplexus bei *Lacerta viridis* Gessn. und *Lacerta agilis* Linn. [On the variations of the spine and extremity plexus of *Lacerta viridis* Gessn. and *Lacerta agilis* Linn.]. Gegenbaurs Morphologisches Jahrbuch 46:563–592. [German]

Congenital – 51 *Lacerta viridis* with variation of sacral position from 28–29, 29–30, and 30–31, with variable lumbar transverse process sacralization or "caudalization" of sacral vertebra, citing sacral variation in *Salamandra maculosa* (Claus 1876, p. 809) and *Necturus* (Parker 1896, pp. 712–713). Last rib was 21 in 85% and 22, in 17%.

15 examples of *Lacerta agilis* with variation of sacral postion from 26–27, 26–27 (with unilateral 28), 27–28, 28–29, 20–30, and 30–31. Last ribs were on vertebra 20 in 67% and on 21 in 20%.

Kühne K. 1915. Über die Variationen der Wirbelsäule, des Brustkorbes und des Extremitätenplexus bei *Lacerta muralis* Dum. u. Bibr. und *Lacerta vivipara* Jacq. [On the variations of the spine, thorax and extremity plexus of *Lacerta muralis* Dum. And Bibr. and *Lacerta vivipara* Jacq.]. Gegenbaurs Morphologisches Jahrbuch 49:407–507. [German]

Congenital – *Lacerta muralis* missing a cervical rib or with variation in number of ribs (three or four on a side) attached to sternum. Variation in sacral position from 27–28 (66.7%), 28–29 (23.5%), 29–30 (22.2%), 30–31 (11.6%), and 31–32 (1%), with variable lumbar transverse process sacralization or "caudalization" of sacral vertebra. Additionally, sacra were sometimes asymmetrically formed of different vertebra. Last ribs were 19 in 97.5%, 20th in 2.5%.

Lacerta vivipara with variation in number of ribs (three or four on a side) attached to sternum. Variation in sacral position from 27–28 (3%), 28–29 (61%), 29–30 (20%), and 30–31 (16%), with variable unilateral sacralization and lumbarization of adjacent vertebrae.

Kumar A, Godwin JW, Gates PB, Garza-Garcia AA, Brockes JP. 2007. Molecular basis for the nerve dependence of limb regeneration in an adult vertebrate. Science 318:772–777.

Trauma – Limb regeneration in *Xenopus* dependent upon relevant nerve remaining intact.

Kuramoto M. 1974. Experimental hybridization between the Brown Frogs of Taiwan, the Ryukyu Islands and Japan. Copeia 1974(4):815–822.

Congenital – Malformed tails in crosses between *Rana okinavana* and *Rana japonica* and also between *Rana okinavana* and *Rana ornativentris*.

Kuroda N. 1919. *Elaphe rufodorsata* from Korea (red-backed ratsnake; "Se-aka-hebi" in Japanese) Dobutsugakuzasshi 31:255. [Japanese]

Congenital – Dicephalic red-backed ratsnake *Elaphe rufodorsata* from Korea.

Kusrini MD, Alford RA, Fitri A, Nasir DM, Rahardyansah S. 2004. Morphological abnormalities of frogs of West Java, Indonesia. Newsletter of the Declining Amphibian Populations Task Force, Froglog 64:1–4.

Congenital – Amelia, micromelia, brachymelia, hemimelia, ectomelia, taumelia, ectodactyly (12%), brachydactyly (9.8%), syndactyly (4.9%), polydactyly, clinodactyly, and scoliosis in West Java, Indonesia: In rice field habitats, 3.6% of 2695 *Fejervarya limnocharis*, 5.26% of 323 *Fejervarya cancrivora*, 0.85% of 117 *Occidozyga lima*, and one of one *Limnonectes kuhlii*. From stream habitats, 8.4% of 262 *Limnonectes kuhlii*, 8.02% of 162 *Bufo asper*, 2.23% of 269 *Rana chalconota*, 0.43% of 235 *Rana hosii*, 5% of 18 *Bufo melanostictus*, 5% of 18 *Rhacophorus javanus*, and 10% of 10 *Fejervarya limnocharis*, 12.2% ectodactyly, 9.8% brachydactyly, and 4.9% syndactyly.

Trauma – In rice field habitats, one third of pathology was related to trauma. From stream habitats, half of pathology was due to trauma – amputation 2.4%, fractured limb 2.4% in stream habitats.

Kutscherowa FN. 1948. Индукция дополнительные конечности у обыкновенной жабы эмбриона. [Induktion einer zusätzlichen Extremität beim Embryo *Bufo vulgaris*.] Doklady Akademija Nauk SSSR 59(4):805–808. [Russian]

Toxicology – Induction of supernumerary forelimb in *Bufo vulgaris*.

Kuwano H. 1902. On a double-headed tortoise. Dobutsugakuzasschi 14:3. [Japanese]

Congenital – Dicephalic *Chrysemys picta*.

Labanick GM. 1984. Anti-predator effectiveness of autotomized tails of the salamander *Desmognathus ochrophaeus*. Herpetologica 40:110–118.

Trauma – Autotomy increased escape for *Desmognathus ochrophaeus*.

Lacépède 1790(1787–88). Histoire naturelle des quadrupèdes ovipares et des serpents. [Natural history of quadrupedal lizards and snakes] 1. Edition. 432 pp.; Paris [French]

Trauma – Double and triple tails in lizards.

Congenital – Dicephalic snake *Lachesis lanceolatus* (p. 482).

Wrote that some travelers of Louisiana assured him that two-headed snakes formed a fixed species that reproduced regularly like other species of snakes. He observed that this is not admissible, and that no traveler had observed a two-headed snake set down two-headed snakes or eggs with two-headed fetuses, but that the frequency of this phenomenon may be dependant on climate. After citing the cases of Aristotle, Aelianus, Aldrovandi, and Lanzoni, he described the trigonocephalus with two heads in the Museo del Re in Parigi. This specimen (of unknown provenance) presents the monstrosity known as atloidic dicephalism: two heads on one single neck, a less common monstrosity than derodymous He also remarks in page 391, that many ancient observations about two-headed snakes referred to amphisbaena, whose enlarged terminal portion of the tail was mistaken for a second head for a long time.

La Clair JJ, Bantle JA, Dumont J. 1998. Photoproducts and metabolites of a common insect growth regulator produce developmental deformities in *Xenopus*. Environmental Science and Technology 32:1453–1461.

Environmental – Deformed North American bullfrogs *Rana catesbeiana*, American green frogs *Rana clamitans*, gray tree frogs *Hyla versicolor*, North American leopard frogs *Rana pipiens*, mink frogs *Rana septentrionalis*, wood frogs *Rana sylvatica*, spring peepers *Hyla crucifer*, Pacific tree frogs *Hyla regilla*, long-toed salamanders *Ambystoma macrodactylum*, and spotted salamander *Ambystoma maculatum* including supernumerary limbs, ectromelia, ectrodactyly, missing limbs, deformed jaws, cranial and facial defects increased near pesticide (oxamyl, dieldrin, malathion, lindane, paraquat)-exposed areas. S-methoprene (mimics juvenile hormone) reacts with environment (olefinic bonds excited to singlet excited state) to interfere with amphibian development – *Xenopus laevis*.

Lada GA. 1999. Polydactyly in anurans in the Tambov Region (Russia). Russian Journal of Herpetology 6:104–106.

Congenital – Isolated hind foot polydactyly in moor frog *Rana arvalis* (six toes), marsh frog *Rana ridibunda* (eight toes), *Bufo bufo*, 41% with seven toes in 1990–1991 (but much less since), Latvian common frog *Rana temporaria*, pool frog *Rana lessonae*, edible frog *Rana esculenta, Rana ridibunda, Rana arvalis*, Japanese tree frog *Hyla arborea*, Asiatic toad *Bufo gargarizans*, Asiatic glass frog *Rana chensinensis*, and Siberian wood frog *Rana amurensis*.

Ladds PW, Mangunwirjo H, Sebayang D, Daniels PW. 1995. Diseases in young farmed crocodiles in Irian Jaya. Veterinary Record 136:121–124.

Metabolic – Cites urate crystals in periarticular gout in crocodiles by Buenviaje et al. 1994.

Ladeiro JM. 1935. Um caso de atlodimia no *Tropidonotus natrix* Linn. [A case of atlodimia in *Tropidonotus natrix* Linn.]. Memórias e Estudos do Museu Zoológico da Universidade de Coimbra 4(2):1–4. [Portuguese]

Congenital – Derodymous *Tropidonotus natrix* snake. Ladeiro classified it as monosomianous altodimous.

LaFortune MJ, Wellehan FX, Heard DJ, Rooney DE, Fiorello CV, Jacobson ER. 2005a. Vacuum assisted closure (Turtle VAC) in the management of traumatic shell defects in chelonians. Journal of Herpetological Medicine and Surgery 15:4–8.

Shell disease – Carapace fractures in *Pseudemys concinna* and *Terrapene carolina*.

LaFortune MJ, Wellehan FX, Terrell SP, Jacobson ER, Heard DJ, Kimbrough JW. 2005b. Shell and systemic hyalohyphomycosis in Fly River turtles (*Carettochelys insculpta*) caused by *Paecilomyces lilacinus*. Journal of Herpetological Medicine and Surgery 15:15–19.

Infection – *Paecilomyces lilacinus* fungal shell erosions in Fly River turtle *Carettochelys insculpta*.

Shell disease – Circular 0.5–3.0 mm circular carapace lesions in *Carettochelys insculpta* caused by *Paecilomyces lilacinus*.

Noted *Paecilomyces lilacinus* is closely related to *Penicillium* and possible misdiagnosis as *Penicillium* suggested for reports claim *Penicillium*.

Lagerlund O, Hanstrom B. 1961. Huggorm med två huvud. [Snake with two heads]. Fauna och Flora 46:175. [Dutch]

Congenital – Dicephalic *Vipera berus*.

Lambiris AJL. 1973. Apparent syndactyly in *Microsaura ventralis*. Journal of the Herpetological Association of Africa 10:9–11.
Congenital – Syndactyly in *Microsaura ventralis*.
Lambiris AJL. 1979. Surgery on a captive salamander. British Journal of Herpetology 5:843–844.
Infection – Contemporary *Salamandra* sp. with ulcerated mandibular lesion attributed to fungal infection.
Lambiris AJ. 1982. Syndactyly in *Bufo maculatus* Hallowell (Amphibia: Anuria: Bufonidae). South African Journal of Zoology 17:122–125.
Congenital – Syndactyly in *Bufo maculatus*.
Lande R. 1978. Evolutionary mechanisms of limb loss in tetrapods. Evolution 32:73–92.
Congenital – 1/53 *Bachia dorbignyi* and 1/37 *B. peruana* had missing toe. 2/10 *panoplia* had only three toes on hind feet. 1/63 *B. intermedia* had four digits; 3/216, four toes, and one, an extra toe. *B. barbouri* had extra or missing digit.
Landois H. 1882. Über eine zweiköpfige Schlange [On a bicephalic snake]. Jahresbericht des Westfälischen Provinzial-Vereins für Wissenschaft und Kunst 11:12–13. [German]
Congenital – Dicephalic snake from North America and refers to Porta's 1650 report in a viper, Dorner's 1873/4 snake and Hennig's 1869 lizard.
Trauma – Cited Porta's 1650 report of lizards with two or three tails.
Landoïs H. 1884. Ein sechsbeiniger Molch. [A six-legged urodele]. Zoologischer Garten 25:94. [German]
Congenital – Supernumerary legs in front of the right hind leg in *Triton taeniatus*.
Lane TJ. 1996. Crocodilians. In Mader DR. ed. Reptile Medicine and Surgery. Pp. 336–340; Philadelphia: Saunders.
Metabolic – Metabolic bone disease and gout, citing Cooper and Jackson, 1981.
Lane TJ, Boyce WM, Reinhard MK, Larsen RE, Poulos PW, King MM, Buergelt CD, Cardeilhac PT. 1984. Disease problems in farm raised hatchling alligators in Florida alligator farms. Proceedings of the I Symposium on Crocodilian farming. IAAAM Proceedings 1(1):9–12.
Metabolic – Metabolic bone disease from feeding boneless turkey (with mineral supplementation) to alligator *Alligator mississippiensis*, as manifest by rubber jaw, spontaneous fractures, multiple folding (green stick) fractures, third cervical vertebrae compression fracture, and reduced long-bone density. Note unique use of "green stick" to describe a folding fracture.
Lannoo M. 2008. Malformed Frogs: The Collapse of Aquatic Ecosystems. 270 pp.; Berkeley: University of California Press.
Environmental – This is a report of his life work, predominantly reporting isolated finds (X-rays illustrated) in isolated wetlands, but does describe hotspots and lists causes as density, lathyrogens, nutrition, UV-B (ozone depletion), temperature extremes, heredity, parasites (Johnson et al. 1999, 2007), acidified water, radiation, heavy metals, retinoids, pesticides, fertilizers, and xenobiotics and noted that malformations have increased in frequency. Skeletal defects from polychlorinated biphenyls. Abnormal jaws from DDT.
52/105 US frog species and 19/188 US salamander species had malformations. Genera species malformation rates were ½ for *Ascaphus*, 10/43 for *Bufo*, 2/2 for *Acris*, 7/10 for *Hyla*, 1/1 for *Osteopilus*, 8/14 for *Pseudacris*, 2/2 for *Gastrophryne*, two third for *Scaphiopus*, ¼ for *Spea*, 1/1 for *Xenopus*, 17/31 for *Rana*, 8/17 for *Ambystoma* (especially unisexual hybrids), one third for *Notophthalmus* and two third for *Taricha*, 1/19 for *Desmognathus*, 2/27 for *Eurycea*, and 5/54 for *Plethodon*.
Northern leopard frog *Rana pipiens* had amelia, ectromelia, micromelia, hemimelia, taumelia (bone triangle), citing Meteyer et al. 2000 report of expansile bubble at end, arthrogryposis with kinks, polymelia, and toxicity from mercuric chloride.
American bullfrog *Rana catesbeiana* had amelia, arthrogryposis with kinks (Lannoo et al. 2003), and pelvic duplication (Johnson et al. 2003).
Green frog *Rana clamitans* had Amelia, ectromelia, and ectodactyly from carbamate insecticide carbofuran and bent back from pyrethroid insecticide fenvalerate.
Mink frog *Rana septentrionalis* had polymelia.
Blanchard's cricket frog *Acris crepitans blanchardi* had ectomelia, citing Meteyer et al. 2000 on an expansile bubble at end, calcified tail.
Bufo americanus ectomelia.
Pacific tree frog *Pseudacris regilla* polymelia.
Northwestern salamander *Ambystoma gracile* abnormalities from organophosphates.
Rana perezi peripheral skeleton and spinal pathology from organophosphates.
Cope's gray tree frog *Hyla chrysoscelis* spinal abnormalities from organophosphates and toxicity from mercuric chloride.
Eastern narrow-mouthed toad *Gastrophryne carolinensis* spinal abnormalities from organophosphates and toxicity from mercuric chloride.
Southern leopard frog *Rana sphenocephala* spinal abnormalities from organophosphates.

Microhyla ornata midtrunk and tail abnormalities spinal abnormalities from carbamate insecticide carbofuran.
Rana hexadactyla kinked tail from carbamate insecticide propoxur.
Rana aerosa limb abnormalities from herbicide diuron.
Triturus cristata malformation during regeneration from fungicide manganese ethylene bisdithiocarbamate.
Ambystoma mexicanum limb abnormalities from tributyltin oxide.
Wood frog *Rana sylvatica* digit and limb abnormalities from rodenticide thiosemicarbazide.
Xenopus laevis teratogenicity from calcium chloride.
Rana tigrina teratogenicity from potassium dichromate.
Bufo arenarum teratogenicity from lead nitrate.
Fowler's toad *Bufo fowleri* toxicity from mercuric chloride.
Red spotted toad *Bufo punctatus* toxicity from mercuric chloride.
Rana grylio toxicity from mercuric chloride.

Lannoo MJ, Sutherland DR, Jones P, Rosenberry D, Klaver RW, Hoppe DM, Johnson PT, Lunde KB, Facemire C, Kapfer JM. 2003. Multiple causes for the malformed frog phenomenon. Pp. 233–262. In Linder GS, Krest D, Sparling D, Little E, eds. Multiple Stressor effects in Relation to Declining Amphibian Populations. American Society for Testing Materials International, West Conshoshocken, Pennsylvania.
Congenital – Among 274 *Bufo americanus, Pseudacris triseriata, Rana clamitans, Rana pipiens, Rana septentrionalis*, and *Rana sylvatica* were four cases of amelia, four polymelia, eight ectromelia/hemimelia, seven micromelia, five taumelia, and seven spongiform bone (expansion of cancellous bone at distal tip of ectromelic limbs), predominantly in ecoregions with mixed forext component in Minnesota. Sutherland (of this group of authors) found 27 malformed green frogs *Rana clamitans* in Trempealeau County, Wisconsin, with 16 unilateral hind limb ectromelia, five bilateral hind limb ectromelia, one hind limb polydactyly, and one unilateral taurmelia. Sixteen of the hemimelic individual had bony expansions which were also contralateral.

Lanzi L. 1911. Contributo allo studio dei monstri doppi dei Teleostei [Contribution to the study of double monsters of the Teleostei]. Archivio italiano di anatomia e di embriologia 9(4):509–569. [Italian]
Environmental – Refers to the work of Hertwig (1892, 1895, 1896, and 1898) reporting that change of temperature of developing eggs of *Rana esculenta* and *Rana fusca* resulted in bifid spines.

Laporta LL, Sawaya P, Hogue A-R. 1983. Bicephalous snake – *Sibynomorphus mikanii* (Schlegel, 1837)- Colubridae. Boletim Fisioliogia Animal 7:79–84.
Congenital – Dicephalic *Sibynomorphus mikanii.*

Larger L. 1913. La contre-évolution ou dégénérescence par l'hérédité pathologique cause naturelle de l'extinction des groupes animaux actuels et fossiles. [The counter-evolution or degeneration by hereditary pathology causing natural extinction of current and fossil animal groups]. Bulletin et Mémoire de la Société Anthropologique de Paris 4:1 [French]
Metabolic – General comments without specifics; e.g., with discussion of acromegaly because of size.

Larsen S. 1993. Tændstikben og andre deformiteter has farvefrøer. [Tail disorders and other deformities of Dendrobatidae.] Nordisk Herpetologisk Forening 36:68–72. [Danish]
Metabolic – *Dendrobates auratus* with arched vertebral column; *D. azureus* with shortened hind legs and reduced animal size.

Lasher DN. 1980. A bicephalic *Crotalus horridus* from Alabama. Herpetological Review 11(4):89–90.
Congenital – Dicephalic *Crotalus horridus.*

Lasserre 1880. Note sur une couleuvre à deux tètes. [On a snake with two heads]. Bulletin de la Société d'Histoire Naturelle Toulouse 14:265–267. [French]
Congenital – Dicephalic *Vipera berus.*

Lataste F. 1879. À propos d'un squelette monstrueux de batracien anoure, (*Al. obstetrians* Wagler) [On a monstrous skeleton of a batrachian anura (*Al. obstetrians* Wagler)]. Revue internationale des sciences 3:49–52. [French]
Congenital – *Bufo vulgaris* with a long coccygeal diapophysis, as was reported by Leydig in *Bufo viridis.*

Latimer K, Rich GA. 1998. Colonic adenocarcinoma in a corn snake (*Elaphe guttata guttata*). Journal of Zoo and Wildlife Medicine 29:344–346.
Neoplasia – Colonic adenocarcinoma in corn snake *Elaphe guttata guttata*, but no indication of bone involvement, despite claim by Diaz-Figueroa and Mitchell (2006).

Latimer KS, Rich GA, Gregory CR. 2000. Osteosarcoma in a California mountain kingsnake (*Lampropeltis zonata*). Proceedings 3rd International Virtual Conference in Veterinary Medicine: Diseases of Reptiles and Amphibians. http:// www.vet.uga/edu/IVCVM/2000/Latimer/Latimer.htm.
Neoplasia – Osteosarcoma in a California mountain kingsnake *Lampropeltis zonata*, presenting as slightly elliptical mass with starburst pattern involving 2–3 vertebrae and proximal portion of eight ribs.

Lauria V. 1960. Editor's (B. Rothman) note to Dichocephaly in a northern water snake. Bulletin Philadelphia Herpetological; Society 8(5):17.

Congenital – Dicephalic Eastern garter snake *Thamnophis s. sirtalis* and northern pine snake *Pituophis m. melanoleucus.*
Lauthier M. 1971. Étude descriptive d'anomalies spontanées des members postérieurs chez *Pleurodeles waltlii* Michah. [Descriptive study of spontaneous posteriour limb anomalies in *Pleurodeles waltlii* Michah]. Annales de Embryologie et Morphologie 4:65–78. [French]
Congenital – Ectodactyly and ectomely in *Pleurodeles waltlii.*
Lavilla EO, Sandoval PE. 1999. A new Bolivian species of the genus *Telmatobius* (Anura: Leptodactylidae) with a humeral spine. Amphibia-Reptilia 20:55–64.
Congenital – Humeral spine in *Telmatobius* is a normal phenomenon.
Lawton M. 1992. Ophtalmology. Pp. 157–169 in Beynon P (ed.), Manual of reptiles. BSAVA, Cheltenham.
Congenital – Cyclopia – cites Bellairs as 1991, although references states 1981.
Lawton MP. 1992. Neurological diseases. Pp. 128–137. in Beynon P, Lawton MP, Cooper JE (eds.), Manual of Reptiles. Cheltenham: British Small Animal Veterinary Association.
Infection – Vertebral abscess, species not named.
Lazell J. 1995. *Plethodon albagula* (western slimy salamander). Foot anomalies. Herpetological Review 26:198.
Environmental – 20% of *Plethodon albagula* (western slimy salamander) from Driftwood area of Hayes County, but not other sites in that county had syndactyly and duplications.
Lebedinsky NG. 1923. Über eine Duplicitas anterior von *Rana fusca* und über die teratogenetische Terminationsperiode der symmetrischen Doppelbildungen der Placentalier.[On anterior duplication in *Rana fusca* and on the teratogenetic termination period of the formations of symmetrical duplication in Placentalia]. Anatomischer Anzeiger 56:257–266. [German]
Congenital – Cites duplications in *Salamandra maculosa* (Braun 1876; Jourdain 1877; Klausner 1890; Przibam 1921) and *Rana esculenta* (Born 1882).
Dicephalic *Rana temporaria* tadpole and *Rana fusca* (Born 1881; Loyez 1897; Spemann 1919).
Trauma – cites *Triton palmatus* with two tails (Josephy 1913).
Lebedinsky NG. 1925. Entwicklungsmechanische Untersuchungen an Amphibien. II. Die Umformung der Grenzwirbel bei *Triton taeniatus* und die Isopotenz allgemein homologer Körperteile des Metazoenorganismus. [The transformation of vertebrae in *Triton taeniatus* and the importance of general homologous body parts of the metazoan organism]. Biologisches Zentralblatt 45(2):94–121. [German]
Congenital – Differential formation of sacral ribs on vertebrae 14 (enlargement on one or both sides) and 15 (loss of left rib or on both sides) of *Triton taeniatus* (Adolphi 1898).
Different position of first hemal arch on second or third caudal vertebra, free ribs on first caudal vertebra or one on the right side of *Triton taeniatus.*
Broadened transverse processes on left side of vertebra eight and on right side of vertebra 9+10 for attachment of pelvis in *Atelopus varius* (Noble 1922). Such asymmetries were also found in *Necturus maculosus, Megalobatrachus maximus, Cryptobranchus alleganiensis, Salamandra atra, S. maculosa, Triton cristatus, Triton taeniatus* and in reptiles, *Lacerta muralis* (Kühne 1915), *L. vivipara* (Kühne 1915), *L. viridis, L. agilis, L. simonyi, L. (Chrysolamprus) ocellata, Grammatophora barbata* (= *Amphibolurus barbatus*), *Haplocercus spinosus, Iguana tuberculata, Moloch horridus, Otenodus (Tupinambis) nigropunctatus, Phrynosoma douglasii,* and *Tropidurus torquatus.*
Lecamp M. 1935. Les formations multiple dans la greffe et la regeneration des members chez le Crapaud accoucheur [Multiple formations in limb transplant and regeneration among birthed toad]. Bulletin Biologique de France et de Belgique Suppl 19:1–106. [French]
Congentital – Surgically transplanted *Ambystoma punctatum* limbs, noting few spontaneous cases of supernumerary limbs.
Lee J. 2005. A Sword-Swallowing Collector Closes an Odd Little Museum. The New York Times 1 January 2005:B6.
Congenital – Dicephalic turtle.
Lee SY, Elinson RP. 2008. Abnormalities of forelimb and pronephros in a direct developing frog suggest a retinoic acid deficiency. Applied Herpetology 5:33–46.
Metabolic – Direct developing frog *Eleutherodactylus coqui* with forelimb reduction attributed to vitamin A deficiency.
Abnormal, absent, or supernumerary forelimbs are rare (Ankley et al. 2004; Guderyahn 2006).
Leeke P. 1912. Regeneration und Selbstverstümmelung. [Regeneration and self-induced injury]. Wochenschrift für Aquarien- und Terrarienkunde (9):447–450, 463–464. [German]
Congenital – *Pelobates fuscus* with four additional hind legs (from Tornier artificially produced); frog with additional hind leg (natural find).
Dicephalic *snake* (after Tornier from Thesing).
Lehmensick R. 1934. Über Panzerverletzungen bei Schildkröten. [On injuries of carapace in turtles]. Zoologischer Anzeiger 105:325–331. [German]

Trauma – *Clemmys leprosa* with fracture healed of marginal bone, in contrast no healing along suture (split in dorsal shield) or because too superficial (superficial bone defects in ventral shield)
Emys orbicularis with defect in ventral shield with little healing.

Leibovitz HE, Culley Jr. DD, Geaghan JP. 1982. Effects of vitamin C and sodium benzoate on survival, growth and skeletal deformities of intensively cultured bullfrog larvae *Rana catesbeiana* reared at two pH levels. Journal of World Mariculture Society 13:322–328.
Metabolic – Diets with 2–8% vitamin C in *Rana catesbeiana* were associated had less scoliosis.

Leighton GR. 1901. The life history of British serpents and their local distribution in the British Isles. XVI+383pp.; London: Blackwood.
Congenital – Dicephalic adder.

Lema T. de. 1957. Bicefalia em serpentes–Descrição de um novo caso [Bicephaly in serpents—Description of a new case]. Iheringia: Zoologia, Porto Alegre 5:5–8. [Portuguese]
Congenital – Derodymous snake *Liophis miliaris* var. *semiaureus* (Cope), commonly known in Brazil as cobra-lisa, cobra-de-banhado, and cobra- dágua [Museu Rio-Grandense de Ciências Naturais (Rio-Grandense Museum of Natural Sciences, Pôrto Alegre, Rio Grande do Sul, Brazil MRCN57)].

Lema T. de. 1958. Notas sôbre os Répteis do Estado do Rio Grande do Sul – Brasil. III. Bicefalia em *Helicops carinicauda* (Wied); Descrição de um caso [Notes on the reptiles of Rio Grande do Sul State—Brazil. III. Bicephaly in *Helicops carinicauda* (Wied); Description of a case]. Iheringia: Zoologia, Porto Alegre 10:25–27. [Portuguese]
Congenital – Dicephalic water snake, *Helicops carinicauda infrataeniata* (Institute of Natural Sciences of the University of Rio Grande do Sul, in Pôrto Alegre, Brazil ICN-Co. 186). Radiographs revealed first vertebrae were longer than the rest and the crania were fused at their base.

Lema T. de. 1961a. Bicefalia em serpentes—Descrição de um novo caso [Bicephaly in serpents—Description of a new case]. Semanas Universitarias Gauchas de Debates Biológicos, 1–4, Porto Alegre, Universidad Federal do Rio Grande do Sul, Porto Alegre (2):34–35. [Portuguese]
Congenital – Derodymous *Liophis miliaris semiaureus* Cope (previously published by de Lema in 1957) whose heads were fused externally near the last labial scales and internally at the 14th vertebra.

Lema T. de. 1961b. Notas sôbre os répteis do Rio Grande do Sul– Brasil. XI. Bicefalia em *Xenodon merremii* (Wagler) (Serpentes: Colubridae) [Note on the reptiles of Rio Grande do Sul—Brazil. XI. Bicephaly in *Xenodon merremii* (Wagler) (Serpentes: Colubridae)]. Iheringia: Zoologia, Porto Alegre 17:15–19. [Portuguese]
Congenital – Derodymous male *Xenodon merremii* male, captured in 1957 in Estado do Rio Grande do Sul, Brazil.

Lema T. de 1982. Descrição de dois espécimens bicéfalos de *Liophis miliaris* (L., 1758) (Serpentes, Colubridae) [Description of two bicephalous specimens of *Liophis miliaris* (L., 1758) (Serpentes, Colubridae)]. Iheringia (Zool) 61:67–79. [Portuguese]
Congenital – Opodymus and a derodymous *Liophis miliaris*, the latter from the Zoobotanical Foundation of Rio Grande do Sul, Brazil MCN 7782. Includes a list of all known dicephalic snakes found in South America until the date of publication of this paper.

Lema T. de 1994. Descrição de um exemplar com bifurcação axial de *Echinanthera cyanopleura* (Cope, 1885) (Serpentes, Colubridae, Xenodontinae) [Description of a specimen of *Echinantera cyanopleura* (Cope, 1885) (Serpentes, Colubridae, Xenodontinae) with anterior axial bifurcation]. Acta Biologica Leopoldina 16(2):113–117. [Portuguese]
Congenital – Derodymous/thoracodymous *Echinantera cyanopleura* (Science and Technology Museum of the Pontifical Catholic University of Rio Grande do Sul MCP 2532) with gibbus at point of fusion of the two vertebral columns.

Leong TM. 2001. Parasitic copepods responsible for limb abnormalities. Newsletter of the Declining Amphibian Populations Task Force, Froglog 46:2–3.
Environmental – 0.06% of copper-cheeked frog *Rana chalconota* had brachymely, polydactyly, and ectomely, which were attributed to parasitism by *Lernaea cyprinacea.*

LePlat G. 1920. Action du milieu sur le développement des larves d'amphibiens. Localisation et différenciation des premières ébauches oculaires chez vertébrés. Cyclopie et anophthalmie. [Effect of milieu on development of amphibian larvae. Localization and differentiation of first ocular primordial among vertebrates. Cyclops and anophthalmia]. Archives de Biologie 30:231–240. [French]
Congenital – Cyclops induction in *Rana fusca, Tr. cristatus*, and *Tr. marmoratus*, citing Stockard on *Ambystoma.*

LeRoy W, Giles 1948 Polydactylism in an Alligator. Copeia 1948(3):214.
Congenital – three extra forefoot toes, four extra left and three extra right hind foot toes in an *Alligator mississippiensis.*

Lesbre F-X. 1927. Traité de Tératologie de l'homme et des animaux domestiques. [Traits and teratology of humans and domestic animals]. 342pp. Paris: Vigot Frères. [French]
Congenital – Double monster in *Rana*, citing O. Schulze.

Meromelia – Multiple forearms in a *Rana*, citing Geoffroy Saint Hilaire.
Polydactyly and abdmelomelia, citing C. Bonnet.
Trauma – Lizard tail duplication, citing C. Bonnet.

Lessona M. 1876–1877. Nota in torno ad un caso di dicefalia nell' *Anguis fragilis* (Linn.). [Note about a case of dicephaly in *Anguis fragilis* (Linn.)] Atti della Reale accademia delle scienze di Torino 12:174–179 [Italian].
Congenital – Derodymous *Anguis fragilis* in Cuneo, Italy, on Aug. 25, 1875, with description of the specimen in vivo, and of the postmortem dissection findings, including detailed measurements.

Leuckart KGFR. 1845. De monstris eorumque causis et Ortu. [Of scrutiny of causes and occurrence of monsters]. Goettingae: Academica Dieterichiana [Latin].
Congenital – Amphibian cyclops.

Levey R, Shambaugh N, Fort D, Andrews J. 2003. Investigations into the causes of amphibian malformations in the Lake Champlain Basin of New England. Vermont Department of Environmental Conservation, Waterbury, Vermont. Pp. 1–239.
Environmental – Ward Marsh, West Haven Vermont northern Leopard frogs (*Rana pipiens*) had 3% brachydactyly and ectodactyly, 3% hemimelia, brachydactyly, and ectodactyly, 8% with hemimely and brachydactyly, 3% polydactyly, 3% micromelia, 3% hemimelia, 3% ectomelia, 5% syndactyly, 31% ectomelia, 11% brachydactyly, and 13% amelia. This compared with 20% of spotted salamander *Ambystoma maculatum* with abnormalities (hemimelia, ectodactyly, brachydactyly, polyphagy), and 0–2.2% (brachydactyly, ectodactyly, syndactyly) in green frogs (*Rama clamitans*) and 0.4–0.7% in *Rana pipiens*.

Leviton AE, Gibbs RH Jr, Heal E, Dawson CE. 1985. Standards in herpetology and ichthyology: Part 1. Standard symbolic codes for institutional resource collections in herpetology and ichthyology. Copeia 1985(3):802–832.
Other – List of museum identifiers for herpetology collections.

LeWin A. 1962. Two-headed snake born. St. Paul Pioneer Press, Aug. 6:15.
Congenital – Dicephalic swamp rattlesnake.

Leydig F. 1872. Die in Deutschland lebenden Saurier. [Germany being saurian]. 262 pp.; Tübingen: H. Laupp'sche Buchhandlung. [German]
Trauma – Formation of double tail (Glückselig 1863; Dugès 1829; Gachet 1833; Calori 1858; Müller 1852) and observations on the cartilage continuation in the new tails of *Lacerta agilis, L. viridis, L. muralis*, and *L. vivipara*.
Double-tailed lizard (Plinius, Gessner 1586; Aldrovandi 1637; Lacépède 1790). Three-tailed lizard (Redi 1684; Seba, Evermann 1858).

Leydig F. 1877. Die anuren Batrachier der deutschen Fauna. [Anuran batrachians of German fauna]. VII+164; Bonn: Max Cohen & Sohn. [German]
Congenital – Supernumerary toe in *Bombinator igneus*.
Rana fusca with one forelimb without toes, while the other forelimb with three toes and additional toe-like outgrowth at base of hand.
Rana arvalis missing left hind foot.
Trauma – Reduction of toes, one specimen of *Bufo vulgaris* with only one toe on forelimb, caused by bivalve *Cyclas cornea*).

Leydig F. 1879. Die Rippenstacheln des *Pleurodeles waltii*. [The spines of ribs of *Pleurodeles waltii*]. Archiv für Naturgeschichte, Jahrgang 45:211–234 [German].
Congenital – *Pleurodeles waltii* with six toes on hind foot and four (instead of normal five) toes on forefeet.

Liceto (Licetus) F. 1634. De monstrorum natura, caussis, et diffenrentiis. [Natural and differentiated monsters] libri duo. 2nd edition 262pp.; Patavii [Padova], Italy: Apud Paulum Frambottum [Latin].
Congenital – Repeated Aristotle's citations. Derodymous snake and illustrated a large-mouthed herp with a tail but no pelvis or hind limbs.

Licht LE, Grant KP. 1997. The effect of ultraviolet radiation on the biology of amphibians. American Zoologist 37(2):137–145.
Environmental – Ultraviolet radiation-induced *Rana pipiens* and *temporaria* "cephalic enlargement" and *Rana sylvatica* tail kinking.

Lichtenstein S. 1923. Ein Fall von spontaner Froschtuberkulose [A case of spontaneous frog tuberculosis). Centralblatt für Bakteriolollgie, Parasitenkunde und Infektionskrankheiten 85:249–252. [German]
Infection – Frog tuberculosis, but no comment on osseous involvement.

Lillywhite HB. 2008. The Dictionary of Herpetology. Malabar, Florida: Krieger Publishing Company, 376 pp.
Dictionary of herpetology terminology, also illustrating similar use of terms (to other scientific fields), but with very disparate meaning:
Diagnosis – Formal character state description distinguishing one taxon from another (herpetology definition), contrasted with use in this annotated bibliography as identified disease state.
Heterotopic – Transplantation to abnormal location or occurring in many habitats (herpetology definition), contrasted with use in this annotated bibliography to indicate spontaneous occurrence in locations usually lacking the enlage.

Lytic – Destruction of structures (herpetologic definition), contrasted with use in this annotated bibliography to holes in bone.

Osteomyelitis – Inflammation of marrow cavity (herpetologic definition), contrasted with use in this annotated bibliography to identify bone infection not limited to the external surface.

Osteitis – Inflammation of bone (herpetologic definition), contrasted with use in this annotated bibliography to identify bone infection, not discriminating use from osteomyelitis.

Porosity – Volume of openings in rock/soil (herpetologic definition), contrasted with use in this annotated bibliography to identify presence of minute surface holes in bone.

Lin Z, Qu Y, Ji X. 2006. Energetic and locomotor costs of tail loss in the Chinese skink, *Eumeces chinensis*. Comparative Biochemnistry and Physiology 143A:508–513.

Trauma – 73% of Chinese skink *Eumeces chinensis* had lost part of tail, but speed reduction was noted only those missing more than 51%.

Caudal autotomy reduces locomotor performance in many lizard species (Ballinger et al. 1979; Downes and Shine 2001; Lin and Ji 2005; Martin and Avery 1998; Pond 1978; Punzo 1982; Shine 2003; Zani 1996).

Lindsey CC. 1966. Temperature-controlled meristic variation in the salamander *Ambystoma gracile*. Nature 209:1152–1153.

Environmental – Northwestern salamander *Ambystoma gracile* apparently independent of temperature, while *Ambystoma mexicanum* had highest number of vertebrae at intermediate temperatures and *Triturus vulgaris*, at higher temperatures.

Liner EA. 1960. A new subspecies of false coral snake (*Pliocercus elapoides*) from San Luis Potosi, Mexico. The Southwestern Naturalist 5(4):217–220.

Trauma – Part of tail of false coral snake *Pliocercus elapoides* broken off during capture.

Lingham-Soliar T. 1995. Anatomy and functional morphology of the largest marine reptile known, *Mosasaurus hoffmanni* (Mosauridae, Reptilia) from the Upper Cretaceous, Upper Maastrichtian of The Netherlands. Philosophical Transactions of the Royal Society of London B 347:155–180.

Fossil – *Mosasaurus hoffmanni* Institut Royal des Sciences Naturelles de Belgique, Brussels, Belgium IRSNB R26 illustrated with focal enlargement at distal third of dental on both superior and inferior surface.

Lingham-Soliar T. 1998. Unusual death of a Cretaceous giant. Lethaia 31:308–310.

Trauma – Rehealed breaks on jaw bones of *Mosasaurus hoffmanni* and toothmarks on giant turtle by this animal.

Fossil – Rehealed breaks on jaw bones of *Mosasaurus hoffmanni* and toothmarks on giant turtle by this animal.

Lingham-Soliar T. 2001. The ichthyosaur integument: Skin fibers, a means for a strong, flexible skin. Lethaia 34:287–302.

Congenital – *Stenopterygius quadriscissus* and *Ichthyosaurus* skin fiber structure contributes to streamlining. Intriguing feather-like appearance is noted.

Fossil – *Stenopterygius quadriscissus* and *Ichthyosaurus* skin fiber structure contributes to streamlining. Intriguing feather-like appearance is noted.

Lingham-Soliar T. 2004. Palaeopathology and injury in the extinct mosasaurs (Lepidosauromorpha, Squamata) and implications for modern reptiles. Lethaia 37:255–262.

Trauma – *Mosasaurus hoffmanni* dentaries with fracture nonunions in IRSNB R25, *Platecarpus ictericus* BMNH R2834 with bony callus. Dentary and maxillary fracture reported in contemporary *Agama atricollis* and road accident related dentary fracture in *Varannus niloticus*. Snout fractures in Indian crocodile *Gavialis gangeticus*.

Fossil – *Mosasaurus hoffmanni* dentaries with fracture nonunions in IRSNB R25, *Platecarpus ictericus* BMNH R2834 with bony callus.

Lipsett JC. 1941. Disproportionate Dwarfism in *Ambystoma*. Journal of Experimental Zoology 86:441–455.

Congenital – Recessive occurrence of blunt heads with abbreviated snouts, shorter limbs, and stubbier digits from hybridization of *Ambystoma tigrinum* and *A. mexicanum*, but first seen in the Berlin strain of *A. mexicanum*. Homozygous state is lethal.

List JC, Smith PW. 1954. A rare snake anomaly. American Midland Naturalist 51(1):312–313.

Congenital – Cites Aristotle double monsters and reports Gulf water snake *Natrix sipedon clarkii* with poster dichotomy.

Liu SK. 2002. Metabolic disease in animals. Seminars in Musculoskeletal Radiology 6(4):341–346.

Metabolic – Deficiency of vitamin D, calcium, phosphorus, and sunlight can cause rickets (actually osteomalacia) in reptiles.

Little RB. 1973. Variation in the plastral scutelation of *Graptemys pulchra* (Reptilia, Chelonia, Emydidae). Association of Southeast Bioloy Bulletin 20:65–66.

Congenital – Greater variability of anterior and left aspect of plastron in *Graptemys pulchra*.

Locatelli P. 1924. Sulla formazione di arti soprannumerari [On the formation of supernumerary limbs]. Bollettino della Società Medico-Chirurgica di Pavia 36:417–419. [Italian]

Trauma – Supernumerary limbs produced experimentally in the *Triton* salamander by the amputation of the hind limbs and also by severing the sciatic nerve to differing extents.

Lock BA. 2006. Behavioral and morphologic adaptations. In Mader DR. ed. Reptile Medicine and Surgery. Pp. 163–179; Philadelphia: Saunders.

Trauma – Lizard tail regrowth occurs in most lizards, with the exception of Agamidae, Chamaeleonidae, Helodermatidae, Lanthanotidae, Xenosauridae, and Varanidae.

Loeffler IK, Stocum DL, Fallon JF, Meteyer CU. 2001. Leaping lopsided: A review of the current hypotheses regarding etiologies of limb malformations in frogs. Anatomical Record 265:228–245.

Trauma/Environmental – *X. laevis* tadpoles with amputated hind limb buds raised in Minnesota pond water in which 7% of surveyed newly metamorphosed feral frogs had malformations, resulted in 6% abnormal redevelopment, contrasted with only one raised in dechlorinated municipal water.

Long Beach Animal Hospital. 2007a. Tortoise bladder stone. http://lbah.com/reptile/cdtbladderstone.html.

Stone – Bladder stone in a desert tortoise.

Long Beach Animal Hospital. 2007b. Iguana bladder stone. http://lbah.com/reptile/cdtbladderstone.html.

Stone – Bladder stone in a green iguana.

Longman HA. 1939. A bicephalous snake. Memoirs of the Queensland Museum 11(3):288.

Congenital – Dicephalic common black snake *Pseudechis porphyriacus.*

Lönnberg E. 1907. En dubbelhöfdad huggorm. [A double-headed snake]. Fauna och Flora 2(3):125–128. [Dutch]

Congenital – Dicephalic viper (Vipera berus) from southern Sweden and reports dicephalic Tunisian snake in the collection of The Naturhistorska Riksmuseet Stockholm and cites Blatchley (1906).

Lopez TJU, Maxson LR. 1990. *Rana catesbeiana* (bullfrog). Polymely. Herpetological Review 21(4):90.

Congenital – Polymely in *Rana catesbeiana* (bullfrog).

López del Castillo C. 1996. Panzernekrosen bei Schildkröten. [Necrosis or gangrene of the shell in tortoises and turtles]. Reptilia (D) 1:64–67. [German]

Shell – Necrosis, a multifaceted syndrome, more common in turtles than in tortoises. Pictures of necrosis in *Emydura albertisi, Cuora galbinifrons, Chelus fimbriatus, Chrysemys picta belli, Cistoclemmys galbinifrons,* and *Mauremys leprosa.* Author suggests infection by fungi (*Basidiolobus, Dermatophyton, Fusarium, Aspergillus, Trochoderma, Trichophyton*) or bacteria (*Aeromonas, Proteus, Klebsiella, Arizona, Bacteroides, Citrobacter, Enterobacter, Escherichia, Pasteurella, Providencia, Serratia, Staphylococcus, Streptococcus, Salmonella, Mycobacterium,* and other).

López del Castillo C. 1998. Metabolic diseases: Gout and articular pseudogout. Reptilia(GB) 3:13–16.

Metabolic – Herman's tortoise *Testudo hermanni* had articular gout, apparently of hips, noting that*Trachemys scripta* fed high protein diet developed gout.

Dabbs' mastigure *Uromastyx acanthinurus* and spur-thighed tortoise *Testudo graeca* with articular pseudogout.

López del Castillo C. 2001. Accidental loss of toes and claws in lizards. Reptilia (GB) 19:71–72.

Trauma – Toe loss from bite in green basilisk *Basiliscus plumifrons* and claw loss in green iguana *Iguana iguana.*

Loren JF. 2007. Airport screeners expect the unexpected. Unusual items discovered in baggage increase over busy holiday periods. Lawrence Journal World November 21, 2007:A2.

Congenital – Dicephalic snake found undeclared in carry on luggage.

Lovich JE, Ernst CH. 1989. Variation in the plastral formulae of selected turtles with comments on taxonomic utility. Copeia 1989(2):304–318.

Congenital – Scute variation in turtles, listing plastal formula and variation for *Chelus fimbriatus, Elseya dentata, latisternum, Emydura krefftii, Phrynops geoffroanus, gibbus, nasutus, williamsi, zuliae, Platemys macrocephala, platycephala, Rheodytes leukops, Caretta caretta, Eretmochelys imbricata, Lepidochelys olivacea, Chinemys kwangtungensis, reevesii, Chrysemys picta, Cuora chriskarannarum, flavomarginata, pani, trifasciata, yunnanensis, Emys orbicularis, Geoemyda spengleri, Graptemys, Heosemys silvatica, Kachuga dhongoka, kachuga, Malaclemys terapin, Mauremys caspica, leprosa, mutica, Ocadia sinensis, Pseudemys concinna, floridana, nelsoni, texana, Pyxidea mouhotii, Rhinoclemmys annulata, areolata, diademata, funerea, nasuta, punctularia, pulcherrima, rubida, Sacalia bealei, Trachemys scripta, Kinosternon alamosae, angustipons, Sternotherus carinatus, despressus, minor, Peltocephalus dumerilianus, Podocnemis erythrocephala, expansa, unifilis, vogli, Platysternum megacephalum, Chersina angulata, Geochelone carbonaria, denticulata, pardalis, sulcata, Gopherus agassizi, berlandieri, polyphemus, Homopus areolatus, boulengeri, fermoralis, signatus, Kinixys belliana, erosa, homeana, Malacochersus tornieri, Psammobates geometricus, oculifera, tentorius, Testudo graeca, kleinmanni.* Knoll (1935) related variation to forest fires and Little (1973), to geographic variation.

Trauma – Scute variation in turtles, listing plastal formula and variation for *Chelus fimbriatus, Elseya dentata, latisternum, Emydura krefftii, Phrynops geoffroanus, gibbus, nasutus, williamsi, zuliae, Platemys macrocephala, platycephala, Rheodytes leukops, Caretta caretta, Eretmochelys imbricata, Lepidochelys olivacea, Chinemys kwangtungensis, reevesii, Chrysemys picta, Cuora chriskarannarum, flavomarginata, pani, trifasciata, yunnanensis, Emys orbicularis, Geoemyda spengleri, Graptemys, Heosemys silvatica, Kachuga dhongoka, kachuga, Malaclemys terapin, Mauremys caspica, leprosa, mutica, Ocadia sinensis, Pseudemys concinna, floridana, nelsoni, texana, Pyxidea mouhotii, Rhinoclemmys annulata, areolata, diademata, funerea, nasuta, punctularia, pulcherrima, rubida, Sacalia bealei, Trachemys scripta, Kinosternon alamosae, angustipons, Sternotherus carinatus, despressus, minor, Peltocephalus dumerilianus, Podocnemis erythrocephala, expansa, unifilis, vogli, Platysternum megacephalum, Chersina angulata, Geochelone carbonaria, denticulata, pardalis, sulcata, Gopherus agassizi, berlandieri, polyphemus, Homopus areolatus, boulengeri, fermoralis, signatus, Kinixys belliana, erosa, homeana, Malacocherus tornieri, Psammobates geometricus, oculifera, tentorius, testudo graeca, kleinmanni*. Knoll (1935) related variation to forest fires and Little (1973), to geographic variation.

Lovich JE, Gotte SW, Ernst CH, Harshbarger JC, Laemmerzahl AF, Gibbons JW. 1996. Prevalence and histopathology of shell disease in turtles from Lake Blackshear, Georgia. Journal of Wildlife Diseases 32:259–265.

Shell – 74% of *Pseudemys concinna* and 35% of *Trachemys scripta* from Lake Blackshear, Georgia had carapace necrosis.

Lowne BT. 1872. Descriptive catalogue of the Teratological Series in the Museum of the Royal College of Surgeons of England. Pp. XX+110, London, R. Hardwicke.

Congenital – Twin fetus of a serpent from Guinea joined by the dorsal region of the vertebral column, with fusion of superior aspect of vertebral column in its superior part with fused heads and necks, medial eyes were fused into a single abnormally wide organ and lower jaws forming a single organ – Catalogue no. 43.

Dicephalic – small snake – Catalogue no. 2.

Small English snake – Catalogue no. 24 a. 27.

Indian water-snake *Hydrophis* – Catalogue no. 26.

Derodymous English snake.

Frog with supernumerary forelimb attached at sternum – Catalogue no. 23.

Frog missing right anterior extremity – Catalogue no.340.

Trauma – lizard with tail duplication – Catalogue no. 28.

Lloyd Morgan C. 1886. Abnormalities in the vertebral column of the common frog. Nature 35:531.

Congenital – Centrum of eighth vertebra of *Rana temporaria* was procoelous; eighth and ninth in another were sacralized on the right, with absent left transverse process.

Loyez M. 1897. Sur un têtrad de *Rana temporaria* bicéphale. [Tetrad of dicephalic *Rana temporaria*]. Bulletin de la Sociétè Zoologique de France 22:146–148. [French]

Congenital – Dicephalic *Rana temporaria.*

Lucan(us) A.M. 1631. Farsaglia. Book IX, verse 719. [Latin] Actually Lucans Pharsalia: or the Ciuill Warres of Rome, between Pompey the great and Julius Caesar. 2nd edition. 310 pp. London: Mathews. [Latin]

Pseudopathology – Mistakes amphisbaena as a two-headed snake because of enlargement of terminal portion of the tail.

Lucas FA. 1886. The sacrum of Menopoma. American Naturalist 20:561–562.

Congenital – Figure of *Menopoma* pelvis in ninth edition of Encyclopaedia Britannica was from an abnormal skeleton, suggesting two sacral vertebrae, as opposed to the normal one. Another specimen showed right ilium attached to one vertebrae and left to two.

Variation in number of dorsals in certain Urodeles: *Menopoma*, 19 or 20; *Necturus*, 18 or 19; *Siren* 41,42, or 43 with total number of vertebrae 101 and 108; *Muraenopsis*, 64 or 65 with total number of vertebrae 105, 107, and 111.

Lucas SG. 2000. Pathological aetosaur armor from the Upper Triassic of Germany. Stuttgarter Beiträge zur Naturkunde Serie B (Geologie und Paläontologie) 281:1–6. [German]

Armor – Three dorsal paramedian scutes of the aetosaur *Paratypothorax andressorum* from the Upper Triassic Lower Stubensandstein of Stuttgart, SW-Germany, are fused together dorsally by a massive bony overgrowth caused most likely by an osteoblastic reaction to an infection of the bone after an injury. First documented pathology in an aetosaur and a rare example of a paleopathology in an osteoderm.

Fossil – Three dorsal paramedian scutes of the aetosaur *Paratypothorax andressorum* from the Upper Triassic Lower Stubensandstein of Stuttgart, SW-Germany, are fused together dorsally by a massive bony overgrowth caused most likely by an osteoblastic reaction to an infection of the bone after an injury. First documented pathology in an aetosaur and a rare example of a paleopathology in an osteoderm.

Lüdicke M. 1964. 5. Ordnung der Klasse Reptilia, Serpentes. [Organization of class reptilia, serpents]. In: Helmcke H, Lengerken H. v, Starck D, Wermuth H. eds. Handbuch der Zoologie, eine Naturgeschichte der Stämme des Tierreiches 7(1). part 6. [Handbook for Zoology, a natural history of the phyla of the animal kingdom]. Lieferung:129–298. Berlin: Walter Gruyter. [German]

Congenital – Dicephalic *Natrix natrix* and *Lampropeltis californiae.*

Lüdicke M. 1971. Ein fünfbeiniger Grasfrosch (*Rana temporaria* L.). [A five-legged glass frog (*Rana temporaria* L.)]. Zoologischer Anzeiger 186: 188–197. [German]

Congenital – Detailed description of a five-legged frog *Rana temporaria* with X-ray, photo of skeleton and drawing of articulation in pelvic: two pairs of ischia, two pairs (one rudiment) of ilia and two completely developed left hind legs.

Lugaro G. 1957. Elenco sistematico dei Rettili italiani conservati nella collezione di studio esistente presso il Museo di Storia Naturale di Milano, con brevi note critiche ed esplicative [Systematic list of Italian reptiles conserved in the existing study collections in care of the Natural History Museum of Milano, with brief critical and explanatory notes]. Atti della Società italiana di scienze naturali 96(1–2):20–36. [Italian]

Congenital – Dicephalic *Coluber viridiflavus*.

Trauma – Review of Italian reptile collections of the Natural History Museum of Milan (Italy) revealed bifid tail in four *Lacerta (Podarcis) muralis* and one with multiple tails.

Luis R, Baez M. 1987. Anomalias morfológicas en los anfibios de las Islas Canarias (Amphibia, Anura) [Morphological anomalies in the amphibians of the Canary Islands (Amphibia, Anura)]. Vieraea 17:295–296. [Spanish]

Congenital – First report in Canary Islands of amphibians with morphological abnormalities was a *Rana perezi* (among 365 normals) with two supernumerary left forelimbs and a *Hyla meridionalis* lacking a tibiotarsus, but with digits attached to distal femur.

Lunel G. 1868. Sur deux cas de polymélie (members surnuméraires), observés chez la *Rana viridis* seu *esculenta* (Lin.). [On two cases of polymelia (supernumerary limbs), observed in *Rana viridis esculenta* (Lin.)]. Mémoires de la Société de Physique et d'Histoire naturelle de Genève 19:305–312. [French]

Congenital – Supernumerary posterior limbs in *Rana esculenta, Rana viridis*, and *Rana clamata*, and anterior in *Pelobates cultripes*, ectomelia in *Rana viridis* and *Rana temporaria*, and polymelia in *Rana viridis*, noting Cisternas' report of supernumerary limb in *Alytes obstetricans*.

Luo S-Q, Plowman MC, Hopfer SM, Sunderman FW. 1993. Mg^{2+} deprivation enhances and Mg^{2+} supplementation diminishes embryotoxic and teratogenic effects of Ni^{2+}, Co^{2+}, Zn^{2+}, and Cd^{2+} for frog embryos in the FETAX assay. Ann Clin Lab Sci 23:121–129.

Environmental – Mg2+ deprivation enhances and Mg2+ supplementation diminishes teratogenic effects (facial abnormalities) of Ni2+, Co2+, Zn2+, and Cd2+ in South African frog *Xenopus laevis*.

Luscher S. 1994. Metabolic bone disease. Bulletin of the Alberta Reptile and Amphibian Society 3(3):12–19.

Metabolic – General article on metabolic bone disease, suggesting that osteoporosis occurs in hibernation from insufficient protein deposits or excess phosphate, citing Ippen et al. 1985.

Lydekker R. 1889. Catalogue of the fossil Reptilia and Amphibia in the British Museum, Part II. Containing the orders Ichthyopterygia and Sauropterygia. 238pp., London: Trustees of the British Museum.

Congenital – *Cimoliosaurus plicatus* cervical vertebrae BMNH 48001 on one side it was divided into two parts, each with its distinct costal facet. This suggests a hemivertebra.

Fossil – *Cimoliosaurus plicatus* cervical vertebrae BMNH 48001 on one side it was divided into two parts, each with its distinct costal facet. This suggests a hemivertebra.

Lydekker R. 1890. Catalogue of the fossil Reptilia and Amphibia in the British Museum, Part IV. Containing the orders Anomodontia, Ecaudata, Caudata, and Labyrinthodontia; and supplement. XXIII + 295pp., London: Trustees of the British Museum.

Congenital – Unpaired, median bone, "Wormian bone," between parietals and frontals in dicynodont therapsids (mammal-like reptiles) [p. 40].

Fossil – Unpaired, median bone, "Wormian bone," between parietals and frontals in dicynodont therapsids (mammal-like reptiles) [p. 40].

Lynch JD. 1965. Apparent syndactyly in a Mexican frog. Journal of the Ohio Herpetological Society 5:57.

Congenital – Snapper with protruding lower jaw.

Syndactyly in *Rana palmipes* and cited Munyer's (1963) report in common fence lizard *Sceloporus undulatus hyacinthinus*.

Environmental – Snapping turtle with deeply indented right side secondary to and painted turtle with three curled marginal scutes attributed to drying.

Shell disease – Painted turtle with asymmetrical broad carapace and irregular scutes, another with irregular neurals and scutes that could not be separately identified, *Chelydra serpentina* with short broad carapace, a painted turtle with carpace of only five scutes, "covering only one third of usual area" and cited Lynn and Ullrich's 1950 report of *Pseudemys scripta troostii* with eight neurals instead of five, six right costals instead of usual five and *Chrysemys picta picta* with irregular shaped shell, painted turtle with seven marginal scutes on right, eight on left, three of which were curled and attributed to drying, five pair of plaston scutes.

Cited Parker's (1901) report in *Clemmys insculpta* and Wandolleck's 1904 report in *Testudo graeca* of distorted carapaces.

Egg – Cites Coker's 1910 report of a loggerhead with dent in egg shell.

Lynn WG. 1937. Variation in scutes and plates in the box-turtle, *Terrapene carolina*. American Naturalist 71:421–426.

Shell disease – Asymmetrical costal scutes in box-turtle, *Terrapene carolina;* Supernumerary scutes in *Malaclemys* (especially females), *centrata* and *pileata littoralis, Graptemys geographica, Chrysemys martinata* and *picta, Pseudemys (Ptychemys) elegans, Chelonia (Chelone) mydas, Caretta (Thalassochelys) caretta, Sternotherus (Aromochelys) odoratus, Chelydra serpentina, Clemmys guttata, Clemmys (Chelopus) insculpta, Terrapene carolina.*

Lynn WG. 1948. The Effects of Thiourea and Phenylthiourea upon the Development of *Eleutherodactylus Ricordii.* Biology Bulletin (Woods Hole) 94:1–15.

Toxicology – Thiourea and phenylthiourea retard limb development in leptodactylid toad *Eleutherodactylus ricordii.*

Lynn WG. 1950. A case of duplication of the tail in *Plethodon.* Herpetologica 6:81–84.

Trauma – Tail duplication in *Plethodon cinereus cinereus,* noting previous reports in *Plethodon cinereus* (Blatchley 1906), *Proteus anguinus* (Przibram 1909), and *Triturus viridescens* (Collins 1932; Dawson 1932; Porter 1934; Smith 1913)

Lynn WG, Ullrich MC. 1950. Experimental production of shell abnormalities in Turtles. Copeia 1950(4):250–262.

Congenital – Eastern painted turtle *Chrysemys picta picta* with hare lip and another with single digit on each hind foot.

Snapping turtle *Chelydra serpentina serpentina* with protruding lower jaw and eyeless upper jaw and another with eyeless stump.

Shell disease – *Pseudemys scripta troostii* with eight neurals (N = 5) and six right and seven left costals (N = 4).

Chrysemys picta picta with plastron bulges to right, wavy border, reduced right and enlarged left abdominal scute.

Abnormal scutes and bony plates in *Clemmys insculpta* (Parker 1901) Distorted carapace in *Testudo graeca* by Wandolleck 1904.

Grant 1937 reported that 10% of *Gopherus* had hard, brittle abnormal shells.

Noted distorted turtle shells reported by Hay 1904, Wandolleck 1904, Coker 1910, Newman 1923, Hildebrand 1930 and 1938, Gressitt 1936 and 1937, Chn 1937, Deraniyagalia 1939, Necker 1940, Mertens 1940, Fox 1941, Smith 1947, Nixon and Smith 1949.

Environmental – Specifically investigated eggs exposed to partial drying with eggs removed from oviduct, covered with moist cotton, removed for 4–5 days during gestation interval days 35–50. Embryos ranged from "almost unrecognizable carapaces and plastons, distorted limbs and eyeless or jawless heads."

Eastern painted turtle *Chrysemys picta picta* with seven right and three left (curved over) marginals and unusually broad neurals, single left costal scute, fusion of plastron pectorals and abdominals and hare lip. Another with long broad carapace with three left and two right costals, eight neurals and enlarged eigth and ninth marginals and only a single digit on each hind foot and a third with carapace one third normal size with only five scutes, and plaston lacking one pair of scutes.

Snapping turtle *Chelydra serpentina serpentina* – deeply indented carapace, another with protruding lower jaw and eyeless upper jaw, supernumerary carapace scutes with seven neurals and 14 marginals, but no nuchal and a third with eyeless stump, short, broad carapace with disorganized scutellation.

Cites Newman's 1923 suggestion of temperature effect and Coker's 1910, of egg distension causing cyclopean embryos (one with fused eyes and nostrils, short upper jaw and bulging lower jaw and a second which lacked eyes, a carapace with only two scutes and an asymmetrical mishappened carapace.

Annotated Bibliography M-N

MacCulloch R.D. 1981. Variation in the shell of *Chrysemys picta belli* from southern Saskatchewan. Journal of Herpetology 15:181–185.

Shell disease – 24 of 128 *Chrysemys picta belli* had scute abnormalities: Six males and 18 females with one or more abnormalities ($P < 0.01$). Two females with 11 marginals on one side; a male with 11 on one side and 12 on the other. Eleven had one or more bridge area carapace abnormalities; ten with extra pleurals; six with divided vertebrals. Average of abnormalities was 3.19 in those with multiple abnormalities.

Cites shell abnormality in 39% of 355 *C. picta* (Zangerl and Johnson 1957), 39% of 51 *C. marginata* (Whillans and Crossman 1977) and 13.4% of 929 *C. picta* x *marginata* (Ernst 1971). The latter referred to kyphosis.

MacGregor JH. 1906. The Phytosauria, with special reference to *Mystriosuchus* and *Rhytidodon*. Memoirs of the American Museum of Natural History 9(2):27–101.

Congenital – Odontoid fused with axis centrum in Phytosauria, but separate in *Erpetosuchus*.

Fossil – Odontoid fused with axis centrum in Phytosauria, but separate in *Erpetosuchus*.

MacGregor JH. 1908. On *Mesosaurus brasiliensis* from the Permian of Brazil. Comissão de Estudos das Minas de Carvão de Pedra do Brazil: 303–306.

Congenital – States that Seeley (1892) was erroneous in representation of two clavicles as fused into a single transverse bar. Refers to Osborn's (1903) claim as coossified radiale and intermedium as an artifact.

Fossil – States that Seeley (1892) was erroneous in representation of two clavicles as fused into a single transverse bar. Refers to Osborn's (1903) claim as coossified radiale and intermedium as an artifact.

Machotka SV. 1984. Neoplasia in reptiles. In: Hoff GL, Frye FL, Jacobson ER. eds. Diseases of Amphibians and Reptiles. Vol 2. Plenum Press, New York. Pp. 519–580.

Neoplasia – Rarity of tumors in reptiles and offers standard classification system. Enchondroma in distal metaphyses of humerus, metacarpals, hyoid, and cervical vertebrae in Indian monitor, *Vasranus dracaena*. Osteosarcoma in emerald lizard *Lacerta viridis*.

Machotka SV, Whitney GD. 1980. Neoplasms in snakes; Report of a probable mesothelioma in a rattlesnake and a thorough tabulation of earlier cases. In Montali RJ, Migaki G. eds. Pathology of Zoo Animals. Smithsonian Institution Press, Washington DC. Pp. 593–602.

Neoplasia – Osteochondrosaarcoma in a black cobra *Naja melanoleuca*. Malignant melanoma of hard palate of pine snake *Pituophis melanoleucus* and of vertebrae in "Everglade snake" *Elaphe obsoleta rossalleni*. Neurofibrosarcoma in Korean viper *Agkistrodon halys brevicaudua*. Osteosarcomata in spinal column of rufous-beaked snake *Ramphiophis rostratus* and African black cobra.

Chondrosarcoma in corn snake *Elaphe guttata*.

Osteochondroma in spinal column of African black cobra.

Mackness B. 2000. Pathology of a limb bone in the estuarine crocodile *Crocodylus porosus*. Herpetofauna 30(2):12–14.

Trauma – Reports (without detail) occurrence of injuries in *Crocodylus niloticus*, *Crocodylus johnstoni*, *Crocodylus porosus*, *Caiman crocodilus*.

Infection – *Crocodylus porosus* humerus with curvature at distal third with radiolucent zone with sequestra and draining sinuses from osteomyelitis complicating trauma.

B.M. Rothschild et al., *Herpetological Osteopathology: Annotated Bibliography of Amphibians and Reptiles*, 219
DOI 10.1007/978-1-4614-0824-6_8, © Springer Science+Business Media, LLC 2012

Mackness B, Sutton R. 2000. Possible evidence for intraspecific aggression in a Pliocene crocodile from north Queensland. Alcheringa 24:55–62.

Other – Pliocene crocodile (Queensland Museum QMF F30566) with semi-discrete roughened knobby protrusions, distinctly demarcated from underlying bone. They interpreted this as infection, but it is a very distinct pathology, whose etiology must be considered unproven.

Fossil – Pliocene crocodile (Queensland Museum QMF F30566) with semi-discrete roughened knobby protrusions, distinctly demarcated from underlying bone. They interpreted this as infection, but it is a very distinct pathology, whose etiology must be considered unproven.

Mackness BS, Cooper JE, Wilkinson C, Wilkinson D. 2011. Palaeopathology of a crocodile femur from the Pliocene of eastern Australia. Alcheringa: An Australasian Journal of Palaeontology 34:515–521.

Trauma – Pliocene crocodile femur (similar to morphology of *Crocodylus porosus*) QMF 30583, identified as probably *Pallimnarchus* or *Crocodylus* with an oblique linear fissure transversing the diaphysis, associated with proliferative new bone formation. They suggest a lateral draining sinus adjacent to a small isolated nodule at the proximal-lateral edge of the fissure, noting bump at opposing edge. Two large puncture wounds were noted in the middle of the fissure, along with several shallow surface depressions. Healed fracture is suspected. If this were seen in an extant animal, a wire injury would have been blamed.

Maddux S. 1996. Unusual turtles survive in spite of physical challenges; Natural oddities on display at fireworks stand. South Bend Tribune (Indiana) 19 July 2004:1.

Congenital – Dicephalic and Siamese twin turtles.

Madej Z. 1965. Variations in the sacral region of the spine in *Bombina bombina* (Linnaeus 1758) (Salientia, Discoglossidae). Acta Biologica Cracoviensia, Serie Zoologica 8:185–197.

Environmental – Number of vertebrae and symmetry of sacrum in *Bombina bombina* depend upon incubation temperature: if nine is normal, then eight with decreased temperature and ten with increased temperature.

10.45% of 555 fire-bellied toad *Bombina bombina* from large water reservoirs (with minimal temperature change) had spinal abnormalities as did 10.70% of 813 yellow-bellied toad *Bombina variegata* from puddles.

Unilateral sacralization of last lumbar vertebra was noted in 26% of *B. bombina*, compared with 1.97% of *B. variegate*. It was bilateral in 2.52% and 2.95%, respectively and affected two vertebrae (1.08% and 1.85%) with abnormal urostyle in 4.6% and 5.43%, respectively.

Maden M. 1982. Vitamin A and pattern formation in the regenerating limb. Nature 295:672–675.

Toxicology – Retinoids produced complete limbs from carpal (but not tarsal) amputations in *Ambystoma mexicanum*. Extra carpals and abnormal radius/ulnar lengths and digits deficiencies were also noted.

Maden M. 1983a. The effect of vitamin A on limb regeneration in *Rana temporaria*. Developmental Biology 98:409–416.

Toxicology – Vitamin A produces serial proximo-distal duplications and mirror images anterior-posterior duplications in *Rana temporaria.*

Maden M. 1983b. The effect of vitamin A on the regenerating axolotl limb. Journal of Embryology and Experimental Morphology 77:273–295.

Toxicology – Vitamin A causes proximal elements to regenerate from distal amputation levels in *Ambystoma mexicanum.*

Maden M. 1993. The homeotic transformation of tails into limbs in *Rana temporaria* by retinoids. Developmental Biology 15379–391.

Toxicology – *Rana temporaria* developed double-posterior limbs on exposure to 10 IU/ml for 2–3 days, but at 20, are inhibited. Missing digits, shortened, fused, or missing long bones also were noted.

Vitamin A and derivative retinoids posteriorize limb regeneration subsequent to axolotl limb amputation (Maden 1983; Kim and Stocum 1986).

Treatment of amputated *Uperodon systoma* tails with retinoids transforms tails into limbs (Mohanty-Hejmadi et al. 1992). In contrast, only inhibition of tail regeneration was noted in *Bufo andersonii, Notophthalmus viridescens, Ambystoma mexicanum, Xenopus laevis*, and *Rana catesbiana* (Niazi and Saxena 1979; Scadding 1987).

Mader DR. 1990a. Metabolic bone disease in captive reptiles. Vivarium (Lakeside) 2(3):12–14.

Metabolic – States that a number of terms often associated with Metabolic Bone Disease (MBD) are sometimes erroneously used interchangeably. These include: osteoporosis, osteomalacia, rickets, fibrous osteodystrophy (osteodystrophy fibrosa), osteopetrosis, and nutritional secondary hyperparathyroidism. He states that these are all distinct conditions or lesions, which relate to the changes in the bones during MBD.

Gopherus agassizi shell hump

Iguana iguana fibrous dystrophy in

Mader DR. 1990b. Metabolic bone disease in captive reptiles. Part 2. Vivarium (Lakeside) 2(4):12–13.

Trauma – Tegu *Tupinambis* sp with fracture from nutritional secondary hyperparathyroidism.

Metabolic – The bones most commonly decalcified by metabolic bone disease are the long bones (legs or ribs), the lower jaw, the flat bone of the skull, and the vertebrae, commonly causing soft, pliable shells in turtles and swollen or broken legs. In severe conditions, the weight of the animal alone is sufficient to cause a pathological fracture

Iguana iguana with nutritionally related fibrous dystrophy.

Tegu (*Tupinambis* sp.) with nutritional secondary hyperparathyroidism-related fracture.

Mader DR. 1993. Use of calcitonin in green iguanas, *Iguana iguana*, with metabolic bone disease. Bulletin of the Association of Amphibian and Reptilian Veterinarians 3(1):5.

Metabolic – Calcitonin treatment of metabolic bone disease in *Iguana iguana* produced recovery in 4–6 weeks, rather than 4–6 months.

Mader DR. 1996. Gout. In Mader DR. ed. Reptile medicine and Surgery. Pp. 374–379; Philadelphia: Saunders.

Metabolic – Recalls that Hippocrates named disease of big toe, hand, or knee of humans as podagra, cheigagra, or Gonagra.

Reptilian risk factors include dehydration, kidney (especially tubular) damage, excess purine-rich foods (e.g., high animal protein as in cat food) and nephrotoxic antibiotics.

Article is based on human gout, but lytic changes in an old world chameleon manus, a "tortoise" neck and an albino gopher tortoise stifle were attributed to gout.

Mader DR. 2000. Reptilian metabolic disorders. In Fudge AM. ed. Laboratory Medicine: Avian and Exotic Pets. Saunders, Philadelphia. Pp. 210–222.

Metabolic – General discussion of gout, using the human model.

Mader DR. 2002. Tail problems in the green iguana. Proceedings of the North American Veterinary Conference 16: 925–926.

Trauma – Tail damage from trauma and autotomy.

Infection – Tail damage from osteomyelitis.

Mader DR. 2006a. Perinatology. In Mader DR. ed. Reptile Medicine and Surgery. Pp. 365–375; Philadelphia: Saunders.

Congenital – Dicephalic California king snake *Lampropeltis getulua*and bearded dragon *Pogona vitticeps.*

Scoliosis and hydrocephalus in Chinese water dragon *Physignathus cocincinus.*

Flipper duplication in *Caretta caretta*

Siamese (at pelvis) twin red-eared slider *Trachemys scripta elegans.*

Mader D. 2006b. Calculi: Urinary. In Mader DR. ed. Reptile Medicine and Surgery. Pp. 763–771; Philadelphia: Saunders.

Stones – Urinary stones in Argentine horned frog *Ceratophrys ornata*, citing: Greek 2000, Blahak 1994, Mader et al. 1999, Kolle et al. 2001, McKown 1998, Homer et al. 1998, Raiti 2004,Frye 1972, Mangone and Johnson 1998, Mayer 2003, 2005.

Calcium pyosphate stone in red-eared slider *Trachemys scripta elegans* and leopard tortoise *Geochelone pardalis.*

Calcified renal abscess in green iguana *Iguana iguana.*

Pseudo-calculus (actually egg) in California desert tortoise *Gopherus agassizii.*

Uroliths in *Uromastyx maliensis* and African spurred tortoise *Geochelone sulcata.*

Mader DR. 2006c. Gout. In Mader DR. ed. Reptile Medicine and Surgery. Pp. 793–800; Philadelphia: Saunders.

Metabolic – 64% of European tortoise have renal disease and 16%, renal tophi, citing Kolle and Hoffman (2002).

Carpal involvement by gout in veiled chameleon *Chamaeleo calyptratus.*

Carpal, elbow, and phalange gout in gecko.

Cervical vertebrae destruction by gout in tortoise.

Mader DR. 2006d. Metabolic bone diseases. In Mader DR. ed. Reptile Medicine and Surgery. Pp. 841–851; Philadelphia: Saunders.

Congenital – Scoliosis in green iguana *Iguana iguana.*

Metabolic – Nutritional bone disease, manifested as fractures and bone enlargement in California desert tortoise *Gopherus agassizii*, Argentine horned frog *Ceratophrys ornata*, spring-tailed iguana *Ctenosaura pectinata.*

Pagetic appearance with cement lines and mosaic appearance is a normal phenomenon or result of infection in reptiles.

Other – Hypertrophic osteoarthropathy (periosteal reaction starting distally and Subsequently developing proximally) in black and white tegu *Tupinambus teguixin.*

Alleged osteopetrosis in Savannah monitor *Varanus exanthematicus.* The periosteal reaction was called solid with solid cortices (although that is not the radiologic appearance) and brittle bones.

Mader DR, Ling GV, Ruby AL. 1999. Cyctic calculi in the California desert tortoise (*Gopherus agassizii*): Evaluation of 100 cases. Proceedings of the Association of Amphibian and Reptile Veterinarians Annual Conference 1999:81–82.

Stone – Single calculi in 73; two, in 21; 3, in 5; and one California desert tortoise *Gopherus agassizii* with five stones, ranging in size from 0.4 to 14 cm (average = 5.4 cm).

Maderson PF, Bellairs A d'A. 1962. Culture methods as an aid to experiment on reptile embryos. Nature 195:401–402.

Trauma – Lack of tail regeneration in surgically treated *Anguis fragilis* and *Lacerta vivipara* embryos.

Mahapatra PK, Mohanty-Hejmadi P. 1994. Vitamin A-mediated homeotic transformation of tail to limbs, limb suppression, and abnormal tail regeneration in the Indian jumping frog *Polypedates maculatus*. Development Growth & Differentation 36:307–317.

Metabolic – Vitamin A-induced ectopic hindlimb in *P. mauclatus, U. systoma, R. temporaria, Bufo melanostictus*, and *Microhyla ornata.*

Mahapatra PK, Mohanty-Hejmadi P, Dutta SK. 2001. Polymely in the tadpoles of *Bufo melanostictus* (Anuria: Bufonidae). Current Science (Bangalore) 80:1447–1451.

Metabolic – Vitamin A-induced ectopic hindlimb in *P. mauclatus, U. systoma, R. temporaria, Bufo melanostictus*, and *Microhyla ornata.*

Mahendra BC. 1936. A case of polymely in the Indian bull-frog *Rana tigrina* Daud. Proceedings of the Indiana Academy of Science (B) 4(6):483–493.

Congenital – Polymely in Indian bull frog *Rana tigrina.* Cites Gervais 1864 supernumerary forelimb in *Pelobates cultripes*; Lunel 1868, in *Rana viridis*, Mazza 1888 in *Rana esamelica*, Sutton 1889 and Bergandale (citation not provided) in *Rana temporaria*, Washburn 1899 in *Bufo columbiensis* and Eigerman and Cox (1901) in *Rana pipiens.* Johnson 1901 reported polymely in *Rana palmipes* and *Rana halecinum*; O'Donoghue 1910 in *Hyla aurea* and *Rana temporaria.*

Mahnert V. 1984. Le (petit) monstre du Parc de la Grange. [The (small) monster of Grange Park]. Musées Genève 243:7–10. [French]

Congenital – Partial dicephalic *Podarcis muralis* and partial head and body duplication in *Elaphe conspicillata.*

Maiorana VC. 1977. Tail autotomy, functional conflicts and their resolution by a salamander. Nature 265:533–535.

Trauma – Tail loss in plethodontid salamander *Batrachoseps attenuatus* reduces reproductive output – perhaps from loss of energy reserves. Reduced breeding and delayed maturity were noted.

Makris SL, Solomon HM, Clark R, Shiota K, Barbellion S, Buschmann J, Ema M, Fujiwara M, Grote K, Hazelden KP, Hew KW, Horimoto M, Ooshima Y, Parkinson M, Wise LD. 2009. Terminology of developmental abnormalities in common laboratory mammals (Version 2). Congenital Anomalies 49:123–246.

Congenital – Approach to a vocabulary for congenital anomalies in mammals. Its application to amphibians and reptiles to be considered. It emphasizes the shifting patterns of "normal" to historical alterations in species manifestations. It notes that the term gnathoschisis has been used for both split mandible and for split maxilla and offers other definitions.

Malacinski GM, Brothers AJ. 1974. Mutant genes in the Mexican axolotl. Science 184:1142-1147.

Congenital – Mutant genes in Mexican axolotl *Ambystoma* mexicanum:

- s – Short toes (reduced number of phalanges) and abnormal rib formation (Briggs 1973; Humphrey 1967, 1974/5; Martin and Signoret 1968)
- ph – phocomelia (Briggs 1973; Humphrey 1974/5)
- y – limb development arrest (Briggs 1973; Humphrey 1973, 1974/5)
- b – slow front limb development (Briggs 1973; Humphrey1974/5)
- h – foot malformation (Briggs 1973; Humphrey1974/5)

Malashichev YB. 2002. Asymmetries in amphibians: A review of morphology and behaviour. Laterality 7(3):197–217.

Congenital – Asymmetries in amphibians, noting that *Bombina* and *Xenopus* are extremely symmetrical compared to other anurans and urodeles.

Malashichev YB, Roger LJ. 2002 Behavioural and Morphological Asymmetries in Amphibians and Reptiles. Proceedings of the 4th World Congress of Herpetology Satellite Symposium:108 pp. Laterality 7(3):195–299.

Congenital – Asymmetries in amphibians, noting that *Bombina* and *Xenopus* are extremely symmetrical compared to other anurans and urodeles.

Malkmus R. 1981. Bemerkungen zu einer *Triturus boscai*-Population in einem Brunnenbecken der Serra de Sintra. [Remarks on a *Triturus boscai* population in a fountain basin in Serra de Sintra]. Boletim da Sociedade Portuguesa de Ciências Naturais 20:25–40. [German]

Congenital – Digital anomalies in *Triturus boscai*: Short digits, lateral outgrowth on digits.

Malz H. 1967. "*Branchiosaurus*," ein problematisches Ur-Amphib aus dem Perm. Natur und Museum 97(10):397–406. [German]
Congenital – *Micromelerpeton* with five instead of four phalanges in anterior extremity.
Fossil – *Micromelerpeton* with five instead of four phalanges in anterior extremity.

Mangold O, Wächter H. 1953. Der Einfluß ungünstiger äußerer Bedingungen während der ersten Entwicklungsphasen auf die Ausgestaltung der Larven von *Triton alpestris*. [The influence of unfavorable external conditions during the first developmental stages in the formation of larvae of *Triton alpestris*]. Naturwissenschaften 40:323–334. [German]
Environmental – Anomalies in O_2-deficient water (after Freytag 155).

Mangone B, Johnson JD. 1998. Surgical removal of a cystic calculi via the inguinal fossa and other techniques applicable to the approach in the desert tortoise, *Gopherus agassizii*. Proceedings of the Association of Amphibian and Reptile Veterinarians Annual Conference 1998:87–88.
Stones – States that calculi are frequently encounted in many species of captive tortoise.
Calculus in desert tortoise, *Gopherus agassizii*.

Manharth A, Lemberger K, Mylniczenko N, Pinkerton M, Pessier AP, Kammeyer P, de Hoog S. 2005. Disseminated phaeohyphomycosis due to an *Exophiala* species in a Galapagos tortoise, *Geochelone nigra*. Journal of Herpetological Medicine and surgery 15:20–26.
Infection – *Phaeohyphomycosis exophiala* infection of Galapagos tortoise *Geochelone nigra* gluteal skin with stones.
Stone – *Phaeohyphomycosis exophiala* infection of Galapagos tortoise *Geochelone nigra* gluteal skin with stones.

Manimozhi A, Baker N, Sekar N. 2006. Twinning in *Python molurus molurus* in captivity. Hamadryad 30(1): 203–204.
Congenital – Identical, not Siamese Indian rock python *Python molurus molurus* twins.

Manner HW, Zapisek WF, Vallee JA. 1960. The effect of body size on the rate of salamander limb regeneration. Anatomical Record 137:378–390.
Trauma – No effect of body size on regeneration in *Triturus viridescens*.

Manteifel YB, Reshetnikov AN. 2002. Avoidance of noxious tadpole prey by fish and invertebrate predators: Adaptivity of a chemical defense may depend on predator feeding habits. Archiv fur Hydrobiologe 153:657–668.
Trauma – Dragonfly nymphs *Aeschna cyanea* capture but prefer to dine on *Rana arvalis* to *Bufo bufo*. More of the latter are released in a damaged state, mostly with tail fin "rupture."

Marcgrav de Liebstadt G. 1648. Historiæ rerum naturalium Brasiliæ, libri octo: quorum tres priores agunt de plantis. Quartus de piscibus. Quintus de avibus. Sextus de quadrupedibus & serpentibus. Septimus de insectis. Octavus de ipsa regione, & illius incolis. Cum appendice de Tapuyis, et Chilensibus. Ioannes de Laet... in ordinem digessit & annotationes addidit, & varia ab auctore omissa supplevit & illustravit. [Natural history of Brazil, book 8: priority among plants; 4, fish; 5, birds; 7, insects; 8, regional. With appendix of Tapuy, Chile, Greece with additional annotations and various omissions supplemented and illustrated]. 293 p. Lugdun. Batavorum [i.e., Leiden]: F. Hackium; Amstelodami: L. Elzevirium. [Latin]
Trauma – Bifid lizard "meiva" (= Taraguira of the Indians) tail on p. 238.

Marcucci E. 1930. Il potere rigenerativo degli arti nei rettili [The regenerative power of limbs in reptiles]. – Archivo zoologico Italiano 14: 227–252. [Italian]
Trauma – *Lacerta muralis*, *L. viridis*, and *Gongylus ocellatus* could regenerate their posterior limbs to differing degrees and *L. muralis* and *L. viridis*, sometimes regenerate their anterior limbs. The more proximal the location of the amputation, the greater extent of regeneration. Regenerated portions were never limblike, but rather occurred as cylindrical, tapering, long, and thin outgrowth that externally resembled a small tail. The outgrowths sometimes had a cartilaginous axis.

Marcucci E. 1931. La rigenerazione nei Rettili [Regeneration in Reptiles]. Archivo zoologico Italiano 16: 455–458. [Italian]
Trauma – Review of literature on regeneration in reptiles, focusing on the discussion of regeneration of limbs. He concluded based on his experiments and the literature, that *Lacerta muralis*, *L. viridis*, *L. agilis*, *L. vivipara*, *Gongylus ocellatus*, and probably other saurians, were able to regenerate not only the tail, but the limbs too. This regeneration is composed of a cartilaginous skeleton, which may also be in part bony, of varying length and shape (usually stump-like).

Marcus LC. 1968. Diseases of snakes and turtles. In Kirk RW. Current Veterinary Therapy III. Small Animal Practice. Saunders, Philadelphia. Pp. 435–442.
Metabolic – Articular gout in turtles.

Marcus LC. 1971. Infectious diseases of reptiles. Journal of the American Veterinary Medical Association 159:1626–1631.
Shell disease – Algae lifts up margins of turtle shields.

Marcus LC. 1980. Bacterial infections in reptiles. In Murphy JB, Collins JT. eds. Reproductive Biology and Diseases of Captive Reptiles. Society for the Study of Amphibians and Reptiles. Contributions to Herpetology No. 1, Oxford, Ohio. Pp. 211–221.

Shell disease – *Beneckea chitinova* produces shell disease in turtles.

Marcus LC. 1981. Veterinary Biology and Medicine of Captive Amphibians and Reptiles. 293 pp.; Lea & Febiger, Philadelphia.

Congenital – Achondroplasia in *Chrysemys (Pseudemys) scripta elegans.*

Metabolic – Decalcification, bowing, and expanded irregular metaphyses from rickets in *Xenopus laevis*); spontaneous fracture of ribs, mandible, leg bones, cranium, vertebrae, digits in green iguana *Iguana iguana*, osteoporosis, and fractures from vitamin D deficiency in African clawed toad *Xenopus laevis* and *Iguana iguana.*

Secondary hyperparathyroidism in *Alligator mississippensis* in AFIP 63–3000–2.

All meat diet produces secondary hyperparathyroidism in *Alligator* and *Phelsuma.*

Gout deposits in hind leg of dwarf crocodile *Osteolaemus tetraspis.*

Pseudomys scripta elegans is a species in which shrimp feeding produces gout (serum uric acid level of 7.8/dl, which is 6 × normal).

Vertebral – Osteoperiostitis in Chinese strip tail snake *Elaphe taenura* is actually proliferative non-marginal syndesmophytes of spondyloarthropathy.

Marelli CA. 1942. Monstruosidad en nuestro sapo común [Monstrosity in our common frog]. In Ministerio de Obras Públicas de la Provincia de Buenos Aires. Memorias del Jardín Zoológico correspondientes al periodo 1940–1941. Agregados al tomo IV: Los Vertebrados exhibidos en los Zoológicos del Plata. Año 1930–1931. Volume 10: Plate CXIII. La Plata. [Spanish]

Congenital – common frog *Bufo arenarum* with two partially functional supernumerary hind limbs, situated below the functional hind limbs when in jumping position. Marelli also stated, in passing, that she thought it was uncommon in cold-blooded animals, in contrast to warm-blooded animals.

Marion KR, Nowak MC. 1980. One-egg twins in a snake, *Elaphe guttata guttata*. Transactions of the Kansas Academy of Science 83(2):98–100.

Congenital – *Elaphe guttata guttata.*

Marolt J, Frank A. 1970. A case of keratoconjunctivitis and endogenous uveitis in an alligator. Deutsche Tierärztliche Wochenschrift 77(8):190–192.

Vertebral – Keratoconjunctivitis and endogenous uveitis in *Alligator mississippiensis* (significant because of association with human pondyloarthropathy – BMR).

Martill DM. 1996. Fossils explained: Ichthyosaurs. Geology Today 12:194–196.

Trauma – Tooth marks attributed to pliosaurs (e.g., *Lipopleurodon*) are frequently present on ichthyosaur vertebrae.

Ichthyosaur bones in coprolites.

Fossil – Tooth marks attributed to pliosaurs (e.g., *Lipopleurodon*) are frequently present on ichthyosaur vertebrae.

Ichthyosaur bones in coprolites.

Martin D. 1996. Blackboard; Turtle Turns Heads, Literally. The New York Times 31 March 1996; 4A:12.

Congenital – Dicephalic turtle.

Martin J, Avery RA. 1997. Tail loss affects prey capture "decisions" in the lizard *Psammodromus algirus*. Journal of Herpetology 31:292–295.

Trauma – Loss of tail in *Psammodromus algirus* reduces sprint performance in prey pursuit and delays "decision" to attack and have lower capture success.

Martin J, Avery RA. 1998. Effects of tail loss on the movements of the lizard *Psammodromus algirus*. Functional Ecology 12:794–802.

Trauma – Escape speed and stride length were reduced by tail loss in *Psammodromus algirus*, as was distance moved and size of home range.

Tail loss in *Scincella lateralis* reduces frequency of movement (Formanowicz et al. 1990) and in *Lacerta monticola*, reduces distance moved and size of home range.

Martin JE, Bell GL Jr. 1995. Abnormal caudal vertebrae of mosasauridae from Late Cretaceous marine deposits of South Dakota. Proceedings of the South Dakota Academy of Science 74:23–27.

Trauma – Wounds in skulls of many mosasaurs and plesiosaurs also fed on one another. *Mosasaurus conodon* at Museum of Geology had "chewed" skull and cervical vertebrae, with tooth of another jammed between quadrate and lower jaw. Pliosaurid plesiosaur paddle had healing around imbedded plesiosaur tooth.

Infection – Seven infectious fusion of mosasaur (possibly *Clidastes*) caudal vertebrae and three in a plioplatecarpine mosasaur.

Fossil – Wounds in skulls of many mosasaurs and plesiosaurs also fed on one another. *Mosasaurus conodon* at Museum of Geology had "chewed" skull and cervical vertebrae, with tooth of another jammed between quadrate and lower jaw. Pliosaurid plesiosaur paddle had healing around imbedded plesiosaur tooth.

Seven infectious fusion of mosasaur (possibly *Clidastes*) caudal vertebrae and three in a plioplatecarpine mosasaur.

Martin JE, Bjork R. 1987. Gastric residues associated with a mosasaur from the Late Cretaceous (Campanian) Pierre Shale in South Dakota. Dakoterra 3:68–72.

Trauma – *Clidastes liodontus* ingested by *Tylosaurus proriger* SDSM 10439, citing Williston's (1898) speculation that mosasaurs may have fed on mosasaurs and noting Russell's 1967 report of smaller mosasaur remains intermingled with a large mosasaur skeleton.

Fossil – *Clidastes liodontus* ingested by *Tylosaurus proriger* SDSM 10439, citing Williston's (1898) speculation that mosasaurs may have fed on mosasaurs and noting Russell's 1967 report of smaller mosasaur remains intermingled with a large mosasaur skeleton.

Martin LD, Rothschild BM. 1989. Paleopathology and diving mosasaurs. American Scientist 77:460–467.

Infection – Infections spondylitis from shark predation in mosasaurs.

Vascular – Presence of avascular necrosis (from bends/decompression syndrome) allowed recognition of diving behavior in Cretaceous mosasaurs and their prey, turtles.

Fossil – Infections spondylitis from shark predation in mosasaurs.

Presence of avascular necrosis (from bends/decompression syndrome) allowed recognition of diving behavior in Cretaceous mosasaurs and their prey, turtles.

Martin J, Salvador A. 1992. Tail loss consequences on habitat use by the Iberian rock lizard, *Lacerta monticola*. Oikos 65:328–333.

Trauma – During summer, Iberian rock lizard, *Lacerta monticola* that lost tails were more often on rocks and used grassy and bare ground less than tailed lizards. Notes long tails in widely foraging lizards and that they use running speed to escape (Vitt 1983).

Martin J, Salvador A. 1995. Effects of tail loss on activity patterns of rock lizards, *Lacerta monticola*. Copeia 1995:984–988.

Trauma – Tailless rock lizard *Lacerta monticola* males were less active during the reproductive period.

Martin J, Salvador A. 1997. Effects of tail loss on time budgets, movements and spacing patterns of the Iberian rock lizard *Lacerta monticola*. Herpetologica 53:117–125.

Trauma – Iberian rock lizard *Lacerta monticola* were minimally affected by tail loss, but tail-less males moved less, had more pauses and spent more time on rocks and less on bushes, with smaller home range size.

Martin R, Signoret J. 1968. Contribution à l'étude de la mutation "short toes" chex l'axolotl. [Contribution to the study of the "short toe" mutation in axolotl]. Annals of Embryology and Morphology 1:141–149. [French]

Congenital – Short toes in *Ambystoma mexicanum*.

Martín de Jesús S, Jiménez Fuentes E, Mulas E. 1989. Un pelomedúsido (Chelonia) con malformaciones patológicas, del Eoceno de Zamora [A pelomedusid (Chelonia) with pathologic malformations, from the Eocene of Zamora]. Studia Geologica Salmanticensia 26:355–364. [Spanish]

Trauma – Eocene Pelomedusidae turtles *Neochelys* aff. *salmanticensis* with trauma-induced asymmetrical xifiplastron and pelvis.

Fossil – Eocene Pelomedusidae turtles *Neochelys* aff. *salmanticensis* with trauma-induced asymmetrical xifiplastron and pelvis.

Martinez A, Soler J. 2000. Anomalies and malformations in European amphibians and reptiles. Reptilia (GB)11:10–12.

Congenital – Dicephalic viperine water snake, *Natrix maura*, common in captivity-bred red-eared slider, *Trachemys scripta elegans*, but rare in diamondback terrapin, *Malaclemys terrapin*.

Polymely in marbled newt, *Triturus marmoratus*.

Polydactylism in Hermann's tortoise.

Amelia in jeweled lizard *Timon lepidus*.

Shell disease – Duplication of vertebral scutes in *Testudo hermanni*, reduction in scales in *Testudo graeca*.

Martínez I, Alvarez R, Herráez L, Herráez P. 1992. Skeletal malformations in hatchery reared Rana perezi tadpoles. The Anatomical Record 233:314–320.

Environmental – High temperature produces scoliosis from hemivertebrae, arched dysplasia, aplasia, and failure of ossification in *Rana perezi*. They noted that increased temperature produces increased food ingestion, which they said can produce malformations.

Martínez Silvestre A. 1994. Manual Clínico de Reptiles [Clinic Manual of Reptiles]. Barcelona: Grass-Iatros Ediciones, 169 pp. [Spanish]

Congenital – Siamese or double monsters are well known in saurians, chelonians, and ophidians, as are bicephalic, two-tailed, and conjoined bodies (thoracodymus, ischiodymus, etc.).

Nuchal plate is frequently absent in *Testudo*. Scale duplication in carapace of chelonians and ventral scales of ophidians, as was illustrated for *Testudo hermanii*. It is often associated with missing eyes (cyclops) and other head malformations, including micrognathia (e.g.,defective upper jaw in *Trachemys scripta elegans*).

Absence and deformation of the tail and extremities has been noted. Amely has been reported in *Tryonix*, *Emydura*, and *Caretta*; polydactyly in crocodilians.

Chelonians are known to have congenital lordosis and kyphosis.

Polydactyly is illustrated in a young *Testudo hermanni*, which had six digits in each fore foot, and eight in each hind foot (normal count is five fingers in each foot).

Trauma – *Trachemys scripta* with a gaping hole in its carapace (chewed, about ¼ the size of the carapace), a half eaten forelimb, and other limb and face damage from of a domestic gerbil.

Fractures were reported in pectoral and humoral plastron plates in *Testudo hermanni*; carapace of *Trachemys scripta* (bitten by a dog), fracture in proximal zone of *Mauremys leprosa* humeral diaphysis (area that coincides with the place hit by the carapace as result of a fall).

Rostral abrasion common in saurian and ophidian specimens, which keep trying to escape their enclosures, or animals that are very curious and active in enclosures that may have sharp corners or abrasive parts

Infection – Subcutaneous abscess in a hind foot of a *Trachemys scripta* produced tarsal osteolysis.

Otic (ear) Abscesses are most common in chelonians, and can cause grave cranial deformation in extreme cases, exemplified by *Testudo graeca*

Metabolic – Metabolic Bone Disease (or Nutritional Osteodystrophy) very common in saurians and chelonians, uncommon in ophidians and varanids because the latter swallow prey whole.

Nutritional osteodystrophy in a very young *Iguana iguana* with femoral fracture and deformed toes and hindlimbs. *Trachemys scripta elegans* with soft, deformed carapace from osteodystrophy.

Renal Osteopathy in reptiles, exemplified by a chameleon with fibrotic formations in the ribcage.

Paravertebral bony mass (miscalled osteitis deformans) in ophidians as described by Frye (1991), in pythons, boas and crotalids, and reported in *Elaphe obsoleta quadrivittata*.

Articular gout is common in ophidians and chelonians, but possible in any reptile

Pseudogout reported in chelonians and saurians.

Neoplasia – Fibrosarcoma in ophidians and osteofibromas in saurians.

Stone disease – *Iguana iguana* with a huge vesicle stone

Shell disease – Ulcerative shell disease, caused by *Citrobacter freundii* in soft-shelled chelonians like *Tryonix* and *Apalone*, illustrated in *Trachemys decorata*.

Ulcerative Shell Disease, caused by *Beneckea chitinovora*, is common in captive aquatic chelonians. Affected animals usually develop secondary infections and osteomyelitis. Cutaneous Mycosis caused usually by *Trichosporon* can affect the carapace of turtles like *Testudo graeca*.

Martínez Silvestre A, Soler J, Solé R, Sampere X. 1998. Polidactilia en *Testudo hermanni* y causas teratogénicas en reptiles. [Polydactyly in *Testudo hermanni* and causes of teratology in reptiles.] Boletín de la Asociación Herpetológica Española:35–38. [Spanish]

Congenital – Polydactylous *Testudo hermanni*: The left foot had eight fingers, the right had nine, and the hands had six fingers each. Malformations were observed in the metacarpus/metatarsus and associated phalanges of all affected fingers, but not in the carpus/tarsus.

Martins d'Alte, J.A. 1937. Repteis bicefálicos [Dicephalic reptiles]. Anais da Faculdade de Ciências do Pôrto 22(1): 53–56. [Portuguese]

Congenital – Dicephalic *Lacerta muralis* Laur. (possibly atlodimous) and *Podocnemis unifilis* Troschell, described and pictured in the work entitled Viagem Filosófica by Dr. A. Rodrigues Ferreira, which remains unpublished but exists as two manuscripts – one in the Bocage museum of Lisbon, and another in the National Library of Rio de Janeiro (Brazil). There exist in addition, two books with drawings and watercolor illustrations from this expedition to northern Brazil (in 1783–1793), drawn by artists that accompanied Dr. Rodrigues. According to Martins, information about this turtle specimen was published subsequently by J. Bethencourt Ferreira in 1923. Lastly, Martins d'Alte reported a *Clemys leprosa* Schwg. in the museum of Lisbon.

Martins d'Alte JA. 1941. Notas teratológicas – VI-VII – [Teratology notes VI-VIII] Anais da Faculdade de Ciências do Pôrto XXVI No 4, p. 245 Porto. [Portuguese]

Congenital – Bifid forelimb in *Pleurodeles waltl*, with six digits (Normal = 4), two atrophied metacarpals, lacking phalanges

Martins D'Alte JA. 1945. Un cas de mélomélie chez la *Chioglossa lusitanica* Boc. [A case of melomely in *Chioglossa lusitanica*]. Boletim da Sociedade Portuguesa de Ciencias Naturales 15:47–51. [French]

Congenital – Melomelia in *Chioglossa lusitanica*.

Definitions – Pygomely – posterior hypogastric origin of accessory limbs
Gastromelia – anterior abdominal origin of accessory limbs
Notomelia – accessory limb insertion in dos
Cephalomelia – accessory limb insertion in head
Melomelia – accessory limb insertion at normal site
Schistomelia – divided hand (Giraldes 1869; Pires de Lima 1927)
in *Pleurodeles waltl* (Michah 1941)
Trimelia – limb divided in three.

Martof B. 1954. Variation in a large litter of garter snakes, *Thamnophis s. siratlis*. Copeia 1954:100–105.

Congenital – 17% of garter snakes *Thamnophis sirtalis* had half-ventrals and an elongated lower jaw was also reported.

Marvin GA. 1995. *Plethodon glutinosus* (slimy salamander). Morphology. Herpetological Review 26:30.

Congenital – Scoliosis in slimy salamander *Plethodon glutinosus*.

Marvin GA. 2010. Effect of caudal autotomy on aquatic and terrestrial locomotor performance in two Desmognathine salamander species. Copeia 2010:468–474.

Trauma – Autotomy in salamanders (Duey and Brodie 1983; Majorana 1977).

Autotomy developed predominantly in terrestrial species (Wake and Dresner 1967).
Tail loss alters behavior, activity, and social status in lizards (Cooper 2003; Gox et al. 1990; Salvador et al. 1996).
Tailless individuals less effective in avoiding predation (Congdon et al. 1974; Ducey and Brodie 1983; Wilson 1992).
When tails contribute to locomotion, loss reduces spring speed (Downes and Shine 2001; Lin et al. 2006; Martin and Avery 1998).
Tail loss without effect on speed (Hamley 1990; Huey et al. 1990; Kelehear and Webb 2006).
Tail loss increases speed in lizards that do not use it for locomotion (Brown et al. 1995; Daniels 1983).
Plethodontid salamander genus *Desmognathus* aquatic locomotion speed is reduced by autotomy.
Autotomized *Eurycea bislineata* have slower speeds (Dowdey and Brodie 1989), while *Eumeces chinensis* requires loss of more than 51% of tail to reduce speed (Lin et al. 2006).

Marx, H. and G. B. Rabb. 1972. Phyletic analysis of fifty characters of advanced snakes. Fieldiana: Zoology 63:1–321.

Congenital – Variation in scales in Colubridae, Elaphidae, Hydrophiidae, Viperinae, Crotalinae, *Azemiops*, and *Atractaspis*.

Maryan B. 2001. A note on a bicephalic death adder. Herpetofauna 31:73.

Congenital – Dicephalic *Pseudonaja affinis* andin a *Acanthophis wellsi* x *Acanthophis pyrrhus* hybrid.

Maskey TM, Schleich HH. 1992. Untersuchungen und Schutzmassnahmen zum Gangesgavial in Südnepal. [Investigations and protective measures for the Ganges gavial in South Nepal]. Natur und Museum (Frankfurt-am-Main) 122:258–267. [German]

Congenital – Deformed and crossed snouts or short jaws, fused and missing extremities in captive gavials, noting "vertebral dystrophies" in more than 20%.

Conjoined twin gavial.

Mason R. 2000. Gentle winter visitors. St. Petersburg Times (Florida) 7 February 2000:3D.

Congenital – Dicephalic turtle.

Massare JA. 1987. Tooth morphology and prey preference of Mesozoic marine reptiles. Journal of Vertebrate Paleontology 7(2):121–137.

Trauma – Mosasaur fragments in *Tylosaurus proriger*, citing Bjork 1981.

Tooth marks on mosasaur SMM 13094 skull.
Hainosaurus sp. Containing turtle bones, citing Dollo 1987, Russell 1967.

Fossil – Mosasaur fragments in *Tylosaurus proriger*, citing Bjork 1981.

Tooth marks on mosasaur SMM 13094 skull.
Hainosaurus sp. containing turtle bones, citing Dollo 1987, Russell 1967.

Massare JA. 2002. Bone microstructure of *Platypterygius americanus* (Reptilia, Ichthyosauria). Journal of Vertebrate Paleontology 22 (supplement to 3):85A.

Metabolic – Suggested that increased bone porosity is relatable to deeper diving behavior for *Platypterygius americanus*. However, this is a different definition than in the clinical studies which actually describe a failure of cortical bone resorption.

Fossil – Mosasaur fragments in *Tylosaurus proriger*, citing Bjork 1981.

Tooth marks on mosasaur SMM 13094 skull.
Hainosaurus sp. containing turtle bones, citing Dollo 1987, Russell 1967.

Massare J, Young H. 2003. Gastric contents of a Jurassic ichthyosaur. Journal of Vertebrate Paleontology 23 (supplement to 3):76A.

Trauma – *Baptanodon (= Ophthalmosaurus)* UW34653 with hooklets.

Fossil – *Baptanodon (= Ophthalmosaurus)* UW34653 with hooklets.

Master TL. 1998. *Dendrobates auratus* (Black and green poison dart frog). Predation. Herpetological Review 29:164–165.

Trauma – Predation on black and green poison dart frog *Dendrobates auratus* by rufous motmot *Baryphthengus marhi.*

Mathur JK, Goel SC. 1974. A note on a tailless embryo of the lizard *Calotes versicolor.* British Journal of Herpetology 5: 420–422.

Congenital – Complete absence of tail in *Calotes versicolor.*

Maticic D, Stejskal M, Vnuk D, Stanin D, Babic T, Pecin M. 2007. Internal fixation of a femoral fracture in a green iguana developing metabolic bone disease – a case report. Veterinarski Arkhiv 77:81–86.

Trauma – Reptile bones take months to heal, in contrast to weeks in mammals (Gillespie 1994)

Metabolic – Comments that long bone fractures are usually the result of trauma or metabolic bone disease. Metabolic bone disease in a green iguana

Mathure JK, Goel SC. 1974. A note on a tailess embryo of the lizard, *Calotes versicolor.* British Journal of Herpetology 5:420–422.

Congenital – Tailess *Calotes versicolor* embryo.

Mathure JK, Goel SC. 1976. Pattern of chondrogenesis and calcification in the developing limb of the lizard *Calotes versicolor.* Journal of Morphology 149:401–420.

Congenital – *Calotes versicolor* normal chondrogenesis patterns. Sesamoids as examples of metaplastic mineralization.

Matuschka F-R, Bannert B. 1987. Cannibalism and autotomy as predator-prey relationship for monoxenous Sarcosporidia. Parasitology Research 74:88–93.

Trauma – Autotomy in Canarian lizard *Gallotia galloti.*

Matz G. 1977. Sur une anomalie de la carapace chez *Testudo* (*Chelonoidis*) *chilensis* Gray (Reptilia, Testudines). [On carapce anomalies in *Testudo* (*Chelonoidis*) *chilensis* Gray (Reptilia, Testudines)]. Bulletin de la Société Zoologique de France 102:497–500. [French]

Shell disease – Abnormal carapace in *Testudo* (*Chelonoidis*) *chilensis.*

Matz G. 1989a. An axial duplication with double body in the lizard *Egernia striolata* (Peters). Herpetopathologia 1:57–59.

Congenital – Vertebral duplication in *Egernia striolata.*

Matz G. 1989b. La duplication axiale chez les reptiles. 2. Lézards et serpents. [Axial duplication in reptiles. 2. Lizards and serpents]. Bulletin de la Société d'Étude Scientifique de l'Anjou, Nouvelle Série 13:183–208. [French]

Congenital – Dicephalism in *Lampropeltis getula californiae*

Review of literature:

Dicephalic *Mabuya striata striata, Trachydosaurus rugosus, Anguis fragilis, Leptotyphiops bicolor, Boa constrictor* (without subspecies, common name cannot be determined), Sri Lanka or rough-scaled sand boa *Eryx conicus*, Burmese python *Python molurus bivittatus*, dwarf or black-tailed boa *Tropidophis melanurus, Coluber constrictor, Coronella austriaca, Dromicus chamissonis, Elaphe climacophora, Elaphe conspicillata, Elaphe longissima, Elaphe obsoleta, Elaphe quatuorlineata sauromates, Elaphe vulpina, Erythrolamprus aesculapi, Helicops carinicauda infrataeniata, Heterodon platyrhinos, Homalopsis buccata, Heterodon platyrhinos platyrhinos, Hydrophis cyanocinctus, Lampropeltis getula californiae, Lampropeltis getula yumensis, Leimadophis poecilogyrus, Leptodeira annulata ashmeadii, Liophis miliaris, Natrix maura, Natrix natrix, Tropidonotus natrix, Natrix septemvittata, Philodryas patagoniensis, Pituophis melanoleucus, Rhabdophis tigrinus, Tachymenis peruviana, Thamnophis sirtalis concinnus, Thamnophis sirtalis sirtalis, Tropidoclonion lineatum lineatum, Xenochrophis piscator, Xenodon merremii, Naja naja, Pseudechis porphyriacus, Agkistrodon halys blomhoffi, Bitis arietans, Bothrops alternatus, Bothrops asper, Bothrops jararaca, Bothops jaracaracussu, Crotalus adamanteus, Crotalus atrox, Crotalus basiliscus basiliscus, Crotalus durissus terrificus, Crotalus horridus, Crotalus viridis oreganus, Crotalus viridis viridis, Sistrurus miliarius streckeri, Vipera ammodytes, Vipera aspis, Vipera berus*, Sri Lanka or rough-scaled sand boa *Eryx conicus, Pituophis melanoleucus* and "large rattler."

Partial dicephalism in *Lacerta agilis* and *Tropidonotus natrix*

Duplicated head and body of *Tiliqua scincoides, Coluber constrictor, Herpetodryas flagelliformis, Tropidonotus natrix, Ptyas mucosus, Crotalus viridis viridis, Viper aspis*

Posterior bifurcation in *Storeria dekayi, Calliophis japonicus, Vipera berus, Natrix s clarki, Thamnophis sirtalis, Natrix sipedon, Herpetodryas flagelliformis, Crotalus viridis, Coluber constrictor*

Head and tail bifurcation in *Tropidonotus sipedon.*

Two bodies in *Egernia striolata, Podarcis muralis, Anguis fragilis*
Body fusion in *Lacerta agilis*
Kyphoscoliosis in *Boa constrictor* (without subspecies, common name cannot be determined)

Matz G. 1994. Teratology. Pp. 106–107. In Bauchot R. ed. Snakes: a natural history. 220pp. New York: Sterliong Publ. Co.
Congenital – Derodymous smooth snake Natrix natrix, bull snake *Pituophis melanoleucus.*
Cycloptic viper *Vipera xanthina.*
Shortening or absence of mandible.

Matz G. 1997 (not 1998). La teratologie des reptiles. [Teratology in reptiles.] Bulletin de la Sociétè Herpètologique de France 82–83:5–14.
Congenital – Anomalies in 23 of 85 *Testudo hermanni* (Kirsche 1972), two with only one supracaudal (Wermuth 1961), 7 of 209 *Testudo h. hermanni.*
Multiple malformations in *Chrysemys, Gopherus*, and *Sternotherus odoratus.*
Cyclops in *Natrix maura.*
Vertebral torsion in *Epicrates c. maurus.*
Incomplete pastron, absent tail in *Emys orbicularis* (Hiller 1990).
Absence of nose, eyes, and tail in *Testudo graeca.*
Duplication of posterior part of body is rarely found in serpents
Amphidichotomy – duplication of posterior portions of body and head, noting general reviews on anomalies by Tuck and Hardy (1970), Smith and Pérez-Higareda (1987), Matz (1989), Payen 1995).
Anomalies *Epicrates c. maurus* were significantly more common than the in 0.5% of 250 *Lepidochelys kempi* newborns and 8.3% of 3656 embryos (Shaver and Chaney 1989).
30 of 72 *Phelsuma madagascariensis grandis* with anomalies.
Body duplication in *Egernia striolata.*
Vertebral deformity in *Epicrates c. maurus.*
Environmental – Temperatures above or below optimum produce anomalies.
Shell disease – Supernumerary scales in *Testudo h. hermanni, Testudo chilensis, Testudo elegans* (Calmonte 1968; Stemmler 1964), *Pseudemys s.* elegans (Mebs 1964), marine turtles (Babcock 1930; Capocaccia 1966; Coker 1910).
Frequent supernumerary scales in *Testudo g. graeca* (Kabisch 1989), *Testudo g. Iberia* (Kabisch 1989), *Testudo sulcata* (Cludsley-Thompson 1970), *Gopherus polyphemus* (Barker and Anyonio 1983), *Pseudemys s. elegans.*
Decreased number of scales in *Tesduod hermanni* (Wermuth 1961).

Mauldin GN, Done LB. 2006. Oncology. In Mader DR. ed. Reptile Medicine and Surgery. Pp. 299–322; Philadelphia: Saunders.
Neoplasia – Plaston involvement by lymphoblastic lymphoma in *Careta caretta*, citing Oros et al. (2001).
Bone involvement by myelogenous leukemia in *Pagona vitticeps*, citing Tocidiowski et al. (2001).
Digit involvement by squamous cell carcinoma in round island skink *Trachydosaurus rugosus*, citing Garner et al. (1994).
Squamous cell carcinoma involvement of mandible in water moccasin *Agkistrodon piscivorus* (citing Machotka 1984) and maxilla, nasal and palate in Burmese python *Python molorus bivittatus* (citing Wilhelm and Emswiller 1977).
Ossifying fibrosarcoma in mandible of green tree python *Morelia (Chondromython) viridis* and in Ball python *Python regius.*

Mautino M, Page D. 1993. Biology and medicine of turtles and tortoises. Veterinary Clinics North America Small Animal Practice 23(6):1251–1270.
Trauma – Shell injury.
Metabolic – Insufficient calcium/phosphorus ratio producing metabolic bone disease.
Shell disease – Septicemic cutaneous ulcerative disease.

Mayor A. 2000. The First Fossil Hunters: Paleontology in Greek and Roman Times. Princeton: Princeton University Press, 361 pp.
Philosophy – Teras, the Greek for monster, means "strikingly abnormal offspring" (p. 195). Mayor took this to mean the abnormalities were not hereditary.
Greek and Roman philosophers were "motivated by an argumentative relationship with tradition, rather than the explanation of evidence" (Page 213).
Aritstotle's method was a search for "regular combinations of characteristics" (page 217), concentrating on predictable, living forms, noting, but excluding unique "monstrosities." Aristotle's History of Animals contains a passage devoted to fishermens' reports of sea monsters, creatures "whose rarity makes them unclassifiable" (page 218). He referred to anomalies only to label them as "monstrous" and "irrelevant" (page 218).
Plato and Pausanias's concept that "animals take different forms in different climates and places" (page 224).

Mayer J. 2005. Comparison of different methods applicable for the reptilian urolith analysis. Journal of Herpetological Medicine and Surgery 15(1):31–35.

Stones – reptilian uroliths are not rare, though most have not been analyzed and none examined for microstructure (Frye 1972; Geisel and Kriegleder 1978; Keymer 1978; Blahak 1994; Homer et al. 1998; McKown 1998; Mader and Ling 1999; Kölle et al. 2001).

Reviews chemical analysis, which is simple, but cannot distinguish various calcium phosphates or hydrates. This contrasts with infrared spectroscopy, which is sensitive and accurate, but requires expensive equipment and requires special expertise. He notes that optical crystallography is inexpensive, but subjective) and that x-ray microanalysis is unambiguous, but can miss minor components. Scanning electron microscopy allows minor components to be recognized on basis of crystal shape, but crystal shape is variable, requiring second technique for verification.

Mazza F. 1888. Caso di melomelia anteriore in una *Rana esculenta* Linn. [Case of anterior melomely in a *Rana esculenta* Linn.] Atti della Società Italiana Scienze naturali e del Museo Civico di Storia Naturale in Milano 31: 145–150. [Italian]

Congenital – Melomelia in a *Rana esculenta.* There were two supernumerary arms attached to the body 4 mm below the left forelimb on separate scapula. Ectodactily in the feet was also present.

De Superville (1744) described *Rana* sp.? with a supernumerary arm to the right of its back (Gervais reported the extra arm on that specimen was to the left). Mazza concluded that anterior polimelia in batrachians is predominantly left sided.

Mazza F. 1892. Caso di dicefalia derodimica in *Anguis fragilis* [Case of derodymous bicephaly in *Anguis fragilis*]. Atti Soc. Ligustica di Scienze Nat. Anno III. [Italian]

Congenital – Derodymus young *Anguis fragilis.*

McArthur S. 1996. Veterinary Management of Tortoises and Turtles. Cambridge, Massachusetts: Blackwell Science, 170 pp.

Metabolic – Nutritional osteodystrophy produces pyramiding of carapace in turtles and soft shells in terrapenes.

Post-hibernation dehydration produces gout.

Vitamin A deficiency produces renal dysfunction which produces secondary gout in red eared terrapene *Pseudemys scripta elegans.*

Shell disease – Excess protein produces horny tissue overgrowth.

Ulcerative shell disease in red eared slider.

McArthur S. 2004. Interpretation of presenting signs. In McArthur S, Wilkinson R, Meyer J, eds. Medicine and Surgery of Tortoises and Turtles. Pp. 273–307; Oxford: Blackwell.

Congenital – Abnormal plate closure in Herman's tortoise *Testudo hermanii.*

Trauma – Rat bite trauma to limbs in post-hibernation animals.

Infection – Infected shoulder in red-eared slider *Pseudemys scripta elegans.*

Infected plaston/carapace in red-eared slider *Pseudemys scripta elegans.*

Star tortoise *Geochelone elegans* with mycotic infection of temporomanidbular joints and knees.

Metabolic – Metabolic bone disease in *Testudo marginata* and Turkish tortoise *Testudo ibera* and with infected plaston/carapace in red-eared slider *Pseudemys scripta elegans.*

Gout in *Testudo graeca graeca* and spur-thighed Moroccan or Moorish tortoise *Testudo graeca* (affecting stifle).

Shell disease – Infected plaston/carapace in red-eared slider *Pseudemys scripta elegans.*

McArthur S, Barrows M. 2004. Nutrition. In McArthur S, Wilkinson R, Meyer J, eds. Medicine and Surgery of Tortoises and Turtles. Pp. 73–85; Oxford: Blackwell.

Metabolic – One third of fresh water chelonians seen in Leeds/London clinic had nutritional osteodystrophy, citing Jackson 1980, compared to Keymer's (1978) 19%.

McAllister CT, Wallach V. 2006. Discovery of dicephalic western diamondback rattlesnake, *Crotalus atrox* (Serpentes: Viperidae), from Texas, with a summary of dicephalism among members of the genus *Crotalus.* Journal of the Arkansas Academy of Science (2006) 60:67–73.

Congenital – Dicephalic snakes include eight families: Leptotyphlopidae, Pythonidae, Boidae, Tropidophiidae, Viperidae, Elapidae, Hydrophiidae, and Colubridae.

Congenital

Dicephalic

Crotalus adamanteus (Anon 1954; Murphy and Shadduck 1978)

Western diamondback rattlesnake *Crotalus atrox* (Engelmann and Obst 1981).

Crotalus basiliscus (Wiley 1930)

Crotalus durissus (Amaral 1927; Belluomini et al. 1974; Matz 1989; Vanzolini 1947)

Crotalus horridus (Anon 1877 1915; Cunninghan 1937; Lasher 1980; Miller 1938; Proctor 1933; Rimkus 1947; Sherer 1995)

Crotalus lutosus (1936)
Crotalus mitchellii (Baird 1856)
Crotalus oreganus (Bridges 1926)
Crotalus scutulatus (Kelly 1909)
Crotalus viridis (Klauber 1956; McMullin 1963; Payen 1991; Reid 2005; Rogers 1970)
Burmese python *Python molurus* (Manimozhi et al. 2006).
Common adder *Vipera berus* (Lagerlund 1951)
Craniodichotomous
European asp *Vipera aspis* (Naulleau 1983).
Western diamondback rattlesnake *Crotalus atrox* with duplication of all but the most caudal viscera and with short left hyoid, noting previous reports by Anonymous 1975 and Muir 1990.
Fused mouths in two western diamondback rattlesnake (Muir 1990).

McCallum ML. 1999. *Rana sphenocephala* (southern leopard frog) malformities found in Illinois with behavioral notes. Transactions of the Illinois State Academy of Science 92:257–264.
Environmental = Low pH was associated with increased frequency of deformities in *Rana sphenocephala* (southern leopard frog).

McCallum ML, Trauth SE. 2000. Curly tail malformity in hatchlings of the alligator snapping turtle *Macroclemys temminckii* (Testudines: Chelydridae), from northeastern Arkansas. Journal of the Arkansas Academy of Science 54:150–152.
Congenital – Curly tail malformity, presenting as tight spiral or coil in hatchings of alligator snapping turtle *Macroclemys temminckii*

McCallum ML, Trauth SE. 2003. A forty-three year museum study of northern cricket frog (*Acris crepitans*) abnormalities in Arkansas: Upward trends and distributions. Journal of Wildlife Diseases 39:522–528.
Environmental = 104 abnormalities among 1464 northern cricket frog *Acris crepitans*, especially from Ozarks highland counties = $p < 0.001$, with mountains higher than coastal plains of Arkansas – $p < 0.01$.
Abnormalities were found in 3.33% between 1957 and 1979; 6.87%, in the 1990s; and 8.48%, in 2000.

McCarty D. 1999. Two-headed rattler. Oregonian 6 September 1999: Local Stories 1. Congenital – Dicephalic rattlesnake.

McCauley RH Jr. 1945. The reptiles and amphibians of Maryland and the District of Columbia. Private printing, Hagerstown, Md., 194 pp.
Trauma – Tail loss in five lined skink *Eumeces fasciatus*, amputation of one or more limbs in spotted/speckled turtle/terrapin *Clemmys guttata*, and loss of foot or part of tail in map turtle *Graptemys geographica*, *Kinosternon subrubrum subrubrum*, and *Sternotherus odoratus*.
McCracken H, Birch CA. 1994. Periodontal disease in lizards – A review of numerous cases. Annual Proceedings of the American Association of Zoo Veterinarians and Association of Reptilian and Amphibian Veterinarians 1994:108–115.
Infection – Facio-oral osteomyelitis in 39 agamids at Melbourne Zoologic Gardens, affecting common bearded dragon *Pogona barbata*, inland bearded dragon *P. vitticeps*, dwarf bearded dragon *P. minor*, Eastern water dragon, *Hydrosaurus pustulatus*, and Jackson's chameleon, *Chamaeleo jacksoni*. Calculus with multifocal jaw osteomyelitis was related to inappropriate diet of "soft, sloppy" food, mainly fruit. Seven cases of osteomyelitis in *P. vitticeps* and *P. barbata*, one in *P. minor*, and three in *C. jacksoni*.
Dental – Agamids, chameleons, and tuataras are unique among lizards in having acrodont teeth ankylosed to their jaws, as opposed to shed/replaced pleurodont teeth.
Calculus with multifocal jaw osteomyelitis was related to inappropriate diet of "soft, sloppy" food, mainly fruit.

McCurley K. 2005. The Complete Ball Python: A Comprehensive Guide to Care, Breeding and Genetic Mutations. China: ECO Herpetological Publishing, 304 pp.
Congenital – Genetics of color in ball python *Python regius*, but no comment on skeletal alterations.

McEwan B. 1982. Bone anomalies in the shell of *Gopherus polyphemus*. Florida Scientist 45:189–195.
Congenital – 50% of *Gopherus polyphemus* with supernumerary carapace bones, compared with 68% of *Gopherus barlandieri*.

McGowan C. 1973. Differential growth in three ichthyosaurs: *Ichthyosaurus communis*, *I. breviceps*, and *Stenopterygius quadriscissus* (Reptilia, Ichthyosauria). Royal Ontario Museum Life Sciences Contributions 93:1–21.
Congenital – *Ichthyosaurus communis*, *I. breviceps*, and *Stenopterygius quadriscissus* had isometric intracranial growth, but with negative allometry in *I. communis* and positive allometry of the orbit in *I. breviceps*. External nares had positive allometry in *I. communis*; maxillary growth, negative in *I. breviceps*, and positive in *Stenopterygius quadriscissus*.
Fossil – *Ichthyosaurus communis*, *I. breviceps*, and *Stenopterygius quadriscissus* had isometric intracranial growth, but with negative allometry in *I. communis* and positive allometry of the orbit in *I. breviceps*. External nares had positive allometry in *I. communis*; maxillary growth, negative in *I. breviceps*, and positive in *Stenopterygius quadriscissus*.

McGowan C. 1989. The ichthyosaurian tailbend: A verification problem facili by computed tomography. Paleobiology 15:429–436.

Congenital – Computerized tomography revealed wedge-shaped centra in *Leptopterygius tenuirostris* and *Eurhinosaurus longirostris*, which identifies tailbend.

Fossil – Computerized tomography revealed wedge-shaped centra in *Leptopterygius tenuirostris* and *Eurhinosaurus longirostris*, which identifies tailbend.

McKown RD. 1998. A cystic calculus from a wild western spiny softshell turtle (*Apalone* (*Trionyx*) *spiniferus hartwegi*). Journal of Zoo and Wildlife Medicine 29(3):347.

Stones – Captive western spiny softshell turtle *Apalone* (*Trionyx*) *spiniferus hartwegi* with apatite (basic calcium phosphate) and struvite (magnesium ammonium phosphate hexahydrare) crystals.

Emphasizes that calculi were not found in any wild-caught reptiles.

Cites calculi in captive green iguana *Iguana iguana* (Frye 1983) and San Esteban Island chuckwalla *Sauromalus varius* (Ryer 1983).

McMullin R. 1963. Two-headed rattler. Outdoor Life 131(3):14.

Congenital – Derdymous prairie rattlesnake.

McVean A. 1975. Autotomy. Comparative Biochemistry and Physiology A 51:497–505.

Trauma – Autotomy in *Lacerta vivipara* (Pratt 1946; Sheppard and Bellairs 1972) and amphibian (Wake and Dresner 1967).

Mebert K. 1993. Untersuchung zur Morphologie und Taxonomie der Würfelnatter *Natrix tessellate* (Laurenti 1768) in der Schweiz und im südlichen Alpenraum. [Research on the morphology and taxonomy of dice snake *Natrix tessellate* (Laurenti 1768) in Switzerland and the southern Alps]. Diploma thesis, University of Zurich. [German]

Congenital – Three types of scale anomalies – ventral scales parted in the middle (cleft), incompletely formed ventral scales which have only unilateral contact with dorsal scales and unpaired which fuse across the axis, predominantly subcaudal in position, noting all more common in introduced, in contrast to naturally occurring populations of dice snake *Natrix tessellate.*

Mebs D. 1964. Eine Anomalie der Carapaxschilder bei *Pseudemys scripta elegans*. [Anomalies in the carapace plates of *Pseudemys scripta elegans*]. Aquaterra 1:47–48. [German]

Congenital – Duplication of keratin plates.

Mebs D. 1965. Harnsteine bei Schildkröten. [Urinary calculi in turtles]. Salamandra 1:47–49. [German]

Stone – Urinary calculi (salts of uric acid) in bladder of *Testudo angulata* and in bladder, urethra, and some renal tubules of *Pseudemys scripta elegans.*

Meche G, Jaeger RG. 2002. Association of male red-backed salamanders with tail-intact versus tail-autotomized females during the courtship season. Journal of Herpetology 36:532–535.

Trauma – *Plethodon cinereus* males prefer autotomized females.

Medel RG, Jiménez JE, Fox SF, Jaksić FM. 1988. Experimental evidence that high population frequencies of lizard tail autotomy indicate inefficient predation. Oikos 53:321–324.

Trauma – *Lacerta lemniscatus, L. monticola*, and *L. altissimus* were equally efficient at escaping predators through tail autotomy, but have very different frequencies of tail breakage/regeneration (49, 68, and 53%, respectively), uncorrected for species longevity.

Cites Núñez an Yáñez (1984) for 51, 24, and 59% at another site.

Meek R. 1985. Aspects of the colony of *Testudo hermanni* in southern Yugoslavia. British Journal of Herpetology 6:437–445.

Trauma – 15% abnormal scutes, secondary to trauma, in *Testudo hermanni.*

Shell disease – 15% abnormal scutes, secondary to trauma, in *Testudo hermanni.*

Meek R, Inskeep R. 1981. Aspects of the field biology of a population of Hermann's tortoise (*Testudo hermanni*) in Southern Yugosalvia. British Journal of Herpetology 6:159–164.

Trauma – 4% of carapace injuries in Hermann's tortoise, *Testudo hermanni.*

Méhely L.von 1902 Über das Entstehen überzähliger Gliedmaszen. [On the formation of supernumerary extremities]. Allattani Közlemények (Zoologische Mitteilungen) 1:19–34. [German]

Congenital – *Pelobates fuscus* with two additional anterior left legs: specimen and bone arrangement figured. *Rana esculenta* with three right anterior legs (after Tornier 1898).

Méhely L.von 1905 Über das Entstehen überzähliger Gliedmaszen. [On the formation of supernumerary extremities]. Mathematischer und naturwissenschaftlicher Anzeiger der ungarischen Akademie der Wissenschaften 3:239–259 (repetition of Méhely 1902). [German]

Congenital – *Pelobates fuscus* with two supernumerary left forelegs and *Rana esculenta* with three supernumerary forelegs.

Meisel J. 1994. 2-headed turtle lacks direction, otherwise healthy. Arkansas Democrat-Gazette (Little Rock, AR) 28 June 1994:1.

Congenital – Dicephalic turtle.

Meister G. 1988. Terrariengeburt einer zweikopfigen lebenden *Boa constrictor*. [Birth of a two-headed living *Boa constrictor* in the terrarium]. Deutsche Aquarien-und Terrarien-Zeitschrift 35(4):137. [German]

Congenital – Dicephalic *Boa constrictor* (without subspecies, common name cannot be determined).

Mendelson JR. 1993. Comparison of the caudal morphology of fragile-tailed colubrid snakes. Abstracts of the Second World Congress on Herpetology, Adelaide, p. 170.

Trauma – High frequency of tail breakage in colubrids (*Scaphiodontophis, Rhadinaea*, and *Natriciteres*).

Merilä J, Forsman A, Lindell LE. 1992. High frequency of ventral scale anomalies in *Vipera berus* populations. Copeia 1992(4):1127–1130.

Congenital – High frequency (92/229) of ventral scale anomalies often associated with duplicated or fused vertebrae and or ribs in *Vipera berus*.

Merrell D. 1969. Natural selection in a lepard frog population. Journal of the Minnesota Academy of Science 35:86–91.

Environmental – 14.8% of Northern leopard frog *Rana pipiens* deformed hindlimbs (toes, absent foot, lower leg or entire leg, unilaterally) in July (attributed to Mud Lake Minnesota, 3.6% (attributed to migratory frogs) in September, but curiously statistically significant difference ($p < 0.01$) between 28 and 30 July samples.

Mertens R. 1924/5. Zwei Mißbildungen bei Schwanzlurchen. [Two malformations of urodeles]. Pallasia 2(1):82–83. [German]

Congenital – *Triturus carnifex cristatus* (Kammolch) specimen lacking right front leg (shoulder girdle formed only on left side).

Salamandra salamandra taeniata specimen with additional four digit hand on right front leg (metacarpals on broadened end of humerus).

Mertens R. 1940. Der Knochenpanzer einer kyphotischen Weichschildkröte. [The bony skull of a kyphotic soft turtle]. Senkenbergiana 22:236–243. [German]

Congenital – Kyphosis in *Kinosternon subrubrum hippocrepis, Deirochelys reticularia*, and *Amyda (Trioinyx) triunguis*.

Messonnier SP. 1996. Common Reptile Diseases and Treatment. 174 pp.; Blackwell Science, Oxford, England.

Metabolic – Metabolic bone disease in iguana and rarely in turtles.

Gout in snakes and tortoises.

Vascular – Digital "avascular necrosis" in iguana.

Shell disease – Shell disease in turtles.

Meteyer CU. 2000. Field Guide to Malformations of Frogs and toads, with Radiographic Interpretations. U.S. Geological Survey, Biological Resources Division, Biological Science Report USGS/BRD/BSR-2000–0005:1–16.

Congenital – Images of malformations, noting dramatic increase in 44 states in 38 species of frogs and 19 species of toads affecting as many as 60% (e.g., American toad, Pickerel frog, Northern leopard frog, bronze frog).

Definitions

Brachydactyly – reduced number of phalanges producing short toe.

Brachygnathia – mandibular micrognathia-short mandible.

Bone bridge – spans space between two margins of bent bone. Linear rays of bone extension, filling margins between bone margins.

Ectodactyly – complete missing toe.

Ectomelia – incomplete limb with missing lower portion.

Hemimelia- short bone with normal distal elements.

Kyphosis – convex thoracic spine.

Microcephaly – small head, snout blunted.

Micromelia – small or short limb.

Phocomelia – incomplete limb with missing proximal portion.

Polydactyly – increased number of metatarsals

Polymelia – additional limbs.

Polyphalangy – duplicate phalangeal sets.

Scoliosis – lateral deviation of vertebrae.

Meteyer CU, Loeffler IK, Fallon JF, Converse KA, Green E, Helgen JC, Kersten S, Levey R, Eaton-Poole L, Burkhart JG. 2000. Hind limb malformations in free-living northern leopard frogs (*Rana pipiens*) from Maine, Minnesota, and Vermont suggest multiple etiologies. Teratology 62:151–171.

Environmental – 86% of northern leopard frogs (*Rana pipiens*) from Maine, Minnesota, and Vermont had amelia polymelia, polydactyly, polyphalangy, phocomelia, ectomelia, ectodactyly, bradydactyly, bone rotations, bone bridging or micromelia with amelia, and reduced segments most common.

Meyer H. von 1860a. *Osteophorus roemeri* aus dem Rotliegenden von Klein-Neuendorf in Schlesien. [*Osteophorus roemeri* from the Rothiegund of Klein-Neuendorf in Silesia]. Palaeontographica 7:99–104. [German]

Congenital – Unpaired, median bone between frontals and nasals, "Inter-Naso-Frontale" or "Naso-Frontale," in eryopoid temnospondyl, *Osteophorus*. P.101

Fossil – Unpaired, median bone between frontals and nasals, "Inter-Naso-Frontale" or "Naso-Frontale," in eryopoid temnospondyl, *Osteophorus*. P.101

Meyer H. von 1860b. Zur Fauna der Vorwelt. Reptilien aus dem lithographischen Schiefer des Jura in Deutschland und Frankreich. [The prehistoric fauna. Reptiles of the Jurassic lithographic shales of Germany and France]. VIII+142 pp.; Frankfurt am Main, pp. 91–94 [German].

Congenital – Supernumerary vertebrae in crocodilians.

Meyer H von. 1861. Reptilien aus dem Stubensandstein ds oberen Keupers. [Reptiles from Stubensandstein of the Upper Keuper]. Palaeontographica 7:1–351. [German]

Infection – Frog larvae with rugose fusion of vertebrae, which became urostyle in adult.

Other – Irregularity of osteoderm ledges in *Sclerosaurus armatus*.

Fossil – Irregularity of osteoderm ledges in *Sclerosaurus armatus*.

Meyer J. 1958. Dichocephaly and polyembryony in the Hognose Snake. Herpetologica 14:128.

Congenital – Dichocephalic hognose snake *Heterodon p. platirhinus*.

Meyer-Rochow VB, Asashima M. 1988. Naturally occuring morphological abnormalities in wild populations of the Japanese newt, *Cynops pyrrhogaster* (Salamandridae; Urodela; Amphibia). Zoologischer Anzeiger 221:70–80.

Congenital – 2% of Japanese firebelly newt *Cynops pyrrhogaster* with supernumerary tails, polydactyly and missing limbs, or digits.

Mienis HK. 1986. A bicephalic *Ptyodactylus hasselquistii* from Israel. Zoology in the Middle East 1:90–92.

Congenital – Dicephalic *Ptyodactylus hasselquistii*.

Miller HE. 1938. Three dangerous serpents of Pennsylvania. Pennsylvania Farmer, Pittsburgh 119:5, 17–18.

Congenital – Dicephalic rattlesnake.

Miller CE. 1968. Frogs with five legs. Carolina Tips 31:1.

Congenital – *Hyla* specimens with with supernumerary limbs located in the pelvic region anterior to the jumping legs. The extra legs were not used for hopping, but had digits and were used in walking and holding on to any supporting medium when at rest.

Miller J. 2002. Snake eyes. News and Observer (Raleigh, NC) 22 September 2002:1.

Congenital – Dicephalic turtle.

Millichamp NJ. 1981. Nutritional diseases of captive reptiles. Proceedings of the Symposium of the Associacion of British Wild Animal Keepers 6:21–32.

Metabolic – General commentary, but no specifics, suggesting that 25% of captive deaths were related to nutritional problems, especially vitamin and mineral deficiencies.

Minnich JE. 1972. Excretion of urate salts by reptiles. Comparative Biochemistry and Physiology A41:535–549.

Metabolic – Chelonians, crocodilians, and tuatara produce ammonia, urea, and uric acid, whereas snakes and lizards are uricotelic. He suggests reptilian urine contains "uric acid dehydrate instead of uric acid."

Minobe H. 1930. A double-headed *Agkistrodon*. Dobutsugaku zasshi [Zoological Magazin Japan] 42 (492):273–274.

Congenital – Dicephalic *Agkistrodon blomhoffi*.

Mishra KD, Majupuria TC. 1979. A report of twin-headed snake, *Spalerosophis diadema*. Geobios 6 Suppl:338.

Congenital – Dicephalic *Spalerosophis diadema*.

Misuri A. 1910a. Ricerche sulla struttura della coda normale e rigenerata nella *Lacerta muralis* Merr. [Research on the structure of the normal and regenerated tails of *Lacerta muralis* Merr.]. Bollettino della Società Zoologica Italiana (serie 2) 11:103–135 [Italian].

Trauma – Multiple tails in lizards are regenerative.

Misuri A. 1910b. Ricerche sulla struttura della coda normale e rigenerata nella *Lacerta muralis* Merr. [Research on the structure of the normal and regenerated tails of *Lacerta muralis* Merr.]. Bollettino della Società Zoologica Italiana (serie 2) 11: 300–316 [Italian].

Trauma – *Lacerta muralis* with bifid tails, one at a 90 degree angle from the longer tail.

Mitchell JC. 1994. The Reptiles of Virginia. 352 pp.; Smithsonian Institution Press, Washington DC.

Congenital – Dicephalic box turtle.

Mitchell MA, Diaz-Figueroa O. 2004. Wound management in reptiles. Veterinary Clinics Exotic Animals 7:123–140.

Trauma – Injuries are common to bony osteoderms of chelonians and crocodilians.

Metabolic – Nutrition diseases are common in captive reptiles.

Neoplasia – Neoplasia is rare in reptiles.

Vascular – Dysecdysis (abnormal shedding) commonly produces avascular necrosis of lizard digits.

Mitchell JC, Fieg M. 1996. Natural history notes: Serpents: *Agkistrodon contortrix mokasen* (northern copperhead): Bicephaly. Herpetological Review 27(4):202–203.

Congenital – Derodymous *Agkistrodon contortrix*, noting dicephalic *Agkistrodon blomhofii* (Inuka 1929 and Nakamura 1938), *A. contortrix* (Strecker 1926) and *Agkistrodon piscirorus* (Johnson 1901).

Mitchell JC, Georgel CT. 2005. *Sceloporus undulatus undulatus* (Eastern fence lizard). Kyphosis and scoliosus. Herpetological Review 36:183–184.

Congenital – Kyphosis and scoliosis in *Sceloporus undulatus undulatus*.

Trauma – Crooked tail in *Uta stansburiana*.

Lizard tail bifurcations are the results of injuries (Blair 1960; Scott 1982; Smith 1946; Tinkle 1967).

Mitchill SL. 1824. Observations demontrant que les serpens à deux têtes sont des monstres et non une race distincte. [Observations demonstrating that two headed serpents are monsters and not a distinct race (species)]. Bulletin des sciences naturelles et de géologie 7:252–253. [French]

Congenital – Dicephalic and ododymus"black or runner snakes" *Boa constrictor.*

"Black or runner snakes" *Boa constrictor* with two bodies.

Mitchill S-M. 1826a. Fait et observations demontrant que les serpens à deux têtes de l'Amerique septentrionale, ou des autres contrées du globe, ne sont pas des individus d'un race distinct, mais des monstres. [Observations made demonstrating two head serpents from North America or other contries of the world ar not of a distinct race (species), but are monsters.] Bulletin des sciences naturelles et de géologie VII: 252–256. [French]

Congenital – Dicephalic, ododymus, and derodymus in "black snake or runner" also called *Boa constrictor.*

Comments that two-headed snakes have been found in North America, West Indies, Polynesia, England, and Italy.

Mitchill SL. 1826b. Facts and considerations showing that the two-headed snakes of North America and other parts of the world are not individuals of a distinct race, but universally monsters. American Journal of Science 10:48–53.

Congenital – Corrects previous literature use of term *Serpens bicips* (representing sentiment that dicephalic snakes were a specific breed).

Derodymous snake from "Islands" and cites Aristotle, Aelian, Lanzoni, Francis Redi, and Aldrovandus as reporting dicephalic snakes, as well as

"Mr. President Clinton" (of Society, not of United States) (1814) and reported "hearing of" a dicephalic *Testudo.*

Dicephalic, diprosopus, and dicaudatus black snake or runner *Coluber* constrictor snakes, the latter with a single jaw. This species was called *Bascanium constrictor* by Johnson (1902).

Reports "intelligence by a most credible source" of a three-headed snake from the Black River near Lake Ontario.

Mizgireuv IV, Flax NL, Borkin LJ, Khudoley VV. 1984. Dysplastic lesions and abnormalities in amphibians associated with environmental conditions. Neoplasma 31:175–181.

Congenital – Microcephaly, reduction of feet and leg bone size and chondrodysplastic tail lesions in *Rana chensinensis* and *amurensis* and *Bufo bufo gargarizans.*

Neoplasia – Enchondroma in *Rana chensinensis,* attributed to teratogenic agent.

Modzelewski EH, Culley DD Jr. 1974. Growth responses of the bullfrog, *Frana catesbeiana,* fed various live foods. Herpetologica 30:396–405.

Metabolic – Metabolic bone disease in bullfrog *Rana catesbeiana.*

Moffat LA, Bellairs D'A. 1964. The regenerative capacity of the tail in embryonic and post-natal lizards (*Laceta viviparta* Jacquin). Journal of Embryology and Experimental Morphology 12:769–786.

Trauma – common lizard *Lacerta vivipera* has fracture planes, as does *Sphenodon* (Howes and Swinnerton 1901), but embryos younger than stage 39 did not regenerate.

Mohan K, Foggin CM, Muvavarirwa P, Honywill J, Pawandiwa A. 1995. Mycoplasma associated polyarthritis in farmed crocodiles (*Crocodylus niloticus*) in Zimbabwe. Journal of Veterinary Research 62:45–49.

Infection – Mycoplasma polyarthritis in *Crocodylus niloticus.*

Mohanty-Hejmadi P, Dutta SK, Mahapatra P. 1992. Limbs generated at the site of tail amputation in marbled balloon frog after vitamin A treatment. Nature 355:352–353.

Toxicology – Amputated tails of marbled balloon frog *Uperodon systoma* exposed to retinoic acid instead regenerate legs. Retionid-induced *Polypedates maculatus* limb formation at ectopic sites.

Moncharmont U. 1949. Descrizione di un arto posteriore soprannumerario in una *Rana esculenta* rinvenuta in natura [Description of a supernumerary hind limb in a *Rana esculenta* found in nature]. Bolletino della Societa dei Naturalisti in Napoli 57:78–91 [Italian].

Congenital – Supernumerary left hind limb and ilium in *Rana esculenta.* The ischium formed a good part of the inferior surface of the acetabular cavity, not only of the normal left limb, but also of the supernumerary.

Montague JJ. 1984. Abnormalities and injuries in New Guinea freshwater crocodiles (*Crocodylus novaeguineae*). Journal of Herpetology 18:201–204.

Trauma – 2.1% of 1073 New Guinea freshwater crocodiles

Crocodylus novaeguineae had deformed tails; 2.9%, deformed trunk; 1.2%, digit amputations; 0.47%, hand or back deformities; and one, a round bump on its snout.

Montalenti G. 1937. "Lusus naturae." Un serpe bicipite ["Freak of nature." A two headed serpent]. Rassegna Faunistica Roma 4(1):42–49. [Italian]

Congenital – Dicephalic *Tropidonotus natrix* snake captured in the Roman countryside.

Montgomery D. 2003. Strange attraction; as sideshows vanish from the midway, a film recalls their glory days. The Washington Post 24 October 2003: C01.

Congenital – Dicephalic turtle.

Moodie RL. 1908. Reptilian epiphyses. American Journal of Anatomy 7:38–467.
Congenital – Variable occurrence of epiphyses – lacking in crocodiles and turtles; present in some lizards – *Amblyrhynchus, Iguana, Sceloporus, Phrynosoma, Chameleon owenii, Heloderma suspectum, Uroplatus, Varanus*, but not *Cnemidophorus*, pterosauria or theropoda.
Moodie RL. 1911. An embryonal plesiosaurian propodial. Kansas Academy of Science 1911:95–101.
Other – Illustrates 1926 (Biologia Generalis) article plesiosaur specimen, without calling it pathologic.
Fossil – Illustrates 1926 (Biologia Generalis) article plesiosaur specimen, without calling it pathologic.
Moodie RL. 1916. Bacteriologic and pathologic evidence in past geological ages. Transactions of the Chicago Pathological Society 1916:84–88.
Trauma – Healed fractures in fossil reptiles,
Other – Periostitis in mosasaur humerus and necrosis (without description) in a mosasaur and in a Jurrassic crocodile.
Fossil – Healed fractures in fossil reptiles,
Periostitis in mosasaur humerus and necrosis (without description) in a mosasaur and in a Jurrassic crocodile.
Moodie RL. 1917. Studies in paleopathology. I. General consideration of the evidence of pathological conditions found among fossil animals. Annals of Medical History 1917:374–381.
Trauma – *Dimetrodon* radius and rib fracture.
Infection – Triassic phytosaur with fractured snout "with many necrotic sinuses." Infected mosasaur humerus and radius (KUVP 1172).
Neoplasia – Mosasaur osteoma
Fossil – *Dimetrodon* radius and rib fracture.
Triassic phytosaur with fractured snout "with many necrotic sinuses." Infected mosasaur humerus and radius (KUVP 1172).
Mosasaur osteoma
Moodie RL. 1918. Paleontological evidences of the antiquity of disease. The Scientific Monthly 1918(9):265–281.
Trauma – *Dimetrodon* left radius and rib fractures.
Triassic Parasuchian *Mystriosuchus plieningeri* with broken snout.
Infection – Crocodylian *Metriohynchus moreli* with palatine cavities apparently from infection and deformed left femoral head, with abnormally small diameter femur and vertegbra with irregularly jagged outer side with deep holes, called necrosis – apparently infectious.
Platecarpus coryphacus humerus with osteomyelitis.
Neoplasia – *Platecarpus* with vertebral mass called an osteoma.
Dental – Caries with hyperlasia and necrosis of mosasaur jaw.
Other – Mosasur with proximal surface "extensive necrosis."
Fossil – *Dimetrodon* left radius and rib fractures.
Triassic Parasuchian *Mystriosuchus plieningeri* with broken snout. Crocodylian *Metriohynchus moreli* with palatine cavities apparently from infection and deformed left femoral head, with abnormally small diameter femur and vertegbra with irregularly jagged outer side with deep holes, called necrosis – apparently infectious.
Platecarpus coryphacus humerus with osteomyelitis.
Platecarpus with vertebral mass called an osteoma.
Caries with hyperlasia and necrosis of mosasaur jaw.
Mosasur with proximal surface "extensive necrosis."
Moodie RL. 1921. Osteomyelitis in the Permian. Science 56:333.
Trauma – Fracture of posterior dorsal spine of *Dimetrodon* with "sinus-filled tumefaction" called osteomyelitis.
Infection – Fracture of posterior dorsal spine of *Dimetrodon* with "sinus-filled tumefaction" called osteomyelitis.
Fossil – Fracture of posterior dorsal spine of *Dimetrodon* with "sinus-filled tumefaction" called osteomyelitis.
Fracture of posterior dorsal spine of *Dimetrodon* with "sinus-filled tumefaction" called osteomyelitis.
Moodie RL. 1922. The paleopathology of the parasuchians. Science 56:417.
Pseudopathology – Says Phytosauria (Parasuchia) nasal prominences are not pathologic, contradicting Abel (1922).
Fossil – Says Phytosauria (Parasuchia) nasal prominences are not pathologic, contradicting Abel (1922).
Moodie RL. 1923a. The Antiquity of Disease. University of Chicago Press, Illinois, 148 pp.
Infection – Microbial invasion of osseous spaces in *Rhamphocephalus*.
Trauma – Fractured *Elaphosaurus* spinal process, *Dimetrodon* radius and *Mystriosuchus* rostrum.
Toe damage in a mosasaur called "rheumatoid arthritis" (actually clearly post-traumatic arthritis; definitely not rheumatoid arthritis).

Neoplasia – Vertebral osteoma in *Platecarpus*.

Other – Palatine "necrosis" in *Metroiorhynchus moreli*. Radial "necrosis" in mosasaur.

Fossil – Microbial invasion of osseous spaces in *Rhamphocephalus*.

Fractured *Elaphosaurus* spinal process, *Dimetrodon* radius and *Mystriosuchus* rostrum.

Toe damage in a mosasaur called "rheumatoid arthritis" (actually clearly post-traumatic arthritis; definitely not rheumatoid arthritis).

Vertebral osteoma in *Platecarpus*.

Palatine "necrosis" in *Metroiorhynchus moreli*. Radial "necrosis" in mosasaur.

Moodie RL. 192b. Paleopathology. An introduction to the study of ancient evidence of disease. Urbana, Illinois: Universityof Illinois Press, 567pp.

Congenital – Plesiosaur *Proneusticosaurus* from Lias of Silesia with pachyostotic/hyperostostic vertebrae and ribs Volz (1902) – ? Normal finding.

Trauma – *Dimetrodon* radius and spinous process fractures; *Edaphosaurus* with incompletely healed radius fracture and a second (phylogeny uncertain) with fractured fibula.

Archelon ischyros from South Dakota with oblique remains of proximal 2/3 of right tibia and fibula, allegedly bitten off (Wieland 1909).

Phytosaur *Mystriosuchus plieningeri* with "deep oval, abnormal depression" anterior to the nostrils (Von Huene 1911). Snout fracture and callus at distal third also illustrated by Moodie (although he pointed to the most proximal portion of the snout). He reported "bone necrosis," but this was not discernable in his image.

Jurassic crocodile *Metriorhynchus moreli* from Oxford Clay of England with maligned femoral fracture with proximal hole with reactive bony lip (puncture wound versus draining sinus?) and healed sacral fracture.

Platecarpus with shortened, flattened, necrotic metatarsal. Illustration suggests amputating bite, but "terminal necrotic sinus" is interpreted by one of us (BMR) as residua from a bite, without evidence of infection.

Rana catesbiana with repaired fracture (Foot 1916) with "new cancellous bone... laid down with an arrangement similar to that seen in Haversian systems."

Infection – Irregular enlargement of fractured *Dimetrodon* vertebral spine with draining sinuses was called oldest case of osteomyelitis (Gilmore 1919).

Niobara mosasaur with "deep sinus in articular surface of one of the arm bones... It is a deep, irregular, rough-sided pocket, surrounded by a lipped surface indicating the existence of an extensive suppuration." Cross section clearly documents periostitis with marrow continuity.

Vertebral – Lower Miocene crocodile *Tomistoma dowsoni* from Egypt with "a thick band of osseous tissue, obviously pathological, firmly binds the vertebrae together," but sharply separated from the bodies (Ruffer 1917).

While certainly not spondylosis deformans, possibility of spondyloarthropathy or diffuse idiopathic skeletal hyperostosos (DISH) must be considered.

Neoplasia – *Platecarpus coryphaeus* with "osteoma... and surrounding lesions at the interarticular surface of the third and forth dorsal vertebrae."

Overlapping two vertebrae, with a thickness of 10 mm and length of 25 mm, he described it as follows: "On the right of the bone the lesion is relatively smooth with lines of growth running circularly around the body of the mass, interrupted anteriorly by an invading mass of rougher bone.

The portion on the left side is quite roughened and raised into a series of irregular ridges. The osteoma is sharply marked off from the body of the vertebra itself and has involved only a portion of the vertebral tissue." On the basis of the histology he illustrated, the fine trabecular pattern leads one of us (BMR) to make the diagnosis of the benign bone tumor known as an osteoblastoma.

Dental – Cites Abel (1912, p. 95) as saying Dollo described dental caries in lower jaw of *Mosasaurus* (No. 1503) at the "Brussels Museum." Moodie further states that "the same jaw shows also alveolar changes indicating pyorrhea." From the phraseology, it is not clear if the latter is Dollo's observation or that of Moodie.

Other – Flattened, irregular eburnated, dense, pitted periosteal reaction on mosasaur humerus, radius, carpus, and metacarpal was called osteoperiostitis (Williston 1898).

Crocodile necrosis (Auer 1909).

Fossil – Plesiosaur *Proneusticosaurus* from Lias of Silesia with pachyostotic/hyperostostic vertebrae and ribs Volz (1902) – ? Normal finding.

Dimetrodon radius and spinous process fractures; *Edaphosaurus* with incompletely healed radius fracture and a second (phylogeny uncertain) with fractured fibula.

Archelon ischyros from South Dakota with oblique remains of proximal 2/3 of right tibia and fibula, allegedly bitten off (Wieland 1909).

Phytosaur *Mystriosuchus plieningeri* with "deep oval, abnormal depression" anterior to the nostrils (Von Huene 1911). Snout fracture and callus at distal third also illustrated by Moodie (although he pointed to the most proximal portion of the snout). He reported "bone necrosis," but this was not discernable in his image.

Jurassic crocodile *Metriorhynchus moreli* from Oxford Clay of England with maligned femoral fracture with proximal hole with reactive bony lip (puncture wound versus draining sinus?) and healed sacral fracture.
Platecarpus with shortened, flattened, necrotic metatarsal. Illustration suggests amputating bite, but "terminal necrotic sinus" is interpreted by one of us (BMR) as residua from a bite, without evidence of infection.
Rana catesbiana with repaired fracture (Foot 1916) with "new cancellous bone… laid down with an arrangement similar to that seen in Haversian systems."
Irregular enlargement of fractured *Dimetrodon* vertebral spine with draining sinuses was called oldest case of osteomyelitis (Gilmore 1919).
Niobara mosasaur with "deep sinus in articular surface of one of the arm bones… It is a deep, irregular, rough-sided pocket, surrounded by a lipped surface indicating the existence of an extensive suppuration." Cross section clearly documents periostitis with marrow continuity.
Lower Miocene crocodile *Tomistoma dowsoni* from Egypt with"a thick band of osseous tissue, obviously pathological, firmly binds the vertebrae together," but sharply separated from the bodies (Ruffer 1917).
While certainly not spondylosis deformans, possibility of spondyloarthropathy or diffuse idiopathic skeletal hyperostosos (DISH) must be considered.
Platecarpus coryphaeus with "osteoma… and surrounding lesions at the interarticular surface of the third and forth dorsal vertebrae."
Overlapping two vertebrae, with a thickness of 10 mm and length of 25 mm, he described it as follows: "On the right of the bone the lesion is relatively smooth with lines of growth running circularly around the body of the mass, interrupted anteriorly by an invading mass of rougher bone.
The portion on the left side is quite roughened and raised into a series of irregular ridges. The osteoma is sharply marked off from the body of the vertebra itself and has involved only a portion of the vertebral tissue." On the basis of the histology he illustrated, the fine trabecular pattern leads one of us (BMR) to make the diagnosis of the benign bone tumor known as an osteoblastoma.
Cites Abel (1912, p. 95) as saying Dollo described dental caries in lower jaw of *Mosasaurus* (No. 1503) at the "Brussels Museum." Moodie further states that "the same jaw shows also alveolar changes indicating pyorrhea." From the phraseology, it is not clear if the latter is Dollo's observation or that of Moodie.
Flattened, irregular eburnated, dense, pitted periosteal reaction on mosasaur humerus, radius, carpus, and metacarpal was called osteoperiostitis (Williston 1898).
Crocodile necrosis (Auer 1909).

Moodie RL. 1926a. Spondylitis deformans in a crocodile from the Pleistocene of Cuba. Annals of Historical Medicine 78:78–82.
Vertebral – "Spondylitis deformans," an erroneous diagnosis was made in crocodile from Pleistocene of Cuba, noting Ruffer reported disease in "two posterior vertebrae of a crocodile from the lower Miocene of Egypt." Actual examination of photos reveals anulus fibrosus, as well as anterior longitudinal ligament ossification. Classic of early marine reptiles is bridging through the disk space, a phenomena that appears to occur whenever vertebral movement is precluded (by osseous bridging either through the outer layers of the vertebral disk or the anterior longitudinal ligament). Syndesmophytes are actually illustrated – indicating spondyloarthropathy as the cuorrect diagnosis.
Fossil – "Spondylitis deformans," an erroneous diagnosis was made in crocodile from Pleistocene of Cuba, noting Ruffer reported disease in "two posterior vertebrae of a crocodile from the lower Miocene of Egypt." Actual examination of photos reveals anulus fibrosus, as well as anterior longitudinal ligament ossification. Classic of early marine reptiles is bridging through the disk space, a phenomena that appears to occur whenever vertebral movement is precluded (by osseous bridging either through the outer layers of the vertebral disk or the anterior longitudinal ligament). Syndesmophytes are actually illustrated – indicating spondyloarthropathy as the correct diagnosis.

Moodie RL. 1926b. Studies in paleopathology. XIII. The elements of the Haversian system in normal and pathological structures among fossil vertebrates. Biologia Generalis 2(1/2):63–95.
Other – Illustrates *Platecarpus* humerus with periostitis and a plesiosaur metapodial with a bump located on the edge at the proximal third.
Fossil – Illustrates *Platecarpus* humerus with periostitis and a plesiosaur metapodial with a bump located on the edge at the proximal third.

Moodie RL. 1930. The dinosaurs of Wyoming. The Wyoming Geological Survey 12:14–114.
Infection – Jurassic crocodile with expanded ilium, draining sinus, and reactive bone around acetabulum.
Fossil – Jurassic crocodile with expanded ilium, draining sinus, and reactive bone around acetabulum.

Mook C. 1921. Individual and age variations in the skulls of recent Crocodilia. – Bulletin of the American Museum of Natural History 44(7):51–66.
Dental – Two *Caiman sclerops* with abnormal mandibular tooth grouping and "skulls themselves have somewhat unusual proportions." Commented, without detail on variation in size and direction of teeth, unequal teeth on

opposite sides of skull, cranial table and suture (especially premaxillo-maxillary on palatal surface) separation of nasal bones from external nares, tooth piercing of opposing surfaces, and prominence of teeth.

Mookerjee H.K. 1933. On the malformation of the vertebral column and the pelvic girdle of *Microhyla rubra.* Anatomischer Anzeiger Jena 75:299–300.

Congenital – Right-sided sacralization of 8th "lumbar" vertebra and lumbarization of right "sacral" vertebra with elongated right ilium in *Microhyla rubra.*

Urostyle abnormality was reported by Thirumulachar 1928.

Moore D. 1988. One for the toad. Townsville Bulletin: Magazine, Saturday, March 19, 1988:1–2.

Congenital – Humble cane toad *Bufo marinus* with supernumerary leg.

Morgan TH. 1827. German *Rana.* Anatomischer Anzeiger 9:703 [German].

Congenital – Forked tail in *Rana* embryos.

Morgan CL. 1886. Abnormalities in the vertebral column of the common frog. Nature 35:53.

Congenital – *Rana temporaria* with procoelous (as opposed to normal amphicoelous) 8th vertebra and a second with only one transverse process on the 9th (for articulation with the ilium). Its centrum was anteriorly convex on right and concave on left, with irregular ill-developed articulation (for urostyle). The left transverse process of the 8th vertebra was abnormally large and articulated with the ilium. A right, but no left posterior zygapophysis was present.

Morgan TH. 1901. Regeneration. New York: Macmillan, 316pp.

Lizards develop double tails secondary to injury and if vertebral injury cuts through 2 or 3 vertebrae, can have more tails (e.g., *Lacerta agilis*) or digits (e.g., *Triton cristatus*).

Morgan TH. 1903. Regeneration of the leg of *Amphiuma means.* Biological Bulletin 5:293–296.

Trauma – Limb regeneration after amputation in *Amphiuma means.*

Morgan TH. 1907. Regeneration – Uebersetzung von M. Moszowski. [Regeneration – Translation of M. Moszowski]. XVI + 437pp.; Leipzig: Wilhelm Engelmann [German].

Congenital – Anomalies of legs in *Triton cristatus.*

Pelobates fuscus with six hind legs and pelvic anomalies (figs. 49–53).

Trauma – Limb regeneration in *Proteus, Amphiuma, Necturus, Triton,* but not in frogs; tail regeneration in lizards, but not in snakes and turtles.

Duplication and triplication of tail in lizards after Tornier (fig. 44A, B).

Morin PJ. 1985. Predation intensity, prey survival, and injury frequency in an amphibian-prey interaction. Copeia 1985(3):638–644.

Trauma – Direct relationship of injury frequency to predator biomass in *Notophthalmus viridescens* predation on *Scaphiopus holbrooki*

Motani R. 2000. Rulers of the Jurassic Seas. Scientific American 26: 52–59.

Environmental – First ichthyosaur bone fragments found in 1927 was *Utatusaurus.* Correlation with deep diving and eye size noted.

Fossil – First ichthyosaur bone fragments found in 1927 was *Utatusaurus.*

Correlation with deep diving and eye size noted.

Motani R, Rothschild BM, Wahl W Jr. 1999. Large eyeballs in diving icthyosaurs. Nature 402:747.

Vascular – Ichthyosaurs proved susceptible to bends, based on examination of proximal femoral and humeral subsidence. Examination of the stylopodials of 13 ichthyopterygian genera reveals a large intergeneric variation in frequencies of avascular necrosis, but a general phylogenetic trend:

Epoch	Genera	Affliction rate
Lower Cretaceous	*Platypterygius*	2/11
Mid/Upper Jurassic	*Nannopterygius*	0/4
	Ophthalmosaurus	14/82
Lower Jurassic	*Stenopterygius*	0/31
	Ichthyosaurus	9/72
	Temnodontosaurus	1/6
Upper Triassic	*Californosaurus*	0/2
	Shonisaurus	0/1
	Shastosaurus	0/5
Middle Triassic	*Mixosaurus*	0/1
	Cymbospondylus	0/1
Lower Triassic	*Grippia*	0/6

Non-parvipelians (i.e., most Triassic ichthyosaurs) rarely have avascular necrosis, if equivocal cases are not considered. Parvipelvians (predominantly post-Triassic ichthyosaurs) manifest variable frequency of decompression syndrome (bends). One interpretation of the ichthyosaur data is that the habit of repetitive diving in ichthyosaurs did not evolve until parvipelvians appeared. The latter are equipped with the true tailbend and a semi-symmetrical caudal fin. This increases the efficiency of continuous swimming, probably a requirement for repetitive diving. *Stenopterygius*, a typical tuna-shaped ichthyosaur, is an exception, in that no bends have been identified in specimens examined to date. It is possible that the genus solely had a physiological adaptation to avoid bends, as do modern whales.

Fossil – Ichthyosaurs proved susceptible to bends, based on examination of proximal femoral and humeral subsidence. Examination of the stylopodials of 13 ichthyopterygian genera reveals a large intergeneric variation in frequencies of avascular necrosis. A general phylogenetic trend:

Epoch	Genera	Affliction rate
Lower Cretaceous	*Platypterygius*	2/11
Mid/Upper Jurassic	*Nannopterygius*	0/4
	Ophthalmosaurus	14/82
Lower Jurassic	*Stenopterygius*	0/31
	Ichthyosaurus	9/72
	Temnodontosaurus	1/6
Upper Triassic	*Californosaurus*	0/2
	Shonisaurus	0/1
	Shastosaurus	0/5
Middle Triassic	*Mixosaurus*	0/1
	Cymbospondylus	0/1
Lower Triassic	*Grippia*	0/6

Non-parvipelians (i.e., most Triassic ichthyosaurs) rarely have avascular necrosis, if equivocal cases are not considered. Parvipelvians (predominantly post-Triassic ichthyosaurs) manifest variable frequency of decompression syndrome (bends). One interpretation of the ichthyosaur data is that the habit of repetitive diving in ichthyosaurs did not evolve until parvipelvians appeared. The latter are equipped with the true tailbend and a semi-symmetrical caudal fin. This increases the efficiency of continuous swimming, probably a requirement for repetitive diving. *Stenopterygius*, a typical tuna-shaped ichthyosaur, is an exception, in that no bends have been identified in specimens examined to date. It is possible that the genus solely had a physiological adaptation to avoid bends, as do modern whales.

Mudge BF. 1876. Notes on the Tertiary and Cretaceous Period of Kansas. Bulletin of the Survey (U.S. Geological and Geographical Survey of the Territories) ii(3):211–221.

Trauma – Healed rib fractures in saurians.

Bite marks in saurian bones from *Galeocerdo* shark.

Fractured saurian vertebrae which have fused and become confluent.

Fossil – Healed rib fractures in saurians.

Bite marks in saurian bones from *Galeocerdo* shark.

Fractured saurian vertebrae which have fused and become confluent.

Mudge BF. 1878. Geology of Kansas. In First Biennial Report of the Kansas State Board of Agriculture: pp. 60–63. Topeka, KS: Kansas State Board of Agriculture.

Trauma – Plesiosaur with five fractured spinous fractures, coalesced in healing.

Serrated *Galeocerdo* shark tooth marks on Cretaceous bones of a possible plesiosaur.

Fossil – Plesiosaur with five fractured spinous fractures, coalesced in healing.

Serrated *Galeocerdo* shark tooth marks on Cretaceous bones of a possible plesiosaur.

Mufti SA. 1969. Tail regeneration in an adult salamander, *Desmognathus fuscus*. American Zoologist 9:613.

Trauma – Time course of tail regeneration in *Desmognathus fuscus* and *Triturus viridescens*.

Mufti SA. 1973. Tail regeneration following amputation in adult *Triturus viridescens*. Pakistani Journal of Zoology 5:31–49.

Trauma – No autotomy, but regeneration takes place in *Triturus viridescens*.

Mufti SA, Simpson SB. 1972. Tail regeneration following autotomy in the adult salamander, *Desmognathus fuscus*. Journal of Morphology 136:297–312.

Trauma – Tail regeneration in *Desmognathus fuscus*.

Muggiasca F, Gandolla E. 1976. I Rettili del Ticino. Descrizione delle singole specie e considerazioni generali [The reptiles of Ticino. Descriptions of single species and general considerations]. 78pp.; Canobbio-Lugano: Aurora S.A [Italian].

Congenital – Dicephalic *Natrix helvetica* embryo (fig. 25).

Muir JH. 1990. Three anatomically aberrant albino *Crotalus atrox* neonates. Bulletin Chicago Herpetological Society 25(3):41–42.

Congenital – Dicephalic and two western diamondback rattlesnakes *Crotalus* with fused mouths, attributed to abnormal cool period.

Cites dicephalic *Crotalus adamanteus* (Klauber 1972; Murphy and Shadduck 1978), *Crotalus durissus terrificus, Crotalus viridis oreganus, Crotalus b. basiliscus, Crotalus horridus, Crotalus v. viridis* (Klauber 1972) and *Crotalus horridus* (Lasher 1980).

Mulder EWA. 2001. Co-ossified vertebrae of mosasaurs and cetaceans: Implications for the mode of locomotion of extinct marine reptiles. Paleobiology 27:724–734.

Infection – *Mosasaurus camperi* MNHNP AC 9649–9775/9776 with 6 fused caudal vertebrae with spongiform bone (Dollo 1882).

Vertebral – cites Martin and Bell's 1995 report of club-tailed *Clidastes.* Dollos's (1882) fused pygal vertebrae in *Plioplatecarpus marshi* were actually abnormal. Dollo also referred to this specimen as *Mosasaurus camperi* which had two areas of fused caudal vertebrae. This MNHNP AC 9649- 9775/9776 specimen is said to closely resemble IRSNB 1496F and 1496HH, also described by Dollo.

Fossil – *Mosasaurus camperi* MNHNP AC 9649–9775/9776 with 6 fused caudal vertebrae with spongiform bone (Dollo 1882).

Cites Martin and Bell's 1995 report of club-tailed *Clidastes.* Dollos's (1882) fused pygal vertebrae in *Plioplatecarpus marshi* were actually abnormal. Dollo also referred to this specimen as *Mosasaurus camperi* which had two areas of fused caudal vertebrae. This MNHNP AC 9649-9775/9776 specimen is said to closely resemble IRSNB 1496F and 1496HH, also described by Dollo.

Mulder EWA. 2003. Co-ossified vertebrae of mosasaurs and cetceans: Implications for the mode of locomotion of extinct marine reptiles. In Mulder WW. On latest Cretaceous tetrapods from the Masstrichtian type area. Doctoral Thesis, Universiteit Amsterdam: 149–159.

Infection – Co-ossified vertebrae in *Plioplatecarpus marshi* MNHNP AC 9649- 9775/9776 were clearly infected, as were MND 20.01.842 and MND 20.01.826. Fused, infected vertebrae in *Mosasaurus camperi.*

Vertebral – Co-ossified vertebrae and "club-tailed" mosasaurs described. Co-ossified infected vertebrae in *Plioplatecarpus marshi.* Those in IRScNB 1 496 and 1497 reveal smooth fusion, predominantly ligamentous.

Fossil – Co-ossified vertebrae in *Plioplatecarpus marshi* MNHNP AC 9649- 9775/9776 were clearly infected, as were MND 20.01.842 and MND 20.01.826. Fused, infected vertebrae in *Mosasaurus camperi.*

Co-ossified vertebrae and "club-tailed" mosasaurs described. Co-ossified infected vertebrae in *Plioplatecarpus marshi.* Those in IRScNB 1 496 and 1497 reveal smooth fusion, predominantly ligamentous.

Müller H. 1852. Eine Eidechse, *Lacerta viridis*, mit zwei über einander gelagerten Schwänzen, welche beide als das Produkt einer überreichen und durch den feinern Bau der wiedererzeugten bemerkenswerthen Reproductionskraft erscheinen. [A lizard, *Lacerta viridis*, with two tails lying above each other]. Verhandlungen der physikalisch-medicinischen Gesellschaft in Würzburg 2 [German].

Trauma – *Lacerta viridis* specimen with divided tail (upper and lower tail without vertebrae, only cartilaginous axis = regenerations).

Müller H. 1864. Ueber die Regeneration der Wirbelsäule und des Rückenmarks bei Eidechsen und Tritonen. [On the regeneration of the vertebral colume and of the nerul cord of lizards and tritons]. Gratulationsschrift der physikalisch-medizinischen Gesellschaft in Würzburg zu der Jubelfeier der Senckenberg'schen Stiftung, Frankfurt a. M.: 62–64 [German].

Trauma – Discussion of the structure of the new formed part of double tail of *Lacerta viridis*, which he described in 1852.

Müller E. 1996. Ueber die Abstossung und Regeneration des Eidechsenschwanzes. [About the rejection and regeneration of the lizard tail]. Jahreshefte des Vereins für vaterländische Naturkunde in Württemberg Stuttgart 52: LXXXV–LXXXVII (not 85–87) [German].

Trauma – Mentions tail loss in lizards.

Muneoka K, Bryant S. 1984. Regeneration and Development of vertebrae appendages. Symposium of the Zoological Society of London In: The Structure, Development and Evolution of Reptiles: A Festschrift in Honour of Professor A. d'A. Bellairs on the occasion of his retirement, WJ Ferguson, ed. London: Academic Press, 52:177–196.

Congenital – Surgical implantation produces supernumerary limbs in *Ambystoma mexicanum.*

Munyer EA. 1963. Syndactylism in the lizard *Sceloporus undulatus hyacinthinus*. Bulletin of the Philadelphia Herpetological Society 11:28.

Congenital – Syndactylism in northern fence lizard *Sceloporus undulatus hyacinthinus* and southern prairie lizard *Sceloporus undulatus consorbrinus* (Loewen 1941).

Murphy TD. 1965. High incidence of two parasitic infestations and two morphological abnormalities in a population of the frog, *Rana palustris* Le Conte. American Midland Naturalist 74:233–239.

Trauma – Faulty forelimb eruption as small limb in 1.2% of *Rana palustris*, attributed to bite by larva of *Hannemania dunni.*

Murphy JB, Shadduck JA. 1978. Reproduction in the eastern diamondback rattlesnake, *Crotalus adamantus* in captivity, with comments regarding a teratoid birth anomaly. British Journal of Herpetology 5:727–733.

Congenital – Derodymous *Crotalus adamantus*.

Muto Y. 1969a. Anomalies in the hindlimb skeletons of toad larvae reared at a high temperature. Congenital Anomalies 9:61–73.

Environmental – *Bubo vulgaris formosus* at 30°C produces hindlimb abnormalities including reduction in size and fusion of tarsals and metatarsals and hyperphalangism.

Muto Y. 1969b. Hindlimb development and malformations of toes in the larvae reared at a high temperature in the toad, *Bufo vulgaris formosus*. Congenital Anomalies 9:1–12.

Environmental – *Bufo vulgaris formosus* raised at 30°C has retardation of hindlimb growth, oligodactylism. 141 of 143 larvae had abnormal feet or an eliminated tallicus. Rare brachydactylia, hypoplasia, and syndactyly were reported.

Muto Y. 1970. Digital malformations in the forelimbs of the toad larvae reared at a high temperature. Congenital Anomalies 10:135–147.

Environmental – Thirty degrees Centigrade produces partial elimination of toe I in *Bufo vulgaris formosus*. 27% of hands had slight brachydactylism; 48%, distinct brachydactylism. 24% of remaining had severe brachydactylism or hypoplasia. Syndactylism (metacarpal fusion) and hypophalangism were rare.

Muto Y. 1971. Skeletal anomalies in the feet of toad larvae reared in aerated water at a high temperature. Congenital Anomalies 11:159–169.

Environmental – *Bufo vulgaris formosus* incubated at 31°C resulted in regressive changes in metatarsal I in 11.5% (fused with tarsal in 69.2%), prehallucal metatarsale in 23.1% (fused with tarsale in; 7.7%, with other metatarsals in 3.8%). He suggested insufficient oxygen supply due to "regressive changes of the spiracle"

Mutschmann F. 2008. Snake Diseases: Prevention and recognizing illness. Frankfort an Main: Serpents Tail, 306 pp.

Trauma – Emerald tree boa *Corallus canicus* with rib fractures.

Infection – *Boa constrictor* (without subspecies, common name cannot be determined) with osteolytic vertebrae.

Metabolic – *P. regius* with calcified deposits adjacent to angulated vertebrae.

Vertebral – *P. regius* with calcified deposits adjacent to angulated vertebrae.

Myers CW. 1967. The pine woods snake *Rhadinaea flavilata* (Cope). Bulletin of the Florida State Museum 10:47–97.

Trauma – Tail breaks in 29% of pine woods snake *Rhadinaea flavilata.*

Najbar B. 1988. [Interesting cases of congenital anomalises of development in *Elaphe guttata guttata* (Linnaeus, 1766) (Reptilia, Serpentes, Colubridae).] Przeglad Zoologiczny 33:419–425. [Polish]

Congenital – *Elaphe guttata guttata* with hydrocephalus, underdeveloped and distorted vertebrae

Nakamura K. 1938. Studies on some specimens of double monsters of snakes and tortoises. Memoirs of the college of Science, Kyoto Imperial University, ser. B, 14:171–192.

Congenital – Derodymus *Agkistrodon halys blomhoffii, Elaphe conspicillata, Elaphe climacophora, Natrix tigrina, Natrix vivakari, Geoclemys reevesii.*

Cites dicephalic snake reports by Redi 1684, Vsevolojsky 1812, Dutrochet 1829, Dorner 1873, Borgert 1896, Cantoni 1912, Strohl 1925, Heasman 1933 and double tortoise by Kuwano 1902.

Craniophagus in *Hemibungarus japonicus.*

Psodymus (bifurcated near sacrum) *Geoclemys reevsii.*

Nandi NC, Raut SK. 1986. A note on a tail abnormality of the spotted house gecko (*Hemidactylus brooli*). Hamadryad 11:22.

Trauma – Bifid tail in common wall lizard *Hemidactylus brooki*, noting Chandra and Mukherjee's (1980) report in *Agama tuberculata.*

Natale GS, Basso NG, Ronco AE. 2000. Effect of Cr(VI) on early life stages of three species of hylid frogs (Amphibia, Anura) from South America. Environmental Toxicology 15:509–512.

Toxicology – Common pllutant chromium CR(VI) produce malformations in *Hyla pulchella, Hyla nana*, and *Scinax squalirostris*.

Naulleau G. 1983. Tératologie chez *Natrix natrix* et *Vipera aspis*. [Teratology of *Natrix natrix* and *Vipera aspis*]. In Constantin V, Matz G. eds. Comptes Rendus du Premier Colloque International de Pathologie des Reptiles et des Amphibiens, Angers, France. Pp. 245–249. [French]

Congenital – Bifurcation of body with two heads in *Natrix natrix* and of neck in *Vipera aspi.*

Environmental – Temperature-related *Vipera aspis* vertebral deformity.

Naulleau G. 1983. Les serpents de France. [The serpents of France]. Revue française de Aquariologie 11(3/4):1–56. [French]

Congenital – Dicephalic *Natrix natrix* and *Vipera aspis*.

Necker WL. 1940. Hump-backed turtles. Chicago Naturalist 3:62.

Congenital – Kyphosis in *Chrysemys picta bellii x marginata* cross.

Néedham M. 1750. De la Langue du Lezard. In: Nouvelles Observations Microscopiques avec de découvertes intéressantes sur la Composition et la Decomposition des Corps organises. [Microscopic observations with interesting discoveries on the development and decomposition of body organization.] Néedham M, ed, pp. 140–144. Paris: Ghuerin and Delatour. [French]

Trauma – Portuguese lizard with double tail, as previously noted by Marchant (1718) and Pliny.

Neiffer DL, Marks SK, Klein EC, Brady NJ. 1998. Shell lesion management in two loggerhead sea turtles, *Caretta caretta*, with employment of PC-7 epoxy paste. Proceedings of the Association of Reptilian and Amphibian Veterinarians 8(4):12–17.

Infection – Fungal hyphae in loggerhead sea turtle *Caretta caretta* carapace erosions.

Shell disease – Multiple small carapace erosions with fungal hyphae in loggerhead sea turtle *Caretta caretta*.

Neill WT. 1941. A dicephalic queen-snake (*Natrix septemvittata*). Copeia 1941(4):266.

Congenital – Dicephalic queen-snake *Natrix septemvittata*.

Neill WT. 1946. An autophagous lizard. Copeia 1946:104.

Trauma – Autotomy in *Gehyra oceanica* which subsequently ate its own tail.

Neill WT. 1960. No title. Bulletin of the Philadelphia Herpetological Society 8(6):6–7.

Congenital – Dicephalic eastern diamondback *Crotalus adamanteus*, noting similar report by Klauber 1956 (also present in his 1972 edition.

Cites dicephalic reports by Mitchill 1826, Meyer 1958, Tripplehorn 1955, Belluomini 1957/8, Nakamura 1938 and comment that what Hyde (1925) called a copperhead was actually a dicephalic Northern water snake *Natrix sipedon sipedon*.

Opisthodichotomic Gulf Salt Marsh Snake *N. s.* clarki described by List and Smith (1954) was actually a Gulf water snake *Natrix sipedon clarkia*.

Neill WT. 1971. The Last of the Ruling Reptiles: Alligators, Crocodiles and their Kin. New York: Columbia University Press, 486 pp.

Trauma – Most salamanders regenerate tails. Newts regenerate hyoids.

Some lizards (e.g., glass lizard) regenerate tails.

Gopher tortoise with post-traumatic carapace hole.

Crocodile with missing foot.

Alligator mcgrewi with healed lateral snout injury.

Cuban ground sloth with tooth marks from Cuban crocodile (Figure 126 on page 150).

Illustrates water snake "ingesting two of its kind" (figure 21, page 56). It has the appearance of a snake with the tails of two larger snakes coming out of its mouth.

Nelson WB, Slavens F. 1975. Two-headed garter snake. Journal of Zoo Animal Medicine 6(3):23.

Congenital – Dicephalic garter snake.

Nelson SR, Foltz FM, Camarata P, Serras MP Jr. 1985. Occurrence of bone fractures and parathyroid hyperplasia in paraphysectomized frogs (*Rana catesbeiana*). Anatomical Record 211(3):311–317.

Metabolic – Paraphysectomized bullfrog *Rana catesbeiana* manifest extensive long leg deformities, presenting as shortening with knob-like deformities from callus formation around spontaneous fractures.

Newman HH. 1906a. The significance of scute and plate "abnormalities" in Chelonia. Biological Bulletin, Lancester 10:68–98.

Congenital – Good description of normal turtle scute/plate appearances.

Newman HH. 1906b. Correlated abnormalites in the scutes and bony plates of chelonia. Science 23:526.

Congenital – Supernumerary or deficient bony plates in *Graptemys geographica* and *Chrysemys marginata*.

Newman HH. 1923. The Physiology of twinning. Chicago: University of Chicago Press. 232 pp.

Congenital – Three-headed snake seen near Lake Ontario (Gemmill 1912) and another was reported by Androvandus from the Pyrenees Mountains.

Predominantly dicephalic twinning in *Triton*, frog, turtle, snake tortoise with supernumerary forearm (Bateson 1894).

Limb duplication in *Ambystoma*.

Dichotomous fission of a rib in association with doubling. When ribs are twinned ribs there are usually twinned costal plates and accessory costal scutes

Situs invertus (Spemann and Falkenburg 1919).

Newman S. 1997. Earthweek; a diary of the planet. Two-headed viper. Syracuse Herald-Journal, Syracuse, NY (25 November 1997): B-3.

Congenital – Dicephalic Russell's viper.

Newth DR. 1958. On regeneration after amputation of abnormal structures. II. Supernumerary induced limbs. – Journal of Embryology and experimental Morphology 6:384–392.

Congenital – Supernumerary induced limbs in *Triturus helveticus* are frequently either highly polydactylous or hypodactylous or simply occur as protrusions.

Trauma – Regeneration of normal urodele limb from amputated abnormal *Triturus helveticus* limb.

Newton ET. 1893. On some new reptiles from the Elgin sandstone. Proceedings of the Royal Society London 52:389–391.

Congenital – Unpaired, median bone ("a single intercalary bone"), "interparietale," parietals and frontals in dicynodont therapsids (mammal-like reptiles), *Gordonia traquairi.*

Fossil – Unpaired, median bone ("a single intercalary bone"), "interparietale," parietals and frontals in dicynodont therapsids (mammal-like reptiles), *Gordonia traquairi.*

Niazi IA, Saxena S. 1978. Abnormal hind limb regeneration in tadpoles of the toad, *Bufo andersoni*, exposed to excess vitamin A. Folia Biologica (Cracov) 26:3–8.

Metabolic – Hypervitaminosis A produces "abnormal hind limb regeneration" (after amputation) in *Bufo andersoni.*

Niazi IA, Saxena S. 1979. Relationship between inhibiting influence of vitamin A and developmental stage of regenerating tail in toad tadpoles (*Bufo andersonii*). Indian Journal of Experimental Biology 17:866–868.

Metabolic – Regeneration in *Bufo andersonii* was inhibited by vitamin A at day 2, but not day 3 or 4.

Nicandro. 1763. Theriacis (redd. latinis versibus I. Gorrhaeus-italicis A. M. Salvini) curante Baudinio. Florentiae ex off. Monchiana. [Theriacis (Rendered Latin verses I. Gorrhaeus-Italian A. M. Salvini) care of Baudinio. Monchian: Florence, Italy]. [Latin]

Pseudopathology – This Greek doctor was the first to report on amphisbaena as a two-headed snake, with a head on either end of the body (after Cantoni 1921).

Nicholas JS. 1955. Vertebrates. In Willier BH, Weiss PA, Hamburger V, eds. Analysis of Development. pp 674–698; Philadelphia: WB Saunders.

Congenital – Salamander with "single huge limb near dorsal midline," citing Hellmich 1930.

Trauma – Regeneration producing double or triple tails in lizards, especially *Lacerta muralis*, citing Bateson 1894; Przibram 1926; Tornier 1906; Raisse 1885. Cites Gachet 1834 for tail regeneration in *Lacerta muralis, agilis, viridis*, and *ocellatus* and *Anolis iguanae* and Woodland 1921 in *Hemidactylus flaviridis.*

Nicholls GE. 1916. The structure of the vertebral column in the anuran *Phaneroglossa*, and its importance as a basis of classification. Proceedings of the Linnean Society of London 1916:80–92.

Congenital – Diplasiocoelous offered as term to describe vertebrae in which only the first 7 amphibian vertebrae are hollow in front; the eight, hollow on both faces and the ninth; doubly concave.

Revision from the diplasiocoelous state to the more primitive procoelous condition was described by Lloyd Morgan (1886) in the common frog and observed by author in a *Rana esculent* and a *Rana tigrina.*

States that vertebral column of common frog is variable in 8%, but all were still diplasiocoelous.

Only vertebral centra abnormality among families Bufonidae, Hylidae, and Cystignathidae was described by Cope (1866) in *Borborocoetes.*

Ten procoelous vertebrae were described by Howe (1886) in a *Rana temporaria.*

Pelobatidae vertebral column has frequent variation, but mostly limited to *Megalophrys.* Adolphi reported 23% of *Pelobates fuscus* with vertebral column abnormalities. A rare variation in Pelobatidae is failure of intervertebral spheres to unite with the center, resulting in the amphicoelous (ancestral) condition persistence to maturity in *P. cultripes* (Dugès 1834) and *P. fuscus* (Stannius 1854). Two other anomalous centra were recorded: *Pelobate fuscus* (Stannius 1854) and *Xenophrys monticola* (*Megalophrys parva*) (Boulenger 1882).

Fusion of sacral vertebra with coccyx occurred occasionally in *P. cultripes* and *M. nasuta* (Boulenger 1908).

Nicholson K. 1999. Years of animal magnetism Retiree closes cages of popular pet shop. The Denver Post 3 June 1999. Denver and the West: 1.

Congenital – Dicephalic turtle.

Nickerson MA. 1966. Bicephalism in three colubrids. British Journal of Herpetology 3:284.

Congenital – Dicephalic Tropidocloniom lincatan lincatan, Thamnophis sirtalis concinnus, and Heterodon platyrhinos.

Niekisch M. 1990. Untersuchungen zur Besiedlungsstrategie der Gelbbauchunke *Bombina variegate variegate* (LINNAEUS 1758) (Anura, Amphibia). [Investigations to the strategy of the yellow-bellied toad *Bombina variegate variegate* (LINNAEUS 1758) (Anura, Amphibia)]. Dissertation Universität Bonn. [German]

Congenital – *Bombina variegate* without left eye; others lacking hind leg and one with enlarged (huge) right hind leg.

Niewiarowsky PH, Congdon JD, Dunham AE, Vitt LJ, Tinkle DW. 1997. Tales of lizard tails: Effects of tail autotomy on subsequent survival and growth of free-ranging hatchling *Uta stansburiana*. Canadian Journal of Zoology 75:542–548.

Trauma – More female *Uta stansburiana* with broken tails survivied, while tail loss had no effect on male survival. Growth rates, however, were lower in the presence of broken tails.

Autotomized individuals had higher survivorship in Nevada (Althoff and Thompson 1994), but lower in Washington.

Social status effects in were noted with tail loss in *Anolis sagrei* (Kaiser and Muschinsky 1994).

Niimi T. 1965. On a dichotomous snake, *Elaphe climacophora* (Boie). Hachu ryo seiruigaku zasshi [Acta Herpetologica Japonica] 1(2):31–32. [Japanese]

Congenital – Derodymus *Elaphe climacophora*

Niimi T. 1969. [Note on the monstered tail of two species of Lacertilia.] Acta Herpetologica Japonica 3(2, 3):14–16. [Japanese with Engl. abstract]

Trauma – *Gekko japonicus* with forked tail; *Takydromus tachydromoides*, with tripartite tail.

Niimi T. 1971. Additional report on the dichotomous snakes. Acta Herpetologica Japonica 4(1–4):5–11.

Congenital – Dicephalic *Elaphe climacophora.*

Nishimura G. 1938. A Manchurian specimen of the double-headed snake. Dobutsugaku Zasshi [Zoological Magazine Japan] 50(3): 135.

Congenital – Dicephalic *Agkistrodon blomhoffi.*

Nixon CW, Smith HM. 1949. The occurrence of kyphosis in turtles. Turtox News 27:1–2.

Congenital – Kyphosis in *Sternotherus odoratus* and *Amyda (Trionyx) ferox.*

Noble GK. 1922. The phylogeny of Salientia. I. The osteology and the thigh musculature: Their bearing on classification and phylogeny. Bulletin of the American Museum of Natural History 46:1–87.

Congenital – Variation in vertebral fusion between species, but not within.

Noble GK. 1931. The biology of the Amphibia. New York, McGraw-Hill. 577 pp.

Trauma – Autotomy and regeneration in plethodontid salamanders. Regeneration occurs as single digit and partial limb region in frog, but in salamander producies forked tail, extra digit or supernumerary limb.

Vertebral – Fusion of first and second cervical vertebrae in *Atelopus* and *Xenopus.*

Noden DM, De Lahunta A. 1985. The embryology of domestic animals. Developmental mechanisms and malformations. Baltimore, London: Williams & Wilkins, 367 pp.

Source of definitions:

Hemimelia – absence of at least half of a limb segment

Meromelia – absence of part of a limb

Phocomelia – absence/reduction of proximal limb segments.

Ectrodactyly – absence of one or more digits.

Thoracopagus – twins conjoined at sternal region

Pygopagus – twins conjoined back to back at pelvis or sacrum

Cephalopagus – twins conjoined at head.

Diprosopus – two faces

Dicaudatus – two tails

Tetrabrachius – two pairs of arms

Tetrascelus – two pairs of legs.

Notomelus – extra limb attached at back.

Nöllert A, Günther R. 1996a. Gelbbauchunke – *Bombina variegata* (Linnaeus 1758). [common spade foot toad *Bombina variegata* (Linnaeus 1758)]. In Günther R. ed. Die Amphibien und Reptilien Deutschlands. Pp. 232–252; Jena: Gustav Fischer. [German]

Congenital – Defects in the vertebral column, malformed extremities in larvae and six legged forms in *Bombina variegata.*

Nöllert A, Günther R. 1996b. Knoblauchkröte – *Pelobates fuscus* (Laurenti 1768). [Yellow-bellied toad – *Pelobates fuscus* (Laurenti 1768)]. In Günther R. ed. Die Amphibien und Reptilien Deutschlands. Pp. 252–274; Jena: Gustav Fischer. [German]

Congenital – Malformation of *Pelobates fuscus* limbs and tail vertebrae.

Trauma – *Pelobates fuscus* with supernumerary foot in Berlin Sapandau collection ZMB 35875.

Nopcsa F. 1923a. Vorläufige Notiz über Pachyostose and Osteosklerose einiger mariner Wirbeltiere. [Preliminary note on pachyostosis and osteosclerosis of some marine vertebrates]. Anatomischer Anzeiger 56: 353–359. [German]

Other – Reference to pachyostosis in *Mesosaurus, Pachypleura,* and *Eidolosaurus*, and to pachyostosis and osteosclerosis in Cretaceous snake *Pachyophis* in the anterior part of trunk explained as counterweight to the enlarged lung.

Fossil – Reference to pachyostosis in *Mesosaurus, Pachypleura* and *Eidolosaurus*, and to pachyostosis and osteosclerosis in Cretaceous snake *Pachyophis* in the anterior part of trunk explained as counterweight to the enlarged lung.

Nopcsa F. 1923b. *Eidolosaurus* und *Pachyophis*: Zwei neue Neocom-Reptilien. [*Eidolosaurus* and *Pachyophis*: Two new Neocomian reptiles]. Palaeontographica 65:97–154. [German]

Other – *Eidolosaurus* and *Pachyophis* with pachyostosis of ribs of the anterior trunk connected with osteosclerosis.

Fossil – *Eidolosaurus* and *Pachyophis* with pachyostosis of ribs of the anterior trunk connected with osteosclerosis.

Nopcsa F. 1923c. Über eine neue Kreideschlange aus Dalmatien. [On a new Cretaceous snake of Dalmatia]. Paläontologische Zeitschrift 5:258, 264–265. [German]

Other – Pachyostosis of ribs in a Cretaceous snake.

Fossil – Pachyostosis of ribs in a Cretaceous snake.

Nopcsa F. 1926. Heredity and evolution. Proceedings of the Zoological Society of London 1926, part 2:633–665.

Congenital – Cites *Chamaesaurus anguinea* with variably one or two hindlimb toes (Duerden and Essex1922), but they actually reported 3, 2, or 1.

Testudo graeca with variably four or five toes no manus (Ghigi 1900).

Vertebral number in *Dryophis prasinus* varies by up to 45 (11%) and those of *Python reticulatus*, by as many as 47 (12%).

Trauma – Longirostral gavials suffer bites or lower ramus excision of ramus and exostoses on top, if deep enough. *Mystriosuchus* suffers similar injuries. (Abel 1922, not 1933 as cited) as may *Pezophaps*.

Some *Mystriosuchus* species have variable rostral bumps/excrescences.

Case (1924) reported a tumor-like bump on a *Leptosuchus imperfecta*, not *Mystriosuchus*, as Nopcsa cited.

Nopcsa F, Heidsieck E. 1934. Über eine pachyostotische Rippe aus der Kreide Rügens. [On a pachyostotic rib in the Cretaceous of Rügen]. Acta Zoologica Academiae Scentiarum Hungaricae 15:431–455. [German]

Other – Pachyostosis of ribs in lizards (*Eidolosaurus, Adriosaurus* [= *Acteosaurus*], *Opetiosaurus* [= *Carsosaurus*]), snakes (*Pachyophis, Symoliophis*), champsosaurs (*Champsosaurus* sp.), mesosaurs (*Mesosaurus brasiliensis, Stereosternum* sp.), nothosaurs (*Pachypleurosaurus edwardsi, Proneusticosaurus carinthiacus* and *P. voltzi*) and plesiosaurs (one specimen of *Pliosaurus ferox*).

Fossil – Pachyostosis of ribs in lizards (*Eidolosaurus, Adriosaurus* [= *Acteosaurus*], *Opetiosaurus* [= *Carsosaurus*]), snakes (*Pachyophis, Symoliophis*), champsosaurs (*Champsosaurus* sp.), mesosaurs (*Mesosaurus brasiliensis, Stereosternum* sp.), nothosaurs (*Pachypleurosaurus edwardsi, Proneusticosaurus carinthiacus* and *P. voltzi*) and plesiosaurs (one specimen of *Pliosaurus ferox*).

Nopsca – common misspelling for Nopcsa

Noriega KC, Rega E, Sumida S. 2002. Indirect evidence of sail presence from healed spinous process fractures in several *Dimetrodon* specimens. Journal of Vertebrate Paleontology 22 (supplement to 3):92A.

Trauma – *Dimetrodon* spinous process fractures, which healed in good anatomic position.

Fossil – *Dimetrodon* spinous process fractures, which healed in good anatomic position.

Norton TM, Ackerman N, Rossi J. 1991. Clinical challenge: Treatment of osteomyelitis in a green iguana (*Iguana iguana*). Journal of Zoo and Wildlife Medicine 22:259–260.

Infection – Osteomyelitis with collapse and malalignment of two lumbar vertebrae in *Iguana iguana*.

Novak SS, Seigel RA. 1986. Gram-negative septicemia in American alligators (*Alligator mississippiensis*). Journal of Wildlife Disease 22:484–487.

Infection – Abscess in *Alligator mississippiensis* which extended into shoulder joint, but no comment on bone.

Nuñez H, Yáñez JL. 1984. Colas de lagartijas de género *Liolaemus*: Autotomía e influencia en la predación. [Tails of the *Liolaemus* lizards: autotomy and the influence of predation] Studies on Neotropical Fauna and Environment 19:1–8. [Spanish]

Trauma – Nuñez and Yáñez report that in the lizard genus *Liolaemus* in central Chile, the tail breaks from the body in variable locations, and not in a consistent, invariable location as previously described in the literature. In fact, the tail can break from very close to its base, to its distal end. In their experiment, the authors had collectors without experience in the field catching the lizards. Sometimes tails were broken off during collection. The sample was divided into three groups: individuals with an intact tail, individuals with regenerated tails, and individuals that suffered tail breakage during capture. They concluded that the tail broke off primarily within the first third of its length, and secondarily within the second third. However, breakage points existed throughout the length of the tail. The authors also observed that the instances of the tail breaking closest to its base corresponded to females, while breakages that occurred more distally corresponded to male individuals. They speculated that this may be due to the hemipenis being in the first tenth of the tail.

Nybelin O. 1942. Zoologiska avdelningen. Berättelse för år 1941.[Zoological Section. Report for year 1941] Göteborgs Musei Årstryck 1942:8–19. [Swedish]

Congenital – Dicephalic *Vipera berus*, from Bohuslän, Sweden

Annotated Bibliography O-R

Obst FJ. 1976. Ein siamesischer Zwilling bei der Vierzehen-Landschildkröte *Agrionemys horsfieldii* (GRAY). [Siamese twins in the four-toed tortoise *Agrionemys horsfieldii* (GRAY)]. Aquarien Terrarien 23(5–6):174–175. [German]

Congenital – Siamese twins of *Agrionemys horsfieldii* from Tadschikistan fused in the posterior region. It has two heads and two pairs of front limbs but only one pair of hind limbs, noting similar cases in *Chelonia mydas* and *Malaclemys terrapin*.

Obst FJ, Richter K, Jacob U. 1984. Lexikon der Terraristik. [Lexicon of Terroristic]. Leipzig, 465 pp. [German]

Trauma – Discussion of fractures and regeneration in amphibians and reptiles.

Metabolic – Osteodystrophia fibrosa, osteoporosis, rickets, and gout in reptiles.

O'Donoghue CH. 1910. Instances of polymely in two frogs. Together with notes on the absence of a right pre-caval vein in two frogs. Zoologischer Anzeiger 35:759.

Congenital – Supernumerary forelimb in tree frog *Hyla aurea* and *Rana temporaria*, citing absence of phylogeny by De Superville 1740 and Royal College of Surgeons 1842, Gervais 1864 in *Pelobates cultripes*, Lunel 1868 in *Rana viridis*, Mazza 1888 in *Rana esamelica*, Sutton 1889 and Bergendal 1889 in *Rana temporaria*, Washburn 1899 in *Bufo columbiensis*, Johnson 1901 in *Rana palmipes and Rana halecinum*, and Eigerman and Cox 1901 in *Rana pipiens*.

Ogden JA, Rhodin AGJ, Conlogue GJ, Light TR. 1981. Pathobiology of septic arthritis contiguous osteomyelitis in a leatherback turtle (*Dermochelys coriacea*). Journal of Wildlife Diseases 17(2):277–287.

Infection – Erosion of distal humerus and replacement of elbow with fibrous ankylosis, with reactive overgrowth of proximal radius and ulna in leatherback turtle *Dermochelys coriacea*. Sclerosis suggests infection. Cites Ackerman et al. (1971) on osteomyelitis in a *Tupinambis teguixin*.

Ôgi K-I. 1954. On the lithium embryo of the frog's egg. Science Reports of the Tôhoku University, 4th series (Biology) 20:163–174.

Toxicology – Lithium-induced spina bifica and cyclopia in *Rana japonica* and *Rana nigromaculata*.

Ohler A. 1991. Entwicklung, Ausprägung und Häufigkeit einer Polydaktylie assoziiert mit Brachymelie bei Larven eines *Rana ribunda*-Pärchens. [Development, expression and frequency of polydactyly associated with brachymely in larvae of one *Rana ridibunda* pair]. Zoologische Jahrbücher, Abteilung für Anatomie und Ontogenie der Tiere 121:127–135. [German]

Congenital – Development and characteristics of polydactyly associated with brachymelia in *Rana ridibunda*: forelimbs with 4–5, hind limbs with 6–8 phalanges.

Ohlmacher HP. 1898. Several examples illustrating the comparative pathology of tumors. Bulletin of the Ohio Hospital for Epileptics 1:223–236.

Trauma – *Rana virescens* spindle shaped femur swelling with trabeculae of bone and cartilage surrounded by large spaces was originally called a medullary osteosarcoma but is more characteristic of a callus after a fracture.

Neoplasia – *Rana virescens* spindle shaped femur swelling, originally called medullary osteosarcoma. Trabeculae are surrounded by large spaces, actually a characteristic of callus after a fracture.

O'Keefe RR, Street HP, Cavigelli JP, Socha JJ, O'Keefe RD. 2009. A plesiosaur containing an ichthyosaur embryo as stomach contents from the Sundance Formation of the Bighorn Basin, Wyoming. Journal of Vertebrate Paleontology 29:1306–1310.

Trauma – Cryptocleidid plesiosaur (?*Pantosaurus*) with ingested ichthyosaur (?*Ophthalmosaurus natans*) embryo.

Fossil – Cryptocleidid plesiosaur (?*Pantosaurus*) with ingested ichthyosaur (?*Ophthalmosaurus natans*) embryo.

B.M. Rothschild et al., *Herpetological Osteopathology: Annotated Bibliography of Amphibians and Reptiles*,
DOI 10.1007/978-1-4614-0824-6_9,

Olivier E 1893. Un crapaud phénomène. [A phenomenal toad]. Revue scientific Bourbonnais et du Centre de la France 6:105. [French]
Congenital – *Bufo vulgaris* with a recurved, saber-like limb.

Olson WM. 2000. Phylogeny, ontogeny, and function: extraskeletal bones in the tendons and joints of *Hymenochirus boettgeri* (Amphibia: Anura: Pipidae). Zoology 103:15–24.
Congenital – More sesamoids in *Hymenochirus boettgeri* than in any other frog.

O'Malley B. 2008. Nutritional problems in reptiles. World Small Animal Veterinary Association, Congress, Dublin, Ireland 20–24 August 2008 (Medicine):1.
Metabolic – Nutritional secondary hyperparathyroidism presenting as fractured limbs, digital and caudal lateral vertebral processes with osteopenia, thin long bone cortices with folding fractures and spinal deformities. This occurs mainly in lizards and chelonians but is uncommon in snakes.

Oringderff NB. 1969. That's right! Two heads! – Sunday Oklahoman, Oklahoma's Orbit, Oklahoma City, Oklahoma, Oct. 12.
Congenital – Dicephalic snake.

Orós J, Rodríguez JL, de los Monteros AE, Rodríguez F, Herráez P, Fernández A. 1997. Tracheal malformation in a bicephalic Honduran milk snake (*Lampropeltis hondurensis*) and subsequent fatal *Salmonella arizonae* infection. Journal of Zoo and Wildlife Medicine 28:331–335.
Congenital – Dicephalic Honduran milk snake *Lampropeltis hondurensis*.

Orós J, Torrent A, de los Monteros AE. 2001. Multicentric lymphoblastic lymphoma in a loggerhead sea turtle (*Caretta caretta*). Veterinary Pathology 38:464–467.
Neoplasia – Lymphoblastic lymphoma presenting as plaston nodules in a loggerhead sea turtle (*Caretta caretta*).

Orska J, Imiolek Z. 1962. Preliminary studies on the effect of temperature on the development of meristic characters in the Urodela. Acta Universitatis Wratislaviensis. Prac Zoologiczne 3:135–155.
Congenital – Cites Gerecht's (1929) report or three *Triturus vulgaris* individuals with 13 vertebrae of 177 from Latvia and Lebedinsky (1925) suggestion that seven Urodela and *Atelopus varius* had an asymmetrical sacrum.
Environmental – Percent of *Triturus vulgaris* with 15 (rather than "normal" 14) trunk vertebrae increased from 13.9% in those incubated at 12°C to 27.3% at 20°C to 51.7% at 28°C. 13 and 16 trunk vertebrae were limited to the cold series. In two cold series specimens, the 14th and 15th vertebrae had unilateral sacral ribs connected to ilium. five old *Ambystoma mexicanum* had 17 trunk vertebrae as did all of the 16.4°C series. In the 7.7°C series, 1/19 had 16 trunk vertebrae. In the 21.4°C series, 9/50 had 17 trunk vertebrae; 7, 16; 1, 16.5 (asymmetrical sacrum); and 1, 15 vertebrae. Asymmetric sacra were found in 13.9% of the low temperature, and only one individual was found in the high temperature series. Conversely, tail vertebrae decreased by mean of 1.5 from cold to warm incubation.

Osborn D, Cooke AS, Freestone S. 1981. History of a teratogenic effect of DDT on *Rana temporaria* tadpoles. Environmental Pollution 25:305–319.
Toxicology – DDT-induced *Rana temporaria* mandibular hole and "jowled" appearance with blunted mandible in, citing Freestone 1981 and noting history of teratogenic effect of DDT on *Rana temporaria.*

Osborn HF. 1903. Reptilian subclasses Diapsida and Synapsida and the early history of the Diaptosauria. Memoirs of the American Museum of Natural History 1(8):449–519.
Congenital – Reports variation in reptiles, but no pathology.
Pseudopathology – Deemed a coossified radiale and intermedium as an artifact.

Osborne CA, Albasan H, Lulich JP, Nwaokorie E, Koehler LA, Ulrich LK. 2008. Quantitative analysis of 4468 uroliths retrieved from farm animals, exotic species, and wildlife submitted to the Minnesota Urolith Center: 1981 to 2007. Veterinary Clinics Small Animal 39:65–78.
Stone – Urate (50%) and unknown composition urolith in gecko.
Urate (97%), mixed (2%), and compound urolith (1%) in iguana.
Urate (92%), calcium carbonate (4%), and compound (4%) urolith in lizards.
Urate (94%), calcium carbonate (3%), and mixed urolith in tortoises.
Urate (58%), calcium phosphate (17%), compound (17%), and mixed (8%) urolith in turtles.

Osgood DW. 1978. Effects of temperature on the development of meristic characters in *Natrix fasciata.* Copeia 1978:33–47.
Congenital – 17% of *Thamnophis sirtalis* and 10% of *Thamnophis sauritus* had half-ventrals.
Environmental – 11% of *Natrix fasciata* have increased frequency of "half-ventrals" with both high and low temperature exposure.

O'Shea M. 1990. A note on a dicephalic Chinese racer (*Elaphe rufodorsata*). In J Coote, ed. Reptiles. Proceedings of the 1988 UK Herpetological Societies Symposium on Captive Breeding. British Herpetological Society, London. Pp 23–24.
Congenital – Dicephalic Chinese racer *Elaphe rufodorsata.*

Ostood DW. 1978. Effects of temperature on the development of meristic characters in *Natrix fasciata*. Copeia 1978(1):33–47.

Environmental – Number of vertebrae in *Natrix fasciata* skeleton increased at high and low temperature.

Otto AG. (Ad. W.) 1816. Seltene Beobachtungen zur Anatomie, Physiologie und Pathologie. Band 1:22 Breslau. [German]

Congenital – Polymely in *Pelobates fusca*, a supernumerary limb to the right of the thorax, noting that St. Hilaire incorrectly wrote that the polymely was posterior when in fact, it was anterior.

In the zoological cabinet of Breslau, there is a *Rana esculenta* with three hind limbs. The supernumerary limb originates in the inside of the normal right limb and is attached along the normal femur. This extra limb is as well formed as the normal limbs and has likewise the muscles. Frog with a supernumerary limb inserted to the right of the pelvis, alongside the normal limb.

Otto AG. 1841. Monstrorum sexcentorum descriptio anatomica. [Descriptive of the anatomy of monsters]. XX + 335pp.; Vratislav: Ferdinand Hirt. [Latin]

Congenital – *Hyla* sp. with reduced right front leg, *Rana temporaria* with reduced right hind leg, *Bufo igneus* both hind legs without toes, *Pelobates fusci* with additional right front leg, and citing polydactyly in *Siren pisciformis* p. 276, Table 26, Fig. 7 (Ercolani 1882).

Trauma – *Gekko* with bifurcated tail; *Lacerta* specimens with bifurcated and trifurcated tails.

Ouellet M. 2000. Amphibian deformities. Current state of knowledge. In Sparling DW, Linder G, Bishop CA eds. Ecotoxicology of Amphibians and Reptiles. SETAC, Pensicola, Florida: 617–661.

Environmental – Limb deformities in 7% of 4286 *Bufo americanus* from agricultural areas of St. Lawrence River Valley of Quebec between 1992 and 1996. Frequency varied from 0–67% in pesticide-exposed sites (0–7.7% in controls): 2.6% of 732 in 54 pesticide-exposed sites versus 1.6% of 693 controls. Ectomelia, ectrodactyly, polymelia, polydactyly, syndactyly, and clinodactyly were noted. There were 41 limb/digit amputations, 14 cases of unilateral anophthalmia, and three deformed jaws.

Ouellet M, Rodriguez J, Bonin J, Lair S, Green DM. 1996–1997. Developmental abnormalities in free living anurans from agricultural habitats. In Anon. The North American Amphibian Monitoring Program Third Annual Meeting: 1p.

Environmental – Limb deformities were noted in 7.0% of 4286 *Bufo americanus* from agricultural areas of St. Lawrence River Valley of Quebec between 1992 and 1996. Frequency varied from 0–67% in pesticide-exposed sites, contrasted with 0–7.7% in controls, 2.6% of 732 in 54 pesticide-exposed sites versus 1.6% of 693 controls. Ectomelia, ectrodactyly, polymelia, polydactyly, syndactyly, and clinodactyly were noted. There were 41 limb/digit amputations, 14 cases of unilateral anophthalmy, and three deformed jaws.

Ouellet M, Bonin J, Rodriguez J, DesGranges J-L, Lair S. 1997. Hind limb deformities (ectromelia, ectrodactyly) in free living anurans from agricultural habitats. Journal of Wildlife Diseases 33:95–104.

Environmental – Correlation of green frogs *Rana pipiens*, American toad *Bufo americanus*, and bullfrog *Rana catesbeiana* hind limb deformity rates between proximity of sites to agricultural habitats (69% versus 0–7.7% in control sites), but forelimb frequency was not affected. This was attributed to pesticide runoff in the St. Lawrence River Valley. Absence of all or part of a limb (ectromelia) or digit (ectodactyly) in *Rana clamitans* was 7% in 1992 and 1993 and in *Rana pipiens* was 66% in 1992, but not 1993. This compares with *Bufo americanus* (695 in pond/corn sites, but only 6% in ditch/corn, soya and 1.4% in pond/barley) in 1992 and *Rana catesbeiana* in which 10% were affected in 1992. All were attributed to agricultural contaminants.

They noted that the carbamate insecticide oxamyl produced curvature and tail deformities in *Rana temporaria* and lordosis in *Bufo boreas boreas* after prolonged UV radiation.

Owen, C. 1746 (not 1742). An essay towards a natural history of serpents: in two parts. London: J. Gray, 240 pp.

Congenital – Dicephalic venomous snake.

Pseudopathology – Denies the dicephalic *Amphisbaena* cited by Pliny, Aelian, Lucan (p. 270), Mataar, Matthiolus, and Hesychius.

Owen R. 1853. Descriptive Catalogue of the Osteological Series contained in the Museum of the Royal College of Surgeons of England. Volume I. Pices, Reptilia, Aves, Marsupialia. London: Taylor and Francis, 635 pp.

Congenital – Ankylosis between 148th and 149th vertebrae and vertebrae 166 and 167 and 184 and 185 are fused on left side, whereas having two ribs on the right in a tiger boa *Python tigris*.

Owen R. 1866. Anatomy of vertebrates: Fish and Reptiles. vol. I. (1984 Indian reprint – New Delhi: International Books and Periodicals Supply Service, 650 pp – reviewed).

Trauma – Regeneration of *Rana temporaria* and *Bombinator igneus* tail and limbs.

Pacces Zaffaroni N, Zavanella T, Arias E. 1989. Spontaneous skeletal malformations of the forelimbs in the adult crested newt. Herpetopathologia 1(1):49–50.

Congenital – 21% of male and 12% of female crested newt *Triturus cristatus carnifex* had skeletal defects, including bent radius and ulnare, fusion of proximal or distal carpal elements, fusion of radiale with prepollicis, intermedium with ulnare and centrale, radiale with prepollicis, fusion of basal carpal three and four, separated intermedium and ulnare. Phalangeal formula deviations (predominantly missing phalanges of digits 1–3, occasionally supernumerary digit 2 or 4 or rudimentary or distorted phalanges were noted in 10 of 152 females and 4 of 61 males.

Pacces Zaffaroni N, Arias E, Zavanella T. 1992. Natural variation in the limb skeletal pattern of the crested newt, *Triturus carnifex* (Amphibia: Salamandridae). Journal of Morphology 213:265–273.

Congenital – *Triturus carnifex* with reduced carpals, alternative phalangeal formula in 34% and 12% of crested newt from northern and central Italy, respectively, and fusions of radiale and prepollicis as part of normal variation. Contrasting Bagnaia and Rosate, Italy, only the frequency of aberrant digital formulas differed between the two locations, 2 and 18%, respectively. Reduced carpals and alternative phalangeal formula were noted. Severe defects were present in 3% of former and 1% of latter. Malformations included supernumerary digits and meromelia.

Pacces Zaffaroni N, Arias E, Lombardi S, Zavanella T. 1996. Natural variation in the appendicular skeleton of *Triturus carnifex* (Amphibia: Salamandridae). Journal of Morphology 230:167–175.

Congenital – Skeletal variability in 36% of northern Italian versus 14% of central Italy crested newt *Triturus carnifex* forelimbs, but no difference in hind limb frequencies of 9–12%.

Pachori, G. N. 1950. On the occurrence of an accessory hind limb in *Rana tigrina* Daud. J. Roy. Asiatic Soc. Bengal, Sci. 16:115–122.

Congenital – Polymelia in *Rana tigrina*, noting previous reports by Lunel 1868, Mahendra 1936, and Kingsley 1991 (= Kingsley 1882??), and Bateso's 1894 report of 50 cases. Supernumerary forelimbs in Salienties, in *Rana nigromaculata* (Wu and Liu 1941) and in *Rana esculenta* (Hvass 1942). Extra acetabulum with hind limb in *Rana tigrina.*

Packard GC, Tracy CR, Roth JJ. 1977. The physiological ecology of reptilian eggs and embryos, and the evolution of viviparity within the class reptilia. Biological Review 52:71–105.

Environmental – Prolonged exposure to temperatures falling several degrees below the optimum produced developmental abnormalities.

Pagenstecher HA. 1863–1864. Über Harnablagerungen bei *Alligator sclerops*, und über Harnausscheidung im allgemeinen. [On deposits of urine (?urate) in *Alligator sclerops* and an excretion of urine (?urate) in general]. Verhandlungen des naturhistorisch medizinischen Vereins Heidelberg 3:129–133. [German]

Metabolic – *Alligator sclerops* with hip gout, deposition of urine crystals in muscles.

Pales L. 1930. Paléopathologie et Pathologie Comparative. [Paleopathology and Comparative Pathology]. VII + 352 pp.; Paris: Masson & Cie. [German]

Trauma – Factures and "scars" in ichthyosaurs, citing *Dimetrodon* with fracture "traces" (Moodie 1923).

Maxillary fracture (Huene 1911) in Triassic phytosaur *Mystriosuchus* (*Belodon* or Parasuchia) *plieningeri*, scarred fracture of maxilla of *Mosasaurus giganteus* and fracture in *Plioplatecarpus marshi* (Abel 1912).

Jaw "wounds" in Triassic phytosaur *Mystriosuchus* (*Belodon* or Parasuchia) *plieningeri*, attributed to tails (Abel 1922).

"Curious lesion" described as atrophy of tibia and fibula of *Archelon ischyros* (Wieland 1909).

Infection – Alveolar-dentary "polyarthritis" or pyorrhea in *Mosasaurus.*

Humeral osteoperiostitis in mosasaur (Moodie 1923).

Osteitis in mosasaurs and in an English crocodile (Moodie 1926).

Arthritis – "Chronic osteoarthritis" in quaternary Egyptian crocodile and of phalanges of mosasaur.

Vertebral – *Tomistoma dowsoni* with a band of osseous tissue between two vertebrae (Ruffer 1917), possible with separation from the vertebral body (unclear if description is of diffuse idiopathic skeletal hyperostosis or of spondyloarthropathy. Ossification of anterior longitudinal ligament and anulus fibrosus of a Pleistocene Cuban saurian (Moodie 1926).

Miocene Egyptian crocodile with "spondylosis."

"Osteoma" in *Platecarpus* (Moodie 1923). However, describes it as ankylosis between two vertebrae. That is definitely not an osteoma. Pales suggest that it is really an exostosis. However, such bridging also requires consideration of spondyloarthropathy.

Neoplasia – "Osteoma" in *Platecarpus* (Moodie 1923). However, describes it as ankylosis between two vertebrae. That is definitely not an osteoma. Pales suggest that it is really an exostosis. However, such bridging also requires consideration of spondyloarthropathy.

"Osteoma" in mosasaur (Abel 1924), but Pales comments on difficult to distinguish osteomas and exostoses, so diagnoses questioned as above.

Dental – Mandibular caries in American and Belgian mosasaurs (Moodie 1923).

Other – "Pachyostosis" in *Pachycanthus*, plesiosaurs and mosasaurs (Moodie 1923). He lists it as "Trias" (Triassic) mosasaurs, although mosasaurs were Cretaceous.

Fossil – Factures and "scars" in ichthyosaurs, citing *Dimetrodon* with fracture "traces" (Moodie 1923).

Maxillary fracture (Huene 1911) in Triassic phytosaur *Mystriosuchus* (*Belodon* or Parasuchia) *plieningeri*, scarred fracture of maxilla of *Mosasaurus giganteus* and fracture in *Plioplatecarpus marshi* (Abel 1912).

Jaw "wounds" in Triassic phytosaur *Mystriosuchus* (*Belodon* or Parasuchia) *plieningeri*, attributed to tails (Abel 1922).
"Curious lesion" described as atrophy of tibia and fibula of *Archelon ischyros* (Wieland 1909).
Alveolar-dentary "polyarthritis" or pyorrhea in *Mosasaurus*.
Humeral osteoperiostitis in mosasaur (Moodie 1923).
Osteitis in mosasaurs and in an English crocodile (Moodie 1926).
"Chronic osteoarthritis" in quaternary Egyptian crocodile and of phalanges of mosasaur.
Tomistoma dowsoni with a band of osseous tissue between two vertebrae (Ruffer 1917), possible with separation from the vertebral body (unclear if description is of diffuse idiopathic skeletal hyperostosis or of spondyloarthropathy. Ossification of anterior longitudinal ligament and anulus fibrosus of a Pleistocene Cuban saurian (Moodie 1926).
Miocene Egyptian crocodile with "spondylosis."
"Osteoma" in *Platecarpus* (Moodie 1923). However, describes it as ankylosis between two vertebrae. That is definitely not an osteoma. Pales suggest that it is really an exostosis. However, such bridging also requires consideration of spondyloarthropathy.
"Osteoma" in mosasaur (Abel 1924), but Pales comments on difficult to distinguish osteomas and exostoses, so diagnoses questioned as above.
Mandibular caries in American and Belgian mosasaurs (Moodie 1923).
"Pachyostosis" in the whale *Pachyacanthus*, plesiosaurs and mosasaurs (Moodie 1923). He lists it as "Trias" mosasaurs (although they were Cretaceous).

Pallaske G. 1961. Gefässwandveränderungen bei D-Hypervitaminose eines Leguan. [Changes of vessel caused by Hypervitaminosis D in a lizard]. Berlin Münchener tierärztliche Wochenschrift 74:132–133 [German]
Metabolic – calcium deposition from hypervitaminosis D in *Iguana iguana*.

Pan Z-C, Ji X. 2001. The influence of incubation temperature on size, morphology and locomotor performance of hatchling grass lizards (*Takydromus wolteri*). Acta Ecologica Sinica 21:2021–2038.
Environmental – Temperature effects on grass lizard *Takydromus wolteri* morphology.

Pancharatna K, Chandran S, Kumbar S. 2000. Phalangeal growth marks related to testis development in the frog *Rana cyanophlyctis*. Amphibia-Reptilia 21:371–379.
Congenital – Growth rings in *Rana cyanophlyctis* correlated poorly with size.

Papenfuss TJ. 1982. The ecology and systematics of the amphisbaenian genus *Bipes*. Occasional Papers of the California Academy of Science 136:1–42.
Trauma – Autotomy in *Bipes biporus*, *B. tridactylus* and *B. canaliculatus* but no regeneration.

Paré JA, Jacobson ER. Mycotic diseases of reptiles. 2007. Infectious Diseases and Pathology of Reptiles: Color Atlas and Text. ER Jacobson, ed. Boca Raton, Fl: Taylor & Francis, pp. 527–570.
Infection – 50% of loggerhead and 25% of Kemp's Ridley turtles had dermatomycotic scutes, sometimes ulcerating underlying bone (Duguy et al. 1998).
Jaw necrosis in *Crocodylus porosus*.
Foot infections in turtles and tortoises (Frye and Dutra 1974; Shreve et al. 2004).
Aspergillus terreus-induced toe gangrene in San Estebar/painted chuckwalla *Sauromalus varius* (Tappe et al 1984).
Chrysosporium occasionally breaks through dermis into bone in squamates and crocodilians.
Fusarium solani-induced tail necrosis in carpet python *Morelia spilota* and by *F. oxysporum* in red bellied black snake in *Pseudechis porphyriacus* (Holz and Slocombe 2000)
Mucormycosis in *Apalone (Trionyx) ferox* (Jacobson et al.1980) *Paecilomyces lilacinus* induced disease in Fly river *Carettochelys insculpta* and hawksbill *Eretmochelys imbricata*.
Shell disease – Fungal ulcerative disease in Bolivian side necked turtle *Platemys platycheles*.
50% of loggerhead and 25% of Kemp's Ridley turtles had dermatomycotic scutes, sometimes ulcerating underlying bone (Duguy et al. 1998).
Fusarium incarnatum (semitectum) induced necrotizing scute in Texas tortoise *Gopherus berlandieri* (Sheve et al. 2004; Rose et al. 2001)
Trichosporon induced carapacial disease (Schildger et al. 1991)

Paré JA, Sigler L, Rosenthal KL, Mader DR. 2006. Microbiology: Fungal and bacterial diseases of reptiles. In Mader DR. ed. Reptile Medicine and Surgery. Pp. 217–226; Philadelphia: Saunders.
Shell – *Fusarium semitectum* scute disease in Texas tortoise *Gopherus berlandieri*.

Paris P. 1912. Curieux cas de teratologie chez une Grenouille. [Curious case of teratology in a frog]. Comptes Rendus de l'Association du Advancement des Sciences 1922:449. [French]
Congenital – Individual with "straighter skull" than normal in a *Rana esculenta*.

Park S. 1987. Same parents for rare Siamese snakes? Herpetological Association of Africa Newsletter (10):20–21.
Congenital – Dicephalic red-lipped herald snake.

Parker GH. 1896. Variations in the vertebral column of *Necturus*. Anatomischer Anzeiger Centralblatt für die Gesamte Wissenschaftliche Anatomie. Amtliches Organ der Anatomischen Gesellschaft 11:711–717.

Congenital – Two of 26 perennibranch salamander *Necturus maculosus* had asymmetrical sacrum. Four of the individuals with symmetrical sacra had first haemal arch on 22nd rather than 23rd vertebra. Nineteen had sacra from the 19th vertebrae, contrated with the 20th in six. In the individuals with asymmetrical sacra, one was composed of #19 and 20, with reduced ribs on the transverse process of #18. A sacral rib and ilium were attached on the left, but only a rudimentary rib on the right. The 20th vertebrae manifested the opposite condition. The second individual with asymmetrical sacrum was formed from the 20th and 21st vertebrae, but illustrated the same pattern. Noted are previous cases in *Bombinator* (Camerano 1880, p. 448; Howes 1890, p. XVI), *Alytes* (Lataste 1879, p. 49), and *Menopoma* (Huxley 1875, p. 752).

Parker GH. 1901. Correlated abnormalities in the scutes and bony plates of the carapace of the sculptured tortoise. American Naturalist 35:17–24.

Trauma – Sculptured tortoise marginal carapace fractures with irregularity of scutes or reduction in number.

Parker HW. 1956. The lizard genus *Aprasia*. Bulletin of the British Museum (Natural History), Zoology 3:363–385.

Congenital – Presacral vertebrae range from 88 to 110 in *Aprasia striolata*, 98–110 in *Aprasia pulchella*, and 96–137 in *Aprasia repens*. Number of vertebrae is higher in areas at 70°F than 64.2°F.

Environmental – Presacral vertebrae range from 88 to 110 in *Aprasia striolata*, 98–110 in *Aprasia pulchella*, and 96–137 in *Aprasia repens*. Number of vertebrae is higher in areas at 70°F than 64.2°F.

Parker M. 1992. More on spindly leg syndrome. British Dendrobatid group Newsletter #14 as reprinted in American Dendrobatid Society Newsletter 12:2–3.

Congenital – Relates spindly leg to a "dietary deficiency" in *Dendrobates truncatus*, as prevented by Cricket Plus dietary supplement – that contains calcium, iron, copper, manganese, zinc, cobalt, iodine, vitamin A, vitamin D3, vitamin E, vitamin K, vitamin C, Vitamins B1, B2, B6, and B12.

Cites Krintler (1988) noting three batches of *Hyla ebraccata* with spindly leg disease. Anomalies included no legs, on or both spindly front legs. He attributed them to poor nutrition, related to "poor weather" and absence of dietary insect availability.

Metabolic – Relates spindly leg to a "dietary deficiency" in *D. truncatus*, as prevented by Cricket Plus dietary supplement – that contains calcium iron, copper, manganese, zinc, cobalt, iodine, vitamin A, vitamin D3, vitamin E, vitamin K, vitamin C, Vitamins B1, B2, B6, and B12.

Cites Krintler (1988) noting three batches of *Hyla ebraccata* with spindly leg disease. Anomalies included no legs, on or both spindly front legs. He attributed them to poor nutrition, related to "poor weather" and absence of dietary insect availability.

Parker WS, Pianka ER. 1975. Comparative Ecology of Populations of the Lizard *Uta stansburiana*. Copeia, Vol. 1975:615–632

Trauma – Broken tails in 40–76% *Uta stansburiana* males and 14–86% of females, especially in more southern locations. Similar observation in *Cnemidophorus tigris*.

Parker WS, Pianka ER. 1976. Ecological observations on the leopard lizard (*Crotaphytus wislizeni*) in different parts of its range. Herpetologica 32:95–114.

Trauma – Low frequency of broken regenerated tails in *Crotaphytus wislizeni*, but slightly higher frequency than in *Phrynosoma platyrhinos* (Pianka and Parker 1975). Similar to *Cnemidophorus tigris* (Pianka 1970), *Uta stansburiana* (Pianka and Parker 1975), and *Phrynosoma platyrhinos*, broken tails are more common in southern populations.

Parona C. 1879. Comunicazioni dai Laboratorj. Dal Laboratorio di Anatomia e Fisiologia Comparate della R. Università di Pavia [Communications of Laboratories. From the Laboratory of Anatomy and Comparative Physiology of the R. University of Pavia]. Bollettino Scientifico dai Professori all'Università di Pavia 1 (1): 64. [Italian].

Congenital – polymely in *Rana mangereccia*. The head of the femur of the accessory limb was attached to the posterior inferior tuberosity of the ilium, and the distal portion of the limbs were fused, with nine toes. Parona concluded that this was a pentamelic frog of the genus *Pigomele* and the variety *pleomele*. He added that four cases of polymely in the genus *Rana* had been found in the Comparative Anatomy Museum of the University of Pavia: His and three already illustrated by Balsamo-Crivelli. The latter eluded and thus were not reported by Cavanna in his then recent and accurate work on polymely in batrachian anurans.

Parona, C. 1880. Sulla pigomelia nei vertebrati [On pygomely in vertebrates]. Bollettino scientifico, Pavia 25(6):91–95. [Italian]

Congenital – Review of the different types of polymely proposed by Geoffroy-St. Hilaire, which was based on the shape, arrangement, degree of development, number of accessory limbs, and most importantly, the point of attachment of the latter to the main body. These subdivisions are: (1) pygomely (2) gastromegaly (3) notomely (4) cephalomegaly, and (5) melomely. The most common is pygomely, which can be defined as the presence of one or two accessory pelvic limbs attached to the hypogastric region, its rear, or in between the normal limbs. The concept of pygomely is not well defined, and at some point, it becomes a true double monster. Some authors suggest that all cases that involve additional normal limbs attaching to two different, more or less complete, trunks or pelvises should be excluded from pygomely. The extra limbs attach to different parts of the pelvic belt, and thus, pygomely can be naturally classified depending on the bones associated with the region of attachment (i.e., iliac, ischiatic, pubic, sacral, coccigeal regions) into: iliomely, ischiomely, pubemely, sacromely, and cocigemely. Compound classifications such as sacro-iliomely can be used when the limbs attach to more than one region:

Species	Author of report	Locality	Peculiarity (extra limbs)	Type of pigomely
Rana viridis	Van der Hoeven	Strassbourg	2 limbs	Pubemely
Rana viridis	Van Deen	Leiden?	2 incomplete limbs	?
Bufo vulgaris	Thomas	Nantes	1 limb	Sacromely
Rana viridis	Duméril	Paris	1 incomplete limb	?
Rana temporaria	Duméril	Paris	1 incomplete limb	Iliomely
Rana clamata	Duméril	Paris	2 limbs	Iliomely
Alytes obstetricans	Cisternas		1 incomplete limb	Sacro-iliomely?
Rana viridis	Balsamo-Crivelli	Pavia	1 left limb	Pubemely
Rana viridis	Balsamo-Crivelli	Pavia	1 incomplete left limb	Iliomely
Rana viridis	Balsamo-Crivelli	Pavia	1 incomplete left limb	Iliomely
Rana viridis	Lunel	Geneva	2 limbs	?
Rana viridis	Fabretti	Perugia	1 incomplete limb	Pubemely
Rana viridis	Fabretti	Perugia	1 incomplete left limb	Pubemely
Rana viridis	Strobel	Parma	1 incomplete right limb	Iliomely
Rana viridis	Strobel	Parma	1 right limb	Iliomely
Rana viridis	Strobel	Modena	1 right limb	Iliomely
Rana temporaria	Strobel	Reggio Emilia	2 incomplete limbs	Iliomely
Rana viridis	Sordelli	Milano	1 incomplete leftlimb	?
Rana viridis	Cavanna	Siena	2 limbs	?
Rana viridis	Cavanna	Firenze*	2 incomplete limbs	Pubemely
Rana viridis	Parona	Pavia	2 incomplete limbs	Iliomely

*Istituto di studi superiori

Parona C. 1883. La pigomelia nei vertebrati. [Pygomelia in vertebrates]. Atti della Società Italiana di Scienze naturali e del Museo Civico di Storia Naturale in Milano 26: 211–316. [Italian]

Congenital – Review of anatomical (monstrous) anomalies (in particular pygomely) and report of pygomelia in a *Rana esculenta* with three legs purchased in the market in Pavia (Italy). The extra hind limb was in between the normal limbs (its femur connected to the lower tuberosity found posterior to the left ilium) and was comprised of the fusion of two feet as evidenced by the presence of nine toes. Internally, despite the duplication of toes and having three tarsal complex bones, it had only one tibia and femur. The second new case had a subtle stump representing an atrophied third hind limb that projected ventrally from the pubic symphysis (lacking any intermediary bones).

Pasquarelli P, Mendes EG, Costacurta L. 1981. Natural polymely in the toad *Bufo icterícus icterícus* Spix, 1824 (Amphibia – Anura). Zoologischer Anzeiger 206:292–296.

Environmental – Polymely in the toad *Bufo icterícus icterícus*. Cause: an unknown (eventually even industrial pollution) stimulus applied below the mid region of somatopleura at a proper embryonic stage.

Paulicki DR. 1883. Ein fünfbeiniger Frosch. [A five-legged frog]. Deutsche Zeitschrift für Tiermedizin und vergleichende Pathologie 12:258–259 [German].

Congenital – *Rana fusca* with an additional posterior extremity.

Pavalko P. 1986. Shell and scute anomalies in some midwestern turtles. Bulletin of the Chicago Herpetological Society 21:36–38.

Congenital – Kyphosis and hump in *Chrysemys picta.*

Shell disease – Kyphosis and hump in *Chrysemys picta.*

Pawar KR, Katdare M. 1984. Toxic and teratogenic effects of fenitronthion, BHC and carbofuran on embryonic development of the frog *Microhyla ornata*. Toxicology Letters 22:7–13.

Toxicology – "Curvature of body axis" caused by pesticides fenitronthion, BHC, and carbofuran in frog *Microhyla ornata.*

Pawar KR, Ghate HV, Katdare M. 1983. Effect of malathion on embryonic development of the frog *Microhyla ornata* (Dumeril and Bibron). Bulletin of Environmental Contamination and Toxicology 31:170–176.

Toxicology – Malathion at 5–10 ppm produced microcephaly in *Microhyla ornata.*

Pawlowska-Indyk A. 1976. Zmiennosc szkieletu osiowego kumaka gorskiego (*Bombina varigata* L.). Zokolic Krynicy (Beskid Sadecki). [Variability of the vertebral column of *Bombina variegata* (L.) of the Krynica region (Beskid Sadecki)]. Acta Universitatis Wratislaviensis. Prace Zoologiczne 7:57–68. [Polish]

Congenital – 70 of 523 *Bombina variegate* had vertebral column abnormalities, including variation in number and structure: 1% with reduced and 2% with increased number of vertebrae; 5% asymmetrical sacrum 4, 5% with asymmetrical sacral duplication, 6% with urostyle abnormalities, 7% with reduction, and 1% with augmentation of number of presacral vertebrae, and 1% with hemivertebrae.

Additionally, abnormalities noted by Madej (1965) in 10% of 555 *Bombina bombina* reduced number of vertebrae in 2%; increased in 1%; asymmetrical sacra in 5%, and abnormal urostyle in 5%, 11% of 813 *Bombina variegata* (1% of decrease, 2% of increase, 5% of sacral asymmetry, and 6% of urostyle abnormality), by Adolphi (1895) in 23% of 65 *Pelobates fuscus* (20% vertebral increase, 11% asymmetrical sacra), (1898) in 10% of 200 *Bufo cinereus* (1% decreased vertebrae), (1893) in 6% of 212 *Bufo variabilis*, and (1895) in 3% of 117 *Rana esculenta*. The latter contrasted with 7% of 205 reported by Zaharesco (1935), who noted 0.5% decreased and 2% increased vertebrae and 1% asymmetrical sacra.

Payen S. 1991. Tératologie chez les reptiles. Étude particulière des monstres à bifurcation axiale. Inventaire des exemplaires conservés dans les museums d'histoire naturelle. [Teratology in reptiles. Particular study of axial bifurcation monsters: Inventory of examples conserved in the museums of natural history]. DVM Dissertation, École Nationale Vétérinaire d'Alfort, Creteil, France, 175 pp. [French]

Congenital – Dicephalic copperhead *Austrelaps superbus*, fer-de-lance *Bothrops atrox*, *Elaphe rufodorsata*, common water snake or Schneider's water snake *Enhydris enhydris*, rainbow boa *Epicrates cenchria maurus*, Bahama boa *Epicrates striatus*, Russell's sand boa or red earth boa *Eryx conicus*, *Eutaenia macrostemma*, common king snake *Lampropeltis getula*, *Leimadophis poecilogyrus*, common Indian wolf snake *Lycodon aulicus*, Cape common wolf snake *Lycophidion capense*, viperine snake *Natrix maura*, diamondback water snake *Natrix rhombifera*, Queen water snake *Natrix septemvittata*, common water snake or northern water snake *Natrix sipedon*, gulf salt marsh snake *Natrix sipedon clarkia*, water snake *Nerodia fasciata*, rough-scaled green snake *Opheodrys aestivus*, San Diego gopher snake *Pituophis melanoleucus annectens*, Pacific gopher snake *Pituophis melanoleucus catenifer*, Great Basin gopher snake *Pituophis melanoleucus deserticola*, Dark Indian python *Python molurus bivittatus*, *Rhabdophis tigrinus*, western aquatic garter snake *Thamnophis couchii*, elegant garter snake or mountain garter snake *Thamnophis elegans*, wandering garter snake *Thamnophis elegans vagrans*, black-tailed boa *Tropidophis melanurus*.

Turkish gecko *Hemidactylus turcicus*, yellow-headed gecko *Gonatodes albogularis*, *Mabuya striata striata*, *Tiliqua scincoides* (Willis 1932), *Trachydosaurus rugosus* (Matz 1989).

Derodymus *Lacerta saxicola* (Darevsky 1960, 1961), *Lacerta vivipara*, asp viper, *Trachydosaurus rugosus* (Matz 1989).

Dicephalic *Mabuya striata striata* (Broadley 1977) and *Varanus varius* with partial body duplication.

Cephaloderopagus sand lizard *Lacerta agilis* (Pleticha 1968; Bellairs 1965), *Mabuya striata punctatissima* and *Vipera aspis*.

Siamese *Trachydosaurus rugosus* (Matz 1989).

Axial duplication in *Anguis fragilis* (Cantoni 1921; Riches 1955; Reichenbach-Klinke and Elkan 1965).

Opisthodichotomous *Phelsuma madagascariensis*, *Podarcis muralis*, Posterior duplication in *Egernia striolata* (Matz 1989).

Trauma – bifid tail in *Stenocercus ornatissimus* and *Xenodon severus* and multiple tail bifurcations in giant green lizard *Lacerta trilineata*.

Payen S. 1995. Axial duplications in lizards. Herpetopathologia 2:171–180.

Congenital – Dicephalic yellow-headed gecko *Gonatodes albogularis albogularis*, *Hemidactylus turcicus*, *Lacerta vivipara*, *Mabuya striata striata*, and apparently *Tiliqua scincoides* (Willis 1932) and *Trachydosaurus rugosus* (Matz 1989).

Derodymus *Trachydosaurus rugosus* (Matz 1989).

Cephaloderopagus *Lacerta agilis* (Pleticha 1968; Bellairs 1965), *Mabuya striata punctatissima*, *Mabuya striata striata*.

Dicephalic *Lacerta saxicola* with axial bifurcation (Darevsky 1960, 1961), *Mabuya striata striata with* duplication of head, forelimbs, and anterior half of body (Broadley 1977) and *Varanus varius* with two forelegs.

Axial duplication (Matz 1989) in *Anguis fragilis* – axial duplication (Cantoni 1921; Riches 1955; Reichenbach-Klinke and Elkan 1965).

Posterior duplication in *Egernia striolata* (Matz 1989)

Opisthodichotomous *Phelsuma madagascariensis* and *Podarcis muralis*.

Siamese twin *Trachydosaurus rugosus* (Matz 1989).

Trauma – Bifid tail in *Stenocercus ornatissimus* and multiple tail bifurcations in *Lacerta trilineata*.

Peabody RB, Brodie ED Jr. 1975. Effect of temperature, salinity and photoperiod on the number of trunk vertebrae in *Ambystoma maculatum*. Copeia 175:741–746.

Environmental – *Aprasia* collected in warm weather had more vertebrae than those collected in cold (Parker 1956).

Number of caudal, but not trunk, vertebrae correlated with temperature in geckos *Ptyodactylus*, *Hemidactylus*, *Tropiocolotes*, *Cearamodactylus*, and *Stenodactylus* (Werner 1964).

More caudal vertebrae were noted in *Triturus vulgaris* raised above and below intermediate temperatures, opposite of findings for *Ambystoma mexicanum* (Orska and Imiolek 1962).

Positive correlation with temperature in *Ambystoma gracile* (Lindsey 1966).

Peale 1830. Tiedemann's Zeitschrift für Physiologie [Tiedemann's Newspaper for Physiology] 4: 123 [German]. Actually Tiedemann 1930.

Pearson PG. 1960. A description of a six-legged bullfrog, *Rana catesbeiana*. Copeia 1960:50–51.

Congenital – Six-legged bullfrog *Rana catesbeiana* with duplicate pelvic girdle from Delaware-Raritan Canal of New Jersey.

Pelgen JL. 1951. A *Rana catesbeiana* with six functional legs. Herpetologica 7:138–139.

Congenital – Supernumerary legs in *Rana catesbeiana*.

Pendlebury GB. 1976. Congenital defects in the brood of a prairie rattlesnake. Canadian Journal of Zoology 54(11):2023–2025.

Congenital – Prairie rattlesnake *Crotalus viridis viridis* with umbilical area fusion, cleft palate (passing through maxillary bone in floor of nose), with partially developed parietal and overdeveloped frontals.

Penner JE. 1972. An unusual specimen of *Ameiva undulata* (Sauria: Teiidae) from Yucatan, Mexico. Herpetological Review 4:202.

Congenital – Absent distal portion of limbs in *Ameiva undulata*.

Pereira AA. 1944. Um caso de *Bothrops jararacuçú* bicefalo [A case of bicephaly in *Bothrops jararacuçú*]. Anais Instituto Pinheiros, São Paulo 7(13):1–2. [Portuguese]

Congenital – Dicephalic *Bothrops jararacussú*, Espírito Santo, Brazil.

Pereira AA. 1950. Um outro caso de bicefalia em serpents (*Bothrops jararaca*) [Another case of bicephaly in snakes (*Bothrops jararaca*)]. Anais Instituto Pinheiros, São Paulo 13(26): 103–106. [Portuguese]

Congenital – Dicephalic *Bothrops jararaca* in the collection of the Instituto Pinheiros (specimen number 5) in the city of Vitória, state of Espírito Santo, Brazil. Necks were juxtaposed and formed a singular, triangular overall shape. They were joined laterally until slightly before the eyes, at which point the heads separated.

Perez-Coll CS, Herkovitz J, Salibvan A. 1988. Embryotoxicity of lead to *Bufo arenarum*. Bulletin of Environmental Contamination and Toxicology 41:247–252.

Toxicology – Lead-induced stunted tail and bifid spine in *Bufo arenarum*.

Pérez García A, Gascó F. 2010. Preservación excepcional de un disco intervertebral atribuido a un reptil marino, descubierto en la Sierra de Albarracín (Cordillera Ibérica, España) en el siglo XIX. [Exceptional preservation of an intervertebral disc attributed to a marine reptile, discovered in the Sierra de Albarracín (Iberian Range, Spain) in the nineteenth century]. Geogaceta 48:75–78.

Vertebral – Thin (~1.5 cm), 11.5 cm diameter disk structure MNCN 58622 attributed to a Jurassic marine reptile found in Sierra de Albarracín, Spain. They interpreted it as an intervertebral disk. If so, it is an important observation, as current thinking suggests that marine reptiles have joint spaces, not disk spaces.

Pérez-Higareda G, Smith HM. 1987. A Two-headed snake, *Bothrops asper*, from southern Veracruz, Mexico. Bulletin of the Maryland Herpetological Society 23(2):72–73.

Congenital – Dicephalic *Bothrops asper*.

Pérez-Mellado V, Corti C, Lo Cascio P. 1997. Tail autotomy and extinction in Mediterranean lizards. A preliminary study of continental and insular populations. Journal of Zoology, London 243:533–541.

Congenital – Tails are shed more easily in high predation continental and "insular" populations.

Autotomy rates are given in ratio of present to absent: *Podarcis bocagei* (24/4), *Lacerta monticola* (14/2), *Psammodromus algirus* (4/1), *Podarcis milensis* (8/1), *Podarcis hispanica* (28/0), *Podarcis lilfordi* (45/20), *Podarcis sicula* (12/5), *Podarcis raffonei* (7/0), *Podarcis pityusensis* (10/0), but there is a discrepancy with the number listed in the methods section of animals examined: 15, 6, 24, 10, 31, 81, 19, 13, and 10, respectively.

Pérez-Higareda G, Smith HM. (Actually Smith HM and Pérez-Higareda G) 1987b. The literature on somatodichotomy in snakes. Bulletin of the Maryland Herpetological Society 23:139–153.

Congenital – Cites Cunningham (1937) gleaning of 150 references and lists 117 additional references. 22% of reports of somatodichotomy were in Colubridae, 18% in Naticidae, and 21% in Crotalidae. Earliest report was by Nakamura (1938) who recognized teratophagus presenting as craniophagus (fusion of cranium), cephaloderophagus (fusion of cranium and anterior trunk), and anakatamesodidymus (separation at both ends). He referred to tyeratodymus as rhinodymus (double nosed), opodymus (two headed) and derodymus (bifurcated head and neck), craniodichotomy (head only), prodichotomy (head and anterior body), proarchodichotomy (up until anus), opisthodichotomy (head and tail duplicated), urodichotomy (only tail duplicated) and amphidichotomy (both anterior and posterior, rarely middle duplicated), and holodichotomy (total duplication).

Peri S, Williams J. 1988. Anomalías osteológicas en *Hyla pulchella pulchella* y *Pseudis paradoxa platensis* (Amphibia: Anura) [Osteological anomalies in *Hyla pulchella pulchella* and *Pseudis paradoxa platensis* (Amphibia: Anura)]. Boletín de la Asociación Herpetológica Argentina 4(1): 4–5 [Spanish].

Congenital – *Hyla pulchella pulchella* in the collections of the Museum of La Plata, Argentina (MLP A 712 and MLP A 714) with dilated diapophysis (sacral style) over the left margin of the last pre-sacral vertebra, while the sacral vertebra possessed the normal diapophysis in its right side. On the opposing side, it had a posteriorly facing, thin, transverse apophysis. A thin, anteroposteriorly expanded cartilage was observed in its end.

Pseudis paradoxa platensis (MLP. A 704 with two supernumerary hind limbs inserted ventrally into the pelvic girdle.

Perkins CB. 1955. Two-headed snake has problems. Statesville Record & Landmark, (Statesville, NC) 29 July 1955: 8.

Congenital – Dicephalic California king snake, named Duplex at San Diego Zoo.

Perkins A. 2002. Two-for-the-show turtle beating odds: Jam and I is a star attraction at the Natural Science Center. News & Record (Greensboro, NC) 22 May 2002:

Congenital – Dicephalic yellowbelly slider turtle.

Perpiñán D, Garner MM, Trupkiewicz JG, Malarchik J, Armstrong DL, Lucio-Forster A, Bowman DD. 2010. Scoliosis in a tiger salamander (*Ambystoma tigrinum*) associated with encysted digenetic trematodes of the genus *Clinostomum*. Journal of Wildlife Diseases 46:579–584.

Congenital – Hemivertebrae-related scoliosis in urodeles (Witzmann 2007).

Infection – 1/202 surviving tiger salamander *Ambystoma tigrinum* with multiple skin nodules developed scoliosis secondary to encysted metacercarial stages of digenean trematodes *Clinostomum* sp. (order Strigeidida, family Clinostgomatidae). Soft tissue swelling dislodged carpus, radius, and ulna with pressure erosions and new bone formation.

Digenean trematode *Ribeiroia ondatrae* disrupts normal limb bud cell growth (Stopper et al. 2002), as have microsporidia (Gamble et al. 2006).

Clinostomum attenuatum or *C. complanatum* have been reported in green frogs *Rana clamitans*, bullfrogs *Rana catesbeiana*, cane toads *Bufo marinus*, red-spotted newts *Notophthalmus viridescens*, cave salamanders *Eurycea lucifuga*, tiger salamanders *Ambystoma tigrinum*, but no musculoskeletal deformities were reported.

Trematode induces scoliosis in San Marcos salamanders *Eurycea nana* (Gamble et al. 2006).

Toxicology – Toxin-induced scoliosis (Iwamuro et al. 2003)

Environmental – Pesticides and habitat modification are suggested as determinants of parasite levels (Belden and Kiesecker 2005; Gray et al. 2007; Kiesecker 2002; King et al. 2007, 2008; Koprivnikar et al. 2007).

Perri T. 1952. Variazioni della competenza arto-formativa negli Anfibi Anuri [Variations of the limb-forming ability in Anuran Amphibians]. Rendiconti (8), Accademia nazionale dei Lincei. Classe di scienze fisiche, matematiche e naturali 12: 753–759 [Italian].

Congenital – Although an experimental paper, it contains a very short mention of the literature on incidence of supernumerary limbs in Anurans in nature. Perri's experiments concentrated on *Hyla arborea* and *Discoglossus pictus*.

Persson O. 1971. Sjukdomar och missbildningar hos saurier. [Diseases and malformations in saurians.] Recip Revlex 4:3–8. [Swedish].

Congenital – Misshapen cervical centrum of a plesiosaur from Oxford Clay (Upper Jurassic) BMNH 48001, possibly *Cimoliosaurus plicatus* represents segmentation defect with differential (laterality) penetration.

Fossil – Misshapen cervical centrum of a plesiosaur from Oxford Clay (Upper Jurassic) BMNH 48001), possibly *Cimoliosaurus plicatus* represents segmentation defect with differential (laterality) penetration.

Petch S. 1990. A case of twins from a single egg in the Chinese ratsnake *Elaphe bimaculata*. Herptile 15(1):13–16.

Congenital – Identical, not Siamese Chinese rat snake *Elaphe bimaculata* twins.

Peterka H. 1941. The anatomy of an anomalous thoracic region in *Rana castebeiana*. Proceedings of the Oklahoma Academy of Science 21:47–49.

Congenital – Six-legged bullfrog with humeri emerging perpendicularly from the thorax.

Peters J. 1969. The snakes of the subfamily Dipsadinae. Misc Publ Univ Michigan Mus Zool 114:1–122

Congenital – Scale asymmetry and variation in *Dipsas brevifacies, Dipsas gaigeae, Dipsas viguieri, Dipsas pavonina, Dipsas vermiculata, Dipsas indica indica, Dipsas indica bucephala, Dipsas indica cisticeps, Dipsas neivai, Dipsas elegans, Dipsas ellipsifera, Dipsas oreas, Dipsas boettgeri, Dipsas latifasciata,Dipsas latifrontalis, Dipsas peruana, Diposas pratti, Dipsas sanctijoannis, Dipsas schunkei, Dipsas albifrons, Dipsas incerta, Dipsas variegata variegata, Dipsas variegata nicholsi, Diposas variegata trinitatis, Sibynomorphus mikani mikani, Sibynomorphus mikani neuwiedi, Sibynomorphus turgidus, Sibynomorphus vagrans, Sibynomorphus ventrimaculatus, Sibon annulata, Sibon anthracops, Sibon dimidiate dimidiate, Sibon dimidiate grandoculis, Sibon sanniola, Sibon argus, Sibon carri, Sibon dunni, Sibon nebulata.*

Peterson SM, Wilson AG Jr, Wilson EM. 1999. *Plethodon idahoensis* (coeur d'alene salamander) skeletal abnormality. Herpetological Review 30:222.

Congenital – Spinal curvature in coeur d'alene salamander (*Plethodon idahoensis*).
Petranka JW, Kennedy CA. 1999. Pond tadpoles with general morphology: Is it time to reconsider their functional roles in aquatic communities? Oecologia 120:621–631.
Trauma – Review of predation by tadpoles on other amphibians, but no distinguishing comments on morbidity versus mortality.
Petzold H-G. 1963. Notizen zur Fortpflanzungsbiologie und Jugendentwicklung zweier Grubenottern (Serpentes: Crotalidae: *Crotalus atrox* und *Agklistrodon p. piscivorus*). [Notes on the reproductive biology and the juvenile development of two adders (Serpentes..)] Bijdragen tot Dierkunde 33:61–69. [German]
Congenital – *Agklistrodon piscivorus* (Eastern Cottonmouth) with double fracture in vertebra 19 and 20.
Petzold H-G. 1967. Ein Nashornleguan, *Cyclura cornuta* mit Gabelschwanz im Tierpark Berlin. [A rhinoceros iguana, *Cyclura cornuta* with forked tail in the Berlin Zoo] Deutsche Aquarien- und Terrarien-Zeitschrift 14: 306–308. [German].
Trauma – Forked-tail in a rhinoceros iguana *Cyclura cornuta.*
Pflaumer C. 1945. Un segundo hallazgo de una culebra con dos cabezas en Chile [A second find of a two-headed snake in Chile]. Revista Chilena de Historia Natural Pura y Aplicada 48:97–102. [Spanish]
Congenital – Dicephalic *Dromicus chamissonis* (Wiegman) captured in Niebla beach in Valdivia, Chile.
Philippeaux JM. 1867a. Sur la régénération des members chez l'Axolotl. [On regeneration of Axolotl limbs]. Comptes Rendus de l'Académie des sciences de l'Institut de France 64: 1204–1205. [French]
Trauma – *Triton cristatus* limb regeneration requires more than stump.
Philippeaux MJ. 1867b. Sur la régénération des members chez l'axolotl (*Siren pisciformis)*. [On regeneration of axolotl limbs]. Annales des Sciences Naturelles. Zoologie et Biologie Animale Series 5, 7:228. [French]
Trauma – *Triton cristatus* limb regeneration. Although *Siren pisciformis* was in the title, it was not mentioned in the text.
Philippeaux JM. 1875. Notes sur les resultants de l'extirpation complète d'un des members anterior sur l'Axolotl et sur la salamander acqu. [Notes on the results of complete removal of anterior limbs of axolotl and of aquatic salamander]. Comptes Rendus et Mémoire de la Société Biologique Paris 6: 113–114. [French]
Trauma – Rudimentary limb after amputation in aquatic salamander.
Piana G.P. 1894. Ricerche sulla polidactilia acquisita determinata sperimentalmente nei Tritoni e sulle code soprannumerarie nelle lucertole [Research on acquired polydactyly determined experimentally in Tritons and on supernumerary tails in lizards]. Ricerche fatte nel Laboratorio di anatomia normale della Reale Universitá di Roma ed in altri laboratory biologici 4:65–71. [Italian]
Congenital – Experimental research on polydactyly in *Triton*
Trauma – Supernumerary tails in lizards. In the latter section, he observed two different vertebral canals formed in cartilaginous cases inside the supernumerary tail, a situation to his knowledge unique. He compared the supernumerary tail of his lizard to that reported in an Axolotl by Sordelli (1882).
Pianka ER. 1967. On lizard species diversity: North American flatland deserts. Ecology 48:333–351.
Trauma – More tail loss in southern than northern *Cnemidophorus, Uta, Phrynosoma* and female *Callisaurus.*
Pianka ER, Pianka HD. 1976. Comparative Ecology of Twelve Species of Nocturnal Lizards (Gekkonidae) in the Western Australian Desert. Copeia 1976:125–142.
Trauma – incidence of tails broken is less than 1% in the knob-tailed *Nephrurus laevissimus*, contrasted with 72.7% in *D. elderi.*
Suggested that arboreal lizards have more broken tails than terrestrial (Werner 1968).
Pieau C. 1966. Effets des rayons x sur le jeune embryon d'orvet (*Anguis fragilis* L.). [Effects of X-rays on the early embryo of slow worm (*Anguis fragilis* L.)]. Comp Rendu Soc Biol 160:59–62. [French]
Environmental – Radiation produced reduced limb development in *Anguis fragilis*.
Pietsch P. 1987. The effects of retinoic acid on mitosis during tail and limb regeneration in the axolotl larva, *Ambystoma mexicanum.* Wilhelm Roux's Archiv of Developmental Biology 196:169–175.
Toxicology – Retinoic acid depresses regeneration of *Ambystoma mexicanum* tails.
Piiper E. 1933. The development of *Rana temporaria* under the influence of cane sugar solution. Proceedings of the Royal Society of London, Series B 112:359–365.
Environmental – 25% of *Rana temporaria* develop double mouth if develop in 8% cane sugar solution.
Pikulik MM. 1985. Zemnowodnye Belorussii. [Amphibians of Bella Russia] 191 pp.; Minsk: Nauka y Teknika. [Russian]
Congenital – Congenital pathologies, but not osseous.
Pilch J Jr. 1981. Life history note. Testudines *Chrysemys concinna texana* (Texas river cooter) Morphology. Herpetological Review 12:81.
Congenital – Dicephalic Texas river cooter *Chrysemys concinna texana.*
Pilliod DS. 1999. *Rana cascadae* (Cascade frog). Predation. Herpetological Review 30:93.
Trauma – Predation on cascade frog *Rana cascadae* by giant water bug *Lethocerus* sp.
Pires de Lima JA. 1927. As Anomalias dos membros nos Portugueses [The anomalies of limbs in the Portuguese]. Colecção "Natura" –Porto. Araújo e Sobrinho Suc. [Portuguese]
Congenital – Schistomely – divided hand.
Pires de Lima, J-A. 1929. Queue bifurquée chez les *Lacertidae.* [Tail bifurcation in *Lacertidae*]. Comptes Rendus de l'Association des Anatomistes (vingt-quatrième reunion: Bordeaux, 25–27 mars 1929: 426–431. [French]

Trauma – General comments, citing Pline (1539), Volante (1923), Vialleton (1911) on fibrocartilaginous tail regeneration and the three-tailed lizard discussion by Taruffi (1886).

Bifid tail in *Lacerta ocellata* and in *Lacerta muralis* (de Kerville (1909) in *Rhacodactylus trachyrhyncus* de Kerville (1909)

Pires de Lima, J.A. 1930. O sardão nas tradições populares. [The lizard in popular traditions]. Trabalhos da Sociedade Portuguesa de antropologia e etnologia 4(3):285–289. [Portuguese]

Trauma – Brief reference to 1929 publication, mentioning again *Lacerta ocellata* Daud. and *Lacerta muralis* Laur. with bifid tails and discussed various traditions, perceptions, and legends about lizards and snakes (including two-tailed lizards) of the Portuguese folk.

Pirlot JM, Welsch M. 1934. Etude anatomique et experimental de quelques tumeurs chez la grenouille rousse (*Rana fusca* L.). [Anatomical and experimental study of several tumors in a red frog (*Rana fusca* L)]. Archives Internationales de Médecine Experimentale 9:341–365. [French]

Neoplasia – Chondromyxoma in *Rana fusca.*

Pitman CR. 1941. About crocodiles. Uganda Journal 8(3):89–114.

Congenital – Related dwarf races of *Crocodylus niloticus* in Aswa/Moroto and Greek rivers in Eastern Uganda to deficiency of permanent water.

Pizzi R, Miller J. 2005. Amputation of a *Mycobacterium marinum*-infected hindlimb in an African bullfrog (*Pyxicephalus adspersus*). Veterinary Records?? 156:747–748.

Infection – *Mycobacterium marinum*-infected tibiofibular fracture in African bullfrog *Pyxicephalus adspersus*

Plateretti V.I. 1777. Su le riproduzioni di gambe e della coda delle salamandre acquatiche. Memoria di Vicenzo Ignazio Plateretti. Scelta di opuscoli interessanti [On the reproduction of limbs and tail in aquatic salamanders. Memoires of Vicenzo Ignazio Plateretti. Selection of interesting papers]. Tomo III, Volume 27:31, 103. Milano, Italia. [Italian]

Trauma – Salamander limbs are regenerated with more or less digits than normal more commonly when a limb is severed repeatedly. Almost six digits (where norm is four) occurred the fourth time the limb had been regenerated.

Platt A. 1910. The works of Aristotle translated into English (under the editorship of Smith JA and Ross WD), Oxford, 240 pp. [Greek translated to English]

Congenital – Noted rarity of dicephalic and cited St. Hilaire III:185, 193 report.

Pleticha P. 1968. Micro-anatomical analysis of a congenital head duplication of a lizard (*Lacerta agilis* L.). *Věstník* Československé zoologické společnosti 32:232–236.

Congenital – Opodymus *Lacerta agilis.*

Pline 1772. Histoire Naturelle traduite en francais. [Natural History Translated into French]. IV Paris. [French].

This appears to be a French translation of Pliny the Elder 1539.

Plinii Secundi C. 1539. Historiae Naturalis. [Natural History]. Lipsiae: Caroli Tauchitii; 500 pp. [Latin]

Trauma – Bipartite and tripartite reptile tails – in books 10 and 11

Pseudopathology – Mistook two mating snakes as dicephalic – in book 10. Mistook amphisbaena as having heads on either end – in book 8.

Plowman C, Montali R, Phillips L, Schlater L, Lowenstine L. 1987. Septicemia and chronic abscesses in iguanas (*Cyclura cornuta* and *Iguana iguana* associated with a *Neisseria* species. Journal of Zoo Animal Medicine 18:86–93.

Infection – Lytic changes in caudal vertebrae of rhinoceros iguana *Cyclura cornuta* from infection. Lytic changes in caudal vertebrae of common iguana *Iguana iguana* from infection.

Plummer M. 1980. Ventral scute anomalies in a population of *Opheodrys aestivus.* Journal of Herpetology 14:199.

Congenital – 15.4% of *Opheodrys aestivus* had ventral scute anomalies. 18/28 had a half ventral inserted between the two normal ventral ends and nine had divided ventral. One of the first group had corresponding vertebral and rib duplication.

Plymale HH, HH, Jacksdon CG Jr, Collier G. 1978. Kyphosis in *Chrysemys scripta yaquia* (Testudines: Emydidae) and other turtles. Southwestern Naturalist 23:457–461.

Congenital – Kyphosis in *Chrysemys scripta yaquia, Caretta caretta, Chelydra serpentina, Kinosternum subrubrum hippocrepis, Sternotheres odoratus, Chrysemys picta, Chrysemys scripta, Pseudomys (Chrysemys) floridana, Clemmy guttata, Deirochelys reticularia, Mallemys centrata, Graptemys (Malaclemys) pseudogeographica, Melanochelys trijuga, Terropene carolina, Terropene ornata, Testudo hermanni, Amyda (Trionyx) ferox, muticus, sinensis, spinifera, triunguis,* and *Trionyx steindachnderi.*

Pol R. 1913. Die Vertebraten-Hypermelie. Studien zur Pathologie der Entwicklung I. [The hypermely of vertebrates. Studies of the pathology of ontogeny]. 1: 71–184. [German]

Congenital – *Rana esculenta* with additional anterior legs and scapula (after Alessandri 1854; Ercolani 1882; Johnson 1901; Lunel 1863; Mazza 1885; Tornier 1898) and 8–7 fingers (Cavanna 1879). *Bufo viridis* (Tornier 1898), *Hyla aurea* (O'Donoghue 1910), *Rana temporaria* (O'Donoghue 1910), *Rana* (Johnson 1901), *Triton taeniatus* (Barfurth 1899) with additional rudimentary scapula and extremity and *Bufo columbiensis* (Washburn 1899)with additional rudimentary scapula and extremity with more fingers, axolotl with rudimentary additional anterior extremity and hyperdactyly (Sordelli 1882; Duméril 1867; Barfurth 1894).

Supernumerary hind legs in *Pelobates fuscus* (Tornier 1901), *Rana esculenta* (Deen 1838; Lunel 1868; Balsamo-Crivelli 1865), *Rana clumata* (Duméril 1867), *Rana halecina* (Kingsly 1882; Johnson 1901), *Rana palustris* (Tuckermans 1885), *Rana temporaria* (Duméril 1865), *Rana fusca* (Bender 1906; Woodland 1885), *Rana viridis* (Duméril 1867), *Ambystoma punctatum* (with hypophalangy: Winslow 1904), *Alytes abstetricans* (Cisternas 1865), *Triton cristatus* (Camerano 1882), and *Triton niger* (Siebold 1828).

Pol R. 1958. Mißbildungen der Extremitäten. Hyperdaktylie (Polydaktylie), Diplocheirie und Diplodie, Hypermelie, Oligodaktylie und Defekte von Röhrenknochen. I. Teil. [Malformation of extremities. Hyperdactyly (Polydactyly), Diplociry and diplody, hypermely, oligodactyly and defects of long bones]. In Schwalbe E, Gruber GB. eds. Die Morphologie der Missbildungen des Menschen und der Tiere. Jena: Fischer Verlag; pp. 683–719. [German]

Congenital – Review of hypermely and hyperdactyly in amphibians: Hypermely more frequent in amphibians than hyperdactyly, in hind legs (Duméril 1865), in front legs (Tornier 1898, 1901). Duméril (1867) reported that A. V. Humboldt found only one specimen of *Rana viridis* with hypermely within 3,000–4,000 specimens used as food in the zoo in Paris.

Scapula – Hypoplasy of *Rana esculenta* scapula (lack of episcapula), large additional scapulae, two accessory fused coracoids on the left side (Alerssandri 1854; Ercolani 1882), another with two accessory fused scapulae, fused episcapulae, additional coracoid on the left side (Ercolani and Lunel 1868), a third with two accessory rudimentary scapulae and an additional coracoid on the right side (Ercolani 1882), a fourth with accessory rudimentary scapula, two additional coracoids and a fused accessory precoracoid on right side (Tornier 1898; Johnson 1901), and a fifth with accessory rudimentary scapula, a thick accessory humerus and 3/2 similar lower arm bones (Cavanna 1879; Washburn 1899). Accessory scapula and two additional coracoids on left side in *Rana esamelica* (Mazza 1888) (*Bufo viridis* (Tornier 1901) and *Hyla aurea* (O'Donoghue 1910)) with accessory rudimentary scapulae, additional anterior extremity. *Triton taeniatus* with additional articulation in shoulder girdle (Barfurth 18990. *Rana palmipes* with paired anterior short additional extremity and additional bone in shoulder girdle (Johnson 1901).

Pelvis – Additional complete pelvic in *Rana temporaria* and *Pelobates fuscus* (Reichenow 1908; Tornier 1901), *Rana viridis* (Duméril 1867) and *Rana halecina* (Johnson 1901), additional hypoplastic in *Rana esculenta* (Van Deen 1838; Lunel 1868), additional additional half pelvis, tarsus with three centralia and hypophalangy (Winslow 1914). *Rana clunata* with additional pair of articulation fossae (acetabulum) in pelvis (Duméril 1865), *Rana halecina* with additional acetabulum ventral in symphysis of pelvis with additional "normal" leg (Kingsley 1882), *Rana temporaria* with small acetabulum dorsal of acetabulum and additional hind leg with foot with five elongated toes (Duméril 1865), *Rana fusca* with small acetabulum dorsal of acetabulum and additional right hind leg with two tarsalia, three metatarsalia, and three toes (middle one with four phalanges, both lateral ones with three phalanges (Bender 1906)), *Bufo vulgaris* with small acetabulum dorsal of acetabulum and additional right hind leg with duplicated foot with four and three toes, respectively (Sutton 1889).

Supernumerary – Supernumerary limbs in toad (Thomas 1861), frog (Balsamo-Crivelli 1865; Guettard 1783; Otto 1816). Frog with extra extremity (D'Alton 1853), fifth leg on right shoulder (De Superville 1744), water frog with six feet (Van Deen 1838).

Axolotl with five toes on front leg (Otto 1841), additional anterior extremity with three or five toes (Gervais 1864; Bassi 1874; Sutton 1889), partial subcutaneous additional humerus, lower arm free (Bergendal 1889; Eigenmann and Cox 1901).

Axolotl with supernumerary forelimb (Barfurth 1894; Sordelli 1882), another with additional rudimentary hand and lower arm and hyperdactyly (Duméril 1867).

Rana fusca with additional rudimentary hind leg attached to body with two lower legs and two feet (Woodland 1908).

Rana palustris additional rudimentary hind leg attached to body with two toes (Tuckerman 1885).

Triton cristatus with division of femur articulating with two lower legs (Camerano 1882).

Triton niger with additional two-toed foot on knee of hind leg (Siebold 1828).

Digit – hyperdactylous frog (Brown-Séquard 1849; Van der Hoeven 1840) and salamander (Van der Hoeven 1840).

Triton with additional toes on front leg (Rayon 1849), *Salamandra cristata* with seven toes on left front leg and six toes on right hind leg (Geoffroy Saint Hilaire 1832), *Bufo columbiensis* with seven or eight toes on the front leg (Cavanna 1879; Washburn 1899).

Poleshaew LW. 1944a. Das Fehlen von Extremitäten bei Anuren und seine Bedeutung für die Entwicklungsmechanik. [The absence of extremities in anurans and its importance for the development of mechanics] Doklady Akademija Nauk SSSR 42(9):418+ [German]

Congenital – Occasional apodic *Rana temporaria* tadpoles, thought to be hereditary.

Poležajew LW. 1944b. Phenomenon of natural apody in anura and its bearing on the mechanics of development. Comptes Rendus (Doklady) de l'Académie des Sciences de l'URSS 42(9):404–407.

Congenital – Occasional apodic *Rana temporaria* tadpoles, thought to be hereditary.

Politzer N. 1926. Die Doppelbildungen bei Urodelen. II. Mitteilung. [Double-headed urodeles]. Roux's Archiv für Entwicklungsmechanik 108:417–462. [German]

Congenital – Dicephalic salamanders.

Pond CM. 1978. The effect of tail loss on rapid running in *Dipsosaurus dorsalis*. American Zoologist 18:612.

Trauma – Tail breakage reduces running speed in *Dipsosaurus dorsalis*. Fully regenerated tails are shorter and thicker and allow faster running.

Pond CM. 1981. Storage. In Physiological ecology – an evolutionary approach to resource use, Townsend CR, Calow P, eds. Oxford: Blackwell, pp. 190–219.

Trauma – Running speed of desert iguana *Dipsosaurus dorsalis* is reduced by 14% with tail loss. Tail loss with regeneration is rare in jubeniles, but observed in 45% of adult males and 34% of adult females in museum collections.

Ponse K. 1941a. Sur l'ectromélie bithoracique et diverses anomalies présentées par les descendants issus d'une ponte biddérienne d'un crapaud femelle. [On bithoracic ectomelia and diverse anomalies in descendents of a female bridge bidderienne toad]. Revue Suisse de Zoologie 48:545–551. [French]

Congenital – Ectomelia in *Bufo bufo*.

Ponse K. 1941b. Sur une mutation dominante chez le Crapaud: l'ectromélie bithoracique. [On a dominant mutation in a toad: Bithoracic ectomelia]. Compt Rendu des Séancesvde la Société de Physique et d'histoire naturelle de Genève 1941:174. [French]

Congenital – Ectomelia as dominant mutation in *Bufo bufo*.

Poole DF. 1961. Notes on tooth replacement in the Nile crocodile *Crocodylus niloticus*. Proceedings of the Zoological Society London 136:131–140.

Dental – Tooth anomalies "frequently occur" in Nile crocodile *Crocodylus niloticus*. Two fully erupted teeth in a single socket (possibly because old root never resorbed). Crown breaks off because of resorption weakening neck of old tooth, with root still imbedded in socket and only partial eruption of the new tooth or part of old root fully enclosed within root of new tooth.

Pooley AC. 1962. The Nile crocodile, *Crocodylus niloticus*. Notes on the incubation period and growth rate of juveniles. The Lammergeyer 2:1–55.

Dental – Variation in which is longest tooth in Nile crocodile *Crocodylus niloticus* is 9th or 10th upper and 2nd, 4th, 10th, or 11th in mandible.

Pope TEB. 1925. Two-headed bull snake (*Pituophis sayi*). Yearly Publication of the Museum Milwaukee 5:161–167.

Congenital – Dicephalic bull snake *Pituophis sayi*, citing dicephalic snakes reported by Aldrovandi 1640, Johnson 1901, Redi 1684, Lacépède 1789, Mitchell 1826, Hyde 1925, Do Amaral 1926, noting Hyde (1925) reported a dicephalic milk snake *Lampropeltis triangulum*, and Redi (1684), a derodymus bull snake *Pituophis sayi.*

Combined cephalad and caudal bifurcation was reported by Johnson in 1901, Wyman in 1862, and Mitchell in 1826. Redi 1684 (Osserv-int. Agli, anim, vienti, etc., p. 2, tav. 1, 1778) described a bull snake with duplicate spinal cord to middle of back.

Fused black snake *Bascanium constrictor* (Johnson 1901) and copperhead (Hyde 1925) heads.

Johnson 1901 described black snake *Bascanium constrictor* and Hyde 1925, in a copperhead, fused heads, and the latter, a dicephalic milk snake *Lampropeltis triangulum.*

Pope CH. 1939. Turtles of the United States & Canada. 348 pp.; New York, NY.: Alfred A. Knopf, Inc.

Shell disease – Illustrated *Kinosternon subrubrum steindachneri* with two lesions on its plastron, although the text makes no mention of the lesions.

Pope CH. 1961. The giant snakes. 289 pp; London: Routledge & Kegan Paul.

Infection – Mouth rot is limited to captive snakes, producing canker mouth and osteomyelitis.

Porras L, Beraducci J. 1980. Dicephalic *Kinosternon.* Herpetological Review 11(2):35.

Congenital – Dicephalic *Kinosternon flavescens.*

Porta JB. (also cited as GD and BD) 1650. (front piece states 1669 but listed as 1658 in catalogue). Neapolitani magiae naturalis libri viginti [Full vigilance of Neapolitan greater nature]. Book II. Devariis animalibus gignendis [Taking notice of cursed birth]. Chapter XVIII. Serpentes pluris capitibus candibusque [Snakes with many heads]. 412 Pp. Rouen: Rothomagi, Sumptibus loannis Berthelin. [Latin]

Congenital – Dicephalic viper.

Suggested that the legend of the *Hydra Lernaea* was born from the fact that similar monsters were seen among serpents.

Trauma – Lizards with 2–3 tails, noting many lizards with double or triple tails.

Porter HC. 1934. A case of nearly perfect cauda bifida in *Triturus viridescens*. Proceedings of the West Virginia Academy of Science 7:37–38.

Vertebral – Basal vertebra of supernumerary tail is fused with ventrolateral edge of normal vertebral tail of red-spotted newt *Triturus viridescens*, which they called cauda bifida.

Potter FE Jr, Sweet SS. 1981. Generic boundaries in Texas cave salamanders, and a redescription of *Typhlomolge robusta* (Amphibia: Plethodontidae). Copeia 1981(1):64–75.

Congenital – Trunk vertebra diapophyses extend beyond lateral zygapophysis margins and presence of parapophysis alar processes as synapomorphies in some individuals of all species of *Eurycea* and in *Typhlomolge tridentifera*. Elongate diapophyses and well-developed alar processes in at least some larger Edwards Plateau hemidactyliines.

Absent normal fusion of fourth and fifth tarsal in a *Typhlomolge rathbuni*. Fourth and fifth tarsal fusion in 21 of 28 *Eurycea tridentifera* (with accompanying tarsal 3 fusion in three), in 3 of 14 *Typhlomolge troglodytes*, 1 of 20 troglobitic populations of *Eurycea neotenes*, but only 6 of 100 tarsi of epigean populations of the latter.

Powell CL II. 1992. Conformation of adult frogs causing spindly leg in tadpoles. American Dendrobatid Society Newsletter 12:4–5.

Congenital – 98% of *Dendrobates auratus* developed spindly leg, in spite of high Spirulina algae food, but noted a genetic component – parent-dependent. Note that B. Ian Hyler had similar experience with *Epipedobates (= Phyllobates) bassleri.*

Poyntz SV, Bellair SA d'A. 1965. Natural limb regeneration in *Lacerta vivipara.* British Journal of Herpetology 3:204–205.

Trauma – Hind limb region regenerates more readily than forelimb in *Lacerta vivipara.* Y-shaped region of bone on distal femur.

Other rare cases of lizard regeneration in the wild (Singer 1961): *Lacerta vivipara* (Arel and Verrier 1930) and *Liolaemus altissimus altissimus*, *Lacerta agilis*, *muralis*, *viridis*, *vivipara* and *Chalcides ocellatus* (Hellmich 1951).

Prado A. 1942. Um novo caso de bicefalia em serpentes [A new case of bicephaly in snakes]. Ciencia (Mexico) 3:254–255. [Portuguese]

Congenital – Derodymus "Cobra de capim" *Leimadophis poecilogyrus* (Wied.), in the collection of the Butantan institute in São Paulo number 10370.

Prado A. 1943. Notas ofiológicas 16. Um novo caso de bicefalia em serpentes [Ophiological notes 16. A new case of bicephaly in snakes]. Memórias do Instituto Butantan, São Paulo 17:7–9. [Portuguese]

Congenital – Prado reported on a young female dicephalic snake, *Leimadophis poecilogyrus* (Wied.), commonly known as "Cobra de capim," collected in October of 1942 in Pedro Leopoldo, State of Minas Gerais, Brazil. The specimen was entered into the collection of the Butantan Institute in São Paulo #10370. Each head possessed a short neck. Based on the size of the snake, Prado noted that it "lived for some time" (was not a neonate). A photograph of the specimen was included.

Prado A. 1945. Serpentes do Brasil [Serpents of Brazil]. São Paulo: Biblioteca Agro-Pecuária de "Sítios e Fazendas." 134 pp., XXII il. [Portuguese]

Congenital – Derodymus cobra de capim *Leimadophis poecilogyrus* (WIED), donated to the Butantan Institute in October of 1942. This specimen came from Pedro Leopoldo, Estado de Minas Gerais.

States that "monstrosities" in snakes were not very common; cited (Amaral 1927; Daniel 1941; Fischer 1868; Johnson 1901; Redi 1684), and also reports (without providing dates) by Aldrovandi, Aristotele, and Dumeril and Bibron.

Prado A. 1946. Ofidios bicéfalos [Bicephalic ophidians]. Anais Paulistas de Medicina e Cirugia, São Paulo 51(6):393–396. [Portuguese]

Congenital – Review (brief) of literature on dicephalic snakes. Dicephalic *Philodryas schottii* (Schlegel) Instituto Butantã number 10592 and "cobra de capim" *Leimadophis poecilogyrus* (Wied) in the Instituto Butantã number 10370, similar to the case reported by Daniel (1941), in Chile by Gunckel (1944) and Pflaumer (1944).

Pratt CW. 1946. The plane of fracture of the caudal vertebrae of certain lacertilians. Journal of Anatomy 80:184–188.

Trauma – Autotomy in *Sphenodon, Lacerta viridis, Anguis fragilis, Ophisaurus, Hemidactylus flaviridis.*

Preziosi R, Diana A, Florio D, Gustinelli A, Nardini G. 2007. Osteitis deformans (Paget's disease) in a Burmese python (*Python molurus bivittatus*) – A case report. The Veterinary Journal 174:669–672.

Infection – Alleged osteitis deformans (Paget's disease) in a Burmese python *Python molurus bivittatus* because of the presence of multifocal calcific vertebral overgrowths producing bulky cotton wool appearance, but this is extrinsic to the vertebrae, as well as illustrating disrupted trabeculae characteristic of osteomyelitis.

Metabolic – Alleged osteitis deformans (Paget's disease) in a Burmese python *Python molurus bivittatus* because of the presence of multifocal calcific vertebral overgrowths producing bulky cotton wool appearance, but this is extrinsic to the vertebrae, as well as illustrating disrupted trabeculae characteristic of osteomyelitis.

Vertebral – Alleged osteitis deformans (Paget's disease) in a Burmese python *Python molurus bivittatus* because of the presence of multifocal calcific vertebral overgrowths producing bulky cotton wool appearance, but this is extrinsic to the vertebrae, as well as illustrating disrupted trabeculae characteristic of osteomyelitis.

Price LI. 1940. Autotomy of the tail in Permian reptiles. Copeia 28:119–120.

Trauma – Transverse cartilaginous vertebral septa in extant *Sphenodon*, Jurassic *Homeosaurus* and *Saphaeosaurus (Sauranodon)*, cervical 7 of Univ Oklahoma 1020 *Captorhinus* and also in *Labidosaurus*. Transverse process is vestigial.

Fossil – Transverse cartilaginous vertebral septa in extant *Sphenodon*, Jurassic *Homeosaurus* and *Saphaeosaurus (Sauranodon)*, cervical 7 of Univ Oklahoma 1020 *Captorhinus* and also in *Labidosaurus*. Transverse process is vestigial.

Prigioni CM, Langone JA. 1985. Anomalías anatómicas registradas en anfibios de la colección herpetológica del Museo Nacional de Historia Natural [Anatomical anomalies registered in amphibians of the herpetology collection of the National Natural History Museum]. Actas de las Jornadas de Zoología del Uruguay 1985:73–75. [Spanish]

Congenital – Ten anomalous anurans from South America, all part of the collections of the National Natural History Museum of Uruguay (MNHN):

1. MNHN 5366: *Hyla microcephala meridiana*. Rio de Janeiro, Brazil. Supernumerary vertebra between the sacrum and urostyle, with lateral asymmetries.
2. MNHN 5365: *H. sanborni*. C. Largo, Uruguay. IX and X vertebrae fused, also has left and right sacral expansions over vertebra VIII.
3. MNHN 701: *Leptodactylus ocellatus*. Montevideo, Uruguay. Polymelous individual with a right supernumerary forelimb, located behind the regular limb. A radiograph showed normal bones in the extra limb.
4. MNHN 394: *Melanophryniscus* sp. Montevideo, Uruguay. Finger IV in right hand has bifid extremity, with possible syndactylia.
5. MNHN 405: *Melanophryniscus* sp. Canelones, Uruguay. Brachydactylous finger III on left hand, equal in length to finger IV. Finger III also exhibited clinodactyly.
6. MNHN 613: *Melanophryniscus* sp. Barra Arroyo Maldonado, Uruguay. Both hind limbs had feet with brachydactylous fingers I and II.
7. MNHN 1153: *Melanophryniscus* sp. Rocha, Uruguay. Left hind limb (foot) with brachydactylous finger IV.
8. MNHN 5182: *Melanophryniscus* sp. Córdoba, Argentina. The specimen lacked a left eye.
9. MNHN 5637: *Melanophryniscus* sp. Rocha, Uruguay. The right hand had the following abnormalities: Brachydactylous finger I, ectrodactylous finger II, syndactyly and brachydactylous fingers III and IV.
10. MNHN 5638: *Melanophryniscus* sp. 50 km from Asunción, Paraguay. Urostyle with two lateral and asymmetric, proximal apophyses.

The *Melanophryniscus* genus had anomalies in 7.86% of their sample, contrasted with Amaro and Sena's 1968 study, which only showed a 2% incidence of anomalies in *Leptodactylus*.

Pritchard PC. 1967. Living Turtles of the World. Neptune City, NJ: T.F.H.

Shell – *Clemmys leprosa* with shell flaking related to a fungal disease.

Pritchard PCH, 2003. Akinesis and plastral scute homologies in *Sternotherus* (Testudines: Kinosternidae). Chelonian Conservation and Biology 4:671–674.

Congenital – Defined true kinesis is the presence of defined active or passive hinging mechanism between shell elements. Note that demands for kinesis associated with adults (e.g., oviposition) are genetically determined while those that are involved in respiration, retraction, or "defense/threat, as in the kinosternids" result from pressures during ontogeny. One *Sternotherus minor minor* and two *Sternotherus odoratus* had akinetic plastons. Of 100 additional *Sternotherus odoratus*, ten had deficiencies of kinesis.

Pritchard PC. 2006. Translation of Bonin F, Devaux B, Dupré A. Turtles of the World. Baltimore: Johns Hopkins University Press, 416 pp.

Shell disease – *Pelusios adansoni* with carapace holes. *Pelusios marani, Lissemys scutata, Chitra vandijki, Kinosternon acutum, Kinosternon flavescens, Mauremys caspica, Rhinoclemmys nasuta, Sacalia bealei, Graptemys gibbonsi, Morenia petersi, Aspideretes gangeticus* with carapace damage.

Pritchard PC. 2008. Evolution and structure of the turtle shell. In: J Wyneken, MH Godfrey, V Beels. Biology of Turtles. Boca Raton, CRC Press, Pp. 45–83.

Congenital – Lordosis or swayback in *Platysternon megacephalum* (big-headed turtle) and *Malacochersus tornieri*.

Kyphosis in trionychids *Apalone ferox*.

Trauma – Retention of sutures may reduce crack propagation induced by trauma.

Environmental – Pyramiding, caused by abnormal humidity (excessively dry) or diet (too rich in protein or abnormal calcium/phosphate ratio), overfeeding dietary fiber, temperature, UV light, with central areolae sharply elevated in vertebral and costal scutes are mild in *Geochelone pardalis*, but extreme in wild *Psammobates tentorius tentorius* females. Only captive *Geochelone carbonaria* manifests pyramiding; also seen in *Astrochelys radiata*.

Shell disease – Pits illustrated in plastron of *Hydromedusa tectifera*, *Chelodina oblonga*, *Heosemys spinosa*; carapace of *Macrochelys temminckii* and *Batagur baska*.

Peculiar geographically circumscribed "pits" in carapace of *Callagur borneoensis* and irregular "resorption" in plastron of *Cycloderma aubryi*.

Scute anomalies: 5–7 Vertebral scutes in *Notochelys platynota*, 5–9 vertebral scutes in *Lepidochelys olivacea*, the latter often with a different number of costals on each side.

Note genetic and trauma (during embryonic stage)-induced scute abnormalities.

Triangular-shaped scutes producing symmetrical zigzag series in carapace of *Geochelone carbonaria* (red-footed tortoise).

Ankylosis occurs in very large old *Apalone ferox*, extinct Mascarene tortoises *Cylindraspis*. If it occurs solely between two shell bones, this may produce asymmetrical carapace distortion, as noted in *Geochelone nigra*.

Other – Ankylosis occurs in very large old *Apalone ferox*, extinct Mascarene tortoises *Cylindraspis*. If it occurs solely between two shell bones, this may produce asymmetrical carapace distortion, as noted in *Geochelone nigra*.

Fossil – Ankylosis occurs in very large old *Apalone ferox*, extinct Mascarene tortoises *Cylindraspis*.

Pritchard JJ, Ruzicka AJ. 1950 Comparison of fracture repair in the frog, lizard and rat. Journal of Anatomy 84:236–261.

Trauma – Femoral fracture healing in *Rana temporaria* and *Lacerta vivipara* were compared to that of rats. They noted "little disability." Dense fibrous tissue laid down between advancing masses of cartilage, with fibrous union only in rat. Cartilage progressively eroded and replaced by subchondral bone.

Frogs started later and progressed more slowly, especially with endochondral ossification. Lizard repair was temperature dependent. Even at 26°C, it was only half that at 32–37°C. Frog was faster than lizard at room temperature and doubled at 26°C, but the latter was slower than lizards. In frog and lizard, whole periosteal blastoma was replaced by cartilage. Thickened periosteum had only sparse blood vessels. Lizard cartilage was hypertrophic, while that in frog was hyaline. Cartilage was eroded from within in the frog. Chondroclasts were rare in frogs in which cartilage nodules formed and were covered externally by new bone. Wide endochondral trabeculae were found.

Pritchett WB, Dent JN. 1972. The role of size in the rate of limb regeneration in the adult newt. Growth 36:275–289.

Trauma – Red-spotted newt *Notophthalmus* (*Triturus*, Diemyctylus) *viridescens* limb regeneration rate was inversely proportional to limb size.

Progscha KH, Lehmann HD. 1970. Angeborene Mißbildungen in einem Wurf von *Sanzinia madagascariensis* (Reptilia, Boidae). [Congenital malformations in a *Sanzinia madagascariensis* (Reptilia, Boidae) litter]. Salamandra 6:108–114. [German]

Congenital – 4 of 12 Madagascar tree boa *Sanzinia madagascariensis* with scoliosis and kyphoscoliosis.

Przibram H. 1909. Experimental-Zoologie. 2. Band: Regeneration: Eine Zusammenfassung der durch Versuche ermittelten Gesetzmässigkeiten tierischer Wieder-Erzeugung (Nachwachsen, Umformung, Missbildung). [Experimental zoology 2. Regeneration. A summary of roles of animal regeneration in growth, transformation, malformations found by experiments]. VIII + 338 pp.; Leipzig und Wien: Franz Deuticke. [German]

Congenital – Dicephalic

Lizards: *Lacerta* (Hennig 1870), *Anguis fragilis* (Lessona 1876),

Turtle: *Emys* (Bateson 1894),

Snakes: *Lycodon aulicus* (Dobson 1873), *Pelias berus* (Dorner 1873; Haldeman 1879), *Tropidonotus septemvittatus* (Kirkland 1871), *Coluber* (Lasserre 1880), *Hydrophis sublaevis* (Shortt 1868), *Ophiobolus getulus* (Yarrow 1878), *Zamenis constrictor, Tropidonotus fasciatus* "Ausland" 1865)

Polydactyly (Fraisse 1885; Philippeaux 1875, 1867; Tornier 1906) in urodeles: *Siredon* (Barfurth 1895; Duméril 1865, 1867; Vulpian 1867), *Triton* (Barfurth 1899; Bonnet 1777–1779; Piana 1894; Reuter 1875; Siebold 1828; Simmermacher 1885), *Pleurodeles* (Giard 1895).

Polymely in anurans: *Alytes* (Lataste 1879), *Bombinator* (Giebel 1867), *Pelobates* (Gervais 1864; Méhely 1905). *Bufo* (Washburn 1899), *Rana* (Bassi 1874; Bergendal 1874; Cavanna 1877; Cisternas 1865; Balsamo-Crivelli 1865; De Betta 1883; Deen 1838; Duméril 1865; Fabretti 1876; Giebel 1867; Johnson 1901; Jourdain 1877; Kingsly 1882; Lunel 1867; Ryder 1878; Strobel 1875; Tornier 1898).

Split of toes in urodeles: *Siredon* (Barfurth 1995).

Trauma – Regeneration of legs and tail in Lissamphibia: *Siren* (Erber 1876; Kammerer 1908), *Proteus* (Fraisse 1885; Goette 1879; Kammerer 1905), *Necturus* (Towle 1901), *Amphiuma* (Morgan 1903; Towle 1901), *Spelerpes* (Towle 1901; Kammerer 1908), *Plethodon, Desmognathus, Manculus* (Towle 1901), *Ambystoma, Siredon* (Barfurth 1895).

Regeneration of jaw: *Lacerta agilis* (Werber 1905), *Tarentola mauritanica* (Werber 1906).

Regeneration of tail in Ophidia (*Chrysopelea, Psammophis*) *Sphenodon* (Werner 1896).

Nine species of Ascalobotae (Tytler 1863), *Phyllodactylus* (Fraisse 1885; Werner 1896), *Ptyodactylus* (Werner 1896), *Tarentola* (Tofohr 1903), Lacertilia: *Ptychozoon, Geckolepis, Gekko, Hemidactylus* (Fraisse 1885; Werner 1896), *Platydactylus* (Fraisse 1885), *Gymnodactylus, Gonatodes, Diplodactylus, Thecadactylus, Gehyra, Leptodactylus, Hoplodactylus* (Werner 1896), Iguanidae: *Liolaemus, Liocephalus, Tropidurus, Uranoscodon, Brachylophus, Ctenosaura, Dipsosaurus, Sclerops* (Werner 1896), *Anolis* (Gachet 1834; Werner 1896), *Iguana* (Boulenger 1888; Gachet 1834; Müller 1864; Ridewood 1905), Agamidae: *Agama* (Werner 1896), *Draco* (Müller 1865), Pygopodidae: *Pygopus, Lialis* (Werner 1896), Scincidae: *Ablepharus, Scincus, Eumeces, Mabuia, Lygosoma, Chalcides* (Werner 1896), Anguidae: *Anguis fragilis* (Fraisse 1885; Lessona 1876; Müller 1864; Werner 1896), *Ophiodes, Ophisaurus ventralis* (Werner 1896), *Ophisaurus gracilis* (Boulenger 1888), *Ophisaurus moguntinus* [fossil] (Lydekker 1888), *Gerrhonotus* (Werner 1896), Teiidae: *Ameïva* (Gachet 1834), *Tejus* (Merian 1705; Tornier 1897), *Gymnophthalmus* (Bouleneger 1888), Anniellidae: *Anniella*, Lacertidae: *Lacerta* (Werner 1896; Fraisse 1885; Giuliani 1878; Müller 1863; Néedham 1750; Perrault 1688), Crocodilia: *Alligator* (Werner in Przibam 1902).

Bifid tail – Urodeles: *Siredon* (Tornier 1900), *Triton cristatus, T. vulgaris* (Tornier 1901), *Rana fusca* (Barfurth 1898, 1900; Vulpian 1862), *Rana catesbeiana* (Ryder 1891).

Lizards (Néedham 1750; Eversmann 1858; Giebel 1862; Laver 1879; Monteil 1880): *Lacerta* (Perrault 1776; Tofohr 1905; Brindley 1898), *Lygosoma telfairi* (Brindley 1898), *Mabuya carinata* (Brindley 1898), *Scincus* (Hirota 1895), *Chalcides sepsoïdes* (Tofohr 1905), *Gongylus ocellatus* (Tofohr 1903), *Trogonophis wiegmanni* (Brindley 1894), geckos: *Platydactylus* (Tofohr 1905), *Hemidactylus gleadovii* (Brindley 1894), *Acanthodactylus, Physignathus lesueurii* (Tofohr 1903).

Triple tail in lizards: *Lacerta* (Marchant 1718; Vincent 1877), *Tejus teju* (Tornier 1897), *Podinema teguixin* (Fraisse 1885), *Calotes cristatellus, Anolis grahami* (Brindley 1898).

Przibram H. 1919. Mitteilungen aus der Biologischen Versuchsanstalt der Akademie der Wissenschaften in Wien (Zoologische Abteilung, Vortsand: H. Przibam). Nr. 44. Die Bruchdreifachbildung im Tierreiche. [Contributions of the Biological Experimental Institute of the Academy of Sciences in Vienna. The injury-induced triple formation in the animal kingdom]. Akademischer Anzeiger Wien nr. 18:252. [German]

Trauma – Short information on triple and multiple formations caused by regeneration (e.g., hyperdactyly) refers to article published 1921.

Przibram H. 1920. Teratologie und Teratogenese. [Teratology and teratogenesis]. Vorträge und Aufsätze über Entwicklungsmechanik der Organismen, Heft 25:1–91. Berlin: Springer Verlag. [German]

Pathology – Gives definition (= every formation different from the standard and not caused by a present sickness) and classification (a. defects, b. excesses, c. alien [=different]) of anomalies.

Congenital – Cites one-eyed salamander.

Supernumerary leg in triton and *Pelobates fuscus* with three additional legs.

Trauma – Lizards with two and three tails.

Przibram H. 1921. Die Bruch-Dreifachbildung im Tierreiche. [The injury-induced triple formation in the animal kingdom]. Wilhelm Roux' Archiv für Entwicklungsmechanik 48: 205–244. [German]

Congenital – Polydactyly in urodeles *Triton taeniatus, Triton cristatus*, and *Triton alpestris* and anuran *Bufo columbiensis*.

Ectodactyly in urodele *Ambystoma punctatum*, anurans *Pelobates fuscus, Rana temporaria*.

Polymely of front legs in urodeles *Triton taeniatus, Triton cristatus, Triton alpestris*, and *Siredon pisciforme*, anurans *Pelobates fuscus, Pelobates* sp., *Bufo viridis, Bufo columbiensis, Rana halecina, Rana palmipes, Rana temporaria*, and *Rana esculenta* and hind legs in urodeles *Triton* (*Molge*) *vulgaris, Triton taeniatus, Triton cristatus, Ambystoma punctatum*, and *Pleurodeles waltlii*, and anurans *Alytes obstetricans, Pelobates fuscus, Bufo vulgaris, Rana halecina, Rana palustris, Rana temporaria, Rana clamata, Rana sylvatica, Bombinator igneus, Bufo lentiginosus*.

Pugener LA, Maglia AM. 2009. Developmental evolution of the anuran sacro-urostylic complex. South American Journal of Herpetology 4:193–209.

Congenital – Sacro-urostylic region of anurans consists of a vertebra and rod-like urostyle, which articulates in saltatorial (jumping) frogs *Acris crepitans* and *Discoglossus sardus* but fuses in hopping-burrowing *Spea multiplicata* and aquatic *Xenopus laevis*. Fusion of additional vertebrae produces a synsacrum (e.g., *Hymenochirus curtipes*) with increased stability.

Anomalous partial or total fusion occurs in *Bombina bombina* (Madej 1965), *Discoglossus pictus* (Sanchiz and Perez 1974), and *Phyllomedusa lemur* (KU 686228).

Anomalous synsacrum-like structures were reported in *Bombina bombina* (Madej 1965), *Rana curtipes* (Smit 1953), *Xenopus gilli* (KU 220990 and 2209991), and *Syncope antenori* (KU 124004 and 124006).

Punzo F. 1982. Tail autotomy and running speed in the lizards *Cophosaurus texanus* and *Uma notata*. Journal of Herpetology 16:329–331.

Trauma – Tail loss reduced running speed in *Cophosaurus texanus* by 32% and in *Uma notata*, by 41%.

Qin Z-F, Zhou J-M, Cong L, Xu X-B. 2005. Potential ectotoxic effects of polychlorinated biphenyls on *Xenopus laevis*. Environmental Toxicology and Chemistry 24:2573–2578.

Toxicology – Polychlorinated biphenyl exposure of *Xenopus laevis* produced forelimb malformations at shoulder in which the forelimbs were fixed in adduction-backward rotation.

Qualls CP, Andrews RM. 1999. Cold climates and the evolution of viviparity in reptiles: Cold incubation temperatures produce poor-quality offspring in the lizard, *Sceloporus virgatus*. Biological Journal of the Linnean Society 67:353–376.

Environmental – Phrynosomatid lizard *Sceloporus virgatus* hatchlings incubated in the cold had greater fluctuating asymmetry.

Sceloporus virgatus tail length at hatching was 22.46 +/− 0.27 mm and 22.55 +/− 0.33 mm at high and low elevation at 15–25°C, contrasted with 26.75 +/− 0.21 mm and 26.55+/− 0.27 mm at 20–30°C incubation.

A number of asymmetrical characters were 6.12 +/− 0.252 mm and 6.07 +/− 0.308 mm, contrasted with 5.23 +/− 0.167 mm and 4.76 +/− 0.160 mm.

Quattrini D. 1952a. Ricerche anatomiche e sperimentali sulla autotomia della coda delle lucertole I – Dinamica dell'autotomia e conseguenze nel tegumento [Anatomic and experimental research on the autotomy of the tail of the lizards I – Dynamics of autotomy and consequences in the integument]. Archivio Zoologico Italiano 37: 131–170. [Italian]

Trauma – In this study of 40 adult male individuals of *Lacerta sicula sicula*, Quattrini found a distal region of the tail that could be entirely or partially lost by autotomy and a proximal region where autotomy could not happen. He noted that there could be a potential plane of autotomy in the proximal region, judging from characters in the integument and the skeleton. The undisputed planes of autotomy began at the seventh vertebra and were always intravertebral, with the autotomy not being able to happen by disarticulation of two contiguous vertebrae. In the skeleton of the proximal part of the tail, there were six vertebrae, plus the proximal part of a seventh. The sixth vertebra could be partially or completely divided in two parts. This was also rarely seen in the fifth, but in none of the rest. Quattrini highlighted that there were no intervertebral autotomy planes. The experiment consisted of hanging weights from a set section of the native tail of the individual lizards, which were secured by their pelvic extremities and the sacral region. The weight was increased until the tail separated (500–1200 g). A much heavier weight (3.5 kg) was necessary to cause tail separation from the proximal region of the tail than that required to cause tail separation from the distal part. When tail separation from the proximal part was achieved, the break was irregular, and disarticulation of two contiguous vertebrae took place, unlike the clean, intravertebral autotomy plane from any part of the distal tail. Quattrini also stated that the regenerated tail could only be lost as a whole, with the exception of the very short and fragile apical end of the tail. He also conducted the experiment in individuals in his sample that had regenerated tails, and it took 2 kg to cause separation of the regenerated tail from the body.

A line illustration was provided of the dorsal, ventral, and right lateral aspects of the fifth, sixth, and seventh tail vertebrae. It showed the typical extent of subdivision for each.

Quattrini D. 1952b. Ricerche anatomiche e sperimentali sulla autotomia della coda delle lucertole II – Muscolatura, adipe sottomuscolare e scheletro [Anatomic and experimental research on the autotomy of the tail of the lizards II – Musculature, adipose tissue under musculature, and skeleton]. Archivio Zoologico Italiano 37: 465–515. [Italian]

Congenital – Quattrini studied 40 individuals of *Lacerta sicula sicula* and 30 of *Lacerta sicula campestris*. Only the distal part of the tail showed a typical segmental anatomical structure. In the skeleton of the proximal part of the tail there are 6 vertebrae, plus the proximal part of a seventh, which can be partially or completely divided in two parts. The chevrons were found to be intervertebral and not coossified with the vertebrae. The first chevron was situated between the third and fourth vertebra, near the caudal limit of the kidney. In this part of the tail, the autotomy cannot happen.

In the distal part of the tail, the skeleton was formed by 41–52 vertebrae, each one subdivided in two parts linked by the periosteum, without interposition of cartilaginous septa. There was a thick cartilaginous septum between vertebra and vertebra, without an articular fissure. The autotomy planes were always intravertebral, and the fracture of the tail thus cannot happen by disarticulation of the vertebrae. In the case of regenerated tails, the cartilaginous tube that forms the skeleton is jointed with the anterior part of the vertebra through which the autotomy took place, and does not show any connection with the cartilaginous septa of the native tail. A line illustration of the vertebral column of the tail of *Lacerta sicula campestris* was included, starting at the sacral vertebrae and progressing distally to the eighth vertebra of the tail.

It showed the ventral and right lateral aspects, labeling the chevrons, intervertebral cartilage, transverse processes of the fifth vertebra and of the sixth vertebra (the latter being partially divided in two), the spinal processes, the proximal and distal parts of the seventh vertebra, the long surface that thickens the seventh vertebra when autotomy takes place, its accessory spinal apophyses, the transverse processes of the eighth vertebra (which are completely divided in two), and the zygapophyses. Quattrini noted that five of the vertebrae of the tail were whole, that the sixth was partially divided in two, and that the seventh and eighth were completely divided in two parts.

Quattrini D. 1953. Autotomia e struttura anatomica della coda dei rettili [Autotomy and anatomical structure of the tail of reptiles]. Monitore Zoologico Italiano 61:36–48. [Italian]

Trauma – Quattrini began his paper by referring the reader to his 1952 paper and then explained that he would present the existing differences between the autotomous and non-autotomous regions of the tail, as well as the differences between regenerated and normal tails of *Lacerta vivipara* and *Lacerta viridis*.

Regarding the non-autotomous part: in *L. vivipara*, the skeleton of this portion of the tail is comprised of six vertebrae plus half of a seventh, through which the primary autotomy plane passes. The other half of the seventh vertebra is therefore always found in the non-autotomous portion of the tail. The first five vertebrae are whole; the sixth is regularly partly subdivided into two pieces. In one instance of a sample of seven individuals, the subdivision was nearly complete. In another, the prime autotomy plane passed through the eighth vertebra, and thus, the skeleton of the non-autotomous part was comprised of seven vertebrae, plus half of the eighth.

In the corresponding part of the skeleton of *L. viridis*, there are also six vertebrae, plus half of the seventh, through which the primary plane of autotomy passes. The sixth vertebra always shows traces of a subdivision into two parts. These traces are, as in *L. vivipara*, observed on the center and over the neural arch and, in *L. viridis*, are particularly evident on the transverse processes, which are split at the apex. None of the individuals sampled by Quattrini had a sixth vertebra completely divided in two parts.

The vertebrae have well-developed zygapophyses, transverse processes, and spinal processes. The first chevron is placed behind the third and fourth vertebrae, next to the caudal end of the kidney, which is contained partly in the tail, ventral to the first three vertebrae. The chevrons are never coossified with the vertebrae, and are always placed intervertebrally. Measured over the tegument, starting at the cloaca, the non-autotomous part in the adult specimens of *L. vivipara* is 9–10 mm (and nearly 12 mm in the specimen that had the primary plane of autotomy going through the eighth vertebra). In *L. viridis* adults, the length of the non-autotomous part varies between 15 and 17 mm. Observed over the tegument, in *L. vivipara*, the primary plane of autotomy is found between 9–10 mm from the cloaca, while in *L. viridis*, it is 15–17 mm. A transverse section through the cloaca always passes through the transverse processes of the first vertebra.

Regarding the autotomous part: in *L. vivipara*, the skeleton is comprised of 37–39 vertebrae, plus the half of the vertebra through which passes the primary autotomy plane (regularly the seventh vertebra of the tail). The length of the vertebrae in adult animals varies between 2 and 2.5 mm, but is a bit smaller near the distal extremity of the tail. Considering also the non-autotomous part, the skeleton of the tail of *L. vivipara* has 44–46 vertebrae. In *L. viridis*, the number of vertebrae varies from 56–61 and their length in adults from 4–6 mm (and like in *L. vivipara*, the smallest are found in the distal end of the tail). Comprehensively, there are 63–68 vertebrae in this part of the skeleton.

The single vertebrae of the autotomous part are always divided in two parts; the part situated anteriorly is always smaller than the posterior. These are united only by periosteum, without cartilage, and between them run diverse planes of autotomy, which therefore are always intravertebral.

The fracture of the tail cannot be verified by disarticulation of the vertebrae. Zygapophyses, chevrons, transverse processes, and spinal processes are well developed in the more anterior vertebrae but are reduced progressively posteriorly. The subdivision surface of each vertebrae passes through the neural arch, the centrum, and the transverse processes (when present). In *L. vivipara*, the fracture surface of single vertebrae passes suddenly by the transverse processes and, in the neural arch, extends after the anterior spinal apophysis, and ends there. The two pieces into which vertebrae are divided are thus united by a subtle bony bridge, which thickens when autotomy is verified.

Regarding the regenerated tails: the skeleton is always represented by a simple cartilaginous tube, which is united anteriorly to the extremity of the small piece of the vertebra through which autotomy came, and it is not in direct rapport with the cartilage present behind the vertebra of the stump.

Quattrini highlighted that in *L. vivipara*, the skeleton of the tail is comprised of fewer vertebrae than in the other genera he had studied (varying from 44 to 46 in *L. vivipara*, and 48 to 58 in *L. sicula sicula* and *L. sicula campestris*, and from 62 to 67 in *L. viridis*). He concluded that the anatomical structures of the tails of *L. viridis* and *L. vivipara* have no notable differences and are nearly identical to that of *L. sicula sicula*, and *L. sicula campestris*, previously described by him in 1952. He also stated that in the proximal part of the tail, the vertebrae are not divided, and thus, autotomy cannot happen, but that in the distal part of the tail, the vertebrae are subdivided into two parts, through which passes the autotomy plane. Lastly, Quattrini found no anatomical disposition to autotomy in regenerated tails.

Quenstedt A. 1867. Handbuch der Petrefactenkunde. [Handbook of Fossil Science]. 2nd ed. Tübingen: Laupp'schen Buchhandlung, 932 pp. [German]

Dental – Crocodilian from Lower Oolite of Schnaitheim with "bad" (p. 131) rotten teeth.

Fossil – Crocodilian from Lower Oolite of Schnaitheim with "bad" (p. 131) rotten teeth.

Rage JC. 1972. Les amphibians et les reptiles du gisement des Abimes de la Fage. [The amphibians and reptiles deposited in Abimes of Fage]. Nouvelles Archives du Musée Guimet d'histoire naturelle, Lyon 10: 79–90. [French]

Congenital – *Bufo* with urostyle transverse apophyses or fusion of two dorsal vertebrae, *Rana* with similar changes, *Ptychadena* with fusion of presacral and sacral vertebrae, all were attributed to fracture healing.

Trauma – *Bufo* with urostyle transverse apophyses or fusion of two dorsal vertebrae, *Rana* with similar changes, *Ptychadena* with fusion of presacral and sacral vertebrae, all were attributed to fracture healing.

Rahmann H, Engels WH, Grüter HG. 1962. Doppelmißbildung einer Vorderextremität bei *Rana temporaria* L. [Double malformation of a front leg of *Rana temporaria* L]. Zoologischer Anzeiger 169: 449–454. [German]

Congenital – Description of *Rana temporaria* with duplication of right forelimb: One right hand with four fingers, the other with five fingers, where the middle finger is split in two.

Raichoudhury DP, Das BK. 1931. Observations on the malformations in the common Bengal toad, *Bufo melanostictus*. Anatomischer Anzeiger 71:120–131.

Congenital – Bengal toad, *Bufo melanostictus*, with crowding of sixth to seventh vertebrae. Coossification of seventh and eighth vertebra and neural arches, hemivertebra, exceeding of normal number of os coccygeum vertebrae by one. Ninth vertebra had well-developed left diapophysis but only a small spinous process on the right with no left posterior zygapophysis. Centra of second and third and sixth to tenth vertebrae were procoelous; fourth, amphicoelous; and fifth, biconvex.

Raidal SR, Shearer PL, Prince R. 2006. Chronic shoulder osteoarthritis in a loggerhead turtle (*Caretta caretta*). Australian Veterinary Journal 84:231–234.

Trauma – Loggerhead turtle (*Caretta caretta*) with shoulder remodeling and bony proliferation. It was called osteoarthritis, but there seems to be subchondral damage. Additionally, the opposite humerus is listed as healed fracture. However, it has the appearance of growth around a wire or filament.

Arthritis – Loggerhead turtle (*Caretta caretta*) with shoulder remodeling and bony proliferation. It was called osteoarthritis, but there seems to be subchondral damage. Given negative cultures, the possibility of a spondyloarthropathy must be entertained.

Rainwater TR, McMurry ST, Platt SG. 1999. Ectomelia in Morelet's crocodile from Belize. Journal of Wildlife Diseases 35:125–129.

Congenital – Ectomelia in Morelet's crocodile.

Raiti P. 2004. Endoscopic-assisted retrieval of a cloacal urolith in an African spurred tortoise (*Geochelone sulcata*). Proceedings of the Association of Amphibian and Reptile Veterinarians Annual Conference 2004:145.

Stone – Urate urolith in an African spurred tortoise *Geochelone sulcata*.

Raiti P, Haramati N. 1997. Magnetic resonance imaging and computerized tomography of a gravid leopard tortoise (*Geochelone pardalis pardalis*) with metabolic bone disease. Journal of Zoo and Wildlife Medicine 28:189–197.

Metabolic – Secondary nutritional metabolic bone disease in leopard tortoise (*Geochelone pardalis pardalis*).

Ramsay EC, Munson L, Lowenstine L, Fowler M. 1996. A retrospective study of neoplasia in a collection of captive snakes. Journal of Zoo and Wildlife Medicine 27:28–34.

Neoplasia – 2.9–3.5% of captive reptiles had neoplasia, without comment on bone involvement.

Ramsay EC, Daniel GB, Tryon BW, Merryman JI, Morris PJ, Bemis DA. 2002. Osteomyelitis associated with *Salmonella enterica ss arizonae* in a colony of ridge-nosed rattlesnakes (*Crotalus willardi*). Journal of Zoo and Wildlife Medicine 33:301–310.

Infection – *Salmonella enterica ss arizonae* osteomyelitis in three original and five subsequent ridge-nosed rattlesnakes (*Crotalus willardi*) among 19 producing rib nodules.

Rand AS. 1954. Variation and predation pressure in an island and mainland population of lizards. Copeia 1954:260–262.

Trauma – Ratio of regenerated tails in *Cnemidophorus lemniscatus* indicates that predation of mainland Roatan, Honduras' specimens was eight times more common than island specimens.

Rand AS 1965. On the frequency and extent of naturally occurring foot injuries in *Tropidurus torquatus* (Sauria: Iguanidae). Papeis Avulsos de Departemento Zoologico 42:225–228.

Trauma – Front legs were naturally "damaged" as frequently as hind legs in *Tropidurus torquatus*, with longest toes most commonly affected. 21/164 were damaged, frequency increasing with size. Only statistically significant between intermediate and large females. Usually loss of one toe (front III or hind IV; rarely all).

Raphael BL, James SB, Cook RA. 1999. Evaluation of vitamin D concentrations in *Uromastyx* spp. With and without radiographic evidence of dystrophic mineralization. Proceedings of the American Association of Zoo Veterinarians (AAZV. Puerto Vallarta, Mexico, AAZV 1999;20–22.

Metabolic – calcium hydroxyapatite and calcium pyrophosphate deposition in skin and periarticular regions of agamid lizards *Uromastyx ornate* (3/8) and *Uromastyx aegyptia* (4/14).

Rauber A. 1877. Die Theorien der excessiven Monstra. [Theory of excessive monsters]. Virchow's Archiv 71:133–206. [German]

Congenital – *Salamandra maculata* Siamese twin. *Tropidonotus natrix* embryo with spiral tail.

Rayer M. 1849a. 3 cas de polydactylie chez un triton [Third case of polydactyly in a triton]. Gazette Médicale (Paris) 1849:901. [French]

Congenital – Polydactyly in a *Rana* forelimb and *Triton* forelimbs, citing *Rana esculenta* hind foot with six digits (Brown-Séquard) However, Ercolani's 1882 citation of page 185 is erroneous.

Rayer P. 1849b. Sur un nouveau cas de monstruosité (atlodyme, Isid. Geoffroy-Saint-Hilaire) observé sur une espèce de reptile ophidien (*Homalopsis schneiderii*) [A new case of monstrosity (altodyme, Isid Geoffroy-Saint Hilaire) observed in an ophidian reptile]. Comptes Rendus des Séances et Mémoires de la Société Biologique (and Gazette Medicale) (1849) 1:185. [French]

Congenital – Dicephalic *Homalopsis schneiderii.*

Raynaud A. 1960. Essais de destruction, par irradiation localisée au moyen des rayons X, des ébauches hypophysaires de l'embryon d'Orvet (*Anguis fragilis* L.). [Testing of destruction through localized radiation X-rays, drafts of the embryonic pituitary Orvet (*Anguis fragilis* L)]. Compte Rendu de l'Academie de Science, Paris 251:2416–2418. [French]

Environmental – Radiation-induced jaw hyperdevelopment.

Raynaud A. 1983. Tératogenèse par l'action des substances chimiques chez les Reptiles, avec mention particulière de l'action de la cytosine-arabinofuranoside chez les embryons de *Lacerta viridis* (Laur.). [Teratogenesis by the action of chemical substances on reptiles, with particular mention of the action of cytosine-arabinofuranoside on *Lacerta viridis* (Laur.) embryos]. In Vago C, Matz G. eds. Proceedings of the First International Colloquium on Pathology of Reptiles and Amphibians, 1982, Angers:251–258. [French]

Toxicology – Cytosine arabinofuranoside-induced variable digit reduction in *Lacerta viridis* and trypan blue-induced limb "anomalies" in *Calotes versicolor.*

Razzetti E., Gentilli A. 2002. Atti del terzo Convegno "Salvaguardia Anfibi", Lugano, 23–24 giugno 2000 [Proceedings of Third Conference Safeguard the Amphibians, Lugano, June 23–24, 2000]: 137–143. [Italian]

Congenital – Posterior polymely in *Rana esculenta* reported by Balsamo-Crivelli 1865, Cavanna 1877, Frabretti 1866, Parona 1879, 1881, Sordelli 1877, Strobel 1876, 1877; in *Rana dalmatina*, by Fabretti 1866; in *Rana temporaria*, by Strobel 1876, 1877; and anterior polymely in *Rana esculenta* by Cavanna 1879, Sordelli 1878.

Razzetti E, Faiman R, Werner YL. 2007. Directional asymmetry and correlation of tail injury with left-side dominance occurs in Serpentes (Sauropsida). Zoomorphology 126:31–43.

Trauma – Within species, individuals with regenerated tails were more often left- footed, and left-foot individuals more often had injured tails. Proportion of injured tails increased with size/age (both in number and rate) and therefore with accumulation over time (*Chalcides ocellatus*-excepted).

Read J. 1997. Comparative abnormality rates of the trilling frog at Olympic Dam. Herpetofauna 27:23–27.

Congenital – 3.5% Trilling frog *Neobatrachus centralis* at Olympic Dam, compared with 1.3% at control sites. Partial ectodactyly (single digit shortening) was present in 81%.

Read JL, Tyler MJ. 1990. The nature and incidence of post-axial, skeletal abnormalities in the frog *Neobatrachus centralis* parker at Olympic Dam, South Australia. Transactions of the Royal Society of South Australia, Adelaide 114:213–217.

Congenital – Frog *Neobatrachus centralis* at Olympic Dam, South Australia with ectodactyly, especially of fourth toe.

Read JL, Tyler MJ. 1994. Natural levels of abnormalities in the trilling frog (*Neobatrachus* [error: *Neobatrachus*] *centralis*) at the Olympic Dam Mine. Bulletin of Environmental Contamination and Toxicology 53:25–31.

Congenital – 0.5–2.2% Abnormalities [predominantly partial ectrodactyly (especially of fourth toe) or distal tarsal bones loss, but also occasional foot element fusion or hand syndactyly] in trilling frog *Neobatrachus [Neobatrachus] centralis* 2–10 km from copper–uranium–gold–silver mine contrasted with 0–1.6% further from the mining sites. Read and Tyler (1990) suggested that the fourth toe ectrodactyly might be a common mutant, rather than environmental.

Environmental – 0.5–2.2% Abnormalities [predominantly partial ectrodactyly (especially of fourth toe) or distal tarsal bones loss, but also occasional foot element fusion or hand syndactyly] in trilling frog *Neobatrachus [Neobatrachus] centralis* 2–10 km from copper–uranium–gold–silver mine contrasted with 0–1.6% further from the mining sites.

Reaves DA. 2004. Turtle turns heads – 2 of them, Moncks Corner brothers find anomaly of nature behind grandparents' house. 15 May 2004:1.

Congenital – Dicephalic yellow-bellied slider turtle.

Rebell G, Roth FJ, Taplan D, Wodinsky J. 1971. Fusariosis in marine turtles. Bacteriological Proceedings 71:121.

Shell disease – Loggerhead turtles, *Caretta caretta*, with ulcerative shell disease from *Fusarium solani.*

Recuero-Gil E, Campos Asenjo O. 2002 (2003). *Triturus marmoratus* (Marbled Newt): Polymely. Herpetological Bulletin 82:31–32.
Congenital – Supernumerary hind limb in Carracedelo *Triturus marmoratus*.
Polymely reported only rarely in Salamandridae: A Catalonian *Triturus marmoratus* (Arias et al. 1999), 7 of 13,815 Japanese *Cynops pyrrhogaster* (Meyer-Rochow and Asashima 1988), one of 557 southern Portuguese *Triturus pygmaeus* (Caetano 1991)
Trauma – regeneration of accessory limbs in *Triturus vulgaris* (Griffiths 1981).

Redi F. 1684a. Osservazioni intorno agli animali viventi che si trovano negli animali viventi [Observations about living animals that are found in living animals]. 253 pp.; Firenze: Piero Matini [Italian].
Congenital – *Tropidonotus* dicephalous snake, derodymus asp, reported personal communication from Foghelio Martino of Hamburg about a dicephalic viper in Rome, and another in Lyon, and reviewed Aristotle, Aelianus, Aldrovandi, and Liceto.
Trauma – Bifid tail in a snake; *Lacerta* with triple tail.
Pseudopathology – Myth of the seven-headed snake related by Greek and Latin (Roman) poets but deemed it an amplified hyperbole.

Redi F. 1684b. Observations de François Rédi sur les animaux vivans trouvés dans les animaux vivans. [Observations of Francesco Redi about living animals that are found in living animals] Collection académique, partie étrangère 4: 464–540. [French]
Congenital – Cites multiple reports of serpents (including a viper) with two heads and described a derodymus viper.
Trauma – Multiple tails in snake and lizard.

Reece RL, Dickson DB, Butler R. 1986. An osteopetrosis-like condition in a juvenile rhinoceros iguana (*Cyclura cornuta*). Australian Veterinary Journal 63:343–344.
Other – Rhinoceros iguana (*Cyclura cornuta*) with greatly swollen left femur and tibia, right tibia and right radius. Amorphous pericortical woven bone trabecular projection from cortex, containing many osteoblast mitoses, with zones of cartilage at periphery. Fibrous tissue penetrated adjacent muscles. They distinguish osteopetrosis as a failure to resorb formed bone. This contrasts with the fowl phenomenon, which they considered an "osteogenic osteoblastoma." The latter has cellular, immature bone proliferation. Fine, radial, sometimes branched "osteophytes" were reported with cartilage islands and fibrous tissue nests. The phenomenon in *Cyclura cornuta* is as yet unexplained.

Reed MA. 1903. The regeneration of a whole foot from the cut end of a leg containing only the tibia. Wilhelm Roux's Archiv für Entwicklungsmechanik 17:150–154.
Trauma – Experimental: Post-traumatic (surgical) foot regeneration in salamander *Spelerpes ruber.*

Reeder J. 2002. Treasure coast boasts a rich, varied history. Palm Beach Post (Florida) 27 October 2002 Sunday: SPECIAL 1.
Congenital – Dicephalic turtle.

Reese AM. 1906. A double embryo of the Florida alligator. Anatomischer Anzeiger 28:229–231.
Congenital – Double embryo in Florida alligator.

Reeves MK, Trust KA. 2004. National Abnormal Amphibian Study FY2002: National Wildlife Refuges in Alaska, Annual Progress Report. U.S. Fish and Wildlife Service Technical Report AFWFO-TR-04–02. 22pp.
Congenital – 3.4–8.6% Abnormalities in frogs including amelia, brachygnathia, bone bridge, micromelia, ectodactylia, brachydactyly, scoliosis, microcephaly.

Reeves MK, Dolph CL, Zimmer H, Tjeerdema RS, Trust KA. 2008. Road proximity increases risk of skeletal abnormalities in wood frogs from National Wildlife refuges in Alaska. Environmental Health Perspectives 116:1009–1014.
Congenital – Wood frog *Rana sylvatica* ectromelia, micromelia, amelia related to proximity to roads. Kenai site also had microcephaly, scoliosis, and kinked tail.

Reichenbach-Klinke H-H. 1956. Knochendegeneration bei einem Rippenmolch (*Pleurodeles waltli* MICHAHELLES) im Zusammenhang mit intermuskulärer [Bone degeneration in the sharp-ribbed newt (*Pleurodeles waltli* MICHAHELLES) in combination with the intermuscular]. Melanophorenanreicherung – Biologisches Zentralblatt 75:407–416. [German]
Congenital – *Pleurodeles waltl* specimens with shortened toes, lost toes, reduced lower and upper part of legs attributed to wrong feeding.

Reichenbach-Klinke H-H. 1961. Krankheiten der Amphibien. [Sickness in Amphibians]. VIII+100pp; Stuttgart: Gustav Fischer Verlag. [German]
Congenital – Dicephalic *Triturus vulgaris, Triturus helveticus*. Body duplication in *Triturus vulgaris, Triturus helveticus*. Supernumerary legs in *Rana graeca, Rana esculenta, Salamandra salamandra, Alytes obstetricans*. Arching of vertebral column, shortening of vertebral column and legs.
Metabolic – Arching of vertebral column, shortening of vertebral column and legs, osteoporosis and rickets caused by deficiency of vitamin D, calcium, phosphorus, and UV light, or exposure to parasites.
Infection – Arching of vertebral column, shortening of vertebral column and legs caused by parasites.
Neoplasia – Chondroma in *Triturus vulgaris*.

Other – Degeneration of extremities in *Pleurodeles waltl.*

Reichenbach-Klinke H-H. 1963. Krankheiten der Reptilien. [Sickness in Reptiles]. VIII + 142pp.; Stuttgart: Gustav Fischer Verlag. [German]

Congenital – Dicephalic turtles, snakes, and lizards. *Anguis fragilis* with two bodies and one head.

Trauma – Healing of fractures and regenerations, duplications of tail in *Sphenodon* and other reptiles: turtles, snakes, and lizards.

Metabolic – Rickets in tortoise.

Reichenbach-Klinke H-H. 1977. Krankheiten der Reptilien. [Sickness in reptiles]. 2. Aufl. 228pp.; Stuttgart: Gustav Fischer Verlag. [German]

Congenital – Dicephalic snakes. *Anguis fragilis* specimen with one head and two bodies. Siamese twin *Pseudemys floridana* and "*Chrysemys scaber*" = *Pseudemys s. scripta.*

Scoliosis in *Trionyx triunguis.*

Metabolic – Rickets in tortoise and crocodiles.

Reichenbach-Klinke H, Elkan E. 1965. The Principal Diseases of Lower Vertebrates. XII + 600pp.; New York: Academic Press.

Congenital – Dicephalic *Triturus vulgaris* x *Triturus helveticus* cross. Anadidymus, arm duplication; katadidymus, leg duplication. Siamese *Anguis fragilis,* joined at head and body; *Pseudemys floridana* with post shield Siamese changes; *Chrysemys scaber* with lateral Siamese changes. Slow worm *Anguis fragilis* with two bodies, but one head. Supranumerary hind limbs in *Rana graeca, Pelobates fuscus, Salamandra salamandra.* Grossly abnormal leg in *Alytes obstetricans. Testudo hermanni* with unpaired scale.

Metabolic – Horse liver diet produces skeletal deformities in *Xenopus.* Rickets in ornamental turtles.

Neoplasia – Maxillary osteoma in *Lacerta viridis.*

Reichenow E. 1908. Beispiele von Abweichungen in der Zahl der Hintergliedmaßen bei *Rana esculenta.* [Examples of deviations in the diameter of bird legs in *Rana esculenta*]. Zoologischer Anzeiger 32:677–682. [German]

Congenital – *Rana esculenta* with two additional hind legs (articulating with pelvic girdle) or lacking hind leg.

Reid F. 1967. Once upon a time. International Turtle & Tortoise Society Journal 1967(May-June):24–30.

Congenital – Derodymus common snapping turtle.

Reid RE. 1984. The histology of dinosaurian bone, and its possible bearing on dinosaurian physiology. Symposium of the Zoological Society of London 52:629–663. In: The Structure, Development and Evolution of Reptiles: A festschrift in honour of Professor A. d'A. Bellairs on the occasion of his retirement, WJ Ferguson, ed. London: Academic Press.

Trauma – Normal fracture callus, hypertrophic growth after fracture, hyperostosis.

Infection – Suppurative periostitis after fracture, osteomyelitis,

Vertebral – Spondylosis deformans in dinosaurs, citing Moodie 1923.

Neoplasia – Osteoma in dinosaurs, citing Moodie 1923.

Fossil – Normal fracture callus, hypertrophic growth after fracture, hyperostosis.

Suppurative periostitis after fracture, osteomyelitis,

Spondylosis deformans in dinosaurs, citing Moodie 1923.

Osteoma in dinosaurs, citing Moodie 1923.

Reid A. 2005. Craniodichotomy in a *Crotalus*: a two headed prairie rattlesnake near Leader, SK. Blue Jay 63(3):139–143.

Congenital – Dicephalic ladder snake *Rhinechis scalaris* from Spain, citing Anonymous. 2002. BBC News. April 4. http://news.bbc.co.uk/1/hi/sci/tech/1910471.stm, and grass snake from China – citing Anonymous. Two-headed snake found. 2004. China View News Xinhuanet October 21. http://www.beijingportal.com.cn/7838/2004/10/21/207@2333847.htm (neither of which could not be located on the Web).

Derodymus prairie rattlesnake *Crotalus viridis* with "abnormally large vertebra."

Reinhardt J. 1873. Anomalier I Krydevirvel hos Krokodilerne. [Anomalies of sacral vertebrae in crocodilians]. Videnskabelige meddelelser frå den Naturhistoriske forening i Kjöbenhavn 42–45: 221–228. [Swedish]

Congenital – *Alligator sclerops acutus* with sacralization of last lumbar vertebra, forming sacrum consisting of three rather than normal two vertebrae. *Crocodylus acutus* with sacralization of first caudal vertebra, forming sacrum consisting of three rather than normal two vertebrae.

Reinhardt. J. 1874. Sur les anomalies des vertebras sacrées chez les crocodilians [Sacral vertebral anomalies in crocodilians]. Journal de Zoologie 3(4): 308–312 [French].

Congenital – Probable *Alligator sclerops* and *Crocodylus acutus* with sacralization of last lumbar vertebra, forming sacrum consisting of three rather than normal two vertebrae. *Crocodylus acutus* with sacralization of first caudal vertebra, forming sacrum consisting of three rather than normal two vertebrae.

Reinhardt F. 1941. Über die Wirkung einseitiger Fütterung auf die Entwicklung der Hinterbeine des Rippenmolches. [About the effect of unilateral feeding on the development of the hind legs of the ribbed newts]. Roux's Archiv für Entwicklungsmechanik 141:340–384 [German].

Metabolic – Modification of rate of growth and development by modification of nutrition.

Reisz RR, Scott DM, Pynn BR, Modesto SP. 2011. Osteomyelitis in a Paleozoic reptile: Ancient evidence for bacterial infection and its evolutionary significance. Naturwissenschaften DOI 10.1007/s00114–011–0792–1

Infection – Paleozoic (275-million-year-old) captorhinid *Labidosaurus hamatus* CMNH 76876 with abscess in lower mandible, apparently starting as a periapical abscess and developing osteomyelitis and fistula.

Dental – Paleozoic (275-million-year-old) captorhinid *Labidosaurus hamatus* CMNH 76876 with infilled tooth sockets in mandible adjacent to abscess. Loss of anterior mandibular teeth was attributed to trauma, with subsequent periapical abscess and osteomyelitis and fistula.

Fossil – Paleozoic (275-million-year-old) captorhinid *Labidosaurus hamatus* CMNH 76876 with infilled tooth sockets in mandible adjacent to abscess. Loss of anterior mandibular teeth was attributed to trauma, with subsequent periapical abscess and osteomyelitis and fistula.

Reuter P. 1873 (not 1875). *Triton* with five legs. Naturwissenschaftliche Gesellschaft zu Chemnitz, 5. Jahresbericht November 3, 1873:26.

Congenital – *Triton* with five legs.

Rew RW. 1916. A frog with symmetrically abnormal hind feet. Proceedings of the Zoological Society London 1916:87–89.

Congenital – Missing first pedal digit in *Rana temporaria.*

Reynolds TD, Stephens TD. 1984. Multiple ectopic limbs in a wild population of *Hyla regilla*. Great Basin Naturalist 44:166–169.

Congenital – 22% Polymelia of hind limb and polydactyly in *Hyla regilla.*

Rhoads P. 2008. The complete Suboc: A comprehensive Guide to the Natural History, Care, and Breeding of the Trans-Pecos Ratsnake. Ada, Oklahoma: Living Art, 291 pp.

Vertebral – Vertebral kinks in Trans-Pecos ratsnake.

Rhodin AGJ, Pritchard PC, Mittermeier RA. 1984. The incidence of spinal deformities in marine turtles, with notes on the prevalence of kyphosis in Indonesian *Chelonia mydas*. British Journal of Herpetology 6:369–373.

Congenital – 0.08% Kyphosis, 0.03% lordosis/scoliosis in *Chelonia mydas*, less than the 0.04% in *Chelonia agassizi*, 0.1–0.4 and 0.3% in *Lepidochelys olivacea* respectively, the 0.2–0.5% in *Caretta caretta*, and the 0.07 and 0.07–0.11% in *Dermochelys coriacea.*

Richardson LR, Barwick RE. 1957. Faulty eruption of the fore-limb in *Hyla aurea*. Transactions of the Royal Society of New Zealand 84:941–942.

Congenital – *Hyla aurea* forelimb encased in skin, but bones were normal.

Richardson LR, Truscoe R. 1963. Vesical calculus in the frog *Hyla aurea*. Transactions of the Royal Society of New Zealand Zoology 3(1):1–3.

Stone – Three *Hyla aurea* with non-concretionary hydroxyapatite urinary calculi, distending bladder to 14 × 5 mm in a 67 mm long animal.

Riches RJ. 1955. Unusual deformity in *Anguis fragilis*. British Journal of Herpetology 1:251.

Congenital – Slow worm *Anguis fragilis* with one head, but two bodies.

Riches RJ. 1967. Notes on a clutch of eggs of the viperine snake (*Natrix maura*). British Journal of Herpetology 4:14–16.

Congenital – Protruding lower jaw in *Natrix maura.*

Richman LK, Montali RJ, Allen ME, Oftedal OT. 1995. Paradoxical pathologic changes in vitamin D deficient green iguanas (*Iguana iguana*). Annual proceedings of the American Association of Zoo Veterinarians 1995: 231–232.

Metabolic – Green iguana *Iguana iguana* with metastatic calcification from low vitamin D levels.

Richter SC, Seigel RA. 2002. Annual variation in the population ecology of the endangered gopher frog, *Rana sevosa* Goin and Netting. Copeia 2002:962–972.

Trauma – Rare toe regeneration in gopher frog *Rana sevosa.*

Ridewood WG. 1902. Abnormal sacra in edible frog (*R. esculenta*) and one common frog (*R. temporaria*). Proceedings of the Linnean Society of London 1902:46–47.

Congenital – Eighth and ninth vertebrae fused with coalescence of centra and neural arches in Rana esculenta, two with stout and two with slender diapophyses.

Rana temporaria with biconcave seventh centrum, biconvex eighth (without corresponding neural arch), concave front of ninth (with unilateral failure of development) of postzygapophysis.

Rieck AF. 1954. The effects of ultraviolet, and of photorecovery, on the developing forelimb of *Ambystoma.* Journal of Morphology 94:367–407.

Environmental – UV light exposure and photo recovery in *Ambystoma maculatum* and *A. opacum* produces ectodactyly and polydactyly.

Rienesl J, Wagner GP. 1992. Constancy and change of basipodial variation patterns: A comparative study of crested and marbled newts – *Triturus cristatus, Triturus marmoratus* – and their natural hybrids. Journal of Evolutionary Biology 5:307–324.

Congenital – Among 190 *Triturus cristatus, Triturus marmoratus* and their natural hybrids, there were 3 tarsal fusions and 2 supernumerary elements. 24 of 327 limbs had variation in phalangeal formula.

Basipodial variance was present in 7/69 sympatric *Triturus cristatus*, 4/46 *Triturus marmoratus* and 4/63 natural hybrids, compared with 1/48 allopatric *Triturus cristatus*, 4/70 *Triturus marmoratus* and 10/78 natural hybrids.

Riley EE, Weil MR. 1986. The effects of thiosemicarbazide on development in the wood frog, *Rana sylvatica*: I. Concentration efforts. Ecotoxicology and Environmental Safety 13:202–297.

Toxicology – Thiosemicarbazide-induced curvature of digits and abnormal limb articulations in wood frog *Rana sylvatica*, but no images/clarification.

Riley EE, Weil MR. 1987. The effects of thiosemicarbazide on development in the wood frog, *Rana* sylvatica. II. Critical exposure length and age sensitivity. Ecotoxicology and Environmental Safety 12:154–160.

Toxicology – Thiosemicarbazide-induced curvature of digits, abnormal limb articulations and acceleration of long bone ossification in wood frog *Rana sylvatica*, but no images/clarification.

Rimkus J. 1948. Better than one. Hunting Fishing 25(5):7.

Congenital – Dicephalic rattler.

Rimpp, K. 1981. Bericht über den Fund mißgebildeter Wechselkröten (*Bufo viridis*). [Report on the Fund malformed toads (*Bufo viridis*)]. Herpetofauna 3 (11):25–28 [German].

Environmental – 53% of European green toad *Bufo viridis* with malformations, attributed to radioactivity in the quarry where they were found: supernumerary front legs or hind legs, absent front leg(s), malrotated hind legs, lower jaw deformation (with part missing), and tumor-like elevation on head.

Neoplasia – European green toad *Bufo viridis* with tumor-like elevation on head.

Rio-Tsonis KD, Washabaugh CH, Tsonis PA. 1992. The mutant axolotl short toes exhibits impaired limb regeneration and abnormal basement membrane formation. Proceedings of the National Academy of Science USA 89:5502–5506.

Trauma – Amputated limbs of short toe *Ambystoma mexicanum* fail to regenerate.

Ritz J, Hammer C, Clauss M. 2010. Body size development of captive and free-ranging tortoises (*Geochelone pardilis*). Zoo Biology 29:517–525.

Metabolic – Renal disease, pyramiding, and fibrous osteodystrophy in captive tortoises (Häfeli and Schildger 1995; McArthur 2004; McArthur and Barrows 2004; Donoghue 2006; Hatt 2008).

Rivera S, Lock B. 2008. The reptilian thyroid and parathyroid glands. Veterinary Clinics: Exotic Animals 11:163–175.

Metabolic – Nutritional secondary hyperparathyroidism produces weak bones, fibrous osteodystrophy, and pathologic fractures. It is caused by decreased dietary calcium, vitamin D3 deficiency, improper dietary calcium/phosphorus intake ratio and manifest as soft shell, carapace deformities, thickening/swelling of long bones and mandible, pathological fractures, and stunted growth.

Parathyroid adenoma in red-footed tortoise *Geochelone carbonaria* resulted in deformed carapace and plaston (Frye and Carney 1975).

Hypothyroidism-induced stunting, loss of dorsal spines in green iguana *Iguana iguana* by Hernandez-Divers et al. 2001. Goiter (hypothyroid) from dietary goitrogens such as bok choy, broccoli, cabbage, cauliflower, kale, mustard seed, rapeseed, soybean sprouts and turnips has been reported in tortoises.

Environmental – Osteomalacia related to inadequate exposure (for diurnal species) to ultraviolet light manifest as soft shell, carapace deformities, thickening/swelling of long bones and mandible, pathological fractures, and stunted growth.

Roberts JM, Verrell PA. 1984. Physical abnormalities of the limbs of smooth newts (*Triturus vulgaris*). British Journal of Herpetology 6:416–418.

Congenital – Supernumerary or bifurcated toes or loss in 1% of smooth newts *Triturus vulgaris*.

Robertson DR. 1969. The ultimobranchial body of *Rana pipiens*. X. Effect of glandular extirpation on fracture healing. Journal of experimental Zoology 172(4):425–441.

Trauma – Normal fracture healing presented as thickening of diaphysis or between fracture ends by 4 weeks.

Metabolic – Removal of ultimobranchial body in *Rana pipiens* reduces skeletal density. Fracture healing presented as thickening of diaphysis or between fracture ends took 8 weeks in absence of ultimobranchial glands and resulted in smaller callus. Increase in density in callus and periosteal region, noted at 12 weeks in controls, did not occur in absence of ultimobranchial glands – who had less trabecular ossification.

Robertson P. 1981. Comparative reproductive ecology of two southeastern skinks. Proceedings of Melbourne Herpetology Symposium, CB Banks and AA Martin, eds. Zoological Board of Victoria, Australia.

Trauma – 53% of adult males and 76% of adult female *Hemiergis decresiensis* had tail injuries, compared to 73% and 72%, respectively, of *Anotis maccoyi*.

Robinson E. 2000. 2-headed snake discovered in Cape-Horn area yard. Unusual reptile sent to live at Washington serpentarium. Seattle Times 7 September 2000: 1.

Congenital – Dicephalic garter snake.

Robinson PT, Sedgwick CJ, Meier JE, Bacon JP. 1973. Internal fixation of a humeral fracture in a Komodo dragon lizard (*Varanus komodoensis*). Veterinary Medicine, Small Animal Clinician 5:645–649.

Trauma – Komodo dragon *Varanus komodoensis* with spiral fracture from fall from artificial rockwork.

Rodhain J. 1949. Tumeurs chondro-osteo-fibreuses multiples chez le lezard *Cyclura cornuta*. [Many chondro-osteo-fibrous tumors in the lizard *Cyclura cornuta*]. Revue Belge de Pathologie et de medicine experimentale 19:317–320. [French]

Neoplasia – Chondro-osteofibroma in rhinoceros iguana.

Rodrigues O. 1968. Um teratódimo deródimo em Jibóia (*Constrictor constrictor constrictor*, Linn., 1766) (Ophidia, Boidae) [A teratodymus derodymus of Jibóia (*Constrictor constrictor constrictor*, Linnaeus, 1766) (Ophidia, Boidae)]. Boletim do Museo Parense Emílio Goeldi; Nova Série, Zoologica 67:1–17. [Portuguese]

Congenital – Derodymus *Constrictor constrictor*. Left head could not digest food because it led to an intestine that had no excretory exit, and all organs in that side of the body were reduced or malformed. Osvaldo Rodrigues Da Cunha also provided a summary of reptile dicephalic studies in Brazil, and discussed embryogenic interpretations. He suggested the fusion of two simultaneously developing embryonic areas in the same egg as a cause of this teratology, as previously suggested by Vanzolini (1947) for a similar case.

In his summary of reptile dicephalic studies in Brazil, Osvaldo Rodrigues Da Cunha stated that accounts of teratology in Brazilian reptiles were quite rare, with snakes appearing most often in the record. Rarer were instances in lizards and turtles, especially regarding cases of dicephaly. He referred the reader to Rosa (1966) for a review of cases in lizards and stated that he was not aware of any cases reported in turtles in the Brazilian literature. However, he also reported an account from a credible person (which he did not name) of a small dicephalic turtle having lived in two aquarium tanks from 1910 to 1912 in the Museum Emilio Goeldi of Para, Brazil. According to him, this turtle was probably *Geoemyda punctularia punctularia* (Daudin 1802), but he was unable to find any records in the museum that could confirm the veracity of this account.

Rodrigues Da Cunha also wrote that the majority of cases of dicephaly were reported from snakes and that cases had been reported sporadically in all of the Americas. The first reports for Brazil were by Do Amaral (1927), Prado (1942, 1943, 1945, 1946), and Vanzolini (1947). Pereira (1944 and 1950), Lema (1957, 1958, 1961), and Belluomini (1957/8) were also reviewed. Belluomini and Lancini (1960) were reported to have presented a case for Venezuela.

Romagnano A, Jacobson ER, Boon, GD, Broeder A, Ivan L, Homer BL. 1996. Lymphosarcoma in a green iguana (*Iguana iguana*). Journal of Zoo and Wildlife Medicine 27:83–89.

Neoplasia – Chronic lymphocytic leukemia in bone marrow of *Iguana iguana*.

Romer AS. 1947. Review of Labyrinthodontia. Bulletin of the Museum of Comparative Zoology 1:1–368.

Congenital – Variation in suture position – variable position of ossification centers, time of their establishment or rate of growth from them.

Labyrinthodont postfrontal, intertemporal, supratemporal, and tabular "all vary in their anteroposterior dimensions, and the intertemporal frequently disappears."

Variation in labyrinthodontia where small tabular does not meet parietal versus tabular and parietal has common sutures, citing Säve- Söderbergh (1935, Fig. 45).

Romer AS. 1956. Osteology of the reptiles. 572pp.; Chicago: University of Chicago Press.

Congenital – True epiphyses were present in *Sphenodon*, lizards and crocodilians.

Rosa CN. 1966. Contribuição ao estudo da bicefalia (deródimos) em lagartos [Contributions to the study of bicephaly (derodymia) in lizards]. Papeis avulsos do Departamento de Zoologia, São Paulo 19(20):245–250. [Portuguese]

Congenital – Derodymus *Hemidactylus mabouia*, preserved in the biology laboratory of the Education Institute "Aurélio Arrobas Martins" in Jaboticabal, Brazil. Fusion of vertebral columns starts at the sixth vertebra, with the right one clearly irregularly shaped, and ends at the 12th, making this section of the vertebral column wider than normal. Irregular ribs were present on the internal sides of the dual necks.

Rose SM. 1964. Regeneration. In Moore JA. The physiology of Amphibia. Pp.545–622; New York: Academic Press.

Trauma – Regeneration is inhibited by hypophysectomy (Schotté and Hall 1952) and by thyroid levels (Schotté and Washburn 1954; but not by thyroid treatment in *Triturus* Schmidt 1958a, b)

Rose FL, Armentrout D. 1976. Adaptive strategies of *Ambystoma tigrinum* inhabiting the Llano Estacado of West Texas. Journal of Animal Ecology 45:713–729.

Trauma – Cannibalistic *Ambystoma tigrinum* has mouth 20% wider at angle of jaw, slightly elongated dentary, premaxillary and maxillary teeth, with up to threefold increase in length of coronoid, palatopterygoid and vomerine teeth, slit-like eyes and slender bodies, protruding lower jaw, and depressed snout.

Rose FL, Koke J, Koehn R, Smith D. 2001. Identification of the etiological agent for necrotizing scute disease in the Texas tortoise. Journal of Wildlife Disease 37:223–228.

Shell – *Fusarium semitectum*-induced scute necrosis in Texas tortoise *Gopherus berlandieri.*

Rosenthal KL, Mader DR. 1996. Microbiology. In Mader DR. ed. Reptile medicine and Surgery. Pp. 117–125; Philadelphia: Saunders.

Infection – *Serratia marcescens* abscess illustrated in ankle and foot of "lizard" as lysis.

Rosenthal K, Divers SJ, Donoghue S, Garner M, Klingenberg JR. 2000. Renal disease. Journal of Herpetological Medicine and Surgery 10(1):34–43.

Metabolic – Renal failure results from relative humidity and water content of diet (reptiles are dependent on visual recognition of water movement to drink), habitat temperature, improper diet (low quantity or poor protein quality, stress hormones, dietary fiber, potassium phosphorus and purine), improper ultraviolet light exposure.

Gout is more prevalent in alligators in winter months when they eat but grow slowly.

Environmental – Renal failure results from relative humidity and water content of diet (reptiles are dependent on visual recognition of water movement to drink), habitat temperature, improper diet (low quantity or poor protein quality, stress hormones, dietary fiber, potassium, phosphorus and purine), improper ultraviolet light exposure.

Rosine W.N. 1952. Notes on the occurrence of polydactylism in a second species of amphibian in Muskee Lake, Colorado. Journal of the Colorado, Wyoming Academy of Science 4(4): 100.
Congenital – leopard frogs *Rana pipiens* with polydactylism, one of which had "a hind limb deformity."
Polydactylism and hind limb duplication in a large percentage of Muskee Lake Colorado tiger salamander *Ambystoma tigrinum*, claiming this is the first record of mass polydactylism in amphibians.

Rosine WN. 1955. Polydactylism in a second species of Amphibia in Muskee Lake, Colorado. Copeia 1955:136.
Congenital – Polydactylism in five salamanders and three leopard frogs *Rana pipiens* from Muskee Lake Colorado in the University of Colorado Museum number 6701–6705, one with a badly deformed hind leg.

Rösler H. 1979. Eine siamesische Zwillingsbildung bei der Gecko-Species *Crossobamon eversmanni* (Reptilia: Sauria: Gekkonidae). [Development of Siamese twins in the gecko species *Crossobamon eversmanni* (Reptilia: Sauria: Gekkonidae)]. Salamandra 15(2):108–110 [German].
Congenital – Development of Siamese twins in egg of *Crossobamon eversmanni*, head of one twin without lower jaw.

Rösler H. 1987. Aufzeichnungen über einige Krankheitserscheinungen bei Geckonen (Sauria: Gekkonidae). 1. Mitteilung: Knochenstoffwechselstörungen. [Descriptions of some illnes in geckos (Sauria: Gekkonidae)]. Aquarien Terrarien Information, Steyr 5(1):7–11 [German].
Congenital – Twin dwarfs.
Metabolic – Rickets common in juvenile geckos; osteomalacia, in adult geckos, citing 23 specimens of 11 different species with deformation of the lower jaw, followed by lordosis and rare tail kyphosis and deformation of extremities. Examples included *Hemidactylus mabouia* with elongated lower jaw, *Pachydactylus c. vansoni* with shortened, widened lower jaw and malformed front legs, *Cosymbotus platyurus* with scoliosis, and *Phelsuma m. grandis* with tail kyphosis and deformation of left hind leg.

Rösler H. 1999 Zur Behandlung einer Rachitis bei Aufzucht von *Paroedura* cf. *bastardia* (Sauria: Gekkonidae). [The treatment of rickets during rearing of *Paroedura* cf. *Bastardia* (Sauria: Gekkonidae)]. Gekkota 1: 193–198. [German]
Metabolic – *Paroedura* cf. *Bastardia* with rickets caused by "wrong nutrition."

Rösler H. 2000. Anomalien bei Geckos (Sauria: Gekkonidae). [Anomalies of geckos (Sauria: Gekkonidae)]. Gekkota 2: 259–262. [German]
Congenital – Dicephalic *Rhacodactylus auriculatus*, *Strophurus s. spinigerus*. Cheilognathoschisis inferior and bicephaly in *Rhacodactylus auriculatus*, brachycephaly and anomalies of skeleton of body and limbs in *Chondrodactylus a. angifer*. Siamese twins in *Crossobamon e. eversmanni*. Polydactyly in *Gehyra oceanica*.

Ross RA, Marzec G. 1990. The reproductive husbandry of pythons and boas. The Institute for Herpetological Research, Stanford. 270pp.
Congenital – Kinked spine in Oenpelli (*Morelia oenpelliensis*) and green tree python *Chondropython viridis*.
Environmental – Kinked spine induced by X-ray in blood ore Sumatran blood python *Python curtus curtus*.
Kinked spine induced by incubating at increased or decreased temperature in Burmese *Python molurus bivittatus* and Ball or royal python *Python regius* (the latter with anophthalmia and maxillary defect.

Rosskopf WJ. 1986. Shell disease in turtles and tortoises. In Kirk RW. Current Veterinary Therapy IX. WB Saunders, Philadelphia. Pp. 751–759.
Congenital – Missing nuchal bone in hatchling desert tortoise, similar to Galapagos tortoise on Hood Island.
Trauma – Crack between pectoral and abdominal scutes from falls.
Infection – Pitting from old, healed osteomyelitis.
Metabolic – Malformed shell from calcium deficiency. Excess protein produces mounded shell bone plates.
Neoplasia – Rare, classified as a chondroma in a desert tortoise.
Shell disease – Ulcerations in diamond-back terrapin *Malaclemys terrapin* from too much time in fresh water.
Ulcerative shell disease from *Beneckea chitinovora* in fresh water turtles, through crustacean intermediary hosts.
Shell disease from *Pseudomonas, Aeromonas, Proteus, Serratia, Klebsiella, Escherichia coli, Staphylococcus aureus*, alpha-hemolytic *Streptococcus* and mycobacteria. Fungal isolates include *Fusarium solanae, Aspergillus, Penicillium, Mucor.*
Maggot-induced hole in carapace of *Galapagos tortoise.*

Rosskopf WJ Jr, Shindo MK. 2003. Syndromes and conditions of commonly kept tortoise and turtle species. Seminars in Avian and Exotic Pet Medicine 12:149–161.
Metabolic – Malnutrition-induced pyramidal shell disease or beak overgrowth.
Infection – Shell rot from algae or bacteria.
Stones – Bladder stones in desert tortoises.
Shell disease – Malnutrition-induced pyramidal shell disease and shell rot from algae or bacteria.

Rosskopf WJ Jr, Woerpel RW. 1981. Repair of shell disease in tortoises. Modern Veterinary Practice 62(12):938–939.
Trauma – Generic discussion of repair of shell fractures in tortoises.

Rosskopf WJ Jr, Woerpel RW, Yanoff S. 1982. Severe shell deformity caused by a deficient diet in a California desert tortoise. Veterinary Medicine, Small Animal Clinician 77:593–594.

Congenital – Wavy carapace in *Gopherus agassizi.*
Metabolic – Wavy carapace in *Gopherus agassizi.*

Rossman DA. 1963. The colubrid snake genus *Thamnophis*: A revision of the *sauritus* group. Bulletin of the Florida State Museum, Biological Sciences Series 7:99–178.
Congenital – 10% of *Thamnophis sauritus* had half-ventrals.

Rostand J. 1932. L'évolution des espèces. [The Evolution of Species]. Paris: Hachett, 205 pp. [French]
Congenital – Considers mutations as part of evolution, without providing any specifics on pathology.

Rostand J. 1947a. Gynogenèse et anomalies digitales chez le Crapaud. [Embryology and anomalies in toads]. Comptes Rendus Hebdomadaires Séances de l'Académie des Sciences Paris 225:417–419. [French]
Congenital – In 7/130 *Bufo bufo.*

Rostand J. 1947b. Gynogenèse et polydactylie chez le Crapaud ordinaire. [Embryology and polydactyly in toads]. Comptes Rendus des Séances de la Société de Biologie et de ses filiales 141:629–630. [French]
Congenital – Polydactyly in *Bufo bufo.*

Rostand J. 1947c. De quelques anomalies digitales chez le Crapaud ordinaire (*Bufo bufo*) [Several digital anomalies in the ordinary toad (*Bufo bufo*)]. La Revue Scientifique, Decembre 1947:1121–1125. [French]
Congenital – *Bufo bufo* polydactyly (4), posterior (16, four of which had syndactyly), anterior (1), and quadripedal (4) ectrodactyly and isolated syndactyly (6).

Rostand J. 1948a. Anomalies digitales chez le Crapaud ordinaire (*Bufo bufo*) [Digital anomalies in the toads (*Bufo bufo*)]. Bulletin intérieur de la Société française de Génétique no. 1, April 1948: 1 p. [French]
Congenital – Posterior limb polydactyly, ectodactyly and anterior limb ectodactyly and syndactyly in *Bufo bufo.*

Rostand J. 1948b. Les mutations des Batraciens Anoures. [Mutations of anuran batrachians]. La Revue scientifique 86e année, fasc 7, n 3295:438–440. [French]
Congenital – Ectomelia, polydactyly (bifurcation of toe), and ectodactyly (four instead of five toes) in *Bufo bufo.*

Rostand J. 1948c. Polydactylie naturelle chez le Crapaud ordinaire (*Bufo bufo*). [Natural polydactyly in toad (*Bufo bufo*)]. Comptes Rendus Hebdomadaires Séances de l'Académie des Sciences Paris 226:1398–1399. [French]
Congenital – Ectodactyly in 27/7407 *Bufo bufo*

Rostand J. 1948d. Ectrodactylie naturelle chez le Crapaud ordinaire. [Natural ectrodactyly in toad] Comptes Rendus Hebdomadaires Séances de l'Académie des Sciences Paris 226:1224. [French]
Congenital – 6/7407 *Bufo bufo* with polydactyly.

Rostand J. 1949a. Hérédité de la polydactylie chez le Crapaud ordinaire (*Bufo bufo*). [Heredity of polydactyly in toad (*Bufo bufo*)]. Comptes Rendus Hebdomadaires Séances de l'Académie des Sciences Paris 228:778–779. [French]
Congenital – 125 of 248 larvae of *Bufo bufo* manifest polydactyly.

Rostand J. 1949b. Les anomalies héréditaires chez les Batraciens. [Hereditary anomalies among batrachians]. La Vie rustique (Bruxelles) 15. 2. 1949:162. [French]
Congenital – Polydactyly in *Bufo bufo.*

Rostand J. 1949c. Sur la descendance d'un male polydactyle de Crapaud ordinaire. [On the descendents of polydactyly in male toads]. Comptes Rendus Hebdomadaires Séances de l'Académie des Sciences Paris 228:1159–1161. [French]
Congenital – Posterior polydactyly in *Bufo bufo* transmitted as a male phenomenon.

Rostand J. 1949d. Polydactylie naturelle chez la Grenouille verte (*Rana esculenta*). [Natural polydactyly in green frog (*Rana esculenta*)]. Comptes Rendus Hebdomadaires Séances de l'Académie des Sciences Paris 228:1666–1667. [French]
Congenital – Polydactyly in 9/49 *Rana esculenta.*

Rostand J. 1949e. Polymélie chez le Crapaud ordinaire (*Bufo bufo*). [Polymelia in toad (*Bufo bufo*)]. Comptes Rendus de la Société de Biologie 143:666–668. [French]
Congenital – Polymely in *Bufo bufo.*

Rostand J. 1949f. Sur diverses anomalies relevées dans une population de Crapauds (*Bufo bufo*). [On diverse anomalies revealed in a population of toads (*Bufo bufo*)]. Comptes Rendus des Séances de la Société de Biologie 143:758–760. [French]
Congenital – Among 44,000 *Bufo bufo* from Alsace, there were 53 polydactyly, 298 ectodactyly, and 100 clinodactyly.

Rostand J. 1949g. La génétique des batraciens. [The genetics of amphibians]. Revue Générale des Sciences pure et appliquées et bulletin de la Société philomathique 56:262–267. [French]
Congenital – Monorhinie in larval, but not adult *Bufo bufo*, noting it can also be caused by lithium. Skeletal effect not addressed.
Polydactyly in triton (Geoffroy St. Hilaire 1832), *Bufo bufo*, and *Rana esculenta.*
Polydactyly in 4/44,000 *Bufo bufo* in Jura and Alsace, noting genetic transmission in half of progeny and 37/256 *Rana esculenta*, transmitted as a recessive trait.
Polydactyly and polymelia in *Ambystoma mexicanum* from Boulder Colorado.
Polymelia is rare in Anourans, found mainly in *Bufo bufo, Rana, Alyte*, and *Discoglossa* (de Falvard 1931).
Polymelia, polydactyly, ectromelia, syndactyly, clinodactyly in *Bufo bufo.*

Ectrodactyly in *Bufo bufo*.

Ectromelia in parthenogenetic *Rana nigromaculata* (Kawamura 1939), bithoracic ectromelia in *Bufo bufo* by Ponse (1941).

Toxicology – Monorhinie in larval, but not adult *Bufo bufo*, noting it can also be caused by lithium. Skeletal effect not addressed.

Rostand J. 1950a. Polydactylie chez la Grenouille rousse (*Rana temporaria*) et clinodactylie chez la Grenouille verte (*Rana esculenta*). [Polydactyly in red frog (*Rana temporaria*) and clinodactyly in the green frog (*Rana esculenta*)]. Comptes Rendus des Séances de la Société de Biologie 144: 19–20 (not 17–19). [French]

Congenital – Polydactyly, bifid, and supernumerary in *Rana temporaria* and clinodactyly in *Rana* esculenta and *Bufo bufo* (noting 100 of 250 offspring had polydactyly).

Rostand J. 1950b. Sur la gynogenèse comme moyen d'exploration du patrimonie héréditaire chez le Crapaud commun (*Bufo bufo*). [On embryology as a means of exploring paternal hereditary among the common toad (*Bufo bufo*)]. Comptes Rendus Hebdomadaires Séances de l'Académie des Sciences Paris 230:2130–2131. [French]

Congenital – *Bufo bufo* with polydactyly and clinodactyly.

Rostand J. 1950c. Sur la descendance des Grenouilles polydactyles. [On the descendants of polydactylous frogs]. Comptes Rendus Hebdomadaires Séances de l'Académie des Sciences Paris231:496–498. [French]

Congenital – Five cases of anterior polydactyly in Trevignon *Rana esculenta*.

Rostand J. 1950d. Essais de chimiotératogenèse chez les Batraciens anoures. [Essay of chemoteratogenesis in batrachians]. Comptes Rendus des Séances de la Société de Biologie 144:915–917. [French]

Toxicology – Brachymelia (short limbs) from paraminobenzenesulfamide in batraciens.

Ectrodactyly from colchicines and trypaflavine in *Rana temporaria*.

Syndactyly from serine and paraminobenzenesulfamide in *Rana temporaria*.

Rostand J. 1951a. La Génétique des Batraciens. [The genetics of batrachians]. Paris: Hermann et Cie, 80 pp. [French]

Congenital – 26/44,000 *Bufo bufo* from Alsace had bilateral polydactyly; 27, unilateral. 9/49 and 37/256 *Rana temporaria* from Bretagne had polydactyly. Polydactyly was also noted in *Rana esculenta, Rana palustris, Discoglossus pictus, Ambystoma mexicanum* (also with polymely), *Salamandra cristata* (also with polymely), *Triturus helveticus* (also with polymely).

Ectodactyly was found in 3 of 44,000 *Bufo*.

Syndactyly was present in 0/44,000 adults, but 10/153 tadpoles, four of which also had ectodactyly.

Ectodactyly and syndactyly were not found in *Rana esculenta*. Clinodactyly was found in 100/44,000 *Bufo*.

Rostand J. 1951b. Sur une nouvelle anomalie génétique (diplopodie) chez le Crapaud ordinaire (*Bufo* bufo). [On a new genetic anomaly (diplopody) in the common toad]. Comptes Rendus Hebdomadaires Séances de l'Académie des Sciences Paris 232:2043–2045. [French]

Congenital – Polydactyly and short limbs in *Bufo bufo*.

Rostand J. 1951c. Génétique des Batraciens anoures: anomalies héréditaires et phénocopies. [Genetics of batrachians: Hereditary anomalies and phenotypes.] Bulletin de la Société d'Histoire naturelle de Toulouse 86:230–236. [French]

Congenital – Polydactyly in 15% of green frog *Rana esculenta*.

Polydactyly, syndactyly, and ectrodactyly in *Rana temporaria*.

Two varieties of thoracic ectromelia in *Bufo bufo*: dominant and recessive.

Ectrodactyly in salamanders.

Syndactyly in *Rana esculenta*.

Clinodactyly, ectrodactyly, and syndactyly in *Bufo bufo*

Rostand J. 1951d. Sur la polydactylie des Batraciens anoures [Polydactyly in batrachians]. Bulletin Biologique de la France et de la Belgique 85:113–136. [French]

Congenital – Polydactyly in *Rana esculenta* and *Rana temporaria, Ambystoma mexicanum*. 26/44,000 *Bufo bufo* with polydactyly (perhaps "weakly" autosomal dominant); polymelia was also present.

Rostand J. 1952a. Sur maturité des oeufs et anomalies chez *Rana temporaria*. [Overmature eggs and anomalies in *Rana temporaria*.] Comptes Rendus des Séances de la Société de Biologie 146:624–625. [French]

Congenital – Supernumerary arm in *Rana temporaria* (Gallien 1944).

Polydactyly, unilateral and bilateral ectomelia and globular replacement of bones of hind limbs in *Rana temporaria*.

Bilateral ectrodactyly [2/7407, 11/1200, and 12/44,000 (Rostand 1951)] in *Rana temporaria*. Syndactyly in *Rana temporaria*.

Rostand J. 1952b. Ectrodactylie and syndactylie chez *Rana temporaria* [Ectrodactyly and syndactyly in *Rana temporaria*]. Comptes Rendus des Séances de la Société de Biologie 146:4–5. [French]

Congenital – Two cases of ectrodactyly among 7407 *Bufo bufo*; 12 among 44,000 (Rostand 1951) and 11/1200 *Rana temporaria* had ectrodactyly. Syndactyly *Rana esculenta*.

Rostand J. 1952c. Polydactylie, polypodie et polymélie chez *Rana esculenta*. [Polydactyly, polypodia, and polymelia in *Rana esculenta*]. Comptes Rendus Hebdomadaires Séances de l'Académie des Sciences Paris 235:322–324. [French]

Congenital – 14.4% of 256 Trevignon *Rana esculenta* with polydactyly in 1949, 12.9% of 177 in 1950, 5.2% of 173 in 1951, 2.5% of 117 in 1952.

Rostand J. 1952d. Sur la variété d'expression d'une anomalie (*P*) chez *Rana esculenta*. [On the variety of expression of the P anomaly in *Rana esculenta*]. Comptes Rendus Hebdomadaires Séances de l'Académie des Sciences Paris 235:583–585. [French]

Congenital – Supernumerary, shortened and globular digits in *Rana esculenta.*

Rostand J. 1952e. Régénération de members normaux chez de *Rana esculenta* présentant l'anomalie P [Regeneration of normal limbs in *Rana esculenta* with anomaly P]. Comptes Rendus des Séances de la Société de Biologie 146:1530–1531. [French]

Trauma – Regeneration of normal limbs in *Rana esculenta* with anomaly P polydactyly implies nongenetic etiology.

Rostand J. 1952f. Sur une certaine anomalie *P* de la Grenouille verte (*Rana esculenta* L.). [On a certain anomaly P in green frog (*Rana esculenta* L.)]. Revue scientifique 1952, Septembre–Octobre:353–365. [French]

Congenital – Polydactyly in 14.4% in 1949, 12.9% in 1950, 5.2% in 1951, and 2.5% in 1952 *Rana esculenta* from Bretagne; 1.4% in 1950, 1.7% in 1951, and 4.6% in 1952 from Champdieu (Loire) and from Trévignon, 73 of 153, detailing longer and shorter limb occurrence. The latter group also had brachymelia in an equal number.

Rostand J. 1953. Ce que nous apprenent les Crapaud et les Grenouilles. [What now is present in toads and frogs]. Les Conférences du Palais de la Découverte, Série A, no 183:5–25. [French]

Congenital – *Bufo bufo* 125/248 with anomalies – polydactyly, ectodactyly, syndactyly, and supernumerary. *Rana esculenta* from Trevignon – 216/458 with polydactyly.

Rostand J. 1955a. Polydactylie associée à la polymélie chez *Alytes obstretricans* Laur. [Polydactyly associated with polymelia in *Alytes obstretricans* Laur.]. Comptes Rendus des Séances de la Société de Biologie149:1916–1918. [French]

Congenital – Polydactyly and polymely in *Alytes obstetricans.*

Rostand J. 1955b. Nouvelles observations sur l'anomalie *P* chez la Grenouille verte (*Rana esculenta* L.) [New observations on anomaly P in green frog *Rana esculenta* L.)]. Rendus des Séances de la Société de Biologie149:2075–2077. [French]

Congenital – 15–35% with polydactyly in 1952 of *Rana esculenta*, compared to 155 of 435 in 1955, 73 of which are the severe variety he refers to as anomaly P.

Rostand J. 1955c. Les Crapauds, les Grenouilles et quelques grands problèmes biologiques. [Toads, frogs and several major biologic problems]. 217pp.; Paris: Gallimard. [French]

Congenital – Polydactyly, ectodactyly and syndactyly in *Bufo*.

Sacral pathologies in *Rana esculenta* and *Rana temporaria* in Bretagne:

1949–14.4% of 256

1950–12.9% of 177

1951–5.2% of 173

1952–2.5% of 117

Trevigno – 47% of 153, with brachymelia in 15.

Rana esculenta from Champdieu (Loire):

1950–1.4% of 70

1951–1.7% of 960

1952–4.6% of 238

Ambystoma mexicanum – 17 larval and two adult 1946–1947 from Muskee Lake at 8300 feet, near Boulder Colorado had polydactyly and supernumerary limbs.

Rana temporaria – 1.3–3% of 4122 from Hennezel (Vosges) had ectodactyly (47 male, nine female).

Rostand J. 1955d. Effets tératogènes des rayons ultraviolets sur les larves de Grenouilles. [Teratogenic effects of ultraviolet rays on frog larvae]. Comptes Rendus des Séances de la Société de Biologie149:905–907. [French]

Environmental – UV exposure of *Rana temporaria* produced 32 with ectomely, four with supernumerary and eight with atrophy.

Rostand J. 1956a. Sur une nouvelle anomalie (*E*) de *Rana temporaria* L.: ectrodactylie, torsion de la queue et pupille Claire. [On a new anomalie (E) in *Rana temporaria* L: Ectrodactyly, tortion and tail and pupil]. Comptes Rendus Hebdomadaires Séances de l'Académie des Sciences Paris 242:3000–3002. [French]

Congenital – Ectodactyly in *Rana temporaria.*

Rostand J. 1956b. Polymorphisme de l'anomalie *E* chez *Rana temporaria* L. [Polymorphism of anomalie E in *Rana temporaria* L.]. Comptes Rendus Hebdomadaires Séances de l'Académie des Sciences Paris 243:973–974. [French]

Congenital – Ectomelia in 13/63 Chaville *Rana temporaria*.

Rostand J. 1957. Grenouilles monstrueuses et radioactivité. [Frog monstrosities and radioactivity]. Comptes Rendus Hebdomadaires Séances de l'Académie des Sciences Paris 245:1175. [French]

Congenital – Amsterdam canal *Rana esculenta* with polydactyly, polymelia and brachymelia.

Rostand J. 1958a. Les anomalies des amphibiens anoures [Anomalies of anuran amphibians]. Paris: Sedes, 100pp. [French]

Congenital – Polydactyly, polymely, and supernumerary limbs in 40% of *Bufo bufo* and in 216/458 *Rana esculenta* from Trevignon.

Ectodactyly was rare (1–2/100) in Hennezel *Rana* (but not *temporaria*) and *Bufo*. Polymely in *Alytes obstetricans* and *Rana esculenta*. Syndactyly in *Bufo*.

Rostand J. 1958b. Sur un nouveau type of polydactylie chez la Grenouille verte (*Rana esculenta* L.). [On a new type of polydactyly in green frog verte (*Rana esculenta* L.)]. Comptes Rendus Hebdomadaires Séances de l'Académie des Sciences Paris 247:336. [French]

Congenital – 1/100 *Rana esculenta* from Puycasquier has anterior polydactyly.

Rostand J. 1958c. Effet tératogène des rayons ultraviolets sur les larves d'amphibiens. [Teratogenic effect of ultraviolet light on amphibian larvae]. Comptes Rendus Hebdomadaires Séances de l'Académie des Sciences Paris 247:972. [French]

Environmental – Exposure of *Rana temporaria* to mercury UV light within the first 2 days of life produced seven ectomely, and one supernumerary leg among 147 larval forms.

Rostand J. 1959. L'anomalie *P* chez la grenouille verte (*Rana esculenta* L.). [Anomaly P in a green frog (*Rana* esculenta L)]. Bulletin Biologique de la France et de la Belgique 93:7–15. [French]

Congenital – *Rana esculenta* with polydactyly.

Rostand J. 1960. Le problème des Grenouilles monstreuses [The problem with frog monsters]. Sciences, novembre-decembre 1960:25–32. [French]

Congenital – Polydactyly in *Rana esculenta, Hyla arborea* and *Rana temporaria.*

Supernumerary *Rana ridibunda.*

Rana esculenta – from Trevignon – 1952 – 47%

1953–1957 – 15–25%

1958 – 77%

1959 – 18%

1960 – 15%

from Lingé – 1%; Champdieu (Loire), 1–5%.

Rostand J. 1962. Sur la distribution de l' anomalie *P* chez la Grenouille verte (*Rana esculenta* L.). [On the distribution of anomaly P in green frog (*Rana esculenta* L.)]. Comptes Rendus Hebdomadaires Séances de l'Académie des Sciences Paris 255:2189. [French]

Congenital – Trevignon *Rana esculenta*: 25%, abnormal in 1960; 80%; and in 1961, 1% in 1962.

Rostand J. 1971. Les etangs a monstres. [The basis of monsters]. Histoire d'une recherche (1947–1970). Paris: Stock, 89p. [French]

Congenital – Supranumerary limbs in *Rana esculenta*, polydactyly in *Rana temporaria*:

Trevignon	Ling
1958 – 80/100	
1959 – 25/	
1961 – 75/	35/
1962 – 01/	60/
1963 –	00/
1964 –	60/
1965 –	70/
1966 –	14/
1970 –	30/

Rostand J, Darré P. 1967. Sur les conditions d'apparition de l'anomalie *P* chez *Rana esculenta*. [On the conditions of appearance of the anomaly P in *Rana* esculenta]. Comptes Rendus Hebdomadaires Séances de l'Académie des Sciences Paris 265:761–762. [French]

Environmental – Incubation of *Rana esculenta* with fish (e.g., eel) produces polydactyly.

Rostand J, Darré P. 1968. Conditions déterminantes de l'anomalie P chez *Rana esculenta*. [Conditions which determine anomaly P in *Rana esculenta*]. Comptes Rendues des Séances de la Société de Biologie et sa Filiales 162:1682–1683. [French]

Environmental – 125 of 185 with abnormalities when *Rana esculenta* incubated with anguilles versus 16 of 26 with tanches.

Rostand J, Darré P. 1969. Action tératogène des dejections de certains poisons sur les larves de *Rana esculenta*. [Teratogenic effect of poisons on *Rana esculenta* larvae]. Comptes Rendues des Séances de la Société de Biologie et sa Filiales 163:2033–2034. [French]

Environmental – Fish responsible for teratogenesis in *Rana esculenta*. Twenty three of 85 abnormal in coenvironment with tanche; 19 of 61 with anguille in May; 73 of 110 versus 41 of 135 in June; and 154 of 112 and 248 of 351 in July, 1969.

Rostand J, Fischer J-L. 1959. Polydactylie chez *Rana temporaria*. [Polydactyly in *Rana temporaria*]. Comptes Rendus Hebdomadaires Séances de l'Académie des Sciences Paris 249:329–330. [French]

Congenital – Polydactyly in *Rana temporaria*.

Rostand J, Tétry A. 1962. La vie. [The Life] 466pp.; Paris: Larousse édition. [French]

Congenital – Derodymus snake and Grenouille (*Rana*) and *Chrysemys scaber*.

Supernumerary forelimb in *Alytes obstetricans* and *Rana pipiens* with supernumerary leg at inferior pelvis.

Polydactyly in *Bufo bufo*.

Rana esculenta with polydactyly, digit bifurcations, phycomelia, and ectromelia and extreme polydactyly and excrescences.

Opisthodichotomic *Pleurodeles*.

Rana temporaria lacking hind limbs and most of pelvis.

Pleurodeles with supernumerary flank arm or parasitic twin.

Xenopus laevis with parasitic duplication of posterior portion of body.

Situs inversus (experimental without details) in *Triton* (Spemann and Falkenburg, no reference provided).

Trauma – *Triton* regeneration, sometimes with polydactyly.

Lizard tail duplication.

Toxicology – Serine and tryptoflavin-induced failure of peripheral forelimb development, syndactyly and ectrodactyly in *Rana temporaria*.

Environmental – Supernumerary forelimb induced in *Rana temporaria* by ultraviolet exposure.

Rostand J, Jacquot M, Darré P. 1967. Nouvelles experiences sur les causes de l'anomalie *P* chez *Rana esculenta*. [New experiences on the causes of anomaly P in *Rana esculenta*]. Comptes Rendus Hebdomadaires Séances de l'Académie des Sciences Paris 264:2395–2397. [French]

Congenital – 80% of *Rana esculenta* larva with anomalies.

Roth M. 1978. Experimental skeletal teratogenesis in the frog tadpole. Anatomischer Anzeiger 143:296–300.

Toxicology – *Lathyrus odoratus* seeds produce achondroplastic-like micromely and curled toes and trident hand in *Rana temporaria* tadpoles.

Roth M. 1988. The 2 stage neuroskeletal pathomechanism of developmental deformitiers of the limb skeleton. Anatomischer Anzeiger 167:271–279.

Toxicology – *Lathyrus odoratus* seeds produce achondroplastic-like stunting and hypothyroidism in *Rana temporaria* tadpoles.

Rothschild BM. 1987a. Decompression syndrome in fossil marine turtles. Annals of the Carnegie Museum 56:253–258.

Vascular – Avascular necrosis has been described in Cretaceous mosasaurs and marine turtles. Avascular necrosis was present in eight families of marine/aquatic turtles from the Cretaceous to recent. Desmatochelyidae, Toxochelyidae, Protostegidae, and Pleurosternidae were especially afflicted in the Cretaceous. Contemporary representation is limited to its extremely infrequent occurrence in Cheloniidae (*Lepidochelys*) and Chelydridae (*Chelydra* and *Macroclemys*).

Fossil – Avascular necrosis has been described in Cretaceous mosasaurs and marine turtles. Avascular necrosis was present in eight families of marine/aquatic turtles from the Cretaceous to recent. Desmatochelyidae, Toxochelyidae, Protostegidae, and Pleurosternidae were especially afflicted in the Cretaceous.

Rothschild BM. 1987b. Avascular necrosis: A familial study. Clinical Research 35:360A.

Vascular – Avascular necrosis described in Cretaceous mosasaurs, present in *Platecarpus*, *Tylosaurus*, *Mosasaurus*, *Plioplatecarpus*, *Prognathodon*, and *Hainosaurus*, but not in *Clidastes*, *Ectenosaurus*, or *Halisaurus*. Occurrence in affected genera was universal, independent of locale.

Fossil – Avascular necrosis described in Cretaceous mosasaurs, present in *Platecarpus*, *Tylosaurus*, *Mosasaurus*, *Plioplatecarpus*, *Prognathodon*, and *Hainosaurus*, but not in *Clidastes*, *Ectenosaurus*, or *Halisaurus*. Occurrence in affected genera was universal, independent of locale.

Rothschild BM. 1988. Avascular necrosis in diving vertebrates: The turtle test. Journal of Vertebrate Paleontology 8(supplement to 3):25A.

Vascular – Avascular necrosis in turtles was limited to marine turtles, with no cases in terrestrial and one instance in a fossil Mediterranean *Trionyx*, which was marine habitat at that time.

Fossil – Avascular necrosis in turtles was limited to marine turtles, with no cases in terrestrial and one instance in a fossil Mediterranean *Trionyx*, which was marine habitat at that time.

Rothschild BM. 1989. Paleopathology and its contributions to vertebrate paleontology: Technical perspectives. Journal of Vertebrate Paleontology 9 (supplement to 3):36A–37A.

Philosophy – As paleopathology has evolved from observational speculation to analysis of testable hypotheses, so too has recognition of its contribution to vertebrate paleontology. In the presence of significant structural and density variation (between matrix and osseous structures), X-rays provide an additional perspective of osseous response to stress and disease. As film techniques are time and cost expensive, fluoroscopy has proven a valuable alternative. Radiologic techniques also allow noninvasive "sectioning" of specimens, illustrating significant internal detail. The object can be "split" on a plane and the two portions rotated to "open" the image. This three-dimensional approach now can be applied to other forms of sequential data to facilitate their three-dimensional representation graphically or with solid representations. Antigen and microstructure may be well preserved in fossils. Molecular preservation with retention of helical structure and sensitivity to collagenase has been demonstrated in 10,000-year-old collagen. Antigen has been extracted from 100 million-year-old bone and documented, in situ, in 11,000-year-old bone. If the appropriate site in the tissue is assessed, if antigen is still present, and if the appropriate antisera are utilized, fixation of the antibody to the specimen can be detected. Minute amounts of DNA can be amplified and analyzed.

Rothschild BM. 1990. Absence of decompression syndrome in recent and fossil Mammalia and Reptilia. Annals of Carnegie Museum 59:287–293.

Vascular – Avascular necrosis was notably lacking in sea snakes, champosaurs, *Claudiosaurus*, seals, sea lions, dolphins, and sea cows.

Fossil – Avascular necrosis was notably lacking in sea snakes, champosaurs, *Claudiosaurus*, seals, sea lions, dolphins, and sea cows.

Rothschild BM. 1991. Stratophenetic analysis of avascular necrosis in turtles: Affirmation of the decompression syndrome hypothesis. Comparative Biochemistry and Physiology 100A:529–535.

Vascular – Avascular necrosis from affected turtle families:

Perseverance of avascular necrosis was 41% in the Cretaceous, 9% in the Eocene, 5% in the Oligocene, and 0.3% in the Holocene. Reduction in frequency of avascular necrosis in the early Eocene was followed by near disappearance, subsequent to the Oligocene. Contemporary representation is limited to its extremely infrequent occurrence in Cheloniidae (*Lepidochelys*) and Chelydridae (*Chelydra* and *Macroclemys*).

Fossil – Perseverance of avascular necrosis was 41% in the Cretaceous, 9% in the Eocene, 5% in the Oligocene, and 0.3% in the Holocene. Reduction in frequency of avascular necrosis in the early Eocene was followed by near disappearance, subsequent to the Oligocene. Contemporary representation is limited to its extremely infrequent occurrence in Cheloniidae (*Lepidochelys*) and Chelydridae (*Chelydra* and *Macroclemys*).

	Number	Avascular necrosis	
Family	Evaluated	Number	Proportion
Pleurosternidae	4	1	.25
Kinosternidae	805	1	.001
Trionychidae	204	2	.001
Protostegidae	6	4	.67
Toxochelyidae	17	5	.29
Dermochelyidae	74	4	.05
Cheloniidae	170	12	.07
Desmatochelyidae	6	2	.33
Baenidae	14	1	.07

Rothschild BM. 2003. Pathology in Hiscock site vertebrates and its bearing on hyperdisease among North American mastodons. In Laub RS. ed. The Hiscock Site: Late Pleistocene and Holocene Paleoecology and Archaeology of Western New York State. Proceedings of the Second Smith Symposium. Bulletin of the Buffalo Society of Natural Sciences 37:171–175.

Congenital – Fusion anomaly, manifest as conjoined frog pelvic bones, was noted in a frog from the Pleistocene Hiscock Site.

Fossil – Fusion anomaly, manifest as conjoined frog pelvic bones, was noted in a frog from the Pleistocene Hiscock Site.

Rothschild BM. 2008a. Scientifically rigorous reptile and amphibian osseous pathology: Lessons for forensic herpetology from comparative and paleo-pathology. Applied Herpetology 10:39–116.

Philosophy – As the study of bone disease in recent and fossil amphibians and reptiles has evolved from observational speculation to analysis of testable hypotheses, so too has recognition of its contribution to our understanding of diseases and organisms. Given the development of a "library" of macroscopic osseous manifestations of a variety of diseases, the power of such examination of skeletons for identification of the etiology of pathology has greatly reduced the need for destructive analysis. High frequency of malformations in amphibians or of spondyloarthropathy in reptiles should stimulate evaluation for environmental factors. Notation of previously unrecognized/undescribed pathology affords unique opportunities. Scientific approach, validated database and phylogeny- independent pathology recognition form the basis for this review of the current knowledge of contemporary and extinct amphibian and reptile osseous pathology. This provides baseline data for forensic herpetologists attempting to identify and interpret osseous lesions, disease, and trauma in a legal context.

Rothschild BM. 2008b. Vertebral pathology in mosasaurs, pp. 90–94. In Everhart, M. ed. Proceedings of the Second Mosasaur Meeting, Fort Hays Studies Special Issue 3, Fort Hays State University, Hays, Kansas.

Trauma – Grooves on vertebrae evidence shark bites, while associated new bone formation allows recognition of predation. The former is common; the latter is rare.

Fused mosasaur vertebrae are attributable to reactive bone from trauma and infection (e.g., related to trauma of shark bites) and perhaps to splinting or a disease documented in contemporary varanids.

Infection – Fused mosasaur vertebrae are attributable to reactive bone from trauma and infection (e.g., related to trauma of shark bites) and perhaps to splinting or a disease documented in contemporary varanids.

Vascular – Recognition of diving disease in mosasaurs was based upon identification of a specific associated pathology, avascular necrosis, invariably present in *Platecarpus*, *Tylosaurus*, *Mosasaurus*, *Plioplatecarpus*, *Prognathodon*, *Hainosaurus*, and an Antarctic mosasaur, and invariably absent from *Clidastes*, *Ectenosaurus*, *Globidens*, *Halisaurus*, and *Kolposaurus*.

Neoplasia – Moodie's (1918) suggestion of osteoma was rejected on the basis of apparent lesion size and location. The only other recognized "tumor" has also been called an osteoma, but it has not been histologically examined to assure it is not simply a hamartoma, as human skull so-called osteoma has been now reclassified.

Vertebral – Three forms of pathology are routinely noted in the vertebrae of mosasaurs: shark bites, avascular necrosis, and fusion, in addition to isolated reports of "tumors."

Fused mosasaur vertebrae are attributable to reactive bone from trauma and infection (e.g., related to trauma of shark bites) and perhaps to splinting or a disease documented in contemporary varanids.

Fossil – Grooves on vertebrae evidence shark bites, while associated new bone formation allows recognition of predation. The former is common; the latter, rare.

Fused mosasaur vertebrae are attributable to reactive bone from trauma and infection (e.g., related to trauma of shark bites) and perhaps to splinting or a disease documented in contemporary varanids.

Recognition of diving disease in mosasaurs was based upon identification of a specific associated pathology, avascular necrosis, invariably present in *Platecarpus*, *Tylosaurus*, *Mosasaurus*, *Plioplatecarpus*, *Prognathodon*, *Hainosaurus*, and an Antarctic mosasaur, and invariably absent from *Clidastes*, *Ectenosaurus*, *Globidens*, *Halisaurus*, and *Kolposaurus*.

Moodie's (1918) suggestion of osteoma was rejected on the basis of apparent lesion size and location. The only other recognized "tumor" has also been called an osteoma, but it has not been histologically examined to assure that it is not simply a hamartoma, as human skull so-called osteoma has been now reclassified.

Three forms of pathology are routinely noted in the vertebrae of mosasaurs: shark bites, avascular necrosis, and fusion, in addition to isolated reports of "tumors."

Rothschild BM. 2008c. Reproducibility between pathologies in recent and fossil crocodilians and crocodilian ecomorphs. Journal of Vertebrate Paleontology 28:134A.

Pathology – Analogous to suggestion that crocodilomorphs have not changed greatly during much of their history, neither has the osseous pathologies with which they were afflicted. Gastralia fractures appear to be

Cretaceous phenomenon, while gout may be a more recent disease. The consistency of pathology in crocodilian ecomorphs may reflect a similar consistency between disease and host habitat.

Trauma – Isolated occurrence of jaw fractures in Triassic phytosaurs *Mystriosuchus*, *Mahajangasuchus*, *Phytosaurus buceros*, and *Leptosuchus imperfecta* and puncture wounds in Cretaceous *Brachychampsa sealeyi* and Eocene *Tilemsisuchus lavocati* suggests intraspecies conflict. That behavioral interpretation is confirmed in contemporary *Alligator mississippiensis* and *Crocodylus americanus*. It contrasts with frequent gastralia fractures in late Paleocene *Leidyosuchus* (= *Borealosuchus*) *formidabilis*, not recognized in contemporary crocodilians.

Gastralia fractures appear to be a Cretaceous phenomenon.

Infection – Alterations compatible with infection are noted in distal elements of *Machimosaurus mosae* and *Leidyosuchus formidabilis*, similar to infections in contemporary *Alligator mississippiensis* (4/88), *Caiman latirostris*, *Crocodylus siamensis*, and *Crocodylus acutus* (2/26).

Neoplasia – Benign neoplasia, in the form of button osteoma and a probable chondroma, is found in an Eocene *Crocodylus*, and an osteochondroma in contemporary *Caiman yacare*.

Metabolic – Isolated cases of articular gout were observed in *Alligator sclerops* and *mississippiensis*; *Tupinambis*; *Crocodylus americanus*, *niloticus*, *novaeguineae*, *porosus*, and *johnstonii*; *Caiman crocodilus*; *Tomistoma schlegelii*, and *Gavialis gangeticus*, but gout has yet to be identified in the fossil record – exclusive of dinosaurs.

Vertebral – Late Paleocene *Leidyosuchus formidabilis* and lower Miocene *Tomistoma dowsoni* and *Leidyosuchus* (=*Borealosuchus*) *formidabilis* have ossification of anulus fibrosus characteristic of the inflammatory arthritis "spondyloarthropathy," as have 2/88 contemporary *Alligator mississippiensis*, 1/26 *Crocodylus acutus*, 1/5 *Crocodylus moreletii*, and 1/22 *Paleosuchus palpebrosus*.

Fossil – Analogous to suggestion that crocodilomorphs have not changed greatly during much of their history, neither has the osseous pathologies with which they were afflicted. Gastralia fractures appear to be Cretaceous phenomenon, while gout may be a more recent disease. The consistency of pathology in crocodilian ecomorphs may reflect a similar consistency between disease and host habitat.

Isolated occurrence of jaw fractures in Triassic phytosaurs *Mystriosuchus*, *Mahajangasuchus*, *Phytosaurus buceros*, and *Leptosuchus imperfecta* and puncture wounds in Cretaceous *Brachychampsa sealeyi* and Eocene *Tilemsisuchus lavocati* suggests intraspecies conflict. That behavioral interpretation is confirmed in contemporary *Alligator mississippiensis* and *Crocodylus americanus*. It contrasts with frequent gastralia fractures in Late Paleocene *Leidyosuchus* (= *Borealosuchus*) *formidabilis*, not recognized in contemporary crocodilians.

Gastralia fractures appear to be Cretaceous phenomenon.

Alterations compatible with infection are noted in distal elements of *Machimosaurus mosae* and *Leidyosuchus formidabilis*, similar to infections in contemporary *Alligator mississippiensis* (4/88), *Caiman latirostris*, *Crocodylus siamensis*, and *Crocodylus acutus* (2/26).

Benign neoplasia, in the form of button osteoma and a probable chondroma, is found in an Eocene *Crocodylus*, and an osteochondroma in contemporary *Caiman yacare*.

Isolated cases of articular gout were observed in *Alligator sclerops* and *mississippiensis*; *Tupinambis*; *Crocodylus americanus*, *niloticus*, *novaeguineae*, *porosus*, and *johnstonii*; *Caiman crocodiles*; *Tomistoma schlegelii*, and *Gavialis gangeticus*, but gout has yet to be identified in the fossil record – exclusive of dinosaurs.

Late Paleocene *Leidyosuchus formidabilis* and Lower Miocene *Tomistoma dowsoni* and *Leidyosuchus* (=*Borealosuchus*) *formidabilis* have ossification of anulus fibrosus characteristic of the inflammatory arthritis "spondyloarthropathy," as have 2/88 contemporary *Alligator mississippiensis*, 1/26 *Crocodylus acutus*, 1/5 *Crocodylus moreletii*, and *1/22 Palaeosuchus palpebrosus*.

Rothschild BM. 2011a. Gout. eMedicine Orthopedics. http://ww.emedicine.com/orthoped/topic124.htm.

Metabolic – Delineation of the nature and distribution of gout and its treatment. Gout was rarely noted in crocodilians, lizards, and turtles and was found in an African bullfrog *Pyxicephalus*.

Rothschild BM. 2011b. Lumbar spondylosis (Spondylosis deformans). eMedicine Obstetrics, Gynecology, Psychiatry and Surgery. http://www.emedicine.com/med/topic2901.htm.

Vertebral – Widely reported in the reptilian literature, spondylosis deformans usually is asymptomatic – unless pressing on a nerve. It may be simply a phenomenon of aging.

Rothschild BM, Bruno MA. 2011. Calcium pyrophosphate deposition disease. eMedicine Radiology. http://www.emedicine.com/radioTOPIC125.HTM.

Metabolic – Delineation of the nature and distribution of calcium pyrophosphate deposition disease and its treatment. It was recognized in Egyptian spiny-tailed lizard *Uromastyx*, *Chrysemys picta elegans*, *Gopherus agassizi*, and *Testudo graeca*.

Rothschild BM, Martin LD. 1987. Avascular necrosis: Occurrence in diving Cretaceous mosasaurs. Science 236:75–77.

Vascular – Prevalence of avascular necrosis in 10–66% of vertebrae of *Platecarpus* (average = 25), 5–15% of *Tylosaurus* (average = 9), but absent in *Clidastes*.

Fossil – Prevalence of avascular necrosis in 10–66% of vertebrae of *Platecarpus* (average = 25), 5–15% of *Tylosaurus* (average = 9), but absent in *Clidastes.*

Rothschild BM, Martin LD. 1990. Shark-induced infectious spondylitis: Evidence in the Cretaceous record. Journal of Vertebrate Paleontology 10 (supplement to 3):39A

Trauma – Evidence of shark attacks on mosasaurs abound in the fossil record. It has, however, been unclear whether this represented scavenging or predation. Healing and presence of a tooth in an infected abscess verify active predation by sharks on mosasaurs.

Fossil – Evidence of shark attacks on mosasaurs abound in the fossil record. It has, however, been unclear whether this represented scavenging or predation. Healing and presence of a tooth in an infected abscess verify active predation by sharks on mosasaurs.

Rothschild BM, Martin LD. 2006. Skeletal Impact of Disease. New Mexico Museum of Natural History and Science, Albuquerque, New Mexico, 226 pp.

Trauma – In depth diagnostic criteria, examples and delineation of pathology literature, especially related to reptiles.

Congenital – In depth diagnostic criteria, examples and delineation of pathology literature, especially related to reptiles.

Infection – In depth diagnostic criteria, examples and delineation of pathology literature, especially related to reptiles.

Metabolic – In depth diagnostic criteria, examples and delineation of pathology literature, especially related to reptiles.

Neoplasia – In depth diagnostic criteria, examples and delineation of pathology literature, especially related to reptiles.

Vascular – In depth diagnostic criteria, examples and delineation of pathology literature, especially related to reptiles.

Vertebral – In depth diagnostic criteria, examples and delineation of pathology literature, especially related to reptiles.

Dental – In depth diagnostic criteria, examples and delineation of pathology literature, especially related to reptiles.

Pseudopathology – In depth diagnostic criteria, examples and delineation of pathology literature, especially related to reptiles.

Other – In depth diagnostic criteria, examples and delineation of pathology literature, especially related to reptiles.

Environmental – In depth diagnostic criteria, examples and delineation of pathology literature, especially related to reptiles.

Fossil – In depth diagnostic criteria, examples and delineation of pathology literature, especially related to reptiles.

Rothschild BM, Storrs GW. 2000. Decompression syndrome in plesiosaurs. Journal of Vertebrate Paleontology 20 (supplement to 3):65A

Vascular – Plesiosaurs proved susceptible to bends, based on examination of proximal femoral and humeral subsidence, as vertebrae of plesiosaurs did not reveal evidence of disease.

The implication is that Plesiosauridae, Elasmosauridae, Polycotylidae, and Pliosauridae pursued more deep and repetitive diving than did Thaumatosauridae and Cryptocleididae.

Fossil – The implication is that Plesiosauridae, Elasmosauridae, Polycotylidae, and Pliosauridae pursued more deep and repetitive diving than did Thaumatosauridae and Cryptocleididae.

Family	Humeri	Femora
Plesiosauridae	7/26	2/33
Cryptocleididae	1/25	16
Thaumatosauridae	14	16
Elasmosauridae	5/37	1/45
Pliosauridae	1/17	2/49
Polycotylidae	5/20	2/13

Rothschild BM, Martin LD, Bell G, Lamb J. 1987. Avascular necrosis in Cretaceous reptiles. Journal of Vertebrate Paleontology 7 (supplement to 3):24A.

Vascular – Decompression syndrome was recognized on the basis of avascular necrosis in 100% of *Platecarpus* and *Tylosaurus*, but absent in any *Clidastes*, allowing recognition of their deep, repetitive and shallow diving behavior, respectively.

Fossil – Decompression syndrome was recognized on the basis of avascular necrosis in 100% of *Platecarpus* and *Tylosaurus*, but absent in any *Clidastes*, allowing recognition of their deep, repetitive and shallow diving behavior, respectively.

Rothschild BM, Martin LD, Schulp AS. 2005. Sharks eating mosasaurs, dead or alive. Netherlands Journal of Geosciences/Geologie en Mijnbouw 84:335–340.

Trauma – Shark bite marks on mosasaur bones, which abound in the fossil record, were reviewed to attempt to distinguish scavenging and predation. Evidence of healing and presence of a shark tooth in an infected abscess document that sharks actively hunted living mosasaurs.

Fossil – Shark bite marks on mosasaur bones, which abound in the fossil record, were reviewed to attempt to distinguish scavenging and predation. Evidence of healing and presence of a shark tooth in an infected abscess document that sharks actively hunted living mosasaurs.

Rothschild BM, Schultze H-P, Pellegrini R. In press. Osseous and articular pathologies in turtles and abnormalities of mineral deposition. In: Morphology and Evolution of Turtles, D Brinkman, J Gardner, P Holroyd (eds.). New York: Springer.

Trauma – In depth examples and delineation of pathology literature, related to turtles. See Chapter 4 for details.

Congenital – In depth examples and delineation of pathology literature, related to turtles. See Chapter 4 for details.

Infection – In depth examples and delineation of pathology literature, related to turtles. See Chapter 4 for details.

Metabolic – In depth examples and delineation of pathology literature, related to turtles. See Chapter 4 for details.

Neoplasia – In depth examples and delineation of pathology literature, related to turtles. See Chapter 4 for details.

Vascular – In depth examples and delineation of pathology literature, related to turtles. See Chapter 4 for details.

Vertebral – In depth examples and delineation of pathology literature, related to turtles. See Chapter 4 for details.

Shell disease – In depth examples and delineation of pathology literature, related to turtles. See Chapter 4 for details.

Pseudopathology – In depth examples and delineation of pathology literature, related to turtles. See Chapter 4 for details.

Other – In depth examples and delineation of pathology literature, related to turtles. See Chapter 4 for details.

Environmental – In depth examples and delineation of pathology literature, related to turtles. See Chapter 4 for details.

Fossil – In depth examples and delineation of pathology literature, related to turtles. See Chapter 4 for details.

Row RW. 1916. A frog with symmetrically abnormal hind feet. Proceedings of the Zoological Society London 1916:87–89.

Congenital – *Rana temporaria* common frog with fourth toe (small pre-hallux in *Rana temporaria*) with four phalanges, but hallux missing bilaterally.

Ruben JA. 1990. Evidence of convergent behavioral patterns in male crocodilians and phytosaurs. Evolutionary Paleobiology of Behavior and Coevolution 1990:427–428.

Trauma – Cites 63% of male *Crocodylus niloticus* >4 meters in length had scars from prior injuries (Cott 1961).

Cites abscessed wounds (Abel 1922) in European *Nicrosaurus* (= *Phytosaurus*; *Machaeroprosopus*) with scars and *Nicrosaurus* University of California Museum of Paleontology A270/27200 with deep puncture along midline of distal third of premaxillary, attributed to large phytosaur.

Infection – Cites abscessed wounds (Abel 1922) in European *Nicrosaurus* (= *Phytosaurus*.

Fossil – Cites abscessed wounds (Abel 1922) in European *Nicrosaurus* (= *Phytosaurus*; *Machaeroprosopus*) with scars and *Nicrosaurus* University of California Museum of Paleontology A270/27200 with deep puncture along midline of distal third of premaxillary, attributed to large phytosaur.

Rubin L, BellairsA d'A, Bryant SV. 1967. Congenital malformations in snakes. British Journal of Herpetology 4:12–13.

Congenital – Two American water snake *Natrix sipedon* with only 75 pre-caudal vertebrae (1/2 normal).

Rübsaamen H. 1949. Missbildungen am Zentralnervensystem von Tritonen durch allgemeinen Sauerstoffmangel bei Normaldruck. [Malformations in the central nervous system of tritons by general lack of oxygen and normal pressure]. Roux's Archiv für Entwicklungsmechanik 143:615–641. [German]

Environmental – Oxygen deprivation producing teratogenesis in *Triturus*.

Rugh R. 1954. The effect of ionizing radiations on amphibian development. Journal of Cell and Comparative Physiology 43:39–67.

Environmental – Tail development in axolotl *Ambystoma* inhibited by X-ray exposure.

Ruffer MA. 1917. A pathological specimen dating from the lower Miocene period. (Extrait de 'Contributions à l'étude des Vertébrés miocène de l'Egypte.) Cairo Survey Department 1917:1–7, Le Caire. [French]

Vertebral – Fused Miocene crocodile vertebrae by syndesmophyte. Miscalled spondylosis deformans, it is actually characteristic of spondyloarthropathy.

Fossil – Fused Miocene crocodile vertebrae by syndesmophyte. Miscalled spondylosis deformans, it is actually characteristic of spondyloarthropathy.

Ruffer MA. 1920. A pathological specimen dating from the lower Miocene period. In Moodie RL. ed. Studies in the Palaeopathology of Egypt. pp. 184–193; Chicago: University of Chicago Press.

Vertebral – Osseous lesions of spondylitis deformans of lumbar/caudal vertebrae of Miocene crocodile *Tomistoma dowsoni.*

Fossil – Osseous lesions of spondylitis deformans of lumbar/caudal vertebrae of Miocene crocodile *Tomistoma dowsoni.*

Russell DA. 1967. Systematics and morphology of American mosasaurs. Bulletin of the Peabody Museum of Natural History, Yale University 23:1–240.

Trauma – Fragments of small mosasaur mixed with skeleton of a larger one, citing Anonymous, 1962, p. 5.

Fossil – Fragments of small mosasaur mixed with skeleton of a larger one, citing Anonymous, 1962, p. 5.

Russell AP, Bauer AM. 1988. Paraphalangeal elements of gekkonid lizards: A comparative study. Journal of Morphology 197: 221–240.

Congenital – Paraphalanges in 57 species of 16 genera of gecko lizards.

Ruth FS. 1961. Seven-legged bullfrog, *Rana catesbeiana.* Turtox News 39:232.

Congenital – Seven-legged bullfrog *Rana catesbeiana.*

Ryan MJ. 1985. The Tungara Frog. Chicago: University of Chicago; 230 pp.

Trauma – Predation on *Physalaemus pustulosus* by bat *Trachops cirrhosus*, opossum, and other frogs.

Ryan JJ. 1986. Malformation congenitale de la queue tortue de blanding. [Congenital malformation of the tail of a Blanding turtle]. Bulletin de la Société Herpétologie de France 38:15. [French]

Congenital – Curly tail in *Emydoidea blandingii.*

Ryder I. 1878. A monstrous frog. The American Naturalist 12:751–752.

Congenital – *Rana palustris* with supernumerary hind limb with five toes.

Ryder I. 1891. Abnormal duplication of urosome in *Rana catesbeiana.* The American Naturalist 25:740–753.

Trauma – *Rana catesbeiana* with dorsiventral urosome bifurcation (dorsoventrally bifurcated tail).

Ryer KA. 1983. Urinary calculi in a green iguana. Veterinary Medicine & Small Animal Clinician 78:607.

Stone – Carbonate, phosphorus, ammonium, and urate calculus in a green iguana *Iguana iguana.*

Annotated Bibliography S

Sachsse W. 1977. Normale und pathologische Phänomene bei Zuchtversuchen mit Schildkröten, hier anhand von *Kinosternon baurii* (Reptilia, Testudines, Kinosternidae). [Normal and pathological phenomena in breeding of tortoise, in this case in *Kinosternon baurii* (Reptilia, Testudines, Kinosternidae)]. Salamandra 13:22–35. [German]

Trauma – Nothing about pathology, only nice stories about injuries and behavior.

Sachsse W. 1983. Inheritance and environment as causes for teratogenesis in amphibians and reptiles. In Constantin V, Matz G. eds. Comptes Rendus du Premier Colloque International de Pathologie des Reptiles et des Amphibiens, Angers, France: 197–205.

Congenital – Non-hereditary causation as occurring if malformed offspring derive from normal parents. Reviews lethal "bent tail" mutation in *Xenopus*.

Sack S. 1821 (not 1831). Nachricht von des Verfassers Abreise von Surinam nach Europa über Nordamerika. [News about the departure of the author to Surinam for Europe through North America]. Beschreibung einer Reise nach Surinam und des Aufenthaltes daselbst in den Jahren 1805, 1806, 1807, so wie von des Verfassers Rückkehr nach Europa. [Description of a travel to Surinam and the stay in Surinam during the years 1805, 1806, 1807, as the return of the author to Europe]. Berlin, pp. 136–153 [German].

Congenital – Double-faced snake.

Sadler R. 2008. Phenotypic plasticity in the evolution of Alligatorinae. Journal of Vertebrate Paleontology 28:135A.

Congenital – Farmed *Alligator mississippiensis* have wider, shorter snouts, ridge below orbit, squamosal flange, and less high sphenoid than wild-caught.

Said-Aliev SA. 1973. The occurrence of abnormality in *Rana ridibunda* Pall. from Tadzhikistan. Izvestiya Akademii Nauk Respubliki Tadzhikistan Otdelenie Biologicheskikh Nauk 1:105–106.

Congenital – *Rana ridibunda* Pall (Lake Frog) with five extremities.

Cited dicephalic snake (Alekperov 1954), eight legged Lichtenstein's toadhead agama *Phrynocephalus interscapularis* (Chemov 1944S).

Trauma – Bifid tails in wall lizard *Eremias velox* (Said-Aliev 1957–1958).

Sailer A, Pyczak C, Hartman U-K. 1997. Siamese twins of *Testudo hermanni boettgeri*. Siamesische Zwillinge bei *Testudo hermanni boettgeri*. Herpetofauna (Weinstadt) 19:12–13.

Trauma – Additional connections between hypapophyses and condylus margins along the entire vertebral column, allegedly produced by touching.

Vertebral – Additional connections between hypapophyses and condylus margins along the entire vertebral column, allegedly produced by touching.

Saka M. 2004. Developmental toxicity of p,p′-dichlorodiphenyltrichloroethane, 2,4,6-trinitrotoluene, their metabolites, and benzo(a) pyrene in *Xenopus laevis* embryos. Environmental Toxicology and Chemistry 23 (4):1065–1073.

Environmental – Prolonged exposure in Kitakyushu, Japan to p,p′- dichlorodiphenyldichloroethylene (more toxic and lethal than DDT) of *Xenopus laevis* embryos. In 1995, 11.9% of montane brown frogs, *Rana ornativentris*, had supernumerary forelimbs. In 1996 supernumerary limbs, polymelia, hemimelia, ectromelia, ectodactyly, syndactyly, polydactyly in 5,6% of 1928 montane brown frogs, 3.0% of 1356 Japanese brown frog *Rana japonica japonica*, and 1.38% of 944 western-Japanese common toad *Bufo japonicus japonicus*.

Salvador A, Martin AJ, Lopez P. 1995. Tail loss reduces home range size and access to females in male lizards, *Psammodromus algirus*. Behavioral Ecology 6:382–387.

Trauma – Tail loss reduces home range size and access to females in male lizards, *Psammodromus algirus*, because of reduction to avoid predation.

B.M. Rothschild et al., *Herpetological Osteopathology: Annotated Bibliography of Amphibians and Reptiles*, 287
DOI 10.1007/978-1-4614-0824-6_10,

Samarasinghe E. 1951. Report on the anatomical variations observed in a polymelous specimen of *Bufo melanostictus*. Ceylonese Journal of Science (B) 24:181–186.
Congenital – Supernumerary limb in *Bufo melanostictus*, noting that De Superville (1740) was probably the first to report supernumerary limb.
Cited polymely in *Rana tigrina* (Mahendra 1936).
"Polycrural" (with two separate pelvic girdles) in *Rana cyanophlyctis* (Deraniyagala 1944).

Sánchez García I, Martínez Silvestre A. 1999. Un caso de duplicacíon axial en *Natrix maura* [A case of axial duplication in *Natrix maura*]. Boletín de la Asociación Herpetológica Española 10:37–38 [Spanish].
Congenital – Derodidymus *Natrix maura*.

Sanchiz FB. 1978. Nuevos restos fósiles de la familia Pelodytidae (Amphibia, Anura) [New fossil remains of the Pelodytidae family (Amphibia, Anura)]. Estudios geológicos (Madrid) 34:9–27 [Spanish].
Congenital – Pelodytid frogs from the Neogene of Spain with neural arch of the first postsacral vertebra dorsally well isolated in comparison to the rest. The second postsacral vertebra had acquired the morphology typical of the first, existing between the two a zone in which the spinal cord must have lacked the normal dorsal osseous protection. A second case had a suture line in the neural arch of VS+1 in comparison to the rest.
Fossil – Pelodytid frogs from the Neogene of Spain with neural arch of the first postsacral vertebra dorsally well isolated in comparison to the rest. The second postsacral vertebra had acquired the morphology typical of t the first, existing between the two a zone in which the spinal cord must have lacked the normal dorsal osseous protection. A second case had a suture line in the neural arch of VS+1 in comparison to the rest.

Sanchiz FB, Pérez PJ. 1974. Frecuencia de anomalías óseas en la población de *Discoglossus pictus* (Anura, Discoglossidae) de Campos (Asturia) [Frequency of osseous anomalies in the population of *Discoglossus pictus* (Anura, Discoglossidae) of Campos (Asturia)]. Boletín de la Estación Central de Ecología (Madrid) 3(6):69–77. [Spanish]
Congenital – Bone anomalies in a *Discoglossus pictus* fauna obtained from *Tyto alba* (owl) pellets in Asturia, Spain, in the collections of Instituto José de Acosta (normal urostyles and ilia only) and Instituto Lucas Mallada (Section of Vertebrate and Human Paleontology): Partial fusion of urostyle to sacrum.
Supernumerary foramen in the first postsacral vertebra, expanded transverse processes in the first postsacral vertebra, and in one case, one of the transverse processes of the first postsacral vertebra expanded enormously and was situated perpendicular to the axis of the urostyle, which made it look remarkably like the sacrum. The sacrum articulation was very deformed, as also noted by Madej (1965). There was a sacral diapophysis in one side only, with the presacral vertebra's transverse process taking its shape on the other side.
Synostosis were observed; phalangeal duplication was reported and figured from a previously unpublished specimen of *Pleurodeles waltl* of the National Natural History Museum in Paris, France.
Trauma – Ilia – Healed fractures, anomalous contours or curvature of the bones, and in one case, a dorsal, hook-shaped exostosis.
Urostyle – Healed fractures, transverse processes in the second postsacral vertebra (the normal condition in *D. pictus* is to only have transverse processes on the first), exostosis in the sacral articulation, anomalous curvatures, partial fusion to the sacrum.
Femur – Medial fracture.
Tibiofibula – Distal and medial healed fractures.
Tibiale and fibulare – Incomplete fusion in between the two bones at half their length in two cases.
Metatarsal – Distal fracture.
Phalanges – Three oblique fractures.
Vertebral – Fused vertebrae centra.

Sassenburg L. 2000. Ratgeber Schildkröten Krankheiten. [Guide to Turtle Diseases]. Ruhmannsfelden: Beder-Verlag. 60 pp. [German].
Metabolic – Rickets in Greek land turtle.
Gout in rosy peep turtle (Rotwangen-Schmucksschildkröte).
Stones – Stones in soft-shell turtle.

Sassenburg L. 2005. Handbuch Schildkröten Krankheiten. [Handbook of Turtle Diseases]. Ruhmannsfelden: Beder-Verlag. 75 pp. [German].
Trauma – Foot amputation in Greek land turtle.
Metabolic – Rickets in Greek land turtle.
Osteodystrophy in Moorish land turtle.
Gout in rosy peep turtle (Rotwangen-Schmucksschildkröte).
Stones – Soft-shell turtle and bladder stone in Spornschildkröte [translates as horn bag spur (bubble stone) turtle] with calcium stones.
Shell disease – Scute repair and shell necrosis in Greek land turtle.

Missing carapace portion in Greek land turtle.

Sassernò A. 1888–89. Ricerche intorno alla struttura della colonna vertebrale del genere *Bombinator* [Research about the structure of the vertebral column of the genus *Bombinator*]. Atti della Reale Accademia delle Scienze di Torino 24: 451–466, Tav. XII [Italian]

Congenital – Dr. Sassernò reported nine cases of abnormal vertebral column (sacral region, some cases represented by multiple individuals) in *Bombinator* from the collections of the Zoology museum of Turin, Italy. He also reported one instance of an individual with transverse apophyses in the atlas. The first case had a normal presacral vertebra, and a sacral vertebra with short apophyses (at the end of the range), but very long (abnormal) apophyses in the coccyx. The second case had a normal presacral vertebra, a sacral vertebra with long apophyses, and a coccyx with rudimentary apophyses. The third case had a normal presacral vertebra, but the apophyses of both the sacral vertebra and coccyx were asymmetrical. Within this case, there were three subcategories: (a) the left sacral apophysis was longer than the right, while the left apophysis of the coccyx was shorter than the right. (b) The reverse of (a). (c) The left apophysis of both the sacral vertebra and the coccyx were longer than the corresponding right. The fourth case had an abnormal presacral vertebra: while one of its apophyses was normal, the other more or less expanded tends to join with the corresponding apophysis of the sacral vertebra, forming part of the sacrum portion that supports the ilium. This case had two subdivisions: (a) the left presacral apophysis was more developed than the right, and (b) the left presacral apophysis was normal, while the right was expanded. The fifth case had a regular presacral vertebra apophysis, and the opposite was more developed like in the previous case, but in addition, joined to the homologous sacral apophysis via a cartilaginous edge. This case had two subdivisions based on where the condition occurred, a right and a left. The sixth case had irregularities in both apophyses of the presacral vertebra, and both the developing apophyses strongly competed with the sacral vertebra to form the sacrum. The seventh case had a normal presacral vertebra, but the apophyses of the sacral vertebra were unequally long. The shortest was joined by a cartilaginous fringe to the corresponding apophysis of the coccyx, which was abnormally dilated and competed with the sacral vertebra to form the sacrum. Both formed a single long apophysis opposite the normally developed apophysis of the sacral vertebra. The other apophysis of the coccyx (normal and little developed) corresponded to the latter. The eighth case had a normal presacral vertebra, but the apophyses of both the coccyx and the sacral vertebra were asymmetric, and irregularly developed so that in a side, the sacral apophyses and those of the coccyx present a regular conformation, while in the other, the apophysis of the coccyx tends to substitute the sacral one, developing very little the latter and very much the former. The ninth case had a normal presacral vertebra and could be referred to the previous case as an extreme example of it, but Sassernò decided to make special mention of it because he found it very interesting. The right apophysis of the sacral vertebra is normally formed, but the left is reduced to the point of resembling the normal left apophysis of vertebrae six, seven, and eight. The left apophysis of the coccyx is extraordinarily developed to functionally replace this reduced apophysis, to the point of resembling perfectly the left apophysis of the sacral vertebra. Meanwhile, the right apophysis of the coccyx is normal and thus reduced. *Bombinator* presents great variability in shape in the last three vertebrae of the column, that this variability prevents usage of this section of the skeleton to differentiate between species of the genus, but that it does not prevent differentiation from other discoglossids, or from bufonids, pelobatids, and hylids.

Two males with normal presacral and sacral vertebrae, but instead of one pair of transverse apophyses, he saw two pairs in the coccyx. One was in the normal place, and the second pair originated almost immediately above the first, possibly related to fusion of a postsacral vertebra to the true coccyx, forming a divertebral coccyx.

Sato T. 1953. [Anatomical notes on a two-headed snake]. Saishu to shiiku [Collecting and breeding] 15: 118–120 [Japanese].

Congenital – Derodymus *Elaphe climacophora.*

Sato T. 2003. *Terminonatator ponteixensis*, a new elasmosaur (Reptilia, Sauropterygia) from the Upper Cretaceous of Saskatchewan. Journal of Vertebrate Paleontology 23:89–103.

Trauma – Could not locate purported healed limb fracture of elasmosaur *Terminonatator ponteixensis.*

Other – Holes, present between metacarpal, were formed by the development of small nodes on the lateral side(s) of the proximal portions. The astragalus has a central penetrating pit with four radiating shallow grooves. Distal tarsal IV has a grove from center to postaxial edge. Diagnosis is unclear as photographs not provided.

Fossil – Could not locate purported healed limb fracture of elasmosaur *Terminonatator ponteixensis.*

Holes, present between metacarpal, were formed by the development of small nodes on the lateral side(s) of the proximal portions. The astragalus has a central penetrating pit with four radiating shallow grooves. Distal tarsal IV has a grove from center to postaxial edge. Diagnosis is unclear as photographs not provided.

Saxena S, Niazi IA. 1977. Effect of vitamin A excess on hind limb regeneration in tadpoles of the toad, *Bufo andersonii* (Boulenger). Indian Journal of Experimental Biology 15:435–439.

Toxicology – Oligodactyly in *Bufo andersonii*.
Saumure RA. 2001. Kyphosis in a musk turtle (*Sternotherus odoratus*) from Ontario, Canada. Chelonian Conservation and Biology 4:159.
Congenital – Kyphosis in musk turtle *Sternotherus odoratus*.
Savage JM. 2002. The Amphibians and Reptiles of Costa Rica: A Herpetofauna Between Two Seas. Chicago: University of Chicago Press, 934 pp.
Trauma – Pseudoautotomy (where tail is twisted off) occurs in plethodontids: *Thamnophis, Desmognathus, Leurognathus, Phaeognathus, Batrachoseps, Plethodon, Chiropterotriton, Lineatriton, Oedipina, Thorius,* many *Bolitoglossa* and *Hemidactylium* and some *Pseudoeurycea*.
Säve-Söderbergh, G. 1935. On the dermal bones of the head in labyrinthodont stegocephalians and primitive Reptilia with special reference to Eotriassic stegocephalians from East Greenland. Meddelelser om Grønland, 98(3), 1–211.
Congenital – Variation in labyrinthodontia where small tabular does not meet parietal versus tabular and parietal have common sutures.
Sawyer GT, Erickson BR. 1985. Injury and diseases in fossil animals. The Intriguing world of paleopathology. Encounters May/June 25–28.
Trauma – Tibia fracture in Paleocene crocodile *Leidyosuchus*.
Alligator with hindfoot traumatic amputation.
Ten-twenty-inch hatchling and Indo-Australian crocodile *Crocodylus porosus* longer than 5 ft had the greatest number of head, tail, and limb injuries.
Apparent stress fracture (diagnosed by BMR) in Paleocene crocodile *Leidyosuchus* metapodial.
Infection – Probable infection causing articular surface destruction with reactive new bone formation at Paleocene crocodile *Leidyosuchus* metapodial phalangeal joint.
Arthritis – Articular surface destruction with reactive new bone formation at Paleocene crocodile *Leidyosuchus* metapodial phalangeal joint.
Shell – Pond turtle with healed carapace puncture.
States that puncture wounds are frequently noted in both living and fossil turtle carapaces.
Fossil – Tibia fracture in Paleocene crocodile *Leidyosuchus*.
Alligator with hindfoot traumatic amputation.
Ten-twenty-inch hatchling and Indo-Australian crocodile *Crocodylus porosus* longer than 5 ft had the greatest number of head, tail, and limb injuries.
Apparent stress fracture (diagnosed by BMR) in Paleocene crocodile *Leidyosuchus* metapodial.
Probable infection causing articular surface destruction with reactive new bone formation at Paleocene crocodile *Leidyosuchus* metapodial phalangeal joint.
Articular surface destruction with reactive new bone formation at Paleocene crocodile *Leidyosuchus* metapodial phalangeal joint.
States that puncture wounds are frequently noted in both living and fossil turtle carapaces.
Sawyer GT, Erickson BR. 1987. Injury and diseases in fossil animals. Field Museum of Natural History Bulletin 58:20–25.
Trauma – Tibial fracture in Paleocene crocodile *Leidyosuchus*.
Leidyosuchus with irregular metatarsal phalangeal joint surface and adjacent bone (apparently infected).
Fossil – Tibial fracture in Paleocene crocodile *Leidyosuchus*.
Leidyosuchus with irregular metatarsal phalangeal joint surface and adjacent bone (apparently infected).
Sawyer GT, Erickson BR. 1998. Paleopathology of the Paleocene crocodile *Leidyosuchus* (=*Borealosuchus*) *formidabilis*. Science Museum of Minnesota Monograph Volume 4:1–38.
Pathology – Among 80 crocodile specimens of *Leidyosuchus* (*Borealosuchus*) *formidabilis* from late Paleocene in collections of the Science Museum of Minnesota.
Congenital – Anomaly of a chevron P79.19.997 in the form of elongation with curvature on one side.
Trauma- Fractures of 67 elements among 80 (minimum number), at least some related to bites and rarely showing signs of infection (e.g., filigree surface reaction). One, expanded proximally, was called a perforating fracture, but etiology would benefit from further study. Most affected were metapodials and phalanges (13%), gastralia (22/22 examined), and osteoscutes (24/4126 examined).
Perforating fractures – only pathology observed in skulls.
Transverse ribbing of unclear etiology – rule out possible stress fractures?
Dorsal rib P89.6.79 and pubis P.73.25.1 (neither illustrated).
Exostoses – isolated – versus enthesitis. One on dorsal vertebra endplate. P76.33.289 is a phenomenon which has yet to be explained in any species. They called some button osteomas, but images do not allow exostoses to be excluded.
Infection – Several digital elements (MCV P95.13.19, MT and phalange P76.28.3, ungal P79.19.1005, phalange P79.19.1008 were illustrated) called spondyloarthropathy, but those illustrated have the joint surface destruction more characteristic of infection.

Osteomyelitis in an atlas, dorsal vertebra, cervical rib, radius, tibia, ilium, and femur, at least one – bite related.

Arthritis – Several digital elements (MCV P95.13.19, MT and phalange P76.28.3, ungal P79.19.1005, phalange P79.19.1008 were illustrated) called spondyloarthropathy, but those illustrated have the joint surface destruction more characteristic of infection.

Vertebral – Spondyloarthropathy, as manifested by vertebral fusion (P70.20.408).

Cited two fused pathological dorsal vertebrae in *Crocodylus* sp. (Gilmore 1946)

Spondylosis deformans was not found.

Other – Transverse ribbing of unclear etiology in dorsal rib P89.6.79 and pubis P.73.25.1 (neither illustrated) – rule out possible stress fractures?

Expanded shafts of radius P74.24.12 and metapodial P71.16.248.

"Periosteopathy" of vertebra P70.20.1775 is circumferential mid-centrum new bone formation of unclear etiology, also found on 49/422 dorsal vertebrae, 1/56 sacrum, 24/268 caudal vertebrae, ½ dorsal ribs, 1/69 humeri, 2/53 radii, 3/44 ulna, 1/33 pubi, 11/284 metapodials (not counting those diagnosed as spondyloarthropathy), 13/414 phalanges (not counting those diagnosed as spondyloarthropathy), and 28 osteoscutes.

They also reported "periosteopathy" in *Alligator mississippiensis* SMM Z69.26, localized to all dorsal vertebra, 14/37 caudal, five metapodials, and four phalanges. While infectious suspected, this phenomenon appears to be unique to crocodilians, or at least to reptiles, and its etiology has not been resolved.

Exostoses – isolated – versus enthesitis. One on dorsal vertebra endplate. P76.33.289 is a phenomenon which has yet to be explained in any species. They called some button osteomas, but images do not allow exostoses to be excluded.

Osteoscutes – 28 with pits (they called erosional cavities) or canals.

Fossil – Among 80 crocodile specimens of *Leidyosuchus* (*Borealosuchus*) *formidabilis* from late Paleocene in collections of the Science Museum of Minnesota.

Anomaly of a chevron P79.19.997 in the form of elongation with curvature on one side.

Fractures of 67 elements among 80 (minimum number), at least some related to bites and rarely showing signs of infection (e.g., filigree surface reaction). One, expanded proximally, was called a perforating fracture, but etiology would benefit from further study. Most affected were metapodials and phalanges (13%), gastralia (22/22 examined), and osteoscutes (24/4126 examined).

Perforating fractures – only pathology observed in skulls.

Transverse ribbing of unclear etiology – rule out possible stress fractures?

Dorsal rib P89.6.79 and pubis P.73.25.1 (neither illustrated).

Exostoses – isolated – versus enthesitis. One on dorsal vertebra endplate. P76.33.289 is a phenomenon which has yet to be explained in any species. They called some button osteomas, but images do not allow exostoses to be excluded.

Several digital elements (MCV P95.13.19, MT and phalange P76.28.3, ungal P79.19.1005, phalange P79.19.1008 were illustrated) called spondyloarthropathy, but those illustrated have the joint surface destruction more characteristic of infection.

Osteomyelitis in an atlas, dorsal vertebra, cervical rib, radius, tibia, ilium, and femur, at least one – bite related.

Several digital elements (MCV P95.13.19, MT and phalange P76.28.3, ungal P79.19.1005, phalange P79.19.1008 were illustrated) called spondyloarthropathy, but those illustrated have the joint surface destruction more characteristic of infection.

Spondyloarthropathy, as manifested by vertebral fusion (P70.20.408).

Cited two fused pathological dorsal vertebrae in *Crocodylus* sp. (Gilmore 1946).

Spondylosis deformans was not found.

Transverse ribbing of unclear etiology in dorsal rib P89.6.79 and pubis P.73.25.1 (neither illustrated) – rule out possible stress fractures?

Expanded shafts of radius P74.24.12 and metapodial P71.16.248.

"Periosteopathy" of vertebra P70.20.1775 is circumferential mid-centrum new bone formation of unclear etiology, also found on 49/422 dorsal vertebrae, 1/56 sacrum, 24/268 caudal vertebrae, ½ dorsal ribs, 1/69 humeri, 2/53 radii, 3/44 ulna, 1/33 pubi, 11/284 metapodials (not counting those diagnosed as spondyloarthropathy), 13/414 phalanges (not counting those diagnosed as spondyloarthropathy), and 28 osteoscutes.

They also reported "periosteopathy" in *Alligator mississippiensis* SMM Z69.26, localized to all dorsal vertebra, 14/37 caudal, five metapodials, and four phalanges. While infectious suspected, this phenomenon appears to be unique to crocodilians, or at least to reptiles and its etiology has not been resolved.

Exostoses – isolated – versus enthesitis. One on dorsal vertebra endplate. P76.33.289 is a phenomenon which has yet to be explained in any species. They called some button osteomas, but images do not allow exostoses to be excluded.

Osteoscutes – 28 with pits (they called erosional cavities) or canals.

Scadding SR. 1987. Vitamin A inhibits amphibian tail regeneration. Canadian Journal of Zoology 65:457–459.
Trauma – Retinol can cause proximodistal duplications in regenerating limbs, but inhibits regeneration of amputated *Notophthalmus viridescens* and *Ambystoma mexicanum*, but not *Xenopus laevis* tails.
Scadding SR. 1990. Effects of tributyltin oxide on the skeletal structures of developing and regenerating limbs of the axolotl larvae, *Ambystoma mexicanum*. Bulletin of Environmental Contamination and Toxicology 45:574–581.
Toxicology – Organotin/antifouling compound tributyltin produces skeletal deletions in axolotl *Ambystoma mexicanum.*
If normal hind limb has nine tarsals, five metatarsals, and 13 phalanges, (2.2.3.4.2 pattern), with variants including loss of a phalange from digit four or absence of a tarsal from fusion of distal tarsal one with tibiale, deletion of two tarsals, reduction in digit five size, and shortening of long bones were induced by tributyltin. If normal forelimb has eight carpals, four metacarpals, and nine phalanges (2.2.3.2 pattern), distal carpal one and radiale are often conjoined and intermedium and centrale, occasionally joined, no alterations were induced by tributyltin.
Scadding SR, Maden M. 1986a. Comparison of the effects of vitamin A on limb development and regeneration in the axolotl, *Ambystoma mexicanum*. Journal of Embryology and Experimental Morphology 91:19–34.
Toxicology – Severity of vitamin A-induced deletins in *Ambystoma mexicanum* correlated with dose, time, and earlier stages of exposure. Extra carpals/tarsals, partial radius-ulna, tibia-fibula, humerus, femur, similar to that noted in *Bufo melanosticus.*
Scadding SR, Maden M. 1986b. Comparison of the effects of vitamin A on limb development and regeneration in *Xenopus laevis*. Journal of Embryology and Experimental Morphology 91:35–53.
Toxicology – Fusion into a single tibia-fibula in *Xenopus laevis*, produced by vitamin A. Failure of limb development, hypomorphism (fewer digits, phocomelia, defective femur, humerus, radius-ulna, tibia-fibula, tibiale, fibulare, duplications with extra carpals, tarsals, serial metatarsals, partial radius-ulna, tibia-fibula, tibiale, fibulare, humerus, femur, digit, or foot).
Scarbo G. 1961. Zoo's Who. Two-headed skink of Australia. Great Bend Daily Tribune 23 July:Young Folks 1.
Pseudopathology – Stump-tailed skink mimics dicephaly.
Schaefer N. 1973. A case of polymely in *Microsaura ventralis*. Journal of the Herpetological Association of Africa 11:1–6.
Congenital – Polymely in *Microsaura ventralis*.
Schall JJ, Pianka ER. 1980. Evolution of escape behavior diversity. American Naturalist 115:551–566.
Trauma – Diversity of escape behaviors reduces tail loss in "bisexual" *Cnemidophorus inornatus, tigris, gularis* (with no gender differences in tail loss) and parthenogenetic *exsanguis* and *tesselatus*. Growth is reduced during tail regeneration. Increased tail breakage at lower latitudes, correlated with number of predators. Southwestern Texas, *Cnemidophorus tigris*, had fewer broken tails than western sites, perhaps reflecting diverse escape behaviors.
Schall JJ, Bromwich CR, Werner YL, Midledge J. 1989. Clubbed regenerated tails in *Agama agama* and their possible use in social interactions. Journal of Herpetology 23:303–305.
Trauma – Clubbed regenerated tails in *Agama agama* as regenerative benefit, versus *A. stellio*, which does not engage in tail whipping.
Schaube MK. 1972. Seasonal variation of newt forelimb regeneration under controlled environmental conditions. Journal of Experimental Zoology 181:281–286.
Trauma – Newt *Notophthalmus viridescens* tail regeneration ceases below 14°C and decreased above 27.8°C (Ellis 1909).
Schauble MK, Nentwig MR. 1974. Temperature and prolactin as control factors in newt forelimb regeneration. Journal of Experimental Zoology 187:335–344.
Trauma – *Notophthalmus viridescens* limb regeneration rate increased with temperature from 10°C to 25°C and prolactin exposure, but no pathology reported.
Environmental – Regeneration in *Notophthalmus viridescens* increased from 10°C to 25°C.
Schauinsland H. 1900. Weitere Beiträge zur Entwicklungsgeschichte der *Hatteria*. Skelettsystem, schallleitender Apparat, Hirnnerven etc. [Additional contributions to the development of *Hatteria*. Skeltal system, sound guiding apparatus, cranial nerves, etc.]. Archiv für mikroskopische Anatomie 56: 747–80. [German]
Congenital – Mainly ontogeny and osteology of *Sphenodon*, described abnormalities in carpus (additional centralia, and fusion of radiale + intermedium in one case, or fusion of intermedium +centralia 1+2, ulnar + carpalia 4+5, and carpalia 2+1 in another case) tarsus (additional centralia, in one case separation of tibiale and intermedium and centrale) of *Sphenodon*.
Schenckius. JG. 1609. Monstrorum historia memorabilis, monstrosa humanorum partuum miracula, stupendis conformationum formulis ab utero materno enata, viuis exemplis, observationibus, & picturis, referens. Accessit analogicum argumentum de monstris brutis, supplementi loco ad observationes medicas Schenckianas edita. [Memorable history of monsters, monstrous miraculous human births, conformed to the formulas from the womb on the mother's stupendous partuition, live examples, observations, and pictures. Analogous argument included animal monsters and medical observations]. 135pp.; Francofurti: Matthiae Beckeri. [Latin]

Congenital – *Bufo* with tail, supernumerary arm in viper, arms in serpent, *Testudo* with polymelia, *Lacerta* with hand deformities, crocodilian with wrist deformity.

Scheremetjewa (or Chevemetieva) EA, Brunst VV. 1938. Сохранение способности к регенерации в средней части лимба тритон и одновременная потеря в проксимальных и дистальных отделов конечностей же. [Preservation of the capacity for regeneration in the middle part of the limb of the newt and its simultaneous loss in the proximal and distal parts of the same limb.] Biulleten' eksperimental'noi biologii I meditsiny. 6:723–724. [Russian]

Trauma – Experimental vascular destruction producing amelia.

Scheyer TM. 2006. Amniote osteoderm and turtle shell bone histology indicate archosauriform origin of turtles. Journal of Vertebrate Paleontology 26:121A.

Congenital – Comparison of *Proganochelys quenstedti* shell bone microstructure with archosaur osteoderms (pareiasaurs, placodonts, and lepidosaurs). Dipole structure with internal structural bone. Parallel-ribbed internal cortex with metaplastic ossification of dermal tissues, different from that in sauropterygian reptiles (e.g., pareiasaurs) which lack a dipole structure, but instead have radial growth.

Fossil – Comparison of *Proganochelys quenstedti* shell bone microstructure with archosaur osteoderms (pareiasaurs, placodonts, and lepidosaurs). Dipole structure with internal structural bone. Parallel-ribbed internal cortex with metaplastic ossification of dermal tissues, different from that in sauropterygian reptiles (e.g., pareiasaurs) which lack a dipole structure, but instead have radial growth.

Schiemenz H, Biella H-J, Günther R, Völkl W. 1996. Kreuzotter – *Vipera berus* (Linnaeus 1758). [Crossed viper – *Vipera berus* (Linnaeus, 1758)]. In Günther R. ed. Die Amphibien und Reptilien Deutschlands. Pp. 710–728; Jena: Gustav Fischer. [German]

Congenital – Dicephalic *Vipera berus*.

Schildger BJ, Frank H, Göbel T, Weiss R. 1991. Mycotic infections of the integument and inner organs in reptiles. Herpetopathologica 2:81–87.

Shell – Bacterial infection of Greek tortoise *Testudo hermanni* carapace with deep necrotic ulcer, deep ulceration of shingle back *Trachydosaurus rugosus* by bacteria and "alteration" of red-eared turtle *Pseudemys scripta elegans* carapace by *Trichosporon*.

Schiötz A. 1891. Kleinere Mittheilungen. [Small reports]. Beilage zu Blätter für Aquarien- und Terrarien-Freunde 2 (18):181. [German]

Congenital – Fused heads in egg of *Coronella laevis*.

Schiøtz A, Volsøe H. 1959. The gliding flight of *Holaspis guentheri* Gray, a West African lacertid. Copeia 1959: 259–260.

Trauma – Gliding flight in West African lacertid *Holaspis guentheri*.

Schipp R, Hemmer H, Flindt R. 1968. Vergleichende licht- und elektronenmikroskopische Untersuchungen an Chordomen von Krötenbastardlarven. [Comparative investigations of chordoma of bastard larvae of toads by light and electron microscopy]. Beiträge zur Pathologischen Anatomie und zur Allgemeinen Pathologie 138:109–133. [German]

Congenital – Hybrid European tadpoles with notochord hypertrophy or hyperplasia.

Schlumberger H. 1958. Krankheiten der Fische, Amphibien und Reptilien. [Illness of Fish, Amphibians and Reptiles]. In Cohrs P, Jaffé R, Meessen H. eds. Pathologie der Laboratoriumstiere. [Pathology of animals in the laboratory]. Vol. 2:714–761; Berlin, Göttingen, Heidelberg: Springer Verlag. [German]

Congenital – Anterior duplicity (anadidymy) in *Rana sylvatica* and *Ambystoma punctatum* Lynn Schwind (1942), *Eleutherodactylus alticola*, snakes, and turtles.

Posterior duplicity (catadidymy) more common in amphibians than anadidymy, the converse of snakes and lizards.

Rana catesbeiana with six full developed hind limbs

Duplication of extremities and polydactyly (Rostand).

Chondrodystrophy in *Ambystoma* (shortening of head and anterior legs).

Kyphosis of carapace of turtles.

Trauma – Healed fractures in reptiles are common: healed ribs in one lizard, *Rana pipiens*, and 10 specimens of Burmese, Indian, or Ceylon *Python molurus* (National Museum of Natural History, Washington, DC), fractures of carapace and bones in the extremities of turtles.

Regeneration of lost parts less common in reptiles (more in lizards than snakes or turtles).

Metabolic – *Bufo bufo* with osteoporosis caused by wrong feeding.

Changes of bone structure caused by food shortage.

Schlumberger HG, Lucke B. 1948. Tumors of fishes, amphibians, and reptiles. Cancer Research 8:657–753.

Trauma – *Rana virescens* spindle-shaped femur swelling with trabeculae of bone and cartilage surrounded by large spaces was originally called a medullary osteosarcoma, but is more characteristic of a callus after a fracture.

Metabolic – Indian monitor *Varanus dracaena* with rickets.

Neoplasia – Indian monitor *Varanus dracaena* enchondromas of fifth and seventh cervical vertebrae, humeri, metacarpals, and hyoid.

Rana virescens spindle-shaped femur swelling with trabeculae of bone and cartilage surrounded by large spaces was originally called a medullary osteosarcoma, but is more characteristic of a callus after a fracture.

Schlüpmann M, Günther R. 1996. Grasfrosch – *Rana temporaria*. [Glass frog – *Rana temporaria*]. In Günther R. ed. Die Amphibien und Reptilien Deutschlands. [The amphibians and reptiles of Germany]. Pp. 412–454; Jena: Gustav Fischer. [German]
Congenital – Five-legged specimens in *Rana temporaria*.

Schlüpmann M, Günther R, Geiger A. 1996. Fadenmolch – *Triturus helveticus* (Razoumowsky, 1789). [Palmate newt – *Triturus helveticus* (Razoumowsky, 1789).]. In Günther R. ed. Die Amphibien und Reptilien Deutschlands. Pp. 143–174; Jena: Gustav Fischer. [German]
Congenital – Deformed tails, missing and additional extremities and phalanges in *Triturus helveticus*.

Schmalhausen J. 1916. French summary of Schlupmann et al. (1966) in Revue zoologique russe 1 (livr. 4, 5).

Schmalhausen J. 1925 Über die Beeinflussung der Morphogenese der Extremitäten vom Axolotl durch verschiedene Faktoren. [On influencing the morphogenesis of the limbs of Axolotl by various factors]. Roux's Archiv für Entwicklungsmechanik 105:483–500. [German]
Trauma – Fusion of regenerating *Siredon pisciformis* tarsal or carpal bones blamed on malnutrition or abnormally high temperature.

Schmelcher D, Hellmich W. 1951. Über eine merkwürdige Larve eines Feuersalamanders. [On a curious larva of a fire salamander]. Die Aquarien- und Terrarien-Zeitschrift 4:300–302. [German]
Congenital – One larva of 24 from the same female of *Salamandra salamandra quadric-virgata* with large and broad head (nearly as broad as long), short abdomen and long tail.

Schmidt KC. 1940. A new turtle of the genus *Podocnemis* from the Cretaceous of Alabama. Geological Series of Field Museum of Natural History 8:1–12.
Trauma – Cretaceous pleurodire turtle *Bothremys barberi* Columbus State University Cretaceous Research Collections, Columbus, Georgia CSUK 90–17–3 with carapace penetration by two holes (one round and one irregular) and a shallow pit in the middle of the left seventh costal, which he suggested could have been caused by an external parasite or disease. He noted similar pits in a Field Museum *Pelusios sinuatus*.
Parallel scrape marks on carapace were illustrated by Schimmer et al. (1997).
Fossil – Cretaceous pleurodire turtle *Bothremys barberi* Columbus State University Cretaceous Research Collections, Columbus, Georgia CSUK 90–17–3 with carapace penetration by two holes (one round and one irregular) and a shallow pit in the middle of the left seventh costal, which he suggested could have been caused by an external parasite or disease. He noted similar pits in a Field Museum *Pelusios sinuatus*.
Parallel scrape marks on carapace were illustrated by Schimmer et al. (1997).

Schmidt AT. 1958a. Forelimb regeneration of thyroidectomized adult newts. I. Morphology. Journal of Experimental Zoology 137:197–226.
Metabolic – Thyroidectomized *Triturus viridescens* have fewer hind limb digits.

Schmidt AT. 1958b. Forelimb regeneration of thyroidectomized adult newts. II. Histology. Journal of Experimental Zoology 139:95–135.
Trauma – Thyroidectomized *Triturus viridescens* have stunted limb regeneration.
Metabolic – Thyroidectomized *Triturus viridescens* have stunted limb regeneration.

Schmidt AJ. 1968. Cellular Biology of Vertebrate Regeneration and Repair. Chicago: University of Chicago Press, 420 pp.
Congenital – Trauma produces supernumerary growths in larval urodeles.
Ultraviolet light applied to *Ambystoma* elbow produces accessory limb (Blum et al. 1958; Butler and Blum 1955).
Chemical induction of accessory limb (Balinsky 1956, 1957).
Trauma – *D.* (*Triturus*) *viridescens* polydactyly from amputated forearm.
Trauma produces supernumerary growths in larval urodeles.
Placement of a ligature around axolotl limbs produces extra limbs (Nassonov 1930).
Environmental – Ultraviolet light applied to *Ambystoma* elbow produces accessory limb (Blum et al. 1958; Butler and Blum 1955).
Chemical induction of accessory limb (Balinsky 1956, 1957).
Induction of accessory tail in *Lygosoma laterale* (Simpson 1964).

Schmidt M. 1985. Missbildungen bei Dendrobatiden als Folge von Inzucht und falscher Ernährung. [Deformations among Dendrobatids as a result of inbreeding and deficient feeding]. Das Aquarium 19 (187):33–35. [German]
Congenital – Absence of anterior legs or thin anterior legs in strawberry frog *Dendrobates pumilio* caused by raising the frogs in a terrarium over many generations, attributed to inbreeding or "wrong food."

Schmidt CW. 1997. Amphibian deformities continue to puzzle researchers. Environmental Science and Technology A 31:324–326.
Congenital – Up to 1995, only two Northern leopard frog, mink frog, and green frogs with supernumerary or ectomely; in 1996, over 200. 5–23% in 1996 in Vermont leopard frogs. Allegedly related to trematode *Manodistomum syntomentera*.

Schmidt RE, Hubbard GB. 1987. Atlas of Zoo Animal Pathology. II. Avian, Reptile and Miscellaneous Species. 192 pp.; Boca Raton, Florida: CRC Press.
Other – Bone necrosis of unclear etiology in African bullfrog *Pyxicephalus*.
Schmidt KP, Inger RF. 1957a. Living Reptiles of the world. 287 pp.; London: Hamisch Hamilton.
Congenital – Dicephalic California king snake *Lampropeltis getula californiae*.
Derodymus diamond – back terrapin *Malaclemys terrapin*.
Schmidt KP, Inger RF. 1957b. Knaurs Tierreich in Farben–Reptilien. [Knaurs' Animal Kingdom in Color – Reptiles]. München, Zürich: Droemersche Verlagsanstalt, 312 pp. [German]
Congenital – Dicephalic California king snake *Lampropeltis getula californiae* pictured on p. 217.
Schmidt RE, Hubbard GE, Fletcher KC. 1986. Systemic survey of lesions from animals in a zoologic collection. 7. Musculo-skeletal system. Journal of Zoo Animal Medicine 17:37–41.
Trauma – Fracture in one of eight reptiles autopsied at the San Antonio Zoological Gardens and Aquarium.
Other – Bone necrosis in one of two amphibians autopsied at the San Antonio Zoological Gardens and Aquarium.
Schneider B. 1980. Zusätzliche gelenkige Verbindungen bei Wirbeln von *Malpolon m. monspessulanus* (Reptilia: Serpentes: Colubridae). [Additional articular connections of vertebrae of *Malpolon m. monspessulanus* (Reptilia: Serpentes: Colubridae)]. Salamandra 16: 268–269. [German]
Congenital – An abnormal extra articulation between successive trunk vertebrae in a colubrid snake *Malpolon m. monspessulanus*.
Schoener TW. 1979. Inferring the properties of predation and other injury-producing agents from injury frequencies. Ecology, Brooklyn 60: 1110–1115.
Trauma – Injury frequency correlated with survival, but predation intensity correlated negatively with survival rates. Both were independent of tail-break frequencies.
Schoener TW, Schoener A. 1980. Ecological and demographic correlates of injury rates in some Bahamian *Anolis* lizards. Copeia 1980:839–850.
Trauma – Tail break was proportional to survival rates in *Anolis*. Male *Anolis sagrei* were more frequently affected than females. Front toe breaks were proportional to conspecific density in *Anolis sagrei* and *Anolis distichus*, but not in the less territorial *Anolis angusticeps*, suggesting intraspecific etiology. Tail-break and back-toe break frequencies were inversely proportional to conspecific density, suggesting predation effect. Percent injured was proportional to body size. Frequencies of breaks in *Anolis carolinensis* were very low.
Schoff PK, Johnson CM, Schotthoefer AM, Murphy JE, Lieske C, Cole RA, Johnson L, Beasley VR. 2003. Presence of skeletal and eye malformations in frogs from north-central United States: Estimations based on collections from randomly selected sites. Journal of Wildlife Diseases 39:510–521.
Congenital – Skeletal malformation rates in randomly selected north central USA wetlands in the over three consecutive years: 1998, 62 sites, 2.3% had skeletal or eye malformations; 1999, 1.6%; in 2000, 1.4%.
Hind limb malformations predominated in all three years, but other abnormalities, involving forelimb, and pelvis were also found. Especially affected were Northern leopard frogs *Rana pipiens*, but also mink frogs *R. septentrionalis*, wood frogs *R. sylvatica*, and gray tree frogs *Hyla* spp. Brachydactyly, polydactyly, ectomely, pelvic asymmetry, ectodactyly, syndactyly, and ectomely were present.
Schönbauer M, Loupal G, Schönbauer-Längle A. 1982. Osteoidchondrosarkom bei einem Wüstenwaran (*Varanus griseus*). [Osteoid chondrosarcoma in one desert monitor (*Varanus griseus*)]. Berliner Münchener tierärztliche Wochenschrift 95(10):193–194. [German]
Neoplasia – Metastasizing osteoid chondrosarcoma of a female desert monitor, *Varanus griseus*.
Schotté OE. 1956. Effects of cortisone and allied adrenal steroids upon limb regeneration in hypophysectomized *Triturus viridescens*. Revue Suisse de Zoologie 63:353–375.
Trauma – Corticosteroids restore limb regeneration in hypophysectomized *Triturus viridescens*.
Schotté OE. 1961. Systemic factors in initiation of regenerative processes in limbs of larval and adult amphibians. – In Rudnick D. ed. Synthesis of Molecular and Cellular Structure. 19th Growth Symposium: 161–192. New York: Ronald Press.
Trauma – Regeneration after amputation in *Rana clamitans*.
Hypophysectomy inhibits regeneration, correctable by ACTH (Schotté and Chamberlain 1955) or cortisone (Schotté and Bierman 1956).
Triturus cristatus, Triturus taeniatus, and *Ambystoma maculatum* regeneration is unaffected by hypophysectomy (Mayo-Smith 1946).
Schotté OE, Hall AB. 1952. Effects of hypophysectomy upon planes of regeneration in progress (*Triturus viridescens*). Journal of Experimental Zoology 121:521–560.
Trauma – Hypophysectomized *Triturus viridescens* develops callus at amputation.
Schotté OE, Harland M. 1943. Effects of denervation and amputation of hind limbs in anuran tadpoles. Journal of Experimental Zoology 93:453–493.
Trauma – Nerve supply necessary for tadpole extremity regeneration.

Schotté OE, Washburn WW. 1954. Effects of thyroidectomy on the regeneration of the forelimb in *Triturus viridescens*. Anatomical Record 120:156.

Trauma – Thyroidectomy of *Triturus viridescens* results in amputation regeneration with coalescence of skeletal parts and long bone stumping.

Schotthoefer AM, Koehler AV, Meteyer CU, Cole RA. 2003. Influence of *Ribeiroia ondatrae* (Trematoda: Digenea) infection on limb development and survival of northern leopard frogs (*Rana pipiens*): Effects of host stage and parasite-exposure level. Canadian Journal of Zoology 81:1144–1153.

Environmental – Limb bud sage of leopard frogs *Rana pipiens* is most susceptible to malformation induction by *Ribeiroia ondatrae*, risk increasing with level of exposure.

Schrader GM, Allender MC, Odoi A. 2010. Diagnosis, treatment and outcome of eastern box turtles (*Terrapene carolina carolina*) presented to a wildlife clinic in Tennessee, USA, 1995–2007. Journal of Wildlife Diseases 46:1079–1085.

Trauma – Trauma in eastern box turtle *Terrapene carolina carolina*, but no comment on bone or shell involvement.

Schreitmüller W, Lederer G. 1930a. Krankheitserscheinungen an Fischen, Reptilien und Lurchen. Beobachtet im Aquarium des zoologischen Gartens zu Frankfurt a. M. [Symptoms of fish, reptiles and amphibians. Observed in the aquarium of the zoo in Frankfurt a. M.]. I. Krankheitserscheinungen an Fischen und Lurchen [Symptoms of fish and amphibians] Berlin: Das Aquarium, 56 pp. [German]

Congenital – *Molge marmoratus* and *Ambystoma mexicanum* with forearm duplication.

Trauma – *Molge alpestris* and *Molge cristata* with tail duplication.

Schreitmüller W, Lederer G. 1930b. Krankheitserscheinungen an Fischen, Reptilien und Lurchen. Beobachtet im Aquarium des zoologischen Gartens zu Frankfurt a. M. [Symptoms of fish, reptiles and amphibians. Observed in the aquarium of the zoo in Frankfurt a. M.]. Krankheitserscheinungen an Reptilien [Symptoms of reptiles]. Berlin: Das Aquarium, 46 pp. [German]

Trauma – Arching of vertebral column, carapace hump, and deformation of tail in *Emys orbicularis*.

Metabolic – Carapace hump in *Emys orbicularis*.

Rickets with arching of vertebral column in Blue tongue skink *Tiliqua scincoides* and in crocodiles caused by wrong food (feeding only raw meat).

Neoplasia – Tumor – with destruction of parts of fingers III, IV, and V of right hand of teju *Tupinambis*.

Shell disease – Abnormal plate formation (additional or/and irregular neural plates) in *Emys orbicularis* and *Testudo graeca*.

Schrötter M, Heckers KO, Rüschoff B, Laufs R, Mack D. 2005. Severe case of spinal osteomyelitis due to *Enterococcus* spp. in a three-year-old rhinoceros viper, *Bitis nasicornis*. Journal of Herpetological Medicine and Surgery 15:53–56.

Infection – Osteolysis and exostoses (reactive bone) diagnosed as spinal osteomyelitis from *Enterococcus* spp. in rhinoceros viper *Bitis nasicornis*, associated with fracture and ankylosis.

Schubert M. 1925. Über einem Fall von Polydaktylie bei *Rana esculenta*. [A case of polydactyly in *Rana esculenta*]. Anzeiger 59: 460–463. [German]

Congenital – Difed left hind extremity with cross sections through leg to show the duplication of lower leg.

Schulp AS, Walenkamp GH, Hofman PA, Rothschild BM, Jagt JW. 2004. Rib fracture in *Prognathodon saturator* (Mosasauridae, Late Cretaceous). Netherlands Journal of Geosciences/Geologie en Mijnbouw 83(4):251–254.

Trauma – Unusual rib bumps on Cretaceous mosasaur *Prognathodon saturator* were caused by stress fractures.

Fossil – Unusual rib bumps on Cretaceous mosasaur *Prognathodon saturator* were caused by stress fractures.

Schulte R. 1983. Bemerkungen zu Froschkrankheiten speziell bei Dendrobatiden. [Observations on frog diseases especially in Dendrobatid]. Herpetofauna 5:15–17. [German]

Other – "Bone corrosion" at tip of snout, nasal region, skull roof, region above eyes, knees, elbows, feet, and legs in black legged poison frog *Phyllobates color*. Spindly legs in Golfo Dulce poison frog *Phyllobates vittatus*, bumblebee poison dart frog *Dendrobates leucomelas*, and Green and black poison dart frog *Dendrobates auratus*.

Schultz CL. 1999. An example of paleopathology in rhynchosaurs from the Triassic of Southern Brazil. Ameghiniana 36(4):20R.

Pathology – Triassic rhynchosaur PV0298T has a 1.5-cm circular right suprangular perforation. PVo232T has a 3-cm irregular jugal excavation, "result of superposition of various smaller circles." These were attributed to "some kind of cyst under the skin."

Fossil – Triassic rhynchosaur PV0298T has a 1.5-cm circular right suprangular perforation. PVo232T has a 3-cm irregular jugal excavation, "result of superposition of various smaller circles." These were attributed to "some kind of cyst under the skin."

Schüßler H. 1925. Larve von *Rana esculenta* L. mit drei Hintergliedmaßen. [Larvae of *Rana esculenta* L. with three legs]. Zoologischer Anzeiger 64:137–138. [German]

Congenital – Larva of *Rana esculenta* with three hind legs, the additional leg and the left side with eight toes.

Schuytema GS, Nebeker AV. 1998. Comparative toxicity of diuron on survival and growth of Pacific treefrog, bullfrog, red-legged frog, and African clawed frog embryos and tadpoles. Archives of Environmental Contamination and Toxicology 34:370–376.

Toxicity – Increased deformities with herbicide diuron concentrations over 20 mg/l in Pacific treefrog *Pseudacris regilla* and African clawed frog *Xenopus laevis*. However, such doses are much higher than in field spray situations.

Schwalbe E. 1906a. Die Morphologie der Mißbildungen des Menschen und der Tiere. I. Teil. Allgemeine Mißbildungslehre (Teratologie). Eine Einführung in das Studium der abnormen Entwicklung. [The morphology of malformations of humans and animals. Part I. An introduction to the study of abnormal development]. XVI + 230 pp.; Jena: Verlag Gustav Fischer. [German]

Congenital – Supernumerary legs in *Rana fusca* and *Rana esculenta* (Fig. 104: three right front legs after Tornier) and Unke (Fig. 105 after Braus); additional phalanges in *Triton* (Fig. 106 after Barfurth).

Trauma – Bifid tail in lizards (Figs. 95, 96, and after Tornier, Fig. 97), in *Triton* (Figs. 98, 99 after Tornier), and in larvae of *Rana fusca* (Figs. 100–103 after Barfurth).

Schwalbe H. 1906b. Die Morphologie der Missbildungen des Menschen und der Tiere. Ein Lehrbuch für Morphologen, Physiologen, Praktische Ärtze und Studierende. I. Allgemeine Missbildungslehre (Teratologie). Eine Einführung in das Stadium der abnormen Entwicklung. Jena: Gustav Fischer, 216 pp. [German]

Trauma – Experimental production of supernumerary legs deriving from inferior pelvis in *Rana esculenta*.

Limb regeneration in frogs and *Lacerta vivipara*.

Tail duplication in *Lacerta agilis, Triton, Rana fusca*.

Polydactyly in *Triton*.

Schwalbe H. 1910. Morphologie der Missbildungen. [Morphology of Deformities]. Book II. Doppelbildungen. [Double bodies]. Jena: Gustav Fischer, 410 pp. [German]

Congenital – Siamese *Rana esculenta*.

Dicephalic *Rana fusca*.

Ocular deformity in *Salamandra maculate*.

Triton taeniatus fused at tail.

Schwaner TD. 1990. Geographic variation in scale and skeletal anomalies of tiger snakes (Elaphidae: *Notechis scutatus-ater* complex) in southern Australia. Copeia 1990(4):1168–1173.

Environmental – Tiger snakes *Notechis scutatus-ater* incubated at higher or lower than preferred resulted in increased anomalies but did not separate scale and skeletal anomaly frequency.

Schwarz E. 1923. Ueber zwei Geschwülste bei Kaltblütern. [On two tumors in cold-blooded animals]. Zeitschrift für Krebsforschung 20:353–357. [German]

Neoplasia – Squamous cell cancer destroying metacarpal and two phalanges of fifth digit of *Tupinambis teguixin*.

Schwartz FJ, Peterson C. 1984. Color and teratological abnormalities of green turtle, *Chelonia mydas*, hatchlings from North Carolina. Florida Scientist 47:65–68.

Congenital – Missing scute, partial prefrontal, and arched upper jaw in green turtle *Chelonia mydas*.

Schwarz CW, Schwartz ER. 1974. The three-toed box turtle in central Missouri: Its population, home range and movements. Missouri Department of Conservation, Terrestrial Series 5:1–28.

Shell – Three-toed box turtle *Terrapene carolina triunguis* with carapace pits.

Schwimmer DR. 2002. King of the Crocodylians. The Paleobiology of *Deinosuchus*. Bloomington, Indiana: Indiana University Press, 220 pp.

Dental – Conjoined dentary tooth pair from single alveolus in *Deinosuchus* as a basis for the generic name Diplocynodon for the tertiary genus, but conjoined third and fourth dentary teeth are common in *Deinosuchus* and *Leidyosuchus*. 15% of *Deinosuchus* high-crowned teeth had wear or breakage.

Shell disease – Bite marks attributed to *Deinosuchus* in Cretaceous side-necked (Pleurodira) turtle *Bothremys barberi*. One had a healing carapace hole.

Fossil – Conjoined dentary tooth pair from single alveolus in *Deinosuchus* as a basis for the generic name Diplocynodon for the tertiary genus, but conjoined third and fourth dentary teeth are common in *Deinosuchus* and *Leidyosuchus*. 15% of *Deinosuchus* high-crowned teeth had wear or breakage.

Bite marks attributed to *Deinosuchus* in Cretaceous side-necked (Pleurodira) turtle *Bothremys barberi*. One had a healing carapace hole.

Schwimmer DR, Stewart JD, Williams GD. 1997. Scavenging by sharks of the genus *Squalicorax* in the Late Cretaceous of North America. Palaios 12:71–83.

Trauma – Cites bite marks on elasmosaurid plesiosaurs (Williston and Moodie 1917; Welles 1943) and mosasaurs (Hawkins 1990; Everhart et al. 1995) without evidence of healing.

Fossil – Lists observed pathologies in Alabama State Museum, Tuscaloosa, Alabama (ALAM); Columbus State University Cretaceous Research Collections, Columbus, Georgia (CSUK); Dallas Museum of Natural History, Dallas, Texas (DMNH); Fort Hays State University Museum, Fort Hays, Kansas (FHSM), University of Kansas, Lawrence, Kansas (KUVP); Los Angeles County Museum, Los Angeles, California (LACM); but none with any evidence of healing:

Phylogeny	Pathology	Location
Turtles		
Toxochelys? (LACM 50974 Campanian)	Serrated bite marks	Humerus
Desmatochelys lowii (KUVP 32401 Turonian)	Parallel serrated cuts	Humerus
KUVP 32405 Turonian	Cut marks, occ serrated	Humerus
Protostega gigas (ALAM PV985.10.2 Campanian)	Bite marks	Humerus
Bothremys barberi (CSUK 90–17–3 Schmidt 1940)	Scrape marks	Carapace
Mosasaurs		
Platecarpus ictericus (LACM 131156 Campanian)	Cuts, most with serration	Ribs
Platecarpus sp. (KUVP 86656 Senonian)	Serrated cut marks	Pubis and ischium
Tylosaurus proriger (FHSM VP-3 Senonian)	Bite marks with serration	Mandible
Mosasauridae (FHSM VP-2156 Senonian)	Serrated cut marks	Ilium
KUVP 1117 Senonian	Bite marks, some serrated	Radius or ulna
DNMH no # Campanian	Bite marks	Vertebra
Plesiosauria, Polycotylidae		
Dolichorhynchops sp. (FHSM VP 12059 Turonian)	Serrated bite marks	Coracoid
Trinacromerum willistoni (KUVP 5070 Turonian)	Bite marks	Pubis
Plesiosauria, Pliosauridae		
Brachauchenius lucasi (FHSM no # Turonian)	Serrated tooth marks	Rib
	With embedded	
	Squalicorax tooth	

Schwind JL. 1942. Spontaneous twinning in the Amphibia. – American Journal of Anatomy 71:117–151.

Congenital – Dicephalic *Rana sylvatica* and several cases of anterior back and head duplication and a supernumerary limb.

Twinning in 27 *Rana sylvatica* from Alpine, New Jersey.

Scott H. 1927. Report on deaths occurring in the Society's Gardens during the year 1926. Proceedings of the Zoological Society London 1927:173–198.

Infection – Tuberculosis in Aesculapian snake, rough-eyed Cayman, and Mexican black iguana, but no comment on bone disease.

Metabolic – Rickets presenting as soft bones in Hilaire's terrapin.

Scott AF. 1982. *Eumeces fasciatus* (five-lined skink) morphology. Herpetological Review 13(2):46.

Trauma – Bifid-tailed five-lined skink *Eumeces fasciatus*, citing bifid and trifid tailed lizards (Smith 1946; Banta 1963; Couch 1969; Goin and Goin 1971; Clark 1973).

Scott PW. 1992. Nutritional diseases. Pp. 138–152. in Beynon P, Lawton MP, Cooper JE (eds.), Manual of Reptiles. Cheltenham: British Small Animal Veterinary Association.

Metabolic – Nutritional osteodystrophy in *Iguana iguana.*

Lumpy shell with pyramiding in turtles from calcium deficiency/rickets.

Sealander JA. 1944. A teratological specimen of the tiger salamander. Copeia 1944:63.

Congenital – Tiger salamander *Ambystoma tigrinum tigrinum* with supernumerary forearm with seven toes.

Seba A. 1734. Locupletissimi rerum naturalium thesauri accurate description, et iconibus artificiosissimis expression, per universam physics historiam. [Cabinet of Natural Curiosities accurate description and artificial images of universal physical history]. Amstelaedami: Janssonio-Waesbergios & J. Wetstenium. [Latin]

Other – Drawing of snake with peculiar constricted distal tail (Vol 2, plate 30).

Pseudopathology – noted general problem existed of accuracy errors (e.g., limbs incorrectly positioned).

Hydra – Seven-headed serpent representation with two forefeet and a long tail, which allegedly belong to Count von Königsmark. The mouths were open and lined with "lion's teeth" including canines. Also noted a similar report by Conrad Gesner and by Aldovandas and cites "page 91 of 8th book on subterranean animals by Athanasius Kircher."

Myhtology – Seven-headed serpent representation with two forefeet and a long tail, which allegedly belong to Count von Königsmark. The mouths were open and lined with "lion's teeth" including canines. Also noted a similar report by Conrad Gesner and by Aldovandas and cites "page 91 of 8th book on subterranean animals by Athanasius Kircher."

Seeley HG. 1898. On the skull of *Mochlorhinus platyceps*, from Bethulie, Orange Free State, preserved in the Albany Museum, Grahamstown. Annals and Magazine of natural History (7) 1:164–176.

Congenital – Unpaired, median bone, "praeparietale," between parietals and frontals in dicynodont therapsid (mammal-like reptiles), *Lystrosaurus* (his *Mochlorhinus*) *platyceps*.

Fossil – Unpaired, median bone, "praeparietale," between parietals and frontals in dicynodont therapsid (mammal-like reptiles), *Lystrosaurus* (his *Mochlorhinus*) *platyceps*.

Seeley HG. 1892 The mesosauri of South Africa. Quarterly Journal of the Geological Society (of London) 48:586–604.

Congenital – Illustration in Fig. 5 of two *Mesosaurus* clavicles fused into a single transverse bar, but not discussed in text.

Fossil – Illustration in Fig. 5 of two *Mesosaurus* clavicles fused into a single transverse bar, but not discussed in text.

Seidel MR. 1979. The Osteoderms of the American alligator and their functional significance. Herpetologica 35:375–380.

Metabolic – Osteoderm role in *Alligator mississippiensis* in heat transmission and release because of vasculature.

Seligmann H. 1997. Tail injury alters activity patterns in *Podarcis muralis* (Reptilia: Lacertidae). Israel Journal of Zoology 44:87.

Trauma – *Podarcis muralis* with regenerated tails moved less often and spent less time in vertical movement.

Seligmann H. 1998. Evidence that minor directional asymmetry is functional. Journal of Zoology 245:205–208.

Congenital – Significant correlation between minimal directional asymmetry and injury, but not with specific side.

Seligmann H. 2000. Evolution and ecology of developmental processes and of the resulting morphology: Directional asymmetry in hindlimbs of Agamidae and Lacertidae (Reptilia: Lacertilia). Biological Journal of the Linnean Society 69:461–481.

Trauma – Notes fourth toe most commonly injured in Agamidae and Lacertidae.

Injuries related to intraspecific combats and from unsuccessful predation attempts.

Correlation between minimal directional asymmetry and injury was stronger in Agamidae than Lacertidae, possibly related to different foraging patterns.

Seligmann H. 2001. Microevolution of proneness to tail loss in lizards. Ph.D. Thesis. Jerusalem: Hebrew University of Jerusalem.

Trauma – Frequency of regeneration correlates with sex in geckos (Werner 1968), geography in teids (Schall and Pianka 1980), altitude in iguanids (Brown and Ruby 1977) and tail shedding with temperature in *Gehyra* (Bustard 1968) and *Stenodactylus* (Werner 1968).

Tail loss reduces bipedal running speed in *Dipsosaurus* (Pond 1978), doubled speed in *Phyllodactylus* (Daniels 1983) and was associated with increased survival in *Uta stansburiana* (Niewiarowsky et al. 1997). Males have higher resistance to tail breakage (Fox et al. 1998).

10% of *Thamnophis sirtalis fitchi* had injured tails, possibly from cattle trampling (Jayne and Bennett 1989) with negative correlation between speed and extent of tail amputation.

Seligmann H, Krishnan NM. 2006. Mitochondrial replication origin stability and propensity of adjacent tRNA genes to form putative replication origins increase developmental stability in lizards. Journal of Experimental Zoology (Molecular Development and Evolution) 306B:433–449.

Congenital – Positive association between low genetic variability and developmental anomalies (Soulé 1979), more in Anguidae than Amphisbaenidae (Gautschi et al. 2002). Increased frequency of scale anomalies and loss of genetic variation in serially bottlenecked populations of the dice snake *Natrix tessellate*.

Seligmann H, Beiles A, Werner YL. 1996a. Tail loss frequencies of lizards and predator specialization. Proceedings of the Sixth International Conference of the Israeli Society for Ecology and Environmental Quality Sciences. In Y Steinberger, ed. Preservation of Our World in the Wake of Change. Jerusalem: ISEEQS Pub. VI A/B:520–522.

Trauma – Lizard tail loss is more common when there is high pressure by predators for which lizards are a small percentage of the diet.

Seligmann H, Beiles A, Werner YL. 1996b. Morphotypes related to tail loss in lizards. Proceedings of the Sixth International Conference of the Israeli Society for Ecology and Environmental Quality Sciences. In Y Steinberger, ed. Preservation of Our World in the Wake of Change. Jerusalem: ISEEQS Pub. VI A/B:44–47.

Trauma – Tail loss causes low social hierarchy (Fox and Baxter 1982) and decreased clutch size (Ballinger and Tinkle 1979; Dial and Fitzpatrick 1981).

Link between tail loss frequency and predation depends on predator specialization; predators more often take atypical animals.

Longer limbs correlated with retention of original tail in *Acanthodactylus pardalis*.

Seligmann H, Beiles A, Werner YL. 2003. Avoiding injury and surviving injury: Two coexisting evolutionary strategies in lizards. Biological Journal of the Linnean Society 78:307–324.

Trauma – Proneness to tail shedding is hereditary, greater in lizards from predator-dense mainland than insular localities (Pérez-Mellado et al. 1997).

Regeneration is greater among paedotypic individuals.

Frequency of regeneration correlates with sex in geckos (Werner 1968), geography in teids (Schall and Pianka 1980), altitude in iguanids (Brown and Ruby 1977), and tail shedding with temperature in *Gehyra* (Bustard 1968) and *Stenodactylus* (Werner 1968).

Tail loss reduces perch height in *Anolis* (Ballinger 1973), bipedal running speed in *Dipsosaurus* (Pond 1978), *Cnemidophorus* (Ballinger et al. 1979), and *Cophosaurus* (Punzo 1982), climbing speed in *Podarcis muralis* (Brown et al. 1995), percent of time moving in *Lacerta monticola* (Martin and Salvador 1997) and *Podarcis muralis* (Seligmann 1997), success of escape from predators in *Scincella* (Dial and Fitzpatrick 1984), growth in *Sceloporus* (Ballinger and Tinkle 1979), and female reproductive investment (related to tail loss and regeneration) in *Coleonyx* (Dial and Fitzpatrick 1981).

Tail loss in *Lacerta monticola* results in shift in habitat use (Martin and Salvador 1992), decrease in home range and female access in *Psammodromus algirus* (Salvador et al. 1995), and decrease in social status in *Uta stansburiana* (Fox and Rostker 1982; Fox et al. 1990).

Loss of tail in *Scincella* is compensated for by cryptic behavior (Formanowicz et al. 1990).

Advantage is given by the clubbed regenerated tails in male *Agama agama* (Schall et al. 1989).

Regenerated tail mimics head shape in *Hemitheoconyx caudicinctus* (Werner 1972), fooling predators.

Seligmann H, Moravec J, Werner YL. 2008. Morphological, functional and evolutionary aspects of tail autotomy and regeneration in the 'living fossil' *Sphenodon* (Reptilia: Rhynchocephalia). Biological Journal of the Linnean Society 93:721–743.

Trauma – Autotomy, usually with regeneration, occurs in lizards, snakes, and amphisbaenians. Left-handed lizards are accident-prone (injured digits more frequently), lose tail more frequently, suggests greater fluctuating asymmetry in tail losers. Tail retainers are more frequently right-sided dominant.

Sphenodon has 36–29 caudal vertebrae – number uncertain, as intact tails difficult to find – and may live 100 years (Dawbin 1982). *Goniourosaurus kuroiwae* may also be long lived, as 91% have regenerated tails (Werner et al. 2006). Percentage of intact tails in *Sphenodon* decreases 1.14% in juveniles and 0.86% in adults per year.

Tail breakage (especially from intraspecific, intrasexual bites) was usually intravertebral in tuatara *Sphenodon*. Tail losers were more left-sided dominant. Only two lizard taxa exceeded regeneration rate of *Sphenodon:* Arabian *Asaccus* (*"Phylodactylus"*) and *Agama bibronii* (Arnold 1984).

Regeneration occurs not from end of stump, but from ablation of approximately the posterior half of terminal half-vertebra – from caudal 10 to 27 – and regenerates as a rod, rather than a tube, and is gender-neutral.

Occasional bifurcated tail in *Sphenodon.*

Homeosaurus is a Jurassic sphenoid with autotomy (Barbour and Stetson 1929).

Autotomy is frequent, but with regeneration in Amphisbaenia (Gans 1978). The only snake with regeneration was *Amphiesma stolatum* (Sharma 1980).

Terminal stumps of *Zootoca* ("*Lacerta*") *vivipara* have terminal vertebral units shorter than a half-vertebra (Bellairs and Bryant 1985), as in *Takydromus septentrionalis* (Boring et al. 1948), in contrast to *Laudakia stellio* which retains ¾ of terminal vertebra.

Tail regeneration in some crocodiles (Dathe 1960).

Seligmann H, Beiles A, Werner YL. 2003. More injuries in left-footed lizards. Journal of Zoology (London) 260: 129–144.

Congenital – Lacertidae asymmetries associate with tail state. Right side is dominant in toads. Most species are right dominant: Agamidae 15/21, Anguidae 4/12, Gekkonidae 27/46, Gymnophthalmidae 9/12, Iguanidae 4/9, Lacertidae 34/48, Scincidae 22/37, Teidae 3/5, *Varanus griseus, Lepidophyma flavimaculatum*, and *Sphenodon punctatus*. Right dominance was significant in *Sphenodon*, Anguidae *Mesaspis gadovii*, Gekkonidae *Stenodactylus sthenodactylus, Gekko swinhonis*, Lacertidae *Acanthodactylus boskianus, A. schmidti, Eremias argus, Lacerta laevis, Meroles anchietae, Mesalina olivieri*, and Sphenodontidae *Sphenodon punctatus*. Left dominance was found in *Calotes versicolor and Ptyodactylus guttatus*.

Trauma – Lacertidae asymmetries associate with tail state. Right side is dominant in toads. Agamidae have more injuries on the right hind limb, "presumably reflecting more use of that side." (p. 130). Injured lizards were more frequently left dominant: Agamidae 16/21, Gekkonidae 31/46, Gymnophthalmidae 5/12, Iguanidae 4/9, Lacertidae 25/48, Scincidae 17/37, Teidae 3/5, *Varanus* and *Sphenodon. Gymnophthalmus underwoodi, Crotaphytus collaris*, and *Sphenops sepsoides* (Scincidae) were significantly more right-sided.

Injured terrestrial *Rhoptropus afer* and *Tropiocolotes nattereri* were left- dominant, in contrast with tree-climbing relatives.

Rate of injured increased with body size, apparently as a corollary of age in left dominant *Ptyodactylus guttatus* and *Laudakia stellio*, but not with *Acanthodactylus beershebensis* or *Mesalina guttulata.*

Semlitsch RD, Moran GB, Shoemaker CA. 1981. Life history notes. Caudata, *Ambystoma tadpoideum* (mole salamander). Morphology. Herpetological Review 12:69.

Congenital – Polydactyly in mole salamander *Ambystoma tadpoideum.*

Sequeira F, Goncalves H, Menses C, Mouta-Faria M. 1999. Morphological abnormalities in a population of *Chioglossa lusitanica*. Boletín de la Asociación Herpetológica Española 10: 35–36.

Congenital – Forelimb polymelia in one, polydactyly in six, ectodactyly in 13 in one of 1738 *Chioglossa lusitanica*.

Trauma – Bifid tail in one of 1738 *Chioglossa lusitanica.*

Sessions SK. 1996–1997. Evidence that trematodes cause deformities, including extra limbs, in amphibians. In Anon. NAAMP III. The North American Amphibian Monitoring Program Third Annual Meeting. 12 unnumbered pages; Patuxent, MD.: US Geological Survey, Biological Resources Division.

Environmental – Parasitic flatworm Manodistomum syntomentera form cysts in Pacific tree frogs *Hyla regilla* and long-toed salamanders *Ambystoma macrodactylum.* 72% of young had missing limbs, ectopic limbs, polydactyly. 40% of 4148 larval and newly metamorphosized and 5% of 1778 adult salamanders had deformities.

Sessions SK, Ruth SB. 1990. Explanation for naturally occurring supernumerary limbs in amphibians. Journal of Experimental Zoology 254:38–47.

Environmental – Trematode (*Ribeiroia ondatrae)* cyst infestation induced deformities in Pacific Tree frogs (*Hyla regilla*). Of 280, 140 had extra limbs, 55 grossly distorted, 13 shortened, nine with extra pelvis, four with missing limb. Among 6000 *Ambystoma macrodactylum*, predominantly hind limb involvement, with polydactyly in 230, forked digits in 12, syndactyly in 17, supernumerary limbs in 30, ectopic hand/ft in 70, short limbs in four, and missing limbs in three.

Sessions SK, Franssen RA, Horner VL. 1999. Morphological clues from multilegged frogs: are retinoids to blame? Science 284:800–802.

Congenital – Multilegged cascade frog *Rana cascadae* from Oregon, wood frog *Rana sylvatica* and green frog *Rana clamitans* from New York and leopard frog *Rana pipiens* from Arizona. Bone triangles also reported.

Seymour RS. 1974. How sea snakes may avoid the bends. Nature 250:489–490.

Vascular – skin permeability to the effect of venous shunting in *Pelamis platurus* and *Pseudemys scripta*.

Seymour RS. 1978. Gas tensions and blood distribution in sea snakes at surface pressure and at simulated depth. Physiological Zoology 51:388–407.

Vascular – Twenty-eight percent of shallow diving *Laticauda colubrina* sea snake cardiac output bypasses lung via intrapulmonary and intraventricular shunting, compared with 70% in deep diving *Hydrophis belcheri*, where blood nitrogen levels are independent of depth. This offers protection from decompression-induced nitrogen bubble formation.

Shacham B. 2005. Tail injury linked to morphological asymmetry in a polymorphic snake. Israel Journal of Zoology 51:77–78.

Congenital – Higher escape frequency in rear-stripped *Psammophis schokari* suggests that symmetry correlates with escape ability.

Shaffer HB. 1978. Relative predation pressure on salamanders (Caudata, Plethodontidae) along an altitudinal transect in Guatemala. Copeia 1978:268–272.

Trauma – Male lizards have more tail loss (Pianka 1967; Vitt et al. 1974).

Female neotropical salamanders have more tail loss than males:

Species	% Male tail loss	% Female tail loss
Pseudoeurycea rex	3	5
Bolitoglossa rostrata	3	11
Pseudoeurycea goebeli	5	9
Bolitoglossa resplendens	10	16
Pseudoeurycea brunnata	2	5
Chiropterotriton bromeliacia	9	30
Bolitoglossa flavimembris	4	8
Bolitoglossa engelhardti	4	6
Bolitoglossa franklini	17	23
Bolitoglossa occidentalis	18	16

Shapiro G. 2003. At Mager & Gougelman, the Eyes Have It. The New York Sun 5 March 2003:1.

Congenital – Dicephalic turtle.

Sharma BD. 1980. A rare case autotomy seen in *Amphiesma stolatum* (Linn., Serpentes: Colubridae. Snake 12:60).

Trauma – Two *Amphiesma stolatum* with missing tails.

Shaver DJ, Chaney AH. 1989. An analysis of unhatched Kemp's Ridley sea turtle eggs. In: CW Caillouet, AM Landry (eds.). Proceedings of the First Symposium on Kemp's Ridley Sea Turtle Biology, Conservation and Management, p. 82–89. Galveston.

Congenital – 8.3% of 1333 *Lepidochelys kempi* embryos that could be staged of the 4165 eggs were deformed: jaw, eyes, carapace, flippers, head, snout. The frequencies varied from 14% in 1980, 27% in 1982, 3% in 1983, 15% in 1984, 11% in 1985, 18% in 1986, and 12% in 1987. Those from 1980 had more jaw deformities, compared to flipper in 1985.

They also stated that 0.5% of 8395 hatching in the 1982–1987 "class" had deformities, including jaws, eyes, flipper, neck, and plastron Jaw pathologies – 2.3% Upper or lower missing, abnormally long, short, or split.

Eye pathologies – 1.8% – Cyclopia or abnormal size.

Flipper pathology – 1.7% – Missing, short, narrow, wide, polydactyly, or clawless.

Head pathology – 1.6% – Absent, dicephalic, enlarged, lumped.

Snout pathology – 0.6% – Abnormally long or short.

Carapace pathology – 1.8% – Saddleback, indentaition, curved, contorted, small, narrowed, or missing posterior component.

Plastron pathology – 0.2% – Incomplete.

Shaw CE. 1954. Captive-bred Cuban iguanas *Cyclura macleayi*. Herpetologica 10: 73–8.

Congenital – Twin Cuban iguanas *Cyclura macleayi.*

Shaw CE. 1955 Dudley's second birthday. Zoonoz 28(12):11.

Congenital – Dicephalic California king snake – second birthday.

Shaw CE. 1956. Dudley-Duplex did it again! Zoonoz 29(12):12.

Congenital – Dicephalic California king snake – third birthday.

Shaw CE. 1958. Dudley Duplex .... Devours two mice at the same time. Zoonoz 31(11):9.

Congenital – Dicephalic king snake devours two mice at the same time.

Shaw CE. 1959. Double trouble. Zoonoz 32(11):10–11.

Congenital – Dicephalic California king snake – sixth birthday.

Shaw CE. 1963. Notes on the eggs, incubation and young of some African reptiles. British Journal of Herpetology 3:63–70.

Congenital – *Varanus e. albigularis* with skewed upper jaw and longer lower jaw.

Shaw CE. 1968. Double trouble again. Zoonoz 41(4):4–6.

Congenital – Dicephalic California king snake *Lampropeltis getula californiae.*

Shaw CE. 1971. A two-headed tale. Zoonooz 44:4–7.

Congenital – four dicephalic California king snakes donated in 17 years to San Diego zoo.

Shaw CE, Campbell S. 1974. Snakes of the American West. 330 pp.; New York: Alfred A. Knopf inc.

Congenital – Dicephalic king snake survives 6 years.

Sheldon MA, Bell GL Jr. 1999. Paedomorphosis in Mosasauroidea (Squamata): Evidence from fossil bone microstructure. Paludicola 2:190–205.

Congenital – Related paedomorphism to giantism.

Fossil – Related paedomorphism to giantism in mosasauroidea.

Sheppard L, Bellairs A. 1972. The mechanism of autotomy in Lacerta. British Journal of Herpetology 4:276–286.

Trauma – Autotomy in *Lacerta dugesii.*

Sherman E, Tock K, Clarke C. 2009. Fluctuating asymmetry in *Ichthyophonus*-sp. Infected newts, *Notophthalmus viridescens*, from Vermont. Applied Herpetology 6:369–378.

Infection – Increased leg length asymmetry in *Notophthalmus viridescens* caused by *Ichthyophonus* infestation.

Environmental – Increased leg length asymmetry in *Notophthalmus viridescens* caused by *Ichthyophonus* infestation.

Shilton CM, Brown GP, Shine R, Benedict S. 2008. Spinal arthropathy associated with *Ochrobactrum anthropi* in free-ranging cane toads (*Chaunus* [*Bufo*] *marinus*) in Australia. Veterinary Pathology 45(1):85–94.

Neuropathic – Spinal changes in 10% of free-ranging cane toads (*Chaunus* [*Bufo*] *marinus*) in Australia. Joint spaces were "indistinct or absent with varying degrees of smooth, moderately firm to rugose bony swelling." Partial or complete loss of joint space was associated with woven bone and irregular hyaline cartilage islands often bridging adjacent vertebrae. Fifteen lesions were associated with inflammation (dense aggregates of degenerated and necrotic macrophages, some heterophils and neutrophils, necrotic bone and cartilage fragments, with articular cartilage fissure or erosion) while nine fusions manifested none.

Vertebral – Spinal pathology in amphibians has been previously rare and limited to scoliosis (Bacon et al. 2006; Ouellet 2000; Speare 1990).

Spinal changes in 10% of free-ranging cane toads (*Chaunus* [*Bufo*] *marinus*) in Australia. Joint spaces (not disc spaces, which are not found in *Bufo*) were "indistinct or absent with varying degrees of smooth, moderately firm to rugose bony swelling." Partial or complete loss of joint space was associated with woven bone and irregular hyaline cartilage islands often bridging adjacent vertebrae. Fifteen lesions were associated with inflammation (dense aggregates of degenerated and necrotic macrophages, some heterophils and neutrophils, necrotic bone and cartilage fragments, with articular cartilage fissure or erosion) while nine fusions manifested none. Some of the toads also had granulomatous gastritis or granulomatous hepatitis or granulomatous cystitis. Culture of six of nine intervertebral spaces revealed *Ochrobactrum anthropi* (formerly known as *Achromobacter* and closely related to *Brucella*).

Notes also the controversial spinal arthropathy of snakes attributed to *Salmonella*.

Shimada K. 1997. Paleoecological relationships of the Late cretaceous lamniform shark, *Cretoxyrhina mantelli* (Agassiz). Journal of Paleontology 71:926–933.

Trauma – Review of previously published evidence of shark predation and scavenging, offering opinion on species of shark involved in the attack reported by Rothschild and Martin (1993).

Shimada K, Hooks GE III. 2004. Shark-bitten protostegid turtles from the Upper Cretaceous Mooreville Formation of Alabama. Journal of Paleontology 78:205–210.

Trauma – *Protostega gigas* FMNH P27452 and FMNH PR58 from the Mooreville Chalk (Upper Santonian to Lower Campanian) in Greene County, Alabama have tooth marks, and FMNH P27452 had five embedded teeth of the Late Cretaceous shark *Cretoxyrhina mantelli.*

Fossil – *Protostega gigas* FMNH P27452 and FMNH PR58 from the Mooreville Chalk (Upper Santonian to Lower Campanian), in Greene County, Alabama have tooth marks and FMNH P27452 had five embedded teeth of the Late Cretaceous shark *Cretoxyrhina mantelli.*

Shimada K, Parris DC. 2007. A long-snouted Late Cretaceous crocodyliform, *Terminonaris* cf. *T. browni*, from the Carlile Shale (Turonian) of Kansas. Transactions of the Kansas Academy of Science 110:107–115.

Congenital – *Terminonaris* cf. *T. browni* FHSM VP-4387 with fused nasals and maxillae, contrasted with the holotype, which exhibits sutures. They suggested that it could represent variation.

Fossil – *Terminonaris* cf. *T. browni* FHSM VP-4387 with fused nasals and maxillae, contrasted with the holotype, which exhibits sutures. They suggested that it could represent variation.

Shimada K, Tsuihiji T, Sato T, Hasegawa Y. 2010. A remarkable case of a shark-bitten plesiosaur. Journal of Vertebrate Paleontology 30:592–597.

Trauma – Elasmosaurid plesiosaur *Futabasaurus suzukii* from the Upper Cretaceous of central Japan with broken lamniform shark *Cretalamna* (= *Cretolamna*) *appendiculata* teeth in humerus and three vertebrae.

Fossil – Elasmosaurid plesiosaur *Futabasaurus suzukii* from the Upper Cretaceous of central Japan with broken lamniform shark *Cretalamna* (= *Cretolamna*) *appendiculata* teeth in humerus and three vertebrae

Shine R, Langkilde T, Wall M, Mason RT. 2005. The fitness correlates of scalation asymmetry in garter snakes *Thamnophis sirtalis parietalis*. Functional Ecology 19: 306–314.

Congenital – 16% of viviparous red-sided garter snakes *Thamnophis sirtalis parietalis* have different number of ribs on either side of body, related to suboptimal maternal thermoregulation during pregnancy. Asymmetric males mated less successfully than symmetric males. They noted that gravid viviparous snake and lizard females maintain higher and less variable body temperatures than non-gravid. Females more often had asymmetric ventral scale patterns.

Shirley M, Schachner E, Shaw C. 2010. Asymmetric skeletal adaptation to a debilitating pathology in the hindlimb of *Poposaurus gracilis* (Archosauria: Poposauroidea). Journal of Paleontology 30:164A.

Trauma – *Poposaurus gracilis* YPM 57100 from Late Triassic of Grand Staircase, Utah had healed right femoral fracture (or at least focal midshaft expansion), associated with more robust left limb size (compared to right).

Fossil – *Poposaurus gracilis* YPM 57100 from Late Triassic of Grand Staircase, Utah had healed right femoral fracture (or at least focal midshaft expansion), associated with more robust left limb size (compared to right).

Shively JH, Songer JJG, Prchal S, Keasey M III, Thoen CO. 1981. *Mycobacterium marinum* infection in Bufonidae. Journal of Wildlife Diseases 17:3–7.

Infection – *Mycobacterium marinum* metatarsal infection in *Bufo cognatus.*

Shortt J. 1868. Notice of a double-headed water snake. Journal of the Linnean Society London, Zoology 9:49–50.

Congenital – Dicephalic *Hydrophis sublaevis.*

Shreve AA, Spahr J, Fountain A, Sleeman JM. 2004. Phaeohyphomycosis in a free-living Eastern box turtle (*Terrapene carolina carolina*). Proceedings of the American Association of Zoo and Wildlife Veterinarians and the Wildlife Disease Association, San Diego, CA:623–625.

Infection – Chromomycosis in hind limb of Eastern box turtle *Terrapene carolina carolina*, but no comment on osseous involvement.

Sicard RE. 1985. Regulation of Vertebrate Limb Regeneration. 185 pp.; New York: Oxford University Press.

Trauma – Review of limb regeneration in amphibia.
Sibtain S.M. 1938. Studies on the caudal autotomy and regeneration in *Mabuya dissimilis* Hallowell. Proceedings of the Indiana Academy of Sciences 8:63–78.
Trauma – Tail regeneration in Scincidae *Mabuya dissimilis*, noting previous reports predominantly dealt with Lacertidae and Gekkonidae.
Siddall ME, Gaffney ES. 2004. Observations on the leech *Placobdella ornata* feeding from bony tissues of turtles. Journal of Parasitology 90:1186–1188.
Infection – Leech *Placobdella ornate* attaching to plastron and carapace of *Chelydra serpentine* and *Chrysemys picta*. Illustrated is pitting at the hypoplastron margin of *Chelydra serpentine.*
Shell disease – Leech *Placobdella ornate* attaching to plastron and carapace of *Chelydra serpentine* and *Chrysemys picta*. Illustrated is pitting at the hypoplastron margin of *Chelydra serpentine.*
Siebold CTE. de 1828. Observationes quaedam de Salamandris et Tritonibus. [Observations on salamanders and tritons]. Diss. Berolini litteris Augusti Petschii: 30 pp. [Latin]
Congenital – Monstrous extremities in urodeles.
Rudimentary melomelia in *Triton*.
Supernumerary digits with interdigital membrane.
Polydactyly (one extra digit) in the posterior foot of two salamanders.
Bifurcation of a digit in a hind limb foot (page 22, Figs. 20, 21).
Sievers E. 1933. Ein Naturfall von Polymelie beim Frosch. (Mit histologischen Beobachtungen). [A natural case of polymely in frogs (with historical observations)]. Anatomy and Embryology 99:710–720 [German].
Congenital – Male of *Rana esculenta* with additional short extremity on left shoulder girdle.
Silly M. 1841. Tératologie. Couleuvre à deux têtes. [Teratology – Smooth snake with two heads]. Compte Rendu Académie des Sciences Paris 13 (16): 831–832 [French].
Congenital Dicephalic smooth snake.
Silverman S. 2006. Diagnostic imaging. In Mader DR. ed. Reptile Medicine and Surgery. Pp. 471–489; Philadelphia: Saunders.
Trauma – Green iguana *Iguana iguana* with fractures.
Infection – Green iguana *Iguana iguana* with septic arthritis.
Metabolic – Green iguana *Iguana iguana* with metabolic bone disease.
Other – Osteolytic lesions in Parson' chameleon *Chamaeleo parsonii.*
Silverman S, Janssen DL. 1996. Diagnostic imaging. In Mader DR. ed. Reptile Medicine and Surgery. Pp. 258–264; Philadelphia: Saunders.
Infection – Septic elbow in green iguana; osteolysis induced by infection in Parson's chameleon jaw.
Silvestri F. 1892. Un caso di dicefalia in un giovane individuo di *Lacerta viridis*. [Dicephaly in a case of a young individual *Lacerta viridis*]. Rivista Italiana di Scienze Naturali e Bollettino del naturalista collettore, allevatore, coltivatore. 12 (9):116
Congenital – Dicephalic *Lacerta viridis*.
Simbotwe MP. 1983. A report on scoliosis in the diurnal gecko *Lygodactylus chobiensis* Fitzsimons 1932 inhabiting Lochinvar National Park, Zambia. Journal of the Herpetological Association of Africa 29:18.
Congenital – Scoliosis in diurnal gecko *Lygodactylus chobiensis*.
Simmermacher, G. 1885. Notiz über einen sechsfüßigen Molch. [Note on a six-legged urodele]; Der Zoologische Garten 26:93. [German]
Congenital – *Triton cristatus* with with toes on of second left femur.
Simpson SB Jr. 1965. Regeneration of the lizard tail. In V Kiortsis, HA Trampush (eds.). Regeneration in animals and related problems. Amsterdam: North Holland, 431–443.
Trauma – Regeneration of tail in *Lygosoma laterale* dependent upon innervation.
Sims J. 1989. Herpetology matters – Bicephalism. Aquarist and Pondkeeper 54(7):69.
Congenital – Dicephalic eastern diamondback *Crotalus adamanteus*, western diamondback *Crotalus atrox*, timber rattlesnake *Crotalus horridus*, prairie rattlesnake *Crotalus viridus viridus*, Northern Pacific rattlesnake *Crotalus viridis oreganus* and Western pygmy rattlesnake *Sistrurus miliarius streckeri*, snapping turtle *Chelydra serpentina*, red-eared terrapin *Pseudemys scripta elegans*, and false map terrapins *Graptemys pseudogeographica* (latter hold record for longevity).
Singer M. 1952. The influence of the nerve in regeneration of the amphibian extremity. Quarterly Review of Biology 27:169–200.
Trauma – Limb regeneration in *Triturus viridescens* is nerve dependent.
Singer M. 1961. Induction of regeneration of body parts in the lizard *Anolis*. Proceedings of the Society for Experimental Biology, NY 107:106–108.

Trauma – *Anolis carolinensis* limb regeneration does not occur unless nerve supply is supplemented (e.g., experimentally from the other side).
Singh T.S. 1948. Polymely in the common Indian Bull-frog *Rana tigrina*. Proceedings of the 34th Indian Science Congress 34(3):186.
Congenital – Bullfrog *Rana tigrina* polymelia, also noted by Mahendra (1936).
Singh LA, Bustard HR. 1982. Congenital defects in the gharial *Gavialis gangeticus* (Gmelin). British Journal of Herpetology 6:215–219.
Congenital – Bent neck, hunchback, bent tail, and cross snout deformities in gharial *Gavialis gangeticus*.
Singh LA, Sagar SR. 1991. Prolonged egg incubation and congenital tail deformities in *Crocodylus palustris*. Journal of the Bombay Natural History Society 89:194–198.
Environmental – Increased daily temperature fluctuation and humidity was associated with bent or curled tails in 35% of *Crocodylus palustris*.
Singh LA, Sagar SR. 1992. Prolonged egg incubation and congenital tail deformities in *Crocodylus palustris* (Reptilia: Crocodilia). Journal of the Bombay Natural History Society 89:194–198.
Environmental – 26% of *Crocodylus palustris* with tail bends and 10.5% with curled tips with increased incubation temperature.
Singh KS, Thapliyal JP. 1973. Twinning in the Checkered water snake, *Natrix piscator*. Herpetologica 29:19–20.
Congenital – Twin checkered water snake *Natrix piscator.* Notes that Hildebrand (1938) summarized twinning in turtles and that Carpenter and Yoshida (1967) reported one in *Agama agama*.
Sit KH, Kanagasuntheram R. 1972. A structural analysis of congenital limb deformities in experimental hyperthyroid tadpoles. Journal of Embryology and experimental Morphology 28:223–234.
Congenital – Phocomelia-hemimelia and digital deformities in hyperthyroid *Bufo melanostictus*.
Metabolic – Phocomelia-hemimelia and digital deformities in hyperthyroid *Bufo melanostictus*.
Skillcorn J. 1988. Toady. Aquarist and Pondkeeper 52(13):19.
Congenital – Supernumerary forelimb with second glenohumeral joint in *Bufo bufo*.
Skulan J. in press. Evidence for congenital anomaly and range of motion in a pathologic mosasaur tarsus. In Rothschild BM. & Shelton S. eds. Paleopathology. Austin, TX: University of Texas Museum of Natural History, (in press).
Congenital – *Platecarpus* tarsus at University of Wisconsin (Madison) Geology Museum with congenital fusion of three component bones and osteoarthritis reaction to resultant abnormal motion.
Trauma – Three rib fractures in *Platecarpus* with pseudoarthroses were also noted.
Fossil – *Platecarpus* tarsus at University of Wisconsin (Madison) Geology Museum with congenital fusion of three component bones and osteoarthritis reaction to resultant abnormal motion.
Three rib fractures in *Platecarpus* with pseudoarthroses were also noted.
Sladden DE. 1932a. Experimental distortion of development in amphibian tadpoles. Part I. Proceedings of the Royal Society of London, Series B 106:318–325.
Metabolic – 10% sugar produced stiffened limbs in *Rana temporaria.*
Sladden DE. 1932b. Experimental distortion of development in amphibian tadpoles. Part II. Proceedings of the Royal Society of London, Series B 112:1–12.
Metabolic – Exposure of *Rana temporaria* embryos to 50% sugar occasionally produced distal limb duplication.
Slotopolsky B. 1921. Beiträge zur Kenntnis der Verstümmelungs – und Regenerationsvorgänge am Lacertilierschwanze [Contributions to the knowledge of mutilation and regeneration processes in the tail of lacertilians]. Vierteljahresschrift der Naturforschenden Gesellschaft in Zürich 66:39–48 [German].
Trauma – Regeneration of lacertilian tails, specifically mentioning *Anguis fragilis*.
Slotopolsky B. 1922. Beiträge zur Kenntnis der Verstümmelungs- und Regenerationsvorgänge am Lacertilierschwanze. [Contributions to the knowledge of mutilation and regeneration processes in the tail of lacertilians]. Zoologische Jahrbücher. Abteilung für Anatomie und Ontogenie der Tiere Abteilung 43:219–322 [German].
Trauma – Autotomy and regeneration in lacertilians *Anguis fragilis, Iguana, Lacerta muralis, viridis, agilis*.
Slyper EJ. 1931. Die Verletzungen und Erkrankungen der Wirbelsäule und Rippen bei den Cetaceen. [Injuries and illness of vertebral column and ribs in cetacea]. Anatomischer Anzeiger 71:176–185 [German].
Trauma – *Plesiosaurus dolichodeirus* two ribs with thickening of healed fracture *Plioplatecarpus marshi* two ribs with healed fractures (see Abel 1912).
Mosasaurus lemonnieri ribs with healed fractures, also chevrons of caudal vertebrae 33 and 35.
Mosasaurus sp., *Platecarpus coryphaeus*, and *Platecarpus* sp. with exostosis on third and fourth thoracal vertebrae, fused in the last genus (Moodie 1928).
Vertebral – *Platecarpus* sp. with third and fourth thoracal vertebrae fused (Moodie 1928).
Hainosaurus bernardi ventral and lateral synostosis of third and fourth thoracal vertebrae (see also Dollo 1885).

Plioplatecarpus marshi fusion of two last lumbar (or caudal) vertebrae (Dollo 1892).
Trinacromerum bentonianum with two anterior thoracal vertebrae fused (Williston 1908)
Fossil – *Plesiosaurus dolichodeirus* two ribs with thickening of healed fracture *Plioplatecarpus marshi* two ribs with healed fractures (see Abel 1912).
Mosasaurus lemonnieri ribs with healed fractures, also chevrons of caudal vertebrae 33 and 35.
Mosasaurus sp., *Platecarpus coryphaeus*, and *Platecarpus* sp., with exostosis on third and fourth thoracal vertebrae, fused in the last genus (Moodie 1928).
Platecarpus sp. with third and fourth thoracal vertebrae fused (Moodie 1928).
Hainosaurus bernardi ventral and lateral synostosis of third and fourth thoracal vertebrae (see also Dollo 1885).
Plioplatecarpus marshi fusion of two last lumbar (or caudal) vertebrae (Dollo 1892).
Trinacromerum bentonianum with two anterior thoracal vertebrae fused (Williston 1908).

Smirina EM. 1972. [Annual layers in bones of *Rana temporaria.*] Zoologiskii Zhurnal 51:1529–1534. [Russian].
Congenital – One osseous line per year in *Rana temporaria.*

Smirina EM. 1994. Age determination and longevity in amphibians. Gerontology 40(2):133–146.
Congenital – Growth layers visualized in mud puppy *Necturus maculosus* parasphenoid, in zygapophyses of bullfrog *Rana catesbeiana* and pterygoid of amphibians.
Transformation mark from formation of resting line after metamorphosis.
Annual layers in phalanges used for age determination in anurans, but in urodeles, regeneration causes confusion.
Errors in age determination derive from endosteal resorption, intraseasonal layers, failure of distinct separation of layers or doubling or tripling on side of bone which grows most rapidly or for longest period.

Smit AL. 1953. The ontogenesis of the vertebral column of *xenopus laevis* (Daudin) with special reference to the segmentation of the metotic region of the skull. Annals of the University of Stellenbosch 29A:79–136.
Congenital – In depth review of vertebral column ontogenesis.

Smith WH. 1882. Report on the reptiles and Batrachians of Ohio. Report of the Geological Survey of Ohio 4: 630–734.
Congenital – Cites serpent monstrosities: Aristotle and also Redi (1684) Dicephalic *Ophibolus triangulus* and frog with supernumerary leg (Kingsley 1878).
Tropidonotus sipedon with two heads and two tails (Wyman 1862).
Rana palustris with supernumerary limbs and cited *Rana palustris* with supernumerary hind leg fused to normal and polydactyly (Ryder 1878).

Smith HM. 1912. Handbook of Lizards; Lizards of the United States and Canada. Handbooks of American Natural History, Vol VI:557 pp. Ithaca, NY: Comstock Publishing Co.
Trauma – Regeneration of tails occurs when fracture is in middle of vertebrae, just posterior to transverse process and reported a forked tail in regeneration.

Smith MA. 1917. A two-headed snake. Journal of the Natural History Society of Siam 2:255–256.
Congenital – Dicephalic water snake *Homalopsis buccata.*

Smith HM. 1946. Handbook of Lizards. Lizards of the United States and of Canada. 557 pp.; Ithaca, NY: Comstock Publishing Co.
Trauma – Fracture of lizard tails occur just posterior to transverse process.
Incomplete fracture produces forked tail (e.g., *Eumeces egregious*).
Broken, regenerated tail in *Ophisaurus.*
Tail loss in *Holbrookia maculata ruthveni, Sceloporus poinsetti, Sceloporus undulatus consobrinus, Sceloporus undulatus garmani* (with regeneration), *Eumeces anthracinus, Eumeces septentrionalis obtusirostris.*

Smith HM. 1947. Kyphosis and other variations in soft-shelled turtles. University of Kansas Publications: Museum of Natural History 1(6):117–124.
Congenital – *Amyda (Trionyx) emoryi* KU 2219 with sharp hump and reported that one seventh had eight neurals and one with eight costals.
Amyda (Trionyx) mutica KU 23230 with prominent curved hump (five times normal shell height), noting 9/16 had seven neurals; seven, had eight. One had seven costal on one side; eight, on the other.
Amyda (Trionyx) spinifera KU 23026 with sharp hump, noting 2/18 had eight neurals and one with eight costals.
Suggested humps occur if costal plates ankylose early.
Cited Gressitt (1937) for kyphosis in *Amyda sinensis steindachneri.*

Smith M. 1954. The British Amphibians and Reptiles. London: Collins.
Trauma – Slow worm *Anguis fragilis* has less tail regeneration ability.

Smith HM. 1957. The role of urotomy in lizards. Herpetologica 13:235–236.

Trauma – Autotomy defined as self-mutilation or autourotomy in *Eumeces laticeps* and *fasciatus* and *Scincella lateralis* which sometimes swallow their tails.

Smith CW. 1984. Morphological anomalies of two geckos, *Hemidactylus frenatus* and *Lepidodactylus lugubris*, and the toad, *Bufo marinus*, on the island of Hawaii. Proceedings of the fifth Conference in Natural Sciences Hawaii National Park. June 5–7, 1984: 41–50. Cooperative National Park Resources Studies Unit, University of Hawaii at Manoa.

Congenital – Hindfoot amelia in *Bufo marinus*.

Zig-zag tail with kinking and multilobed, bulbous multi-tipped tails in two mourning gecko *Lepidodactylus lugubris*. Zig-zag garter snake *Thamnophis radix haydeni*.

Trauma – Bifurcated tails in two house geckos *Hemidactylus frenatus*.

Smith GR. 1996. Tail loss in the striped plateau lizard, *Sceloporus virgatus*. Journal of Herpetology 30:552–555.

Trauma – Tail loss resulted in slower growth of striped plateau lizard *Sceloporus virgatus*.

Smith ZN. 2003. Basra? What about Boston? Chicago Sun-Times 11 September 2003:46.

Congenital – Dicephalic turtle.

Smith HM, Chiszar D. 1988. The earliest records of Ophidian dicephaly in the Western Hemisphere. Bulletin of the Chicago Herpetological Society 23(8):121–123.

Congenital – Suggested that earliest report of a dicephalic snake was by Herra 1725–1756.

Dicephalic probable *Liophis perfuscus* from Barbados Island (Edwards 1751).

Amphisbaena with two necks, which was probably *Tropidonotus fasciatus spidedon, Elaphe obsoleta obsoleta*, or *Lampropeltis triangulum triangulum* (which Smith and Chiszar favor). They cite McIntyre (1977) for *Elaphe subocularis* and Parek (1977) in *Crotaphopeltis hotamboeia*.

Smith HM, Fitzgerald KT. 1983. Trauma-induced developmental vertebral displacement (Rhoecosis) in a garter snake. Herpetological Review 14:69–72.

Trauma – Western plains garter snake *Thamnophis radix haydeni* with displaced 54th vertebra (called "amyoplasia"), secondary to postembryonic trauma, producing kypholordosis (= mild kyphosis) and scoliosis. They used the term "rhoecosis" from the Greek rhoikos, for crooked. They related the peculiar sine-wave-like appearance to an illusion from vertebral column torsion.

Vertebral – Western plains garter snake *Thamnophis radix haydeni* with peculiar sine-wave-like appearance to an illusion from vertebral column torsion.

Smith HM, James LF. 1958. The taxonomic significance of cloacal bursae in turtles. Transactions of the Kansas Academy of Science 61:86–96.

Vascular – Cloacal bursae, apparently for respiration, were present in Chelydridae, Platysternidae, Emydidae, Pelomedusidae (African genera *Pelomedusa* and *Pelusios*, which no longer have them, excepted) and Chelidae, but were absent in Kinosternidae, Dermatemyidae, Testudinidae, Cheloniidae, Dermochelyidae, Carettochelyidae, and Trionychidae.

Smith HM, Nickon DC. 1961. Preliminary experiments on the role of the cloacal bursae in hibernating turtles. Natural History Miscellanea of the Chicago Academy of Sciences 178:1–8.

Vascular – Importance of cloacal respiration in *Chrysemys*, contrasted with buccal respiration in *Pseudemys scripta*.

Smith HM, Pérez-Higareda G. 1988 (not 1987). The literature on somatodichotomy in snakes. Bulletin of the Maryland Herpetological Society 23(4):139–153 (for 1987).

Congenital – Somatodichotomy in snakes, citing Cunningham 1937s 150 references and listed 117 additional references. Twenty two percent of reports were in Colubridae, 18% in Natricidae, and 21% in Crotalidae. Earliest report was by Nakamura (1938) who recognized teratophagus presenting as craniophagus (fusion of cranium), cephaloderophagus (fusion of cranium and anterior trunk), and anakatamesodidymus (separation at both ends). He called tyeratodymus as rhinodymus (double-nosed), opodymus (two-headed), and derodymus (bifurcated head and neck). Craniodichotomy (head only), prodichotomy (head and anterior body), proarchodichotomy (up until anus), opisthodichotomy (head and tail duplicated), urodichotomy (only tail duplicated), and amphidichotomy (both anterior and posterior, rarely middle duplicated) and holodichotomy (total duplication).

Terminology – Nakamura (1938) who recognized teratophagus presenting as craniophagus (fusion of cranium), cephaloderophagus (fusion of cranium and anterior trunk), and anakatamesodidymus (separation at both ends). He called tyeratodymus as rhinodymus (double-nosed), opodymus (two-headed), and derodymus (bifurcated head and neck). Craniodichotomy (head only), prodichotomy (head and anterior body), proarchodichotomy (up until anus), opisthodichotomy (head and tail duplicated), urodichotomy (only tail duplicated), and amphidichotomy (both anterior and posterior, rarely middle duplicated) and holodichotomy (total duplication).

Smith AR, Lewis JH, Crawley A, Wolpert L. 1974. A quantitative study of blastemal growth and bone regression during limb regeneration in *Triturus cristatus*. Journal of Embryology and Experimental Morphology 32:375–390.

Trauma – Bone regression occurs within 20 days of amputation in *Triturus cristatus*.

Smith DD, Laposha NA, Powell R, Parmerlee JS Jr. 1985. *Crotalus molossus* (blacktail rattlesnake). Anomaly. Herpetological Review 16:78–79.

Congenital – Bent tails in black tail rattlesnake *Crotalus molossus*.

Trauma – Black tail rattlesnake *Crotalus molossus* with short tail, lacking rattle.

Smit AL. 1953. The ontogenesis of the vertebral column of *Xenopus laevis* (Daudin) with special reference to the segmentation of the metotic region of the skull. Annals of the University of Stellenbosch 29:79–136.

Congenital – Variable number (3–4) of caudal vertebrae form urostyle in *Xenopus laevis*.

Smits AW, Brodie ED Jr. 1995. Lack of respiratory "cost" of tail autotomy in the lungless salamander, *Oedipina uniformis*. Comparative Biochemistry and Physiology 111A:115–161.

Trauma – No effect of autotomy on respiration in the lungless salamander *Oedipina uniformis*.

Smyth M. 1974. Changes in the fat stores of the skinks *Morethia boulengeri* and *Hemiergis peronii* (Lacertilia). Australian Journal of Zoology 22:135–145.

Trauma – Tail loss in Australian skink *Morethia boulengeri* and *Hemiergis peronii* females made them (especially *Hemiergis peronii*, which lacks fat bodies) reduced clutch size.

Snodgrass JW, Hopkins WA,Broughton J, Gwinn D, Baionno JA, Burger J. 2004. Species-specific responses of developing anurans to coal combustion wastes. Aquatic Toxicol 66:171–182.

Environmental – Green frog *Rana clamitans* and wood frog *Rana sylvatica* larvae exposure to coal combustion wastes-contaminated sediment [accumulated significant ($P < 0.05$) concentrations of at least six trace elements (As, Cd, Fe, Se, Sr, and V)] had malformation rates were ≤4%, independent of species.

Snyder RC. 1949. Bipedal locomotion of the lizard *Basiliscus basiliscus*. Copeia 1949:129–137.

Trauma – Removal of one third of *Basiliscus basiliscus* tail impeded bipedal running.

Solino (aka Solinus) CG. 1587. Polyhistor, Entreateth of the First Foundation of the City of Rome (ex recens Giov. Franc. Muratori) Augustae Taurinorum ex. off. Regia. [Latin]

Pseudopathology – Described amphisbaena as a two-headed snake; in reality, a snake with an enlarged terminal portion of the tail.

Sordelli F. 1876. Descrizione di una *Rana* polimelica del Museo Civico di Milano con alcune considerazioni sulla Polimelia e sulla Polidattilia negli Articolati [Description of a polymelous *Rana* of the Civic Museum of Milano with some considerations on Polymely and on Polydactyly in the articulated]. Atti della Società Italiana Scienze naturali e del Museo Civico di Storia Naturale in Milano 19:392–398 [Italian].

Congenital – five-legged *Rana viridis* Roesel in the collections of the Museo Civico di Milano.

Sordelli F. 1878. Nuovo caso di Polimelia nella Rana mangereccia [New case of Polymely in Rana mangereccia]. Atti della Società Italiana di Scienze naturali e del Museo Civico di Storia Naturale in Milano 21:48 [Italian].

Congenital – Anterior polymely in *Rana esculenta* from the cabinet of the Collegio Sant'Alessandro. Feet attached to the thoracic belt through a shared scapula that connected to the normally developed left scapula.

Sordelli F. 1882. Di un Axolotl polimelico e della più frequente causa di tale anormalita nei Batraci Urodeli [On a polymelous Axolotl and on the very frequent cause of this anomaly in urodelan batrachians]. Atti della Società Italiana di Scienze Naturali e del Museo Civico di Storia Naturale in Milano 25:251–259 [Italian].

Congenital – *Siredon lichenoides* with supernumerary right forelimb, inserted over the humerus of the normal limb, at about two third from its articulation with the shoulder.

Sordelli F. 1887. Atlante Zoologico. Storia Naturale dei Mammiferi, Uccelli, Rettili, Anfibi e Pesci. [Zoological Atlas. Natural History of the Mammals, Birds, Reptiles, Amphibians, and Fishes]. Milan: U Hoepli Ed., 180 pp. [Italian].

Trauma – Duplication of distal tail of *Chamaeleo vulgaris*.

Sosnowskij IP. 1956. Zweiköpfige Schlange. [Two-headed snake]. Priroda 3:119 [Russian].

Congenital – Partial head duplication in *Natrix natrix*, although image claims a boa.

Souder, W. 2000. A plague of frogs. Hyperion, New York, New York, USA, 299 pp.

Trauma – Rare cause of missing limbs.

Environmental – Review of attention to environmental factors in epidemic of missing and incomplete limbs and jaws and supernumerary and malformed limbs in *Rana pipiens, Rana septentrionalis, Hyla regilla, Rana clamitans*, and *Rana palustris*, starting with serendipitous observation in Minnesota River and development of the controversy related to parasite-teratogen in the water.

Soulé ME. 1979. Heterozygosity and developmental stability: Another look. Evolution 33:396–401.

Congenital – Asymmetry of scales was proportional to heterozygosity in side-blotched lizard *Uta stansburiana sensu lato*.

Sower SA, Reed KL, Babbitt KJ. 2000. Limb malformations and abnormal sex hormone concentrations in frogs. Environmental Health Perspectives 108:1085–1090.

Environmental – Bullfrogs *Rana catesbeiana* and green frogs *Rana clamitans* in central and southern New Hampshire showed malformed frogs at 81% of the sites sampled (13 of 16 sites). Normal frogs had significantly higher concentrations (nearly threefold) of *in vitro* produced androgens and of brain gonadotropin-releasing hormone than malformed frogs.

Sowerby JdeC, Lear E. 1872. Tortoises, terrapins and turtles drawn from life by James de Carle Sowerby. London: Henry Sotheran, Joseph Baer & Co. Series of unnumbered plates.

Shell disease – Two drawings of *Emys concentrica* with pitted carpaces, one with elevated margins, the second solely excavations, often curved.

Spadola F, Insacco G. 2009. Newborn dicephalic *Podarcis sicula.* Acta Herpetologica 4:99–101.

Congenital – *Podarcis sicula* with duplication of anterior half of body.

Cites Holfert 1999 and Broadley 1972, related to dicephalic *Rhacodactylus auriculatus* and *Trachylepis (Mabuya) striata.*

Spallanzani L. 1768. Prodromo di un' opera da imprimersi sopra le riproduzioni animali. [Precursor of work to be printed on animal reproduction]. Modena: Montanari Printing House, 102 pp. [Italian]

Trauma – Experimented on tadpoles, aquatic salamanders, and snails. He observed a regeneration of the severed tail in tadpoles – tail, limbs and even the jaw – in the aquatic salamanders Regenerated limbs in salamanders may have more or less digits than normal (after Taruffi 1886).

Spallanzani L. 1769. Reproductions of the legs in the Aquatic salamander. In Spallanzani L, ed. An Essay on Animal Reproductions. [Translated from the Italian] pp. 68–82. London: Becket and Hondt.

Trauma – Partially cut regenerated limbs of large salamanders may be straightened and deformed, with shortened, thicker femora and tibiae, with variable number of toes. This is in contrast to fractures which produce callus, but healing does not necessarily produce a usable limb.

Speare R. 1990. A review of diseases of the Cane Toad, *Bufo marinus*, with comments on biological control. Australian Wildlife Research 17:387–410.

Trauma – Healing femoral fracture and foot gangrene (Cicmanec et al. 1973).

Infection – 7.6% of *Bufo marinus* carry *Salmonella.*

Environmental – Two *Bufo marinus* with five legs (Moore 1988), attributed to insecticide use.

Speare R, Berger L, O'Shea P, Ladds PW, Thomas AD. 1997. Pathology of mucomycosis of cane toads in Australia. Journal of Wildlife Diseases 33:105–111.

Infection – *Mucor amphibiorum* infection of cane toad *Bufo marinus* "affecting bone," but no details.

Spemann H, Falkenberg H. 1919. Über asymmetrische Entwicklung und Situs inversus viscerum bei Zwillingen und Doppelbildungen. [On asymmetrical development and situs inversus viscera in twins and double formations]. Archiv für Entwicklungsmechanik 45: 371–422 [German].

Trauma – Twins with loss of some legs and dicephalic larvae of *Triton taeniatus* produced by dividing eggs.

Špinar Z. 1972. Tertiary frogs from Central Europe. 286 pp.; Prague: Academia.

Congenital – Variation in frontopareitale, nasale (especially *Palaeobatrachus laubei*), squamosum, columella, prooticum, parasphenoid, and pterygoid, 50% of seventh vertebra does not fuse with eighth and ninth.

Occasionally nine presacral vertebrae in *Palaeobatrachus.*

Frequent unilateral sacral wing enlargement from normal of seven to nine vertebrae, occasionally six to eight, in *Palaeobatrachus*, asymmetrical eighth and tenth transverse processes, transverse seventh fused with eighth, transverse eighth composing most of sacral wing. Synsacrum with unilateral enlargement in *Palaeobatrachus luedeckei* and *diluvianus.*

Palaeobatrachus grandipes eighth vertebral postzygapophysis composing most of synsacrum and asymmetrical sacral wings.

Xenopus laevis with abnormal sacral wing, off urodele on one side, sacral vertebra on the other.

Palaeobatrachus grandipes without synsacrum fenestra.

Half wing with stunted appendage.

Trauma – Fractured *Palaeobatrachus* fibulare (calcaneus) bowed, fused to tibiale.

Fracture fibulare (calcaneum with bowing and fusion to tibiale) in *Palaeobatrachus.*

Fossil – Variation in frontopareitale, nasale (especially *Palaeobatrachus laubei*), squamosum, columella, prooticum, parasphenoid, and pterygoid, 50% of seventh vertebra does not fuse with eighth and ninth.

Occasionally nine presacral vertebrae in *Palaeobatrachus.*

Frequent unilateral sacral wing enlargement from normal of seven to nine vertebrae, occasionally six to eight, in *Palaeobatrachus*, asymmetrical eighth and tenth transverse processes, transverse seventh fused with eighth, transverse eighth composing most of sacral wing. Synsacrum with unilateral enlargement in *Palaeobatrachus luedeckei* and *diluvianus.*

Palaeobatrachus grandipes eighth vertebral postzygapophysis composing most of synsacrum and asymmetrical sacral wings.

Xenopus laevis with abnormal sacral wing, off urodele on one side, sacral vertebra on the other.

Palaeobatrachus grandipes without synsacrum fenestra.

Half wing with stunted appendage.

Fractured *Palaeobatrachus* fibulare (calcaneus) bowed, fused to tibiale.

Fracture fibulare (calcaneum with bowing and fusion to tibiale) in *Palaeobatrachus.*

Spix X. 1824. Animalia sive species novae Testudinidium et Ranavum, etc. [New turtle and frog species] in Fol Monachii Tab 20, f2, p 24. [Latin].
Congenital – Natural polymelia in the toad *Bufo ictericus ictericus*.
Squires ZE, Bailey PC, Reina RD, Wong BB. 2008. Environmental deterioration increases tadpole vulnerability to predation. Biology Letters 4:392–394.
Environmental – Increasing salinity of brown tree frog *Litoria ewingii* environment decreases activity and increases vulnerability to dragonfly nymph *Hemianax papuensis*.
Stannius H. 1856 (not 1954). Die Amphibien. [Amphibians]. In C Siebold, H Stannius. Handbuch der Zootomie. Die Wirbelthiere. [Handbook of Zoology: The Vertebrates], 2nd book. Berlin: Verlag von Veit & Co., 270 pp.
Congenital – Fusion of sacrum and coccyx in *Ceratophrys dorsata*, *Atelopus varius*, *Aglossa*, and *Breviceps*.
Starnes SM, Kennedy CA, Petranka JW. 2000. Sensitivity of embryos of southern Appalachian amphibians to ambient solar UV-B radiation. Conservation Biology 14:277–282.
Environmental – Exposure of *Ambystoma maculatum* in southern Appalachian Mountains to ambient UV-B resulted in embryo deformities in less than 10%. Exposed *Hyla chrysoscelis* and *Pseudacris triseriata* had 8.2% and 8.3% affected respectively, contrasted with unexposed (UV blocked) of 2.9% and 0.1%, respectively. *Rana sylvatica* values were 4.3 versus 2.0.
Stebbins RC. 1954. Natural history of the salamanders of the plethodontid genus *Ensatina*. University of California Publications in Zoology 54(2):47–123.
Trauma – *Ensatina eschscholtzii xanthoptica* recent tail loss in 46 of 603 (7%). Irregularities in tail indicate past breaks, noting that they are shifted posteriorly with animal growth.
Regenerated salamander toes are shorter and broader with less perfect joints and cartilaginous in nature, occasionally with multiple slender regenerates.
Steehouder T. 1990. Overbultvorming in het schild bij de opkweek van landschildpadden. [Plaston malformation in growing tortoises]. Lacerta 48:89–95. [Dutch]
Congenital – Plastron deformities in *Testudo hermanii* and *graeca*.
Stemmler O. 1963. In Reichenbach-Klinke HH. Krankheiten der Reptilien. Stuttgart:Gustav Fischer. [German]
Toxicology – Vitamin A induction of supernumerary limbs.
Stemmler-Gyger O. 1964. Zwei bemerkenswerte Exemplare der Sternschildkröte. [Two notable examples of star turtles]. Die Aquarien- und Terrarien-Zeitschrift 17:116–118.
Shell disease – supernumerary scutes in *Testudo elegans* with elongated carapace.
Stender E. 1911. Abnormitäten bei Reptilien. *Lacerta*. [Abnormalities in reptiles. *Lacerta*]. Gratisbeilage zur Wochenschrift für Aquarien und Terrarienkunde 19:74–75 [German].
Trauma – *Lacerta agilis* and *L. vivipara*, each one specimen with double tail (X- ray pictures); *Vipera berus* with two heads.
Stepanov SA, Arkhangel'sky MS, Ivanov AV, Uspensky GN. 2004. Paleopathology of ichthyopterygians. Ark. Pathol. 2: 29–30.
Congenital – Ichthyosaurids of family Undorosaurinae show changes in the structure of vertebral bone tissue due to the aging.
Trauma – Ichthyosaur *Otschevia panderi* with an extra joint (pseudoarthrosis) at fracture site in the middle of right clavicula.
Metabolic – Plesiosaur *Platypterygius* sp. with osteoporosis and osteophytes on joint surfaces of two anterocaudal vertebrae due to trauma or inflammation.
Vertebral – Plesiosaur *Platypterygius* sp. with osteoporosis and osteophytes on joint surfaces of two anterocaudal vertebrae due to trauma or inflammation.
Fossil – Ichthyosaurids of family Undorosaurinae show changes in the structure of vertebral bone tissue due to the aging.
Ichthyosaur *Otschevia panderi* with an extra joint (pseudoarthrosis) at fracture site in the middle of right clavicula.
Plesiosaur *Platypterygius* sp. with osteoporosis and osteophytes on joint surfaces of two anterocaudal vertebrae due to trauma or inflammation.
Plesiosaur *Platypterygius* sp. with osteoporosis and osteophytes on joint surfaces of two anterocaudal vertebrae due to trauma or inflammation.
Stephens N, Holder N, Maden M. 1985. Motorneuron pools innervating muscles in vitamin A-induced proximal—distal duplicate limbs in the Axolotl. Proceedings of the Royal Society of London, series B 224:341–354.
Toxicology -Vitamin A exposure of axolotls produces full and partial limb (predominantly forelimb) duplication.
Sternberg CH. 1909. The Life of a Fossil Hunter. Henry Holt and Company, 286 pp. Vascular – Image (facing page 115) of *Protostega gigas* with apparent avascular necrosis of the proximal humerus.
Vertebral – Ankylosed *Tylosaurus dyspelor* (AMNH) vertebrae.
Fossil – Image (facing page 115) of *Protostega gigas* with apparent avascular necrosis of the proximal humerus.
Ankylosed *Tylosaurus dyspelor* (AMNH) vertebrae.

Sternberg CH. 1922. Explorations of the Permian of Texas and the chalk of Kansas. 1918. Kansas Academy of Sciences Transactions 30:119–120.
Trauma – Polycyotylid plesiosaur ingested by *Tylosaurus*, or at least inside its rib cage.
Fossil – Polycyotylid plesiosaur ingested by *Tylosaurus*, or at least inside its rib cage.
Stetter MD. 1995. Noninfectious medical disorders of amphibians. Seminars in Avian and Exotic Pet Medicine 4:49–55.
Metabolic – Metabolic bone disease in amphibia, manifest as osteopenia, scoliosis, long-bone bowing, and pathological fractures.
Gout, manifest as radiodense mass on top of third phalanges in "Marine toad."
Steward JW. 1961. A case of dicephalism in the adder. British Journal of Herpetology 3:18.
Congenital – one sixth *Vipera b. berus* littermates with dicephalic head and neck.
Stewart MM. 1999. Status and conservation of Midwestern amphibians. Copeia 1999(2):536–538.
Congenital – Deformed frogs in Minnesota.
Stivers JD. 1914. Two-headed snake one of big family: Mother and 57 youngsters are captured on State Hospital grounds—freak snakelet very much alive. Middletown Daily Times-Press, Middletown, NY (1 August 1914):8.
Congenital – Derodymus garter snake.
Stojanov A. 2000. Hornpanzeranomalien bei den Landschildkröten (*Testudo graeca ibera* and *Testudo hermanni boettgeri*) in Bulgarien. [Carapace anomalies in the tortoise species (*Testudo graeca ibera* and *Testudo hermanni boettgeri*) in Bulgaria.] Historia Naturalis Bulgarica 11:97–105 [German].
Congenital – Dicephalic *Testudo graeca*
Malformations in the horny shields of Testudo graeca ibera and Testudo hermanni boettgeri.
Stojanov A. 2005. New data of abnormalities on the bony and horny shell of tortoises from Bulgaria. Acta Zoologica Bulgarica 57:25–30.
Infection – Infectious horny shell in a Hermann's tortoise and *T. graeca ibera* with shell "infractions" and osteolysis from suppurated infections.
Metabolic – Calcium/phosphorus disturbance producing abnormally large plastron fontanels and highly porous bone especially at rib ends and peripheral plates.
Shell disease – Infectious horny shell in a Hermann's tortoise and *T. graeca ibera* with shell "infractions" and osteolysis from suppurated infections.
Stolk A. 1958. Tumours of Reptiles. 4. Multiple osteomas in the lizard *Lacerta viridis*. Beaufortia 7:1–9.
Neoplasia – Multiple chondromas of the tail a rising in the vertebral cartilage in an extant green lizard *Lacerta viridis*. Cited possible hemangioma of caudal vertebrae in Late Jurassic sauropod dinosaur *Diplodocus*, subsequently identified as actually DISH (Rothschild and Berman (1991).
Fossil – Cited possible hemangioma of caudal vertebrae in Late Jurassic sauropod dinosaur *Diplodocus*, subsequently identified as actually DISH (Rothschild and Berman 1991).
Stone PA, Dobie JL, Henry RP. 1992. Cutaneous surface area and bimodal respiration in soft- shelled (*Trionyx spiniferus*), stinkpot (*Sternotherus odoratus*), and mud turtles (*Kinosternon subrubrum*). Physiological Zoology 65:311–330.
Vascular – Aquatic respiration contributed 38% of oxygen and 85% of carbon dioxide exchange in soft-shell turtle (*Trionyx spiniferus*), 26% and 56% in stinkpot turtles (*Sternotherus odoratus)*, and 14% and 46% in mud turtles (*Kinosternon subrubrum*) proportional to cutaneous surface area.
Stopper GF, Heckler L, Franssen RA, Sesions SK. 2002. How trematodes cause limb deformities in amphibians. Journal of Experimental Zoology 294:252–263.
Environmental – Trematode (*Ribeiroia ondatrae)* is widespread across North America and cyst infestation causes hind limb anomalies by physical interference (perturbation of positional relationships of cells in limb buds) in Pacific tree frogs *Hyla regilla*, western toads *Bufo boreas*, wood frogs *Rana sylvatica*, and northern leopard frogs *Rana pipiens* (88% of the latter two). Mirror-image duplications, "bony triangles," and hylomorphisms were reported. They raise the question of hind limb vulnerability to cercarial attack because of environmental exposure, while the forelimb develops within the branchial chamber and is thus protected.
Storer TI. 1925. A Synopsis of the Amphibia of California. University of California Publications Zoology 27:1–342 [page 58 is pertinent].
Congenital – *Bufo boreas halophilus* with supernumerary leg attached to posterior end of body (Crosswhite and Wyman 1920).
Trauma – Two *Triturus torosus* with bifid tails.
50% of *Bufo boreas halophilus* around golf course in Los Angeles County, California had mutilated limbs with toe, hindfoot, and tibia loss – attributed to lawn mowers.
Streckenbach, P. 1979. Doppelmißbildung beim madagassischen Taggecko. [Double monsters in Madagascar day gecko]. Elaphe, Berlin 1979 (4):45 [German].
Congenital – Siamese *Phelsuma madagascariensis* with two bodies (one stunted), eight legs and two tails.

Strecker JK. Jr. 1926. Ophidian freaks. Two-headed snakes. Contributions Baylor University Museum (6):10.

Congenital – Dicephalic copperhead *Agkistrodon mokasen* and DeKay's snake *Storeria dekayi* and spotted garter snake *Thamnophis marcianus*.

Derodymous gray pilot snake *Elaphe obsoleta confina* (Baird and Girard, no citation).

Stringer EM, Ramer JC, Bowman MR, Heng HG, Bradway DS. 2009. Phaeohyphomycosis of the carapace in an Aldabra tortoise (*Geochelone gigantea*). Journal of Zoo and Wildlife Medicine 40(1):160–167.

Infection – Carapace flaking with bone involvement by fungus *Exophiala oligosperma* in Aldabra tortoise *Geochelone gigantea*.

Shell disease – Carapace flaking with bone involvement by fungus *Exophiala oligosperma* in Aldabra tortoise *Geochelone gigantea*.

Strobel P. 1876a. Cenno su tre casi di polimelia nelle rane [Short account on three cases of polymely in frogs]. Atti della Società Italiana di Scienze naturali e del Museo Civico di Storia Naturale in Milano 18:405–409 [Italian].

Congenital – Three cases of polymely in *Rana esculenta* in the collections of the University of Parma, Italy. One had a supernumerary limb to the left of the ventre, with two divergent digits in the other two limbs. Another had a supernumerary limb that attached to the left of the anus, with only two digits. A third had a supernumerary limb that attached to the right of the pubis. It lacked metatarsals and digits. Two were previously mentioned by De Mortillet in 1865. The supernumerary hand had four fingers like the normal hands, but the fingers were divergent, with the external digits at right angles from the base of the hand.

Strobel P. 1876b. Ulteriori cenni sulla polimelia nelle Rane [Further short accounts on polymelia in frogs]. Atti della Società Italiana di Scienze naturali e del Museo Civico di Storia Naturale in Milano 19: 385–391 [Italian].

Congenital – Two polymeric *Rana esculenta*, in the collections of the anatomy cabinet of the University of Modena, Italy. One was pentameric and had osteomalacia: the diaphyses of the cylindrical bones of the extra arm were only comprised of a subtle cord of apparently fibrous tissue. The second *Rana esculenta* had a supernumerary hind limb that attached above and to the right of the left iliofemoral articulation. The tibia-fibula of the second was discreetly thick, and thus Strobel thought it resulted from the fusion of two. The tarsal bones were only two, but each of them apparently resulted from the fusion of two tarsals.

Polymeric *Rana temporaria* from the collections of the Spallanzani Museum in Reggio dell'Emilia, Italy, with two posterior supernumerary connect to the left side of the body, above the thigh of the normal left hind limb, to which they are attached to the knee. The left normal hind limb foot had only four digits.

Posterior polymelia is more frequent than anterior; left polymely, much less rare than right.

Polymelia accompanied by partial or total fusion of supernumerary limbs is rarer than polymely with simple and distinct limbs.

Strohl J. 1919. A propos des experiences de Tornier concernant la genèse de formations pathologiques chez les animaux, suivi de quelques rèsultants obtenus par l'élevage experimental de diverse espèces de Dorades monstrueuses. [Experiences of Tornier with breeding generations of pathologic phenomenon in animals, following several results of experiments on diverse species and fish monsters]. L'année biologique 23:72–73 [French].

Congenital – Mentions fish monsters, with only passing comment of amphibian monsters.

Strohl J. 1925. Les serpents a deux têtes et les serpents doubles. A propos d'un cas de bicéphalie chez un hydrophyide '*Hydrophis spiralis*' (Shaw) et d'un cas de bifurcation axiale postérieure (déradelphie) chez un '*Vipera berus*' L. [Two head and double-bodied serpents. A case of bicephalism in a hydrophyid '*Hydrophis spiralis*' (Shaw) and a case of posterior axial body duplication (déradelphie) in a '*Vipera berus*' L]. Annales des Sciences Naturelles. Zoologie et biologie animale Paris 8:105–132 [French].

Congenital – Dicephalic *Coronella getula* (Yarrow 1878), *Coronella triangulum* (Blatchley 1906), *Coronella austriaca* (Malaquin 1923), *Coluber catenifer, Coluber melanoleucus* (Johnson 1901), *Coluber longissimus, Distira brugmansii* (*Hydrophis sublaevis*) (Schortt 1868), *Cuban boa Epicrates angulifer, Eutaenia sirtalis, Heterodon simus* (Johnson 1901), *Heterodon platirhinos* (Blatchley 1906), *Homalopis buccata* (Flower 1899), *"Hydrophis spiralis," Lycodon aulicus* (Dobson 1873; Wall 1905), *Lachesis lanceolatus* (Lacépède 1789), *Hydrus platurus* (Pelamis bicolor), *Naja tripudians* (Dobson 1873), *Tropidonotus fasciatus fasciatus* (Jan 1863; Johnson 1901), *Tropidonotus fasciatus sipedon* (Bancroft 1769; Wyman 1862; Johnson 1901), *Tropidonotus natrix* (Redi 1684; Vsevolojsky 1812; Cantoni 1921; Dusseau 1865; Lasserre 1880; Fraucher 1903. *Tropidonotus piscator (quincunciatus)* (Wall 1905), *Tropidonotus septemvittatus* (Kirkland 1871), *Tropidonotus sirtalis* (Johnson 1901), *Tropidonotus tigrinus, Tropidonotus vagrans* (Johnson 1901), *Vipera aspis* (Carlo Bonaparte 1832–1841; Edoardo de Betta 1864), *Vipera berus* (Dorner 1873; Bogert 1897; Lönnberg 1907; Stender 1911; Enzio Reuter 1921), *Zamensis constrictor* (Mitchill 1826; Johnson 1901; Wyman 1863), *Zamenis hippocrepis*.

Derodymous *Hydrophis spiralis*.

Tricephalism allegedly described in seventeenth-century serpents by Ulysse Aldrovande and by Mitchill (1925).

Body and partial skull duplication in *Vipera berus*.

Posterior axial body duplication "*Vipera berus*" L.

Tropidonotus sipedon with two heads and two tails.

Stuart GA. 1908. Abnormal tail in a lizard, *Hemidactylus gleadovii*. Journal of the Bombay Natural History Society 18:688–689.
Trauma – Trifid tail in *Hemidactylus gleadovii.*
Stuart JN. 1996. Additional records of kyphosis in freshwater turtles. Bulletin of the Chicago Herpetological Society 31:60–61.
Congenital – Kyphotic *Chrysemys picta belli* (among 429 captures) and *Trionyx spiniferus emoryi* (among 20). *Trachemys scripta troostii* with abnormal large yolk mass causing distension of carapace (Cagle 1950).
Sturn A, Brattstorm BH. 1958. A serial abnormality in the painted turtle. Herpetologica 13:277–278.
Congenital – *Chrysemys picta* with nine instead of five central scutes.
Sues H-D, Olsen PE, Carter JG, Scott DM. 2003. A new crocodylomorph archosaur from the Upper Triassic of North Carolina. Journal of Vertebrate Paleontology 23:329–343.
Trauma – Triassic crocodylomorph *Dromicosuchus grallator* with head and neck, bite marks attributed to rauisuchian archosaur.
Fossil – Triassic crocodylomorph *Dromicosuchus grallator* with head and neck, bite marks attributed to rauisuchian archosaur.
Sullivan RM. 1979. Dermal armor pathology in the tail of a specimen of the fossil lizard *Helodermoides tuberculatas*. Herpetologica 35(3):278–282.
Trauma – Regenerating osteoderms in Oligocene lizard *Helodermoides tuberculatas* produced a solid mass.
Fossil – Regenerating osteoderms in Oligocene lizard *Helodermoides tuberculatas* produced a solid mass.
Surlève-Bazeille J-E. 1970. Observations sur l'infrastructure des members de larves de Grenouille verte (*Rana esculenta* L.) atteintes de l'anomalie *P*. [Observations on the infrastructure of limbs of larval green frog (*Rana esculenta* L.) with anomalyP]. Comptes Rendus Hebdomadaires des Séances de l'Académie des Sciences, Paris, Série D 270: 193–195 [French].
Congenital – Anomaly *P* in *Rana esculenta.*
Surlève-Bazeille J-E, Cambar R. 1969. Nouvelles recherches sur les modalités de transmission de l'anomalie *P* chez la Grenouille verte (*Rana esculenta* L.). [New research onf transmission modality of anomaly P in green frog (*Rana esculenta* L.)]. Comptes Rendues des Séances de la Société de Biologie 163:2034–2037 [French].
Environmental – Relates *Rana esculenta* supernumerary limbs to contact with certain fish.
Surlève-Bazeille J-E, Cambar R, Calas A. 1969a. Premières observations histologiques sur les membres des larves de Grenouille verte (*Rana esculenta* L.) atteints de l'anomalie *P*. [First histologic observations among green frog (*Rana esculenta* L.) larvae with anomaly P]. Comptes Rendues Hebdomadaires des Seances de l'Académie des Sciences, Paris, Série D 269:2240–2242 [French].
Congenital – Pelvic agenesis and teratoma in *Rana esculenta.*
Surlève-Bazeille J-E, Cambar R, Mauget R. 1969b. Essai de transmission de l'anomalie *P* à diverses espèces d'amphibiens. Premiers resultants obtenus sur *Rana temporaria*. [Essay on transmission of anomaly P to diverse species of amphibians. First results obtained on *Rana temporaria*]. Actes de la Société Linnéenne de Bordeaux 106, série A (2):1–7 [French].
Congenital – Essay on transmission of anomaly P in diverse species of amphibians.
3–6% *Rana temporaria* have polydactyly with either fish present.
Surlève-Bazeille J-E, Cambar R, Calas A. 1970. Observations sur l'infrastructure des members de larves de grenouille verte (*Rana esculenta* L.) atteintes de l'anomalie P. [Observations on the infrastructure of larvae of green frog (*Rana esculenta)* L. larvae with anomaly P]. Comptes Rendues Hebdomadaires des Seances de l'Académie des Sciences Paris 270:193–195 [French].
Congenital – Polydactyly in *Rana esculenta.*
Surya Naryana Rao A. 1990. Double headed common sand boa *Eryx conicus* (Schneider). Journal of the Bombay Natural Historical Society 87:462.
Congenital – Dicephalic common sand boa *Eryx conicus.*
Susebach E. 1935. Eine doppelköpfige, eine spiraling aufgerollte und dunkelgefärbte Larve von schwarzem Feuersalamanderweibchen (Briefliche Mitteilungen). [A double headed, a spiral enrolled and dark colored larva of female black fire salamander]. Blätter zur Aquarien- und Terrarienkunde, Braunschweig, 46:147 [German].
Congenital – dicephalic and spiral enrolled tail of fire salamander *Salamandra maculosa.*
Sutton JB. 1885. Tumours in animals. Journal of Anatomy and Physiology 19:415–475.
Congenital – Bufo vulgaris, Rana palustris, Rana esculenta, Rana temporaria with supernumerary hind limbs.
Trauma – *Lacerta viridis* with bifid tail.
Metabolic – Indian monitor *Varanus dracoena* with rickets.
Neoplasia – *Varanus dracoena* with enchondromas of fifth and seventh cervical vertebrae, humeri, metacarpals, and hyoid.

Sutton JB. 1889. Supernumerary limbs in frogs and toads. Transactions of the Pathological Society of London 40: 461–463.

Congenital – Supernumerary hind limbs in *Bufo vulgaris, Rana palustris*, and *Rana esculenta.*

Supernumerary forelimb on left side with three digits in *Rana temporaria*, in woodcut of *Rana esculenta* (Tuckerman 1886) and in frog from Teratological collection of the Museum of the Royal College of Surgeons.

Sutton JB. 1890. Evolution and Disease. 285 pp.; London: Walter Scott.

Congenital – Derodymous snake.

Supernumerary forelimb from sternum in frog in Royal College of Surgeons Collection, partial from supernumerary coracoid in *Rana temporaria.*

Supernumerary hind limb with foot duplication in *Bufo vulgaris*, duplicate hind limb attached to symphysis pubis in *Rana palustris*, two supernumerary hind limbs in *Rana esculenta.*

Trauma – Bifid tail in green lizard *Lacerta viridis.*

Sutton JB. 1892. Evolution and Disease. New York: Scribner's Sons. 314 pp.

Congenital – Supernumerary limbs are not rare among amphibians.

Dicephalic snake.

Common toad with supernumerary leg with polydactyly.

Rana palustris with supernumerary leg attached to symphysis pubis.

Rana esculenta with posterior dichotomy, noting other side suppressed.

Metabolic – Confined snakes develop soft-skull bones and lose teeth.

Gout – Guanine gout in frogs and lizards.

Dental – Confined snakes develop soft-skull bones and lose teeth.

Swanson A. 1997. Two-bodied bearded dragon. Reptiles Magazine 5(4):8.

Congenital – Opisthodichotomy in bearded dragon *Pogona vitticeps* and king snake.

Swanson S, van Breukelen F, Kreiser B, Chiszar D, Smith HM. 1997. A double-bodied midland water snake and additions to the literature on ophidian axial bifurcation. Bulletin of the Chicago Herpetological Society 32:80–83.

Congenital – Opisthodichotomous *Nerodia sipedon* pleuralis fused ventrally just past the umbilicus, also citing Payen's (1991) work, which did not distinguish between craniodichotomy and prodichotomy, calling both dicephalic.

Swarts C. 2003. Rare two-headed tortoise found in South Africa. National Geographic News 30 May 2003: 1.

Congenital – Dicephalic tortoise, probably angulate tortoise *Chersina angulata.*

Sykes JM, Trupkiewicz JG. 2006. Reptile neoplasia at the Philadelphia Zoological Garden, 1901–2002. Journal of Zoo and Wildlife Medicine 37:12–29.

Neoplasia – Increase in frequency of reptilian neoplasia from 2.3 to 3.8% in older literature, compared with 12.4–23% in more recent literature.

Musculoskeletal primary neoplasm in a snake, but no comment on bone involvement.

Myelomonocytic leukemia in a Solomon Island ground skink *Eugongylus albofasciatus*, but no comment on bone involvement.

Lymphosarcoma in a hognose *Heterodon platirhinos*, a river jack *Bitis nasicornis* and a Carolina pygmy rattlesnake *Sistrurus miliarius miliarius*, but no comment on bone involvement.

Hematopoietic neoplasia is frequently reported in the lizard literature (e.g., San Diego Zoo), although they had only case and Sacramento zoo had none in lizards.

Philadelphia zoo results:

	1901–1967	1968–1979	1980–1991	1992–2002	Total
Chelonia	3.3%		2.4%		6/511,
				only in turtles, not tortoises	
Lizard		0.8	4.7	5.9	19/882
Snake	1.0	2.7	5.2	9.2	53/2202
Crocodilian					0/89

Annotated Bibliography T-Z

Tabin CJ. 1992. Why we have (only) five fingers per hand: Hox genes and the evolution of paired limbs. Development 116:289–296.

Congenital – General review of formation of vertebrate limb pattern, reviewing the five distinct Hox (homeobox gene clusters)-encoded domains across the limb bud.

Takahashi H. 1957. The malformed development of toad larvae treated with estradiol. Annotationes Zoologicae Japonenses 30:199–203.

Metabolic – Estradiol produces bent femur and reduction in tibia and fibula length of *Bufo vulgaris formosus.*

Takeishi M. 1996. [On the frog, *Rana ornativentris*, with supernumerary limbs found at Yamada Greenery Area in Kitakyushu City, Fukuoka prefecture, Japan.] Bulletin of the Kitakyushu Museum of Natural History 15:119–131. [Japanese with English abstr.]

Congenital – Supernumerary forelimbs in 8 of 59 (11.9%) *Rana ornativentris.*

Takeuchi T. 1958. Effects of boric acid on the development of the eggs of the toad, *Bufo vulgaris formosus*. Science Reports of Tohoku University. Fourth Series. Biology 24(1):33–43.

Toxicology – Lithium-induced cyclopia in anura and urodela (Bellamy 1919; Kawakami 1953; Kawakami and Kawakami 1951; Lehmann 1937, 1938; Ôgi 1954).

Glycolysis inhibitors (sodium fluoride, sodium iodoacetate, sodium monoiodoacetate) rarely produce cyclopia in *Ambystoma* and *Rana esculenta* (Tamini 1943).

Talavera RR. 1987. Anomalies in the vertebral column ontogeny of field reared *Pelobates cultripes*. 395–398. In van Gelder JJ, Strijbosch H, Bergers PJ eds. Proceedings of the 4th Ordinary General Meeting of the Societas Europaea Herpetologica; Faculty of Sciences, Nijmegen.

Congenital – Seven (8.54%) *Pelobates cultripes* at Gosner stages 38 and 19 (31.15%) postmetamorphics had one or more vertebral anomalies: 9 (1.22%) (instead of normal 8) presacral vertebrae or decrease to 7 (2.44% in tadpoles and 13.11% in post metamorphics) from atlas fusing to second vertebrae in 7 adults or eight has asymmetric sacral diaphysis in 2 tadpoles and 1 adult. Atlas was fused to second vertebra in 11.47%. Sacrum fused to urostyle in 18.04% of postmetamorphics. Asymmetrical sacral diaphyses and double sacrum (7.32%) of tadpoles and 8.2% of postmetamorphics, with urostyle fused to sacrum. Transverse processes on urostyle in 4.92% of postmetamorphics. Abnormal urostyle fused to sacrum in one adult. Abnormal presacral vertebrae transverse processes in 2.44% of tadpoles and 3.28% of metamorphics, one tadpole with elongate symmetrical transverse processes, abnormal posterior presacral transverse process. Failure of dorsal fusion of neural arches of atlas and second vertebra (1.6% of postmetamorphics).

Vertebral – Atlas fusing to second vertebrae in *Pelobates cultripes* in 11.47%.

Talukdar SK. 1977. A case of anterior dichotomy in a colubrid snake, *Xenochrophis piscator* (Schneider) (Reptilia: Serpentes: Colubridae). Science Culture 43(12):538–539.

Congenital – Dicephalic *Xenochrophis piscator*, noting reports in wolf snake *Lycodon aulicus* (Dobson 1873), Russell's viper *Vipera russelli* (Acharji 1945), European grass snake (Buckland) *Hydrophis cyanocinctus* (Shortt 1868), *Natrix* by Daudin (citing Nicholson), *Homolopsis buccata* (Flower 1899), American hog-nosed snake *Heterodon simus* in *Madras Times* January 13, 1897, and Johnsons (1901) reporting cases as far back as 640 CE.

B.M. Rothschild et al., *Herpetological Osteopathology: Annotated Bibliography of Amphibians and Reptiles*, DOI 10.1007/978-1-4614-0824-6_11, © Springer Science+Business Media, LLC 2012

Talukdar SK. 1981. Notes on an aberrant specimen of colubrid snake, *Enhydris enhydris* (Schneider). Bulletin of the Zoological Survey of India 4(2):141–142.

Congenital – Derodymous *Enhydris enhydris*, *Lycodon* (Wall 1905), Wolf snake *Lycodon aulicus* (Dobson 1873), Russell's viper *Vipera russelli* (Acharji 1945), *Hydrophis* (Shortt 1868), water snake *Homolopsis buccatta* (Flower 1889), and *Xenochrophis piscator* (Talukdar 1977).

Talvi T. 1987. Üks imelik konn. [A strange frog.] Eesti Loodus 1987:599–601,621,623 [Estonian].

Congenital – Immobile, anterior supernumerary limb derived from shoulder girdle in *Rana temporaria*.

Trauma – Immobile, anterior supernumerary limb derived from shoulder girdle in *Rana temporaria*, with review of potential causes, suggesting super regeneration following an injury.

Talvi T. 1992a. Amphibians and reptiles of Estonia: List, geographic relationships and current situation. KorsóS Z, Kiss I, eds. Proceedings of the Sixth Ordinary General Meeting S.E. Hungary, Budapest 1991; pp 429–432.

Congenital – Oligodactyly and polymely in *Rana temporaria* (Talvi 1987).

Talvi T. 1992b. Jäsemete väärarendid ja nende tekkepõhjused kahepaiksetel. [Malformations of limbs and their origin in amphibians.] Schola Biotheoretica 18:56–58 [Estonian].

Congenital – Polydactyly in tiger salamander (Bishop 1947) and polymelia in anurans (Borkin and Pikulik 1986).

Metabolic – Hyperthyroid-induced congenital limb abnormalities (Sit and Kanagasuntheram 1972).

Environmental – UV light-induced supernumerary limbs in urodeles (Butler and Blum 1963) and other anomalies in *Rana esculenta* (Dubois 1979).

Carcinogen-induced supernumerary limbs induced by carcinogens in *Triturus viridescens* (Breedis 1952) and abnormal limb regeneration in newts (Tsonis and Eguchi 1982).

Radiation-induced abnormalities in *Rana nigromaculata* (Kawamura and Nishioka 1978).

Toxicology – Vitamin A-induced dysmorphogenesis (Johnson and Scadding 1991), abnormalities in *Bufo andersonii* (Niazi and Saxena 1978) and polydactyly in *Rana temporaria* (Rostrand 1951) and hind limb fish-related teratogenic changes (Rostand and Darre 1969).

Talvi T. 1992c. Naturally occurring limb abnormalities in anurans of the Estonia. In: Ordinary general meeting of the SEH 19–23 August 1991 Barcelona, Spain, p. 124, Budapest, 1992.

Congenital – Oligodactyly in 3.3% of *Rana temporaria* and 1 of 150 *Bufo calamita.*

Polydactyly in 1.01% of *Rana temporaria* and 1.33% of *Bufo calamita.*

Supernumerary limb with articulating socket on clavicle of a *Rana temporaria.*

Tanaka K. 2011. Phenotypic plasticity of body size in an insular population of a snake. Herpetologica 67:46–57.

Environmental – Dwarfism in Japanese four-lined snake *Elaphe quadrivirgata* on Yakushima Island (diet exclusively lizards), compared to the main Japanese Islands (diet predominantly frogs) as phenotypic response, presumably to food limitation.

Tank PW, Holder N. 1981. Pattern regulation in the regenerating limbs of urodele amphibians. The Quarterly Review of Biology 56(2):113–142.

Trauma – Supernumerary limbs by dedifferentiation when graft and stump are in greatest disharmony in urodele amphibians.

Tanke DH, Currie PJ. 1998. Head-biting behavior in theropod dinosaurs: Paleopathological evidence. GAIA 1998:167–184.

Trauma – Cranial and body injuries in extant and fossil crocodilians (Brazaitis 1981; Buffetaut 1983; Cott 1961; Erickson 1996; Gilmore 1946; Sawyer and Erickson 1985, 1998; Webb and Manolis 1989; Webb and Messel 1977).

Shell disease – Virtually, every Cretaceous Baenidae in Dinosaur Park, Alberta, Canada had pathologic scars with brown versus pale yellow. In sun-bleached specimens where the carapace becomes tan or gray, the lesions are bright white.

Fossil – Cranial and body injuries in extant and fossil crocodilians (Brazaitis 1981; Buffetaut 1983; Cott 1961; Erickson 1996; Gilmore 1946; Sawyer and Erickson 1985, 1998; Webb and Manolis 1989; Webb and Messel 1977).

Virtually, every Cretaceous Baenidae in Dinosaur Park, Alberta, Canada had pathologic scars with brown versus pale yellow. In sun-bleached specimens where the carapace becomes tan or gray, the lesions are bright white.

Taruffi C. 1880. Nota storica sulla polimelia delle rane [Historical note on the polymely of frogs]. Atti della Società Italiana di Scienze Naturali e del Museo Civico di Storia Naturale in Milano 23: 112–122 [Italian].

Taruffi reviewed all cases of polymely in the genus *Rana* from antiquity to this day. He discussed the conclusions of Strobel (1876) and Cavanna (1877, 1879) regarding (among others): frequency of polymely in wild versus domestic animals, manner of attachment of supernumerary limbs. Summarized below, abnormalities in the feet could not be read, because the journal was too tightly bound. Please refer to the corresponding annotations in this volume for details on the feet.

Species	Polymely (above umbilical)	Polymely (below umbilical)	Author and year of publication
Frog	Not present Supernumerary limb	Supernumerary limb Not present	Vallisnieri (1706) De Supperville (1744)
Alytes obstetricans	Not present	Supernumerary limb	Cisternas (1865)
Palliates cultripes	Supernumerary limb	Not present	Gervais (1864)
Rana	Not present	Supernumerary limbs	Ducret (1869)
R. clamata	Not present	Supernumerary limbs	Duméril (1865)
R. esculenta	Supernumerary limbs	Not present	Alessandrini (1854) Bassi (1874) Lunel (1869) Sordelli (1878) Strobel (1876)
	Not present	Supernumerary limbs	Alessandrini (1854), Balsamo-Crivelli (1865) Cavanna (1877) Duméril (1865) Fabretti (1975) Sordelli (1877) Strobel (1876) Thomas (1865)
	Not present	Rudimentary	Cavanna (1877)
R. temporaria	Not present	Supernumerary limb	Duméril (1865)
R. viridis	Not present	Supernumerary limbs	Lunel (1869)

Taruffi C. 1881–94. Storia della Teratologia [History of Teratology]. Bologna [Italian].
See specific volumes below.

Taruffi C. 1881. Storia della Teratologia I [History of Teratology I]. 371 pp. Bologna [Italian].
Congenital – Teratologic cases through time, anecdotal, and scientific, from antiquity to this day – including those mentioned in many mythologies (among them Indian, Egyptian, Nubian, Greek). A good example are the Gorgons of Greek myth, which had reptile features including snakes for hair. He also provides a historic overview and explains how newborn human frog-resembling monsters had been believed to result from hybridization until the mid eighteenth century.
He discussed theories on the lack of certain teratologies in batrachians (Baer 1828; Dareste 1877; Saint-Hilaire 1837).

Taruffi C. 1882. Storia della Teratologia II [History of Teratology II]. 578 pp. Bologna [Italian].
Congenital – Theories on the origins of teratologies (Bruch 1867) and the merits of true double monster and hyper-regenerational causes of teratologies in batrachians.
Dicephalic ophidians survive longer than saurians.
Stated that true posterior body duplicity (a common teratology in mammals) had not been observed in reptiles, despite Tiedmann (1831) and Wiese (1812) having reported to have observed it in lizards.
Trauma – Bifid double tails are found in ofidians, saurians, and batrachians, citing Ambrosini as published in Aldrovandi (1642).
Pseudopathology – Reported that Aldrovandi (1642) received a donation of a live lizard that had two heads, with one being in place of the tail. This account was not only published by Aldrovandi (1642; p. 659, figs. V, VI), but also independently by Beltrami and Rigal (1831) and by Torquato Bembo of Padova, son of Cardinal Bembo (no publication year or title were provided).

Taruffi C. 1885. Storia della Teratologia III [History of Teratology III]. 589 pp. Bologna [Italian].
Congenital – Review of the different types of teratology and noted how many animal cases were known of each. He remarked that of 24 cases of pleuromely (thoracomely) animals known, batrachians accounted for nearly half (11 in total) as also noted by Van Deen (1838), Van der Hoven (1840), Balsamo-Crivelli (1865), Ercolani (1882, 2 cases), Fabretti (1875) and Parona (no date given), Alessandrini (1854), and Strobel (1876).
Corrects last volume, relating Braun's report of a dicephalic *Salamandra maculosa* and Born's of a dicephalic *Rana fusca*.

Supernumerary limb in *Rana* (Ercolani 1882), noting one attached to the abdomen (gastromelia) and citing 11 cases of pleuromely and 9 cases of ileopolymely in *Rana* (Balsamo-Crivelli 1865; Duméril 1865; Fabretti 1875; Lunel 1868; Otto 1816, 1841; Strobel 1876 and Parona).

Femoral duplication was very rare in *Rana* (Otto 1816), *Triton* (Camerano 1882), and axolotl (in which humeral duplication also occurred) (Sordelli 1882).

Polymelia in *Rana* (Bassi 1874; Cavanna 1879; Gervais 1864).

Claimed polydactyly in urodels is acquired, not congenital and that new digits can be added (hyperregeneration), as shown by Siebold (1828), Duméril (1865), and Sordelli (1882).

Duplication of the foot in a single leg is rare; the only case known to him, Ercolani's (1882) *Rana esculenta*.

Trauma – Claimed polydactyly in urodels is acquired, not congenital and that new digits can be added (hyperregeneration), as shown by Siebold (1828), Duméril (1865), and Sordelli (1882).

Taruffi C. 1886. Storia della Teratologia IV [History of Teratology IV]. 534 pp. Bologna [Italian].

Congenital – Reviewed anecdotal reports of anomalous snakes (e.g., Giulio Capitolino).

Reviewed Aristotle's "De generatione animalium" book IV, mentioning his explanation for the occurrence of a dicephalic snake.

Dicephalic lizards (Bibron cited by Saint-Hilaire 1836) and snakes (Aristotle, Aelianus, Aldrovandi 1642; Dutrochet 1829; Edwards 1751; Lacépède 1789; Lowne 1872; Porta, Lanzoni 1690; Redi 1684; Valentin 1704).

Double-faced reptiles and amphibians by [Catesby (as reviewed by Edwards 1754), Sack (1831) and Born (1881), but incorrectly cited and not found].

Supernumerary and abnormal limbs in frogs (Balsamo-Crivelli 1865; Bassi 1874; Cavanna 1877, 1879; Cisternas 1865; D'Alton 1853; Duméril 1865, Ercolani 1882; Fabretti 1875, Gervais 1864; Lunel 1869; Otto 1816, 1841; Parona 1883; Sordelli 1877, 1878; Strobel 1876; Van Deen 1838; Van der Hoeven 1840; Vallisneri 1706, and Thomas, as cited by Duméril (1865)).

Trauma – Review of regenerative powers (Plateretti 1777; Spallanzani 1768).

Multiple-tailed lizards (Alrovandi 1642; Pliny the Elder 1539; Porta 1589; Redi 1684; Saint-Hilaire 1838; Vallisneri 1733; Wiese 1812). List of all.

Taruffi C. 1894a. Storia della Teratologia VII [History of Teratology VII]. 598 pp. Bologna. [Italian].

Congenital – Hemivertebrae in a serpent (Albrecht 1883).

Vertebral anomalies in *Rana* (Bourne 1884; Howes7 1886).

Serpent embryo with celosoma (absent sternum), in the herpetology laboratory collections of the Museum of Paris (Dareste 1877, p. 123).

Taruffi C. 1894b. Storia della Teratologia VIII [History of Teratology VIII]. 534 pp. Bologna. [Italian].

Congenital – Correction to tome III, cites pleuromely in *Rana* (Mazza 1888) and axolotl (Sordelli 1882).

Polymelia in *Rana* (Sordelli 1876; Strobel 1876), but Fauvelle's 1886 interpretation of hyperdactyly makes doubtful that the cause of polymely in the axolotl reported by Sordelli (1883) is congenital.

Trauma – Correction to tome IV, cites Otto (1816) as having observed multiple-tailed lizards.

Fauvelle's 1886 interpretation of hyperdactyly makes doubtful that the cause of polymely in the axolotl reported by Sordelli (1883) is congenital.

Tasnádi-Kubacska A. 1962. Paläopathologie – Pathologie der vorzeitlichen Tiere [Paleopathology – Pathology of prehistoric animals]. 269 pp.; Jena: Gustav Fischer Verlag. [German].

Trauma – Bone fractures and callus formation in the pelycosaurs, *Edaphosaurus* (fibula) and *Dimetrodon* (radius and neural spine) (Permian); in the marine crocodile, *Metriorhynchus* and *Teleosaurus* (femur, ribs, and sacral vertebra); in mosasaurs, *Mosasaurus* (ribs); ichthyosaur, *Eurhinosaurus longirostris* (ribs); plesiosaurs, *Thaumatosaurus victor*, *Plesiosaurus*, *Plioplatecarpus marshi* (ribs) (Jurassic and Cretaceous); and in Pleistocene frog, *Rana mugiens* (pelvic), *Rana mehelyi* (ilium) and extant *Rana temporaria* (leg bone).

Rostral exostoses in Triassic phytosaurs, *Phytosaurus*, and *Mystriosuchus* were healed injuries and two gavials with shortened, bitten snouts.

Refers to literature on pathology of *Proneusticosaurus*, *Mosasaurus*, *Phytosaurus*, *Metriorhynchus*, *Plioplatecarpus*, *Archelon*, and *Crocodylus* citing specific papers. Two gavials with shortened, bitten snouts.

Mutilation of extremities in *Alligator mississippiensis* (Werner, Clarke, Reese).

Loss of three toes on hind leg and loss of front leg in *Caiman niger* (Schomburgk).

Cretaceous *Archelon ischyros* with shortened fibula and tibia (Moodie).

Infection – Osteomyelitis of *Mosasaurus* radius.

Gavials with shortened, bitten snouts, and abscess formation.

Vertebral – Spondylosis and necrosis? in caudal vertebrae of Jurassic sauropod dinosaurs, *Diplodocus* and *Apatosaurus*.

Other – Pachyostosis in Early Cretaceous snake, *Pachyophis woodwardi* Nopsca.

Fossil – Bone fractures and callus formation in the pelycosaurs, *Edaphosaurus* (fibula) and *Dimetrodon* (radius and neural spine) (Permian); in the marine crocodile, *Metriorhynchus* and *Teleosaurus* (femur, ribs, and sacral vertebra); in mosasaurs, *Mosasaurus* (ribs); ichthyosaur, *Eurhinosaurus longirostris* (ribs); plesiosaurs *Thaumatosaurus victor, Plesiosaurus, Plioplatecarpus marshi* (ribs) (Jurassic and Cretaceous); and in Pleistocene frog, *Rana mugiens* (pelvic), *Rana mehelyi* (ilium), and extant *Rana temporaria* (leg bone).

Rostral exostoses in Triassic phytosaurs, *Phytosaurus*, and *Mystriosuchus* were healed injuries and two gavials with shortened, bitten snouts.

Refers to literature on pathology of *Proneusticosaurus, Mosasaurus, Phytosaurus, Metriorhynchus, Plioplatecarpus, Archelon* and *Crocodylus* citing specific papers.

Cretaceous *Archelon ischyros* with shortened fibula and tibia (Moodie).

Osteomyelitis of *Mosasaurus* radius.

Spondylosis and necrosis? in caudal vertebrae of Jurassic sauropod dinosaurs, *Diplodocus* and *Apatosaurus*.

Pachyostosis in Early Cretaceous snake, *Pachyophis woodwardi* Nopsca.

Tassava RA, Huang Y. 2005. Tail regeneration and ependymal outgrowth in the adult newt, *Notophthalmus viridescens*, are adversely affected by experimentally produced ischemia. Journal of Experimental Zoology 303A:1031–1039.

Trauma – Tail regeneration is dependent upon vascular supply.

Tatarinov LP. 1998. [On a rare aberration, the absence of the occipital condyle in the the ocephal, *Viatkosuchus sumini* (Reptilia, Theriodontia).] Paleontologicheskii Zhurnal 1:101–113 [Russian].

Congenital – Absence of the occipital condyle in the the ocephal, *Viatkosuchus sumini*.

Taylor EH. 1954. Further studies on the serpents of Costa Rica. University of Kansas, Science Bulletin 36:673–801.

Trauma – Crushed *Erythrolamprus bizonus, Dipsas tenuissima* KUMNH 31950 and *Scaphiodontophis venustissimus* missing part of tail.

Taylor JA. 1984. Ecology of the lizard, *Ctenotus taeniolatus*. Interaction of life history, energy storage and tail autotomy. PHD Thesis, University of New England, Armidale, Australia.

Trauma – *Ctenotus taeniolatus* has no abdominal fat bodies, so stores majority of lipids in the tail. Survival at 5°C or 10°C was for 66 days without tail, contrasted with more than 90 days for intact individuals. Autotomy levels of 40–70% was somewhat age related, similar numbers to those observed in *Hemiergis decresiensis* (Robertson 1981). Frequencies were high in species with abdominal fat bodies (*Lampropholis delicata* and *Leiolopisma platynota*). Starved or cold geckos retain their tails for a longer period (Daniels 1981).

Increased mortality over winter in *Lacerta vivipara* with partially regenerated tails (Bauwens 1981).

Removal of tail in *Hemiergis peronii* results in smaller clutch size (Smyth 1974).

Taylor MA, Cruickshank AR. 1993. Cranial anatomy and functional morphology of *Pliosaurus brachyspondylus* (Reptilia: Plesiosauria) from the Upper Jurassic of Westbury, Wiltshire. Philosophical Transactions of the Royal Society London B 341:399–418.

Congenital – Large orbits used to suggest it was a visual hunter.

Taylor EH, Elbel RE. 1958. Contribution to the herpetology of Thailand. University of Kansas, Science Bulletin 38:1033–1189.

Trauma – Tail fragility.

Taylor SK, Green DE, Wright KM, Whitaker BR. 2001. Bacterial diseases. In: Amphibian Medicine and Captive Husbandry. KM Wright, BR Whitaker, eds. Malabar, Florida: Krieger Publishing Company; pp. 159–179.

Environmental – Malformations in Vermont wetlands attributed to chemical toxicants in eastern gray tree frog *Hyla versicolor* (1.7%), *Pseudacris crucifer* (0.7%), *Rana pipiens* (1.9%), *Rana catesbeiana* (1.6%), *Rana clamitans* (2.7%), and *Rana sylvatica* (0.3%). Malformed elements (68/83) were more common than missing elements, but statistically correlated with pollution sources rather than parasitic infection.

Taylor B, Skelly D, Demarchis LK, Slade MD, Galusha D, Rabinowitz PM. 2005. Proximity to pollution sources and risk of amphibian limb malformation. Environmental Health Perspectives 113:1497–1501.

Environmental – Rate of nontraumatic limb malformations in Vermont wetlands varied from 0% to 10. 2%. 3/235 eastern gray tree frog *Hyla versicolor* with malformed and 1 with missing limb; 3 each/895 *Pseudacris crucifer*; 2 missing and 3 malformed *Rana catesbeiana*; 9 missing and 28 malformed limbs in *Rana clamitans*; 10 missing, 28 malformed and 2 supernumerary in *Rana pipiens*; and 3/937 malformed relating proximity to agriculture and laws.

Terni T. 1915. Studio anatomico di una coda doppia di *Gongylus ocellatus*, Wagl. [Anatomical study of a double tailed *Gongylus ocellatus*, Wagl.] Archivio Italiano di Anatomia e di Embriologia 14 (2): 290–314 [Italian].

Trauma – *Gongylus ocellatus* with two tails, both in the sagittal plane (one dorsal to the other after the split).

Terni T. 1925. Esperimenti biologici della natura. [Biological experiments of nature]. Archivio di antropologia criminale, psichiatria e medicina legale 45: 363–378 [Italian].

Trauma – Reported a two-tailed lizard and referred to Italian popular writer (Ugo Ojetti) who had written a fictitious and humorous novel (entitled "Un amuleto," in: "Le vie del peccato." Castoldi – Milano, Italy) about the misadventures of someone who kept an amulet made of a multi-tailed lizard.

Themido AA. 1944. Anomalias e monstruosidades (catálogo descritivo das colecções do Museu Zoológico de Coimbra). [Anomalies and monstruosities (descriptive catalog of the collections of the Zoological Museum of Coimbra)]. Memorias e Estudos do Museu Zoológico da Universidade de Coimbra (153): 1–24 [Portuguese].

Congenital – Dicephalic *Tropidonotus natrix* L., collected by Themido in 1934.

Trauma – Bifid tails in *Lacerta ledpida lepida* Daudin, collected in Serra da Estrêla in 1886; *Lacerta bocagei bocagei* Seoane, collected in Coimbra in 1886; another *Lacerta bocagei bocagei* Seoane, collected in Coimbra as well (no date provided); *Psammodromus hispanicus hispanicus* Fitzinger, collected in Serra de Aire in 1887.

Thiemann GW, Wassersug RJ. 2000. Patterns and consequences of behavioral responses to predators and parasites in *Rana* tadpoles. Biological Journal of the Linnean Society 71:513–528.

Environmental – Exposure of *Rana clamitans* to caged fish predators increased susceptibility to trematode infection. It reduced tadpole activity, increasing proximity to cercariae.

Thiesmeier B, Günther R. 1996. Feuersalamander – *Salamandra salamandra* (Linnaeus, 1758). [Fire salamander – *Salamandra salamandra* (Linnaeus, 1758)]. In Günther R. ed. Die Amphibien und Reptilien Deutschlands. pp. 82–104; Jena: Gustav Fischer [German].

Congenital – Frequent abnormalities of body structure in *Salamandra salamandra.*

Thirumalachar B. 1928. The variation in the urostyle of *Microhyla rubra* (Jerd.). The Half-Yearly Journal of the Mysore University (India) 2(1):49–51.

Congenital – Meristematic variation in the urostyle of *Microhyla rubra* in the form of a ball and socket joint in the urostyle.

Thomason S. 1997. Helena's turtle man Mike Gugliotta ships all over the world. Birmingham News (Alabama) 13 August 1997: 8E.

Congenital – Dicephalic turtle.

Thompson d'AW. 1984. History of Animals. In The Complete Works of Aristotle, J Barnes, ed. The Revised Oxford Translation. Bollingen, Princeton University Series, pp. 774–993. [Greek translated to English].

Trauma – Lizard and serpent tails regrow.

Thompson K. 2007. Diseases of bones. In: G Grant Maxie, ed., Jubb, Kennedy and Palmer's Pathology of Domestic Animals. 5th ed. Edinburgh: Saunders Elsevier, 1–129.

Pathology – General review of pathology and terminology, including types of fractures and dysplasias, but no comments on reptiles or amphibians.

Thomson JA. 1935. Biology for Everyman. Vol.1. 756 pp.; New York: E.P. Dutton & Co.

Congenital – Thirty one dicephalic grass snakes, adders, and two dozen other cases including cobra and cited dicephalism, derdromus two bodies and Siamese (Mitchill 1826) and derderomus snake (Redi 1684).

Thorbjarnarson J, Wang X. 2010. The Chinese Alligator: Ecology, Behavior, Conservation and Culture. Baltimore: John Hopkins Press, 265 pp.

Environmental – Temperature-dependent egg incubation sex determination in American alligator Thornton CS. 1951. Beryllium inhibition of regeneration. III. Histological effects of beryllium on the amputated forelimbs of *Ambystoma* larvae.

Trauma – Midshaft amputation of humerus in *Psammodromus hispanicus* and *Ambystoma opacum* followed by erosion of that bone (histologically), with regeneration at forearm level inhibited by beryllium.

Thulborn T, Turner S. 1993. An elasmosaur bitten by a pliosaur. Modern Geology 18:489–501.

Trauma – Elasmosaur bitten by pliosaur.

Fossil – Elasmosaur bitten by pliosaur.

Tiedemann F. von 1831. Beschreibung einiger seltener Thiere – Monstra. [Description of some rare animals - Monsters]. Zeitschrift für Physiologie 4:121–124 [German].

Congenital – Young lizard(s) with doubled posterior part of the body.

Tietge J. 1997. Results of two national meetings on amphibian malformations. In: Anonymous, ed., Society of Environmental Toxicology and Chemistry 18th Annual Meeting, Pensacola, Florida, p. 90.

Environmental – Malformed amphibians are increasing, suggesting possible role for parasites, "xenobiotic chemicals" and UVB and proposes national reporting center.

Tihen JA. 1959. An interesting vertebral anomaly in a toad, *Bufo cognatus*. Herpetologica 15(1):29–30.

Congenital – Sacral vertebra in anurans may be fused with immediate preceding vertebrae or with os coccygeum. Dilated diapophyses may be fused with presacral or sacral vertebrae.

Bufo cognatus with postzygapophyses on posterior border of neural arch, but without articular surfaces. Shortened os coccygeum was also noted.

Timehri 1890. On the Upper Demerara River: About and above the great falls. Journal of the Royal Agricultural and Commercial Society of British Guiana Series 2, volume 4: 96–133.

Trauma – Chalcis (*Chalcides flavescens*) tail breaks off easily.

Tinkle DW. 1967. The Life and Demography of the Side-Blotched Lizard, *Uta stansburiana*. Miscellaneous Publications Museum of Zoology, University Michigan 132:1–182.

Trauma – Complete loss of limb or toe in a side-blotched lizard *Uta stansburiana* from gangrene and noted that Rand 1965 reported that 30–40% of *Tropidurus torquatus* had lost a toe.

Constriction at base of *Uta stansburiana* tail, divided toe, and four forked tails, one of which was triply so.

Blair 1960 reported that 15% and 24% of 1-year-old female and male *Sceloporus olivaceus* had tail breaks, compared with 35% and 50% of 3-year-olds. Cagle (1946) found regenerated tails in 30% of *Hemidactylus garnoti* adults, compared with 10% of "young."

Vertebral – Six instances of abnormalities among 3,500 *Uta stansburiana*, including crocked tail related to "defect in vertebral formation."

Tinkle DW. 1976. Comparative Data on the Population Ecology of the Desert Spiny Lizard, *Sceloporus magister*. Herpetologica 32:1–6.

Trauma – Tail breaks in 22% of yearling desert spiny lizard *Sceloporus magister.*

Tinkle DW, Ballinger RE. 1972. *Sceloporus undulatus*: A Study of the Intraspecific Comparative Demography of a Lizard. Ecology 53:570–584.

Trauma – Higher rate of *Sceloporus undulatus* tail breakage in more southern populations.

Tir, N. S. 1903. *Lacerta ocellata* Daud. Trudy. All in Russian, see Khosatzky.

Tischlinger H. 1993. Ueberlegungen zur Lebensweise der Pterosaurier anhand eines verheilten Oberschenkelbruches bei *Pterodactylus kochi* (Wagner). [Some observations on the mode of life of the pterosaurs based on a healed femur fracture of *Pterodactylus kochi* (Wagner)]. Archaeopteryx 11:63–71 [German].

Trauma – Healed fracture of right femur in pterosaur, *Pterodactylus kochi*, with considerable callus formation compared with comparable femur fractures and callus formation in extant Philippine sailing lizard, *Hydrosaurus pustulatus*, and the water agama, *Physignathus cocincinus*.

Fossil – Healed fracture of right femur in pterosaur, *Pterodactylus kochi*, with considerable callus.

Titheridge M, Berg P. 1979. Fibrous osteodystrophy in a young common iguana. Herpetile 4(1):15–16.

Metabolic – Common iguana unable to support body weight on legs; condition responding to vitamin D supplementation.

Tocidlowski ME, McNamara PL, Wojcieszyn JW. 2001. Myelogenous leukemia in a bearded dragon (*Acanthodraco vitticeps*). Journal of Zoo and Wildlife Medicine 32:90–95.

Neoplasia – Myelogenous leukemia in bearded dragon *Acanthodraco vitticeps* with swollen right elbow and inability to move hind limbs with general right humerus lysis and soft tissue swelling of elbow and stifle. Cortices were thin.

Tofohr, O. 1903. Gabelschwänzige Eidechsen. [Two-tailed lizards]. Blätter für Aquarien- und Terrarienkunde 14: 319–320 [German].

Trauma – Double-tailed *Platydactylus mauritanicus* from regeneration after a fracture and cited double-tailed rolling lizard *Gongylus ocellatus*, French finger *Acanthodactylus*, western green lizard *Lacerta viridis*, Tuesday wall lizard *Lacerta muralis*, and Australian water guan *Physignathus lesuerii.*

Tofohr, O. 1905. Doppelschwänzige Eidechsen. [Two-tailed lizards]. Blätter für Aquarien- und Terrarienkunde 16 (24. Heft): 233–235 [German].

Trauma – Bifid-tailed meadow lizard *Chalcides sepsoides*, three specimens of meadow lizard *Lacerta serpa* and *Lacerta muralis.*

Tonapi G.T. 1958. An abnormal *Rana tigrina* Daud. Science & Culture 23: 663–664.

Congenital – Absent hind limb and most of tail in one and malformed hind limbs in two other *Rana tigrina.*

Environmental – Cited description by Rugh (1954) and Yoshida et al. (1956) of monstrosities in localities "where nuclear fission experiments were carried out."

Tornier G. 1896. Über Hyperdactylie, Regeneration und Vererbung mit Experimenten. [On hyperdactyly, regeneration and heritage from experiments]. Roux' Archiv für Entwicklungsmechanik 3:469–476 [German].

Trauma – Mechanical injuries as explanation for hyperdactyly demonstrated by experiments with *Triton cristatus*.

Tornier G. 1897a. Über Schwanzregeneration und Doppelschwänze bei Eidechsen. [On tail regeneration and double tails in lizards]. Sitzungsberichte der Gesellschaft naturforschender Freunde zu Berlin 1897:59–64 [German].

Trauma – *Varanus flavescens* with double tail.

Tornier G. 1897b. Über Regeneration und Hyperdactylie. [On regeneration and hyperdactyly]. Archiv für Anatomie und Physiologie, physiologische Abtheilung 1897:52–55 [German].

Trauma – Four *Lacerta* specimens with divided regenerated tail.

Tornier G. 1897c. Über experimentell erzeugte dreischwänzige Eidechsen und Doppelgliedmaßen von Molchen. (Vorläufige Mittheilung.). [On experimentally reproduced three-tailed lizards and double extremities of urodeles (preliminary report)]. Zoologischer Anzeiger 20:356–361 [German].

Trauma – Regeneration of two or three ends of the tail in lizards, *Lacerta agilis* and *L. viridis*; regeneration of double hind leg in *Triton cristatus*.

Tornier G. 1897d. Über Operationsmethoden, welche sicher Hyperdactylie erzeugen, mit Bemerkungen über Hyperdactylie und Hyperpedie. [On surgical methods which produce hyperdactyly with certainty, with remarks on hyperdactyly and hyperpedy]. Zoologischer Anzeiger 20:362–365 [German].

Congenital – Hyperdactyly in *Triton cristatus*.

Tornier G. 1897e. Über experimentell erzeugte dreischwänzige Eidechsen und Doppelgliedmaßen von Molchen. (Vorläufige Mittheilung.). [On experimentally produced three-tailed lizards and double extremities in urodeles (preliminary report)]. Zoologischer Anzeiger 20:356–361 [German].

Trauma – Regeneration of two or three ends of the tail in lizards, *Lacerta agilis* and *L. viridis*; regeneration of double hind leg in *Triton cristatus*.

Tornier G. 1898. Ein Fall von Polymelie beim Frosch mit Nachweis der Entstehungsursachen. [A case of polymely in a frog with proof of causes of origin]. Zoologischer Anzeiger 21: 372–379 [German].

Congenital – Anatomy of the normal and additional, abnormal pectoral girdle with two additional arms on the right side of *Rana esculenta*.

Trauma – *Rana esculenta* regeneration after fracture.

Tornier G. 1900. Über Amphibiengabelschwänze und einige Grundgesetze der Regeneration. [On forked tails of amphibians and some basic laws of regeneration]. Zoologischer Anzeiger 23:233–256 [German].

Trauma – Double tail in *Triton vulgaris, Pelobates fuscus* larva, axolotl, and *Lacerta*.

Tornier G. 1901a. Neues über das natürliche Entstehen und experimentelle Erzeugen überzähliger und Zwillingsbildungen. [New results on the natural appearance and the experimental production of supernumerous and twin formations]. Zoologischer Anzeiger 24:488–504 [German].

Congenital – Induction of forked legs in *Bombinator igneus, Molge cristata*, and *M. taeniata*; induction of extremities at their girdles in *Bufo viridis, Pelobates fuscus, Molge cristata* and *M. taeniata*; formation of additional vertebrae in frogs.

Trauma – Doubled tail in *Lacerta*.

Tornier G. 1901b. Uberzählige Bildungen und die Bedeutung der Pathologie für die Biontotechnik. [Supernumerary formations and the importance of pathology for the bio-ontological techniques]. Verhandlungen des V. Internationalen Zoologischen Kongress zu Berlin: 468–500 [German].

Congenital – Dicephalic headed snake (schematic drawing).

Pelobates fuscus with bifurcation of hind leg and with two left hind legs.

Bufo mauritania with two left front legs (second leg with only one digit).

Rana esculenta with three left front legs.

Snakes with one-sided vertebra duplication.

Rana mugicus with malformation of vertebrae resulting in s-shaped arching of vertebral column.

Trauma – Double tail in *Hatteria punctata*.

Description of different kinds of tail duplication.

Tornier G. 1905. An Knoblauchkröten experimentell entstandene überzählige Hintergliedmaßen. [Experimentally induced supernumerous posterious extremities in spadefoots]. Roux' Archiv für Entwicklungsmechanik 20:76–124 [German].

Congenital – Supernumerary hind legs in 12 specimens of *Pelobates fuscus*.

Tornier G. 1906. Experimentelles und Kritisches über tierische Regeneration. [Experimental and critical remarks on regeneration in animals]. Teil 6–10. Sitzungsberichte der Gesellschaft Naturforschender Freunde zu Berlin 1906: 264–287 [German].

Congenital – Reason for anomalies is the small space in the egg membrane (compared with the amnion in mammals).

Tornier G. 1911. Über die Art, wie äussere Einflüsse den Ausbau des Tieres abändern. [On the art of how external influences change the formation of animals]. Verhandlungen der Deutschen Zoologischen Gesellschaft 21:45–91 [German].

Congenital – Tail deformations (shortening, arching) in larvae of axolotl.

Tornier, G. 1919. Mißbildete Embryonen bei *Salamandra mac.* (Briefliche Mitteilung an den Herasugeber). [Malformed embryos of *Salamandra mac.* (Report by letter to the editor)]. Blätter zur Aquarien- und Terrarienkunde, Braunschweig, 30: 161 [German].

Congenital – *Salamandra maculosa* malformed embryos, including cyclops.

Tóth T, Gál J, Ujvári B, Madsen T. 2005. Severe malformation in neonate *Vipera ursinii rakosiensis*. Amphibia-Reptilia 26:388–390.

Congenital – Hungarian meadow viper *Vipera ursinnii rakosiensis* Siamese twins, with fused maxillary bones. One was male, the other a female, demonstrating that these embryos from separate eggs grew together during embryologic development.

Towle EW. 1901. On muscle regeneration in the limbs of *Plethodon*. Biological Bulletin 6:289–299.
Trauma – Limb regeneration in *Necturus maculosus, Amphiuma means, Spelerpes ruber* and *guttolineata, Plethodon cinereus, Desmognathus ochrophaeus, Diemyctylus viridescens, and Manculus quadrigitatus.*
Wiedersheim, (1875 I, p. 95) claimed no regeneration in *Proteus* and *Siren lacertina*, while Erber (1876 4, p. 34) claimed the regeneration occurred in the latter.
Environmental – Regeneration in *Triton* occurs from 48°F to 57°F (Higginbottom 1847).

Townsend CH. 1928. A double turtle. Bulletin of the New York zoological Society 31:35–36.
Congenital – Derodymous *Chrysemys picta, with* depressed area between the united plastrons.

Townsend S. 1977. A short note on a specimen of *Geocrinia victoriana* with five legs. Victorian Naturalist 94:56.
Congenital – Supernumerary limb in of *Geocrinia victoriana.*

Trefz M, Hiller A. 1992. Erfolgreiche Heilung von Panzernekrosen bei der Rotbauch-Spitzkopfschildkröte, *Emydura australis subglobosa*. [Successful healing of necrosis of the shell in the red-bellied, side-necked turtle, *Emydura australis subglobosa*]. Herpetofauna (Weinstadt) 14(81):33–34 [German].
Shell – Red-bellied short-necked turtles (*Emydura australis subglobosa*) with carapace necrosis; successful treatment with oxytetracycline and multivitamin medicine; in addition, local treatment with m-cresol fulson-acid-formaldehyde-polycondensate.

Triplehorn C.A. 1955. Notes on the young of some North American reptiles. Copeia 1955(3): 248–249.
Congenital – *Storeria dekayi* with one head but two bodies.

Troiano JC, Román LH. 1996. Diseases encountered in genus *Caiman* intensive breeding. In: Crocodiles. Proceedings of the 13th Working Meeting of the Crocodile Specialist Group. IUCN – The World Conservation Union, Gland Switzerland, pp. 328–341.
Trauma – Rostral loss in *Caiman crocodylus yacare* from aggressive behavior.
Partial tail loss in *Caiman latirostris.*
Metabolic – *Caiman crocodylus yacare* spine curvature, softening of jaw and limb bones with deformations and fractures from metabolic bone disease produced by low calcium or excess phosphorus diets or ingestion of excess phthalic acid.

Trutnau L. 1977. Behandlung von Darmkrankheiten bei Krustenechsen (*Heloderma horridum* und *H. suspectum.* [Treatment of enteritis in gila monster (*Heloderma horridum* and *H. suspectum*]. Deutsche Aquarien- und Terrarien-Zeitschrift 30:100–102 [German].
Vertebral – Enteritis in gila monster *Heloderma horridum* and *H. suspectum* is of significance as a potential source of spondyloarthropathy – BMR.

Tscharntke T, Zucchi H. 1975. Doppelschwänzige Zauneidechse. [Double-tailed sand lizard]. Kosmos, Stuttgart 71:274–275 [German].
Trauma – *Lacerta agilis* (Zauneidechse) with short regenerated tail in addition to the normal tail.

Tscherepach P. 1969. Ein sensationelles Bilddokument: Zwillingsbildung auch bei Landschildkröten. [A sensational image document: Twinning also in tortoises]. Deutsche Aquarien- und Terrarien-Zeitschrift 16:33 [German].
Congenital – Duplication of anterior portion of tortoise, with shared leg.

Tschudi JJ. v. 1857. Monographie der schweizerischen Echsen.[Monography of lizards of Switzerland]. Neuchâtel, 42 pp [German].
Trauma – Double- or triple-tailed lizards as regeneration.

Tschudi F. v. 1865. Das Thierleben der Alpenwelt. Naturansichten und Thierzeichnungen aus dem Schweizerischen Gebirge. [The Animal Life of the Alps. Views of nature and animal drawings from the Swiss mountains]. Leipzig: JJ Weber, 560 pp. [German].
Trauma – Double-tailed lizard.

Tsonis PA, Eguchi G. 1981. Carcinogens on regeneration: Effects of N-methyl-N'-nitro-N-nitrosoguanidine and 4-nitroquinoline-1-oxide on limb regeneration in adult newts. Differentiation 20:52–60.
Toxicology – Carcinogens *N*-methyl-*N'*-nitro-*N*-nitrosoguanidine and 4-nitroquinoline-1-oxide produce absent radius and ulnar, subregeneration or superregeneration of carpals and digits and disorganization of skeletal elements in Japanese newt *Triturus* (*Cynops*) *pyrrhogaster.*

Tsonis PA, Eguchi G. 1982. Abnormal limb regeneration without tumor production in adult newts directed by carcinogens, 20-methylchloranthene and benzo(α)pyrene. Development and Growth Differentiation 25:183–190.
Toxicology – Direct application of 20-methylchloranthene and benzo(a)pyrene to limb buds produced subregeneration and superregeneration of carpal, digits, ulna, and radius absence and accessory limb formation in Japanese newt *Cynops* (*Triturus*) *pyrrhogaster*.

Tsonis PA, Eguchi G. 1985. The regeneration of newt limbs deformed in nature. Cellular and Molecular Life Sciences 41:918–919.
Trauma – Regeneration in a wild newt population revealed symmetrical reduplication, supernumerary digits, and deformed wrists, independent of the original deformity. They suggested traumatic origin of the anomalies.

Tuck RG. 1973/4. Acranio Dichotomous queensnake, *Regina septemvittata* (Say) (Serpentes: Natricidae) from Frederick County, Maryland. Bulletin of the Maryland Herpetological Society 9(2):22–23.
Congenital – Dicephalic queensnake *Regina septemvittata* USNM 164350.

Tuck RG, Hardy JD. 1970. An amphidichotomous northern black racer, *Coluber c. constrictor*, from Maryland and a resume of recent records of axial bifurcation in snakes. Bulletin of the Maryland Herpetological Society 6(3):37–48.

Congenital – Amphidichotomous (Siamese twin) *Coluber constrictor* USNM 162645.

Dicephalic California king snake *Lampropeltis getula californiae* (Shaw 1956, 1959), European grass snake *Natrix (Tropidonotus) natrix* (Ladeiro 1935), *Amphiesma vibakari* (*Natrix vibakari*) (Nakamura 1938), *Dromicus chamissonis* (Pflaumer 1945; Prado 1946), *Elaphe climacophora* (Nakamura 1938), *Elaphe conspicillata* (Nakamura 1938), *Elaphe quatuorlineata sauromates* (Alekperov 1954; Amrakh 1944; Iki 1946), *Helicops carinicauda infrataeniata* (Lema 1958), *Heterodon platyrhinos platyrhinos* (Meyer 1958), *Lampropeltis getula californiae* (Lüdicke 1964; Schmidt and Inger 1957; Shaw 1956, 1959), *Lampropeltis triangulum (doliata) triangulum* (Lüdicke 1964), *Leimadophis poecilogyrus* (Lüdicke 1964; Prado 1942, 1943, 1946), *Leptodeira annulata ashmeadi* (Belluomini and Lancini 1960), *Liophis miliaris (miliaris semiaureus)* (Lema 1957), *Natrix* (*Tropidonotus*) *natrix* (Ladeiro 1935; Lüdicke 1964; Martins d'Alte 1937; Themido 1944), *Natrix rhombifer rhombifer* (Oringderff 1969), *Philodryas patagoniensis (schottii)* (Prado 1946), *Regina* (*Natrix*) (Goldberg 1967; Neill 1941), *Rhabdophis* (*Natrix*) *tigrina* (Nakamura 1938), *Tachymenis peruviana* (Lüer 1944a, b; Prado 1946), *Thamnophis sirtalis sirtalis* (Cohen 1937), *Xenodon merremi* by (Lema 1961), *Pseudechis porphyriacus* (Longman 1939), *Agkistrodon halys blomhoffii* (Nakamura 1938), *Bothrops alternatus* (Berst 1945; Prado 1946), *Bothrops atrox* (Daniel 1941, 1955; Dupouy 1958; Lüdicke 1964; Pflaumer 1945; Prado 1942, 1946), *Bothrops jararaca* (Pereira 1950), *Bothrops jararacussu* (Pereira 1944), *Crotalus durissus terrificus* (Klauber 1956; Lüdicke 1964; Vanzolini 1947), *Crotalus horridus horridus* (Anonymous 1967; Harris 1968), *Vipera berus* (Curry-Lindahl 1963; Lagerlund and Hanstrom 1961; Nubelin 1942; Steward 1961), *Vipera russelli* (Deraniyagala 1958), *Natrix sipedon clarkii* (List and Smith 1954), *Storeria dekayi* (Tripplehorn 1955), *Thamnophis sirtalis sirtalis* (Martof 1954), *Calliophis* (*Hemibungarus*) *japonicus* (Nakamura 1938), *Coluber constrictor constrictor* (current paper), and *Crotalus viridis viridis* (Klauber 1956).

Tuckerman F. 1886. Supernumerary leg in a male frog, *Rana palustris*. Journal of Anatomy and Physiology 20:516–519.

Congenital – Ectromelic supernumerary leg in *Rana palustris*, but with ankylosed inferior tibia and fibula articular expansion and replacement of astragalus and calcaneus by a single bone.

Accessory acetabulum.

Turley SD, Eaton-Poole L, Pinkney AE, Osborn MA, Burton DT. 2003. Evaluation of the potential impact of water and sediment from National Wildlife Refuge sites using a modified frog embryo teratogenesis assay – *Xenopus* (FETAX). Pp. 79–95. In: Multiple Stressor effects in Relation to Declining Amphibian Populations. West Conshochocken, Pennsylvania: ASTM International.

Environmental – Axial (curvature) and craniofacial (reduced upper jaws) malformations in *Xenopus* from exposure to Missisquoi (Swanton, Vermont) water exposure, but no missing or supernumerary limbs.

Turner FB, Medica PA, Jennrich RI, Maza BG. 1982. Frequency of broken tails among *Uta stansburiana* in southern Nevada and a test of the predation hypothesis. Copeia 1982:835–840.

Trauma – Broken tails were present in 6% of *Uta stansburiana* hatchlings (correlated with predator density and varied from 3% in 1971 to 10% in 1966), 30% of yearlings, and 51% of 2-year-olds.

Twitty VC, Delanney LE. 1939. Size regeneration in salamander larvae under complete starvation. Journal of Experimental Zoology 81:399–414.

Trauma – Starvation in *Ambystoma* did not affect tail growth.

Tyler MJ, Leong AS-Y, Godthelp H. 1994. Tumors of the ilia of modern and Tertiary Australian frogs. Journal of Herpetology 28:528–529.

Neoplasia – 1/49 spotted grass frogs *Limnodynastes tasmaniensis* with tumor, but also describes 4 with expansile tumors, 3 in posterior half of shaft; 1, in anterior. They were radiolucent with cortical thinning and scalloping with focal internal calcification. Histology confirmed osteochondroma in 3, but cortical disruption in one suggests probable chondrosarcoma.

Other – Oligo-Miocene Two Trees and Upper Site, Riversleigh Station, Northwest Queensland frog with ilial shaft distortion.

Fossil – Oligo-Miocene Two Trees and Upper Site, Riversleigh Station, Northwest Queensland frog with ilial shaft distortion.

Tytle T, Grimpe R, Pickering D, Putnam L. 1984. Life history notes. Sauria. *Phelsuma madagascariensis kochi* (Koch's day gecko). Herpetological Review 15:49.

Congenital – Siamese twin *Phelsuma madagascariensis kochi* (Koch's day gecko) jointed at back of head.

Tytler LC. 1864. Observations on a few species of geckos alive in the possession of the author. Journal of the Asiatic Society of Bengal 33:535–548.

Trauma – Tail detachment in geckos.

Tytler RC. 1865. Observations on a few species of Geckos alive in the possession of the author. Journal of the Asiatic Society of Bengal 33:535–548.
Trauma – Conspecific-derived tail loss in *Gekko chameleon.*
Uehlinger V. 1969. Une mutation récessive (pd) déterminant la polydactylie chez *Xenopus laevis* D. (Batraciens, Anoures). [A recessive mutation (pd) for polydactyly among *Xenopus laevis* D. (Batrachia, Anura)]. Journal of Embryology and Experimental Morphology 21: 207–218 [French].
Congenital – Recessively transmitted polydactyly in *Xenopus laevis.*
Uehlinger V, Reynaud J. 1965. Une anomalie héréditaire (kinky tail) chez *Xenopus laevis* D. [A hereditary (kinky tail) in *Xenopus laevis* D]. Revue de Suisse Zoologie 72:680–685 [French].
Congenital – Recessively transmitted kinky tail tip in *Xenopus laevis.*
Ulber T. 1986. Ein Beitrag zur "Knickschwanz"-Problematik bei *Calotes* Cuvier, 1817. [A contribution to tail-bending in *Calotes*]. Sauria 8(3):21–22 [German].
Vertebral – Tail-bending and malformation of vertebral column in *Calotes* and *Phelsuma* in captivity; *Calotes emma* (Thailand) and *Lygodactylus* (Kenya), in nature.
Ullrey DE, Bernard JB. 1999. Vitamin D: Metabolism, sources, unique problems in zoo animals, meeting needs. In ME Fowler, RE Miller, eds. Zoo and Wild Animal Medicine, Current Therapy 4. Philadelphia: Saunders, p. 63–78.
Metabolic – Review of physics and physiology of vitamin D metabolism.
Metabolic bone disease in green iguana *Iguana iguana* manifest as increased flexibility of long bones, apparent thickening of limbs (does not comment on bone thickening), fibrous osteodystrophy, cortical thinning, widening osteoid seams, and poor trabecular mineralization. Soft tissue mineralization was noted in iguana not illuminated by UV lamps.
For Savanna monitor lizard *Varanus exanthematicus* and *Iguana iguana* conversion of provitamin D_3 is temperature dependent.
Day geckos *Phelsuma madagascariensis* and Komodo dragons (*Varanus komodoensis*) require UV light for normal calcium metabolism and normal bone structure.
Ultsch GR. 1985. The viability of Nearctic freshwater turtles submerged in anoxia and normoxia at 3 and 10°C. Comparative Biochemistry and Physiology 81A:607–611.
Vascular – Mud and musk (Kinosternidae) and soft-shelled (Trionychid) turtles have extrapulmonary gas exchange as do emydid turtles (Smith and Nickon 1961), including *Chrysemys picta* (Ultsch and Jackson 1982). *Sternotherus* and *Trionyx* have thin skins with favorable surface to volume ratios. The latter is related to size for *Sternotherus* and shape, in *Trionyx.* Both also have buccopharyngeal respiration, as does *Chelydra serpentina.*
Suggests that all freshwater turtles have effective extrapulmonary gas exchange.
Ultsch GR, Jackson DC. 1982. Long-term submersion at 3°C of the turtle *Chrysemys picta bellii*, in normoxic and severely hypoxic water. I. Survival, gas exchange and acid-base status. Journal of experimental Biology 96:11–28.
Vascular – Extrapulmonary respiration in emydid turtle *Chrysemys picta bellii.*
Underhill DK. 1966a. An incidence of spontaneous caudal scoliosis in tadpoles of *Rana pipiens* Schreber. Copeia 3:582–583.
Congenital – Recessive inheritance of scolosis in *Rana pipiens*, with rates ranging from 28% in Maryland to 5% in North Carolina and 40% in a "French pond."
Underhill DK. 1966b. An incidence of spontaneous caudal scoliosis in tadpoles of *Rana pipiens* Schreber. Copeia 1966:582–583.
Congenital – Caudal scoliosis or lateral tail curvature in *Rana pipiens.*
Urban EK. 1965. Quantitative study of locomotion in teiid lizards. Animal Behavior 13:513–529.
Trauma – Tail loss in *Cnemidophorus* reduces speed.
Ursprung E, Ringler M, Jehle R, Hödl W. 2011. Toe regeneration in the neotropical frog *Allobates femoralis.* Herpetological Journal 21:83–86.
Trauma – Toe regeneration in aromobatid frog *Allobates femoralis.*
Vaglia JL, Babcock SK, Harris RN. 1997. Tail development and regeneration throughout the life cycle of the four-toed salamander *Hemidactylum scutatum.* Journal of Morphology 233:15–29.
Trauma – Larva of four-toed salamander *Hemidactylum scutatum* that experience tail injury have lower survival during the aquatic phase, from predation by the red spotted newt *Notophthalmus viridescens.* Partial compensation for tail loss occurs by elongation of vertebral centra anterior to the injury.
Valente AL, Parga ML, Espada Y, Lavin S, Alegre F, Marco I, Cuenca R. 2007. Ultrasonographic imaging of loggerhead sea turtles (*Caretta caretta*). Veterinary Record 161:226–232.
Vascular – Oxygen storage in loggerhead sea turtles (*Caretta caretta*) was related to "large amounts of solid tissues and a great blood supply."
Vallisneri A. 1706. V. Galleria di Minerva 5:285 [Italian].
Five-legged frog that also possessed a left hind limb with seven digits (same specimen published in 1715 and 1733, after Taruffi 1880).

Vallisneri A. 1715. Nuove osservazioni fisico-med. [New physico-medical observations] Venezia: 203 [Italian].
Five-legged frog that also possessed a left hind limb with seven digits (same specimen published in 1706 and 1733, after Taruffi 1880).
Vallisneri A. 1733a. Opere fisico-mediche [Physico-medical works]. Volume 2: 551 pp. Venezia [Italian].
Congenital – Five-legged with supernumerary leg next to the end of the coccyx, on the right side, and left hind foot had seven digits.
Vallisneri A. 1733b. Opere fisico-mediche [Physico-medical works]. Volume 3: 676 pp, [but pages 410+ were not available for review]. Venezia [Italian].
Congenital – Five-legged frog that also possessed a left hind limb with seven digits (pp. 306–307).
Trauma – Multi-tailed lizards by Pliny the Elder (1539), Porta (1589), Redi (1684), and Aldrovandi (1642) in page 448 of volume 3 (after Taruffi, 1886).
Valmont-Bomare JC. 1775. Dictionnaire raisonné universel d'histoire naturelle: contenant l'histoire des animaux, des végétaux et des minéraux, et celle des corp célestes, des météores, et des autres principaux phénomenes de la nature. [Reasoned universal dictionary of natural history: Containing the history of animals, vegetables and minerals and celestial bodies, meteorites, and other principal phenomena of nature]. Vol 5, 621 pp.; Paris: Brunet [French].
Trauma – *Lacerta vivipara (Lacertus terrestris)* with two and three tails.
Van Deen I. 1838 Anatomische Beschreibung des monströsen, sechsfüssigen Wasser-Frosches (*Rana esculenta*). [Anatomical description of monstrous, six-legged water frog (*Rana esculenta*)]. 24 pp.; Leiden: S. und J. Luchtmans [German].
Congenital – *Rana esculenta* with additional pair of hind legs on additional pelvis ventral to left leg. Additional phalange on front leg and on hind leg in other specimens.
Van der Hoeven J. 1840. Fragments zoologiques sur les Batraciens. [Zoologic questions on batrachians]. Mémoires de la Société d' Histoire Naturelle Strassbourg 3:1–12 [French].
Congenital – *Rana esculenta* with two supernumerary hind limbs which is attached to the pubic symphysis.
Rana with five digits in forelimb and *Salamandra subviolacea* with one accessory digit in a forelimb (after Taruffi 1885).
Van der Hoeven J. 1932. Eenige aanteekeningen over de kenmerken van het geslacht *Bombinator* van Merrem, en over de soorten, die daartoe behooren. [Some office omen for guidance on the characteristics of the *Bombinator* of Merrem genus and species]. In HC van Hall, W Vrolik, GJ Mulder. Bydragen tot de natuurkundige Wetenschappen. [Contributions to the Studies of Natural Sciences]. Amsterdam: Erven H. Gartman, pp. 77–82 [Dutch].
Congenital – *Bombinator obstetricans fuscus* with unique triangular-shaped transverse processes of the sacral vertebrae.
Van der Meulen J. 1995. Een dijbeenbrejk bij een *Litoria infrafrenata*. [A fracture of the tibia in a *Litoria infrafrenata*]. Terra 31:80–81 [Dutch].
Trauma – Spontaneously healed tibial fracture in *Litoria infrafrenata*.
Van Gelder JJ, Rijsdijk G. 1987. Unequal catchability of male *Bufo bufo* within breeding populations. Holarctic Ecology 10:90–94.
Trauma – Thirty of 10,000 *Bubo bubo* were missing part of legs – caused by mowing.
Van Gelder JJ, Strijbosch H. 1995. Adult common toads (*Bufo bufo*) with mutilated legs. Alytes 13(3):105–108.
Trauma – Thirty of 10,000 *Bubo bubo* were missing part of legs – caused by mowing.
Van Hoepen ECN. 1913. Bijdragen tot de Kennis der Reptielen van de Karoo-formatie. 1. De Schedel van *Lystrosaurus latirostris* Owen sp. [Contribution to the knowledge of reptiles of the Karoo formation. 1. The skull of *Lystrosaurus latirostris* Owen sp.]. Annals of the Transvaal Museum 4:1–46 [Dutch].
Congenital – Unpaired, median bone between parietals and frontals in the dicynodont therapsid (mammal-like reptiles), *Lystrosaurus*.
Fossil – Unpaired, median bone between parietals and frontals in the dicynodont therapsid (mammal-like reptiles), *Lystrosaurus*.
Vanni S. 1979. Note di erpetologia della Toscana: *Salamandrina terdigitata, Rana graeca, Coluber viridiflavus, Natrix natrix* [Herpetological notes of Tuscany: *Salamandrina terdigitata, Rana graeca, Coluber viridiflavus, Natrix natrix*]. Atti della Società Toscana di Scienze Naturali Residente in Pisa, Memorie (Serie B) 86:103–123 [Italian].
Congenital – Dicephalic (two) European whip snakes *Coluber viridiflavus viridiflavus* Lacépède 1789 in the collections of the provincial museum of Natural History of Livorno number 2479, one grass snake *Natrix natrix lanzai* Kramer 1971 in the collections of the zoology museum "La Specola" number 23780 and *Crotaphopeltis hotamboeia* Laurenti (1768) captured in 1978 in Bardere Ier (Somalia) also in the collections of the zoology museum "La Specola."
Spectacled salamander *Salamandrina terdigitata* Lacépède 1788 with polydactyly and syndactyly in the hind limb and swelling of both forearms.

Vanni S, Nistri A. 1987. Brevi note su alcuni esemplari anomali conservati nella collezione erpetologica del Museo Zoologico 'La Specola' dell'Università di Firenze (Reptilia) [Brief note on some anomalous specimens conserved in the herpetological collections of the zoological museum 'La Specola' of the University of Florence (Reptilia)]. Atti del Museo Civico di Storia Naturale di Grosseto 11/12: 85–87 [Italian].

Congenital – Derodymous Hermann's tortoise (*Testudo hermanni robertmertensi* Wermuth (1952)) collected near Cavriglia Arezzo, Italy, catalog number 7707 M.F., also number 762 in the Italian Central Vertebrate Collection established by E.H. Giglioli. Marginals are larger than those of normal specimens of like age with variation in count and separation between scutes.

Van Sluys M, Vrcibradic D, Rocha CF. 2002. Tail loss in the syntopic lizards *Tropidurus itambere* (Tropiduridae) and *Mabuya frenata* (Scincidae) in southeastern Brazil. Studies on Neotropical Fauna and Environment 37:227–231.

Trauma – Effects of autotomy relate to habit/habitat (terrestrial or fossorial), ease of tail shedding, frequency of intraspecific aggression and interspecific predation, character of predator, age and sex of sample, and species longevity, citing Bellairs and Bryant (1985).

Autotomy recognized in 23% of *Tropidurus itambere* (as tree branch sit and wait foragers) and 82% of *Mabuya frenata* (as active foragers).

Frequency (% of #) of tail regeneration:

Gekkonidae –	*Coleodactylus amazonicus*	4 of 22
	C. septentrionalis	40 of 35
	Gonatodes humeralis	25 of 20
	Gymnodactylus geckoides	48 of 219
	Hemidactylus mabouia	50 of 40
	H. palaichthus	32 of 20
	Lygodactylus klugei	39 of 361
	Phyllopezus pollicaris	42 of 135
	Thecadactylus rapicauda	66 of 64
Gymnophthalmidae		
	Arthrosaura reticulate	44 of 39
	Leposoma parietale	9 of 21
	L. percarinatum	19 of 24
	Neusticurus ecpleopus	20 of 50
	Prionodactylus oshaugnessy	46 of 24
	Vanzosaura rubricauda	62 of 130
Polychrotidae	*Anolis fuscoauratus*	9 of 40
Scincidae	*Mabuya agilis*	83 of 36
	M. frenata	82 of 216
	M. heathi	59 of 255
	M. macrorhyncha	78 of 106
Teiidae	*Ameiva ameiva*	34 of 303, 26 of 25, 32 of 22
	Cnemidophorus littoralis	42 of 100
	C. ocellifer	44 of 476
	Kentropyx pelviceps	43 of 23
	Tupinambis merianae	17 of 52
	T. rufescens	24 of 422
Tropiduridae	*Liolaemus lutzae*	26 of 299
	Tropidurus etheridgei	19 of 19
	T. hispidus	38 of 431, 57 of 55
	T. itambere	23 of 547
	T. montanus	31 of 147
	T. nanuzae	28 of 108
	T. semitaeniatus	60 of 403
	T. torquatus	34 of 121

Van Valen L. 1974. A natural model for the origin of some higher taxa. Journal of Herpetology 8:109–121.

Congenital – Supernumerary limbs (De Superville 1740) in 29 of 86 *Hyla regilla* from Montana (Hebard and Brunson 1963); similar number from New Jersey (Anonymous 1964).

Polymelous frogs, toads, and rarely salamanders: Bateson (1894), Duméril (1865), Ercolani (1881), Gemmill (1906), Ghorab (1959), Kingsly (1883), Mahendra (1936), Pol (1958), Przibram (1921), Rostand (1951), Voitkevich (1959); Adler (1958), anonymous (1944, 1945, 1962), Banta (1966), Bonnet and Rey (1935), Brandt (1932, 1933, 1934, 1935), Canella (1935), Chalaux (1952), Charles (1944), Cooper (1958), Cunningham (1955), Freytag (1941, 1952), Gaggero (1960), Heatwole and Suarez-Lazu (1965), Hobson (1958), Hvass (1943), Kahn (1926), Mertens (1925), Moncharmant (1949), Pearson (1960), Perri (1951, 1952), Peterka (1941), Rahmann et al. (1962), Rostand (1956), Ruth (1961), Rylkova (1924), Samarasinghe (1951), Schussler (1925), Storer (1925, p. 58), Witschi and Chang (1954), and Wu and Liu (1941).

Polymely in 20 of 60 leptodactylids from Ecuador (Copping, cited by Rostand 1958); 25 of 250 *Rana catesbeiana* from Ohio (Anonymous 1954; Hauver 1958), 350 individuals from Mississippi (Volpe 1967), and 600 polymelous *Rana esculenta ridibunda* from Russia (Voitkevich 1948, 1957, 1958, 1959, 1961, 1963, 1965).

Polydactyly *Ambystoma tigrina* up to 90% from Colorado (Bishop 1947, 1949; Bishop and Hamilton 1947; Rosine 1955), 1/48 *Chioglossa lusitanica* from Portugal, 1/48 (Dubois and Thireau 1972) *Rana e. esculenta* from France up to 85% (Rostand 1949, 1952, 1958, 1959, 1962; Rostand and Darre 1969) and Netherlands (Voitkevich 1961), *Rana esculenta* from France (Dubois and Thireau 1972), 3/23 *Rana iberica* from Portugal (Dubois and Thireau 1972), 3/40 *Rana temporaria* (Rostand and Fischer 1959), 5/22 *Salamandra salamandra* from France (Joly 1966), 4/47 *Triturus cristatus* from France (Dubois and Thireau 1972), 2/43 *Triturus helveticus* from France 2/43 (Dubois and Thireau 1972), and probably in Switzerland and Germany (Rostand et al. 1967).

Vanzolini PE. 1947. Notas sôbre um deródimo de *Crotalus durissus terrificus* (Laur) [Notes about a derodymous *Crotalus durissus terrificus* (Laur).]. Papeis avulsos do Departamento de Zoologia, São Paulo 8(24):273–283 [Portuguese].

Congenital – Derodymous *Crotalus durissus terrificus* in the collections of the Department of Histology and Embryology of the Faculty of Medicine of the University of São Paulo, Brazil, with review of literature

Varela – Lasheras I, Bakker AJ, van der Mije SD, Metz JA, van Alphen J, Galis F. 2011. Breaking evolutionary and pleiotropic constraints in mammals. On sloths, manatees and homeotic mutations. EvoDevo 2:11 doi: 10.1186/2041-9139-2-11

Congenital – Asymmetrical thoraco-lumbar and "caudal boundry" transitional vertebrae are common in Anolis, citing personal communication by Andre Pires da Silva.

Vasilyev DB, Solovyev YN, Mitin VN. 2003. [Bone tumors of reptiles.] Voprosy Onkologii (St. Petersburg) 49:81–84 [Russian].

Neoplasia – Review of tumors in reptiles:

Multiple enchondromas in *Varanus dracaena*

Chondroosteofibroma in *Cyclura cornuta*

Osteosarcoma in *Varanus* sp. and *Rhamphiophis rostratus*

Osteochondroma in *Naja melanoleuca* and *Varanus bengalensis*

Osteogenic sarcoma in Burmese python *Python molurus bivittatus* and *Python reticulatus*

Ossifying fibroma in *Iguana iguana*

Chondrosarcoma of second degree of anaplasia in *Iguana iguana*

Vaughn PP. 1955. The Permian reptile, *Araeoscelis* restudied. Bulletin of the Museum of Comparative Zoology Harvard 113:305–468.

Trauma – Permian MCZ 1262 *Araeoscelis* has a regenerated tail manifest as a tubular fragment of bone with irregular transverse ridges.

Fossil – Permian MCZ 1262 *Araeoscelis* has a regenerated tail manifest as a tubular fragment of bone with irregular transverse ridges.

Vaughn PP. 1970. Alternation of neural spine height in certain Early Permian tetrapods. Bulletin of Southern California Academy of Sciences 69:80–86.

Congenital – Structural modification with alternating height vertebral neural spines for efficient vertebral column dorsiflexion and lateral flexion in *Captorhinus*.

Fossil – Structural modification with alternating height vertebral neural spines for efficient vertebral column dorsiflexion and lateral flexion in *Captorhinus*.

Vega CS, Maisch MW. (in press) Pathological features in Upper Permian and Middle Triassic dicynodonts (Synapsida, Therapsida). Pathologic feature in Upper Permian and Middle Triassic dicynodonts (Synapsida, Therapsida). In: Early Evolutionary History of the Synapsida. Vertebrate Paleobiology and Paleoanthropology Book Series. Springer.

Trauma – *Stachleckeria* GPIT/RE/8001 from Middle Triassic of Brazil with possible scapula tendon avulsion.

Infection – *Stachleckeria* GPIT/RE/8001 from Middle Triassic of Brazil with possible infectious arthritis of knee.

Dermal – *Geikia locusticeps* GPIT K114 lesion, holotype of *Pelanomodon tuberosus*, from Upper Permian of Tanzania with concave circular pit with raised rim on maxilla, probably caused by inclusion/epidermal cyst.

Pathology – *Stachleckeria* GPIT/RE/8001, from Middle Triassic of Brazil with lesion of distal femur which appears to be a large and multiple small pressure erosions, with aneurysm or hydatid cysts to be considered.

Fossil – *Stachleckeria* GPIT/RE/8001 from Middle Triassic of Brazil with possible scapula tendon avulsion.

Stachleckeria GPIT/RE/8001 from Middle Triassic of Brazil with possible infectious arthritis of knee.

Geikia locusticeps GPIT K114 lesion, holotype of *Pelanomodon tuberosus*, from Upper Permian of Tanzania with concave circular pit with raised rim on maxilla, probably caused by inclusion/epidermal cyst.

Stachleckeria GPIT/RE/8001 from Middle Triassic of Brazil with lesion of distal femur which appears to be a large and multiple small pressure erosions, with aneurysm or hydatid cysts to be considered.

Vega-Dias C, Schultz CL. 2003. A paleopathology in *Jachaleria candelariensis* Araújo and Gonzaga 1980 (Synapsida, Dicynodontia) from the Upper Triassic of Southern Brazil. Ameghiniana 40(4):74R.

Pathology – Upper Triassic *Jachaleria candelariensis*, Universidade Federal do Rio Grande do Sul – Instituto de Geociências UFRGS-PVo151T, lacking acromial process with deep semilunar depression with oval shaped excavation.

Fossil – Upper Triassic *Jachaleria candelariensis*, Universidade Federal do Rio Grande do Sul – Instituto de Geociências UFRGS-PVo151T, lacking acromial process with deep semilunar depression with oval shaped excavation.

Veith M, Viertel B. 1993. Veränderungen an den Extremitäten von Larven und Jungtieren der Erdkröte (*Bufo bufo*): Analyse möglicher Ursachen. [Modifications of the extremities of larvae and young adults of the common toad (*Bufo bufo*): Analysis of possible causes]. Salamandra 29(3–4):184–199 [German].

Trauma – Damaged hind limbs in common toad *Bufo bufo* related to pollution of breeding pond or predation by predatory leeches *Erpobdella octoculata.*

Vella D. 2007. Management of freshwater turtle shell injuries. Lab Animal 38(1):13–14.

Trauma – Turtle shell injury results from falling, being dropped, or stepped on.

States that turtles are avid climbers.

Vellard J, Penteado J. 1931. Un cas de bicéphalie chez un serpent (*Liophis almadensis*). [A case of bicephaly in a snake (*Liophis almadensis*)]. Bulletin de la Société de Zooologie de France 56(4):360–362 [French].

Congenital – Dicephalic *Liophis almadensis* and *Testudo ibera.*

Vergnaud-Grazzini C. 1966. Les Amphibiens du Miocène de Beni-Mellal. [The Miocene amphibians of Beni-Mellal]. Mémoires du Service Géologique Maroc 27 (198): 43–69 [French].

Congenital – Miocene *Discoglossus* with gradations of fusion of postsacral vertebrae with urostyle. One Maroccan fossil with fusion through centra and a second that also had neural arch fused.

Fossil – Miocene *Discoglossus* with gradations of fusion of postsacral vertebrae with urostyle. One Maroccan fossil with fusion through centra and a second that also had neural arch fused.

Vergner I. 1990. Beobachtungen zur Vermehrung von Phelsumen im Terrarium. [Observation on the reproduction of Phelsumes in the terrarium]. Herpetofauna, Weinstadt 12(65):25–34 [German].

Trauma – Deformations in tail of *Phelsuma laticauda laticauda* and *P. madagascariensis grandis*, perceived as genetic because occurring in related lineages?

Verhoef-de Fremery R. 1989. Abnormalities occurring during development of *Ambystoma mexicanum* and *Xenopus laevis*. Herpetopathologia 1:35–39.

Congenital – Kinked tail in *Xenopus laevis.*

Vershinin VL. 1989. [Morphological anomalies in urban amphibians.] Ekologiya (Sverdlovsk, Russia) 3:58–66 [Russian].

Environmental – Material was collected in 1980–1981 in large city area and included all amphibians living there. Part of the material was from years 1977, 1978, 1982, and 1983. Examined were moorfrog *Rana arvalis* – 194 adults and 1,784 juvenile offsprings, lake frog *Rana ridibunda* – 9 adults and 518 juvenile offsprings, common frog *Rana temporaria* – 100 adults and 495 juvenile offsprings, Siberian salamander *Salamandrella keyserlingii* – 266 adults and 313 juvenile offsprings, and common newt *Triturus vulgaris* – 275 adults and 28 juvenile offsprings.

Area types

II – Multistoried area

III – Low-storied area

IV – Forest park area K – suburb area

Types of anomalies

Type 1. Abdominal edema as result of kidney disease or mutation that leads to death during metamorphosis period.

Type 2. Abnormal extremities or their parts regeneration in chemically changed environment often observed in tailed amphibians, esp. in *Triturus vulgaris*. Here, relates cases of asymmetric polydactyly, ectodactyly, clinodactyly, polymely, and ectomely.

Type 3. Developmental anomalies induced by unfavorable changes in environment – i.e., cold, chemical pollutants etc. Skeleton deformations, abdominal herniations, symmetric, or asymmetric non-bending extremities probably could be put in this group.

Type 4. Genetic abnormalities (in Dubois opinion (1979) many of them are symmetrical): unusual skin pigmentation, absence of eyelids and iris (mono- and bilateral), symmetric poly- and ectodactyly.

Type 5. Tumors (found only in tailed amphibians).

Vershinin VL. 1995. Types of morphological anomalies of amphibians in urban regions. In: Kuzmin SL, Dodd CK. Jr, Pikulik MM. eds. Amphibian populations in the Commonwealth of Independent States: current status and declines. pp. 91–98; Moscow: Pensoft.

Environmental – Reviews etiology for anomalies as mutation, partial neoteny, parasites, habit isolation, hybridization, ontogenetic deviation, anomalous regeneration, virus, temperature change, pollution, pH change, pesticides, and radioactive contaminants.

Ekaterinburg City Urals from 1977–1993:

	Salamandrella *keyserlingii*			Triturus vulgaris			Rana arvalis				Rana temporaria	Rana ridibunda
	F	C	S	M	S	F	M	F	C	S	S	M
Polydactyly	2(1)			2								
Ectodactyly	13(5)	1		3			13(7)	2(2)	1(1)		4(1)	2(2)
Syndactyly	4						1					
Clinodactyly				3	1		1	1(1)			1(1)	
Hemimely							1(1)	2(1)	2(2)	3(2)	1(1)	
Brachymely	2(1)			1								
Taumely	1										1(1)	
Polymely	2											
Ectomely							9(6)	1(1)			1(1)	
axial skeleton deformity	4	2	1				3(2)	3(1)	1(1)			
mandibular hypoplasia							2(2)	5(5)	5(5)	1(1)		
Population	682	63	6	223	87	121	3,592	5492	2996	1057	2,390	276
Affected	26	4	1	9	1	1	27	14	9	7	8	2
Juveniles	300	8	0	67	17	57	3,422	5309	2939	1004	2,239	953
Affected	7						16	11	9	3	5	2
Species totals												
Total	31/751			11/456			57/13137				8/2869	2/1078
Juveniles	7/308			0/148			41/12674				8/2474	2/953
Population	24/441			11/308			16/463				0/395	0/123

= number affected; () = juveniles in that group.

M = multistory house; S = single story house; F = forest park; C = "control" – which was not identified.

There was no significant frequency variation by region for *Salamandrella keyserlingii* or *Triturus vulgaris*.

Verzár F., Weiss P. 1930. Untersuchungen über das Phänomen der identischen Bewegungsfunktion mehrfacher benachbarter Extremitäten. [Investigations on the phenomenon of identical function of movement of multiple neighbored extremities]. Pflügers Archiv für die gesamte Physiologie 223: 671–684 [German].

Congenital – *Rana esculenta* from Mezötur, Hungary with three anterior legs on the left side explained as heterophory (= dislocation of part of the anlage of the extremity); synchrone and identical function of the additional legs with the normal leg.

Vetter H. 2005. Leopard and African Spurred Tortoise: *Stigmochelys pardalis* and *Centrochelys sulcata*. Lanesboro, Minn: Serpent's Tail. 192 pp.

Environmental – UV radiation produces shell disease in leopard tortoise *Stigmochelys pardalis*, as manifest by holes.

Other – A deformed jaw was present in African spurred tortoise *Centrochelys* sulcata.

Vial Y. 1968. Naissances chez les reptiles. [Origin of reptiles]. Bêtes et Nature, Saint-Ouen (52): 27–29 [French].

Congenital – Derodymous couleuvre (smooth) viper *Natrix viperina.*

Vialleton L. 1911. Eléments de morphologie des vertébrés: Anatomie et embryologie comparées, paléontologie, et classification. [Elements of vertebrate morphology: Anatomy and comparative embryology, paleontology and classification]. XIV + 790 pp. Paris, O. Doin et fils. [French].

Trauma – Tail regenerates as cartilaginous rod.

Vickaryous MK, Russell AP, Currie PJ. 2001. Cranial ornamentation of ankylosaurus (Ornithischia: Thyreophora): Reappraisal of developmental hypotheses. In: K Carpenter, ed. The Armored Dinosaurs. pp. 318–340, Bloomington, IN: Indiana Univ Press.

Trauma – Exostoses of dermal cranial bones, sesamoids and ectopic extracranial bones of armored dinosaurs.

Fossil – Exostoses of dermal cranial bones, sesamoids and ectopic extracranial bones of armored dinosaurs.

Vickaryous M, Reisz R, Modesto S, Head J. 2010. Tail autotomy in the fossil record: New information about voluntary tail loss in captorhinid reptiles. Journal of Vertebrate Paleontology 30:181A.

Trauma – Autotomy in Lower Permian *Captorhinus* and *Labidosaurus* (UCLA 3491) related to intravertebral joints.

Autotomy used in intraspecific conflict, as well as for predators.

Autotomy developed only twice in amniotes, in captorhinids, and in squamata/spenodontea.

Autotomy requires that detached portion be nonessential to daily base, tail is frequent site of attack, that it acts as a deterrent (e.g., wiggles), and there are adaptations to minimize trauma and promote wound healing. Scab blocks bacterial invasion.

Regeneration in leopard gecko produces a cartilaginous (not bony) cone.

Fracture plane is separated by periosteum.

There is no fracture plane in mesosaurs (Modesto 1999), *Tanystropheus* (Reshesta 2000), and *Araeoscelis* (regenerates but no fracture plane (Vaughn 1955)), but does occur in captorhinids in the Permian. The fracture planes in *Captorhinus aguti* and *Captorhinus laticeps* (USM 15024) affect vertebrae 11–14 and are bordered by rounded ridge. The fracture plane does not pass through the neural arch, but the latter is fractured during autotomy. Fusion begins at top of arch, proceeding ventrally in gecko.

Fossil – Autotomy in Lower Permian *Captorhinus* and *Labidosaurus* (UCLA 3491), related to intravertebral joints.

Autotomy developed only twice in amniotes, in captorhinids, and in squamata/spenodontea.

There is no fracture plane in mesosaurs (Modesto 1999), *Tanystropheus* (Reshesta 2000), and *Araeoscelis* (regenerates but no fracture plane (Vaughn 1955), but does occur in captorhinids in the Permian. The fracture planes in *Captorhinus aguti* and *Captorhinus* laticeps) USM 15024 affect vertebrae 11–14 and are bordered by rounded ridge. The fracture plane does not pass through the neural arch, but the latter is fractured during autotomy. Fusion begins at top of arch, proceeding ventrally in gecko.

Vieira de Dandrade D, Shinya AA. 1993. Natural occurrence of shell abnormalities in hatchling red-footed tortoises (*Geochelone carbonaria*). Herpetological Review 24(3):89.

Congenital – Nineteen percent of red-footed tortoises (*Geochelone carbonaria*) with carapace or plastron anomalies and 9% with supernumerary scutes.

Viertel B, Veith M. 1992. Predation by leeches and regeneration, a factor in larval development of *Bufo bufo* (L.). In KorsóS Z, Kiss I. eds. Proceedings of the Sixth Ordinary Meeting S.E. Hungary, Budapest 1991; pp. 479–484.

Trauma – One percent of *Bufo bufo* have missing limb segments caused by the leech *Erpobdella octoculata.*

Environmental – Hind limb "damage" in *Bufo bufo* suggesting teratogenic substances from a nearby refuse dump, hybridization, mutation or effect of preditors should be investigated. The leech *Erpobdella octoculata* was the "external damaging factor."

Viglucci A. 1985. Two-headed snake slithers to stardom. The Miami Herald 6 June 1985: D-1.

Congenital – Derodymus ball python *Python regius.*

Vinegar A. 1974. Evolutionary implications of temperature-induced anomalies of development in snake embryos. Herpetologica 30:72–74.

Environmental – Burmese, Indian, or Ceylon python *Python molurus* normal egg development at 30.5°C, developed kinked vertebral column, and curved tail if incubated at 27°C.

Viner T, Heckert A, Carrano M. 2008. Multiple pathological elements in a phytosaur (Diapsida: Archosauria) skeleton from the Upper Triassic of Arizona, USA: Evidence of extreme osteomyelitis. Journal of Vertebrate Paleontology 28:156A.

Trauma – Upper Triassic *Smilosuchus gregorii* with bumps on right ulnar, the distal one resembling a stress fracture.

Infection – Upper Triassic *Smilosuchus gregorii* with left humeral hole with elevated, sclerotic rim; deltopectoral crest bony overgrowth; left fibular periosteal reaction and draining sinus; small excoriated area; two bumps on right ulnar. The distal one resembles a stress fracture; the lateral, ill-defined. Right femur periosteal reaction with holes.

Metapodial or distal radius with smooth, indented elliptical lesion with elevated margin and periosteal lesion on ? femur cross section. The elliptical holes with reactive surrounding bone suggest fungal infection.

Fossil – Upper Triassic *Smilosuchus gregorii* with bumps on right ulnar, the distal one resembling a stress fracture.

Upper Triassic *Smilosuchus gregorii* with left humeral hole with elevated, sclerotic rim; deltopectoral crest bony overgrowth; left fibular periosteal reaction and draining sinus; small excoriated area; two bumps on right ulnar. The distal one resembles a stress fracture; the lateral, ill-defined. Right femur periosteal reaction with holes.

Metapodial or distal radius with smooth, indented elliptical lesion with elevated margin and periosteal lesion on ? femur cross-section. The elliptical holes with reactive surrounding bone suggest fungal infection.

Virchow R. 1878. Ein grosser Blasen-(Kloaken?)-Stein von einer Meeresschildkröte. [A larger bladder (? Cloacal) stone of one marine turtle]. Virchows Archiv 73:629–630 [German].

Stone – Phosphatic bladder stone in *Chelonia mydas*.

Virey J-J. 1819. Observations et description de deux individus sexdigitaire [Observation and description of two individuals with six digits]. Journal complémentaire du dictionaire des sciences médicales 4:327–330 [French].

Congenital – Polydactyly, six digits in the hind limbs in *Rana*.

Vitt LJ. 1981. Tail autotomy and regeneration in the tropical skink, *Mabuya heathi*. Journal of Herpetology 15:454–457.

Trauma – Male predominance of tail loss in tropical skink *Mabuya heathi* – 22% versus 0 under 40 mm; 10 versus 25 (only exception) between 41 and 50 mm; 57 versus 29 between 51 and 60 mm; 75.5 versus 72 between 61 and 70 mm; 94 versus 59 between 71 and 80 mm, and 0/1 versus 82% of 17 females over 81 mm long.

9.4% had multiple tail breaks – and these were distal to the original break.

Vitt LJ. 1983. Tail loss in lizards: The significance of foraging and predator escape modes. Herpetologica 39:151–162.

Trauma – No difference between tail loss rate in foraging and "sit and wa

Tail alterations (* indicates gender significant at $p < 0.05$):

Species/length (mm)	% Regenerated Male	Female	Multiple breaks (%)
Gekkonidae			
Gymnodactylus geckoides			0
<	5	0	
31–40	44	45	
>	70	66	
Phyllopezus pollicaris			3.5
<	6	0	
41–50	31	25	
51–60	18	36	
61–70	13	50	
>	73	71	
Lygodactylus kluget			0
<	8*	0*	
19–23	25	9	
24–28	35	29	
>	39	55	
Hemidactylus mabouia			5
<	0	0	
31–50	33	33	
>	71	57	
Iguanidae			

(continued)

Species/length (mm)	% Regenerated Male	% Regenerated Female	Multiple breaks (%)
Tropidurus torquatus			2.5
<	5*	0*	
51–70	22	21	
71–90	25	40	
>	50	62	
Platynotus semitaeniatus			6
<	14*	36*	
61–80	48	63	
>	79	100	
Polychrus acutirostris			0
<	0	–	
91–110	6	0	
111–130	5	0	
>	–	0	
Scincidae			
Mabuya heathi			16
<	22*	0*	
41–50	10	25	
51–60	57	29	
61–70	75	72	
71–80	94	59	
>	–	82	
Teiidae			
Ameiva ameiva			5
<	0	50	
81–120	17	19	
>	42	43	
Cnemidophorus ocellifer			3.4
<	21	27	
51–70	29	38	
71–90	60	47	
>	50	–	
Gymnophthalmus multisculatus			3.7
<	36	40	
31–36	68	47	
>	72	72	
Tupinambis teguixin			11
<	10	10	
201–300	13	29	
>	33	18	

Vitt LJ, Ohmart RD. 1974. Reproduction and Ecology of a Colorado River Population of *Sceloporus magister* (Sauria: Iguanidae). Herpetologica 30:410–417.

Trauma – Increases in male *Sceloporus magister* attributed to fighting.

Vitt LJ, Congdon JD, Hulse AC, Platz JE. 1974. Territorial aggressive encounters and tail breaks in lizard *Sceloporus magister*. Copeia 1974:990–993.

Trauma – Territorial aggression produces tail breaks more common in male than female *Sceloporus magister.*

Vitt LJ, Congdon JD, Dickson NA. 1977. Adaptive strategies and energetics of tail autotomy in lizards. Ecology 58:326–337.

Trauma – Tail autotomy in 65% of *Coleonyx variegatus*, 62% of *Eumeces skiltonianus*, 52% of *Eumeces gilberti*, and 74% of *Gerrhonotus multicarinatus*.

Regeneration was most rapid in short-lived, early-maturing species and with multiple breeding opportunities in a single season (e.g., *Coleonyx variegatus.*

Vizotto LD. 1975. Bicefalia em *Philodryas olfersii* (Lichtenstein, 1823) (Ophidia, Colubridae) [Bicephaly in *Philodryas olfersii* (Lichtenstein, 1823) (Ophidia, Colubridae)]. Naturalia, São Jose do Rio Preto, (São Paulo) 1:69–76 [Portuguese].

Congenital – Dicephalic female *Philodryas olfersii* (Lichtenstein 1823) in the collections of the Laboratory of Zoology of the Faculty of Philosophy, Science and Letters of São Jose do Rio Preto number LZ-FFCL-R. 108. The atlas was partially bisected, appearing triangular in shape when observed from above. Reduced individual vertebral length was observed from the axis to the sixth vertebra, as compared to the rest.

Vizotto LD, Taddei VA, Cais A. 1977. Pseudomelia em *Phrynohyas hebes* (Cope, 1882) (Anura – Hylidae) [Pseudomely in *Phrynohyas hebes* (Cope, 1882) (Anura – Hylidae)]. Naturalia 3:35–43 [Portuguese].

Congenital – Pseudomely and scoliosis in *Phrynohyas hebes* in the collections of the Department of Zoology of the Instituto de Biociências, Letras e Ciências Exatas, in São José do Rio Preto, SP, Brazil. There were two supernumerary elongate bones set in a "V" shape, whose apex was below and next to the pubis. The internal branch, after thickening conspicuously, projected forward and united itself with the left ilium. The other branch projected from the middle region, and became the external, apparent incomplete supernumerary limb observed at first glance. At the presacral vertebra (number VIII), the vertebral column was dislocated, bent to the left (scoliosis). This vertebra also lacked transverse apophyses and articulated imperfectly to both ilia.

Vogt T. 1922. Zur Reptilien und Amphibienfauna Südchinas. [On South China reptiles and amphibians]. Archiv für Naturgeschichte 88A:135–146 [German].

Congenital – Kyphosis in *Trionyx sinensis.*

Voigt E. 1960. Über einen mutmaßlichen fossilen Harnstein (Urolith) aus der Oberen Kreide. [About an alleged fossil urinary (uroliths) from the Upper Cretaceous]. Mitteilungen aus dem Geologischen Staatsinstitut Hamburg 29:85–95 [German].

Stone – Cretaceous (Santonian) apatite stone presumably from sea reptile.

Fossil – Cretaceous (Santonian) apatite stone presumably from sea reptile.

Voitkevich AA. 1958. Peculiar reduction of crus in natural limb duplication of frog. Transactions of the Academy of Science USSR. 113(4):841–844.

Congenital – Reduction of hind limb crural and distal bones in *Rana ridibunda* during natural "duplication."

Volante F. 1922. Contributo alla conoscenza della superrigenerazione nei rettili [Contribution to the knowledge of superregeneration in reptiles]. Giornale della Accademia di medicina di Torino 85: 309–314 [Italian].

Trauma – Shorter and more general version of essentially the same data published by Volante in 1923. Six *Lacerta muralis* with multiple tails were studied of which four had two caudal apices and the remaining two had three, all of different shapes and sizes.

Volante F. 1923. Contributo alla conoscenza della superrigenerazione nei Rettili [Contribution to the knowledge of superregeneration in reptiles]. Monitore Zoologico Italiano 34:185–200 [Italian].

Trauma – Four *Lacerta muralis* with two tails; two *Lacerta muralis* with three tails.

Volpe EP. 1967. Understanding Evolution. 3rd ed. 220 pp.; Dubuque, Iowa: WC Brown,

Congenital – In 1958, 350 multilegged bullfrogs were found in Tunica,

Mississippi, but only in one season. They were utilized in this book as hypothetical examples to illustrate various genetic models.

Volz W. 1902. *Proneusticosaurus*, eine neue Sauropterygier-Gattung aus dem unteren Muschelkalk Oberschlesiens. [*Proneusticosaurus*, a new sauropterygia genus from the lower Muschelkalk of Upper Silesia]. Palaeontographica 49:121–162 [German].

Other – *Proneusticosaurus madelungi* with pachyostosis (proximal end of last thoracal and first and second sacral ribs thickened like a tumor).

Fossil – *Proneusticosaurus madelungi* with pachyostosis (proximal end of last thoracal and first and second sacral ribs thickened like a tumor).

von Fejérváry GJ. 1916. Beiträge zur Kenntnis von *Rana mehelyi* BOLKAY. [Contributions to the knowledge of *Rana mehelyi* BOLKAY.]. Mitteilungen Jahrbuch der königlich-ungarischen geologischen Reichsanstalt 23(3):133–155 [German].

Trauma – Fracture of os ilium in *Rana mehelyi.*

Voris HK. 1975. Dermal scale-vertebra relationships in sea snakes (Hydrophiidae). Copeia 1975:747–757.

Congenital – Ratio of dorsal scales to vertebrae is 2:1 in Boidae and some Typhlopidae

Ratio of ventral scales to vertebrae is usually 1:1 in snakes (Leptotyphlopidae, Uropeltidae, Aniliidae, Xenopeltidae, Boidae, Colubridae, Elaphidae, Viperidae) (Alexander and Gans 1966), *Acanthophis antarcticus, Bungarus caeruleus, Furina annulata, Hoplocephalus stephensii, Naja naja, Pseudechis guttatus, Enhydris alternans, Enhydris smithi, Erpeton tentaculatus, Cerberus rynchops, Myron richardsonii,*

Natrix piscator, Laticauda laticaudata, Laticauda colubrina, Laticauda semifasciata, Laticauda schistorhynchus, Aipysurus eydouxii, Aipysurus fuscus, Aipysurus laevis, Aipysurus duboisii, Aipysurus foliosquama, Aipysurus apraefrontalis, Emydocephalus iljimae, Hydrelaps darwiniensis, Ephalophis greyi, Ephalophis mertoni, Acalyptophis peronii (with much variation), *Lapemis hardwickii* (with much variation), except for aquatic *Acrochordus granulatus* (Acrochordidae) and *Lapemis hardwickii* (Hydrophiidae). *Acrochordus granulatus*, and *Acrochordus javanicus* had five times as many ventral scales as vertebrae.

Ratios approaching 2:1 of vertebra versus ventral scales were found in *Thalassophina viperina, Thalassophis anomalus, Kolpophis annandalei, Astrotia stokesii, Pelamis platurus, Enhydrina schistosa, Microcephalophis cantoris, Hydrophis nigrocinctus, Hydrophis spiralis, Hydrophis melanosoma, Hydrophis belcher, Hydrophis elegans, Hydrophis cyanocinctus, Hydrophis klossi, Hydrophis stricticollis, Hydrophis torquatus, Hydrophis ornatus, Hydrophis lapemoides, Hydrophis mammillaris, Hydrophis caerulescens, Hydrophis fasciatus, Hydrophis brooki*, with intermediate values in *Hydrophis kingi, Hydrophis obscurus, Hydrophis major, Hydrophis ornatus.*

Thalassophina viperina, Microcephalophis gracilis, Acalyptophis peronii, Lapemis hardwickii, Hydrophis curtus, Hydrophis kingi,Hydrophis brooki and *Hydrophis major* had 1:1 ratio for only the anterior portion.

Hydrophis cyanocinctus, Hydrophis elegans, Thalassophina anamolus, K. jerdoni, Kerilia annandalei, Pelamis platurus, Astrotia stokesii, Enhydrina schistosa had no relationship between ventral scale and number of vertebrae.

Ratio of ventral annuli to vertebrae is 2:1 in amphisbaenians, except for *Leposternon*, which lacks a regular correlation (Gans 1971).

Regina leberis had ratio reduced because of four to five extra ventrals at the neck and two to three extra vertebrae just anterior to the vent.

Rest of sea snakes have more ventral scales than vertebrae, with great variability in ratio. Ventral counts varied also with geographical area.

Loss of relationship is associated either with burrowing (e.g., Typhlopidae) or highly aquatic mode of life (e.g., Acrochordidae). *Laticauda semifasciata* and *Hydrelaps darwiniensis* have the lowest coefficient of variation. Coefficient of variation for ventrals ranged from 2.84 to 5.64 for lizards, 1.49 to 3.00 for snakes and Amphisbaenidae. Aquatic snakes had coefficients of variation similar to lizards.

Vorobyeva EI. 2000. [The problem of polydactyly in amphibians.] Russian Journal of Herpetology 6:95–103 [Russian].

Congenital – Limnophylus, larval salamander *Salamandrella keyserlingii* versus steam or reophylous larval salamander *Ranodon sibiricus* with sixth toe as normal condition, compared to *Ichthyostega* and *Acanthostega.*

Vsevolojsky N. de 1812. Notice sur un serpent à deux têtes (*Coluber natrix*) vivant. [Notice on a living two-headed serpent (*Coluber natrix*)]. Mémoires de la Société Impérial de Naturaliste de Moscou 3:284–288 [French].

Congenital – Dicephalic *Coluber natrix*, that Lacépède called *Coluber torquatus* and cited Edwards and Catesby (1769) reporting a dicephalic *Bandanensis biceps* and the most famous example as *Hydra* or snake with seven heads (Cabinet de Seba, vol 1, page 158, pl CII, figure 1).

Vulpian A. 1867. Sur la reproduction des members chez l'axolotl dans le cas de polydactylie acquise. [On limb reproduction in axolotl, in the case of acquired polydactyly] Bulletin de la Société philomatique de Paris 6(IV):117 [French].

Congenital – Polydactyly in axolotls.

Wadsworth JR. 1954a. Neoplasms of snakes. University of Pennsylvania, Bulletin Veterinary extension quarterly 133:65–74.

Neoplasia – Osteochondrosarcoma of spinal column in African black cobra *Naja melanoleuca.*

Wadsworth JR. 1954b. Some neoplasms of captive wild animals. Journal of the American Veterinary Medical Association 125:121–123.

Neoplasia – Osteochondroma and Osteochondrosarcoma of African black cobra.

Wadsworth JR. 1956. Serpentine tumors. Journal of the American Veterinary Medical Association No. 51:326–328.

Neoplasia – Quotes Hill 1954 – spinal column osteosarcoma in rufous-beaked snake and Wadsworth 1954 for spinal column osteochondrosarcoma in black cobra.

Wagner G. 1913. On a peculiar monstrosity in a frog. Biological Bulletin 25:313–317.

Congenital – *Rana pipiens* with three supernumerary arms extending from sternum – *Thoracopagus parasiticus* and cited Sutton (1892, p. 112, fig. 60) illustrating supernumerary leg.

Waite ER. 1929. The reptiles and amphibians of South Australia. British Science Guild (South Australian Branch), Adelaide.

Trauma – Tail duplication – even to four partial tails – in *Phyllodactylus marmoratus*, bifid tail in *Diplodactylus vittatus*.
Tail regeneration in *Pygopus lepidopodus*.
The short-tailed *Nephrurus asper* lacks autotomy splits.

Wake DB. 1966. Comparative osteology and evolution of the lungless salamanders, family Plethodontidae. Southern California Academy of Sciences. Memoir 4:1–111.
Congenital – *Thorius* manus variation (0–1)-2-3-(1–2), pes (0–1)-2-3-(2–3)-(1–2) phalanges.
Bolitoglossa rufescens manus (0–1)-(1–2)-(2–3)-(1–2), pes (1-(1–2)-(2–3)-(2–3)-(1–2).
Oedipina poelzi manus (0–1-(1–2)-(2–3)-(1–2), pes (0–1)-2-(2–3)-(2–3)-(1–2).

Wake DB, Dresner IG. 1967. Functional morphology and evolution of tail autotomy in salamanders. Journal of Morphology 122:265–306.
Trauma – Only salamanders regenerate a virtually normal tail.
Primitive salamanders with thick tails and accident-based posterior tail loss include *Pseudotriton ruber, Desmognathus fuscus, Eurycea bislineata.*
Thin-tailed salamanders whose tail can break at any point include *Chioglossa lusitanica, Batrachoseps pacificus, and Plethodon cinereus.*
Those with constriction-based (at base of tail) tails include *Hemidactylum scutatum, Ensatina eschscholtzii, Bolitoglossa subpalmata, Bolitoglossa yucatana*, and *Thorius dubitus* (which lacks visually conspicuous constriction).
Tail breakage is rare in *Aneides lugubris, A. ferreus*, and *A. aeneus, Hydromantes.*
Museum collections revealed 23% of slender-tailed and 9.5% of constricted tail species (chi square, $p < 0.001$) had broken tails:
Thick-based tails – *Desmognathus fuscus* 33%
- *Desmognathus monticola* 11%
- *Desmognathus wrighti* 7%
- *Eurycea bislineata* 8%

Slender-based tails – *Plethodon cinereus* 13%
- *Plethodon jordani* 28%
- *Plethodon ouachitae* 20%

Constricted-based tails – *Ensatina eschscholtzii* 10%
- *Hemidactylum scutatum* 4%
- *Chiropterotriton chiropterus* 7%
- *Chiropterotriton multidentatus* 5%
- *Pseudoeurycea leprosa* 7%
- *Bolitoglossa altamazonica* 11%
- *Bolitoglossa morio* 19%
- *Bolitoglossa platydactyla* 7%
- *Bolitoglossa rostrata* 11%
- *Bolitoglossa rufescens* 6%
- *Bolitoglossa subpalmata* 7%
- *Thorius macdougalli* 5%
- *Thorius narisovalis* 18%

Recent tail loss in 4.9% of 169 *Ensatina* and 14 others with partially regenerating tails (Gnaedinger and Reed 1948), 46 of 603 (7%) found by Stebbins (1954).

Walker T. 1995. Two-headed turtle. Courier-Mail 4 December 1995: 1.
Congenital – Dicephalic turtle.

Walker ML, Dorr JA, Pisani GR. 2008. Observation of aberrant growth in a Timber rattlesnake (*Crotalus horridus*). Transactions of the Kansas Academy of Science 111:156–158.
Congenital – Timber rattlesnake *Crotalus horridus* with 18 rattle segments without tapering, instead of the normal 2–6 with tapering. Suggests hypothyroidism may have resulted in increased shedding frequency for age (Rivera and Lock 2008).
Metabolic – Timber rattlesnake *Crotalus horridus* with 18 rattle segments without tapering, instead of the normal 2–6 with tapering. Suggests hypothyroidism may have resulted in increased shedding frequency for age (Rivera and Lock 2008).

Wall F. 1905. Double-headed snakes. Journal of the Bombay Natural History Society 16:386–388, and 752–753.
Congenital – Dicephalic *Coluber natrix, Hydrophis subloevis (cyanocinctus), Homolopsis buccata, Heterodon simus.*

Wallach JD. 1969. Medical care of reptiles. Journal of the American Veterinary Medical Association 155:1017–1047.
  Trauma – Monostotic fibrous osteodystrophy from vitamin D deficiency alleged in *Iguana iguana* with pneumonia, but actually appears to be a fracture.
  Infection – Oral infections producing osteomyelitis in snakes. *Salmonella* carriage in turtles, tortoises, and anoles.
  Metabolic – Rickets producing enlarged articulations, soft, deformed shells, citing (Reichenbach-Klinke and Elken 1955).
    Monostotic fibrous osteodystrophy from vitamin D deficiency alleged in *Iguana iguana* with pneumonia, but actually appears to be a fracture.
  Stone – Combined calcium phosphate-urate calculus in *Iguana iguana.*
    Low calcium diets produce calculi in turtles and lizards.

Wallach JD. 1970. Nutritional diseases of exotic animals. Journal of the American Veterinary Medicine Association 156:583–599.
  Metabolic – Hypervitaminosis D mentioned.

Wallach JD. 1971. Environmental and nutritional diseases of captive reptiles. Journal of the American Veterinary Medicine Association 159:1632–1643.
  Metabolic – Enlarged articulations and soft shells in reptiles with rickets.
    Fibrous osteodystrophy in a green iguana.
  Vertebral – "Spinal deformities" in cobra with rickets.
  Stones – Irregular stones in urinary bladder of Galapagos tortoise.

Wallach JD. 1975. The pathogenesis and etiology of ulcerative shell disease in turtles. Journal of Zoo Animal Medicine 6:11–13.
  Shell disease – *Beneckea chitinovora* producing loosened plates around sutures and raw punctate 1–12 mm ulcers in sliders, red-eared turtle, musk turtles, soft shell turtles, side-neck turtles, and painted turtles.

Wallach JD. 1977a. Management and Nutritional Problems in Captive Reptiles. In Kirk RW. ed. Current Veterinary Therapy VI. pp. 784–785; Philadelphia: W.B. Saunders Company.
  Metabolic – Mentions rickets in reptiles.
    Intraperitoneal injections of D-serine produce gout in alligators.
  Toxicology – Intraperitoneal injections of D-serine produce gout in alligators.

Wallach JD. 1977b. Ulcerative shell disease in turtles: Identification, prophylaxis and treatment. International Zoo Yearbook 17:170–171.
  Shell disease – *Candida albicans* shell disease, but unresponsive to medication.
    Beneckea (Bacillus) chitinovora isolated from *Trionyx spinifer, Chrysemys scripta elegans, Chrysemys picta picta, Sternotherus minor peltifer*, and *Podocnemis unifilis.*

Wallach JD. 1978. Reptiles. Feeding and nutritional diseases. In Fowler ME. ed. Zoo and Wild Animal Medicine. pp 123–128, Philadelphia: Saunders.
  Metabolic – Gout in chelonians, crocodilians, and herbivorous lizards.
    Suggests high protein diets, hypovitaminosis A, and dehydration etiological factors for gout in reptiles.

Wallach JD. 1995. New records of dicephalic snakes in museum collections. Herpetological Review 26(3):127–129.
  Congenital – Dicephalic snakes (11 taxa in addition to the 99 of Cunningham (1937) and Smith and Pérez-Higardo (1988)) in *Bothrops atrox, Coluber constrictor, Elaphe obsoleta, Lampropeltis getula californiae, Pituophis catenifer catenifer, Thamnophis atratus atratus, T.* cf. *gigas, T. sitalis sirtalis, and T. elegans vagrans.*

Wallach V. 2003. See live mutant snakes right here. Natural New England (16):50–55.
  Congenital – Dicephalic condition in 1 of every 50,000 to 100 births, noting Aristotle's report; occurrence in corn snake, 38 grass snakes, 23 king snakes, 22 garter snakes, 13 water snake, 12 black racer, and 11 corn snakes. The eight families represented include worm snakes, pythons, boas, tropidophiidae, vipers, elapids, sea snakes, and colubrids.
    Edward Bancroft illustrated first New World dicephalic snake in 1769, Benjamin Franklin possessed Park's drawing in 1787, that Johnson (1902) and Strohl (1925) identified it as water snake, Brimley (1920) as a black snake and Babcock, as a milk snake, as confirmed by Smith and Chiszar (1988). It was originally called *Amphisbaenia.*
    Dicephalic garter snake (Jackson in 1847 and 1870) in the catalog of the Anatomical Museum of the Boston Society for Medical Improvement, black racer (Wyman 1863), bisexual black racer (Wells 1849), hognose (Hyde 1925), king (Hyde 1925), garter (Hyde 1925), and rattlesnakes. (Hyde 1925).
    Garter snake with two heads that share a single lower jaw.
    Prodichotomous – duplication of head and neck = derodymous.
    Craniodichotomy and prodichotomy in milk snakes (Mitchell 1826).
    Amphidichotomous water snake (Wyman 1863), milk snake (Mitchell 1826; Wyman 1863).

Wallach V. 2004. Dicephalism in snakes: Two heads are not always better than one! Reptiles 12(2):60–69.

Congenital – Dicephalic Florida pine snake *Pituophis melanoleucus mugitus*, gopher snake, Texas rat snake *Elaphe obsoleta lindheimeri*.
California king snake *Lampropeltis getulus californicus*, Hieroglyphic slider turtle *Pseudemys concinna hieroglyphica*, bearded dragon *Pogona viticeps* and summarized Cunningham (1937).
Wallach V. 2005. A new dicephalic *Lampropeltis t. triangulum* from Maine, USA. Podarcis 6: 36–39.
Congenital – Derodymous common milk snake *Lampropeltis t. triangulum*, noting previous reports by Smith and Chiszar (1988), Smith (1882, p. 690), Howey (1883, p. 161) and Johnson (1902, p. 528) and that the only one that reproduced was *Lampropeltis getula californiae* "Thelma and Louise" at the San Diego Zoo until 2000.
Wallach V. 2006. It's another double-header – and it learns. Reptiles 14(3):10.
Congenital – Derodymous eastern milk snake *Lampropeltis t. triangulum*.
Wallach V. 2007. Axial Bifurcation and Duplication in Snakes. Part I. A Synopsis of Authentic and Anecdotal Cases. Bulletin of the Maryland Herpetological Society 43(2):57–95.
Congenital – Reviewed 950 cases of dicephalism, axial bifurcation, and somatodichotomy in snakes, taken from the primary and secondary literature, museum specimens, press clippings, news reports, the internet, videos, photographs, and postcards. A checklist of 169 species in 93 genera was presented in addition to the distribution of records within the eight families. Data are analyzed according to their temporal discovery and their occurrence by geographic region, number of cases by country, and within the USA by state.
Wallach JD, Hoessle C. 1968. Fibrous osteodystrophy in green iguanas. Journal of the American Veterinary Medical Association 153:863–865.
Metabolic – Swollen soft bone and costochondral swelling from rickets in *Iguana iguana*.
Wallach JD, Hoff GL. 1982. Metabolic and Nutritional Diseases of Reptiles. In Hoff GL, Davis JW eds. Noninfectious Diseases of Wildlife. Ames: Iowa State University Press; 155–167.
Metabolic – Vitamin D deficiency producing rickets, manifest by enlarged articulations, malformed limbs, and malformed shells, especially in *Iguana iguana*.
Fibrous osteodystrophy from nutritional secondary hyperparathyroidism.
Gout in *Alligator mississippiensis*.
Walley HD. 1997. On the occurrence of bifurcation in the scincid lizard *Eumeces anthracinus*. Bulletin Maryland Herpetological Society 33:178–180.
Trauma – One of 350 *Eumeces anthracinus pluvialis* had bifurcated tail.
Tail loss in scincid lizard *Eumeces anthracinus* occurs as fracture along an intravertebral vertical cartilaginous septum (Arnold 1984).
Wallis K. 1927. Zur Knochenhistologie und Kallusbildung beim Reptil (*Clemys leprosa* SCHWEIGG). [To the bone histology and the callus formation in a reptile (*Clemys leprosa* SCHWEIGG)]. Zeitschrift für Zellforschung und mikroskopische Anatomie 6:1–26 [German].
Trauma – *Clemys leprosa* from Tunesia with healed fracture of carapace: histology.
Wallis K. 1928. Über den Knochenkallus beim Kaltblüter (Eidechse). [On bone callus in a cold blood animal (lizard)]. Zeitschrift für Zellforschung 7:257–289 [German].
Trauma – Callus formation at break of fibula in a lizard, noting timing of the healing process.
Walterstorff W. 1887 (not 1885–1886). Ueber fossile Frösche, insbesondere das Genus *Palaeobatrachus*. [About fossil frog of the genus *Palaeobatrachus*]. Jahrbuch und Abhandlungen des Naturwissenschaftlichen Vereins zu Magdeburg 1886;2:1–97 [German].
Congenital – *Palaeobatrachus* usually had three sacral vertebrae.
Extreme sacral variation was noted in *Bombinator*.
Fossil – *Palaeobatrachus* usually had three sacral vertebrae.
Extreme sacral variation was noted in *Bombinator*.
Wanderka HA. 1974. Verhütung von Wachstums- und Entwicklungsstörungen bei Salamanderlarven und -jungtieren durch Wirkstoff- und Organ-Kombinationspräparate. [Prevention of disorders in growth and development of salamander larvae and juveniles by substance and organic combination compounds]. Der praktische Tierarzt 6:324–328 [German].
Metabolic – Addition of Vigantol or D 64 to the food improves growth of captive salamanders.
Wanderka H. 1984. Behandlung rachitischer Prozesse bei tropischen Wasserschildkröten mit zytoplasmatischen Substanzen. [Treatment of rickets processes in tropical aquatic turtles with cytoplasmic substances]. Der praktische Tierarzt 65:894–903 [German].
Metabolic – Successful treatment of tropical rachitic turtles with NeyChondrin, NeyArthros, NeyGeront, Neythymun and Neyimmun.
Wandolleck B. 1904. Eine bucklige *Testudo hermanni* L. [On humped *Testudo hermanni* L.]. Zoologisches Jahrbuch für Systematik 20:151–166 [German].
Congenital – Kyphosis in *Testudo hermanni*.

Wang Y, Evans S. 2010. Polydactyly in a mesozoic salamander from China. Journal of Vertebrate Paleontology 30:183–184.

Congenital – Polydactyly in 12 (during presentation, although abstract says 11)/200 Middle/Late Jurassic salamander *Chunerpeton tianyiensis*, affecting manis and or pes. Additional individuals with polymelia or reduction/addition of/to phalanges.

Fossil – Polydactyly in 12 (during presentation, although abstract says 11/200 Middle/Late Jurassic salamander *Chunerpeton tianyiensis*, affecting manis and or pes. Additional individuals with polymelia or reduction/addition of/to phalanges.

Washabaugh CH, Tsonis PA. 1995. Effects of vitamin D metabolites on axolotl limb regeneration. Development, Growth & Differentiation 37:497–503.

Metabolic – Similar effect of 1–25-dihydroxyvitamin D2 on limb bud mesenchymal cells in axolotl as produced by retinoids = limb fusions and truncations – on regenerating parts.

Washabaugh CH, Rio-Tsonis KD, Tsonis PA. 2005. Variable manifestations in the short toes (s) mutation of the axolotl. Journal of Morphology 218:107–114.

Congenital – Short toes in *Ambystoma mexicanum*.

Washburn FL. 1899. A peculiar toad. American Naturalist 33:139–141.

Congenital – Supernumerary forelimb in *Bufo columbiensis*.

Watkins-Colwell GJ. 1995. A case of prodichotomy in a wild-caught *Elaphe obsoleta obsoleta*. Bulletin of the Maryland Herpetological Society 31(2):97–99.

Congenital – Prodichotomous black rat snake *Elaphe obsoleta obsoleta*.

Watson DMS. 191a. On some features of the structure of the therocephalian skull. Annals and Magazine of Natural History 98, 11:65–79.

Congenital – Unpaired, median bone "preparietale," between parietals and frontals in gorgonopsian therapsids (mammal-like reptiles).

Watson DMS. 1913b. On *Micropholis Stowi* Huxley, a temnospondylous amphibian from South Africa. Geological Magazine (5) 10:340–346.

Congenital – Unpaired, median bone between frontals and nasals, "interfrontal," in microsaur *Ricnodon* and in *Sclerocephalus roemeri*. Unpaired, median bone between the internasal processes of the premaxillae in microsaur *Micropholis*.

Fossil – Unpaired, median bone between frontals and nasals, "interfrontal," in microsaur *Ricnodon*, and in *Sclerocephalus roemeri*. Unpaired, median bone between the internasal processes of the premaxillae in microsaur *Micropholis*.

Webb GA. 1984. A supernumerary limb in the spotted grass frog, *Limnodynastes tasmaniensis*. Herpetofauna, Sydney 15(2):41–42.

Congenital – Supernumerary limb in spotted grass frog, *Limnodynastes tasmaniensis*.

Webb GJW, Manolis SC. 1983. *Crocodylus johnstoni* in the McKinlay River Area, N.T. V. Abnormalities and injuries. Australian Wildlife Research 10:407–420.

Trauma – Eighty one percent of *Crocodylus johnstoni* has tail injuries, especially common in species which congregate in high densities. Most are puncture marks from other crocodiles. One of 449 had mandible almost completely broken through. Limb injuries were present in 14% of injured animals; amputation, 70%, 54% one digit, 8 with 2, 6% with 3 and 2% with 6. One animal had a 0.6 cm jaw protrusion. One had protruding lower jaw curving dorsally around premaxilla. Fresh injuries were more common in animals caught in the dry season.

Injury rates in	Total	Tail only	Tail amputation only:
Crocodylus			
johnstoni	56%	46%	7.3%
niloticus	21%		
porosus	7%	4%	2%
acutus	14%		
crocodylus	–	25%	30%

Webb GJ, Manolis SC. 1989. (not 1985). Crocodiles of Australia. French's Forest, New South Wales, Australia: Reed Books Ply, Ltd., 160 pp.

Congenital – Siamese twin salt water crocodile.

Salt water crocodile with absent or kinked/curved tail attributed to high incubation temperature.

Trauma – Upper jaw of salt water crocodile ripped off by male during breeding season.

Metabolic – Stunting in fresh water crocodile from food shortage.

Environmental – Salt water crocodile with absent or kinked/curved tail attributed to high incubation temperature.

Webb GJ, Manolis C. 1998. Crocodiles of Australia. 160 pp.; New Holland Publishers, Australia.

Congenital – High temperature produces spinal abnormalities and strongly coiled tails, premature skull ossification, and protruding jaws in freshwater crocodiles.

Environmental – Hundred percent of fresh water crocodile progeny are female at temperature below 30°C; 70%, between 31 and 32; above 32, 100% female. However, in the field, some nests have 100% males, specifically in those taking around 74 days to incubate – at 31.5° constant temperature.

Webb GJW, Messel H. 1977. Abnormalities and injuries in the estuarine crocodile, *Crocodylus porosus*. Australian Wildlife Research 4:311–319.

Trauma – Tail injuries in 56% and limb injuries in 29% (64% amputated digits) of injured of 1,345 caught, marked and released *Crocodylus porosus*. Four had amputated forelimbs and one, an amputated hind limb. One had snout swelling and seven had misaligned jaws. Injury rates were higher in Andranangoo Creek than other areas. Injuries apparently occurred first at immediate post-hatching stage and then at 70 cm snout-vent length.

Webb GJ, Sack GC, Buckworth R, Manolis SC. 1983. An examination of *Crocodylus porosus* nests in two northern Australian freshwater swamps, with an analysis of embryo mortality. Australian Wildlife Research 10:571–605.

Congenital – Abnormal *Crocodylus porosus* embryos from northern Australian freshwater swamps:

Kinked tail – Nineteen with misalignment of caudal vertebrae, 1 with shortened lower jaw – in hot and cold nests.

Kinked back – Twenty three with misalignment of thoracic and lumbar vertebrae (four with kinked tails, one with skewed lower jaw, one with protruding lower jaw) from hot and cold nests.

Misalignment of cervical vertebrae – Seven from hot nests.

Shortened abdomen – Two with missing thoracic or lumbar vertebrae from hot and cold nests.

Central cranial "platform" forming a pronounced dome – Seven, one with skewed lower jaw and 1 with kinked back from hot nests.

Jaw abnormalities – Five with short lower jaw; one with skewed jaw, kinked back and cranial "bump," on with short upper jaw; one with short lower jaw, kinked tail and twisted spine.

Absent tail – Eight to seven from a single nest.

Webb GJ, Beal AM, Monolis SC, Demsey KT. 1987. The effects of incubation temperature on sex determination and embryonic development rates in *Crocodylus johnstoni* and *C. porosus*. In: Webb GJ, Monolis SC., Whitehead PJ. (eds.). Wildlife management: Crocodiles and alligators. Pp. 507–531; London, Sydney, Australia, Surrey Beaty.

Congenital – Crocodile runt syndrome, curved and coiled tails and "abnormalities of head, limbs and spine."

Weber B. 1995. Tumor-like lesions in the skeleton of iguanidae - enchondroma or osteofibrosis. Proceedings of the 5th International Symposium on Pathology of Reptiles and Amphibians, Alphen a Rijn:329–332.

Neoplasia – Spindle-shaped enlargements of long bone diaphyses in *Iguana iguana, Cyclura cornuta*, and *Ctenosaura acanthura* interpreted as enchondromas or osteofibrosa cystica from nutritional hyperparathyroidism.

Wechsler A. 2003. Doing a double take over a heads-up find; Poestenkill family adopts turtle with two noggins, and it's getting the attention of wildlife experts. The Times Union (Albany, NY) 20 May 2003:1.

Congenital – Dicephalic turtle.

Weems RE. 1974. Middle Miocene sea turtles (*Syllomus, Procolpochelys, Psephorus*) from the Calvert Formation. Journal of Paleontology 48:278–303.

Congenital – Splitting of first neural and extra postneurals and costals, extra eighth rib and extra longitudinal dorsal keels in *Syllomus crispatus aegyptiacus*.

Syllomus crispatus aegyptiacus USNM 24872 has a bony ridge on the femur between the trochanters and the caput.

Shell scute and bony element variation (Lynn 1937; Lynn and Ullrich 1950; Newman 1906; Parker 1901; Zangerl and Johnson 1957; Zangerl and Turnbull 1955).

Neoplasia – *Syllomus crispatus aegyptiacus*, with a "tumorous bony growth" in USNM 24872 (identified as an osteoma by one of the authors – bmr). Eighth cervical had neural spine enclosed in a body growth, and amorphous lumps were reported on carapace of USNM 24876.

Shell disease – Shallow, well-rounded pits were also reported on outer shell surface of *Syllomus crispatus aegyptiacus* USNM 24872.

Procolpochelys grandaeva USNM 24889 with circular depression on right third and fourth costals. He suggested that it was similar to "basal configuration" of *Balanus concavus* or a similar barnacle.

Fossil – Splitting of first neural and extra postneurals and costals, extra eighth rib and extra longitudinal dorsal keels in *Syllomus crispatus aegyptiacus*.

Syllomus crispatus aegyptiacus USNM 24872 has a bony ridge on the femur between the trochanters and the caput.

Shell scute and bony element variation (Lynn 1937; Lynn and Ullrich 1950; Newman 1906; Parker 1901; Zangerl and Johnson 1957; Zangerl & Turnbull 1955).

Syllomus crispatus aegyptiacus, with a "tumorous bony growth" in USNM 24872 (identified as an osteoma by one of the authors – bmr). Eighth cervical had neural spine enclosed in a body growth and amorphous lumps were reported on carapace of USNM 24876.

Shallow, well-rounded pits were also reported on outer shell surface of *Syllomus crispatus aegyptiacus* USNM 24872.

Procolpochelys grandaeva USNM 24889 with circular depression on right third and fourth costals. He suggested that it was similar to "basal configuration" of *Balanus concavus* or a similar barnacle.

Weichmann U. 1989. Extreme Höcker auf dem Panzer einer Landschildkröte. [Extreme deformation of the carapace of a tortoise]. Deutsche Aquarien- und Terrarien-Zeitschrift 42(10):1 [German].

Metabolic – Greek tortoise with 1-cm high humps on vertebralia and costalia caused by too much and too "good" food.

Weishampel DB, White NM. 2003. Dinosaur Papers 1676–1906. 524 pp.; Washington: Smithsonian Books.

Vertebral – Published plates from Cuvier 1812, listed as Crocodiles fossiles. Pl I and Pl II. Figure 8 in plate I and Fig. 10 in plate II appear to represent vertebral fusion of atlas and axis. Figure 10 appears to be congenital fusion, while Fig. 8 could be simple adherence of adjacent bones.

Fossil – Published plates from Cuvier 1812, listed as Crocodiles fossiles. Pl I and Pl II. Figure 8 in plate I and Fig. 10 in plate II appear to represent vertebral fusion of atlas and axis. Figure 10 appears to be congenital fusion, while Fig. 8 could be simple adherence of adjacent bones.

Welles SP. 1943. Elasmosaurid plesiosaurs with description of new material from California and Colorado. Memoirs of the University of California 13(3):125–254.

Vertebral – Plesiosaur *Hydrotherosaurus alexandrae* UC Museum of Paleontology 22912 with compression of last cervical vertebra.

Fossil – Plesiosaur *Hydrotherosaurus alexandrae* UC Museum of Paleontology 22912 with compression of last cervical vertebra.

Wells C. 1964. Pathological epipodials and tarsus in Stretosaurus macromerus from the Kimmeridge Clay, Stretham, Cambridgeshire. Quarterly Journal of the Geological Society of London 120: 299–304.

Arthritis – *Stretosaurus macromerus* Sedgwick Museum, Cambridge, England J-35,990C-O with "extensive pathologic changes that affect the tibia, tibiale, intermedium, first distal tarsal, and (fused) second and third distal tarsal." He called it osteoarthritis, but it may be taponomic.

Pseudopathology – *Stretosaurus macromerus* Sedgwick Museum, Cambridge, England J-35,990C-O with "extensive pathologic changes that affect the tibia, tibiale, intermedium, first distal tarsal, and (fused) second and third distal tarsal." He called it osteoarthritis, but it may be taponomic.

Fossil – *Stretosaurus macromerus* Sedgwick Museum, Cambridge, England J-35,990C-O with "extensive pathologic changes that affect the tibia, tibiale, intermedium, first distal tarsal, and (fused) second and third distal tarsal." He called it osteoarthritis, but it may be taponomic.

Wenker CJ, Bart M, Guscetti F, Hatt J-M, Isenbugel E. 1999. Periarticular hydroxyapatite deposition disease in two red-bellied short-necked turtles (*Emydura albertisi*). Annual Proceedings of the American Association of Zoo Veterinarians 1999: 23–26.

Metabolic – Hydroxyapatite deposition disease in red-bellied short-necked turtles *Emydura albertisi*.

Werber J. 1905. Regeneration der Kiefer bei der Eidechse *Lacerta agilis*. [Regeneration of jaw bone of lacerta *Lacerta agilis*]. Roux's Archiv für Entwicklungsmechanik 19:248–258 [German].

Trauma – Regeneration of jaw bone of *Lacerta agilis*.

Werber J. 1906. Regeneration der Kiefer bei Reptilien und Amphibien. [Regeneration of jaw bone of reptiles and amphibians]. Roux's Archiv für Entwicklungsmechanik 19:248–258 [German].

Trauma – Regeneration of jaw bone of *Lacerta agilis*, citing Werber (1905).

Werber EI. 1917. Experimental studies on the origin of monsters. II. Regarding the morphogenesis of duplicities. Journal of Experimental Zoology 24:409–435.

Congenital – General commentary on monsters, without specific discussion related to herpetology.

Wermuth H. 1961. Anomalien bei einer griechischen Landschildkröte (*Testudo hermanni hermanni*). [Anomalies in the Greek tortoise]. Sitzungsberichte der Gesellschaft der naturforschenden Freunde zu Berlin 1:139–142 [German].

Congenital – Anomalies in horn shields, reduced (four) number of front toes and four on right and three on left hind leg of Greek tortoise.

Werner F. 1910. Selbstverstümmelung und Regeneration des Eidechsenschwanzes. [Self-mutilation and regeneration of the lizard tail]. Wien Wissen für Alle Naturhistorische Beilage 1910(4):17–18 [German].

Trauma – Autotomy with regeneration in *Coronella austriaca, Ophisaurus moguntinus.*

Werner WE Jr. 1959. Amphibians and reptiles of the Thousand Islands region, New York. Copeia 1959:170–172.

Congenital – Kyphosis in *Chemys picta marginata.*

Werner YL. 1961. The vertebral column of the geckos (Gekkonoidae), with special consideration of the tail. PhD. Thesis. The Hebrew University of Jerusalem, 203 pp. [Hebrew].

Congenital – Variation in number of presacral vertebrae in *Gekko japonicus* (25–26 and *Tropiocolotes tripolitanus* (24–26), and variable number of caudal vertebrae in *Hemidactylus flaviridis, Ceramodactylus doriae* (46–48), and *Stenodactylus elegans* (29–32).

In-depth reporting on individuals of species, summarized in 1964, 1965, and 1968 articles.

Werner YL. 1964a. Frequencies of regenerated tails, and structure of caudal vertebrae, in Israeli desert geckoes (Reptilia: Gekkonidae). Israel Journal of Zoology 13:134–136.

Trauma – Twenty eight percent of climbing geckos *Hemidactylus turcicus* and *Ptyodactylus hasselquistii* had regenerated tails. Thirty nine percent and 23%, respectively, had lost part of tails somewhere between tail base and tail tip, similar to predominantly ground dwelling *Alsophylax blanfordii* and *Tropiocolotes steudneri.* This contrasted with rare tail (from its base) regeneration in *Stenodactylus sthenodactylus* (12%) and *Ceramodactylus doriae* (4%), with missing tails in 12% and 9%, respectively.

Restriction of autotomy to tail base was found in ground dwelling geckos *Palmatogecko, Ptenopus* (Brain 1958), and arboreal broad-tailed geckos *Uroplatus* (Wellborn 1933) and *Phyllurus* (Holder 1960).

Werner YL. 1964b. Intraspecific and temperature-correlated variations of vertebral numbers in some Near-East geckos (Reptilia, Gekkonidae). Proceedings of the Israel Zoological Society 13:131–133.

Environmental – Vertebral number inversely correlated with temperature (Ben-Tuvia 1963; Gabriel 1944).

V-shaped curve of vertebral numbers in snakes (Fox 1948; Klauber 1941) and partially correlated in saurian *Aprasia* (Parker 1956).

84 of 86 *Ptyodactylus hasselquistii* had 27 vertebrae (caudal range 26–31).

32 of 34 *Hemidactylus turcicus* had 28 vertebrae (caudal range 32–40).

10 of 14 *Tropicolotes steudneri* had 26 vertebrae (caudal range 29–30).

11 of 18 *Ceramodactylus doriae* had 26 vertebrae (caudal range 42–50).

36 of 47 *Stenodactylus sthenodactylus* had 27 vertebrae; 9, 28.

18 of 28 west of Gulf of Sytre had 28 vertebrae; 28 of 33 east had 27.

Werner YL. 1965. The comparative caudal osteology of some gekkonid lizards from Israel. Israel J Zool 14:286–301.

Trauma – Almost all caudal vertebrae in *Tropiocolotes steudneri, Cryptodactylus kotschyi orientalis*, and *Alsophylax blanfordii* are autotomous (but not functional in the last 10 or 15).

Only the first few caudal vertebrae of *Stenodactylus sthenodactylus* and *Ceramodactylus doriae* are autotomous.

Palmatogecko rangei only detaches tail at base.

Autotomy in *Hemidactylus turcicus* in all but last three caudal vertebrae; in *Ptyodactylus hasselquistii guttatus*, all but the last two.

Werner YL. 1967. Regeneration of the caudal axial skeleton in a gekkonid lizard (*Hemidactylus*) with particular reference to the 'latent' period. Acta Zoologica (Stockholm) 48:103–125.

Trauma – Replacement by unsegmented cartilaginous tube in *Hemidactylus flaviridis.*

Slow worm *Anguis fragilis* has less tail regeneration ability (Smith 1954).

Werner YL. 1968. Regeneration frequencies of geckos of two ecological types (Reptilia: Gekkonidae). Vie et Milieu Series C 19:199–221.

Trauma – *Agamura* does not lose tail.

Fifty percent of *Hemidactylus flaviridis* (Woodland 1920), 33% of *Tarentola mauritanica* (Furieri 1956), and 40% of *Hemidactylus mabouia* (Brain 1958) (all climbing species) had regenerated tails. Low percentage was found in 19% of ground dwelling *Palmatogecko rangei*, all at the tail base, similar to that noted in *Chondrodactylus angulifer* and *Ptenopus garrulous* and 9% in *Diplodactylus williamsi* (Bustard 1964).

There was no gender bias.

Only the first or first and second post-pygal vertebrae in *Stenodactylus sthenodactylus* and *Ceramodactylus doriae* are autotomous.

Tail condition (break at tail base/distally)

Tail condition	% Intact	% Missing	% Regenerated
Hemidactylus turcicus	39	32 (12/20)	29 (9/20)[a]
Ptyodactylus hasselquistii	49	22 (14/8)	29 (7/22)[a]
Stenodactylus sthenodactylus 82	8 (base)	10 (base)	
Ceramodactylus doriae 75	21 (11/10)	4 (base)	

[a]Overall – significantly more regeneration than other genera, but not significantly different at rate of regeneration at tail base.

No fracture planes or autotomy in *Amphisbaena alba* (Gans 1962, 1964; Gans and Alexander 1962) or in *A. occidentalis occidentalis*, although *A. o. townsendi* has a fracture plane and autotomy (Gans 1961).

Almost all caudal vertebrae in *Tropiocolotes steudneri* and some in *Alsophylax blanfordii* are autotomous (Werner 1965).

Many ground dwelling (especially psammophile) geckos have reduced autotomy.

Arboreal Australian leaf-tailed geckos *Phyllurus cornutus* and rupicolous *Phyllurus platurus* have only two to three fracture planes at base of tail (Holder 1960).

Malagasy arboreal flat-tailed *Uroplatus fimbriatus* has autotomy at seventh and eighth caudal vertebrae only (Wellborn 1933).

The short-tailed *Nephrurus asper* lacks autotomy splits (Holder 1960; Waite 1929).

Environmental – Trend towards reduced autotomy in *Stenodactylus sthenodactylus* as temperature dropped from 31°C to 25.5°C.

No autotomy in *Ceramodactylus doriae* at 21°C.

Werner YL, Moravec J. 1998. Preliminary observations on tail regeneration in *Sphenodon* (Reptilia: Rhynchocephalia). Israel Journal of Zoology 44:90.

Trauma – 13/13 tailless *Sphenodon* with regenerated calcified tube with local irregular pseudosegmental calcified thickenings.

Werner YL, Takahashi H, Yasukawa Y, Ota H. 2006. Factors affecting foraging behavior, as seen in a nocturnal ground lizard, *Goniurosaurus kuroiwae kuroiwae*. Journal of Natural History 40:439–459.

Trauma – Tail loss in ground lizard *Goniurosaurus kuroiwae kuroiwae* affects behavior, especially locomotion – reduced mobility in males, slightly increased mobility in females.

Westenhöfer M. 1926. Vergleichend-morphologische Betrachtungen über die Entstehung der Ferse und des Sprunggelenks der Landwirbeltiere mit besonderer Beziehung auf den Menschen. [Comparative morphological considerations of the formation of heel and ankle articulation of the land vertebrates, with special relation to humans]. Archiv für Frauenkunde und Konstitutionsforschung 12:305–352 [German].

Trauma – Fracture of femur in lizard figured.

Wexo JB. 1981a. Snakes. 20 pp., San Diego: Wildlife Education.

Congenital – Dicephalic *Lampropeltis getula californiae.*

Wexo JB. 1981b. Las serpientes. [Snakes]. Zoobooks: 20 pp., San Diego: Wildlife Education.

Congenital – Dicephalic *Lampropeltis getula californiae.*

Wexo JB. 1987. Snakes. Zoobooks 4(10):1–20. San Diego: Wildlife Education.

Dicephalic *Lampropeltis getula californiae.* One photograph.

Wheldon J. 2006. Unearthed, the dragon with double firepower. The London Daily Mail 20 December 2006: 45.

Congenital – Dicephalic turtle.

145-million-year-old derodymus fossil reptile *Sinohydrosaurus lingyuanensis* or *Hyphalosaurus lingyuanensis* (Choristodera family which evolved crocodile-like characteristics) from China reported by Eric Buffetaut.

Fossil – 145-million-year-old derodymus fossil reptile *Sinohydrosaurus lingyuanensis* or *Hyphalosaurus lingyuanensis* (Choristodera family which evolved crocodile-like characteristics) from China reported by Eric Buffetaut.

Fossil – 145-million-year-old derodymus fossil reptile *Sinohydrosaurus lingyuanensis* or *Hyphalosaurus lingyuanensis* (Choristodera family which evolved crocodile-like characteristics) from China reported by Eric Buffetaut.

Whillans TH, Crossman EJ. 1977. Morphological parameters and spring activities in a central Ontario population of midland painted turtle, *Chrysemys picta marginata* (Agassiz). Canadian Field-Naturalist 91:47–57.

Congenital – 19.5% of midland painted turtle *Chrysemys picta marginata* had misshapen plastson; 6.2%, abnormal carapace, and 16%, "unusual appendages."

Whitaker A. 1924. Society Club features: two-headed snake bracelet. Daily Magazine of the Oakland Tribune, Oakland, CA (23 October 1924).

Congenital – Derodymus snake.

Whitaker R. 1971. Notes on Indian snakes – 1. Journal of the Bombay Natural History Society 68:461–463.

Congenital – Dicephalic dog-faced water snake *Cerberus rynchops.*

White CP. 1925. Regeneration of the lizard's tail. Journal of Pathology and Bacteriology 28:63–68.

Trauma – Autotomy in *Lacerta vivipara* with tail regeneration as continuous, unsegmented tube.

White M. 1989. Tortoises and soft shell illness. Rephiberary 138:6.

Metabolic – Soft shell disease produced by calcium-deficient diet.

White JB, Murphy GG. 1972. A kyphotic eastern spiny soft-shelled turtle, *Trionyx spinifer spinifer.* Journal of the Tennessee Academy of Science 47:61.

Congenital – Kyphotic eastern spiny soft-shelled turtle *Trionyx spinifer spinifer* with nine (normal is ten) thoracic vertebrae. Fourth is long and arched, noting articulation of fourth, fifth, and sixth ribs with anterior, middle, and posterior regions of the fourth (? If fourth and fifth vertebrae were fused congenitally).

Whited JL, Tabin CJ. 2009. Limb regeneration revisited. Journal of Biology 8:5 (doi: 10.1186/jbiol 1 05).

Trauma – Reports Spallanzani (1769) as first scientific treatise on limb regeneration.

Whiteaves J. 1902 A Canadian two-headed snake. Ottawa Natur. 16:148.

Congenital – Dicephalic garter snake *Eutainia sirtalis sirtalis.*

Whitehead WE. 1944. Another abnormal frog. Turtox News 22:183.

Congenital – Six-legged bullfrog tadpole.

Wiedersheim, R. 1875. *Salamandrina perspicillata* und *Geotriton fiscus.* [*Salamandrina perspicillata and Geotriton fiscus*] 17 plates and 3 wood cuts Genua 1875: p. 179 [German].

Congenital – *Salamandrina perspicillata* and *Geotriton fiscus* with foot anomalies after Goette (1879, p. 22) Barfurth 1898.

Wieland GR. 1906. The osteology of *Protostega.* Memoirs of the Carnegie Museum 2: 279–298.

Vascular – Front flipper of *Protostega* is 3–1/2 times the length of the dorso-sacral series of vertebrae, greater than any other turtle. This led to interpretation as a rapid swimmer in pursuit of prey.

Fossil – Front flipper of *Protostega* is 3–1/2 times the length of the dorso-sacral series of vertebrae, greater than any other turtle. This led to interpretation as a rapid swimmer in pursuit of prey.

Wieland GR. 1909. Revision of the Protostegidae. American Journal of Science (4) 27:101–130.

Congenital – "Shark tooth pertaining to a scavenger species related to *Lamna* was found with the type" of *Dermochelys. Archelon ischyros* with obliquely bitten, healed tibia and fibula. Type of *Dromocyon vorax* has healed lower jaw fracture.

Fossil – "Shark tooth pertaining to a scavenger species related to *Lamna* was found with the type" of *Dermochelys. Archelon ischyros* with obliquely bitten, healed tibia and fibula. Type of *Dromocyon vorax* has healed lower jaw fracture.

Wiesner CS, Iben C. 2003. Influence of environmental humidity and dietary protein on pyramidal growth of carapaces in African spurred tortoises (*Geochelone sulcata*). Journal of Animal Physiology and Animal Nutrition 87:66–74.

Environmental – African spurred tortoises *Geochelone sulcata* with pyramidal osseous carapace overgrowth attributed to dry conditions and nutritionally dense diet producing high growth rate. Dehydration reduces pressure at gap areas and the tissue around the gap collapses, becomes ossified and the collapsed areas become fixed.

Shell disease – Pits and humps in Aldabra giant tortoise *Geochelone dussumieri.*

Wijnen A. 1999. Doppelköpfigkeit. [Dicephaly]. Elaphe Magazine 7(4):29 [German].

Congenital – Derodymous *Acrantophis dumerili.*

Wilbur HM, Semlitsch RD. 1990. Ecological consequences of tail injury in *Rana* tadpoles. Copeia 1990:18–24.

Trauma – Tail loss did not affect *Rana catesbeiana* predation rate by newt *Notophthalmus viridescens.*

Wilcox C, Perrin M. 1996. Junior journalists. St. Petersburg Times (Florida) 18 April 1996: 3.

Congenital – Only dicephalic turtle in Florida.

Wilder H.H. 1904. Duplicate twins and double monsters. American Journal of Anatomy 3:386–472.

Trauma – Comments on regeneration of lizard tails and salamander limbs.

Wilder H.H. 1908. The morphology of cosmobia: Speculation concerning the significance of certain types of monsters. American Journal of Anatomy 8:355–440.

Congenital – Dicephalic *Storeria.*

Wiles M, Rand TG. 1987. Integumental ulcerative disease in a loggerhead turtle *Caretta caretta* at the Bermuda Aquarium: Microbiology and histopathology. Diseases of Aquatic Organisms 3:85–90.

Shell disease – Ulcerative plastron disease in *Caretta caretta* attributed to mixed bacterial-fungal infection.

Wiley GO. 1930. Notes on the neotropical rattlesnake (*Crotalus terrificus basiliscus*) in captivity. Bulletin of the Antivenin Institute of America 3(4):100–103.

Congenital – Dicephalic neotropical rattlesnake *Crotalus terrificus basiliscus.*

Wilhelm RS, Emswiller BB. 1977. Intraoral carcinoma in a Burmese python. Veterinary Medicine, Small Animal Clinician 72:272.

Vascular – Necrosis of dorsal palate in Burmese python *Python molurus bivitattus.*

Wilhelm BC, O'Keefe FR. 2010. A new partial skeleton of a cryptocleidoid plesiosaur from the Upper Jurassic Sundance Formation of Wyoming. Journal of Vertebrate Paleontology 30:1736–1742.

Vertebral – Lateral compression of last caudal vertebra of Upper Jurassic cryptocleidoid plesiosaur *Pantosaurus* striatus USNM 536965.

Fossil – Lateral compression of last caudal vertebra of Upper Jurassic cryptocleidoid plesiosaur *Pantosaurus* striatus USNM 536965.

Wilhoft DC. 1980. Kyphosis in the snapping turtle *Chelydra serpentina.* Herpetological Bulletin of the New York Herpetological Society 15:15–26.
Congenital – *Chrysemys, Lissemys,* and *Chelydra serpentina* kyphosis.
Wilkins J. 1998. Little shop horrors; It's not your usual museum; this one takes a gruesome look at the darkside. The San Diego Union-Tribune 12 May 1998: E-1.
Congenital – Dicephalic turtle.
Wilkinson R, Hernandez-Divers S, Lafortune M, Calvert I, Gumpenberger M, McArthur S. 2004. In McArthur S, Wilkinson R, Meyer J, eds. Medicine and Surgery of Tortoises and Turtles. pp. 187–238; Oxford: Blackwell.
Congenital – Tarsal absence in eastern box turtle *Terrapene carolina.*
Metabolic – Spur-thighed tortoise *Testudo graeca* and *Chelydra serpentina* with demineralization due to secondary nutritional hyperparathyroidism.
Increased mineralization from secondary renal hyperparathyroidism in African spurred tortoise *Geochelone sulcata.*
Carapace bone loss from secondary hyperparathyroidism in Hermann's tortoise *Testudo hermanni.*
Willft DC. 1980. Kyphosis in the snapping turtle *Chelydra serpentina.* Herpetologica 15(2):15–26.
Congenital – Kyphosis in common snapping turtle *Chelydra serpentina.*
Willis RA. 1932. A monstrous twin embryo in a lizard, *Tiliqua scincoides.* Journal of Anatomy 66:189–201.
Congenital – *Tiliqua scincoides* embryos partially fused back to back with soft tissue components of two heads and two divergent coiled tails.
Williams K.L. 1957. Yolk retraction as a possible cause of kyphosis in turtles. Herpetologica 13:236.
Congenital – Kyphosis from early vertebral column fusion in turtles.
Williams JS, Bartels WS. 1994. Paleoecological implications of ulcerative shell disease in fossil turtles from the Eocene of Southwestern Wyoming. Abstracts with Programs Geological Society of America 26(5):68.
Shell disease – Shell-rot in trionychids, emydids, and *Dermatemys.* Bacterial infection in *Beneckea chitinovora* from shell injury.
Bridgerian Eocene trionychids (14), 7 dermatemydidaes, and 17 emydids, with disease in only 1 dermatemydidae, but 10 emydids. All five *Echmatemys* septaria had extensive rot, compared with three of seven *E. wyomingensis* with lesions.
Fossil – Bridgerian Eocene trionychids (14), 7 dermatemydidaes, and 17 emydids, with disease in only 1 dermatemydidae, but 10 emydids. All five *Echmatemys* septaria had extensive rot, compared with three of seven *E. wyomingensis* with lesions.
Williamson TE. 1996. ?*Brachychampsa sealeyi,* sp. Nov., (Crocodylia, Alligatoroidea) from the Upper Cretaceous (lower Campanian) Menefee Formation, northwestern New Mexico. Journal of Vertebrate Paleontology 16:421–431.
Trauma – Heavily pitted later surangular of alligatoroid ?*Brachychampsa sealeyi,* sp. nov. Dentary of NMMNH 25050 has partially healed subcircular depression with lip of remodeled bone attributed to conspecific bite. Actually, it is a semicircular ¼- to 1/3-cm groove.
Fossil – Heavily pitted later surangular of alligatoroid ?*Brachychampsa sealeyi,* sp. nov. Dentary of NMMNH 25050 has partially healed subcircular depression with lip of remodeled bone attributed to conspecific bite. Actually, it is a semicircular ¼- to 1/3-cm groove.
Williston SW. 1898. Mosasaurs. The University Geological Survey of Kansas 4, Part V:83–221.
Trauma – Exostoses on lower jaws, vertebrae (especially caudal), and paddles (especially digits).
Platecarpus with distal tail spine fractures with pseudoarthroses.
Infection – *Platecarpus* coryphaeus with fused forearm, carpus, and metacarpal "united by exostoses" (infectious in origin).
Fossil – Exostoses on lower jaws, vertebrae (especially caudal), and paddles (especially digits).
Platecarpus with distal tail spine fractures with pseudoarthroses
Platecarpus coryphaeus with fused forearm, carpus, and metacarpal "united by exostoses" (infectious in origin).
Williston SW. 1908. North American plesiosaurs: *Trinacromerum.* Journal of Geology 16:715–136.
Vertebral – Two "united" plesiosaur *Trinacromerum bentonianum* thoracic vertebrae.
Fossil – Two "united" plesiosaur *Trinacromerum bentonianum* thoracic vertebrae.
Williston SW. 1914. Water Reptiles of the Past and Present. 251 pp.; Chicago: University of Chicago Press.
Trauma – Mosasaur jaw injuries and bitten off tail.
Vertebral – "Hand and foot" and vertebral fusions in mosasaurs.
Fossil – Mosasaur jaw injuries and bitten off tail.
"Hand and foot" and vertebral fusions in mosasaurs.
Williston SW, Moodie RL. 1917. *Ogmodirus martini,* a new plesiosaur from the Cretaceous of Kansas. Kansas University Science Bulletin 10:61–73.
Trauma – Cretaceous plesiosaur *Ogmodirus martini* neural spines deeply scarred by fish teeth and scratches on humeri.

Fossil – Cretaceous plesiosaur *Ogmodirus martini* neural spines deeply scarred by fish teeth and scratches on humeri.

Wilson CB. 1897. Experiments on the early development of the amphibian egg under the influence of Ringer and salt solutions. Roux's Archiv für Entwicklungsmechanik 5:615–636.

Environmental – 0.6% saline egg exposure hinders hind limb development in *Ambystoma, Rana* and *Chlorophilus*.

Wilson LD. 1968. A fracture plane I the caudal vertebrae of *Pliocercus elapoides* (Serpentes: Colubridae). Journal of Herpetology 1:93–94.

Trauma – Tails breaking off in *Pliocercus elapoides hobartsmithi* (Liner 1960) and *P. e. laticollaris* and *Scaphiodontophis venustissimus* (Taylor 1954).

Wilson BS. 1992. Tail injuries increase the risk of mortality in free-living lizards (*Uta stansburiana*) Oecologia 92:145–152.

Trauma – Tail injuries increase the risk of mortality (especially of males) in free-living lizard *Uta stansburiana*. 62% of survivors had complete tails.

Wilson RS, Booth DT. 1998. Effect of tail loss on reproductive output and its ecological significance in the skink *Eulamprus quoyii*. Journal of Herpetology 32:128–131.

Trauma – Tail loss results in reduced escape ability (Dial and Fitzpatrick 1984; Wilson 1992), energy reserve loss (Dial and Fitzpatrick1981; Smyth 1974), social status loss with reduced home range and success in mating (Fox et al. 1990; Salvador et al. 1995), reduction in growth rate (Ballinger and Tinkle 1979), and if tail lipid deposit important to vitellogenesis, reduced reproductive output (Dial and Fitzpatrick 1981; Smyth 1974; Taylor 1984).

Tail loss in lizards lacking abdominal fat bodies results in reduction in clutch size (Smyth 1974; Taylor 1984).

Forty seven percent of sphenomorphine eastern water skink *Eulamprus quoyii* museum specimens had tail loss, frequency increasing with size, and peaking at 66.1% for females and 82.4% in males (frequency in males significantly different from females only at the larger sizes).

Tail loss resulted in clutch size reduction of 75% in *Eulamprus quoyii*. This compares with 50–100% in *Ctenotus taeniolatus* (Taylor 1984) and 55% in *Hemiergis peronii* (Smyth 1974), which lack abdominal fat bodies, and contrasts with 14% in *Morethia boulengeri* (Smyth 1974) and no reduction in *Lampropholis guichenoti* (Taylor 1984) and *Coleonyx brevis* (Dial and Fitzpatrick 1981), which have abdominal fat bodies. The latter, however had 14% reduction of egg mass and 19% reduction of egg energy. *Batrachoseps attenuatus* regeneration of tail is associated with short-term reduced reproductive effort.

Wiman C. 1910. Ichthyosaurier aus der Trias Spitzbergens. [Ichthyosaurs of Triassic Spitsbergen]. Bulletin of the Geological Institute of Upsala 10:124–148 [German].

Other – Pachyostosis in ichthyosaur *Pessosaurus*.

Fossil – Pachyostosis in ichthyosaur *Pessosaurus*.

Wiman C. 1915. Über die Stegocephalen aus der Trias Spitzbergens. [On stegocephalian amphibian from the Triassic of Spitsbergen.] Bulletin of the Geological Institute of Upsala 13:1–34 [German].

Congenital – Median bone (his "Centroparietale") between parietale and postparietale in Triassic trematosaurid amphibian, *Aphaneramma rostratum*.

Fossil – Median bone (his "Centroparietale") between parietale and postparietale in Triassic trematosaurid amphib, *Aphaneramma rostratum*.

Wiman C. 1916. Neue Stegocephalenfunde aus dem Posidonomyaschiefer Spitzbergens. [New finds of stegocephalians in the *Posidonomya* shales of Spitsbergen]. Bulletin of the Geological Institute of Upsala 13:209–222 [German].

Congenital – *Lonchorhynchus* with calcified and weakly ossified cartilage compared with ethmoidal cartilage.

Fossil – *Lonchorhynchus* with calcified and weakly ossified cartilage compared with ethmoidal cartilage.

Wiman C. 1920. Some reptiles from the Niobrara group in Kansas.. Bulletin of the Geological Institute of Upsala 18:9–18.

Congenital – Anomalous "emarginate" coracoid and right scapula in *Clidastes sternbergii*.

Fossil – Anomalous "emarginate" coracoid and right scapula in *Clidastes sternbergii*.

Wise SE. Jaeger RG. 1998. The influence of tail autotomy on agonistic behavior in a territorial salamander. Animal Behavior 55:1707–1716.

Trauma – Tailless red-backed salamander *Plethodon cinereus* was subjected to more aggression by interlopers.

Wise SE, Verret FD, Jaeger RG. 2004. Tail autotomy in territorial salamanders influences scent marking by residents and behavioral responses of intruders to resident chemical cues. Copeia 2004:165–172.

Trauma – Tailless red-backed salamander *Plethodon cinereus* marks more scents, but experiences less aggression from tailed intruders.

Winslow GM. 1904. Three cases of abnormality in urodeles. Tufts College Studies 1:387–410.

Congenital – *Ambystoma punctatum* with bifid toe.

Trauma – *Plethodon glutinosus* with bifid tail.

Wisniewski PJ. 1984. Limb degeneration in two species of Asian newt. Bulletin of the British Herpetological Society 10:54.

Other – Degeneration of bones of toes in *Paramesotriton chinensis, Pleurodeles waltl* and *Cynops ensicauda popei.*

Wissdorf H, Roder B, Horn H-G. 1984. Multiple Zungenbeinfrakturen als Ursache der Nahrungsverweigerung bei einem Bengalenwaren (*Varanus bengalensis* B., Daudin 1803). [Multiple fractures of the hyoid bone as a cause of food inacceptance in a Bengal monitor (*Varanus bengalensis* B., Daudin 1803).] Verhandlungen und Berichte über Erkrankungen der Zootiere 26:313–319 [German].

Trauma – Hyoid fractures in Bengal monitor (*Varanus bengalensis*).

Wissman M, Parsons B. 1994. Metabolic bone disease in green iguanas. A new, effective treatment offers hope to herps with this condition. Reptiles 1(3):68–72.

Metabolic bone disease – Nutritional secondary hyperparathyroidism or fibrous osteodystrophy in *Iguana iguana*, presenting as weak bones, swollen limbs with spontaneous fractures, rubbery and misshapen lower jaws, and tail "breaks" and folding fractures. Curiously, notes response to calcitonin treatment. As the hormone that contributes to bone density and structure, a very intriguing observation.

Witschi E. 1920. Über merogenetische Entwicklung äquipotentieller Fragmente. [On a merogenetic development of equipotential fragments]. Festschrift zum 60. Geburtstag von Friedrich Zschokke, Artikel 36:15pp. Basel: von Kober, CF. Spittlers Nachfolger [German].

Environmental – *Rana temporaria* with hypermely (additional anterior legs and one specimen with additional hind leg, caused by maximum increased temperature during development.

Witschi E. 1922. Überreife der Eier als kausaler Faktor bei der Entstehung von Mehrfachbildungen und Teratomen. [Overripe eggs as cause for the development of multiple formations and teratomes]. Verhandlungen der Naturforschenden Gesellschaft Basel 34: 33–40 [German].

Environmental – Overripe eggs as cause for polymely and duplication of head to addition of an amorphous lump of tissue in frogs. In overripe eggs, the cell connection gets loose so that single regions react independent as in experiments where parts are disconnected from other parts.

Witschi E. 1952. Overripeness of the eggs as a cause of twinning and teratogenesis. A review. Cancer Research 12:763–785.

Environmental – Egg overripeness produces Siamese twins, polymelia, polydactyly, duplication of lower body region, and double tails in *Rana pipiens*. Thirty four twins and 9 polymelic frogs were found in 7 of 25 overripe cultures, contrasted with none of 100 normal cultures. Oxygen deprivation produces teratogenesis in *Ambystoma* (Detwiler and Copenhaver 1941) and *Triturus* (Rübsaamen 1949).

Witschi E, Chang CY. 1954. Hereditary Polymelia combined with edema and sex reversal in the toad. Anatomical Record 118:370.

Congenital – Polymely in 293 of 1,758 edematous (of 1,903 toads) from one female.

Witzmann F, Scholz H, Müller J, Kardjilov N. 2010. Sculpture and vascularization of dermal bones, and the implications for the physiology of basal tetrapods. Zoological Journal of the Linnean Society 160:302–340.

Vascular – Metaplastic dermal bone makes large-scale gas exchange, unlikely in most basal tetrapods, except for the newt-like lepospondyls (Schoch and Carroll 2003).

Thermal regulation by well-vascularized skin in crocodiles (Seidel 1979).

Woitkewitsch AA. 1948. Phenomenon nasledstvennogo izvrashcheniya realizatsii formoobrazovatelnoy potentsii. [Phenomenon of hereditary development of the embryology of deviation]. Doklady Akademija Nauk SSSR 60:305–308 [Russian].

Congenital – *Rana ridibunda* with supernumerary legs.

Woitkewitsch AA. 1955a [Magnification of limb bud in frogs.] Priroda (12):100–102 [Russian].

Congenital – Three hundred and four species of *Rana ridibunda* (lake frog) with additional extremities found in natural water basin reservoirs were studied. 295 of them had an additional right lower extremity (24 variant with connection to main extremity in thigh region, thigh and leg region, or even entire lower extremity length).

Woitkewitsch AA. 1955b. [Gesetzmäßigkeiten bei der Entwicklung von zusätzlichen Extremitäten des Seefrosches.] [Regularities of the development of additional extremities of the marsh frog]. Archives of Anatomy and Histology 33(2):41–50 [German translation from Russian].

Congenital – Pictures of all morphological types of marsh frog supernumerary limbs.

Woitkewitsch AA. 1955c. [Peculiarities of innervation of additional extremities formed in frogs in the wild]. Doklady Akademii Nauk SSSR 116(5):884–887 [Russian].

Congenital – Three hundred and four species of *Rana ridibunda* (lake frog) with additional extremities found in natural water basin reservoirs were studied. 295 of them had an additional right lower extremity (24 variant with connection to main extremity in thigh region, thigh and leg region, or even entire lower extremity length).

Woitkewitsch AA. 1958a. [Eigenartige Unterschenkelreduktion einer duplizierten Extremität eines Frosches in natürlichen Bedingungen]. [A peculiar reduction of the shinbone in the doubling of the limb in the frog in nature]. Doklady Akademija Nauk SSSR. (German Biological Sciences, translation from Russian) 128:28–31 [German].

Congenital – Phocomelia and polydactyly in frogs.

Woitkewitsch AA. 1958b. [Peculiar reduction of crus in natural limb duplication of frog.] Doklady Akademija Nauk SSSR 118(4):841 [Russian].

Congenital – *Rana ridibunda* shows reduction of the crural and distal bones during natural duplication of hind limb.

Woitkewitsch AA. 1958c. A peculiar reduction of the shin-bone in the doubling of the limb in the frog in nature. Doklady Akademija Nauk SSSR (Biological Sciences, translation) 128:28–31.

Congenital – *Rana ridibunda* shows reduction of the crural and distal bones during natural duplication of hind limb.

Woitkewitsch AA. 1958d. Spatial relations between true limbs and accessory limbs developing under natural conditions. Doklady Akademija Nauk SSSR (Biological Sciences, translation) 120:373–376.

Congenital – Variation in lower extremity duplication in frogs.

Woitkewitsch AA. 1959. Natürliche Mehrfachbildungen an Froschextremitäten. [Natural multiple formations of frog extremities]. VEB Gustav Fischer Verlag, Jena: 82 pp [German].

Congenital – Short historical of missing and additional hind legs (polymely, polydactyly, polyphalangy).

Trauma – Polymely not hereditary, occurrence from "external influences."

Woitkewitsch AA. 1961. Le dévelоppement des extrémités surnuméraires chez les amphibians. [The development of supernumerary extremities among amphibians]. Bulletin Biologique de la France et de la Belgique 95:569–600 [French].

Congenital – Variation of position of supernumerary limbs and speculation on etiology in *Rana ridibunda.*

Woitkewitsch AA. 1963. [Symmetrical disorders of the blastemata of frog hind legs.] Doklady Akademija Nauk SSSR 15:1479–1482 [Russian].

Environmental – Five hundred and forty six lake frog *Rana Ridibunda* with supernumerary limbs found in natural water basin reservoirs. Nineteen had not only supernumerary numeral right extremity but also analogous left extremity. Extremities fused into a single complex in three individuals.

Woitkewitsch AA. 1965. [Mass formation of additional hind extremities in the lake frog.] Zhurnal Obshchei Biologii 26:56–62 [Russian, Engl. abstr.].

Congenital – No additional supernumerary extremity cases reported. In 1952, male and female adult frogs with extranumeral right extremity were captured. In lab, 472 eggs were fertilized; of them, 378 tadpoles developed and no cases of extranumeral extremities were found.

Environmental – No additional supernumerary extremity cases reported. In 1952, male and female adult frogs with extranumeral right extremity were captured. In the lab 472 eggs were fertilized; of them 378 tadpoles developed and no cases of extranumeral extremities were found.

Wolff É. 1948. La science des monsters. [The science of monsters]. 265 pp.; Paris: Gallimard [French].

Trauma – Experimental surgical triton derodymous or posterior duplication.

Wolff ED, Fowler DW, Bonde JW. 2007. A possible case of necrotizing dermatitis in the crocodilian *Diplocynodon*, from the Oligocene of the Isle of Wight, United Kingdom. Historical Biology 19:203–207.

Infection – *Diplocynodon*, probably *hantoniensis* with a 1 cm irregular excavated area MSU-VP-Path-10–01. Surrounding area had what they called a porous periosteum. They speculated on fungal disease. While the lesions have a classic bacterial infectious appearance, the lesions lack the characteristics of what has actually been documented in individuals with known/proven fungal disease. They suggested similar finding in BMNH R.5220.

Fossil – *Diplocynodon*, probably *hantoniensis* with a 1 cm irregular excavated area MSU-VP-Path-10–01. Surrounding area had what they called a porous periosteum. They speculated on fungal disease. While the lesions have a classic bacterial infectious appearance, the lesions lack the characteristics of what has actually been documented in individuals with known/proven fungal disease. They suggested similar finding in BMNH R.5220.

Wolterstorff W. 1885/86. Über fossile Frösche insbesondere das Genus *Palaeobatrachus.* [On fossil frogs especially of the genus *Palaeobatrachus*]. I. Theil. Jahresberichte und Abhandlungen des Naturwissenschaftlichen Vereins Magdeburg 1885–86: 1–93 [German].

Congenital *Palaeobatrachus* with abnormalities in sacral region.

Fossil – *Palaeobatrachus* with abnormalities in sacral region.

Wolterstorff W. 1887. Ueber fossile Frösche insbesondere das Genus *Palaeobatrachus.* [On fossil frogs especially of the genus *Palaeobatrachus*]. II. Theil. Jahresberichte und Abhandlungen des Naturwissenschaftlichen Vereins Magdeburg 1886–87: 1–96 [German].

Congenital – Six digits in hind limb in *Palaeobatrachus gigas.*

Fossil – Six digits in hind limb in *Palaeobatrachus gigas*.

Wolterstorff W. 1917. Zoologische Beobachtungen um Büden, bei Magdeburg. [Zoological observations at Büden near Magdeburg]. Blätter zur Aquarien- und Terrarienkunde 28:120–122 [German].

Trauma – Garlic toad *Pelobates fuscus* with cut tail and hind legs due to cannibalism.

Wolterstorff W. 1927. Feuersalamander mit überzähliger Gliedmaße auf dem Rücken. [Fire salamander with supernumerary extremities on the back]. Blätter für Aquarien- und Terrarienkunde 37:211 [German].

Congenital – *Salamander maculosa* with additional leg on dorsum.

Wolterstorff W. 1935. Zusatz [Comment]. Blätter zur Aquarien- und Terrarienkunde, Braunschweig, 46: 147 [German].

Congenital – Dicephalic *Salamander maculosa* larva.

Wolterstorff W, Freytag GE. 1942a (not 1941). Eine zweiköpfige Feuersalamanderlarve. [A two-headed larva of fire salamander]. Wochenschrift für Aquarien- und Terrarienkunde 38:367–368 [German].

Congenital – *Salamandra salamandra* larva with two heads and anterior body; author refers to three other dicephalic specimens.

Wolterstorff W, Freytag GE. 1942b. Rückgratsverkrümmung beim Kammolch (*Triturus cristatus* Laur.). [Arching of vertebral column in crested newt Kammolch (*Triturus cristatus* Laur.)]. Zoologischer Anzeiger 138:90–92 [German].

Vertebral – *Triturus cristatus* with dorsal arching of vertebral column.

Wolterstorff W, Greytag 1942. Rückgratverkrümmung beim Kammolch (*Triturus cristatus* Laur.). [Spinal kink in crested newt (*Triturus cristatus*)]. Zoologische Anzeiger 138:90–92 [German].

Congenital – Spinal kinks in crested newt *Triturus cristatus*.

Woodland W. 1908. A curious instance of polymely in the common frog. Zoologischer Anzeiger 32:354–357.

Congenital – One of three or four hundred *Rana temporaria* had supernumerary legs; two additional hind legs connected by a stalk of integument to the body over a cartilagenous plate (= fused femora).

Woodland WN. 1920. Some observations on caudal autotomy and regeneration in the gecko (*Hemidactylus flaviridis* Rüppel) with notes on the tails of *Sphenodon* and *Pygopus*. Quarterly Journal of Microscopical Science 65:63–100.

Trauma – Autotomy in glass snake *Ophisaurus ventralis*, American "chameleon" *Anolis principalis* and *Uromastyx spinipes*.

Hemidactylus flaviridis only sheds that part of tail necessary for escape. The regenerated, occasionally bifid tail, cannot be shed in parts, but can be broken off, contrasted with non-autotomous *Calotes versicolor.*

Regenerated tails in *Sphenodon punctatus*.

Work TM, Balazs GH. 2010. Pathology and distribution of sea turtles landed as by catch in the Hawaii-based North Pacific pelagic longline fishery. Journal of Wildlife Medicine 46:422–432.

Vascular – Cites Bjorndal et al. (2003) that the pelagic phase in sea turtles is dominated by immature turtles.

Oxygen storage in sea turtles was related to "large amounts of solid tissues and a great blood supply" (Valente et al. 2007).

Worrest RC, Kimeldorf DJ. 1975. Photoreactivation of potentially lethal, UV-induced damage to boreal toad (*Bufo boreas boreas*) tadpoles. Life Sci 17:1545–1550.

Environmental – UV light produces spine curvature in *Bufo boreas boreas*, noting report by Zimskind and Schisgall (1955) in *Rana pipiens* and *Rana catesbeiana.*

Worrest RC, Kimeldorf DJ. 1976. Distortions in amphibian development induced by ultraviolet-B enhancement (290–315 nm) of a simulated solar spectrum. Photochemistry and Photobiology 24:377–382.

Environmental – Concave spinal curvature from UV exposure in *Bufo boreas boreas*.

Worthington RD. 1974. High incidence of anomalies in a natural population of spotted salamanders, *Ambystoma maculatum*. Herpetologica 30:216–220.

Environmental – Thirty five percent of Hattiesburg Mississippi spotted salamanders *Ambystoma maculatum* from a shallow pond have deviant trunk vertebral counts (other than 14), and 88% have ectomely with syndactylism, brachydactylism, oligodactylism, polydactylism, and bifurcated toes. Author considered possibility of heat effect.

Wotton RM, Heckerman RO, Mangery PW. 1953. A report on a naturally occurring supplementary hind limb in a frog, *Rana pipiens*. Proceedings of the Pennsylvania Academy of Science 27:286–290.

Congenital – Cites Curtis (1940) stating that "supplementary limb formation seldom has been reported outside the laboratory."

Accessory hind limb in *Rana pipiens*. Thick, short ilia, pelvis twisted to left, sacral transverse process at acute angle to ilia, fused tibiale and fibulare, two metatarsals, three phalanges, and femur consisting of prominent head (articulating with secondary acetabulum anterior to normal position) with short neck, but no shaft.

Wright JWA. 1878. A very wonderful freak of nature. Mining and Science Press, Febr. 16, 36(7):7.

Congenital – Derodymous *Pituophis wilksei* (gopher snake), referred to here as a bull snake, a pine snake, or a gopher snake.

Wright BA. 1940. Cephalic deformities in embryos of the Massasauga rattlesnake (*Sistrurus c. catenatus* Raf.). Transactions of the Illinois Academy of Science 32(2):221–222.

Congenital – Massasauga rattlesnake *Sistrurus catenatus catenatus* lacking frontal and parietal bones and cerebellum and cerebrum.

Wright J. 1960. No title. Bulletin of the Philadelphia Herpetological Society 8(6):6.

Congenital – Dicephalic timber rattler *Crotalus horridus horridus.*

Wright K. 2001. Nutritional secondary hyperparathyroidism in the Ramsey Canon leopard frog (*Rana subaquavocalis*). In Baer CK, Willette MM eds. Proceedings of the American Association of Zoo Veterinarians, American Association of Wildlife Veterinarians, Association of Reptilian and Amphibian Veterinarians, National Association of Zoo and Wildlife Veterinarians Joint Conference, Orlando, Florida Sept 18–23, 2001. American Association of Zoo Veterinarians: 34–35.

Metabolic – Ramsey Canon leopard frog *Rana subaquavocalis* with osteopenia from nutritional (calcium deficiency) secondary hyperparathyroidism.

Wright D. 2005. Serpentarium a hot spot. Post and Courier 5 November 2005: B8.

Congenital – Dicephalic diamond back terrapin.

Wright KM. 2006. Overview of amphibian medicine. In Mader DR. ed. Reptile Medicine and Surgery. Philadelphia: Saunders, pp. 941–971.

Congenital – Spindly leg in masked tree frog *Smilisca phaeota.*

Infection – Red leg-like disease produced by *Chlamydophila psittaci* in African clawed frog *Xenopus laevis* and Solomon Island eyelash frog *Ceratobatrachus guentheri.*

Metabolic – Metabolic bone disease in White's tree frog *Pelodryas caerulea.*

Gout in waxy tree frog *Phyllomedusa sauvagei.*

Wright K. 2008. Two common disorders of captive bearded dragons (*Pogona vitticeps*): Nutritonal secondary hyperparathyroidism and constipation. Journal of Exotic Pet Medicine 17:267–272.

Metabolic – Secondary hyperparathyroidism in bearded dragons *Pogona vitticeps.*

Wright KM, Whitaker BR. 2001. Nutritional disorders. In Wright KM, Whitaker BR eds. Amphibian Medicine and Captive Husbandry. pp. 73–87; Malabar, Florida: Krieger Publ.

Congenital – Spindly leg defined as skeletomuscular underdevelopment in Espinosa poison frog *Epipedobates espinosai, E. anthonyi*, black-legged poison frog *Phyllobates bicolor*, Phyllomedusine frog *Phyllomedusa* cf *tarsius*, New Granada cross-banded frog *Smilisca phaeota.*

Metabolic – Metabolic disease is a general term. It has been reported in African clawed frog, *Xenopus laevis*, bullfrog *Rana catesbeiana*, Tschudi's African bullfrog *Pyxicephalus adsperus*, ornate horned frog *Ceratophrys ornata*, South American bullfrog *Leptodactylus pentadactylus*, Australian giant frog *Litoria infrafrenata*, Pine Barrens Tree frog *Hyla andersonii*, bufonidae, axolotl *Ambystoma mexicanum* and Mexican caecilian *Dermophis mexicanus*, with mandibular deformity, splayed limbs, scoliosis, abnormally shaped and radiolucent mandibles, thin long bone cortices, osteopenia and angular limb deformities, as illustrated in New Granada cross-banded frog *Smilisca phaeota.*

Amphibians utilize lipoproteins for transport of 25-hydroxycholecalciferol (vitamin D), in contrast with bony fishes, reptiles, birds, and mammals. Thus, diet has a greater effect in amphibia.

Hypervitaminosis A in ornate horned toad *Ceratophrys ornata* maintained on rodents (which have high vitamin A levels) as a possible source of metabolic bone disease.

Scoliosis was reported in the giant monkey frog *Phyllomedusa* cf *tarsius*. Spindly leg was reported in Espinosa poison frog *Epipedobates espinosai, E. anthonyi*, black-legged poison frog *Phyllobates bicolor*, and hylid frogs such as *Phyllomedusa* cf. *tarsius* and *Smilisca phaeota.* They attributed it to flaked fish foods as sole diet, perhaps from B vitamin leaching.

Contrasted with bony fishes and higher vertebrates, amphibians transport 25-hydroxycholecalciferol (vitamin D) utilizing lipoproteins. Vitamin D absorption and utilization may be blocked by high levels of vitamin A in mice and rats (Douglas, Pennino, Dierenfeld 1994).

Hypervitaminosis A as possible cause of metabolic bone disease (Bruce and Parkes 1950), such as ornate horned toad *Ceratophrys ornata.*

Metabolic bone disease (mandibular deformity, scoliosis, long bone fracture) in African clawed frog *Xenopus laevis* (Bruce and Parkes 1950), bullfrog *Rana catesbeiana* (Modzelewski and Culley 1974), Tschudi's African bullfrog *Pyxicephalus adspersus*, ornate horned frog *Ceratophrys ornata*, South American bullfrog *Leptodactylus pentadactylus*, Australian giant frog *Litoria infrafrenata*, Pine Barrens Treefrog *Hyla andersonii*, various bufonids, axolotl *Ambystoma mexicanum* and Mexican caecilian *Dermophis mexicanus.* Radiologically, radiolucent mandibles, thin cortical bone, expanded marrow cavity, osteopenia, angular limb deformities, pathological fractures (e.g., in bullfrog *Rana catesbeiana*).

Scoliosis in giant monkey frog *Phyllomedusa* cf *tarsius*, corrected by replacement of B vitamins – ? from algal-leaching of food.
Wu HW, Liu CK. 1941. A case of polymely in the frog, *Rana nigromaculata*. Sinensia 12:75–80.
Congenital – Supernumerary forelimb attached at sternum of *Rana nigromaculata*.
Wu S. 1994. [The mortal cause and control method on the shell-rot of *Trionyx sinensis*]. Freshwater Fisheries 24:35–37. [Chinese]
Shell disease – Oval shell lesions in *Trionyx sinensis*.
Wurmbach H. 1927. Histologische Untersuchungen über die Heilung von Knochenbrüchen bei Amphibien. [Histologic study of healing fractures of amphibians]. Zeitschrift für Wissenschaftliche Zoologie 129:253–358 [German].
Trauma – Healing fractures in *Triton alpestris, Salamandra amculosa, Rana Temporaria*, and *Rana esculenta* humeri and ulnae
Wyman J. 1862. Two specimens of monstrosity in serpents. Proceedings of the Boston Society of Natural History 9:193–194.
Congenital – Dicephalic water adder *Tropidonotus sipedon*, noting Aristotle reported a dicephalic snake. Derodymous black snake *Coluber constrictor.*
Wyman M, Crosswhite E. 1920. Journal of Entomology and Zoology (Pomona College Claremont, California) 12:78 (Note often reverse citing of these authors).
Congenital – Abnormal toad, evidently *Bufo boreas halophilus* with fused leg bones and two normal feet derived from posterior sacral region.
Wyneken J, Mader D. 2004. Clinical applications: Reptile skeletal system. Proceedings of the North American Veterinary Conference 18:1358–1360.
Trauma – Fractures from nutritional secondary hyperparathyroidism.
Yangpraphakon P, Ousavaplangchai L, Kanchanakpangka S. 1994. A Color Atlas of Diseases of the Crocodile. 76 pp.; Thailand: Style Creative House Co. Ltd.
Congenital – Facial malformations (e.g., absent maxilla), cheiloschisis (cleft palate), arthrogryposis, poly and syndactyly, and "double monsters," but provided no denominator, precluding population frequency assessment. Their (Samutprakarn) crocodile farm raises *Crocodylus siamensis, Crocodylus porosus*, and *Tomistoma schlegelii*, but they did not indicate which had what problems.
Infection – Infection, but provided no denominator, precluding population frequency assessment. Their (Samutprakarn) crocodile farm raises *Crocodylus siamensis, Crocodylus porosus*, and *Tomistoma schlegelii*, but they did not indicate which had what problems.
Metabolic – Feeding chicken liver or intestines to Crocodiles raises Ca/PO4 from normal ratio of 1.2–1.5 to 1:20–50 producing osteodystrophy fibrosa or osteomalacia. Their (Samutprakarn) crocodile farm raises *Crocodylus siamensis, Crocodylus porosus*, and *Tomistoma schlegelii*, but they did not indicate which had what problems.
Yarrow HC. 1878. A two-headed snake. American Naturalist 12:470.
Congenital – Dicephalic *Ophiobolus getulus* USNM 7276.
Yeomans L. 1988. Care and breeding of the Honduran milk snake - (*Lampropeltis triangulum hondurensis*) - and a case of dicephalism. Herpetile 13:5–8.
Congenital – Derodymous Honduran milk snake *Lampropeltis triangulum hondurensis*.
Yeomans L. 1990. A note on a dicephalic Honduran milk snake (*Lampropeltis triangulum hondurensis*). In J Coote, ed. Reptiles. Proceedings of the 1988 UK Herpetological Societies Symposium on Captive Breeding:24–26. British Herpetological Society, London.
Congenital – Dicephalic Honduran milk snake *Lampropeltis triangulum hondurensis*.
Yntema CL. 1960. Effects of various temperatures on the embryonic development of *Chelydra serpentina*. Anatomical Record 136: 305–306.
Environmental – Common snapping turtle *Chelydra serpentina* eggs kept at 15°C during early somite formation and then moved to room temperature developed malformed heads, microphthalmy, and reduced or absent tails, legs or carapace, contrasted with no such abnormalities at 20°. Unusually low incubation temperatures resulted in reduced *Chelydra serpentina* tails.
Yntema CL. 1962. Duplication and innervation in anterior extremities of *Ambystoma* larvae. Journal of Experimental Zoology 149(2):127–145.
Trauma – Polydactyly and polymely in *Ambystoma punctatum* larva from transplantations from extra-limbed animals to normal hosts.
Yom-Tov Y, Geffen E. 2010. Recent spatial and temporal changes in body size of terrestrial vertebrates: Probable causes and pitfalls. Cambridge Philosophical Society Biological Reviews, doi: 10.1111/j.1469–185X.2010.00168.x.
Environmental – Five-fold variation in body size of Japanese four-lined rat snake *Elaphe quadrivirgata*.

Yoshinaga T. 1901. *Agkistrodon* (*Gloydius*) *blomhoffii* from Japan. Dobutsugakuzasshi (Zoological Magazine): 13:369 [Japanese].
Congenital – Derodymous Japanese pit viper *Agkistrodon* (*Gloydius*) *blomhoffii* ("Mamushi" in Japanese).
Youngprapakorn P, Ousavaplangchai L, Kanchanakpangka S. 1994. A Color Atlas of Diseases of the Crocodile. Thai Association for Trade in Reptiles and Amphibians, Bangkok, Thailand, 76 pp.
Congenital – Short lower jaw, docked tail, absent maxilla, nasal, zygomatic arch and part of frontal bones, absent upper jaw, cleft palate (uranoschisis), cleft chin, amelia, polydactyly, supernumerary limb off umbilicus in crocodiles.
Crocodiles from the Samutprakarn Crocodile Farm and Zoo, but not named as to species.
Trauma – Jaw fractures from fighting in crocodile.
Crocodiles from the Samutprakarn Crocodile Farm and Zoo, but not named as to species.
Infection – Purulent arthritis in a crocodile elbow and knee.
Crocodiles from the Samutprakarn Crocodile Farm and Zoo, but not named as to species.
Metabolic – Osteodystrophia fibrosa producing rubber jaw in crocodile.
Accumulation of white chalky urate (gout) in crocodile extremity joints and their capsules.
Crocodiles from the Samutprakarn Crocodile Farm and Zoo, but not named as to species.
Zaffaroni NP, Zavanella T, Arias E. 1989. Spontaneous skeletal malformations of the forelimbs in the adult crested newt. Herpetopathologia 1:49–50.
Congenital – Thirty five percent of male and 20.5% of female *Triturus cristatus carnifex* in Italy had supernumerary phalanges, polydactyly, carpal fusion, phalangeal "distortion," syndactyly, forked digit, carpal and digit duplication, and a bent radius and ulnar.
Zakharov VM. 1989. Future Prospects for Population Phenogenetics. Soviet Scientific Reviews. Section F. Physiology and General Biology Review 4(part 3):1–79.
Congenital – Variation in skull plate pattern in sand lizard *Lacerta agilis*, but less fluctuating asymmetry in *Lacerta strigata* incubated above or below 25°C.
Zamaraev VN. 1974. Does regenerative capacity decrease with age? Soviet Journal of Developmental Biology 4:495–503.
Trauma – Younger animals (not commenting specifically on herpetology) regenerate better than older animals.
Zammit M, Kear BP. 2010. Healed bite marks on a Cretaceous ichthyosaur. Palaeontologica Polonica doi: 10.4202/1pp.2010.0117.
Trauma – Gouge (22.58 mm × 15.56 mm) on ventrolateral surface of left dentary just anterior to tooth row of *Platypterygius* from Lower Cretaceous of Australia. Abrasions and new bone formation on corresponding dentary and at anterior margins.
Fossil – Gouge (22.58 mm × 15.56 mm) on ventrolateral surface of left dentary just anterior to tooth row of *Platypterygius* from Lower Cretaceous of Australia. Abrasions and new bone formation on corresponding dentary and at anterior margins.
Zammit M, Norris RM, Kear BP. 2010. The Australian Cretaceous ichthyosaur *Platypterygius australis*: A description and review of skeletal remains. Journal of Vertebrate Paleontology 30:1726–1735.
Arthritis – Cretaceous ichthyosaur *Platypterygius australis* Queensland Museum FM F48022 left humerus with proximal articular surface defect. Several sharply defined defects at proximal joint surface of right femur. Erosive arthritis or avascular necrosis is a diagnostic consideration (BMR).
Avascular – Cretaceous ichthyosaur *Platypterygius australis* Queensland Museum FM F48022 left humerus with proximal articular surface defect. Several sharply defined defects at proximal joint surface of right femur. Erosive arthritis or avascular necrosis is a diagnostic consideration (BMR).
Fossil – Cretaceous ichthyosaur *Platypterygius australis* Queensland Museum FM F48022 left humerus with proximal articular surface defect. Several sharply defined defects at proximal joint surface of right femur. Erosive arthritis or avascular necrosis is a diagnostic consideration (BMR).
Zangerl R. 1935. IX. *Pachypleurosaurus edwardsi*, Cornalia sp. Osteologie – Variationsbreite - Biologie. [*Pachypleurosaurus edwardsi*, Cornalia sp. Osteology – variation - biology]. Abhandlungen der Schweizerischen Palaeontologischen Gesellschaft 56:1–80 [German].
Other – Macroscopically thickening of bone, microscopically by increased density of ribs and very rarely scapulae, noting thickening of proximal rib endings pachyostotic in middle and only osteosclerotic in upper bone beds. He raises the question of pachyostotic pliosaurian versus "ailing" of the phylum.
Fossil – Macroscopically thickening of bone, microscopically by increased density of ribs and very rarely scapulae, noting thickening of proximal rib endings pachyostotic in middle and only osteosclerotic in upper bone beds. He raises the question of pachyostotic pliosaurian versus "ailing" of the phylum.
Zangerl R, Johnson RG. 1957. The nature of shield abnormalities in the turtle shell. Fieldiana: Geology 10:341–362.

Shell disease – Supernumerary scutes in 39% of 355 box turtle *Chrysemys picta.*

Zangerl R, Turnbull W. 1955 *Procolpochelys grandaeva* (Leidy), an early carettine sea turtle. Fieldiana-Zool 37:345–382.

Congenital – Supernumerary peripheral carapace element in Miocene *Procolpochelys grandaeva.* Division of neural plate into three parts occurs in *Procolpochelys.*

Variation, usually with transverse division of neural, vertebral, and pleural plates in *Lepidochelys.* Complete pleural division in *Lepidochelys* is always associated with vertebral shield division. Division of neural plate into three parts is rare in *Lepidochelys.*

Lengthwise division of neural plates in *Caretta caretta.* Of four Atlantic loggerheads, one had neural division (fragmentation) and one had reduction in neural size.

Fossil – Supernumerary peripheral carapace element in Miocene *Procolpochelys grandaeva.* Division of neural plate into three parts occurs in *Procolpochelys.*

Zani PA. 1996. Patterns of caudal-autotomy evolution in lizards. Journal of Zoology of London 240:201–220.

Trauma – Frequency of tail loss determined by ease of loss and by predation pressure (Jaksic and Greene 1984; Medel et al. 1988; Rand 1954; Turner et al. 1982).

Frequency of tail loss by species:

Family – genus – species	%
Corytophanidae – variable caudal autotomy	
Basiliscus plumifrons	9
Corytophanes cristatus	0
Phrynosomatidae – variable caudal autotomy	
Callisaurus draconoides	0
Phrynosoma cornutum	0
Polychrotidae – variable caudal autotomy	
Dactyloa punctata	0
Norops biporcatus	10
Norops fuscoauratus	9
Norops humilis	22
Norops limifrons	24
Norops nitens	0
Hoplocercidae – caudal autotomy occurs	
Enyaloides laticeps	0
Tropiduridae – variable caudal autotomy	
Tropidurus flaviceps	0
Tropidurus hispidus	57
Tropidurus umbra	0
Uranoscodon superciliosus	28
Crotaphytidae – variable caudal autotomy	
Crotaphytus bicinctores	50
Iguanidae – variable caudal autotomy	
Dipsosaurus dorsalis	51
Sauromalus obesus	0
Gekkonidae – caudal autotomy occurs	
Coleodactylus amazonicus	4
Coleodactylus amazonicus	40
Gonatodes concinnatus	29
Gonatodes humeralis	25
Hemidactylus palaichthus	32
Hemidactylus turcicus	40
Pseudogonatodes guianensis	33
Thecadactylus rapicauda	60
Teiidae – variable caudal autotomy	
Ameiva ameiva	26

Family – genus – species	%
Ameiva festiva	21
Kentropyx pelviceps	43
Gymnophthalmidae – caudal autotomy occurs	
Alopoglossus angulatus	33
Alopoglossus atriventris	0
Arthrosaura reticulate	44
Iphisa elegans	67
Leposoma parietale	9
Leposoma percarinatum	18
Neusticurus ecpleopus	46
Prionodactylus oshaughnessyi	46
Scincidae – variable caudal autotomy	
Eumeces fasciatus	39
Scincella lateralis	11

Opluridae – Caudal autotomy occurs
Chamaeleonidae – No caudal autotomy
Eublepharidae – Caudal autotomy occurs
Lacertidae – Caudal autotomy occurs
Xantusiidae – Caudal autotomy occurs
Cordylidae – Caudal autotomy occurs
Anguidae – Variable caudal autotomy
Xenosauridae – No caudal autotomy
Helodermatidae – No caudal autotomy
Varanidae – No caudal autotomy
Sphenodontidae
Sphenodon – Caudal autotomy occurs

Zavadil V. 1989. Šestinohá rosni'ka [A six-legged frog.] Živa 37:81 [Czech].

Congenital – Supernumerary legs in *Hyla arborea.*

Zavanella T, Zaffaroni NP, Arias E. 1984. Abnormal limb regeneration in adult newts exposed to the fungicide maneb 80. Archives of Environmental Contamination and Toxicology 13:735–745.

Trauma – After amputation, crested newt *Triturus cristatus carnifex* regenerated with "slight" radial and ulnar reduction, absence of metacarpals, supernumerary carpals, fused or missing phalanges, absence of one to three distal phalanges, and incurved distal phalanges were observed.

Toxicology – Fungicide Maneb 80 (manganese ethylenebisdithiocarbamate) produced severe malformations after amputation in crested newt *Triturus cristatus carnifex*, but "slight" radial and ulnar reduction, absence of metacarpals, supernumerary carpals, fused or missing phalanges, absence of one to three distal phalanges, and incurved distal phalanges were observed without fungicide exposure. Fungicide-induced abnormalities included large preaxial elbow cartilaginous nodules, humeral spike-like outgrowths, and shortened or rudimentary radius and ulnar.

Zeidler E. 2010. Cricket frogs talk… with their feet. Kansas Herpetological Society 37th Annual Meeting Program 6–7 November 2010, Topeka Zoo, Topeka, Kansas.

Trauma – Bifid tail and facial disfiguration in common snapping turtle.

Neoplasia – Tumors in common snapping turtle, but no comment on bone involvement.

Shell disease – Shell deformity in common snapping turtle.

Zentek J, Dennert C. 2004. Häufige Fütterungsfehler und Hinweise zur Diätetik bei Reptilien. [Frequent feeding mistakes and suggestions of diet of reptiles]. Gemeinschaftsveranstaltung Deutsche Veterinärmedizinische Gesellschaft und ATF: Kleine Heimtiere: Diagnostik und Therapie von Erkrankungen des Verdauungsapparates – Aktueller Wissensstand. 4 pp. [German].

Metabolic – Diseases caused by diet: rickets caused by deficiency of Ca, P, and vitamin D; osteodystrophy fibrosa, by deficiency of Ca and surplus of P; gout, by surplus of purin and lack of water.

Zhang X, Chen Y, Tian L. 1997. [Wild deformed frogs found in Hebei Province.] Chinese Journal of Zoology 32:49 [Chinese].

Congenital – Supernumerary leg, ectomelia, and extra foot in *Rana nigromaculata.*

Zhi-Chong P, Xiang Ji. 2000. [The influence of incubation temperature on size, morphology, and locomotor performance of hatchling grass lizards (*Takydromus wolteri*).] Acta Ecologica Sinica 21:2031–2038 [Chinese].

Environmental – Grass lizards *Takydromus wolteri* incubated at 24°C and 27°C had larger head size/length than those at 33°C.
Židek L, Židek P. 1985. Dvojhlavé mlád'a užovky hladkej. [Two-headed young in the smooth snake.] Živa 33:228 [Slovak].
Partial Congenital – Head duplication in smooth snake *Coronella austriaca.*
Zingg A. 1966 (not 1968). Zur Fortpflanzung von *Dispholidus typus* (Reptilia, Colubridae). [On reproduction by *Dispholidus typus* (Reptilia, Colubridae)]. Salamandra 4:37–43 [German].
Environmental – Kyphoscoliosis in more than half of *Dispholidus typus* incubated at 28°C.
Zug GR, Hedges SB, Sunkel S. 1979. Variation in reproductive parameters of three Neotropical snakes, *Coniophanes fissidens, Dipsas catesbyi* and *Imantodes cenchoa.* Smithsonian Contributions to Zoology 300:1–20.
Trauma – Six percent of female and 10% of male *Imantodes cenchoa,* 1% of *Dipsas catesbyi* and 30% of *Coniophanes fissidens* had broken tails. Curiously, there was a Bolivian sample of *Dipsas catesbyi* in which 10% were affected.
Zwart P. 1963. Studies on renal pathology in Reptiles. Utrecht: Stichting Pressa Trajectina, 118 pp.
Metabolic – Renal disease in reptiles.
Urate tophus in *Chlamydosaurus kingii* in Fig. 41, labeled (?) acute pyelonephritis.
Zwart P. 1974. Maladies des reptiles. [Illnesses of reptiles]. Part II. Zoo Anvers 40:14–22,63–70 [French].
Metabolic – Osteomalacia in gecko manifests as skeletal decalcification and in iguana as palpable nodules.
Zwart P. 1989. Lymphangitis due to gout in reptiles. Herpetopathologia 1:123–124.
Metabolic – Visceral gout reported in black-headed python *Aspidites melanocephalus,* New Guinea olive-colored python *Liasis papuanus,* and Bengal monitor *Varanus bengalensis.*
Zwart P. 1992. Serpents. In: La Consultation des Nouveaux Animaux de Compagnie [Consultation on new country animals], K Gabrisch, P Zwart, eds. Maisons-Alfort: Editions du Point Veterinaire, pp. 287–311 [French].
Trauma – Rib fractures in snakes.
Vertebral – Alleged block vertebrae in *Thamnophis radix haydeni,* presenting as focal scoliosis.
Alleged temperature-dependent congenital vertebral deformities in *Boa constrictor* (without subspecies, common name cannot be determined), ball, or royal python *Python regius, Bothrops atrox, Thamnophis radix* (Zwart, van Moppes 1983).
Environmental – Alleged temperature-dependent congenital vertebral deformities in *Boa constrictor* (without subspecies, common name cannot be determined), ball, or royal python *Python regius, Bothrops atrox, Thamnophis radix* (Zwart, van Moppes 1983).
Zwart P. 1992a. Lezards. [Lizards]. In: La Consultation des Nouveaux Animaux de Compagnie. [Consultation on new country animals], K Gabrisch, P Zwart, eds. Maisons- Alfort: Editions du Point Veterinaire, pp. 313–333 [French].
Metabolic – Osteodystrophy fibrosa presenting as vertebral fracture in *Cordylus warreni.*
Articular gout in lizards.
Zwart P. 1992b. Amphibiens. [Amphibians]. In: La Consultation des Nouveaux Animaux de Compagnie [Consultation on new country animals], K Gabrisch, P Zwart, eds. Maisons-Alfort: Editions du Point Veterinaire, pp. 351–362 [French].
Congenital – Arthrogryposis in *Dendrobatus pumilio.*
Metabolic – Osteomalacia in amphibians (Bruce and Parkes 1950).
Zwart P. 1995. Amphibian. In Gabrisch K, Zwart P. (eds) Krankheiten der Heimtiere. [Illnesses of Domestic Animals]. 3rd ed. Hannover: Schlütersche Verlagsanstalt, pp. 859–892 [German].
Congenital – Arthrogyposis with reduced anterior limbs in *Dendrobates leucomelas, D. auratus,* and other *Dendrobates* species
Metabolic – Rickets in amphibians.
Zwart P, Gröne A. 2006. Pathologies of tails in reptiles. European Association of Zoo and Wildlife Veterinarians (EAZWV) 6 scientific meeting, May 24–28 – 2006, Budapest Hungary, 8 pp.
Environmental – Sharp dorsal flexion in geckos related to heating lamp.
Other – Garter snake *Natrix natrix* with necrotic tail vertebrae.
Zwart P, Harshbarger JC. 1991. A contribution to tumors in reptiles. Description of new cases. Proceedings of the International Colloquium of Pathological Medicine of Reptiles and Amphibians Sept 27–29, German Veterinary Association, Bad Nauheim, Germany 4:219–224.
Neoplasia – Double-crested basilisk, *Basiliscus plumifrons,* with "generalized" histiocytic sarcoma; water dragon, *Physignathus lesuerii,* with "generalized" lymphoma (all without comment on actual bone involvement); and Timor python *Python timorensis* with osteosarcoma affecting three vertebrae and *C. constrictor* with "localized" osteosarcoma.
Hermann's tortoise, *Testudo hermanni,* with cornifying squamous cell carcinoma that caused pathologic fracture of femur and necrotic adjoining carapace.

Zwart P, van Moppes MC. 1983. Kongenitale Wirbeldeformierungen bei Schlangen, insbesondere bei Strumpfbandnattern (*Thamnophis radix*). [Congenital deformities of vertebrae in snakes, especially in *Thamnophis radix*]. Der Zoologische Garten, Neue Folge, Jena 53:432–440 [German].

Congenital – Symmetrical shortening or lengthening of vertebrae in boa, *Constrictor constrictor*; asymmetrical lordosis in Strumpfbandnatter (garter snake) *Thamnophis radix*; kyphosis and scoliosis in boa *Constrictor constrictor;* and "erratic" missing or added vertebral parts in Strumpfbandnattern *Thamnophis radix*, boa *Constrictor constrictor*, and *Bothrops atrox.*

Vertebral – Bridge building between Strumpfbandnatter (garter snake) *Thamnophis radix* vertebrae.

Environmental – Symmetrical shortening or lengthening of vertebrae in boa, *Constrictor constrictor*; asymmetrical lordosis in Strumpfbandnattern (garter snake) *Thamnophis radix*; kyphosis and scoliosis in boa *Constrictor constrictor*; and "erratic" missing or added vertebral parts in Strumpfbandnattern (garter snake) *Thamnophis radix*, boa *Constrictor constrictor* and *Bothrops atrox;* and vertebral bridging in Strumpfbandnatter (garter snake) *Thamnophis radix* attributed to abnormal temperature during embryonal development.

Zwart P, van de Watering CC. 1969. Disturbance of bone formation in the common iguana (*Iguana iguana* L): Pathology and etiology. Acta Zoologica et Pathologica Antverpiensia 48:333–356.

Trauma – *Iguana iguana* fracture with balloon-shaped callus.

Metabolic – Spontaneous fracture of ribs, mandible, leg bones, cranium, vertebrae, digits; ballooning probably related to vitamin D deficiency.

Spiny-tailed iguana with thickening of extremities probably related to vitamin D deficiency.

Zwart P, Goedegebuure SA, Lambrechts L. 1994. Relationship between intestinal tract and mineralization of skeleton in reptiles, with reference to normal development of the carapace in young tortoises. International Symposium zu Erkrankungen der Zootiere Kristiansand 36:315–319.

Metabolic – Severe hyperparathyroidism in *Geochelone pardalis.*

Decalcified chameleon skeletons.

Slight elevated rims of horny shields in Greek (Hermann's) tortoise *Testudo hermanni.*

Zwart P, Lambrechts L, DeBatist P, Bijnens B, Claessen H, Mennes SS, van Riel C. 1997. Excessive growth in a Hermann's tortoise (*Testudo hermanni*) and its consequences for the development of the carapace. Verhandlungen und Berichte über Erkrankungen von Zootieren, Neue Folge 2:61–64.

Shell disease – Thickened (up to 2 cm), loose-textured carapace with large fat-filled spaces in a Hermann's tortoise *Testudo hermanni.*

Zwart P, Hetzel U, Dik KJ. 2001. Osteitis deformans and concomitant inclusion body disease in a Boa (*Boa constrictor*). Erkrankungen der Zootiere 40:61–55.

Infection – Thickening of *Boa constrictor* (without subspecies, common name cannot be determined) vertebrae (with irregular proliferations) and ribs producing ankylosis of costovertebral joints and calcific mass bridging vertebrae. New bone contained multiple cavities with spikes of necrotic bone. Scoliosis with local osteolysis. Fifty percent of bone marrow was necrotic, with gradual transition from normal. They note that marrow as present, not fibrosis and blood vessels were not increased. They acknowledged this was not Paget's disease but chose to use the term "osteitis deformans" as a "more general name." They illustrated retrovirus like particles.

Metabolic – Thickening of *Boa constrictor* (without subspecies, common name cannot be determined) vertebrae (with irregular proliferations and ribs producing ankylosis of costovertebral joints and calcific mass bridging vertebrae. New bone contained multiple cavities with spikes of necrotic bone. Scoliosis with local osteolysis. Fifty percent of bone marrow was necrotic, with gradual transition from normal. They note that marrow as present, not fibrosis and blood vessels were not increased. They acknowledged this was not Paget's disease, but chose to use the term "osteitis deformans" as a "more general name." They illustrated retrovirus like particles.

Zweifel RG. 1980. Aspects of the biology of a laboratory population of kingsnakes. pp. 141–152 In Murphy JB, Collins JT. (eds.) Reproductive biology and diseases of captive reptiles. Ithaca: Society for the Study of Amphibians and Reptiles.

Environmental – King snake *Lampropeltis getula* incubated at $32.8 \pm 0.8°C$ had tail deformities.

Zwilling E. 1955. Teratogenesis. In: Willer BH, Weiss PA, Hamburger V. eds. Analysis of Development. P. 699–719. Philadelphia: Saunders.

Toxicology – Enlargement of structures in *Rana esculenta* and *Ambystoma tigrinum* from sodium thiocyanate exposure (Ranzi et al. 1946).

Articles Frequently Cited in Discussion of Amphibian and/or Reptile Pathology but which do not Discuss Actual Amphibian or Reptile Abnormalities

Abel O. 1912. Kämpfe, Verletzungen und Kampfanpassungenbei fossilen Wirbeltieren. Verhandlungen der Kaiserlich-Koniglichen Zoologisch-Botanischen Gesellschaft in Wien 62:53–55.

Abbasi SA, Soni R. 1984. Teratogenic effects of chromium (IV) in environment as evidenced by the impact of t larvae of amphibian *Rana tigrina*: Implications in the environmental management of chromium. International Journal of Environmental Studies 23:131–137.

Achariyo LN, Singh LAK, Pattanak SK. 1990. Occurrence of congenital blindness in Gharial *Gavialis gangeticus* (Reptilia: Crocodilia). Journal of the Bombay Natural History Society 88: 117–118.

Ahlfeld F. 1880. Die Mißbildungen des Menschen. Leipzig: Grunow.

Albertini G, Lombardi M. 1977. Metamorfosi ritardata e gigantismo in una larva di *Rana esculenta* L. raccolta presso Chiaravalle (Ancona). Studi trentini di Scienze naturali, Acta Biologica 54:179–196.

Aldrovandi U (but Latin could be V), Cornelius J, Baro TD, Bernia MA. 1647. Quadrupedum omnium bifulcisrum. 429pp. London: Joannis Martyn.

Allen ME, Crissey SD, Demeter BJ. 1986. The effect of diet on growth and bone development in leopard gecko. Ann Proc Am Assoc Zoo Vet 1986:44–45.

Alonzo G. 1903. Ricerche sullo sviluppo della coda e degli arti nel *Gongylus ocellatus*. Atti Acc Gioen 16(5):48.

Alverdes K. 1927. Das Verhalten transplantierter Beinanlagen bei Froschlarven. Zeitschrift für mikroskopisch-anatomische Forschung 2:1

Antunes A. 1963. Classificação dos monstros de Isidoro Geoffroy Saint Hilaire, ampliada e adaptada. Boletim do Departamento de Anatomia Patológica, Faculdade de Veterinária da Universidade de São Paulo 1 (5):4–12; 1 (6):4–11.

Ariel E, Ladds PW, Roberts BL. 1997. Mycobacteriosis in young freshwater crocodiles (*Crocodylus johnstoni*). Australian Veterinary Journal 75: 831–833.

Baer KE von. 1828. Scholien und Corollarien zu der Entwickelungsgeschichte des Hühnchens im Eie.

Baer KE von. 1844 (not 1845). Über die doppelleibigen Missgeburten oder organische Verdoppelungen in Wirbeltieren. Mémoires de l'académie impériale des sciences de St. Pétersbourg, 6. ser. (2) 4: 79–178. Oder 6(2):1–100.

Baevsky YB, Belyaev DK, Utkin LG. 1958. Observations on ovary eggs of sable. Transactions of the Academy of Science USSR. 120(2): 439–440.

Baird SF. 1856. Appendix to the report of the Secretary. H—List of additions to the museum of the Smithsonian Institution in 1855. Annual Reports of the Smithsonian Institution 10:57–61.

Baldus R. 1901. Die Intervertebralspalten v. Ebner's und die Querteilung der Schwanzwirbel bei *Hemidactylus mabouia* Mr. Inaug. Diss. Leipzig.

Balinsky BI. 1957. New experiments on the mode of action of the limb inductor. Journal of Experimental Zoology 134:239–273.

Balinsky Bi. 1956. A new theory of limb reduction. Proceedings of the National Academy of Sciences, USA 42:781–785.

Baljet B, Heyke GCM. 1992. Zur Geschichte der Klassifikationssysteme der Doppelmißbildungen unter besonderer Berücksichtigung des Klassificationssystems von Louis Bolk (1866–1930). Annals of Anatomy 174: 361–368.

Baljet B, Oostra R-J. 1998. Historical aspects of the study of malformations in the Netherlands. American Journal of Medical Genetics 77: 91–99.

Baljet B, Oostra R-J. 1998. Historical aspects of the study of malformations in the Netherlands. American Journal of Medical Genetics 77: 91–99.

B.M. Rothschild et al., *Herpetological Osteopathology: Annotated Bibliography of Amphibians and Reptiles*,
DOI 10.1007/978-1-4614-0824-6_12, © Springer Science+Business Media, LLC 2012

Baljet B, Oostra RJ. 1999. Digital data and the 19th century teratology collection. Journal of Visual Communication in Medicine 22:186–194.

Banchi A. 1904. Sviluppo degli arti addominali del *Bufo vulgaris* innestati in sede anomala. Monitore Zoologico Italiano 25: 396–399.

Banchi A. 1905. Sviluppo di arti pelvici di *Bufo vulgaris* innestati in sede anomala. Archivio Italiano di anatomia e di embriologia 4: 671–693

Banta AM, Gartner RA. 1914. Accessory appendages in *Ambystoma* larva. Proceedings of the Society of experimental medicine 11:177–178.

Barfurth D. 1891. Regeneration. Ergebnisse der Anatomie und Entwickelungsgeschichte 1:103–140.

Barfurth D. 1891. Zur Regeneration der Gewebe. Archiv für mikroskopische Anatomie 37:406–491.

Barkowius IC. 1828. Monstra Animalium Duplicia per Anatomen Indagata. Habito respectu ad Physiologiam Medicinam Forensem et Artem Obstetriciam. Lipsiae: Supptibus Leopoldi Vossii, 142pp.

Barrett S, Schlater L, Montali R, Sneath R. 1994. Iguana-associated salmonellosis-Indiana, 1990. Morbidity and Mortality Weekly Report 41:38–39.

Bataillon E. 1920. Problèmes d'embryonogénie expérimentale. Jumeaux vrais et monsters doubles. Bulletin de l'Association Philomathique d'Alsace et de Lorraine 6(1):1–4

Bateson W. 1890. On some cases of abnormal repetition of parts in animals. Proceedings of the Zoological Society 1890:579–588.

Batzer DP, Sharitz RR. 2006. Ecology of Freshwater and Estuarine Wetlands. Berkeley: University of California Press, 568 pp.

Bedriaga J von. 1879. Über die Begattung bei einigen geschwänzten Amphibien. Zoologischer Anzeiger 5: 265–268, 357–359. ursprüngliches Zitat (ohne Titel): II: 664.

Bedriaga J von. 1879. Verzeichnis der Amphibien und Reptilien Vorderasiens. Bulletin de la Société impériale des naturalistes de Moscou 54(3):22–52.

Bennett RA. 2004. Radiografia en reptiles: tecnicas e interpretacion. Proceedings of the North American Veterinary Conference, 1468–1470.

Bernström J. 1943. Till kännedom on huggormen *Vipera berus* (Swedish with German summary). Göteborgs Kungl Vetenskaps- och Vitterhets-samhälles Handlingar 6 B 10:1–34.

Berrill M, Bertram S, Wilson A, Louis S, Brigham D, Stromberg C. 1993. Lethal and sublethal impacts of pyrethroid insecticides on amphibian embryos and tadpoles. Environmental Toxicology and Chemistry 12:525–539.

Bertacchini P. 1899. Morfogenesi e teratogenesi negli anfibi anuri. (Ia Serie: Blastoporo e doccia midollare.) Internationale Monatsschrift für Anatomie und Physiologie 16: 140–154.

Birge WJ, Black JA. 1979. Effects of copper on embryonic and juvenile stages of aquatic animals. In Nriagu JO, ed. Copper in the Environment. New York, John Wiley and Sons, pp. 590–631.

Birkhahn H. 1992. Ny viden om aminosyreforsyningen hos farbefrøer (dendrobatider). Nordisk Herpetologisk Forening May/June 1992 35(3):38–43.

Blaringhem L. 1911a. Les transformations brusques des êtres vivants. Paris: Ernest Flammarion, 353 pp.

Blaringhem L. 1911b. Anomalie di conformazione. In: Trattato di Anatomia Patologica di Aschof. Italian translation, Torino.

Blum HF, Butler EG, Chang JJ, Mawe RC, Schmidt SE. 1957. Studies on regression and regeneration in the urodele forelimb after localized ultraviolet radiation. Journal of Cellular and Comparative Physiology 49:153–170.

Bodemer CW. 1959. Observations on the mechanism of induction of supernumerary limbs in adult *Triturus viridescens*. Journal of Experimental Zoology 140:79–99.

Boettger O. 1890. Neue Schlange aus Ostindien. Berichte Senckenbergische Naturforschende Gesellschaft Frankfurt 1890:297–298.

Bolk R. 1906. Dubbelmonstra, hun Classificatie en Ontstaan. Geneeskundige Bladen uit Klinieken Laboratorium 12 (Nr. 9+10):235–281.

Bonaparte C. 1929. Analyse des travaux de l'Acad. des Sciences. Analysis of the work of the Academy of Sciences.

Bonnet C. 1781. Oeuvres d'histoire naturelle et de philosophie. Vol. VII. Contemplation de la Natur. Neufchâtel: Samuel Fouche.

Bonnet C. 1779–83. Collection Complete de Oeuvres de Charles Bonnet. Neuchatel, 10 volumes.

Boulenger GA. 1890. Fauna of British India: Reptilia and Batrachia. 541 pp.; London: Taylor & Francis.

Bourne AG. 1886 Anatomischer Anzeiger 1 (21):7–8.

Bourne Quart Jour Micro. Sci, NS 24:83

Boyer TH, Boyer DM. 1996. Turtles, tortoises and terrapins. In: Reptile Medicine and Surgery, DR. Mader, ed. W.B. Saunders, Philadelphia, PA.

Brandt W. 1940a. Extremitätentransplantation an *Triton taeniatus*. Ein experimenteller Beitrag zum Determinationsproblem. Archiv für mikroskopische Anatomie und Entwicklungsmechanik 103:517.

Brandt W. 1940b. Experimental production of functioning reduplication – a triple and a functioning quintuple hindlimb in the frog. Journal of experimental Biology 17:396–401.
Braun M. 1893. Ueber die künstliche Erzeugung von Doppel-, Halb- und Zwergbildungen bei Thieren. Naturwissenschaftliche Wochenschrift 8(27):265–270.
Braus H. 1904. Einige Ergebnisse der Transplantation von Organanlagen bei Bombinatorlarven. Verhandlungen der Anatomischen Gesellschaft 26: 53–66.
Braus H. 1905. Experimentelle Beiträge zur Frage der Entwicklung peripherer Nerven. Anatomischer Anzeiger 26: 433–479.
Braus H. 1906. Vordere Extremität und Operculum bei *Bombinator*-Larven. Ein Beitrag zur Kenntnis morphogener Correlation und Regulation. Gegenbaurs Morphologisches Jahrbuch 35:139.
Braus H. 1908. Entwicklungsgeschichtliche Analyse der Hyperdaktylie. Münchener medizinische Wochenschrift 55 (8): 386–390.
Braus H. 1908. Pfropfung bei Tieren. Verhandlungen des naturhistorisch-medizinischen Vereins Heidelberg, neue Folge 8:525–539. Not relevant. Embryonal transplantations in *Bombinator*
Brehm AE. 1902. La vita degli animali. Torino [Tierleben, translated from German by professor M. Lessona, Torino (Italy) 1902]
Bretscher A. 1949. Die Hinterbeinentwicklung von *Xenopus laevis* DAUD. Und ihre Beeinflussung durch Colchinin. Revue Suisse de Zoologie 56:33–96.
Brock JA, Nakamura RM, Miyahara AY, Chang EM. 1976. Tuberculosis in Pacific green sea turtles, *Chelonia mydas*. Transactions of the American Fisheries Society 105:4:564–566
Broomhall SD, Osborne W, Cunningham R. 2000. Comparative effects of ambient ultraviolet-B (UV-B) radiation on two sympatric species of Australian frogs. Conservation Biology 14:420–427.
Browne C, Dumont JN. 1980. Cytotoxic effects of sodium selenite on tadpoles (*Xenopus laevis*). Archives of Environmental Contamination and Toxicology 9:181–191.
Brož O. 1947. Mnohočetne "chondromy" v kuzi *Triton taeniatus*. – Věstnik čsl. Zoologické Spolećnosti 11:89–91.
Brož M., Horn V, Husák S. 1964. K etioilogii tzv. Nekrózy prstů u mladých ohařů. Veterinářstí 1964(1):23–25.
Brož M., Horn V, Cupák M, Husák S. 1965. Nové poznatky o hereditární neurotrofické osteopatii mladých ohařů krátkosrstých. Veterinářství 1965 (15): 230–233.
Brühl CB. Das Skelet der Krokodilinen. Vienna: Braumüller, 48 pp.
Brunst V. 1932. Über die Bedeutung der Funktion für die definitive Skeletbildung der Extremitäten bei der Regeneration. Roux's Archiv für Entwicklungsmechanik 125:640–662.
Brunst V. 1950a. Influence of x-rays on limb regeneration in urodele amphibians. Quarterly Review Biology 25:1–29.
Brunst V. 1950b. Influence of local x-ray treatment on the development of extremities of the young axolotl (*Siredon mexicanum*). Journal of experimental Zoology 114:1–49.
Brunst V. 1960. Reaction of limb regenerates of adult axolotl (*Siredon mexicanum*) to x-irradiation. Radiation Research 12:642–656.
Brunst V, Figge FH. 1951. The development of secondary tails in young axolotls after local X-ray irradiation. Journal of Morphology 89:111–133.
Bryant SV, Iten LE. 1976. Supernumerary limbs in amphibians: experimental production in *Notophthalmus viridescens* and a new interpretation of their formation. Developmental Biology 50:212–234.
Bube A, Burkhardt E, Weiss R. 1991. Systemische Chromomykose (*Chromoblastomykose*) bei der Aga Kröte (*Bufo* marinus). Internationales Colloquium für Pathologie und Therapie der Reptilien und Amphibien 1991:75–82.
Butler EG. 1955. Regeneration of the urodele forelimb after reversal of its proximo-distal axis. Journal of Morphology 96:265–282.
Byrnes EF. 1898. On the regeneration of limbs in frogs after the extirpation of limb rudiments. Anatomischer Anzeiger 15:
Byrnes EF. 1904. Regeneration of the anterior limbs in the tadpoles of frogs. Roux's Archiv für Entwicklungsmechanik 18:
Cagle FR. 1954. Observations on the life cycles of painted turtles (genus *Chrysemys*). American Naturalist 52 (1):225–235.
Canton JH, Sloff W. 1982. Toxicity and accumulation studies of cadmium (Cd^{2+}) with freshwater organisms of different trophic levels. Ecotoxicology and Environmental Safety 6:113–128.
Carpenter RL. 1932. Spinal-ganglion responses to the transplantation of differentiated limbs in *Ambystoma* larvae. Journal of experimental Zoology 61:149-
Catão-Dias JL, Nichols DK. 1999. Neoplasia in snakes at the National Zoological Park, Washington, DC (1978–1997). Journal of Comparative Pathology 120:89–95.
Chalkley DT. 1954. A quantitative histological analysis of forelimb regeneration in *Triturus viridescens*. Journal of Morphology 94:21–70.
Choi MN. 1932. Production of a limb by grafting of a nose anlage. Folia Anatomica Japonica 10:
Cholodkovsky N. 1896. Anomalies des grenoulles. Travaux de la Société naturale de St. Petersbourg 24:202-

Cicmanec JL, Ringler DH, Beneke ES. 1973. Spontaneous occurrence and experimental transmission of the fungus, *Fonsecaea pedrosoi*, in the marine toad, *Bufo marinus*. Laboratory Animal Science 23:43–47.

Cole E. 1988. Letters. Snake Keeper 2(9):17.

Colucci VL. 1890. Sulla rigenerazione parziale dell' occhio nei Tritoni. Memorie della Reale Accademia delle Scienze dell'Istituto di Bologna, Ser. V, 1:593–629.

Cook AS. 1970. The effect of p,p'-DDT on tadpoles of the common frog (*Rana temporaria*). Environmental Pollution 1:57–71.

Cooper JE. 1975. Exophthalmia in a rhinoceros Viper (Bitis nasicornis). Veterinary Record 97:130–131.

Crawshaw, GJ. 1997. Disease in Canadian amphibian populations. In Green DM. ed. Amphibians in decline: Canadian studies of a global problem. Volume 1, Herpetological conservation. Pp. 258–270; St. Louis, Missouri: Society for the Study of Amphibians and Reptiles.

Cree A, Daugherty CH. 1990. Tuatars sheds its fossil image. New Scientist 20 October 1990:22–26.

Crump ML. 1983. Opportunistic cannibalism by amphibian larvae in temporary aquatic environments. American Naturalist 121:281–287.

Ctesias C. 5th century BC. Storia di Persia: De rebus indicis.

Cunningham AA, Langton TES, Bennett PM, Lwein JF, Drury SEN, Gough RE, MacGregor SK. 1996. Pathological and microbiological findings from incidents of unusual mortality of the common frog (*Rana temporaria*). Philosophical Transactions of the Royal Society London: Biological Sciences 351:1539–1557.

Dareste C. 1889. Nouvelles recherches sur le mode de formation des monstres doubles. Comptes Rendus Hebdomadaires des séances de l'Académie des sciences 104:715–717.

David L. 1932. Das Verhalten von Extremitätenregeneraten des weißen und pigmentierten Axolotl bei heteroplastischer, heterotroper und orthotoper Transplantation und sukzessiven Regeneration. Roux's Archiv für Entwicklungsmechanik 126:457–511.

Davidoff MV. 1884. Über dieVarietäten des Plexus lumbo-sacralis von *Salamandra maculosa*. Morphologisches Jahrbuch 9: 401–413.

De Barros GC, Langenergger CH, Langenergger J, Peixoto PV. 1988. Surto de micobacteriose em riação de rãs (*Rana catesbeiana*) causado por *Mycobacterium marinum*. Pesquisa Veterinária Brasileira 8:75–80.

Dent JN. 1962. Limb generation in larvae and metamorphosing individuals of the South African clawed toad. Journal of Morphology 110: 61–78.

Detwiler SR. 1924. The effects of bilateral extirpation of the anterior limb rudiments in *Ambystoma* embryos. Journal of comparative Neurology 37:1–14.

Detwiler SR. 1924. Experiments on the transplantation of limbs in *Ambystoma*. The innervation and function of limbs transplanted after the outgrowth of peripheral nerves. American Journal of Anatomy 33:407–419.

Detwiler SR. 1925. Coordinated movements in supernumerary transplanted limbs. Journal of Comparative Neurology 38:461–493.

Detwiler SR. 1930. Observations upon the growth, function, and nerve supply of limbs when grafted to the head of salamander embryos. Journal of Experimental Zoology 55:319–379.

Diethelm G, Stauber E, Tilllson M, Ridgley S. 1996. Tracheal resection and anastomosis for an intratracheal chondroma in ball python. Journal of the American Veterinarian Medical Association 209:786–788.

Dobberstein J. 1963. Mißbildungen. 275–283. In Frei W. ed. Allgemeine Pathologie. Berlin und Hamburg: Paul Parey, 345 pp.

Dobberstein JW, Frei A, Hemmert-Halswick, Hjärre A. 1955. Allgemeine Pathologie für Tierärtze und Studierende der Tiermedizin. Berlin: Paul Parey, 345 pp.

Dolent J. 1846. Monstres. Paris: Lemerre 199pp,

Dowdey TG, Brodie ED Jr. 1989. Antipredator strategies of salamanders: Individual and geographical variation in responses of *Eurycea bislineata* to snakes. Animal Behavior 38:707–711.

Droin A, Reynaud J, Uehlinger V. 1968. Folded jaw (fj), une mutation létale recessive affectant le développement de la mâchoire chez *Xenopus laevis*. Revue Suisse de Zoologie 75:531–538.

Duellman WE. 1960. Variation, distribution, and ecology of the Mexican teiid lizard *Cnemidophorus calidipes*. Copeia 1960:97–101.

Duguy R, Moriniere P, Le Millinaire C. 1998. Facteurs de mortalité observes chez les tortues marines dans le Golfe de Gascogne. Oceanological Acta 21:383–388.

Duméril AM, Bibron G. 1841. Erpétologie générale ou histoire naturelle complète des Reptiles. 8. Paris Roret. i–vii + 1–792.
Dunson WA. 1967. Relationship between length and weight in the Spiny Softshell Turtle. 1967 (2):483–485.
Dürken B. 1912. Über frühzeitige Ezstirpation von Extremitätenanlagen beim Frosch. Ein experimenteller Beitrag zur Entwicklungsphysiologie und Morphologie der Wirbeltiere unter besonderer Berücksichtigung des Nervensystems. Zeitschrift für wissenschaftliche Zoologie 99:189–355.
Dutra FR, Largent EJ. 1950. Osteosarcoma induced by beryllium oxide. American Journal of Pathology 26:197–210.
Egar MW. 1988. Accessory limb production by nerve-induced cell proliferation. Anatomical Record 221:550–564.
Eisler R. 1986. Polychlorinated biphenyl hazards to fish, wildlife, and invertebrates: A synoptic Review. Biological Report 8(1.7) Washington DC: United States Department of the Interior, 85 pp.
Eisler R. 1987. Polycyclic aromatic hydrocarbon hazards to fish, wildlife and vertebrates: A synoptic review. U.S. Fish and Wildlife Service Biological Report 85(1.11). Contaminant Hazard Reviews Report No. 11:1–81.
Elkan E. 1974. Malignant melanoma in a snake. Journal of Comparative Pathology 84:51–57. Malignant melanoma in vertebrae of 'Everglade snake' *Elaphe obsoleta rossalleni.*
Elkan E. 1983. Giant-cell arteritis in a terrapin. Veterinary Record 112:183.
Elze K, Bergmann A, Eulenberger K, Selbitz H-J, Engelmann WE. 1977. Auswertung des Krankheitsgeschehens bei Reptilien im Zoologischen Garten Leipzig (1957–1976). Verhandlungsbericht des Internationalen Symposiums über die Erkrankungen der Zootiere 19:45–55.
Ercolani G. 1882. Della polidactylia e della polimelia nell'uomo e nei vertebrati. Memorie della Accademia delle scienze dell'Istituto di Bologna (Serie 4) 3:727–751.
Erdheim J. 1914. Rachitis und Epithelkörperchen. Denkschriften der Akademie der Wissenschaften Wien 90:363–368.
Farinelli-Ferruzza N. 1957. Transformazione di coda in arto nei trapianti xenoplastici di bottone codale di *Triton cristatus* su *Discoglossus pictus*. Acta Embryologiae et Morphologiae Experimentalis 1:171–187.
Farnsworth R, Brannian JR, Fletcher K, Klassen S. 1986. A vitamin E-selenium responsive condition in a green iguana. Journal of Zoo Animal Medicine 17:42–43.
Fatio V. 1872. Faune des vertébrés de la Suisse, Vol. III. Histoire naturelle des reptiles et des batraciens. 603pp.; Gèneve et Bâle: H. Georg.
Faust RJ, Bayless MK. 1996. Nile Monitors: Ecology, morphology and husbandry. Reptiles 4:68–83.
Fejérváry GJ v. 1925. Die phyletische Bedeutung des Prähallux und vergleichend-osteologische Notizen über den Anuren-Tarsus. Annales Musei Nationalis Hungarici 22:1–30.
Feldman F. 1996. Soft tissue mineralization: Roentgen analysis. Current Problems in Diagnostic Radiology 15:161–240.
Fellers GM, Green DE., Longcore JE. 2001. Oral Chytridiomycosis in the mountain yellow-legged frog (*Rana muscosa*). Copeia 2001(4):945–953.
Figiel CR Jr, Semlitsch RD. 1991. Effects of nonlethal injury and habitat complexity on predation in tadpole populations. Canadian Journal of Zoology 69:830–834.
Filatow D. 1933. Über die Bildung des Anfangsstadiums bei der Extremitätenentwicklung. Wilhelm Roux' Archiv für Entwicklungsmechanik 127: 776–802.
Finnie EP. 1972. Lymphoid leucosis in an Indian python. J Path 107:295–297.
Fitch S. 1949. Study of Snake Populations in Central California. American Midland Naturalist 41:513–579.
Fitzgerald SD, Janovitz EB, Burnstein T, Axthelm MK. 1990. A caudal coiling syndrome associated with lymphocytic epaxial perineuritis in newborn boa constrictors. Journal of Zoo and Wildlife Medicine 21:485–489.
Fox H, Turner SC. 1967. A study of the relationship between the thyroid and larval growth in *Rana temporaria* and *Xenopus laevis*. Archive de Biologie (Liège) 78:61–90.
Fraas E. 1896. Die schwäbischen Triassaurier, nach dem Material der kgl. Naturaliensammlung in Stuttgart zusammengestellt. Stuttgart.
Frank W. 1970. Mykotische Erkankungen der Haut und der inneren Organe bei \ Amphibien und Reptilien. Internationales Symposium Erkrankungen Zootier 12:231–255.
Frank W. 1975. Haltungsprobleme und Krankheiten der Reptilien. Diagnose und Behandlung. Tierärztliche Praxis 3:343–364.
Freda J. 1991. The effects of aluminium and other metals on amphibians. Environmental Pollution 71:3??
Frei W. 1963. Allgemeine Pathologie fur Tierärzte, 5th ed. Parey, Berlin.
Freytag D. 1950. Über einige entwicklungsphysiologisch bemerkenswerte Anomalien bei Schwanzlurchen (Urodelen). Deutsche Aquarien- und Terrarien-Zeitschrift 3: 188–189.
Freytag GE. 1982. Eine albinotische Larve des Feuersalamanders (*Salamandra salamandra*) aus dem Harz und andere Feuersalamander-Weißlinge. Salamandra 18(1/2):89–92.
Friedmann FF. 1903. Spontane Lungentuberkulose bei Schildkröten und die Stellung des Tuberkelbazillus im System. Zeitschrift Tuberkulose 4:439–457.

Frye FL. 1974. Surgery in captive reptiles. In Kirk RW. ed. Current Veterinary Therapy. Vol. V: 180–198. Philadelphia: Saunders.

Frye FL. 1984. Hemangioendothelioma in a *Boa constrictor*. Journal of Zoo Animal Medicine 15:78–83.

Frye FL, Dutra FR. 1974. Hypothyroidism in turtles and tortoises. Veterinary Medicine, Small Animal Clinician 69:990–993.

Galis F. 1999. Why do almost all mammals have seven cervical vertebrae? Developmental constraints, Hox genes, and cancer. Journal of Experimental Zoology (Molecular Development and Evolution) 285:19–26.

Gallien, L. 1944. Recherches experimentales sur l'organogenese sexuelle chez les Batraciens anoures. Bulletin Biologique de France et Belgique 78 : 257-359.

Gegenbaur C. 1862. Untersuchungen zur vergleichenden Anatomie der Wirbelsäule bei Amphibien und Reptilien. 72pp.; Leipzig: Wilhelm Engelmann.

Gegenbaur C. 1878 (not 1870). Grundzüge der vergleichenden Anatomie. 2nd ed. Leipzig: Wilhelm Engelmann, 655 pp.

Gegenbaur C. 1880. Kritische Bemerkungen über die Polydaktylie als Atavismus. Morphologisches Jahrbuch 6:584–596.

Gegenbaur C. 1888. Ueber Polydaktylie. Morphologisches Jahrbuch 14: 394–406

Geoffroy St.-Hilaire I. 1836. Histoire générale et particulière des anomalies de l'organisation chez l'homme et les animaux ou, Traité de teratology, ouvrage comprenant des recherches sur les caractères, la classifaction, l'influence physiologique et pathologique, les rapports généraux, les lois et les causes des monstrousités, des variétés et vices de conformation. Volume II:571 pp. Paris: Ballière.

Gerlach L. 1882. Die Entstehungsweise der Doppelmissbildungen bei den höheren Wirbelthieren. VIII+233pp.; Stuttgart: Ferdinand Enke.

Gerlach L. 1886. Ueber die Entstehungsweise von Doppelmissbildungen. Münchner medizinische Wochenschrift.

Geyer M, Borchardt T, Müller-Schrobsdorff F, Schreiyäck C, Müller C, Wietelmann A, Fiehn C, Müller-Ladner U, Dinser R. Joint regeneration in newts? Yes, they can! Ann Rheum Dis 2010;69:A22 doi:10.1136/ard.2010.129593q

Giard A. 1896. Y a-t-il antagonisme entre la "Greffe" et la "Régénération." Comptes Rendues des Séances de la Société de Biologie Paris

Giard A. 1897. Sur la régénération hypotypiques. Comptes Rendus hebdomadaires des Séances et mémoires de la Société de Biologie Paris 4:315–317.

Giraldes 1869. Leçons cliniques sur les maladies chirurgicales des enfants. Paris: Audrien Delahaye, 862 pp.

Giuliani M. 1878. Sulla struttura del midollo spinale, e sulla riproduzione della coda della *Lacerta viridis*. Memoria delle Classe di scienze fisiche, matematiche e naturali della Reale Accademia dei Lincei.

Glaesner L. 1924. Über drei Doppelbildungen von *Chelone mydas*. Zoologischer Anzeiger 60:185–194.

Glick B. 1931. The induction of supernumerary limbs in *Ambystoma*. Anatomical Record 48:407-

Godlewski ME. Jr. 1905. Versuche über den Einfluß des Nervensystems auf die Regenerationserscheinungen der Molche. Bulletin de l'Académie de Science de Cracovie, Classe Math. Science Nat., Jahrgang 1904:492–504.

Goette A. 1869. Untersuchungen über die Entwicklung des *Bombinator igneus*. Archiv für mikroskopische Anatomie 1869: 90–125.

Goldschmidt R. 1921. Ein Beitrag zur Analyse der Doppelmißbildungen. Roux's Archiv für Entwicklungsmechanik 47:654–666.

Golec W, Kasprzak W. 1977. Über einige Erkrankungen bei Reptilien im Zoo Poznań. Verhandlungsbericht des Internationalen Symposiums über die Erkrankungen der Zootiere 19:65–67.

Goodman J. 1881. A Sermon Preached Before the Right Honourable The Lord Mayor and Aldermen of London at the Guild-Hall Chapel December 18, 1881. Amen Carver, London: R Rayfer. 41 pp.

Goss RJ. 1956. The relation of bone to the histogenesis of cartilage in regenerating forelimbs and tails of adult *Triturus viridescens*. – Journal of Morphology 98:89–124.

Goss RJ. 1956. The unification of regenerates from symmetrically duplicated forelimbs.. – Journal of experimental Zoology: 133:191–210.

Goss RJ. 1958. Skeletal regeneration in amphibians. – Journal of Embryologyand experimental Morphology 6:638–644.

Goss RJ, Stagg MW. 1958. Regeneration of lower jaws in adult newts. – Journal of Morphology 102:289–309.

Grant C. 1937. Orthogenic variation. Proceedings of the Indian Academy of Science 46:240–245.

Grant KP, Licht LR. 1995. Effects of ultraviolation on life-history stages of anurans from Ontario, Canada. Can J Zool 73(2):2292–2301.

Gräper L. 1922. Extremitätentransplantationen an Anuren. Roux's Archiv für Entwicklungsmechanik 101:284+

Gräper L. 1926. Zur Genese der Polydactylie. Roux's Archiv für Entwicklungsmechanik 107:1+

Graveson AC, Armstrong JB. 1994 In vivo evidence that the premature death (p) mutation of *Ambystoma mexicanum* affects an early segregating subpopulation of neural crest cells. Journal of Experimental Zoology 269:327–335.

Gray BS. 2002. Observations on the herpetofauna of the Asbury Woods Greenway in Pennsylvania. Bulletin of the Chicago Herpetological Society 37(2):21–24.

Grillitsch B, Grillitsch H. 1989. Teratological and ontogenetic alterations to external oral structures in some anuran larvae (Amphibia: Anura: Bufonidae, Ranidae). Fortschritte der Zoologie 35:276–282.

Grobler JH. 1972. Observations on the amphibian *Pyxicephalus adspersus* Tschudi in Rhodesia. Arnoldia (Rhodesia) Miscellaneous Publication 6:1–4.
Grosse W-R. 1994. Der Laubfrosch. Neue Brehm Bücherei 615: Wittenberg Lutherstadt: A. Ziemsen Verlag.
Gruber W. 1859. Missbildungen. Erste Sammlungen. Mémoires de l'Académie Impériale des Sciences de St.-Pétersbourg, 7. Série 2(2):1–80.
Gruber W. 1871. Über einen bemerkenswerther Fall von Polydactylie. Bull Acad Impériale des Sciences de Saint-Pétersbourgh 15:359–368.
Grünberg W, Kovaciny Jelinek CL, Preisinger A. 1977. Harn und Harnsteine bei Tier und Mensch. Fortschritte der Urologie und Nephrologie 9:125–138.
Grünberg W, Kutzer E, Otte E. 1963. Abszeßähnliche Nekrosen bei Schlangen aus Zoologischen Gärten. Berliner und Münchener Tierärztliche Wochenschrift 76:90–95.
Grundmann E. 1900. Ueber Doppelbildungen bei Sauropsiden Anatomische Hefte 14(1):257–300.
Guérin 1835 Magasin de Zoologis Class 3, Pls. 7, 8 – Actually Cocteau 1935.
Günther AG. 1858. Catalogue of the Batrachia Salientia in the Collection of the British Museum. 160 pp.; London: Taylor and Francis.
Gurwitsch A. 1896. Ueber die formative Wirkung des veränderten chemischen Mediums auf die embryonale Entwicklung. Roux's Archiv für Entwicklungsmechanik 3:219–260.
Guyénot E, Dinichert-Favarger J, Galland M. 1948. L'exploration du territoire de la patte antérieure du *Triton* (asymétrie, duplicature, orientation des regenerates). Revue Suisse de Zoologie, Supplément 2:1–120.
Häfeli H-P. 1971. Zur Fortpflanzungsbiologie des Alpensalamanders (*Salamandra atra* LAUR.). Revue Suisse de Zoologie 78(2):235–293.
Häfeli W, Schildger BJ. 1995. Reptilien. In Göltenboth R, Klös HG,m eds. Krankheiten der Zoo- und Wildtiere. Berlin: Blackwell Wissenschafts, pp. 530–534.
Hamerton AE. 1929. Report on the deaths occurring in the Society's Gardens during the year 1928. Proceedings of the Zoological Society London 1929:49–59.
Hare KM, Daugherty CH, Cree A. 2002. Incubation regime affects juvenile morphology and hatchling success but not sex, of the oviparous lizard *Oligosoma suteri* (Lacertilia: Scincidae). New Zealand Journal of Zoology 29:221–229.
Hargitt, 1899 – Some interesting egg-monstrosities. Zoological Bulletin 2 (5):225–227.
Harrison RG. 1907. Experiments in transplanting limbs and their bearing upon the problems of the development of nerves. Journal Experimental Zoology 4:239–282.
Harrison RG. 1915. Experiments on the development of the limbs in amphibian. Proceedings of the national Academy of Sciences 1(11):539–544.
Harrison RG. 1918. Experiments on the development of the fore limb of *Ambystoma*, a self differentiating, equipotential system. Journal of experimental Zoology 25:78-
Harrison RG. 1921. On relations of symmetry in transplanted limbs. Journal of experimental Zoology 32:1–136.
Harshbarger JC, Shumway SE, Bane GW. 1976. Variably differentiating oral neoplasms, ranging from epidermal papilloma to odontogenic ameloblastoma, in cunners [(*Tautogolabrus adsperus*) Osteichthyes; Perciformes: Labridae]. Progress in Experimental Tumor Research 20:113–128.
Hay ED. 1958. The fine structure of blastema cells and differentiating cartilage cells in regenerating limbs of *Ambystoma* larvae. – Journal of Biophysical and biochemical Cytology 4:583–585.
Hay AW, Watson G. 1976. The plasma transport proteins of 25-hydroxycholecalciferol in fish, amphibians, reptiles and birds. Comparative Biochemistry and Physiology 53B:167–172.
Heimes P, Waitzmann M. 1993. Die Äskulapnatter (*Elaphe longissima* [Laurenti, 1786]) in Deutschland (Reptilia, Serpentes: Colubridae). Zoologische Abhandlungen und Berichte aus dem Museum für Tierkunde und Völkerkunde in Dresden 47:157–192.
Hennig E. 1913. Über pathologische und verwandte Erscheinungen bei fossilen Tieren. Berliner klinische Wochenschrift 50(2), Nr. 36: 1664–1668.
Henry CJ, Ulijaszek SJ. 1996. Long-term Consequences of Early Environment: Growth, Development and the Lifespan Developmental Perspective. 37th Symposium Volume of the Society for the Study of Human Biology. Cambridge: Cambridge University Press. 253 pp.
Hertwig O. 1892. Urmund und Spina bifida. Eine vergleichend-morphologische, teratologische Studie an missgebildeten Froscheiern. Archiv für mikroskopische Anatomie 39: 353–503.
Hertwig O. 1895. Beiträge zur experimentellen Morphologie und Entwicklungsgeschichte. I. Die Entwicklung des Froscheies unter dem Einfluss schwächerer und stärkerer Kochsalzlösungen. Archiv für Mikroskopische Anatomie 44:285.
Hertwig O. 1896. Ueber den Einfluss verschiedener Temperaturen auf die Entwickelung der Froscheier. Sitzungsberichte der Könglich-Preussischen Akademie der Wissenchaften zu Berlin 6:

Hertwig O. 1896. Experimentelle Erzeugung tierischer Missbildung. Festschrift Gegenbaur II.

Hertwig O. 1898. Ueber den Einfluss der Temperatur auf die Entwickelung von *Rana fusca* und *Rana esculenta*. Archiv für Mikroskopische Anatomie 51:319–381.

Heyer WR, McDiarmid RW, Weigmann DL. 1975. Tadpoles, predation, and pond habitats in the tropics. Biotropica 7:100–111

Hiffelsheim M. IV. Tératologie comparée: Recherches sur le développment des monstres doubles; extrait d'un mémoire de M. Valentin. Comptes Rendues des séances de la société et de biologie et de ses filiales 4:99–103.

Highfield AC. 1990. Artificial incubation techniques in relation to *Testudo graeca* and *T. hermanni* with notes on embryonic anoxia as a possible factor in hatchling mortality. Testudo:2–4.

Hodler F. 1949. Untersuchungen über die Entwicklung von Sacralwirbel und Urostyl bei den Anuren. Revue Suisse de Zoologie 56: 747–790.

Hoffmann CK. 1890. Reptilien. II. Eidechsen und Wasserechsen. In HG Bronn's Klassen und Ordnungen des Their-Reichs, wissenschaftlich dargestellt in Wort und Bild. Leipzig: Winter'sche-Verlagshandlung, pp. 443–1619.

Hoge AR, Federsoni PA, Jr. 1976/77. Observacoes sobre uma ninhada de Bothrops atrox (Linnaeus–1758) – (Serpentes: Viperidae: Crotalinae). Memoìrias do Instituto de Butantan 40/41: 19–36.

Holman 1992. Amphibians from a second century Roman well at Tiddington, Warwickshire. Bulletin of the British Herpetological Society 39:5–7.

Honrubia MP, Herráez MP, Alvarez R. 1993. The carbamate insecticide ZZ-Aphox induced structural changes of gills, liver, gall-bladder, heart, and notochord of *Rana perezi* tadpoles. Archives of Environmental Contamination and Toxicology 25:184–191.

Horié H. 1939. Appearance of thumb pads in the metamorphosing *Rana temporaria* by injecting the male hormone. Proceedings Imperial Academy Tokyo 15:362-

Howard JL. 1986. Current Veterinary Therapy: Food Animal Practice. Philadelphia: Saunders, 1008 pp.

Hubbard WD. 1927. Crocodiles. Copeia 1927:115–116.

Hubert J. 1964. Essais de fissuration de l'oeuf de lizard vivipare (*Lacerta vivipara*). Compte Rendu de la Société Biologique, Paris 158: 523–525.

Hübner, H. 1911. Zur Kasuistik der tierischen Doppelmissbildungen. Frankfurter Zeitschrift für Pathologie 8:135–150.

Huey RB, Pianka ER. 1981. Ecological consequences of foraging mode. Ecology 62:991–999.

Huchzermeyer FW, Agnagna M. 1994. A survey of parasites and pathology of African dwarf crocodiles Osteolaemus tetraspis in the Congo Republic. In Crocodiles: Proceedings of the 12th Working Meeting of the Crocodile Specialist Group of the Species Survival Commission of IUCN – The World Conservation Union, convened at Pattaya, Thailand, 2–6 May 1994 Vol. 2: 309–313.

Imhof G. 1905. Anatomie und Entwicklungsgeschichte des Lumbalmarkes bei den Vögeln. Archiv für Mikroskopische Anatomie und Entwicklungsgeschichte 65:498–610.

Ippen R. 1972. Ein Beitrag zu den Spontantumoren bei Reptilien. Verhandlungsbericht des XIV. Internationalen Symposiums über die Erkrankungen der Zootiere. Akademie-Verlag, Berlin:409–418.

Iten LE, Bryant SV. 1973. Forelimb regeneration from different levels of amputation in the newt, *Notophthalmus viridescens*: Length rate and stages. Wilhelm Roux' Archiv 173:263–282.

Jackson CG Jr, Landry MM, Jackson MM. 1971. Reproductive tract anomaly in a box turtle. J Wildlife Dis 7:175–177.

Jackson CG Jr, Fulton M. 1970. A turtle colony epizootic apparently of microbial origin. Journal of Wildlife Disease 6:466–468.

Jackson OF. 1983. Induced skeletal deformities in snake embryos. Herptile 8:137–139.

Jackson OF. 1990. Tortoises and hibernation. Veterinary Record 131:244.

Jacobson ER, Seely JC, Novilla MN. 1980. Lymphosarcoma associated with virus-like intranuclear inclusions in a California king snake (Colubridae: *Lampropeltis*). Journal of the National Cancer Institute 65:577–579.

Jacobson ER, Ginn PE, Troutman JM, Farina L, Stark L, Klenk K, Komar N. 2005. West Nile virus infection in farmed American alligators (*Alligator mississippiensis*) in Florida. Journal of Wildlife Diseases 41:96–106, 2005.

Jasmin AM, Baucom J. 1967. *Erysipelothrix insidiosa* infections in the Jarofke D, Lange J. 1993. caiman (*Caiman crocodilus*) and the American crocodile (*Crocodylus acutus*). American Journal Veterinary Medicine Clinical Pathology 1:173–177.

Ji X, Braña F. 1999. The influence of thermal and hydric environments on embryonic use of energy and nutrients, and hatchling traits in the wall lizards (*Podarcis muralis*). Comparative Biochemistry and Physiology 124A:205–213.

Ji X, Zheng X-Z, Xu Y-G, Sun R-M. 1995. Several aspects of the thermal biology of the skink *Eumeces chinensis*. Acta Zoologica Sinica 41:268–285.

Jickeli CF. 1902. Die Unvollkommenheit des Stoffwechsels als Veranlassung für Vermehrung, Wachstum, Differenzierung, Rückbildung und Tod der Lebewesen im Kampf ums Dasein. XVI + 353 pp. Berlin: R. Friedländer & Sohn.

Johnson DM, Akre BG, Crowley PH. 1975. Modeling arthropod predation: Wasteful killing by damselfly naiads. Ecology 56:1081–1093.

Josephy H. 1913. Über eine Doppelbildung bei einer Tritonenlarve. Archiv für Entwicklungsmechanik 35: 589–597.

Jubb KFV, Kennedy PC. 1963. Pathology of Domestic Animals. 23–25; London: Academic Press.

Kaestner S. 1899. Neuer Beitrag zur Causuistik der Doppelbildungen bei Hühnerembryonen.Archiv für Anatomie und Physiologie 1899:28–33.

Kaestner S. 1907. Pathologische Wucherungen, Divertikel- und Geschwulstbildungen in frühen Embryonalstadien. Archiv für Anatomie und Physiologie 1907:250–266.

Kammerer P. 1905. Über Abhängigkeit des Regenerationsvermögens der Amphibienlarven von Alter, Entwicklungsstadium und spezifischer Größe. Archiv für Entwicklungsmechanik 19: 148–180.

Kawamura T, Nishioka M. 1978. Abnormalities in the descendants of *Rana nigromaculata* produced from irradiated eggs or sperm. Scientific Report of the Laboratory for Amphibian Biology Hiroshima University 3:1–187.

Kim HT, Olson WM, Hall BK. 2002. Effects of hind limb denervation on development of sesamoids in *Hymenochirus boettgeri*. Integrative Comparative Biology 42: 1256.

Kiortis V. 1953. Potentialités du territoire patte chez la *Triton* (adultes, larves, embryons). – Revue suissede Zoologie 60:301–410.

Klauber LM. 1940. A statistical study of the rattlesnakes. VII. The rattle Occ. Pap. San Diego Soc. Nat. Hist. 6:1–62.

Klein F. 1863. Beiträge zur Osteologie der Krokodilschädel. Jahreshefte des Vereins für vaterländische Naturkunde Württembergs, Stuttgart 16 Jahrgang 19:70–100.

Klima M. 1994. Anpassungen an die aquatische Lebensweise. Handbuch der Säugetiere Europas. Band 6, Meeressäuger.

Kluge AG. 1981. The life history, social organization, and parental behavior of *Hyla rosenbergi* Boulenger, a nest-building gladiator frog. Miscellaneous Publications of the Museum of Zoology of the University of Michigan 160:1–170.

Koether F. 1927. Ueber Duplicitas anterior, posterior und posterior partim cruciata bei *Triton*. Archiv für Entwicklungsmechanik 110: 578–625.

Kolbow H. 1928. Experimentell verursachte Bildung von Armen aus ursprünglichem Beinmaterial bei *Triton*. Roux's Archiv für Entwicklungsmechanik 113:12–38.

Kopsch F. 1897. Über eine Doppel-Gastrula bei *Lacerta agilis*. Sitzungsberichte der Akademie der Wissenschaften zu Berlin 29:646–650.

Kotlik P. 1991. Patologický průbeh líhnutí *Elaphe guttata guttata*. Akvárium Terárium 34(12):24.

Kuramoto M. 1984. Systematic implications of hybridization experiments with some Eurasian Treefrogs (Genus *Hyla*). Copeia 1984(3):609–616.

Küster E. 1905. Ueber Kaltblütertuberkulose. Münchener Medizinische Wochenschrift 10 January:1.

Lallier R. 1954. Chlorure de lithium et biochimie du développement de l'œuf d'amphibien. Journal of Embryology and experimental Morphology 2:323–339.

Landois 1894 Several instances of double formations among domestic animals. Westfälischer Provinz-Verein für Wissenschaft und Kunst, 22. Jahresbericht 1893–4:61–62 (ein sechsbeiniges Hausschwein), 64–66 (Ziegen-Missbildungen aus Westfalen).

Lanza B, Poggesi M. 2003. Eine neue, äußerst seltene Zehenanomalie bei der Gattung *Tarentola* Gray, 1825 (Reptilia, Gekkonidae). Studi Trentini di Scienze Naturali, Acta Biologica 79(2002):243–244.

Lawler, S. P., D. Dritz, T. Strange, and M. Holyoak. 1999. Effects of introduced mosquitofish and bullfrogs on the threatened California red-legged frog. Conservation Biology 13:613–622.

Lereboullet A. 1863. Recherches sur les monstruosités du brochet observées dans l'oeuf et sur leur mode de production. Annales des Sciences naturelles. Zoologie et biologie animale 20:177–271.

Leviton AE, Gibbs RH Jr, Heal E, Dawson CE. 1985. Standards in herpetology and ichthyology: Part 1. Standard symbolic codes for institutional resource collections in herpetology and ichthyology. Copeia 1985(3): 802–832.

Leydig F. 187?. Bau der Zehen der Batrachier und die Bedeutung des Fersenhöckers. Morphologisches Jahrbuch 2: 191 (additional toe in *Bombinator igneus*)

Lizana M, Pedraza EM. 1998. The effects of UV-B radiation on toad mortality in mountainous areas of central Spain. Conservation Biology 12:703–707.

Licht P, Bennett AF. 1972. Copeia 1972:702–707.

Linné 1785–90. Amoenitates academicae. Erlangae.

Liosner LD. 1939. Byulleten' Eksperimental'noi Biologii i Meditsiny 8:14–17.

Liosner LD. 1939. A contribution to the morphogenetic role of the skeleton in regeneration. II. Communication. Bulletin de Biologie et de Médecine Expérimentale 8:14–17.

Lissitzky EF. 1910. Durch experimentelle Eingriffe hervorgerufene überzählige Extremitäten bei Amphibien. Archiv für Mikroskopische Anatomie 75:587–633.

Lomovskaya EG. 1948. [A Limb regeneration in the axolotl after X-irridation of half of the limb.] Doklady Akademija Nauk SSSR 61:157–160.[Russian]

López del Castillo C. 2001. Accidental loss of toes and claws in lizards. Reptilia (GB) 19:71–72.

Lourdais O, Shine R, Bonnet X, Guillon M, Naulleau G. 2004. Climate affects embryonic development in a viviparous snake, *Vipera aspis*. Oikos 104: 551–560.

Lucké B. 1938. Carcinoma in the leopard frog; its probable causation by a virus. Journal of Experimental Medicine 68:457–468.

Lyvere DB. 1966. Repair of the shell of a Galapagos tortoise. Modern Veterinary Practice 47:76.

Macdonald JR. 1951. Pathological vertebrates from South Dakota. Bulletin of the Geological Society of America 62:1539.

Maden M. 1982. Supernumerary limbs in amphibians. American Zoologist 22:131–142.

Mader DR. ed. 1996. Reptile medicine and Surgery. XX+512pp.; Philadelphia: W.B.Saunders. Not ossified plastron in specimen of *Gopherus agassizi* (after Häfeli and Zwart 2000).

Malashichev YB, Roger LJ. 2002 Behavioural and Morphological Asymmetries in Amphibians and Reptiles. Proceedings of the 4th World Congress of Herpetology Satellite Symposium:108 pp. Laterality 7(3):195–299.

McCann C. 1932. Notes on Indian batrachians. Journal of the Bombay Natural History Society 36:152–180.

Marcucci E. 1915. Capacità rigenerativa degli arti nelle larve di Anuri e condizioni che ne determinano la perdita. Archivo Zoologico Italiano 8: 89–119.

Marcucci E. 1932. Trapianti di pelle e rigenerazione in *Lacerta muralis*. Archivo Zoologico Italiano 17(3–4): 435–446.

Marelli C.A. 1931. La Plata. (Agregados al tomo IV. Los Vertebrados exhibidos en los Zoológicos del Plata. Año 1930–1931.)

Martin, L. 1880. Histoire des monsters depuis l'antiquité jusqu'à nos jours. Grenoble: Jerome Million, 322 pp.

Mathure JK. 1979. An examination of the reptile fauna of Wyperfeld National Park using pitfall trapping. Victorian Naturalist 96(3):98–101.

May RM. 1933. Reaction neurogéniques de la moelle. À la greffe en surnombre, ou à l'ablation d'une ebauche de patte postérieure chez l'embryo de l'anoure, *Discoglossus pictus*, Otth. Bulletin Biologique de la France et de la Belgique 67:327-

Mayer C. 1854 Über krankhafte Knochen vorweltlicher Thiere. Nova Acta Leopoldiana 25 (2): 673–689.

McEachern MJ. 1991. Keeping and breeding corn snakes. Advanced Vivarium Systems, Lakeside. 60pp.

McGowan C. 1974. A revision of the longipinnate ichthyosaurs of the Lower Jurassic of England, with descriptions of two new species (Reptilia: Ichthyosauria). Royal Ontario Museum Life Sciences Contributions 97:1–37.

McIntyre D.C. 1977. First report of a double embryo in *Elaphe subocularis*. Journal of the Northern Ohio Association of Herpetologists 3(2):29.

McNease L, Joanen T. 1981. Nutrition of alligators. Proceedings of the 1st Alligator Production Conference, Gainesville, Florida, pp. 15–28.

Mehl MG. 1913. *Angistorhinus*, a new genus of Phytosauria from the Trias of Wyoming. Journal of Geology 21:186–191 (not 129–165).

Mertens K 1959. Über die Schädel einiger bemerkenswerter Warane Australiens. Vierteljahrschrift der Naturforschenden Gesellschaft in Zürich, Jahrgang 104:115–128

Meunier 1869. Sur l'accouplement de la grenouille et du crapard. Cosmos, Revue Encylclopédique Hebdomadaire des Progrès des Sciences 3° Series, 4:210–211.

Meunier 1869. Une page de la vie de Geoffroy Saint-Hilaire. Cosmos, Revue Encylclopédique Hebdomadaire des Progrès des Sciences 3° Series, 4:653–657.

Milewski AW. 1918. Ueber Tornier's experimentelle Untersuchungen über Entstehen pathologischer Verbildungen bei Tieren sowie weiteres über experimentelles Erzielen von monströsen Goldfischarten. Archiv für Entwicklungs-Mechanik der Organismen 44:472–498.

Minucci S, Saint-Jeannet J-P, Toyama R, Scita G, DelUca LM, Taira M, Levin AA, Ozato K, Dawid IB. 1996. Retinoid X receptor-selective ligands produce malformations in *Xenopus* embryos. Proceedings of the National Academy of Sciences 93:1803–1807.

Mitrophanow, P. 1898. Teratogenetische Studien. Archiv für Entwicklungsmechanik 1: 347–379.

Møller AP. 1999. Developmental stability is related to fitness. American Naturalist 153: 556–560.

Moodie RL. 1916a. The growth of bone in Cretaceous time. Science 43:35–36.

Moodie RL. 191b. Mesozoic pathology and bacteriology. Science 43:425–426.

Moodie R. 1926. Recent advances in paleopathology. Annals of Medical History 8:327–330 (Book review).

Moodie RL. 1927. Tumors in the Lower Carboniferous. Science 66:541.

Moody SM. 1980. Phylogenetic and historical biogeographical relationships of the genera of the family Agamidae (Reptilia: Lacertilila). Ph.D. Thesis. (University of Michigan).

Moreau JL. 1808. Les écarts de la nature ou description des principales monstruosités dans l'homme et dans les animaux: Précédée d'un discours sur la physiologie et la classification des monstruosities, avec figures coloriées par N.F. Regnault. Paris: Fournier Frères.

Moreau de la Sarthe 1808. Les écarts de la nature ou description des principales monstruosités dans l'homme et dans les animaux. Brussels: Demat.
Morgan TH. 1908. Experiments in grafting. American Naturalist 42(493):1–11.
Nassonov NV. 1930. Die Regeneration der Axolotlextremitäten nach Ligaturanlegung. Roux's Archiv für Entwicklungsmechanik 121:639–657.
Nasanow NW. 1936. [On the peculiarities and causes of the development of additional formations in amphibia]. Doklady Akademija Nauk SSSR (2)11(5): 207–211. [Russian].
Nasanow NW. 1941. Entwicklung zusätzlicher Gebilde durch Transplantation von Knorpel unter die Haut erwachsener Urodelen. Moscow Nauk SSSR 1941:
Nazdrowicz NH, Bowman JL, Roth RR. 2008. Population ecology of the eastern box turtle in a fragmented landscape. Journal of Wildlife Management 72:745–753.
Nash JTC. 1915. Evolution and Disease. VIII+73pp.; New York: William Wood a. Co.
Neill WT, Allen ER. 1954. Algae on turtles: some additional considerations. Ecology 35(4):581–584.
Newman HH. 1904. The biologie of Twinning
Newman HH. 1917a. The biology of twins (mammals). Chicago: University of Chicago Press. 186 pp.
Newman HH. 1917b. On the production of monsters by hybridization. Biological Bulletin 32(5): 306–321.
Newth DR. 1958. New (and better?) parts for old. – New Biology 26:47–62.
Nishioka M. 1972. Abnormalities of the Offspring of Nucleo-cytoplasmic Hybrids between *Rana nigromaculata* and *Rana brevipoda*. Scientific Report of the Laboratory for Amphibian Biology 1:1–94.
Nishioka M. 1978. Abnormalities in the descendants of *Rana japonica* produced from irradiated eggs or sperm. Scientific Report of the Laboratory for Amphibian Biology 3:189–384.
Noden DM, De Lahunta A. 1985. The embryology of domestic animals. Developmental mechanisms and malformations. 367pp.; Baltimore, London: Williams & Wilkins.
Noris KS. 1953. The ecology of the desert iguana *Dipsosaurus dorsalis*. Ecology 34:265–287.
Northway R. 1970. Veterinary Medicine, Small Animal Clinician 65:944
Oostra R-J, Baljet B, Verbeeten BWJM, Hennekam RCM. 1998. Congenital anomalies in the teratological collection of museum Vrolik in Amsterdam, the Netherlands. V: Conjoined and acardiac twins. American Journal of Medical Genetics 80: 74–89.
Pagan O, Müller M. 1990. Disecdysis in snakes: non-infectious and infectious aetiopathogenesis, with special reference to the first viral dermatitis. Schweizerisches Archiv für Tierheilkunde 132:452.
Page LA. 1966. Diseases and Infections in Snakes: A Review. Bulletin of the Wildlife Disease Association 2:111–126.
Parker WS, Pianka ER. 1973. Notes on the ecology of the iguanid lizard, *Sceloporus magister*. Herpetological 29:143–152.
Perri T. 1948. La determinazione di arti soprannumerari in *Bufo vulgaris*. [The determination of supernumerary limbs in *Bufo vulgaris*]. Rendiconti (8), Accademia nazionale dei Lincei. Classe di scienze fisiche, matematiche e naturali 5: 442–447. [Italian] [This paper covers only experimentally induced cases].
Perri T. 1951. La determinazione die arti soprannumerari in "*Bufo vulgaris*." [The determination of supernumerary limbs in "Bufo vulgaris"]. Nova VII. Atti Accad Nazion Lincei (8) Rendiconti 10:64–68. [Italian] [This paper covers only experimentally induced cases].
Perri T. 1956. Induzione di arti soprannumerari negli Anfibi (particolarmente negli Anuri) mediante sostanze citolitiche. [Induction of supernumerary limbs in amphibians (particularly in anurans) by cytolytic substances] Experientia 12(4):125–135. [Italian] [This paper covers only experimentally induced supernumerary limbs]
Petranka JW, Hopey ME, Jennings BT, Baird SD, Boone J. 1994. Breeding habitat segregation of wood frogs and American toads: The role of interspecific tadpole predation and adult choice. Copeia 1994:691–697.
Petranka JW, Rushlow AW, Hopey ME. 1998. Predation by tadpoles of *Rana sylvatica* on embryos of *Ambystoma maculatum*: Implications of ecological role reversals by *Rana* (predator) and *Ambystoma* (prey). Herpetologica 54:1–13.
Philippen H-D. 1986. Rätselhafte Phänomene bei diesjährigen Nachzuchten von Schildkröten – Versuch der Lokalisierung der Ursache. Die Schildkröte (Neue Folge) 1(1/2):35–39.
Piatt J. 1942. Transplantation of aneurogenic forelimbs in *Ambystoma punctatum*. Journal of experimental Zoology 91:79–101.
Piatt J. 1952. Transplantation of aneurogenic forelimbs in place of the hindlimb in *Ambystoma*. Journal of experimental Zoology 120:247–285.
Pierce BA, Montgomery J. 1989. Effects of short-term acidification on growth rates of tadpoles. Journal of Herpetology 23:97–102.
Praschag R. 2005. Land-schildkröten. 2nd ed. Ulmer: Graz, 95pp.
Prescott W. 1937 Die Eroberung von Peru, Kultur und Untergang der Inkas. 479pp.; Josef Belt, Wien.
Prowazek S. 1901. Zur Regeneration des Schwanzes der urodelen Amphibien. Arbeiten aus den Zoologischen Instituten der Universität Wien und der Zoologischen Station Triest 13 (2):81–124.

Przibram H. 1911. Experiments on asymmetrical forms as affording a clue to the problem of bilaterality. Journal of experimental Zoologie 10(3):255-

Przibram H. 1927. Deutungen spiegelbildlicher Lurcharme. (Zur Verständigung mit R. G. Harrison u. a.). Roux's Archiv für Entwicklungsmechanik 109:411–448.

Puckett WO. 1936. The effects of x-radiation on limb development and regeneration in *Ambystoma*. Journal of Morphology 59:173–213.

Puppe A. 1925. Experimentell erzeugte Polydaktylie und Bruchdreifachbildung bei *Rana fusca*. Roux's Archiv für Entwicklungsmechanik 104(2):125–137.

Puşcariu F, Papillian VV, Gabor M, Deac C. 1973. Toxicité et effets morphopathologiques des aflatoxines chez les tetrads. Archiv Roumaine de Pathologie Experimental et Microbiologie 32:255–267.

Quattrini D. 1954. Piano di autotomia e rigenerazione della coda nei sauri. Archivio Italiano di Anatomia e di Embriologia 59: 225–282.

Rabaud E. 1908. L'évolution tératologique. Rivista di scienza, 'Scientia' 3:45–64.

Rabaud E. 1914. La tératogenèse: étude des variations de l'organisme. Encyclopédie Scientific; Paris: Doin, 361 pp.

Radovanović M. 1935. Anatomische Studien am Schlangenkopf. Jenaische Zeitschrift für Naturwissenschaften 69:321–422.

Rauber A. 1878. Ueber Doppelbildungen bei Wirbelthieren. Virchow's Archiv 72:443.

Rauber A. 1878. Die Theorien der excessiven Monstra. Virchow's Archiv 73:551–594.

Rauber A. 1879. Formbildung und Formstörung in der Entwickelung von Wirbelthieren. Morphologisches Jahrbuch 5: 661–705.

Rauber A. 1880. Formbildung und Formstörung in der Entwickelung von Wirbelthieren. Morphologisches Jahrbuch 6: 1–48.

Reed MA. 1903. The regeneration of a whole foot from the cut end of a leg containing only the tibia. Wilhelm Roux's Archiv für Entwicklungsmechanik 17:150–154.

Regnault NF. 1808. Les écarts de la nature ou description des principales monstruosités dans l'homme et dans les animaux. Paris: Fournier Frères.

Reig AO. 1957. Los Anuros del Matildense. In Stipanicic PN, Reig AO. El complejo porfirico de la Patagonia extraandria y su fauna de Anuros. Acta geologica Lillona (Tucuman) 1: 231–297

Rhodin AG, Medem F, Mittermeier RA. 1980. The occurrence of neustophagia among Podocnemine turtles. British Journal of Herpetology 6:175–176.

Rievel H. 1907. Knochenpathologie der Tiere. In: Lubarsch O., Ostertag, R. (eds.) Ergebnisse der Allgemeinen Pathologie und Pathologischen Anatomie des Menschen und der Tiere. 11. Jahrgang, II. Abteilung: 590–708.

Rogers LJ. 1966. The nitrogen excretion of *Chelodina longicollis* under conditions of hydration and dehydration. Comparative Biochemistry and Physiology 18:249–260.

Rokhlina ML. 1936. Influence des glandes à sécrétion interne sur la formatoin de la coque de l'oeuf. Bulletin de Biologie et de Médecine Expérimentale 11:221–222.

Rosenthal K. 2004. A new look at treatment of calcium: Phosphorus disorders in iguanas. Proceedings of the North American Veterinary Conference 18:1348.

Rosskopf WJ, Howard EB, Gendron AP. 1981. Granulocitic leukemia in a tortoise. Modern Veterinary Practice 62:701–702. (now Small Animal Medicine Digest)

Rostand J. 1932. L'évolution des espèces. Paris: Hachett, 205 pp.

Rostand J. 1950. La chimiotératogenèse chez les Batraciens anoures. Revue scientifique 88:155–157.

Ruben LN. 1960. An immunobiological model of implant-induced urodele supernumerary limb formation. American Naturalist 94:427–434.

Rund G. 1926. On the production of double limbs in amphibians. Journal of experimental Zoology 44:419-

Ruthven AG. 1922. The amphibians and reptiles of the Sierra Nevada de Santa Marta, Columbia. University of Michigan Museum of Zoology Miscellaneous Publication 8:1–69.

Sachsse W. 1977. Normale und pathologische Phänomene bei Zuchtversuchen mit Schildkröten, hier anhand von *Kinosternon baurii* (Reptilia, Testudines, Kinosternidae). Salamandra 13:22–35.

Samour JH, Hawkey CM, Pulger, et al. Clinical and pathologic findings related to malnutrition and husbandry in captive giant tortoises [Geochelone spp.]. Vet Rec 1986; 118:299.

Schauinsland H. 1906. Die Entwickelung der Wirbelsäule nebst Rippen und Brustbein. In: O. Hertwigs Handbuch der Entwickelungslehre der Wirbeltiere 3(2):339–593.

Schoch RR, Carroll RL. 2003. Ontogenetic evidence for the Paleozoic ancestry of salamanders. Evolution & Development 5:314–324.

Schultz TW, Dumont JN, Clark BR, Buchanan MV. 1982. Embryotoxic and teratogenic effects of aqueous extracts of tar from a coal gasification electrostatic precipitator. Teratogenesis, Carcinogenesis, and Mutagenesis 2:1–11.

Schultze B. 1855. Ueber anomale Duplicität der Axenorgane. Virchow's Archiv 7:479–531.

Schultze EA, Mason GL, Clyde VL. 1999. Lymphosarcoma with leukemic blood profile in a Savannah monitor (*Varanus exanthematicus*). Journal of Zoo and Wildlife Medicine 30:158–164.

Schumacher J. 2003. Fungal diseases of reptiles. Veterinary Clinics of North America Exotic Animal Practice 6(2): 327–35, vi.
Schuytema GS, Nebeker AV. 1999. Comparative toxicity of ammonium and nitrate compounds to Pacific treefrog and African clawed frog tadpoles. Environmental Toxicology and Chemistry 18:2251–2257.
Schwalbe E. 1910. Missbildungen und Variationslehre. Sammlung Anatomischer und Physiologischer Vorträge und Aufsätze 9:1–35.
Schwartz DA, LoGerfo JP. 1988. Congenital limb reduction defects in the agricultural setting. American Journal of Public Health 78:654–658.
Scott H. 1926. Report on deaths occurring in the Society's Gardens during the year 1925. Proceedings of the Zoological Society London 1926:231–244.
Seligman H. 2002. Behavioral and morphological asymmetries in hindlimbs of *Hoplodactylus duvaucelii* (Lacertilia: Gekkonomorpha: Gekkota: Diplodactylinae). Laterality 7:277–283.
Serres M. 1832. Recherches d'anatomie transendante et pathologique. Theorie des formations et des deformations organiques, appliqué à l'anatomie de Ritta-Christina et de la duplicité monstreuse, accompagné de 20 planches. Paris: JB Bailliere, 315pp.
Shine R. 2003. Locomotor speeds of gravid lizards: Placing 'costs of reproduction' within an ecological context. Functional Ecology 17:526–533.
Sibley WK von. 1889. Ueber Tuberculose bei Wirbelthieren. Archiv für pathologische Anatomie und Physioilogie und für klinische Medicin 116:104–115.
Simpson SB Jr. 1964. Analysis of tail regeneration in the lizard *Lygostoma laterale*. I. Initiation of regeneration and cartilage differentiation. The role of ependyma. Journal of Morphology 114:425–436.
Sinitsin NP. 1956. A method for the grafting of a second head on a frog. Bulletin of experimental Biology and Medicine 42(3):812–814.
Sison TM, Padilla MA, Vizmanos MZ, Follosco M. 1990. Isolation and identification of fungi found in necrotic skin lesions of captive marine turtles (*Eretmochelys imbricata*). Philippine Journal of Veterinary Medicine 27:35–36.
Skowron S, Roguski H. 1957. Regeneration from implanted dissociated cells. I. Regeneration potentialities of limb and tail cells. Folia Biologica 6:163–173.
Smith DD, Powell R. 1983. *Acris crepitans blanchardi* (Blanchard's cricket frog): anomalies. Herpetological Review 14:118–119.
Spemann H. 1903. Entwickelungsphysiologische Studien am Triton-Ei, III. Roux' Archiv für Entwicklungsmechanik 16:551–631.
Spemann H, Falkenberg H. 1919. Über asymmetrische Entwicklung und Situs inversus viscerum bei Zwillingen und Doppelbildungen. Archiv für Entwicklungsmechanik 45: 371–422.
Stannius H. 1854. Handbuch der Zootomie. Anatomie der Wirbelthiere. Berlin, 1854.
Stewart HL. 1976. Comparative aspects of certain cancers. In: Cancer, FF Becker, ed. Vol. 4 pp. 303–374. Plenum publ. Co., New York.
Stewart C, Burne RH, Smith GE, Duckworth WL, Pearson W. 1902. Descriptive and Illustrated Catalogue of the Physiological Series of Comparative Anatomy Contained in the `Museum of the Royal College of Surgeons of England. Volume II, 2nd edition. London: Taylor and Francis, 518 pp.
Stockard CR. 1910. The influence of alcohol and other anaesthetics on embryonic development. American Journal of Anatomy 10: 369–392.
Stockard CR. 1921. Developmental rate and structural expression: an experimental study of twins, double monsters and single deformities, and the interaction among embryonic organs during their origin and development. American Journal of Anatomy 28:115–277.
Strohl J. 1918. La tératogénèse. L'année biologique 23:72.
Suedmeyer WK, Turk JR. 1996. Lymphoblastic leukemia in an inland bearded dragon, *Pongo* vitticeps. Journal of the Association of Reptilian and Amphibian Veterinarians 6:10–12.
Swett FH. 1926. On the production of double limbs in amphibians. Journal of experimental Zoology 44:419–473.
Takahashi H. 1958. Time relation between the administration of estradiol and the occurrence of abnormalities in limbs and gonads of frog larvae. Journal of the Faculty of Science Hokkaido University – Series VI: Zoology 14:101–110.
Tåning ÅV. 1952. Experimental study of meristic characters in fishes. Biological Reviews 27:169–193.
Tappe JP, Chandler FW, Liu SK, Dolensek EP. 1984. Aspergillosis in two San Esteban chuckwallas. Journal of the American Veterinary Medical Association 185:1425–1428.
Taylor SK, Green DE, Wright KM, Whitaker BR. 2001. Bacterial diseases. In: Amphibian Medicine and Captive Husbandry. KM Wright, BR Whitaker, eds. Malabar, Florida: Krieger Publishing Company; pp. 159–179.
Terni T. 1919. Sulla correlazione fra ampiezza del territorio di innervazioni e grandezza delle cellule gangliari —2° Ricerche sui ganglî spinali che innervano la coda rigenerata nei Saurî (*Gongylus ocellatus*). Archivio Italiano di Anatomia e di Embriologia 17 (4): 507–543.

Tenri T. 1922. La rigenerazione del simpatico nella coda regenerate dei Sauri. Monitore Zoologico Italiano Siena 33:58–62.

Thornton CS. 1938. The histogenesis of the regenerating forelimb of larval *Ambystoma* after exarticulation of the humerus. Journal of Morphology 62:219–241.

Thornton CS, Shields TW. 1945. Five cases of atypical regeneration in the adult frog. Copeia 1945:40–42.

Tiedermann H. 1966. The molecular basis of differentiation in early development of amphibian embryos. In Moscona AA, Monroy A. eds. Current Topics in Developmental Biology. New York: Academic Press, pp. 85–112.

Tinkle DW. 1973. A population analysis of the sagebrush lizard, *Sceloporus graciosus* in Southern Utah. Copeia 1973:284–296.

Todd TJ. 1823. On the process of reproduction of the members of the aquatic salamander. Quarterly Journal of Science, Literature and Arts 16:84–96.

Töndury G. 1955. Entwicklungsstörungen durch chemische Faktoren und Viren. Verhandlungen der Gesellschaft Deutscher Naturforscher und Ärzte 98. Versammlung:119–126.

Tornier G. 1896a. Hyperdactylie und Regenerationsexperimente und eine neue Vererbungstheorie. Sitzungsberichte der Gesellschaft naturforschender Freunde zu Berlin 1896:24–25.

Tornier G. 1896b. Ueber eine experimentell erzeugte Doppelgliedmaße. Sitzungsberichte der Gesellschaft naturforschender Freunde zu Berlin 1896:144–145.

Tornier G. 1908. Ueber experimentelles Hervorrufen und Naturentstehn von Mopsköpfen, Cyclopen und andere vorgebürtliche Kopfverbildungen bei Wirbelthieren. Sitzungsberichte der Gesellschaft Naturforschender Freunde zu Berlin 1908: 298–315.

Tornier, G. 1919. Mißbildete Embryonen bei *Salamandra mac.* (Briefliche Mitteilung an den Herasugeber). Blätter zur Aquarien- und Terrarienkunde, Braunschweig, 30: 161.

Travis J. 1989. The role of optimizing selection in natural populations. Annual Review of Ecology and Systematics 20:279–296.

Trutnau L. 1970. Ein Beitrag zur Behandlung erkrankter Krustenechsen (*Heloderma horridum* und *H. suspectum*). Aquarien- und Terrarien-Zeitung 23(8): 251–252.

Tschumi PA. 1954. Kokurrenzbedingte Rückbildungen der Hinterextremität von *Xenopus* nach Behandlung mit einem Chloraethylamin. Revue Suisse de Zoologie 61:117–270.

Tschumi PA. 1964. Die Evolution der Gliedmassen in entwicklungsphysiologischer Perspektive. Revue Suisse de Zooologie 71:215–236.

Tur J. 1903. Sur un cas de diplogénèse très jeune dans le blastoderme de *Lacerta ocellata* Daud. Bibliographie Anatomique fasc. 3, 12:83–88.

Tyler MJ, Davies M, Walker KF. 1985. Abrasion injuries in burrowing frogs (Amphibia, Anura) from the Northern Territory, Australia. Zoologischer Anzeiger 214:54–60.

Ubisch L. v. 1922. Das Differenzierungsgefälle in Amphibienkörpern. Verhandlungen der Deutschen Zoologischen Gesellschaft 27: 44–46.

Ubisch L. v. 1923. Das Differenzierungsgefälle in Amphibien und seine Auswirkungen. Wilhelm Roux'Archiv für Entwicklungsmechanik der Organismen 52:641–670.

Uelhinger V. 1966. Description chez *Xenopus laevis* D. d'une mutation dominante 'Screwy' (S), létale a l'état homozygote. Revue Suisse de Zoologie 73:527–534.

Valentin GG. 1851. Ein Beitrag zur Entwicklungsgeschichte der Doppelmissgeburten. Archiv für Physiologische Heilkunde 10:1–40.

van Westrienen AFAS. 1911. Die vergleichende Teratologie der dicephalen Doppelbildungen. Nieuwe Verhandelingen van het Bataafsch Genootschap 2/6/3: 239 p. (Proefschrift ter verkrijging von den Grad von Doctor ill de Genecskunde). Rotterdam, Hengel.

Vershinin VL. 1995. Status of amphibian populations in anthropogenic landscapes of the Ural, Siberia and the Far East. In: KuzminSL, Dodd CK. Jr, Pikulik MM. eds. Amphibian populations in the Commonwealth of Independent States: current status and declines. Pp. 88–90; Moscow: Pensoft.

Viertel B. 1985. Unvollständige Metamorphose bei *Rana temporaria, Bufo bufo* und *Bufo calamita* (Anura: Ranoidea) unter besonderen Bedingungen. Salamandra 21:313–315.

Vinegar A. 1973. The effects of temperature on the growth and development of embryos of the Indian python (*Python molurus*). Copeia 1973:171–173.

Vinegar MB. 1975a. Life History Phenomena in Two Populations of the Lizard Sceloporus undulatus in Southwestern New Mexico. American Midland Naturalist 93:388–402.

Vinegar MB. 1975b. Comparative aggression in *Sceloporus virgatus, S. undulatus consobrinus* and *S. u. tristichus* (Sauria: Iguanidae). Animal Behavior 23:279–286.

Voitkevich AA. 1957. Innervation peculiarities in naturally originated extra limbs of frogs. Transactions of the Academy of Science USSR. 116(5): 884–887.

Voitkevich AA. 1960. Heterochromia during supernumerary morphogenesis. Transactions of the Academy of Science USSR. 131(5): 1198–1201.

Vorobyeva EI.(ed.) 1995. The Siberian Newt (*Salamandrella keyserlingii* Dybowski, 1870). Ecology, Behaviour, Conservation. 239pp., Nauka, Moscow.
Wagler J. 1830. Natürliches System der Amphibien mit vorangehender Classification der Säugethiere und Vögel. VI + 354 pp.; München, Stuttgart, Tübingen: J. G. Cotta'sche Buchhandlung.
Wallach JD. 1966. Hypervitaminosis D in green iguanas. Journal of the American Veterinary Medical Association 149:912–914.
Wallach V. 1998. The lungs of snakes. Pp. 93–295 In Gans C, Gaunt AS. (eds.) Biology of the Reptilia. Volume 19 (Morphology G), Visceral organs. Ithaca: Society of the Study of Amphibians and Reptiles.
Wallach JD, Hoessle C. 1966. Visceral gout incaptive reptiles. Journal of the American Veterinary Medicine Association 151:897–899.
Watson DMS. 1914. On the skull of a pareiasaurian reptile, and on the relationships of that type. Proceedings of the Zoological Society London 1914:155–180.
Weiss P. 1925a. Unabhängigkeit der Extremitätenregeneration vom Skelett. Roux's Archiv für Entwicklungsmechanik 104:359–394.
Weiss P. 1925b. Die seitliche Regeneration der Urodelenextremität. Archiv für mikroskopische Anatomie und Entwicklungsmechanik 104:395–408.
Weiss P. 1927. Potenzprüfung am Regenerationblastem. I. Extremitätenbildung aus Schwanzblastem im Extremitätenfeld bei Triton. Roux's Archiv für Entwicklungsmechanik 111:317–340.
Wellborn V. 1934. Vergleichende osteologische Untersuchungen an Geckoniden, Eublephariden und Uroplatiden. Sitzungsberichte der Gesellschaft Naturforschender Freunde zu Berlin, Jahrgang 1933:126–199.
Wells LJ. 1945. A case of iliothoracopagus (dicephalus tribradius tripus) with a consideration on the "budding" and "fission" theories. Anatomical Record, Philadelphia 92:1–22.
Wendelstadt H. 1901. Ueber Knochenregeneration. Experimentelle Studie. Archiv für mikroskopische Anatomie 57:798–822.
Wermuth H. 1950. Variationsstatistische Untersuchung der Rassen- und Geschlechtsmerkmale bei der Blindschleiche (*Anguis fragilis* Linné). Deutsche Zoologische Zeitschrift 1(2): 81–121.
Werner F. 1896. Über die Schuppenbekleidung des regenerierten Schwanzes bei Eidechsen. Sitzungsbericht der Akademie Wien 105:122–146.
Werner Yl. 1972. Observations on eggs of eublepharid lizards, with comments on the evolution of the Gekkonoidea. Zoologische Mededelingen 47:211–224.
Wetzel G. 1900. Drei abnorm gebildete Eier von *Tropidonotus natrix*. Anatomischer Anzeiger 18:425–440.
Wexo JB. 1992. Snakes. Zoobooks: 20pp., San Diego: Wildlife Education. Dicephalic *Lampropeltis getula californiae*. One photograph. Removed from the 1987 edition.
Wexo JB. 2003. Snakes. Zoobooks Vol 9:15. San Diego, Wildlife Education Ltd. Photograph removed from the 1987 edition.
Williston S.W. 1904. The relationships and habits of the mosasaurs. Journal of Geology 12:43–51.
Williston SW. 1910. *Cacops, Desmospondylus*: new genera of Permian vertebrates. Bulletin of the Geological Society of America 21: 249–284.
Williston SW. 1911. American Permian vertebrates. 145pp.; Chicago: University Chicago Press.
Wilson LD, Townsend JH, Johnson JD. 2010. Conservation of Mesoamerican Amphibians and Reptiles. Eagle Mountain, Utah: Eagle Mountain Publishing, 812 pp.
Windle B. 1889. On the origin of double monstrosity. Journal of Anatomy and Physiology 23 (new series 3): 390–400.
Witschi E. 1934. Appearance of accessory "organizers" in overripe eggs of the frog. Proceedings of the Society of Experimental Biology and Medicine 31:419–420.
Woitkewitsch AA. 1957. [Innervation peculiarities in naturally originated extra limbs of frogs] [Innervationsbesonderheiten zusätzlicher Extremitäten, gebildet bei Fröschen unter natürlichen Bedingungen.] Doklady Akademija Nauk SSSR 116(5):884-888. [Russian].
Woitkewitsch AA. 1958. [Gesetzmäßigkeiten der räumlichen Beziehung zwischen zusätzlichen und Hauptextremitäten bei Duplizierung unter natürlichen Bedingungen.] Doklady Akademija Nauk SSSR 120(2):438 [Russian].
Woitkewitsch AA. 1960. H [Heterochromia during supernumerary morphogenesis.] Doklady Akademija Nauk SSSR (Biological Sciences, translation) 131:235–238 [Russian].
Wolff É. 1977. Les malformations expliquées par l'embryologie. Revue du Palais de la Découverte 6(53):26–40.
World Health Organization. (1973 and continuing) Bibliography on Congenital Defects in Animals. Document UPH/CMT/73.1. Veterinary Public Health Unit, Division of Communicable Diseases, World Health Organization, Geneva, Switzerland, 92 pp.
Xiao Y, Shi P, Tang R. 1986. Malignant tumor of *Vipera russelli siamensis* from Fujian. Acta Herpetologica Sinica 5:214–216. [Chinese]
Yntema CL. 1970. Observations on females and eggs of the Common Snapping Turtle, *Chelydra serpentina*. American Midland Naturalist 84(1):69–76.

Yoshida M, Suigino H, Swoya H. 1956. Mie. Medical Journal 6:41–45
Zavanella T, Pacces Zaffaroni N, Arias E. 1984. Abnormal limb regeneration in adult newts exposed to the fungicide maneb 80. A histological study. Journal of Toxicology and Environmental Health 13:735–745.
Zwart P. 1974. Maladies des reptiles. part I. Zoo Anvers 39:152–158.
Zwart P, Harshbarger JC. 1972. Hematopoietic neoplasms in lizards: Report of a typical case in *Hydrosaurus amboiensis* and of a probable case in *Varanus salvator*. International Journal of Cancer 9:548–553.
Zwilling E. 1961. Limb morphogenesis. In Abercrombie M, Brachet J. eds. Advances in Morphogenesis. Vol. 1: 301–330; New York: Academic Press.

Cited articles which could not be located in the issue as cited:

Allen ME, Oftedal OH, Horst RL. 1995. Remarkable differences in the response to dietary vitamin D among species of reptiles and primates: Is ultraviolet B light essential? In MF Holick, EG Jung, eds, Biologic Effects of Light. Berlin: Walter de Gruyter, pp. 13–30.
Anon. 1914. Snake stories wax quite numerous as season wanes. Gleaning from our contemporaries concerning various reptile monstrosities – Brother Drake awards blue ribbon to the Times-Press. Middletown Daily Times-Press (Middletown, NY) 10 August 1914: 5.
Anon. 1967. Two-headed rattlesnake. The Baltimore Sun (Baltimore, Maryland) 21 September 1967.
Anon. 1972. Man hires 2-headed snake. Bucks County Courier Times (Levittown, PA) 10 November 1972:A1.
Anon. 1992. Lifestyles: North Hills People. North Hills News Record (Warrendale, PA) 26 November 1992:B1.
Aristotle. 1812. On the generation of animals, pp. 241–448 in Taylor T. (transl), The works of Aristotle, translated from the Greek. Vol. VII. 520pp. (Thomas Taylor), London.
Arnault de Nobleville LD, Salerne F. 1756. Suite de la matière médicale de Geoffroy. Règne animali. [Set of Geoffroy's medical material. Animal kingdom]. T. 12.
Baird SF, Henry J. 1856. Appendix to the report of the secretary. Annual Report of the Board of Regents of the Smithsonian Institution in 1855. Pp. 47–68.
Baird SF. 1856. Appendix to the report of the Secretary. H—List of additions to the museum of the Smithsonian Institution in 1855. Annual Reports of the Smithsonian Institution 10:57–61.
Bedriaga J von. 1879. Beiträge zur Kenntnis des Rippenmolches (*Pleurodeles waltii*). [Contributions to the knowledge of the sharp-ribbed newt (*Pleurodeles waltii*)]. Verhandlungen der Kaiserlichen Gesellschaft der Wissenschaften zu Moskau Bull Soc Nat Mosc 179–201.
Beeld. 1977. Two-headed tortoise. Herpetologica 13:41–42.
Boring AM, Chang L-F, Chang W-H. 1948. Autotomy and regeneration in the tails of lizards. Proceedings of the Zoological Society London 1:351–353.
Cambar R, Haget A. 1949. Sur un cas de polydactylie chez la Grenouille verte (*Rana esculenta* L.). Société sciences physiques et naturelles de Bordeaux, 24 novembre 1949:??
Catesby M.1731–1743. Histoire naturelle de la Caroline Florida et Bahama. [Natural History of Carolina, Florida, and Bahama]. 180 pp.; London: Horace's Head.
Clark DR. 1971. The strategy of tail autotomy in the ground skink, *Lygosoma laterale*. Journal of Experimental Biology 176:295–302.
Coote J. 1979. Notes on nutritional problems in snakes. Herpetile 4(3):31–32.
Cope ED. 1866. On the structure and distribution of the genera of the Arciferous anura. Journal of the Academy of Science in Philadelphia 1866:6.
Cowles RB. 1939. Mother Nature's myths – debunked. The Los Angeles Times, Los Angeles, CA (26 March 1963): I-4–5, I-11.
Cunningham B. 1932. A two-headed fer-de-lance. Field and Stream 75
Devaux B. 1992. Conférence à Sanibel. Tortue 21:7–9.
Divers S. 1995. Metabolic bone diseases in an Argentine horned frog (*Ceratophrys ornata*). Reptilia 3(8):47–49.
Dollo L. 1887. Le hainosaure et les nouveaux vertebras fossils du Musee de Bruxelles. Rev Questions Sci 1° series 22:70–112.
Dorner 1872. College of Surgeons Museum, Teratology Catalogue Nos. 24–27. [Erroneous citation. Actually Lowne 1872].
Dutrochet H. 1829. Observations sur une vipère à deux têtes. [Observations on a viper with two heads]. Annales des travaux de l'Académie royale des Sciences.
Ferguson MJW. 1989. Anomalies congénitales chez l'alligator d'Amérique. In Ross CA, et al. eds. Crocodiles, alligators et caimans. 240pp.; Paris: Bordas edition.
Fredericq L. 1893. L'autotomie ou la mutilation dans le règne animal. Bull de l'Acad Roy de Belgique (Ser 3) 1893;26:180.

Freytag GE. 1954. Molche mit Wirbelsäulenverkrümmungen. – Deutsche Aquarien- und Terrarien-Zeitschrift 1:184–185. Cited as such in Freye 1957.
Froriep R. 1804–1861. FROPIEP'S NOTIZEN aus dem Gebiete der Natur- und Heilkunde, gesammelt und mitgetheilt..... Jahrgang 1856–1857. Jena, Verlag von Friedrich Mauke, 1856–1857. [Cited as volume 15, but there are only 8 volumes].
Gaupp P. 1990. Die Schlange mit zwei Köpfen. [The snake with two heads]. Bild, Leipzig (9 August 1990): p.2.
Gervais P. 1864. Cas de polymélie chez un batracien du genre *Pelobates* (*cultripes*). Mémoire de l'Académie de Montpelier Section de Science 6:66.
Geoffroy St Hilaire 1869. Cosmos, Paris S. 3. v. p. 136, &c & #x2013; actually Meunier 1869.
Girard A. 1897. Sur l'autotomie parasitaire, etc. Comptes Rendues hebdomadaires des Séances et mémoires de la Société de Biologie Paris 4:
Gunckel H. 1944. Culebra de dos cabezas. Diario Austral, Temuco, Apr.24.
Haldemann SS. 1879. Double headed Snakes. American Naturalist 13:535.
Hanley CS, Hernandez Divers S. 2003. Practical gross pathology of reptiles. Seminars in Avian and Exotic Pet Medicine 12:71–80.
Hubner H. 1911. Zur Kasuistik der tierischen Doppelmissbildungen. [On causes of double malformations in animals]. Journal of Experimental Zoology 24:455.
Kovalenko EE, Danilov IG. 1991. [Rare variants of abnormal structure of vertebral column in Anura.] Vestnik Sankt-Peterburgskogo Universiteta Seriya 3 Biologiya 3(17):10–24, 117.
Labbé A. L'année biologique. Comptes Rendus Annuels des GTravaux de biologie Générale l 7:600.
Lacépède 1758. Histoire naturelle de l'Orénoque – Lyon.
Lacépède 1840. Oeuvres.. T II p. 309. Paris.
Lerich. 1951. Journal of Occupational Therapy 5(3).
Lissitzky EF. 1911. [Zur Frage der experimentellen Erzeugung zusätzlicher Extremitäten bei niederen Wirbeltieren.] Kharkov. Veterinarnyi Institut. Sbornik trudov 10(3):4–1.
Lovering, 1978. Pac Rural Press 15(13):194.
Lovering, 1978. Mining and Sci Press 36(13):194.
McArthur S. 1997. Medicine and Surgery of Tortoises and Turtles. In Wilkinson R, Meyer J. Oxford, UK; Malden, MA: Blackwell.
Merian MS. 1705. Metamorphosis Insectorum Surinamensium. Of te verandering der Surinaamsche insecten. Waar in de Surinaamsche rupsen en wormen met alle des zelfs veranderingen, na het leven afgebeeld en beschreeven worden, zynde elk geplaast op die gewassen, bloemen en vruchten, daar sy op gevonden zyn; waar in ook de generatie der kirkvorschen, wonderbaare padden, hagedissen, slangen, spinnen en mieren werden vertoond en beschreeven, alles in America na het leven en levensgroote geschildert en beschreeven. 60 pp., Amsterdam.
Mertens R. 1936. Eine bemerkenswerte Variation des Schildkrötenpanzers. Mitteilung, Isis 1934/36:15–19.
Meyer-Rochow VB, Koebke J. 1986. A study of the extra extremity in a five-legged *Rana* frog. Zoologischer Anzeiger 217:1–13.
Morrow T. 1963. By the way. Chicago Daily Tribune, Chicago, IL (6 March 1963):14.
Neill WT. 1960. Bulletin of the Philadelphia Herpetological Society 8:6.
Pammenes. 5th Century BC. Peri Therion.
Pliny, the Elder. 1539. – Plinii, the Younger is correct.
Pollack ED, Leibig V. 1989. An induced developmental disorder of limbs and motor neurons in *Xenopus*. American Naturalist 289.
Proger LW. 1972. Descriptive catalogue of the Pathological Series in the Hunterian Museum of the Royal College of Surgeons of England. Part II. London: Livingstone [actually Dobson 1972].
Przibram H. 1926. Regeneration und Transplantion bei Tieren. In Handbuch der normalen und pathologischen Physiologie 14(pt 1):1080–1113.
Quinn J. 1974. Two-headed snake; Reginald the Dominant and Llewllyn share zoo pad. The Los Angeles Times, Los Angeles, CA (19 April 1974): C-1.
Ratner JB. [Überzählige vordere Extremitäten beim Frosch (*Rana esculenta*).] Wissenschaftliche Arbeiten Kiew, Veterinary Inst. 2:145.
Reuter E. 1921. En huggorm med twenne väl utvecklade hufond. [a 'couleuvre' with well developed double head]. Meddelser Societ Fauna flora fennica Häft 43:241.
Rimkus J. 1947. Double trouble. Pacific Game News 18(8):23.
Rylkowa EW. 1924. [Zur Frage über die Herkunft und Entwicklung der Polymelie bei Amphibien.] Russian Zoological Journal 4:14–45.
Šanda A, Pivník L. 1963. Spisy fakulty Veterinářství 1963 (3):371.
Sova Z. 1961. Veterinářství 1961 (12):451.
Taylor T. 1812. On the generation of animals, pp. 241–448 in Taylor T. (transl), The works of Aristotle, translated from the Greek. Vol. VII. 520pp. (Thomas Taylor), London.

Thomas A. 1865, see Dumeril 1865.
Tonapi GT. 1958. An abnormal *Rana tigrina* Daud. Quarterly Journal of Science 23:pp. Apparently improper citation for other Tonapi 1958.
Valentin GG. 1852. Comptes Rendues des séances de la société et de biologie et de ses filiales 4: 92–102.
Werber J. 1906. Regeneration der Kiefer bei Reptilien und Amphibien. Archiv für Entwicklungsmechanik 21: 1–14.
Werthemann 1952 Revue Suisse de Zoologie 34:291.
Wisniewski PJ. 1979. Notes on leg-loss in common toad. Regional Anesthesia and Pain Medicine 1(2): 43–47.
Wolterstorff W, Freytag GE. 1942. Über eine zweiköpfige Bastardlarve. – Wochenschrift für Aquarien.
Wyman 1920. Journal of Entomology and Zoolgy 12.

Articles cited in publications apparently not archived and not available for review

[Aldrovandus] (after Fraisse 1885). U. 1645. Patritii Bononiensis de quadrupedibus digitatis oviparis. Lib. II. Bononiae.
Arnold SJ. 1988. Quantitative genetics and selection in natural populations: Microevolution of vertebral numbers in the garter snake *Thamnophis elegans*. In Proceedings of the Second International Conference on Quantitative Genetics, BS Weir, MM Goodman, EJ Eiser, G Namtung, eds. Pp. 619–636. Raleigh North Carolina: North Carolina State University. Sunderland Massachusetts: Sinauer Associates Inc.
Baptistae Portae I. 1650. Neopolitani magiae naturalis libi viginti. Rhotomagi. Lib.II De variis animalibus gignendis. Cap XVIII; Serpentes pluris capitibus caudibusque.
Berst A. 1945. Una serpiente bicéfala en Argentina. El Orden, Santa Fé, Argentina:6,29. Sept. 24.
Bonnet C. 1777. Sur la reproduction des members de la Salamandre aquatique. Observations sur la physique etc. par Rozier. 1. Mémoire vol. 10: Paris.
Bonnet C. 1781. Sur la reproduction des members de la Salamandre aquatique. Oeuvres de Ch. Bonnet, Neuchâtel, Tome V(1):284–358 pl. 5–7.
Bosc LAG. 1817. Nouveau dictionnaire d'histoire naturelle. [New dictionary of natural history.] Vol. 17. or ?18?: 1–542.
Bougon. 1902. Corriere della Domenica IV ann. N. 1, 5. January 1902.
Brown RI, 1995. Metabolic bone disease in captive iguanas. International Reptile Breeders Association: Monitor 1(3):3–4.
Caub JW. v. 1517. Ortus Sanitatis: De herbis et plantis, de animalibus et reptilibus, de auibus et volatilibus, de piscibus et natatilibus, de lapidibus et in terre venis nascentibus, de virinis et earum speciebus, tabula medici9nalis cum directorio generali per omes tract. Argentorati (Strassburg).
Caub von ??..Denamalibus ch 60.
Comerano Atti R Ac Sci Torino 15:8.
Cook J, Jarchow J. 1991. Locomotion, neurological and limb problems in tortoises. Tortoise Trust Newsletter 5(6).
Ctesias C. 5th century BC. Storia di Persia: De rebus indicis. Daniels CB. 1981. The role of caudal lipid in the adaptive strategy of tail autotomy in the gekkonid lizard *Phyllodactylus marmoratus* (Gray 1844). Honors Thesis. Zoology Department, University of Adelaide, Australia.
Deem JM. 1999. Mummy Tombs, Mummy News (25 August 1999).
Fertard B. 2004. Frattura del carapace in una tartaruga. [Carapace fracture in a tortoise]. Summa 21(6): 46–48. [Italian]
Foot JS. 1916. A contribution to the histology of the femur. Smithsonian Contributions to knowledge 35(3), pl 1, fig 5
Fraisse P, Carriere F. 1879. Ueber Regeneration von Organen und Geweben bei Amphibien und Reptilien und den Landpulmonaten. In: Tageblatt des Naturforschervereins in Baden-Baden. [Daily journal of the Naturalists' Association in Baden-Baden]. Jahresbericht über die Gesammte Medicin [Annual report on the whole medicine 1:4].
Girard F. 1997. Presentation des affections metaboliques osseuses (les plus communes) chez les sauriens et leur traitment. [Presentation of the (most common) metabolic osteal illnesses in Sauria and their treatment.] Bulletin de la Société d'Herpétologie de France 79:60–62.
Gould GM, Pyle WL. 1984. Les curiosités médicales. 796pp.; Monaco: SIP edition.
Gregory J. 2005. Heads up, you don't see this two often. Queensland Courier Mail (31 January 2005):3.
Henderson I. 1989. Spinal lumps on a corn snake. Snake Keep 3(8):10–11.
Henderson I. 1989. Spinal lumps on a corn snake. Part 2. Snake Keep 3(10):19–20.
Hennig C. 1869. Eine zweiköpfige Eidechse. [A two-headed lizard] Cosmos. Revue Encyclopédique Hebdoadaire des Progrès des Sciences et de leur Applications aux Arts et a l'Industrie, Paris:21.
Iki BI. 1946. [Two-headed snake.] Kommunist, Azerbaijan, Oct. 26 [In Russian].
Jaisson P. 1969. Anomalies naturelles chez les amphibians anoures. Mémoire présenté à la Faculté des sciences pour obtenir le diplôme d'études supérieures de Sciences (deuxième sujet). June 1969.

Kendrick RE. 1995. Metabolic bone disease in the green iguana. NTHS (National Technical Honor Society) News 13(12):7.
Klauber LM. 1956. Rattlesnakes. Their habits, life histories, and influence on mankind. 2 vols.: 1476pp., Berkeley a. Los Angeles: University California Press.
Köhler G. 1992b. Blasenstein bei einem Schwarzen Leguan, *Ctenosaura similis*. Iguana Rundschreiben, Hanau 2:18–19.
Kranich D. 2006. The fine art of finding 2-headed eastern milksnakes. The Maine Herpetological Society Newsletter (March 2006) 14(2): 2–3, 8.
Kuzmin SL, Godina LB. Urodstva i bolezni. [Anomalies and diseases]. In Vorobyeva EI, ed. Sibirskii uglozub: (Salamandrella keyserlingii Dybowski, 1870): ekologiya, povedenie, okhrana. [The Siberian newt (Salamandrella keyserlingii Dybowski, 1870): Ecology, behavior, conservation]. Nauka, Moskva: 160–164 [Russian].
Lanzoni G. 1690. De vipera duplicâ capite praeditâ. Ephemera Naturae Curios (decade II) 9: 318.
Lauthier M. 1970. Organogenèse de la ceinture pelvienne et des membres postérieurs chez *Pleurodeles waltlii* Michahelles; étude d'anomalies spontanées des membres postérieurs. [Organogenesis of the pelvic girdle and hind limbs in *Pleurodeles waltlii* Michahelles; study of abnormal spontaneous hindlimb]. Thèse de 3e cycle. Paris. Poleshaew LW. 1948. Verlust und Wiedergewinnung der Regenerationsfähigkeit der Extremität bei Anuren. Trudy Instituta citologii, gistologii I embriologii 2(4).
Licerus F. 1665. De Monstris. Edita novissima. Amstrdam: Sumptibus Andreae Frisii.
Meckel JF. 1815. De duplicitate monstrosa commentarius. Halle and Berlin.
Moodie RL. 1925. Studies in Paleopathology. XIII. The elements of the Haversian system in normal and pathological structures among fossil vertebrates. Biologia Generalis 1925 (was in press at time of 1926 article, so unknown pagination).
Parona C. 1881. Nuovi casi di pigomelia nei vertebrati [Pigomely in vertebrates]. Giornale d'Anatomia, Fisiologia e Patologia degli Animali 1888: 1–14 Pisa [Italian].
Pooley AC. 1986. Observations on nutritional requirements of crocodilians. Proceedings of the Crocodile Workshop, Sun City, Bophutatswana, pp. 27–29.
Redi F. 1690. Mélanges des Curieux de la Nature, de Vienne pour l'annèe 1690 pag. 138.
Reuss CF. v. 1784. Ausführliche und accurate Beschreibung nebst genauer Abbildung einiger vorhin fabelhafter Geschöpfe welche in der heutigen Naturgeschichte berühmter Schriftsteller gänzlich verändert und ins Licht gestellet sind. [Detailed and accurate description along with the precise figure of some fabulous creatures that are just now in the modern natural history of famous writers completely changed and in the right light]. W. Nauck, Leipzig.
Rogers. 1970. Two heads are better than one: He could be named "Double trouble." Winkler County News 8 October 1970.
Schmalhausen J. 1915. Entwicklung der Extremitäten der Amphibien und deren Bedeutung für die Frage nach dem Ursprung der Extremitäten der terrestren Wirbeltiere. [Development of the limbs of amphibians and their importance for the question of the origin of the extremities of strict terrestrial vertebrates]. Wissenschaftliche Notizen der Moskauer Universität [Scientific Notes of the Moscow University], Naturwissenschaftliche 37:1–263 [Russian].
Tur J. 1903. Fall von früher Doppelmissbildung bei *Lacerta ocellata*. [Case of earlier double malformation in *Lacerta ocellata*]. Arbeiten des zootomischen Laboratoriums der kaiserlichen Universität Warschau 30:1–8.
Tur J. 1904. Doppelmissbildung von *Mabuia multifasciata* Kubl. Schriften aus dem zootomischen Laboratoriums der kaiserlichen Universität Warschau, Heft 33(Warsch. Univ. iswjäst. No. 7):1–36.
Wanser 1896. Über Missbildungen bei Hydramnion. [On malformations in Hydramnion]. Inaugural Dissertation Tübingen [German].
Wiese. 1812. De monstris animalium. Diss. Berlin.
Zaharesco V. 1935. Recherches anatomiques et morphogégeniques sur les variations numériques de la colonne vertebral chez la grenouille (*Rana esculenta* L.) Jassay Annales Scie 20:370–406 [French].

Inaccessible articles (Solitary, non-circulating holding):

D'Alton, E. 1848. De monstrorum duplicium origine atque evolutione: Cam tabu la acri incise 1. Halle.
Da Cunha PL. 1929. Tratado de Anatomia Pathologica [Treatise on Pathologic Anatomy]. 2nd edition, X + 746 pp.; Rio de Janeiro: Pimenta de Mello & Co. (Bibliotheca Scientifica Brasileira vol. 136) [Portuguese].
Geoffroy Saint-Hilaire I. 1827?. Histoire générale et particulière des anomalies de l'organisation chez l'homme et des animaux: Ouvrage comprenant des recherches sur les charactères, la classification, l'influence physiologique et pathologique, les reppots généraux, les lois et le... Bruxelles. Tom. III, page 132.
Congenital–Dicephalic lizard, captured in the small French province of Roussilon in October of 1829, according to De Betta (1883).
Jackson JBS 1847. Descriptive catalogue of the monstrosities in the cabinet of the Boston Society for medical improvement.

Parona, C. 1878–79. La pigome lia studiata nell'uomo e negli altri vertebrati. [Pygomely studied in man and the other vertebrates] Annali scientifici del Reale Istituto tecnico di Pavia:33–111 (1–79) [Italian].

Pires de Lima JA. 1920. Nova série de observações portuguesas de anomalies dos membros. [New series of Portuguese observations of limb anomalies] Annaes Paulistas de Medicina e Cirurgia 11(12):269 [Portuguese].

Rebello GP. 1914. Noções de Teratologia. [Teratology Notions] XII + 226pp.; [Salvador], Livraria Catilina [Portuguese].

Shacham B. 2004. Polymorphism in the schokari sand snake (*Psammophis schokari*) in the coastal sand dunes of Israel. MSc. Thesis. The Hebrew University of Jerusalem.

Schauinsland H. 1784. Ausführliche und accurate beschreibung: Nebst genauer abbildung einiger vorhin fabelhafter gescho¨pfe welche in der heutigen naturgeschichte beru¨hmter schriftsteller ganzlich verandert und ins licht gesteller sind. Mit einigen kupfern erlautert. Leipzig: W. Nauck, 144 pp.

Schwalbe E. 1914. Anomalie di conformazione. In Trattato di Anatomia patologica de Aschoff. Trad Ital. Torino.

Valentin(i) MB. 1704. Amphiteatrum zootomicum. Accedit Methodus secandi cadav. humana ut & Ars dealbandi ossa pro sceletopoeia, cum osteologia. [Amphiteatrum zootomicum. Besides, the method of cutting the dead body. that the human. The art and the bones of the goddess for the band SKELETON poeia, with osteologia] Francofurti ad Moenum. [Latin] [Part II, page 172, Table 85, Fig. 9 allegedly reports a dicephalic snake].

Wright RJ, Lay M, Balazs GH. 1977. An investigation of the green turtle in the feeding pastures of Ka'u, Hawaii. Report prepared for the Marine Option Program, University of Hawaii at Hilo, 23 pp.

Inadequate citation to allow locating article:

Anon. 1897. [no title]. The Washington Star, Washington, DC (date unknown).

Bartle et al. 1991 – Allegedly mentioned dicephalic and two-tailed *Xenopus*.

Beltrami. 1831. title?. Accademia delle Scienze di Parigi. [Paris Academy of Science] Böttcher.

Cornaglia – Allegedly mentions *Pachypleura*.

Cuvier 1825 – General comment on supernumerary limbs, without naming phyla.

Frye FL. Pathologic conditions in Captive Reptiles – Allegedly mentions renal failure with hyperparathyroidism produces loosening of turtle shell plates kyphoscoliosis in Iguana achondroplastic dwarfism in red-eared slider turtle *Chrysemys (Pseudemys) scripta elegans* and multiple osteocartilaginous exostoses in Varanus.

Howey FL. 1883. Copy book B. Department Report U.S. National Museum.

Leriche 1951 – Orvosi hetilap 92:33 – Allegedly mentioned polydactyly in *Rana temporaria.*

Light 1925

Marchand 1718. Mémoires de l'Académie des Sciences. Année p. 24.

Mateus 1944 – Allegedly mentioned pathology.

Michler 1857 – Allegedly mentioned pathology.

Rayon 1849 – Allegedly mentioned extra toes in *Triton.*

Rostand J. 1949+ Sur les anomalies relevées dans un population de Crapaud ordinaire. [On anomalies in a population of ordinary toad]. Comptes Rendus des Séances de la Société de Biologie – after volume 144.

Sasserno

Schulze O

Seevat 1925 – Allegedly mentioned dicephalic *Testudo ibera* with forelimb duplication.

Seurat 1925

Sordelli F. 1877 – Allegedly mentioned dicephalic *Tropidonotus fasciatus.*

Sturm J. 1797–1828. Deutsche Fauna. Nürnberg.

Sutton JB. 1895 – mentions *Bufo vulgaris, Rana palustris* and *Rana esculenta* with supernumerary hind limb, *Rana temporaria* with supernumerary forelimb, and *Lacerta viridis* with bifid tail.

A List of Fossil Taxa

Araeoscelis	Fossil diapsid reptile
Archelon	Fossil turtle
Bothremys barberi	Fossil turtle
Brachauchenius lucasi	Fossil plesiosaur
Captorhinus	Fossil primitive reptile
Captorhinus aguti	Fossil primitive reptile
Captorhinus laticeps	Fossil primitive reptile
Chunerpeton tianyiensis	Fossil salamander
Deinosuchus	Fossil crocodile
Desmatochelys lowii	Fossil turtle
Dimetrodon	Fossil pelycosaur
Diplodocus	Fossil sauropod dinosaur
Dolichorhynchops	Fossil plesiosaur
Dromicosuchus grallator	Fossil crocodile
Edaphosaurus	Fossil pelycosaur
Eurhinosaurus longirostris	Fossil ichthyosaur
Futabasaurus suzukii	Fossil plesiosaur
Geikia locusticeps	Fossil mammal-like reptile
Hainosaurus bernardi	Fossil varanid
Helodermoides tuberculatas	Fossil lizard
Jachaleria candelariensis	Fossil mammal-like reptile
Labidosaurus	Fossil primitive reptile
Leidyosuchus	Fossil crocodile
Leidyosuchus (Borealosuchus) formidabilis	Fossil crocodile
Lystrosaurus latirostris	Fossil mammal-like reptile
Lystrosaurus (Seeley's Mochlorhinus) platyceps	Fossil mammal-like reptile
Mesosaurus	Fossil mesosaur
Metriorhynchus	Fossil crocodile
Micropholis stowi	Fossil microsaur amphibian
Mosasaurus lemonnieri	Fossil mosasaur
Otschevia panderi	Fossil ichthyosaur
Pachyophis woodwardi	Fossil snake
Palaeobatrachus diluvianus	Fossil frog

(continued)

DOI 10.1007/978-1-4614-0824-6_13,

Palaeobatrachus grandipes	Fossil frog
Palaeobatrachus laubei	Fossil frog
Palaeobatrachus luedeckei	Fossil frog
Pelanomodon tuberosus	Fossil mammal-like reptile
Phytosaurus	Fossil phytosaur
Platecarpus ictericus	Fossil mosasaur
Platecarpus coryphaeus	Fossil mosasaur
Platypterygius	Fossil plesiosaur
Plesiosaurus dolichodeirus	Fossil plesiosaur
Plioplatecarpus marshi	Fossil plesiosaur
Procolpochelys grandaeva	Fossil turtle
Proganochelys quenstedti	Fossil turtle
Prognathodon saturator	Fossil mosasaur
Proneusticosaurus	Fossil nothosaur
Proneusticosaurus madelungi	Fossil nothosaur
Protostega gigas	Fossil turtle
Pterodactylus kochi	Fossil flying reptile
Psephorus	Fossil turtle
Rana mugiens	Fossil frog
Rana mehelyi	Fossil frog
Rienodon	Fossil microsaur amphibian
Sclerocephalus roemeri.`	Fossil temnospondyl amphibian
Smilosuchus gregorii	Fossil phytosaur
Stahleckeria	Fossil mammal-like reptile
Syllomus crispatus aegyptiacus	Fossil turtle
Tanystropheus	
Teleosaurus	Fossil crocodile
Terminonaris cf. T. browni	Fossil crocodile
Terminonator ponteixensis	Fossil elasmosaur
Thaumatosaurus victor	Fossil plesiosaur
Toxochelys	Fossil turtle
Trinacromerum willistoni	Fossil plesiosaur
Trinacromerum bentonianum	Fossil plesiosaur
Tylosaurus dyspelor	Fossil mosasaur
Tylosaurus proriger	Fossil mosasaur

List of Scientific and Common Names

Scientific name	Common name
Abronia	Alligator lizard
Ablepharus	Snake-eyed skink
Acalyptophis peronii	Spiny-headed sea snake
Acanthodactylus beershebensis	Be'er Sheva fringe-fingered lizard
Acanthodactylus boskianus	Bosk's fringe-toed lizard
Acanthodactylus erythrurus	Spiny-footed lizard
Acanthodactylus schmidti	Schmidt's fringe-toed lizard
Acanthodraco vitticeps	Bearded dragon
Acanthophis antarcticus	Southern death adder
Acanthophis pyrrhus	Desert death adder
Acanthophis wellsi	Pilbara death adder
Acanthophis madagascariensis	Madagascar ground boa
Acanthosaurus crucigera	Horn-headed lizard
Acontias	Lance skink
Acontias meleagris	Cape legless skink
Acrantophis dumerili	Dumeril's boa
Acris crepitans blanchardi	Blanchard's cricket frog
Acris gryllus	Southern cricket frog
Acrochordus granulatus	Little file snake
Acrochordus javanicus	File snake, Elephant trunk snake
Aelodon priscus = Crocodylus priscus	Fossil
Agama anchietae	Anchieta's agama, Western rock agama
Agama agama	Red-headed rock agama
Agama bibronii	Bibron's agama
Agama mutabilis	Desert agama

(continued)

Scientific name	Common name
Agama stellio	Roughtail rock agama
Agama s. str.	Agama
Agama tuberculata	Kashmir rock agama
Agamodon anguliceps	Angled worm lizard
Agamura	Spider gecko
Agkistrodon bilineatus	Taylor's cantil
Agkistrodon (Gloydius) blomhoffi	Japanese pit viper ("Mamushi" in Japanese)
Agkistrodon contortrix	Southern copperhead
Agkistrodon halys	Chinese water mocassin
Agkistrodon halys brevicaudus	Korean viper
Agkistrodon mokasen	Copperhead
Agkistrodon piscivorus	Cottonmouth
Aglossa	Aglossa frog
Agrionemys horsfieldii	Russian tortoise
Ahaetulla	Whip snake
Ahaetulla prasina	Malayan whipsnake
Aipysurus apraefrontalis	Short-nosed seasnake
Aipysurus duboisii	Reef shallows seasnake
Aipysurus foliosquama	Leaf-scaled seasnake
Aipysurus laevis	Olive seasnake
Alligator mcgrewi	Fossil alligator
Alligator mississippiensis	American alligator
Alligator sclerops	Jacare
Allobates femoralis	Aromobatid frog
Alopoglossus angulatus	Northern teiid
Alopoglossus atriventris	Keel-bellied shade lizard
Alsophylax blandfordii	Blandford's ground gecko
Alytes obstetricans	Midwife frog
Amblyrhynchus	Marine iguana

(continued)

B.M. Rothschild et al., *Herpetological Osteopathology: Annotated Bibliography of Amphibians and Reptiles*,
DOI 10.1007/978-1-4614-0824-6_14, © Springer Science+Business Media, LLC 2012

Scientific name	Common name
Ambystoma = *Siredon*	Axolotl
Ambystoma argus = *Ambystoma maculatum*	
Ambystoma gracile	Northwestern salamander
Ambystoma jeffersonianum	Jefferson salamander
Ambystoma macrodactylum	Long-toed salamander
Ambystoma maculatum	Spotted salamander
Ambystoma mexicanum	Mexican axolotl
Ambystoma opacum	Marbled salamander
Ambystoma punctatum	Violet salamander
Ambystoma tadpoideum	Mole salamander
Ambystoma texanum	Small mouthed salamander
Ambystoma tigrinum	Tiger salamander
Ameiva	Jungle-runner, Whiptail lizard
Ameiva ameiva	Jungle runner, Dwarf Tegu, Giant ameiva
Ameiva exsul	Puerto Rican ground lizard
Ameiva festiva	Central American ameiva, Middle American ameiva, Tiger ameiva
Ameiva undulata	Rainbow ameiva
Amietophrynus maculatus	See *Bufo maculatus*
Amietophrynus mauritanicus	See *Bufo mauritania*
Amietophrynus pantherinus	See *Bufo pantherinus*
Amietophrynus regularis	See *Bufo regularis*
Amps kingi	Amphisbaenian worm lizard
Amphibolurus (*Grammatophora*) *barbata*	Bearded dragon
Amphiesma stolatum	Buff striped keelback snake
Amphiesma vibakari (*Natrix vibakari*)	Asian keelback snake
Amphisbaena	Amphisbaena
Amphisbaena alba	White-bellied worm lizard
Amphisbaena antillensis	Cope's worm lizard
Amphisbaena caeca	Puerto Rican worm lizard
Amphisbaena dubia	Uncertain worm lizard
Amphisbaena occidentalis	Western worm lizard
Amphisbaena spurrelli	Spurelli's Worm Lizard
Amphiuma means	Two-toed amphiuma
Amyda mutica	Leather turtle
Amyda (*Trionyx*) *sinensis*	Chinese soft-shelled turtle
Amyda (*Trionyx*) *spiniferus*	Eastern spiny softshell turtle
Anaxyrus americanus	See *Bufo americanus*

(continued)

Scientific name	Common name
Anaxyrus boreas	See *Bufo boreas*
Anaxyrus cognatus	See *Bufo cognatus*
Anaxyrus columbiensis	See *Bufo columbiensis*
Anaxyrus debilis	See *Bufo debilis*
Anaxyrus fowleri	See *Bufo fowleri*
Anaxyrus hemiophrys	See *Bufo hemiophrys*
Anaxyrus punctatus	See *Bufo punctatus*
Anaxyrus terrestris	See *Bufo lentiginosus*
Anaxyrus woodhousii woodhousii	See *Bufo woodhousii woodhousii*
Ancistrodon contortrix	Southern copperhead
Ancistrodon piscivorus	Cotton-mouth moccasin
Ancyclocranium	Amphisbaenian worm lizard
Aneides aeneus	Green salamander
Aneides ferreus	Clouded salamander
Aneides lugubris	Arboreal salamander
Anelytropsis	Blind lizard
Anguis fragilis	Slow worm
Anilius scytale	South American red pipe snake, False coral snake
Anisolepis	Tree lizard
Anniella	Legless lizards
Anolis	Anole, See *Chamaeleolis*, *Chamaelinorops*
Anolis angusticeps	Cuban twig anole
Anolis carolinensis carolinensis	Northern green anole
Anolis distichus	Bark anole
Anolis fuscoauratus	Slender anole
Anolis grahami	Jamaican turquoise anole
Anolis iguanae	Common green iguana
Anolis principalis	American "chameleon"
Anolis sagrei	Cuban anole
Anomalepis	Worm snake
Anops	Worm lizard
Anotis maccoyi	Elf skink
Apalone ferox	Florida softshell turtle
Apalone spinifer	Eastern spiny softshell turtle
Aprasia pulchella	Granite worm lizard
Aprasia repens	Sand plain worm lizard
Aprasia striolata	Lined worm lizard
Aptycholaemus	Tree Lizard
Arizona elegans occidentalis	Glossy snake
Arthrosaura reticulata	Reticulated creek lizard
Asaccus ("*Phylodactylus*")	Southwest Asian leaf-toed gecko

(continued)

Scientific name	Common name
Ascalabotes = Tarentola	
Ascaphus	Tailed frog
Ascaphus truei	Costal tailed frog
Aspideretes gangeticus	Ganges softshell turtle
Aspidites melanocephalus	Black-headed python
Aspidonectes spirifer	Soft shell turtle
Aspidites	Python
Astrotia stokesii	Stoke's sea snake
Asyntaxia caudalis	Claimed as species, but actually a term for cauda bifida
Atelopus chiriquiensis	Lewis' subfoot toad
Atelopus limosus	Sapo limosa, Limosa Harlequin frog
Atelopus varius	Costa Rican Variable Harlequin toad
Atelopus zeteki	Panamanian golden frog
Atheris nitschei	Bush viper
Atractaspis	Burrowing asp
Aulura barbouri	Barbour's worm lizard
Austrelaps superbus	Copperhead
Azemiops	Fea's viper
Bachia dorbignyi	Dorbigny's bachia
Bachia (Chalcides) flavescens	Cylindrical skink
Bachia intermedia	Noble's bachia
Bachia panoplia	A spectacled lizard
Bachia peruana	Peruvian worm lizard
Baikia	West African worm lizard
Baptemys garmanii	Fossil turtle
Bascanium constrictor	Black racer
Basiliscus basiliscus	Brown basilisk, Common ballisk
Basiliscus plumifrons	Green-crested basilisk
Bassiana duperreyi	Eastern three-lined skink
Batagur baska	Northern river terrapin
Batagur borneoensis	See *Callagur borneoensis*
Batrachoseps	Slider salamander
Batrachoseps attenuatus	California slender salamander
Batrachoseps major	Garden slender salamander
Batrachoseps nigriventris	Black-bellied slender salamander
Batrachoseps pacificus	Channel Islands slender salamander
Batrachoseps relictus	Relictual slender salamander
Bipes biporus	Mexican mole lizard, Five-toed worm lizard, Ajolote lizard

(continued)

Scientific name	Common name
Bipes canaliculatus	Four-toed worm lizard
Bipes pallasii	Glass lizard
Bipes tridactylus	Three-toed worm lizard
Bitis arietans	Puff adder
Bitis nasicornis	Rhinoceros viper
Blanus cinereus	Iberian worm lizard
Blanus strauchi	Anatolian worm lizard
Boa	Boa snake
Boa constrictor	Boa constrictor
Boa c. occidentalis	Argentinian boa
Boa (Corallus) enhydris	Cook's boa, Amazon tree boa
Boa cookie	Cook's tree boa
Boaedon f. fulliginosus	Common house snake
Boiga dendrophila melanota	Gold-ringed cat snake, Mangrove snake
Bolitoglossa altamazonica	Amazon climbing salamander
Bolitoglossa engelhardti	Mushroom tongue salamander
Bolitoglossa flavimembris	Yellow-legged mush-roomtongue salamander
Bolitoglossa franklini	Mushroom tongue salamander
Bolitoglossa mexicana	Mexican climbing salamander
Bolitoglossa morio	Cope's climbing salamander
Bolitoglossa occidentalis	Southern banana salamander
Bolitoglossa platydactyla	Broadfoot climbing salamander
Bolitoglossa resplendens = Bolitoglossa lincolni	Lincoln's web-footed salamander
Bolitoglossa rostrata	Long-nosed climbing salamander
Bolitoglossa rufescens	Northern banana salamander
Bolitoglossa subpalmata	La Palma salamander
Bolitoglossa yucatana	Yucatan mushroomtongue salamander
Bombina bombina	Fire-bellied toad
Bombina orientalis	Oriental fire-bellied toad
Bombina variegate	Yellow-bellied toad
Bombinator igneus	See *Bufo igneus*
Bombinator pachypus	Appenine firebelly toad
Borborocoetes	Chilean frog
Bothrochilus	Bismarck ringed python
Bothrops asper	Terciopelo, Central American lancehead
Bothrops atrox	Fer-de-lance, Mapanare

(continued)

Scientific name	Common name
Bothrops insularis	Golden lancehead snake
Bothrops jararaca	Jararaca, Yarara
Bothrops jararacuçú	Jararacussu
Bothrops jararacussu	Jararacussu
Bothrops moojeni	Caiçaca, Brazilian lancehead snake
Brachycephalus	Brazilian golden frog
Brachycephalus ephippium	Pumpkin toadlet
Brachylophus	Crested iguana
Brachylophus fasciatus	Fiji banded iguana, South Pacific banded iguana
Brachyophidium rhodogaster	Mountain burrowing snake
Bronchocela cristatella	See *Calotes cristatellus*
Breviceps	Rain frog
Bronia	Worm lizard
Brycon guatemalensis	Machaca
Bufo americanus	American toad
Bufo andersonii	Marbled balloon frog
Bufo arenarum	Common Argentine frog
Bufo asper	Asian giant toad, Boreal toad
Bufo boreas halophilus	California toad
Bufo bufo	Common toad, See also *Bufo vulgaris*
Bufo calamita	Natterjack toad
Bufo cinereus	See *Bufo bufo*
Bufo cognatus	Great Plains toad
Bufo columbiensis	Oregon spotted frog
Bufo crucifer	Striped toad
Bufo debilis debilis	Eastern green toad
Bufo ephippium = Brachycephalus ephippium	Spix's Saddleback Toad
Bufo fowleri	Fowler's toad
Bufo gargarizans	Asiatic toad
Bufo hemiophrys	Dakota toad, Canadian toad
Bufo ictericus	Yellow Cururu toad
Bufo igneus = Bombina bombina	Firebelly toad
Bufo japonicus japonicus	Western-Japanese common toad
Bufo lentiginosus	Wesern Woodhouse's toad, Rocky mountain toad
Bufo maculatus	Flat-backed toad
Bufo marinus	Cane toad, Marine toad
Bufo mauritania	Mauritanian toad; Berber toad
Bufo melanostictus	Common Indian toad
Bufo pantherinus	Western leopard toad
Bufo peltocephalus	Cuban toad

(continued)

Scientific name	Common name
Bufo punctatus	Red-spotted toad
Bufo regularis	Egyptian maculated toad
Bufo variabilis	Variable toad
Bufo viridis	Green frog, Green toad
Bufo vulgaris = Bufo bufo	see *Bufo bufo*
Bufo woodhousii fowleri	Fowler's toad
Bufo w. woodhousii	Woodhouse's toad
Bungarus caeruleus	Common krait, Indian krait
Cadea blanoides	Cuban spotted amphisbaenian
Cadea palirostrata = Amphisbaena palirostrata	Cuban Sharp-nosed Worm Lizard
Caiman crocodilus	Common caiman
Caiman latirostris	Broad-nosed caiman
Caiman sclerops	Common caiman, See also *Caiman crocodilus*
Calliophis (*Hemibungarus*) *japonicus*	Oriental coral snake
Calabaria reinhardtii	Calabar ground python
Callagur borneoensis	Painted terrapin
Callisaurus draconoides	Zebra-tailed lizard
Calotes cristatellus	Green crested lizard
Calotes emma	Emma Gray's forest lizard, Forest crested lizard
Calotes versicolor	Oriental garden lizard, Ceylon lizard
Calumma parsonii	See *Chamaeleo parsonii*
Calyptocephalus (*Calyptocephalella*) *gayi*	Green, web-footed Chilean toad
Calyptocephalella	See *Calyptocephalus*
Caretta caretta	Loggerhead turtle
Carettochelys insculpta	Fly River turtle
Carphodactylus laevis	Chameleon gecko
Casarea	boa
Causus rhombeatus	Common adder, Rhombic night adder
Centrochelys sulcata	African spurred tortoise
Ceramodactylus (*Stenodactylus*) *doriae*	Arabian whip-tailed gecko
Ceratobatrachus guentheri	Solomon Island eyelash frog
Ceratophrys	Horned frog
Ceratophrys americana (*Odontophrynus americanus*)	Brazilian horned frog
Ceratophrys aurita	See *Ceratophrys dorsata*
Ceratophrys (*Proceratophrys*) *boiei*	Alphabet frog
Ceratophrys dorsata (*aurita*)	Brazilian horned frog
Ceratophrys ornata	Argentine horned frog
Cerberus rynchops	Dog-faced water snake

(continued)

Scientific name	Common name
Chalarodon	Malagasy terrestrial iguanian lizard
Chalcides (Bachia) flavescens	Chalcis
Chalcides ocellatus	Ocellated skink, Eyed skink, Gongilo
Chalcides sepsoïdes	Meadow lizard
Chamaeleo calyptratus	Veiled chameleon
Chamaeleo chamaeleo	Common chameleon
Chamaeleo (Trioceros) deremensis	Usambara three-horned chameleon, Giant three-horned chameleon
Chamaeleo dilepis	Flap-necked chameleon
Chamaeleo (Trioceros) jacksonii	Jackson's chameleon
Chamaeleo (Trioceros) johnstoni	Johnston's chameleon
Chamaeleo (Trioceros) montium	Mountain chameleon
Chamaeleo oweni (Trioceros oweni)	Owen's chameleon
Chamaeleo (Furcifer) pardalis	Panther chameleon
Chamaeleo parsonii (Calumma)	Parson's chameleon
Chamaeleo (vulgaris) Chameleon	Common chameleon
Chamaeleolis (Anolis)	*Anolis*
Chamaelinorops (Anolis)	*Anolis*
Chamaesaura aenea	Transvaal grass lizard
Chamaesaura anguina	Cape grass lizard
Chamaesaura anguinea see Chamaesaura anguina	Cape Grass Lizard, Cape Snake Lizard, Highland Grass
Chamaesaura macrolepis	Large-scale grass lizard
Charina bottae	Silver snake, Worm snake, Northern rubber boa
Chaunus (Bufo) marinus (Rhinella marina)	Cane toads
Chelodina longicollis	Australian side-necked turtle
Chelodina oblonga (rugosa)	Narrow-breasted snake-necked turtle, Oblong turtle
Chelodina rugosa (oblonga)	Northern Australian snake-necked turtle
Chelone corticata = Caretta caretta	Loggerhead turtle
Chelonia mydas	Green turtle
Chelonoidis (Testudo) chilensis	Tortuga argentina
Chelus fimbriatus	Matamata terrapin
Chelydra serpentina	Common snapping turtle, Hard shell snapping turtle, Alligator terrapin
Chersina angulata	Angulate tortoise, Bowsprit tortoise
Chinemys kwangtungensis (nigricans)	Kwangtung river turtle
Chinemys nigricans	See *Chinemys kwangtungensis*
Chinemys (Mauremys) reevesii	Reeve's turtle, Chinese pond turtle, Chinese three-keeled pond turtle, Golden turtle, Japanese coin turtle
Chioglossa lusitanica	Golden-striped salamander
Chirindia ewerbecki	Mbanja worm lizard
Chirocolus (Heterodactylus) imbricatus	Microteiid lizard
Chiropterotriton bromeliacia (Dendrotriton bromeliacius)	Cuchumatuna bronelid salamander
Chiropterotriton chiropterus	Common splayfoot salamander
Chiropterotriton multidentatus	Toothy splay foot salamander
Chitra vandijki	Myanmar narrow-headed softshell turtle
Chlamydosaurus kingii	Frilled lizard
Chlorophilus	Swamp tree frog
Chondrodactylus angulifer	Giant ground gecko
Chondrophython (Morelia) viridis	Green tree python
Christinus marmoratus	Marbled gecko
Chrysemys concinna texana	Texas river cooter
Chrysemys elegans	Cumberland turtle
Chrysemys (Pseudemys) floridana	Common cooter
Chrysemys marginata	Painted turtle
Chrysemys neurrie	Painted turtle
Chrysemys picta	Northern painted turtle
Chrysemys scaber (Pseudemys s. scripta)	Yellow-bellied terrapin
Chrysemys s. elegans	Red-eared slider turtle
Chrysemys (Trachemys) scripta	Red-eared tortoise
Chrysolamprus (Lacerta) ocellatus	Ocellated lizard

(continued)

Scientific name	Common name
Chrysopelea	Flying snake, Golden tree snake, Ornate flying snake, Golden flying snake
Cimoliasaurus plicatus	Fossil plesiosaur
Cistoclemmys (Cuora) galbinifrons	Flowerback box turtle
Clemmys (Mauremys) caspica caspica	Caspic water turtle
Clemmys (Mauremys) caspica leprosa	Spanish terrapin
Clemmys (Mauremys) caspica rivulata	Western Caspian turtle
Clemmys cumberlandensis = Chrysemys elegans	Red-eared Slider
Clemmys guttata	Spotted turtle
Clemmys (Chelopus or Glyptemys) insculpta	Wood turtle
Clemmys (Mauremys) leprosa	Mauric water turtle
Cnemaspis	Gecko
Cnemidophorus exsanguis	Chihuahuan spotted whiptail
Cnemidophorus gularis	Texas spotted whiptail
Cnemidophorus inornatus	Little striped whiptail
Cnemidophorus lemniscatus	Rainbow whiptail, South American whiptail
Cnemidophorus littoralis	Brazilian whiptail lizard
Cnemidophorus motaguae	Rainbow lizard, Giant whiptail lizard
Cnemidophorus ocellifer	Spotted whiptail lizard
Cnemidophorus sexlineatus	Six-lined racerunner
Cnemidophorus tesselatus	Colorado checkered whiptail
Cnemidophorus tigris	Western whiptail
Coleodactylus amazonicus	Brazilian pigmy gecko
Coleodactylus septentrionalis	Ilha Maracá gecko
Coleonyx variegatus	Western banded gecko
Colostethus collaris	Dendrobatid frog
Colostethus palmatus	Palm rocket frog
Colostethus trinitatis	Yellow-throated frog, Mannophryne
Coluber	Racers
Coluber biceps = Amphisbaena (Not a species name; latin for "two-headed" snake)	Grass snake
Coluber bicephalus	Grass snake
Coluber catenifer	Pacific gopher snake, Coast gopher snake, Western gopher snake
Coluber constrictor	Black racer
Coluber c. constrictor	Northern black racer

(continued)

Scientific name	Common name
Coluber doliatus triangulus	Milk snake
Coluber flagellum	Coachwhip snake
Coluber florulentus	Flowered racer
Coluber jugularis	Caspian whipsnake, Persiand large whipsnake
Coluber longissimus	Aesculapian snake
Coluber melanoleucas = Pituophis sayi	Pine snake
Coluber natrix	Common ringed snake, Grass snake
Coluber torquatus (natrix)	Grass snake, Ring snake
Coluber viridiflavus	European whip snake
Coniophanes fissidens	Yellowbelly snake
Conolophus	Galapagos land iguanas
Constrictor constrictor	Boa constrictor
Cophosaurus texanus	Greater earless lizard
Corallus caninus	Emerald tree boa
Corallus (Boa) enhydris	Cook's boa, Amazon tree boa
Corallus (Enhydris) enhydris	Amazon tree boa
Corallus enydris cooki	Garden tree boa
Cordylus warreni	Warren's girdled lizard
Coronella	Milk snake
Coronella austriaca	Smooth snake
Coronella getula	Chain snake
Coronella laevis	Smooth snake
Coronella triangulum	Milk snake
Corucia zebrata	Prehensile-tailed skink
Corytophanes	Helmeted iguana or helmeted basilisk
Corytophanes cristatus	Helmeted basilisk
Cosymbotus platyurus	Philippine house lizard, Flat-tailed gecko
Crocodilurus amazonicus	Jacarerana lizard
Crocodylus acutus	American crocodile
Crocodylus americanus	Sharp-nosed crocodile
Crocodylus biporcatus	*Crocodylus porosus*
Crocodylus cataphractus	African slender-snouted crocodile
Crocodylus intermedius	Orinoco crocodile
Crocodylus johnstoni	Johnston's crocodile
Crocodylus moreletii	Morelet's crocodile
Crocodylus novaeguineae	New Guinea crocodile
Crocodylus niloticus	Nile crocodile
Crocodylus palustris	Marsh crocodile
Crocodylus porosus	Estuarine crocodile, Salt water crocodile
Crocodylus rhombifer	Cuban crocodile
Crocodylus siamensis	Siamese crocodile
Crocodylus tetrapis	Dwarf crocodile

(continued)

Scientific name	Common name
Crocodylus vulgaris	*Crocodylus niloticus*
Crossobamon eversmanni	Comb-toed gecko
Crossodactylus gaudichaudii	Gaudichaud's frog
Crotalus	Rattlesnakes
Crotalus adamanteus	Eastern diamondback rattlesnake
Crotalus atrox	Texas rattlesnake, Western diamondbacked rattlesnake
C. b. basiliscus	Mexican west coast rattlesnake, Mexican green rattler
Crotalus durissus terrificus	South American rattler, Cascavel
Crotalus exsul	Red rattlesnake, Red diamond rattlesnake
Crotalus h. horridus	Canebrake rattlesnake
Crotalus lucasensis	San Lucan rattlesnake
Crotalus lutosus	Great Basin rattlesnake
Crotalus mitchellii	Speckled rattlesnake
Crotalus molossus	Black-tailed rattlesnake
Crotalus oreganus	Western rattlesnake
Crotalus ruber	Red diamond rattlesnake
Crotalus terrificus	South American rattlesnake, Tropical rattlesnake
Crotalus vergrandis	Uracoan rattlesnake
Crotalus viridis viridis	Prairie rattlesnake
Crotalus v. oreganus	Northern Pacific rattlesnake
Crotalus willardi	Ridge-nosed rattlesnake
Crotaphopeltis hotamboeia	Herald snake
Crotaphytus bicinctores	Great Basin collared lizard
Crotaphytus collaris	Eastern collared lizard, Common collared lizard
Crotaphytus wislizeni	Leopard lizard
Cryptobranchus alleganiensis	Hellbender
Cryptobranchus japonicus	Japanese giant salamander
Cryptodactylus kotschyi	Kotschy's gecko
Ctenoblepharis	Tree iguana
Ctenodon nigropunctatus	Northeastern spinytail iguana
Ctenonotus cuvieri	Puerto Rico-i tarajos anolisz
Ctenosaura acanthura	Northeastern spinytail iguana
Ctenosaura multispinis	Northeastern spiny-tail iguana
Ctenotus taeniolatus	Striped skink, Copper-tailed skink
Cuora amboinensis	Box turtle
Cuora chriskarannarum	Pan's box turtle, Southern Vietnamese box turtle
Cuora flavomarginata	Yellow-margined box turtle
Cuora galbinifrons	Indochinese box turtle, Vietnamese box turtle, Flowerback box turtle, See *Cistoclemmys galbinifrons*
Cuora pani	Pan's box turtle
Cuora trifasciata	Chinese three-striped box turtle
Cuora yunnanensis	Yunnan box turtle
Cupriguanus	Large-head leguan
Cyclemys dentate	Asiatic leaf terrapin, Brown stream terrapin
Cycloderma aubryi	Aubry's flapshell turtle
Cyclodus scincoides = *Tiliqua scincoides*	Blue-tongued cyclodus
Cyclura	Rock iguana
Cyclura cornata	Rhinoceros iguana
Cyclura macleayi	Cuban lizard, Ground iguana
Cyclura pectinata	Mexican spiny-tailed iguana
Cylindrophis maculatus	Linne's earth snake, Ceylonese cylinder snake, Sri Lankan pipe snake
Cylindrophis rufus	Malaysian pipe snake, Red-tailed pipe snake
Cynisca	Worm lizard
Cynops ensicauda popei	Riu Kiu salamander
Cynops pyrrhogaster	Japanese fire belly newt
Cystignathus ocellatus = *Leptodactylus bolivianus*	Leptodactylus, Criolla white-lipped frog
Dactyletra	Tongueless Frogs
Darevskia	White-bellied lizard
Dasypeltis	Egg eater snake
Deirochelys reticularia	Chicken turtle
Dendrobates auratus	Black and green poison arrow/dart frog
Dendrobates azureus	Blue poison frog
Dendrobates fantasticus	Red-headed poison frog
Dendrobates granuliferus	Granular poison frog
Dendrobates histrionicus	Harlequin poison frog
Dendrobates imitator	Imitator poison frog
Dendrobates lehmanni	Lehmann's poison frog

(continued)

Scientific name	Common name
Dendrobates leucomelas	Bumblebee poison dart frog
Dendrobates pumilio	Strawberry poison-dart frog
Dendrobates pumilius	Blue-jeans poison dart frog
Dendrobates quinquevittatus	Madeira poison frog
Dendrobates reticulatus	Red-backed poison frog
Dendrobates speciosus	Splendid poison frog
Dendrobates tinctorius	Dyeing poison frog
Dendrobates truncatus	Yellow-striped poison frog
Dendrobates variabilis	Splash-back poison frog
Dendrophidion dendrophis	Tawny forest racer
Dendrotriton bromeliacius	See *Chiropterotriton bromeliacia*
Dendrophryniscus brevipollicatus	Coastal tree toad, Brazilian toad
Dermatemys	Freshwater turtle
Dermatonotus muelleri	Mueller's narrow-mouthed frog
Dermochelys coriacea	Leatherback turtle
Desmognathus fuscus	Northern dusky salamander
Desmognathus monticola	Seal salamander
Desmognathus ochrophaeus	Mountain dusky salamander
Desmognathus monticola punctatus	Seal salamander
Dermophis mexicanus	Mexican burrowing caecilian
Diadophis	Ringneck snake
Diadophis s. similes	San Diego ringneck snake
Dibamus argenteus	White blind skink
Diemyctylus viridescens	Spotted newt
Diplodactylus	Australian gecko
Diplodactylus vittatus	Wood gecko
Diplodactylus williamsi	Eastern spiny-tailed gecko, Soft-spined gecko
Diploglossus monotropis	Galliwasp
Diplometopon	Zarudnyi's worm lizard
Diporophora bilineata	Two-lined dragon
Diposaurus dorsalis	Desert iguana
Dipsadoboa unicolor	Günther's green tree snake
Dipsas albifrons	Slug-eater snake, Sauvage's snail-eater snake
Dipsas boettgeri	Boettger's snail-eater snake
Dipsas brevifacies	Snail-eating thirst snake
Dipsas catesbyi	Catesby's snail eater snake, Ornate snail-eating snake
Dipsas elegans	Caracolera subtropical

(continued)

Scientific name	Common name
Dipsas ellipsifera = *Dipsas elegans*	Caracolera subtropical
Dipsas gaigeae	Gaige's thirst snail-eater snake
Dipsas incerta	Cope's snail-eater snake, Jan's snail-eater snake
Dipsas indica	Amazonian snail-eater snake, Big-headed snail-eating snake
Dipsas latifasciata	Striped snail-eater snake
Dipsas latifrontalis	Venezuela snail-eater snake
Dipsas neivai	Variegated snail-eater snake
Dipsas oreas	Ecuador snail-eater snake
Dipsas pavonina	Northern snail-eater snake
Dipsas peruana	Many-scaled snail-eater snake, Striped snail-eater snake
Dipsas pratti	Pratt's snail eater
Dipsas sanctijoannis	Tropical snail-eater snake
Dipsas schunkei	Shunk's snail-eater snake
Dipsas tenuissima	Taylor's snail-eater snake
Dipsas variegata	Variegated snail-eater snake
Dipsas vermiculata	Vermiculate snail-eater snake
Dipsas viguieri	Bocourt's snail-eater snake
Dipsochelys dussumieri	Aldabra giant tortoise
Dipsosaurus dorsalis	Desert iguana
Discoglossus pictus	Painted frog
Discoglossus tricolor = *Epipedobates tricolor*	Phantasmal poison frog, Three-colored poison frog
Dispholidus typus	Boomslang
Distira brugmansii (*Hydrophis sublaevis*)	Sea snake
Draco	Flying dragons
Dromicus chamissonis = *Philodryas chamissonis*	Chilean Green Racer
Duttaphrynus andersonii	See *Bufo andersonii*
Duttaphrynus melanostictus	See *Bufo melanostictus*
Duttaphrynus pantherinus	See *Bufo pantherinus*
Dryophis prasinus	Gunther's whip snake, Oriental whipsnake, Asian vine snake, Jade vine snake
Duberria	Slug-eater snake
Echinantera cyanopleura	Corredeira-do-mato, Argentinian snake
Echmatemys euthenta	Fossil turtle
Egernia	Spiny-tailed skinks

(continued)

Scientific name	Common name
Egernia striolata	Tree-crevice skink
Elachistodon westermanni	Indian egg-eating snake
Elaphe bimaculata	Chinese rat snake
Elaphe climacophora	Japanese rat snake
Elaphe conspicillata	Japanese forest rat snake
Elaphe dione	Dione rat snake
Elaphe g. guttata	Corn snake
Elaphe longissima	Aesculapian snake
Elaphe obsoleta lindheimeri	Texas rat snake
Elaphe o. obsoleta	Black rat snake
Elaphe obsoleta quadrivitta	Yellow rat snake
Elaphe obsoleta rossalleni	Everglades rat snake
Elaphe quadrivirgata	Japanese four-lined ratsnake
Elaphe quatuorlineata sauromates	Four-lined snake
Elaphe rufodorsata	Chinese racer, Red-backed ratsnake
Elaphe schrencki schrencki	Russian rat snake
Elaphe subocularis	Trans-Pecos ratsnake
Elaphe taenura	Chinese stripe-tailed snake
Elaphe vulpina	Western fox snake
Elaps frontalis	See *Micrurus altirostris*
Eleutherodactylus alticola	Jamaican peak frog
Eleutherodactylus cuneatus	Riparian frog
Eleutherodactylus ricordii	Ricordi's robber frog
Elseya dentate	Northern snapping turtle
Elseya latisternum	Saw-shelled turtle
Elusor macrurus	Mary River turtle, Mary River tortoise
Emeces gilberti	Gilbert's skink
Emoia	Gecko
Emprostomelophorus tetrascelus	frog with two supernumerary legs
Emyda granosa	*Cryptopus granosus*
Emydocephalus ijimae	Ijim's turtlehead sea snake
Emydoidea blandingii	Blanding's turtle
Emydura albertisi	Red bellied side neck turtle
Emydura australis subglobosa	Red bellied side-necked turtle
Emydura krefftii	Australian side-necked turtle
Emydura macquarii	Murray River turtle
Emydura signata	Brisbane short-necked turtle
Emydura subglobosa	Red-bellied short-necked turtle
Emydoidea blandingi	Blanding's turtle
Emys concentrica	Concentric terrapin
Emys europaea	European pond turtle

(continued)

Scientific name	Common name
Emys orbicularis	European pond turtle
Enhydrina schistose	Beaked sea snake
Enhydris alterans	Reuss' water snake
Enhydris (Corallus) enhydris	Amazon tree boa, Common water snake, Schneider's water snake
Enhydris hardwicki	Hardwick's water snake
Enhydris smithi	Smith's water snake
Ensatina eschscholtzii	Monterey ensatina
Enhydrina schistose	Beaked seasnake
Enyalioides	Dwarf iguana
Enyaliosaurus	Club tailed iguana
Enyalius	Spinytail iguana
Enyalius catenatus	Wied's fathead anole
Enyalioides laticeps	Guichenot's dwarf iguana, Amazon wood lizard
Enygrus	Python
Ephalophis greyi	Northwestern mangrove seasnake
Ephalophis mertoni	Sea snake
Epicrates angulifer	Maja, Cuban boa
Epicrates cenchria	Rainbow boa, Slender boa
Epicrates colubrinus = Eryx colubrinus	Kenyan sand boa
Epicrates jaculus = Eryx jaculus	Javelin sand boa
Epicrates maurus	Tree boa, Brown rainbow boa
Epicrates striatus	Dominican red mountain boa
Epidalea calmita	See *Bufo calmita*
Epipedobates anthonyi	Anthony's poison arrow frog
Epipedobates bassleri	Pleasing poison frog
Epipedobates bilinguis	Ecuador poison frog
Epipedobates boulengeri	Marbled poison frog
Epipedobates espinosai	Espinosa poison frog
Epipedobates femoralis	Brilliant-thighed poison frog
Epipedobates parvulus	Ruby poison dart frog
Epipedobates pictus	Spot-legged poison frog
Epipedobates pulchripectus	Beautiful-Breasted Poison-Arrow Frog, Beautiful-Breasted Poison Frog, Blue-Breasted Poison Frog
Epipedobates silverstonei	Silverstone's poison frog
Epipedobates tricolor	Phantasmal poison frog, Three-colored poison frog
Epipedobates trivittatus	Simply natural dart frog
Egernia depressa	Pigmy spiny-tailed skink

(continued)

Scientific name	Common name
Epicrates angulifer	Cuban boa
Epicrates cenchria maurus	Rainbow boa
Epicrates striatus	Bahama boa
Eremias argus	Mongolia racerunner
Eremias velox	Wall lizard
Eretmochelys imbricata	Hawksbill turtle
Erpeton tentaculatus	Tentacled snake
Chrysemys picta	Painted turtle
Erythrolamprus aesculapii	Aesculapian false coral snake
Erythrolamprus bizonus	Bright-ringed snake, Double-banded coral snake mimic
Eryx	Old World sand boa
Eryx conicus	Russell's earth-snake, Red earth boa, Red-tailed boa, Russell's sand boa, Sri Lanka boa, Rough-scaled sand boa, Common sand boa
Eryx johnii	Indian sand boa
Eublepharis macularius	Leopard geckos
Eugongylus albofasciatus	Solomon Island ground skink
Eulamprus quoyii	Ensatina
Eumeces anthracinus	Coal skink
Eumeces brevilineatus	Short-lined skink
Eumeces chinensis	Chinese skink
Eumeces fasciatus	Common five-lined skink, Blue-tailed skink
Eumeces gilberti	Gilbert's skink
Eumeces laticeps	Broadhead skink
Eumeces septentrionalis	Northern prairie skink
Eumeces skiltonianus	Western skink
Eunectes	Anaconda
Eunectes murinus	Green anaconda, common anaconda, water boa
Eunectes notaeus	Yellow anaconda
Eurycea nana	San Marcos salamander
Eurycea bislineata	Northern two-lined salamander
Eurycea longicauda	Longtail salamander
Eurycea lucifuga	Cave salamander
Eurycea neotenes	Texas salamander
Eurycea tridentifera	Comal blind salamander
Eutaenia elegans lineolata = *Thamnophis elegans*	Elegant garter snake
Eutaenia macrostemma = *Thamnophis eques*	Mexican garter snake
Eutaenia sirtalis	Garter snake
Eutropis multifasciata	East Indian brown mabuya

(continued)

Scientific name	Common name
Fejervarya (*Limnonectes*) *brevipalmata*	Short-webbed Frog, Pegu Wart Frog, Peters' Frog
Fejervarya cancrivora	Crab-eating frog
Fejervarya (*Limnonectes*) *keralensis*	Verrucose frog
Fejervarya (*Limnonectes*) *limnocharis*	Indian Cricket Frog, Boie's Wart Frog, Field Frog, Indian Rice Frog, Rice Frog, Paddy Frog, Cricket Frog, Terrestrial Frog, White-lined Frog
Fejervarya (*Tomopterna*) *rufescens*	Parambikulam frog, Malabar wart frog
Furcifer pardalis	Panther chameleon, See *Chameleo pardalis*
Furina annulata	Bandy-bandy, Black-and-white-ringed snake
Gallotia galloti	Canarian lizard
Gambelia	Leopard lizard
Gastrophryne	Narrowmouth toads
Gastrophryne carolinensis	Eastern narrow-mouthed toad
Gastrophryne carolinensis	Narrow-mouthed toad
Gavialis gangeticus	Ganges gavial
Geckolepis	
Gehyra oceanica	Oceanic gecko
Gekko	Gecko
Gekko gekko	Tokay gecko
Gekko japonicus	Schlegel's Japanese Gecko
Gekko swinhonis	Peking gecko
Gekko verticillatus	Taukte lizard
Gehyra variegata	Variegated dtella, Tree dtella
Geocalamus	Kenya worm lizard
Geochelone carbonaria	South American red-legged tortoise, Red-footed tortoise
Geochelone denticulate	South American yellow-footed tortoise, Brazilian giant tortoise
Geochelone dussumieri	Aldabra giant tortoise
Geochelone elegans	Star tortoise
Geochelone elephantopus vanderburghi	Galapagos giant tortoise
Geochelone (*Testudo*) *gigantea*	Aldabra tortoise
Geochelone nigra	Galapagos tortoise
Geochelone pardilis	Leopard tortoise
Geochelone pardalis babcocki	African leopard tortoise
Geochelone platynota	Burmese star tortoise

(continued)

Scientific name	Common name
Geochelone radiate	Radiated tortoise
Geochelone sulcata	African spur-thighed tortoise, See also *Centrochelys sulcata*
Geoclemys reevesii	Reeves' turtle
Geocrinia victoriana	Eastern smooth frog, Victorian smooth froglet
Geoemyda punctularia punctularia	Spotted-legged turtle
Geoemyda spengleri	Black-breasted leaf turtle, Vietnamese leaf turtle
Geophis nasalis	Coffee earth snake
Geotriton fuscus = *Speleomantes ambrosii*	Ambrosi's cave salamander, Spezia cave salamander
Gerrhonotus m. multicarinatus	Red-backed alligator lizard
Gerrhonotus taeniatus	Bromeliad arboreal alligator lizard
Gerrhosaurus	Plated lizard
Gerrhosaurus major	Great-plated lizard
Gerrhosaurus validus	Great plated lizard, Giant plated lizard
Gloydius (*Agkistrodon*) *blomhoffi*	Japanese pit viper ("Mamushi" in Japanese)
Glyptemys insculpta	Wood turtle, See *Clemmys insculpta*
Gonatodes albogularis	Yellow-headed gecko
Gonatodes concinnatus	Collared forest gecko
Gonatodes humeralis	Bridled forest gecko
Gongylophis conicus	Rough-scaled sand boa, Rough-tailed sand boa
Gongylus ocellatus	Walzenskink
Goniurosaurus kuroiwae kuroiwae	Kuroiwa's ground gecko, Tokashiki gecko
Gopherus	Gopher tortoise
Gopherus agassizii	Desert tortoise
Gopherus berlandieri	Texas tortoise
Gopherus polyphemus	Gopher tortoise
Graptemys flavimaculata	Yellow-blotched map turtle
Graptemys geographica	Common map turtle, Northern map turtle
Graptemys gibbonsi	Pascagoula map turtle
Graptemys nigrinoda	Ringed map turtle, Black-knobbed map turtle
Graptemys pseudogeo graphica kohnii	Mississippi map turtle
Graptemys (*Malaclemys*) *pseudogeographica ouachitensis*	Ouachita map terrapin
Graptemys ouachitensis	Ouachita map terrapin
Graptemys pulchra	Alabama map turtle

(continued)

Scientific name	Common name
Grammatophora (*Amphibolurus*) *barbata*	Bearded lizard
Gymnodactylus	Brazilian gecko
Gymnodactylus geckoides	Naked-toed gecko
Gymnophthalmus multisculcatus	Microteiid lizard
Gymnophthalmus underwoodi	Underwood's spectacled tegu
Hadrianus corsoni	Fossil turtle
Halbrookia	Lesser earless lizard
Haplocercus spinosus	Lizard-rosette
Hatteria punctata	Tuatara, Great fringed lizard of New Zealand
Helicops carinicauda infrataeniata	Wied's Keelback
Helminthophis albirostris	Whitenose blind snake
Heloderma horridum	Beaded lizard
Heloderma suspectum	Gila monster
Hemachatus haemachatus	Rinkhals
Hemibungarus (*Calliophis*) *japonicus japonicus*	Japanese coral snake
Hemidactylium	Spiny salamander
Hemidactylum scutatum	Four-toed salamander
Hemidactylus brooki	Common wall lizard
Hemidactylus flaviridis	Yellow-bellied house gecko, Indian wall lizard
Hemidactylus flaviviridis	House gecko
Hemidactylus frenatus	House gecko
Hemidactylus flaviridis	House gecko
Hemidactylus garnotii	Indo-Pacific gecko
Hemidactylus gleadovii	Asian spiny gecko, Brook's house gecko
Hemidactylus mabouia	Amerafrican house gecko
Hemidactylus palaichthus	Antilles leaf-toed gecko
Hemidactylus turcicus	Turkish gecko
Hemiergis decresiensis	Three-toed earless skink
Hemiergis peronii	Lowlands earless skink, Four-toed mulch skink
Hemitheconyx caudicinctus	African fat-tailed gecko
Heosemys silvatica	Cane turtle
Heosemys spinosa	Spiny turtle
Herpetodryas carinatus	Saint Vincent blacksnake
Herpetodryas flagelliformis	Coach-whip snake
Heterodactylus imbricatus	See *Chirocolus imbricatus*
Heterodon contortrix	Hog-nosed snake
Heterodon platirhinos	Spreading adder, Eastern hog-nosed snake
Heterodon simus	American hog-nosed snake, Southern hog-nosed snake

(continued)

Scientific name	Common name
Heteronotia binoei	Bynoe's gecko, Prickly gecko
Heteronotia binoei	Binoe's prickly gecko
Hieremys annandalii	Yellow-headed temple turtle
Hyla caerulea = *Hila caerulea* = *Litoria caerulea*	White's tree frog
Holarchus octolineatus	Striped Kukri snake, Grace's kukri snake
Holaspis guentheri	Neon blue-tailed tree lizard
Holbrookia maculata ruthveni	Bleached earless lizard
Holbrookia propinqua	Keeled earless lizard
Homalopsis	Water snake
Homalopsis buccata	Fish snake, Puff-faced water snake
Homalopsis schneiderii	New Guinea bockadam, Dog-faced water snake
Homeosaurus maximiliani	Fossil rhynchocephalan
Homopus areolatus	Common padloper, Parrot-beaked tortoise
Homopus boulengeri	Boulenger's cape tortoise, Donner-weer tortoise
Homopus femoralis	Karoo cape tortoise
Homopus signatus	Speckled tortoise, Speckled padloper tortoise
Hoplocephalus stephensii	Stephen's banded snake
Hoplocercus	Weapontail lizard
Hoplodactylus	New Zealand gecko
Hoplodactylus duvaucelii	Duvaucel's gecko
Hydrelaps darwiniensis	Black-ringed sea snake, Port Darwin sea snake
Hydrodynastes gigas	Brazilian false water cobra
Hydrophis	Sea snake
Hydrophis belcheri	Faint-banded sea snake
Hydrophis (Pelamis) bicolor	Yellowbelly sea snake, Pelagic sea snake
Hydrophis brooki	Brooke's sea snake
Hydrophis caerulescens	Dwarf seasnake
Hydrophis curtus	Shaw's sea snake
Hydrophis cyanocinctus	European grass snake
Hydrophis elegans	Elegant sea snake
Hydrophis fasciatus	Striped sea snake
Hydrophis inornatus	Plain seasnake
Hydrophis kingi	King's Sea Snake, Spectacled Sea Snake
Hydrophis klossi	Kloss' sea snake
Hydrophis lapemoides	Persian Gulf sea snake, Arabian gulf sea snake
Hydrophis major	Olive-headed sea snake

(continued)

Scientific name	Common name
Hydrophis mammilaris	Bombay sea snake, Broad banded sea snake
Hydrophis melanosoma	Black-banded sea snake, Robust sea snake
Hydrophis nigrocinctus	Daudin's sea snake
Hydrophis obscurus	Russell's sea snake
Hydrophis ornatus	Ornate reef sea snake
Hydrophis sirtalis	Sea snake
Hydrophis spiralis	Yellow sea snake
Hydrophis stricticollis	Collared sea snake
Hydrophis sublaevis (Distira brugmansii)	Sea snake
Hydrophis sublaevis (cyanocinctus)	Sea snake
Hydrophis torquatus	West coast black-headed sea snake
Hydrosaurus	Sail-finned lizard
Hydrosaurus pustulatus	Philippine sail-tailed water dragon agamid, Eastern water dragon
Hydrosaurus varanus	Sail-finned lizard
Hydrus platurus (Pelamis bicolor)	Bicolor sea snake
Hyla andersonii	Pine Barrens treefrog
Hyla arborea	European treefrog
Hyla arbora	Japanese treefrog
Hyla aurea	Golden bell frog
Hyla chrysoscelis	Cope's gray treefrog
Hyla cinerea	Green treefrog
Hyla chrysoscelis	Cope's gray treefrog
Hyla crucifer	Spring peeper
Hyla ebraccata	Hourglass treefrog
Hyla femoralis	Pine woods treefrog
Hyla gratiosa	Barking treefrog
Hyla meridionalis	Stripeless tree frog
Hyla microcephala meridiana	Yellow cricket tree frog
Hyla nana	Tree frog
Hyla parviceps	Tree frog
Hyla pulchella	Tree frog
Hyla regilla	Northern Pacific tree frog
Hyla sanborni	Tree frog
Hyla squirella	Squirrel treefrog
Hyla versicolor	Eastern gray tree frog
Hylodes asper	Brazilian torrent frog
Hylodes dactylocinus	Leptodactylid frog
Hydromedusa tectifera	Argentine snake-necked turtle
Hymenochirus boettheri	South African dwarf frog
Hynobius (Salamandrella) keyserlingii	Siberian salamander

(continued)

Scientific name	Common name
Hynobius leechii	Korean salamander
Hynobius nigrescens	Japanese black salamander
Hypsiglena torquata achrorhyncha	Night snake
Hypsiglena ochrorhynchus	Coast night snake
Hypsilophus tuberculatus	Green iguana
Ichthyostega	Fossil early tetrapod
Iguana ardiosé	Slate-coloured iguana
Iguana delicatissima	Lesser Antillean iguana
Iguana iguana	Green iguana
Iguana tuberculata	Common iguana of West Indies
Imantodes cenchoa	Blunt-headed tree snake
Iphisa elegans	Glossy shade lizard
K. jerdoni	Jerdon's Sea Snake
Kachuga dhongoka	Three-striped roofed turtle, Batagur dhongoka
Kachuga kachuga	Red-crowned roofed turtle, Batagur kachuga
Kentropyx pelviceps	Forest whiptail lizard
Kerilia annandalei = *Kolpophis annandalei*	Bighead sea snake
Kinixys belliana	Bell's hinge-back tortoise
Kinixys erosa	Forest hinge-back turtle, Serrated hinge-back tortoise
Kinixys homeana	Kuhl's tortoise
Kinosternon acutum	Tabasco mud turtle
Kinosternon alamosae	Alamos mud turtle
Kinosternon angustipons	Central American mud turtle
Kinosternon baurii	Striped mud turtle
Kinosternon flavescens	Yellow mud turtle
Kinosternon subrubrum	Eastern mud turtle
Kolpophis annandalei	Bighead sea snake
Lacerta agilis	Sand lizard
Lacerta armeniaca	Caucasian parthenogenetic rock lizard
Lacerta altissimus = *Liolamus altissimus*	Dusky lizard
Lacerta bipes = *Anguis bipes* = *Scelotus bipes*	
Lacerta bocagei = *Podarcis bocagei*	Bocage's wall lizard
Lacerta caeruleocephala	
Lacerta chloronotus = *Lacerta viridis*	
Lacerta dahli	Caucasian rock lizard
Lacerta dugesii	Madeiran wall lizard

(continued)

Scientific name	Common name
Lacerta laevis	Lebanon lizard
Lacerta lemniscatus = *Ctemidophorus lemniscatus*	Jungle runner
Lacerta lepida	Ocellated lizard, Jeweled lizard
Lacerta major Lacerta cinereus maculatus	Great spotted Jamaican lizard
Lacerta mixta	Rock lizard
Lacerta monticola	Iberian rock lizard
Lacerta muralis	Wall lizard
Lacerta (*Chrysolamprus*) *ocellata*	Ocellated bronze skink
Lacerta rostombekovi	Caucasian rock lizard
Lacerta saxicola	Caucasian rock lizard
Lacerta serpa	Meadow lizard
Lacerta sicula	Italian wall lizard
Lacerta simonyi	El Hierro giant lizard, Simony's lizard
Lacerta strigata	Caspian green lizard, Striated lizard
Lacerta taurica fiumana	Cimean wall lizard
Lacerta teguixin = *Tupinambis teguixin*	Black tegu, Golden tegu
Lacerta trilineata	Giant green lizard
Lacerta unisexualis	Parthenogenetic rock lizard
Lacerta valentini	Rock lizard
Lacerta (*Zootoca*) *vivipara*	Viviparous lizard
Lacerta viridis	Green lizard
Lachesis = *Lachesis*	Bushmaster snake
Lachesis lanceolatus = *Bothrops alcatraz*	Alcatrazes lancehead
Lachesis muta	South American bushmaster
Laemanctus	Serrated casquehead iguana
Lapemis curtus	Shaw's sea snake
Lapemis hardwickii	Spine-bellied sea snake
Lampropeltis alterna	Gray-banded king snake
Lampropeltis californiae	California king snake
Lampropeltis calligaster	Prairie king snake
Lampropeltis doliata triangulum	Aberrant Western milk snake
Lampropeltis getula yumensis	California king snake
Lampropeltis hondurensis	Honduran milk snake
Lampropeltis mexicana	Mexican kingsnake
Lampropeltis (*doliata*) *triangulum*	Milk snake
Lampropeltis t. triangulum	Common milk snake, Eastern milk snake
Lampropeltis zonata	California mountain kingsnake

(continued)

Scientific name	Common name
Lamprophis fuliginosus	Corn snake
Lampropholis	Metallic skinks
Lampropholis delicata	Plague skink
Lampropholis guichenoti	Common garden skink, Pale-flecked garden sunskink
Laticauda	Sea krait
Laticauda colubrina	Banded sea snake
Laticauda laticaudata	Common sea krait, Blue-lipped sea krait
Laticauda schistorhynchus	Flat-tail sea snake
Laticauda semifasciata	Black-banded sea krait
Laudakia stellio	Stellion, hardim, star lizard, starred agama
Leimadophis poecilogyrus	Cobra de capim
Leiolopisma telfairii	Telfair's skink
Leiolopisma unicolor	Ground skink
Leiocephalus	Curly-tailed lizard
Leiocephalus schreibersi	Curly-tailed lizard
Leiolopisma telfairii	Telfair's skink
Leiosaurus	Southern anole
Lepidobatrachus llanensis	Escuerzo frog
Lepidochelys kempii	Kemp's Ridley sea turtle, Atlantic Ridley turtle
Lepidochelys olivacea	Olive Ridley Turtle, Pacific Ridley turtle
Lepidodactylus lugubris	Mourning gecko
Lepidophyma flavimaculatum	Tropical night lizard
Lepidosternum affine	Smallhead worm lizard
Leposoma parietale	Common forest lizard
Leposoma percarinatum	Muller's tegu
Leposternon	South American worm lizard
Leptobrachium (Megalophrys) carinense = Brachytarsophrys carinense	Karin hills frog, Broad-headed short-legged toad
Leptodactylus melanotus = Leptodactylus latrans	Llanos toad-frog, Creole frog, Butter frog
Leptodactylus ocellatus	Common thin-toed frog
Leptodactylus pentadactylus	South American bullfrog
Leptodactylus pentadactylus labyrinthicus	Labyrinth White-lipped Frog
Leptodeira annulata	Banded cat-eyed snake
Leptodeira annulata ashmeadii	Banded cat-eyed snake
Leptodira hotamboeia	Herald snake, Red-lipped snake
Leptopterygius trigonodon	Ichthyosaur (Fossil)
Lepidosternum affine = Amphisbaena microcephalum	Smallhead worm lizard
Leposternon microcephalum	Smallhead worm lizard

(continued)

Scientific name	Common name
Leposternon polystegum	Bahia worm lizard
Leposternon scutigerum	Shielded worm lizard
Leposternon wuchereri	Wucherer's Worm Lizard
Leptosuchus imperfecta	Fossil parasuchian reptile
Leptotyphlops	Thread snake
Leptotyphlops bicolor	African worm snake
Lerista	Scincid lizard
Leurognathus	
Lialis	Snake lizard
Liasis papuanus	Guinea olive colored python
Liasis amethystinus	Scrub python
Liasis childreni	Children's python
Lichanura	Rosy boa
Lichanura roseofusca	Rosy boa
Lichanura roseofusca (trivirgata) roseofusca	Desert rosy boa
Limnodynastes tasmaniensis	Spotted grass frog
Limnonectes (Fejervarya) brevipalmata	Short-webbed Frog, Pegu Wart Frog, Peters' Frog
Limnonectes (Fejervarya) keralensis	Verrucose frog
Limnonectes kuhlii	Kuhl's creek frog, Large-headed frog
Limnonectes (Fejervarya) limnocharis	Indian Cricket Frog, Boie's Wart Frog, Field Frog, Indian Rice Frog, Rice Frog, Paddy Frog, Cricket Frog, Terrestrial Frog, White-lined Frog
Lineatriton	Mexican plethodontid salamander
Liolaemus	Tree iguanid
Liolaemus altissimus	Dusky lizard
Liolaemus chiliensis	Chilean tree iguana
Liolaemus fuscus	Brown tree iguana
Liolaemus lemniscatus	Lagartija
Liolaemus leopardinus	Leopard tree iguana
Liolaemus lutzae	Lutz's tree iguana
Liolaemus monticola	Peak tree iguana
Liolaemus nigromaculatus	Zapallaren tree iguana
Liolaemus nigroviridis	Black-green tree iguana
Liolaemus nitidus	Shining tree iguana
Liolaemus platei	Braided tree iguana
Liolaemus schroederi	
Liolaemus tenuis	Jeweled lizard, Slender lizard, Thin lizard
Leiolopisma platynota	Red-throated cool-skink, Red-throated skink
Liophis almadensis	Almaden ground snake

(continued)

Scientific name	Common name
Liophis miliaris	Cobra-lisa, Cobra-de-banhado, Cobra-dágua
Liophis perfuscus	Barbados racer, Tan ground snake
Liotyphlops albirostris	White-nosed blind snake
Lissemys punctata	Indian flapshell turtle
Lissemys scutata	Burmese flapshell turtle
Lissemys punctata punctata	Indian flapshell turtle, Spotted flapshell turtle
Lithobates pipiens	Northern leopard frog
Lithobates sylvaticus	Wood frog
Litoria aurea	Australian hylid frog
Litoria caerulea	Green treefrog
Litoria ewingii	Brown tree frog
Litoria fallax	Eastern dwarf treefrog
Litoria genimaculata	Green-eyed treefrog
Litoria infrafrenata	Australian giant frog, White-lipped treefrog
Litoria nannotis	Waterfall frog, Torrent tree frog
Litoria rheocola	Common mist frog
Lophognathus temporalis	Striped water dragon
Lophura amboinensis	Malaysian sail-finned lizard
Loveridgea	Tanzania worm lizard
Lucasius	Australian gecko
Lygodactylus	Dwarf gecko
Lycodon	Wolf snake
Lycodon aulicus	Indian wolf snake, Common wolf snake
Lycophidion c. capense	Cape wolf snake, Cape common wolf snake
Lygodactylus chobiensis	Chobe dwarf gecko, Okavango dwarf gecko
Lygodactylus klugei	K'uge's dwarf gecko
Lygosoma laterale	Ground skink
Lygosoma telfairi	Round Island skink
Mabuya agilis	Agile mabuya
Mabuya carinata	Keeled Indian mabuya, Many-keeled grass skink, Golden skink
Mabuya dissimilis	Striped grass mabuya, Striped grass skink
Mabuya frenata	Cope's Mabuya
Mabuya heathi	Brazilian mabuya
Mabuya macrorhyncha	Hoge's mabuya
Mabuya multifasciata	Doubled lizard, Indonesian skink
Mabuya (*Trachylepis*) *striata*	Striped skink
Mabuya striata punctatissimus	Montane speckled skink

(continued)

Scientific name	Common name
Mabuya striata striata	Common striped skink, African striped skink
Machimosaurus mosae	Fossil crocodile
Macroclemys temminckii	Alligator snapping turtle
Malaclemys centrata	Diamondback terrapin
Malaclemys (*Graptemys*)	
pseudogeographica ouachitensis	Ouachita map turtle
Malaclemys pileata littoralis	Texas diamondback terrapin
Malaclemys terrapin	Diamondback terrapin
Malacocherus tornieri	Pancake tortoise
Malpolon m. monspessulanus	Montpellier snake
Manculus quadrigitatus	Fossil
Manouria emys	Burmese brown tortoise
Mauremys caspica	Caspian turtle, See *Clemmys caspica*
Mauremys leprosa	Mediterranean pond turtle, See *Clemmys leprosa*
Mauremys mutica	Asian yellow pond turtle
Mauremys reevesii See *Chinemys reevesii*	
Megalobatrachus japonicus	Giant Japanese salamander
Megalobatrachus maximus	Giant salamander
Megalophrys (*Leptobrachium*) *carinense* = *Brachytarsophrys carinense*	Karin hills frog, Broad-headed short-legged toad
Megalophrys nasuta	Long-nosed horned frog, Asian horned frog, Asian leaf frog
Megalophrys parva	Lesser stream horned frog
Melanochelys trijuga thermalis	Indian black turtle, Ceylon pond turtle
Melanophidium	Shield tail snake
Melanophryniscus	South American redbelly toad
Melanosuchus niger	Black caiman
Menopoma alleghaniense = *Cryptobranchus alleganiensis*	Hellbender
Menobranchus lateralis = *Necturus maculatus*	Mud puppy
Meroles anchietae	Shovel-snouted lizard
Mesalina olivieri	Olivier's sand lizard
Mesalina guttulata	Small spotted lizard
Mesaspis gadovii	Gadow's alligator lizard
Mesobaena	Indian worm lizard
Microcephalophis cantoris	Günthers sea snake

(continued)

Scientific name	Common name
Microcephalophis gracilis	Slender sea snake
Microhyla ornata	Ornate narrow-mouthed frog
Microhyla rubra	Red narrow-mouthed frog
Microlophus bilineatus	Galapagos lava lizard
Microsaura ventralis	Dwarf chameleon
Micruroides euryxanthus	Western coral snake, Arizona coral snake
Micrurus	Eastern coral snake
Micrurus fulvius	Eastern coral snake
Molge alpestris	Alpine newt
Molge cristata = Triturus cristata	Great crested newt
Molge marmoratus = Triturus marmoratus	Marbled newt
Molge taeniata	Smooth newt
Molge vulgaris = Lissotriton vulgaris	Common smooth newt
Molochus horridus	Horny devil, Thorny dragon, Mountain devil, The Moloch
Monopeltis	Worm lizard
Morelia oenpelliensis	Oenpelli python
Morelia spilota macdowelli	Costal Queensland carpet python
Morelia spilota variegate	Northwest carpet python
Morelia (Chondropython) viridis	Green tree python
Morenia petersi	Indian eyed turtle
Morethia boulengeri	Boulenger's skink
Morunasaurus	Spiny-tailed lizard
Myron richardsonii	Richardson's mangrove snake
Mystriosuchus plieningeri	Fossil phytosaur
Naja	Cobra
Naja bungarus	King cobra
Naja hannah	Hamadryad, King cobra
Naja melanoleuca	Black cobra, Forest cobra
Naja naja sputatrix	Javanese cobra
Naja nigricollis	Spitting cobra
Naja tripudians	Cobra di capello
Narudasia festiva	Festive gecko
Natriciteres	Marsh snake
Natriciteres variegata sylvatica	Forest marsh snake
Natrix fasciata	Water snake
Natrix helvetica	Grass snake
Natrix laurenti	Laurenti's grass snake
Natrix maura	Viperine snake
Natrix natrix	Grass snake
Natrix piscator	Checkered water snake

(continued)

Scientific name	Common name
Natrix rhombifera	Diamondback water snake
Natrix septemvittata	Queen snake, Queen water snake
Natrix sipedon	Common water snake, Northern water snake
Natrix sipedon clarkia	Gulf water snake, Gulf salt marsh snake
Natrix sipedon sipedon	Northern water snake
Natrix tessellate	Dice snake
Natrix (Rhabdophis) tigrina	Tiger keelback
Natrix viperina	Smooth viper
Natrix vibakari (Amphiesma vibakari)	Asian keelback
Necturus maculosus	Mud puppy
Neobatrachus centralis	Trilling frog
Nephrurus laevis	Smooth knobtail gecko
Nephrurus asper	Rough knob-tailed gecko
Nephrurus laevissimus	Knob-tailed lizard
Nerodia clarkii compressicauda	Mangrove salt marsh snake
Nerodia fasciata	Southern water snake, Banded water snake
Nerodia rhombifer	Diamondback water snake
Nerodia sipedon insularum	Lake Erie water snake
Nerodia s. sipedon	Common water snake, Northern water snake
Neusticurus ecpleopus	Common stream lizard
Niveoscincus metallicus	Metallic skink
Norops biporcatus	Green tree anole
Norops fuscoauratus	Brown-eared anole, Slender anole
Norops humilis	Ground anole
Norops limifrons	Slender anole
Norops nitens = Anolis nitens	Yellow tongued forest anole
Notechis scutatus	Tiger snake
Notochelys platynota	Malayan flat-shelled turtle
Notophthalmus viridescens	Red-spotted newt, Eastern newt
Ocadia sinensis	Chinese striped-neck turtle
Occidozyga lima	Floating spotted frog
Odontophrynus americanus	See *Ceratophrys americana*
Oedipina poelzi	Gamboa worm salamander
Oedipina uniformis	Common worm salamander
Oedua ocellata	Eyed gecko
Oligosoma suteri	Egg-laying skink, Suter's skink
Omolepida	Giant slender bluetongue

(continued)

Scientific name	Common name
Opheodrys aestivus	Rough green snake, Rough-scaled green snake, Common grass snake
Ophisaurus ventralis	Eastern glass snake
Ophibolus doliatus triangulus	Milk snake, Garter snake, House snake
Ophibolus getulus	Common chain snake, Common king snake
Ophibolus triangulus	Milk snake, Chicken snake, House snake
Ophiodes striatus	Glass lizard
Ophiomorus	Snake skink
Ophisaurus	Glass lizard
Ophisaurus apodus	European legless lizard
Ophisaurus compressus	Island glass lizard
Ophisaurus gracilis	Burma glass lizard, Asian glass lizard
Ophisaurus moguntinus	Fossil anguid lizard
Ophisaurus ventralis	Eastern glass snake
Ophryoëssa superciliosa	Sword-tailed iguana
Ophryoessoides	Keeled iguana
Oplurus	Malagasy iguanian lizard
Oplurus cuvieri	Cuvier's Madagascar swift, Madagascan spiny-tailed iguana
Oplurus torquatus	Madagascar spiny-tailed iguana
Osteocephalus oophagus	Amazonian treefrog
Osteolaemus tetraspis	Broad fronted crocodile, West African dwarf crocodile
Osteopilus	West Indian treefrog
Otodus (Tupinambis) nigropunctatus	Black tegu
Otocryptis wiegmanni	Brown-patched kangaroo lizard, Sri Lankan kangaroo lizard
Pachycalamus brevus	Short worm lizard
Pachydactylus	Thick-toed gecko
Pachydactylus c. vansoni	Van Son's thick-toed gecko
Pachytriton breviceps	Paddletail salamander, Tsitou newt
Pagona vitticeps	Central bearded dragon, Inland bearded dragon
Palaeobatrachus	Fossil frog
Palaeosuchus niloticus	Nile crocodile
Palmatogecko rangei	Web-foot geckos
Pantherophis guttatus	Corn snake
Paramesotriton chinensis	Chinese warty newt

(continued)

Scientific name	Common name
Paramesotriton deloustali	Tam Dao warty newt, Tam Dao salamander
Paroedura cf. *bastardia*	Madagascar spiny tailed gecko, Malagasy forest gecko
Pelamis bicolor (Hydrophis platurus)	Yellow-bellied sea snake
Pelamis platurus	Yellowbelly sea snake
Pelias (Vipera) berus	Common European adder
Pelobates cultripes	Western spadefoot toad
Pelobates fuscus	Common spadefoot toad
Pelodryas caerulea	White's tree frog
Pelomedusa	African helmeted turtle
Peltocephalus dumerilianus	Big-headed Amazon river turtle, Big-headed side neck turtle
Peltophryne lemur	Puerto Rican crested toad
Peltophryne peltocephala	See *Bufo peltocephalus*
Pelusios adansoni	Adanson's mud turtle
Pelusios marani	Gabon mud turtle
Pelusios subniger	Blackish terrapin
Peromelus monoscelus	with only one posterior limb
Petrosaurus	California rock lizard
Petuophis melanoleucus catenifer	Pacific gopher snake
Phaeognathus	Red Hills salamander
Phelsuma abbotti chekei	Cheke's day gecko
Phelsuma abbotti sumptio	Assumption Island day gecko
Phelsuma cepediana	Blue-tailed day gecko
Phelsuma comorensis	Comoran gecko
Phelsuma dubia	Dull day gecko
Phelsuma guentheri	Rick Island gecko, Round Island gecko
Phelsuma laticauda laticauda	Gold dust day gecko
Phelsuma lineata chloroscelis	Lined day gecko
Phelsuma madagascariensis	Madagascar giant day gecko, Madagascar day gecko
Phelsuma m. grandis	Giant Madagascar day gecko
Phelsuma pusilla pusilla	Madagascar gecko
Phelsuma astriata astriata	Seychelles day gecko, Seychelles small day gecko, Stripeless day gecko
Phelsuma standingi	Standing's day gecko
Phelsuma sundbergi sundbergi	Chacoan monkey tree frog

(continued)

Scientific name	Common name
Phenacocaurus	Anole
Philodryas olfersii	Lichtenstein´s green racer
Philodryas patagoniensis	Patagonia green racer
Philodryas schottii	Schott's green racer
Phryne vulgaris = Bufo bufo	
Phrynohyas hebes	Golden-eyed tree frog
Phrysoma	Horned Lizard
Phrynosoma douglasii	Tapaja, Pygmy short-horned lizard
Phymaturus	Mountain lizard
Phryniscus cruciger	Rancho grande harlequin frog or striped little frog
Phrynhoidis asper	See *Bufo asper*
Phrynocephalus interscapularis	Toadhead agama
Phrynops geoffroanus	Geoffroy's sideneck turtle
Phrynops gibbus	Gibba turtle
Phrynops nasutus	Common toad-headed turtle
Phrynops williamsi	William's side-necked turtle
Phrynops zuliae	Zulia toad-headed turtle, Zulia toad-headed sideneck turtle
Phrynosaura	Tree iguana
Phrynosoma cornutum	Coast horned lizard, Texas horned lizard
Phrynosoma douglasi	Short-horned lizard, Tapaja, Pygmy short-horned lizard
Physalaemus pustulosus	Tungara toad
Phyllobates bicolor	Black-legged poison frog
Phyllobates lugubris	Lovely poison frog
Phyllobates terribilis	Golden poison frog
Phyllobates vittatus	Golfo Dulce poison frog
Phyllodactylus marmoratus	Marbled gecko
Phyllodactylus marmoratus	Marbled gecko
Phyllomedusa	Leaf frog
Phyllomedusa hypocondrialis	Orange-legged leaf frog
Phyllomedusa sauvagei	Waxy tree frog
Phyllomedusa sauvagei	Chacoan monkey tree frog
Phyllomedusa cf. *tarsius*	Brown belly leaf frog
Phyllomedusa tomopterna	Tiger leg tree monkey
Phyllopezus pollicaris	Chaco leaf-toed gecko
Phyllorhynchus decurtatus perkinsi	Spotted leaf-nosed snake
Phyllurus cornutus	Australian leaf-tailed gecko

(continued)

Scientific name	Common name
Phyllurus platurus	Australian "rupicolous" gecko
Phyllyllotes bassleri	Pleasing poison frog
Physalaemus pustulosus	Túngara frog
Physignathus	Water dragon
Physignathus cocincinus	Thai water dragon, Chinese water dragon
Physignathus lesuerii	Eastern water dragon
Phyton sebae	African rock python
Pipa	Surinam toad
Pipa americana	Surinam toad
Pituophis catenifer	Gopher snake, Blow snake, Pine snake, Western bull snake
Pituophis catenifer annectens	San Diego gopher snake
Pituophis conifer	Gopher snake
Pituophis deserticola	Great Basin gopher snake
Pituophis melanoleucus	Eastern pine snake, Bull snake
Pituophis melanoleucus affinis	Gopher snake
Pituophis melanoleucus catenifer	Pacific gopher snake
Pituophis melanoleucus deserticola	Great Basin gopher snake
Pituophis m. melanoleucus	Northern pine snake
Pituophis sayi = Coluber melanoleucas	Bull snake
Pituophis wilksei	Bull snake, Pine snake, Gopher snake
Pixicephalus	African bullfrog, Burrowing bullfrog
Pixicephalus delalandii	Pixie frog
Platydactylus guttatus	Fan-fingered gecko
Platemys macrocephala	Big-headed Pantanal Swamp turtle
Platemys platycephala	Twist-necked turtle, South American flat-headed turtle
Platydactylus mauritanicus	*Tarentola mauritanica* Moorish gecko
Platynotus	Lava lizard
Platynotus semitaeniatus	Striped lava lizard
Platyplecturus madurensis	Shield-tailed snake
Platysternum megacephalum	Big-headed turtle
Plecturus	African burrowing snake
Plethodon albagula	Western slimy salamander
Plethodon cinereus	Red-backed salamander
Plethodon glutinosus	Northern slimy salamander

(continued)

Scientific name	Common name
Plethodon idahoensis	Coeur d'Alene salamander
Plethodon jordani	Jordan's salamander
Plethodon neomexicanus	Jemez Mountains salamander
Plethodon ouachitae	Rich Mountain salamander
Pleurodeles waltl	Spanish newt, Ribbed newt
Plica	Tree runner lizard
Pliocercus	False coral snake
Pliocercus elapoides	False coral snake
Podarcis bocagei	Bocage's wall lizard, Lagarta galega
Podarcis hispanica	Iberian wall lizard
Podarcis lilfordi	Lilford's wall lizard
Podarcis milensis	Milos wall lizard
Podarcis muralis	Common wall lizard
Podarcis pityusensis	Ibiza wall lizard
Podarcis raffonei	Aeolian wall lizard
Podarcis sicula	Italian wall lizard
Podinema teguixin	Red tegu or teju
Podocnemis erythrocephala	Red-headed Amazon River turtle, Red-headed Amazon side-necked turtle
Podocnemis expansa	Arrau river turtle, South American river turtle
Podocnemis unifilis	Terekay turtle, Yellow-spotted Amazon River turtle
Podocnemis vogli	Savanna side-necked turtle
Podonemis	South American river turtle
Pogona barbata	Common bearded dragon, Eastern bearded dragon
Pogona minor	Dwarf bearded dragon
Pogona vitticeps	Inland bearded dragon, Central bearded dragon
Polychroides	= *Polychrus*
Polychrus	Bush anole
Polychrus acutirostris	Brazilian bush anole
Polyodontophis subpunctatas	
Polypedates maculatus	Common Indian tree frog, Chunam tree frog
Prionodactylus oshaugnessyi	White-striped eyed lizard
Proceratophrys boiei	See *Ceratophrys boiei*
Proteus anguineus	Olm, Proteus
Prototretus = *Stenocercus*	Whorl-tail iguana
Prototretus pectinatus = *Stenocercus*	pectinatus, collared lizard

(continued)

Scientific name	Common name
Psammobates geometricus	Geometric tortoise
Psammobates oculifera	Serrated tortoise
Psammobates tentorius	Berger's cape tortoise, African tent tortoise
Psammodromus algirus	Large psammodromus
Psammodromus hispanicus	Spanish psammodromus
Psammophis	Sand snake, Sand racer
Psammophilus dorsalis	Peninsular rock agama
Psammophis dromophis = *Psammophis lineatus,*	Lined olympic snake, Striped swamp snake
Psammophis notostictus	Karoo Sand Snake
Psammophis phillipsii	Olive Grass Racer, Phillips' Sand Snake; Olive Grass Racer, Phillips' Sand Snake
Psammophis schokari	Schokari sand snake, Rear-stripped snake
Psammophis sibilans sibilans	Short-snouted grass snake
Psammophylax	Skaapsteker
Pseudacris crucifer	Spring peeper
Pseudacris feriarum	Upland chorus frog
Pseudacris regilla	Pacific treefrog, Pacific chorus frog
Pseudacris triseriata	Chorus frog
Pseudechis guttatus	Spotted black snake
Pseudechis porphyriacus	Red-bellied snake
Pseudemys concinna	Eastern river cooter
Pseudemys (*Ptychemys*) *elegans*	Red-eared slider turtle
Pseudemys floridana	Florida terrapin, Florida cooter
Pseudemys (*Chrysemys*) *floridana*	Florida cooter, Coastal cooter
Pseudemys nelsoni	Western chorus frog, Florida red-bellied cooter
Pseudemys ornata callirostris	ornata callirostris
Pseudemys scripta elegans	Elegant terrapin
Pseudemys scripta scripta	Yellow-bellied terrapin, Serrated terrapin
Pseudemys scripta troostii	Illinois slider turtle
Pseudemys texana	Texas river cooter
Pseudemys troostii	Cumberland slider
Pseudepidalea variabilis	See *Bufo variabilis*
Pseudepidalea viridis	See *Bufo viridis*
Pseudis paradoxa	Paradox frog
Pseudoeurycea brunnata	Brown False Brook Salamander
Pseudoeurycea goebeli	Goebel's false brook salamander

(continued)

Scientific name	Common name
Pseudoeurycea leprosa	Mexican salamander Tlaconete leproso
Pseudoeurycea rex	Royal false brook salamander
Pseudogonatodes guianensis	Amazon pigmy gecko
Pseudonaja affinis	Dugite
Pseudopus	European glass lizard
Pseudotrapelus	Agama
Pseudotriton rubor	Red salamander
Ptenopus garrulous	Barking gecko
Ptenopus	Chirping gecko
Ptyas mucosus	Oriental ratsnake, common ratsnake
Ptychadena	Ridged frog
Ptychemys (Pseudemys) elegans	= Trachemys scripta elegans
Ptychozoon homalocephalum	Flying gecko, Fringed gecko
Ptychozoon kuhli	Kuhl's flying gecko
Pygodactylus gronovii	Striped worm lizard
Ptyodactylus guttatus	Fan-fingered gecko
Ptyodactylus hasselquistii	Kramer's yellow fan-fingered gecko, Fan foot gecko, Egyptian fan-footed gecko
Pygomeles	Short skink
Pygopus	Australian scaly-foot
Pygopus lepidopodus	long Kur-ring gai, Common scaly foot lizard
Pyidea mouhotii = Pyxidea mouhotii	Keeled box turtle
Python curtus curtus	Sumatran blood python
Python molurus bivittatus	Burmese python, Indian python, Ceylon python, Dark Indian python
Python regius	Ball python, Royal python
Python reticulatus	Reticulated python, Regal python
Python sebae	African rock python
Python spilotes	Carpet snake, Diamond python, Western Australian carpet python
Python tigris	Tiger boa
Python timorensis	Timor python
Pyxicephalus adspersus	Tschudi's African bullfrog
Pyxis arachnoides	Spider tortoise
Raja clavata	Thornback ray
Rhamphiophis rostratus	Rufous-beaked snake
Rana adspersa	Numskull frog
Rana aerosa	
Rana alticola = Clinotarsus alticola	Assam Hills Frog

(continued)

Scientific name	Common name
Rana amurensis	Siberian wood frog
Rana arvalis	Swamp frog, Moor frog
Rana aurora draytonii	Northern red-legged frog, California red-legged frog
Rana berlandieri	Rio Grande leopard frog
Rana blairi	Plains leopard frog
Rana boylii	Yellow-legged frog, Foothill yellow-legged frog
Rana brevipoda	Japanese frog
Rana cascadae	Cascades frog
Rana catesbeiana	North American bullfrog
Rana chalconota	Copper-cheeked frog, Schlegel's Java frog
Rana chensinensis	Inkiapo frog, Asiatic glass frog
Rana clamitans	Green frog
Rana clumata	Green frog
Rana cyanophylictis	Skipper frog, Water skipping frog
Rana esculenta = Rana viridis	Edible frog
Rana esamelica	Indian bull frog
Rana fusca	Brown gras frog
Rana graeca	Greek stream frog, Balkan stream frog
Rana grylio	Pig frog
Rana halecina	Common European green frog, Shad frog
Rana halecinum	Indian bullfrog
Rana heckscheri	River frog
Rana hexadactyla	South Indian green frog
Rana hosii	Poisonous rockfrog, Hose's rock frog
Rana iberica	Iberian frog
Rana japonica	Japanese brown frog
Rana japonica japonica	Japanese brown frog
Rana lessonae	Pool frog
Rana luteiventris	Columbia spotted frog
Rana mehelyi	Fossil frog
Rana muscosa	Mountain yellow-legged frog
Rana mugicus	= Rana catesbeiana
Rana mugiens	Bullfrog
Rana nigromaculata	Dark-spotted frog
Rana okinavana	Kampira falls frog, Yaeyama harpist frog, Harpist brown frog
Rana ornativentis	Montana brown frog
Rana palmipes	Amazon River frog
Rana palustris	Pickerel frog

(continued)

Scientific name	Common name
Rana perezi	Perez's frog, Iberian pond frog
Rana pipiens	Northern leopard frog
Rana pretiosa	Oregon spotted frog
Rana ridibunda	Lake frog, Marsh frog
Rana septentrionalis	Mink frog
Rana sevosa	Gopher frog
Rana sphenocephala	Southern leopard frog, Florida leopard frog
Rana subaquavocalis	Ramsey Canyon leopard frog
Rana sylvatica	Wood frog
Rana temporaria	Latvian common frog
Rana tigrina	Indian bullfrog, Indus Valley bullfrog
Rana uticularia	Southern leopard frog
Rana virescens	Spring frog, Leopard frog
Rana virgatipes	Carpenter frog
Rana viridis = *Rana esculenta*	Edible frog
Ranodon sibiricus	Chiricahua leopard frog
Regina liberis	Yellow-bellied or Leather snake
Regina septemvittata	Queen snake
Rhabdophis tigrinus	Tiger keelback, yamaka-gashi Asian tiger snake
Rhabdophis (Natrix) tigrina	Oriental tiger snake
Rhacheosaurus gracilis	Fossil crocodile
Rhacodactylus	Caledonian lizard
Rhacodactylus auriculatus	Gargoyle gecko
Rhacodactylus trachyrhynchus	Rough-snouted gecko
Rhacoëssa	Type of Australian gecko
Rhacophorus javanus	Java treefrog
Rhadinaea decorata	Adorned graceful brown snake, Litter snake
Rhadinaea flavilata	Pine woods snake
Rhamphiophis rostratus	Rufous beaked snake
Rheodytes leukops	Fitzroy River turtle
Rhinechis scalaris	Ladder snake
Rhinella arenarum	See *Bufo arenarum*
Rhinella crucifer	See *Bufo crucifer*
Rhinella ictericus	See *Bufo ictericus*
Rhinella marina	See *Bufo marinus*, *Chaunus marinus*
Rhineura	Worm lizard
Rhinocheilus lecontei	Long-nosed snake
Rhinoclemmys annulata	Brown land turtle, Brown wood turtle
Rhinoclemmys areolata	Mojina, Furrowed wood turtle

(continued)

Scientific name	Common name
Rhinoclemmys diademata	Maracaibo wood turtle
Rhinoclemmys funerea	Black River turtle
Rhinoclemmys nasuta	Large-nosed wood turtle
Rhinoclemmys punctularia	Spot-legged turtle, South American wood turtle
Rhinoclemmys pulcherrima	Ornate wood turtle, Mexican wood turtle, Painted wood turtle, Central American wood turtle
Rhinoclemmys rubida	Mexican spotted wood turtle
Rhinophis	Rhinocerps earth snake
Rhinophrynus dorsalis	Mexican burrowing frog
Rhoptropus afer	Namib day gecko
Sabnis - ? Trionyx sabnis	
Sacalia bealei	Beal's-eyed turtle
Salamandra atra	Alpine salamander
Salamandra cristata	Southern crested newt
Salamandra maculosa	Spotted salamander
Salamandra salamandra	Common fire salamander
Salamandrella (Hynobius) keyserlingii	Siberian salamander
Salamandrella keyserlingii	Siberian newt
Salamandrina perspicillata	(perspicillata) Spectacled salamander
Salamandrina terdigitata	Southern Spectacled salamander
Salvadora grahamiae	Mountain patchnose snake
Sanzinia	Boa
Sanzinia madagascariensis	Madagascar tree boa
Sator grandavus	Island sator
Sauresia sepsoides	Serpentine four-toed galliwasp
Sauromalus	Chuckwalla
Sauromalus obesus	Western chuckwalla, Desert chuckwalla
Sauromalus showi	Chuckwalla
Sauromalus varius	San Esteban Island chuckwalla lizard, Painted chuckwalla
Scaphiodontophis venustissimus	Common neckbandsnake
Scaphiopus	North American Spadefoot, Southern spadefoot
Sceloporus	Spiny Lizard
Sceloropus biseriatus	Western swift
Sceloporus cyanogenys	Blue spiny lizard
Sceloporus graciosus	Common sagebrush lizard
Sceloporus jarrovii	Yarrow's spiny lizard
Sceleporus magister	Desert spiny lizard

(continued)

Scientific name	Common name
Sceloporus merriami	Canyon lizard
Sceloporus occidentalis biseriatus	San Joaquin fence lizard
Sceloporus olivaceus	Rusty lizard
Sceloporus poinsettii	Crevice spiny lizard
Sceloporus scalaris	Bunch grass lizard
Sceloporus undulatus	Common fence lizard, Eastern fence lizard
Sceloporus undulatus consorbrinus	Southern prairie lizard
Sceloporus undulatus hyacinthinus	Northern fence lizard
Sceloporus virgatus	Striped plateau lizard
Scelotretus	Lined gecko
Scinax squalirostris	Striped-snouted tree frog
Scincus	Skink
Scincella lateralis	Little brown skink
Scincella lateralis	Ground skink
Sclerosaurus armatus	Fossil anapsid reptile
Sepedon haemachates	Ringhals or Ring-necked spitting cobra
Seps chalcides	Algerian cylindrical snake
Sibon annulata	Ringed snail sucker snake
Sibon anthracops	Cope's snail sucker snake, Ringed snail-eater snake
Sibon argus	Argus snail sucker
Sibon carri	Carr's snail sucker
Sibon dimidiate	dimidiate auger shell
Sibon dunni	Dunn's snail sucker snake
Sibon nebulata	Cloudy snail-eating snake
Sibon sanniola	Pygmy snail sucker snake, Pygmy snail-eating snake, Yucatan snail sucker snake
Sibynomorphus mikani	Slug-eating colubrid snake
Sibynomorphus turgidus	Bolivian tree snake
Sibynomorphus vagrans	Dunn's tree snake
Sibynomorphus ventrimaculatus	Boulenger's tree sn*ake*
Sibynophis	Many-toothed snake
Sinocephalus poensis	
Sinomicrurus japonicus	See *Calliophis japonicus*
Siredon lichenoides = Siredon mexicanum	Baird's tiger salamander
Siredon mexicanum	Mexican axolotl
Siredon pisciformis = Siredon mexicanum	
Siren lacertina	Greater siren
Sistrurus catenatus catenatus	Eastern Massasauga rattlesnake

(continued)

Scientific name	Common name
Sistrurus miliarius miliarius	Carolina pygmy rattlesnake
Sistrurus miliarius streckeri	Western pygmy rattlesnake
Sitana ponticeriana	Fan-throated lizard
Smilisca phaeota	Masked tree frog, New Granada cross-banded frog
Sonora episcopa	Great Plains ground snake
Spalerosophis diadema	Diadem snake, Royal snake
Spea	Western spadefoot
Spelerpes ruber	Red salamander
Spelerpes guttolineatus	Three-lined salamander
Sphenodon punctatus	Tuatara
Sphenops capistratus	Wedge-snouted skink
Sphenops sepsoides	Wedge-snouted skink
Spherotheca (Tomopterna) rufescens	Parambikulam frog, Malabar wart frog
Staurois latopalmatus	Rock skipper frog
Staurois parvus	Rock skipper
Staurotypus triporcatus	Three-keeled terrapin
Stellio = Agama	Horned lizard
Stenocercus	Whorl-tail iguana
Stenocercus ornatissimus	Lesser ornate whorltail iguana
Stenodactylus	Dwarf sand gecko
Stenodactylus doriae	See *Ceramodactylus doriae*
Stenodactylus elegans	Egyptian gecko
Stenodactylus sthenodactylus	Dwarf sand gecko
Stenorhina degenhardti	Degenhardt's scorpion-eater, southern scorpion-eater
Sternotherus	Musk turtle
Sternotherus carinatus	Razorback musk turtle
Sternotherus depressus	Flattened musk turtle
Sternotherus minor	Loggerhead musk turtle
Sternotherus odoratus	Common musk turtle
Stenorhina degenhardti	Tropical American opisthoglyph snake
Stigmochelys pardalis	Leopard tortoise
Storeria dekayi	Dekay's snake, Northern brown snake
Storeria occipitomaculata	Redbelly snake
Storeria o. occipitomaculata	Northern red-bellied snake
Strophurus s. spinigerus	Soft spiny-tailed gecko, Southwest spiny-tailed gecko
Tachydromus sexlineatus	Asian grass lizard, Six-striped long-tailed lizard, Long-tailed grass lizard

(continued)

Scientific name	Common name
Takydromus wolteri	Amur grass lizards
Trachylepis (*Mabuya*) *striata*	African striped skink
Tachymenis chilensis	Chilean slender snake
Tachymenis peruviana	Peru slender snake
Takydromus septentrionalis	Northern grass lizard, China grass lizard
Takydromus tachydromoides	Japanese grass lizard
Takydromus wolteri	Grass lizard
Tantilla eiseni	Western blackhead snake
Tarbophis fallax	Mediterranean cat snake, European cat snake
Taricha granulosa	Rough-skin newt
Tarentola mauritanica	Moorish wall gecko, Moorish gecko
Tarocja tprpsa	California newt
Taudactylus eungellensis	Eungella day frog, Eungella torrent frog
Tejus teju	Teju lizard
Telmatobius	Andean water frog
Terrapene coahuila	Coahuilan box turtle
Terrapene carolina bauri	Florida box turtles
Terrapene carolina carolina	Eastern box turtle
Terrapene carolina triunguis	Three-toed box turtle
Terrapine ornata	Ornate box turtle
Terrapene triunguis	Box turtle
Testudo angulata	Angulate tortoise, Bowsprit tortoise
Testudo carbonaria	South American red-footed tortoise
Testudo (*Chelonoidis*) *chilensis*	Chaco tortoise, Southern wood tortoise
Testudo denticulate	Rain forest tortoise
Testudo denticulate	Denticulated tortoise
Testudo elegans	Starred tortoise
Testudo elephantopus	Galápagos tortoise
Testudo gigantea	Giant tortoise
Testudo [*Geochelone*] *gigantean*	Aldabra tortoise
Testudo giantea elephantiae	Galápagos tortoise
Testudo graeca	Mediterranean spur-thighed turtle, Morish tortoise, Greek tortoise
Testudo hermanni	Greek tortoise, Hermann's tortoise
Testudo hermanni hermanni	Italian tortoise, Western Hermann's tortoise
Testudo horsfieldi	*Horsfield's tortoise*
Testudo ibera	Mediterranean spur-thigh tortoise, Greek tortoise
Testudo kleinmanni	Egyptian tortoise

(continued)

Scientific name	Common name
Testudo marginata	Marginated tortoise
Testudo radiate	Radiated tortoise
Testudo sulcata	African spurred tortoise
Testudo tabulate (*Testudo horsfieldii*)	Brazilian tortoise
Thalassophina anamolus	Anomalous sea snake
Thalassophina viperina	Schmidt's sea snake
Thallassochelys caretta	
Thalassophis anomalus	Anomalous sea snake
Thomnophis	Garter snake
Thamnophis atratus atratus	Santa Cruz garter snake
Thamnophis couchii	Western aquatic garter snake, Sierra garter snake
Thamnophis cyrtopsis	Black-necked garter snake
Thamnophis elegans	Elegant garter snake, Mountain garter snake, Western terrestrial garter snake
Thamnophis elegans vagrans	Wandering garter snake
Thamnophis cf. *gigas*	Giant garter snake
Thamnophis hammondii	Two-striped garter snake
Thamnophis marcianus	Spotted garter snake, Checkered garter snake
Thamnophis ordinoides	Northwestern garter snake
Thamnophis radix	Plains garter snake
Thamnophis radix haydeni	Zig zag garter snake
Thamnophis sauritus	Eastern ribbon snake
Thamnophis sirtalis fitchi	Common garter snake, Valley gartersnake
Thamnophis sirtalis infernalis	Pacific garter snake, California red-sided garter snake
Thamnophis sirtalis sirtalis	Eastern garter snake
Thrasops occidentalis	Western black tree snake
Thecadactylus	Turniptail geckos
Thecadactylus rapicauda	Turnip-tailed gecko
Thorius	Minute salamander
Thorius dubitus	Acultzingo pigmy salamander
Thorius macdougalli	MacDougall's minute salamander
Thorius narisovalis	Upper Cerro minute salamander
Thrasops jacksoni	Jackson's tree snake, Black tree snake
Thrasops flavigularis	Yellow-throated bold-eyed tree snake
Tiliqua scincoides	Blue-tongued skink
Timon lepidus	Jeweled lizard
Tomistoma schlegelii	False gavial, False gharial

(continued)

Scientific name	Common name
Tomopterna (*Spherotheca* or *Fejervarya*) *rufescens*	Parambikulam frog, Malabar wart frog
Tomuropeltis	Horntail worm lizard
Trachyboa	Eyelash boa
Trachyboa boulengeri	Northern eyelash boa
Trachydosaurus rugosus	Shingle-back skink
Trachemys scripta	Yellow-bellied turtle, Common slider, See *Chrysemys scripta*
Trachemys scripta elegans	Red-eared slider
Trachydosaurus rugosa	Shingle-backed skink, sleepy lizard
Trachydosaurus rugosus	Shingle-backed skink, sleepy lizard
Tretanorhinus variabilis	Caribbean water snake
Tribolonotus	Crocodile skink
Trioceros deremensis	See *Chameleo deremensis*
Trioceros jacksonii	See *Chameleo jacksonii*
Trioceros johnstoni	See *Chameleo johnstoni*
Trioceros montium	See *Chameleo montium*
Trioceros oweni	See *Chameleo owenii*
Trigonocephalus	pit viper
Trimeresurus	Asian pit vipers, Asian lanceheads, Asian lance-headed vipers
Trimorphodon vandenburghi	Western Lyre snake, California Lyre snake
Trionyx	Soft shell turtle
Trionyx ferox	Florida soft shell turtle
Trionyx gangeticus	Ganges soft shell turtle
Trionyx muticus	Smooth softshell, Pancake turtle
Trionyx (*Amyda*) *sinensis*	Chinese soft-shelled turtle
Trionyx (*Amyda*) *spiniferus*	Spiny softshell turtle
Trionyx (*Amyda*) *spinifera*	Spiny softshell turtle
Trionyx spinifer spinifer	Eastern spiny soft-shelled turtle
Trionyx steindachneri	Wattle-necked soft-shelled turtle
Trionyx triunguis	African soft shell turtle, Nile turtle
Triton cristatus	Southern crested newt
Triton helveticus	Palmate newt
Triton marmoratus	Marbled newt
Triton niger	Northern dusky salamander
Triton noachicus	Fossil newt
Triton palmatus	= *Triton helveticus*
Triton taeniatus	Smooth newt, Smooth Eft, Great water newt

(continued)

Scientific name	Common name
Triton vulgaris	Smooth newt, Lille vandsalamander
Triturus alpestris	Alpine newt
Triturus boscai	Bosca's newt, Iberian newt
Triturus carnifex	Italian crested newt
Triturus cristatus carnifex	Crested newt
Triturus helviticus	Palmate newt
Triturus marmoratus	Marbled newt
Triturus orientalis	Little red-bellied newt
Triturus pygmaeus	Southern marbled newt
Triturus pyrrhogaster	Japanese common or firebelly newt
Triturus sinensis	= *Paramesotriton chinensis*
Triturus torosus	Taricha torosa
Triturus viridescens	Yellow-bellied, Eastern newt
Triturus vulgaris	Smooth newt
Trogonophis wiegmanni	Checkerboard worm lizard
Tropiocolotes tripolitanus	Tripoli dwarf gecko
Tropidoclonion lineatum lineatum	Lined snake
Tropidoleptus invertebrate	Fossil
Tropidonotus fasciata fasciata	Banded water snake
Tropidonotus fasciata sipedon	Midland water snake
Tropidonotus n. natrix	Ring snake, Grass snake
Tropidonotus piscator (*quincunciatus*)	Asiatic water snake
Tropidonotus quincunciatus	Checkered keelback snake
Tropidonotus septemvittata	Queen snake
Tropidonotus sipedon	Northern water snake
Tropidonotus tigrinus	Tiger keelback snake
Tropidonotus vagrans	Wandering garter snake
Tropidophis melanurus	Dwarf or black-tailed boa
Tropidurus etheridgei	Etheridge's lava lizard, Western collared spiny lizard
Tropidurus flaviceps	Amazon thornytail, Tropical thornytail iguana
Tropidurus hispidus	Peters' lava lizard
Tropidurus itambere	Saxicolous lizard
Tropidurus montanus	Brazilian endemic tropidurid lizard
Tropidurus nanuzae	Rodrigues' lava lizard
Tropidurus semitaeniatus	Striped lava lizard
Tropidurus torquatus	Amazon lava lizard

(continued)

Scientific name	Common name
Tropidurus umbra	Blue-lipped tree lizard
Tupinambis merianae	Argentine black and white tegu, Argentine tegu
Tupinambis (Otenodus) nigropunctatus	Iacruaru, Gold tegu, Salipenter
Tupinambis rufescens	Red teju
Tupinambis teguixin	Common tegu, Columbian tegu, Golden tegu
Tropiocolotes	dwarf or pygmy gecko
Tropiocolotes nattereri	Dwarf gecko, Sand gecko
Tropiocolotes steudneri	Steudner's gecko
Tylotriton verrucosus	Emperor newt
Typhlomolge rathbuni	Texas blind salamander
Typhlomolge robusta	Blanco blind salamander
Typhlops	Blind snake
Typhlops mucroso	Burrowing snake
Uma notata	Colorado desert fringe-toed lizard
Uperodon systoma	Marbled balloon frog
Uranoscodon	Driving lizard
Uranoscodon superciliosus	Diving lizard, Brown tree climber, Mop head iguana
Uracentron	Thorny-tailed iguana
Uromastyx	Spiny-tailed lizard
Uromastyx acanthinura	Bell's dab-lizards
Uromastyx aegyptia	Egyptian spiny-tailed lizard
Uromastyx hardwickii	Hardwick's dab-lizards, Indian spiny tail lizard
Uromastyx maliensis	Sudanese spiny-tail lizard
Uromastyx ornata	Ornate spiny-tailed lizard
Uromastix spinipes	= *Uromastyx aegyptia*
Uropeltis	Pipe snake, Shield-tailed snake, earth snake
Uroplatus	Flat-tail gecko
Uroplatus fimbriatus	Common leaf-tailed gecko
Urosaurus	Tree or Brush Lizard
Urostrophys vautieri	
Uta stansburiana	Common side-blotched lizard
Utatsusaurus	Fossil ichthyosaur
Varanus albigularis	Rock monitor, White-throated monitor
Varanus bengalensis	Bengal monitor
Varanus bengalensis nebulosus	Long-necked monitor lizard, Bengal monitor
Varanus dracaena	Indian monitor
Varanus exanthematicus	Savannah monitor
Varanus e. albigularis	African savannah monitor, White-throated monitor

(continued)

Scientific name	Common name
Varanus flavescens	Yellow monitor, Calcutta oval-grain lizard
Varanus gouldii	Sand monitor
Varanus grayi = V. olivaceous	Gray's monitor
Varanus griseus	Desert monitor
Varanus indicus	Pacific varanid
Varanus komodoensis	Komodo dragon
Varanus niloticus	Nile monitor
Varanus ocellatus = exanthematicus	Savannah monitor
Varanus olivaceous	Gray's monitor
Varanus salvadorii	Crocodile monitor
Varanus salvator	Water monitor
Varanus spenceri	Spencer's monitor, Spencer's goanna, Plain goanna
Varanus varius	Lace monitor
Vanzosaura rubricauda	Redtail Tegu
Viatkosuchus sumini	Fossil mammal-like reptile
Vibrissaphora boringii	Omei moustache toad
Vipera ammodytes	Long-nosed viper, Nose-horned viper
Vipera aspis	Aspis viper, Sri Lanka green viper
Vipera (Pelias) berus	Common adder
Vipera russelli	Russell's viper
Vipera ursinii	Meadow viper
Vipera ursinii rakosiensis	Hungarian meadow viper
Vipera xanthina	Tit polonga
Xantusia vigilis	Desert night lizard
Xenobates agassizi	Desert tortoise
Xenochrophis piscator	Chequered keelback, Asiatic snake
Xenodon merremii	Boipeva, Wagler's snake
Xenophrys monticola	Concave-crowned horned toad
Xenopus laevis	African clawed frog
Xenopus tropicalis	Western clawed frog
Xenopeltis unicolor	Asian sunbeam snake
Xerobates agassizi	California desert tortoise
Zamenis constrictor	American black snake
Zamenis mucosus	Sri Lankan rat snake
Zamenis (Coluber) florulentus	Flowered racer
Zamenis hippocrepis	Horseshoe snake
Zootoca ("Lacerta") vivipara	Viviparous lizard, Common lizard
Zygaspis	African worm lizard

Pathologies Index

V

Z

Scientific Index

B

C

D

H

R

U

V

X

Z

Common Index

C

H

N

R

S

T

Y

Z

MIX
Papier aus verantwortungsvollen Quellen
Paper from responsible sources
FSC® C105338

If you have any concerns about our products,
you can contact us on
ProductSafety@springernature.com

In case Publisher is established outside the EU,
the EU authorized representative is:
Springer Nature Customer Service Center GmbH
Europaplatz 3, 69115 Heidelberg, Germany

Printed by Libri Plureos GmbH
in Hamburg, Germany